	DIVISION	SUBDIVISION	CLASS		
Protista—"Algae"	Chrysophyta* (*Chrysophytes*)		Tribophyceae (*Yellow-green algae*) Chrysophyceae (*Golden algae*) Bacillariophyceae (*Diatoms*) Prymnesiophyceae (*Prymnesiophytes*) Eustigmatophyceae (*Eustigmatophytes*) Raphidophyceae (*Raphidophytes*)		
	*Consists of a diverse assemblage that can be considered in several Divisions.				
	Pyrrhophyta		Dinophyceae (*Dinoflagellates*)		
	Euglenophyta		Euglenophyceae (*Euglenoids*)		
	Cryptophyta		Cryptophyceae (*Cryptophytes*)		
Plantae—"Algae"	Chlorophyta (*Green Algae*)		Chlorophyceae (*Chlorophytes*)		Volvocales Tetrasporales Chlorococcales Ulotrichales Ulvales Chaetophorales Oedogoniales Zygnematales
			Bryopsidophyceae (*Bryopsidophytes*)		Cladophorales Siphonocladales Codiales Dasycladales
			Charophyceae (*Charophytes*)		Charales
			Prasinophyceae (*Prasinophytes*)		
	Phaeophyta (*Brown Algae*)		Phaeophyceae	Phaeophycidae	Ectocarpales Chordariales Sporochnales Desmarestiales Sphacelariales Cutleriales Tilopteridales Dictyotales Dictyosiphonales Scytosiphonales Laminariales
				Cyclosporidae	Fucales
	Rhodophyta (*Red Algae*)		Bangiophyceae		Porphyridiales Bangiales Erythropeltidales Rhodochaetales
			Florideophyceae		Acrochaetiales Nemaliales Cryptonemiales Gigartinales Rhodymeniales Palmariales Ceramiales
Plantae—"Embryophytes" (Bryophytes)	Bryophyta		Hepaticae (Hepatopsida)	Jungermanniae	Calobryales Jungermanniales Metzgeriales
				Marchantiae	Sphaerocarpales Monocleales Marchantiales
			Anthocerotae (Anthoceropsida)		Anthocerotales
			Musci (Bryopsida)	Sphagnidae	Sphagnales
				Andreaeidae	Andreaeales
				Tetraphidae	Tetraphidales
				Polytrichidae	Polytrichales
				Buxbaumiidae	Buxbaumiales
				Bryidae	Dicranales Archidiales Fissidentales Pottiales Grimmiales Funariales Eubryales Isobryales Hookeriales Hypnobryales

P L A N T S
An Evolutionary Survey

R. F. Scagel
R. J. Bandoni
J. R. Maze
G. E. Rouse
W. B. Schofield
J. R. Stein

Department of Botany
The University of British Columbia

Wadsworth Publishing Company
Belmont, California
A Division of Wadsworth, Inc.

Biology Editor: Jack Carey

Production: Greg Hubit Bookworks

Copy Editor: Nick Murray

Cover photograph: © Pat O'Hara

© 1984, 1982 by Wadsworth, Inc. All rights reserved. No part of this book may be reproduced, stored in a retrieval system, or transcribed, in any form or by any means, electronic, mechanical, photocopying, recording, or otherwise, without the prior written permission of the publisher, Wadsworth Publishing Company, Belmont, California 94002, a division of Wadsworth, Inc.

Printed in the United States of America

1 2 3 4 5 6 7 8 9 10—86 85 84 83 82

Library of Congress Cataloging in Publication Data
Main entry under title:

Plants, an evolutionary survey.

   Includes index.
   1. Botany. 2. Botany—Classification. 3. Plants—Evolution. 4. Phylogeny (Botany) I. Scagel, Robert Francis, 1921–
QK47.P625   1984   581   83-24369

ISBN 0-534-02802-0

PREFACE

In 1965, the first edition of *An Evolutionary Survey of the Plant Kingdom* appeared. It was designed primarily for a one-year (two semesters or three quarters) course at second-year university level, surveying the plant kingdom with emphasis on comparative morphology and evolution. Although it has now gone through several reprintings, the first edition has not been revised during this period except for minor changes. However, in 1969 a condensed revised version appeared under the title *Plant Diversity: An Evolutionary Approach*. This condensed text was designed primarily for a one-term (one semester or one quarter) course in plant morphology, or for the plant diversity section of an introductory botany or biology course. In 1982 the nonvascular plant chapters were revised and expanded in *Nonvascular Plants: An Evolutionary Survey*. The wide use of *An Evolutionary Survey of the Plant Kingdom* by second-year students, as well as in more advanced undergraduate courses and as a reference, has confirmed our philosophy that a collaboration of specialists in one text can make a unique contribution to students of the plant kingdom, especially at a more advanced level, providing a more satisfactory coverage of each of the groups than is often achieved in a book written by one author.

Plants: An Evolutionary Survey has been written, with one exception, by the same authors, and is essentially a major revision of the subject matter of *An Evolutionary Survey of the Plant Kingdom* and *Plant Diversity*, with much the same overall objectives and scope of the earlier publications.

Since 1965, there have been significant advances in botanical research, which have necessitated some major changes in classification systems, among both nonvascular plants and embryophytes. An outline of the classification system followed in this text appears inside the front and back covers. In addition to these major changes, we have updated all chapters to take into account much new information that has been reported in the literature appearing since 1965.

This text is intended to lead the average interested student beyond what is expected at the second-year university level. At the same time, it provides additional reference material to stimulate further interest, curiosity, and enthusiasm, and to make the text more useful as a general reference for the more advanced student and teacher. A list of pertinent references appears at the end of each chapter. The text assumes that the student has already had a first-year university course in botany or the equivalent and has acquired an understanding of fundamental biological processes, such as mitosis and meiosis.

We have standardized our terminology throughout the text as far as is practical. We have also tried to maintain continuity in treatment so that the overall trends in phylogeny and evolution are not lost or interrupted. By the same token, we have attempted to apply the findings of paleobotany directly to living representatives in the various plant groups in order to gain perspective on these two complementary aspects of botanical science. A comprehensive glossary, including terms used throughout the text, is included at the end of the text preceding the index.

Individual contributions, which have been primarily in our respective fields of specialization, are as follows: bacteria and fungi (including lichens)—J. R. Stein; slime molds—Robert J. Bandoni; bryophytes and lower vascular plants—W. B. Schofield; algae—Robert F. Scagel and Janet R. Stein; gymnosperms and paleobotany—Glenn E. Rouse; vascular plant anatomy and flowering plants—J. R. Maze. However, since we have consolidated our efforts and consulted closely with each other in the interests of uniformity and continuity, it is impossible to define completely the limits of each person's contribution. In the process of writing and editing this work we have come to know one another better and to understand and appreciate each other's discipline and point of view to a greater extent.

Many of the original drawings prepared for *An Evolutionary Survey of the Plant Kingdom, Plant*

Diversity: An Evolutionary Approach, and *Nonvascular Plants: An Evolutionary Survey* by John C. Andrews, Lesley R. Bohm, P. Drukker Brammall, Margaret (Dean) Jensen, Frank Lang, Ernani Meñez, and Muriel Schofield have been reproduced in *Plants: An Evolutionary Survey*. The quality of these original drawings is a feature that has been universally well received by students and teachers using the earlier texts. In addition, some new drawings have been added; these were prepared by Lesley R. Bohm, P. Drukker Brammall, Maureen Douglas, Lindsay J. Brooke, Scott Renyard, and Muriel Schofield. The following figures were prepared by Lesley R. Bohm: 5-15; 9-12C, D, 17T, 18G, H; 10-4, 5, 6, 8, 9, 10, 11, 12; 12-4, 5, 6, 7, 8, 9, 10, 11, 12, 13, 15, 16, 17, 18, 19, 20, 21, 22B, 25, 26, 27, 28, 29, 30, 31, 32; 13-6A; 14-4; 15-2C,D, 3A, B, 5B-I, 10, 11, 17A; 16-2D,G, 3, 4, 6D, 7A, 8; 18-3C, 4, 7; 19-4A, D-F, 5, 6, 7B-D, 7A, 10A-F, 11, 12, 13C-F, 14B-F, 15, 16A-D, F, 17A-B, 18H; 20-1, 3, 4B, 5, 8, 9; 21-3B, 7, 9, 10; 22-6, 16, 17, 20, 24, 25; 23-2, 6B, 9, 13, 14; 24-3, 9, 23B, 26, 27, 28, 30; 26, 6, 7, 8, 10, 14, 15, 23, 25, 27, 28, 29, 41; 27-18, 19, 20, 21, 22, 23, 24, 25,26, 27, 28, 31, 32. P. Drukker Brammall prepared the following figures: 12-1, 2, 3; 13-1, 2A, B, 3A, B; 15-6, 19E; 16-1. Lindsay J. Brooke prepared the following figures: 26-17, 18, 24. Maureen Douglas prepared the following figures: 7-3, 28. Muriel Schofield prepared the following figures: 16-5, 9, 13; 18-1, 3A, B. Scott Renyard prepared the following figures: 13-14B, 5, 9, 10; 14-2C; 26-1, 4, 13, 16, 19, 20, 21, 35A, B, 39.

New photographs, including transmission and scanning electron micrographs, have been added to this revised text to complement the drawings.

For constructive criticism and valuable recommendations concerning preparation of this text, and for reading and commenting on an early draft of the manuscript, or portions thereof, we are especially indebted to the following reviewers:

Chester Bosworth, Southern Connecticut State University; Gilbert Brum, California State Polytechnic University, Pomona; Dennis Clark, Arizona State University; Lois J. Cutter, University of North Carolina at Greensboro; William E. Dietrich, Indiana University of Pennsylvania; Patricia Gensel, University of North Carolina at Raleigh; Allen Graham, Kent State University; Robert B. Kaul, University of Nebraska; P. W. Kirk, Old Dominion University; Norma Lang, University of California, Davis; Estelle Levetin, University of Tulsa; L. W. Macior, Akron University; M. W. Miller, University of South Alabama; Paul Nighswonger, Northwestern State College; Florence Neely, Augustana College; Nancy Nicholson, Miami University (Ohio); Donald Pfister, Harvard University; Harry K. Phinney, Oregon State University; Don R. Reynolds, Natural History Museum, Los Angeles; Rudolf Schmid, University of California, Berkeley; Judith E. Skog, George Mason University; Shirley Sparling, California Polytechnic State University, San Luis Obispo; Anne A. Susalla, St. Mary's College (Notre Dame, Indiana); Sanford S. Tepfer, University of Oregon; Robert Tolbert, Moorhead State University; Henry S. Webert, Nicholls State University; Jonathon J. Westfall, University of Georgia; Richard A. White, Duke University.

Portions of the manuscript were also read and constructively criticized by Drs. Fred R. Ganders, J. W. D. Garbary, Lynda J. Goff, Gayle I. Hansen, Sandra Lindstrom, M. Neushul, R. E. Norris, F. J. R. Taylor, Nancy J. Turner, and M. J. Wynne; to each of these persons we express our sincere appreciation. Although their comments and criticisms have been most helpful and have contributed significantly to our preparation, the authors take full responsibility for the final product.

We are indebted to individuals and publishers who loaned original illustrations and photographs, and who gave permission to copy or redraw figures; these are acknowledged in the appropriate captions.

Finally, we wish to acknowledge the assistance provided by the Department of Botany of the University of British Columbia, and especially to thank Mr. J. Reid for photographic assistance and Mrs. J. C. Oliveira for technical assistance.

CONTENTS

1	INTRODUCTION	1
	The Process of Evolution	1
	Natural Order	6
	Features Used in Classification	12
	References	13
2	MONERANS (Monera—Bacteria and Bluegreen Algae)	14
	Division Schizomycophyta (Bacteria)	14
	Division Archaebacteria	25
	Division Cyanophyta (Bluegreen Algae)	27
	Relationships of the Monera	36
	References	37
3	NONVASCULAR EUKARYOTES	38
	Habitat	39
	Cell Structure	40
	Reproduction	40
	Nonphotosynthetic Forms	46
	Photosynthetic Forms	49
	References	54
4	FUNGI	55
	General Characteristics	55
	Division Oomycota	64
	Divisions Chytridiomycota and Hyphochytridiomycota	69
	Division Zygomycota	72
	Division Eumycota	75
	Subdivision Ascomycotina	78
	Subdivision Basidiomycotina	92
	Form Class Fungi Imperfecti	111
	Relationships of Fungi	114
	References	117
5	PROTISTANS—MYXOMYCOTA (Slime Molds)	118
	Class Myxomycetes (True Slime Molds or Acellular Slime Molds)	118
	Class Dictyosteliomycetes (Dictyostelids)	127
	Class Plasmodiophoromycetes (Endoparasitic Slime Molds)	132
	Class Labyrinthulomycetes (Labyrinthulids)	139
	Relationships of Myxomycota	139
	References	141
6	PROTISTAN ALGAE	142
	Division Chrysophyta (Chrysophytes)	142
	Division Pyrrhophyta (Class Dinophyceae) (Pyrrhophytes)	159
	Division Euglenophyta (Class Euglenophyceae) (Euglenophytes)	165
	Division Cryptophyta (Class Cryptophyceae) (Cryptophytes)	169
	Relationships of the Protistan Algae	172
	References	172
7	PLANTS—CHLOROPHYTA (Green Algae)	174
	Class Chlorophyceae	175
	Class Bryopsidophyceae	200

Class Charophyceae	208
Class Prasinophyceae	214
Ecology of Chlorophyta	214
Relationships of the Chlorophyta	216
References	217

8 PLANTS—PHAEOPHYTA (Brown Algae) 219

Cell Structure	219
Classification and Morphological Diversity	225
Reproduction and Life Histories	235
Lines of Evolution	241
Relationships of Brown Algae	242
Distribution and Ecology	242
Importance and Uses	247
References	251

9 PLANTS—RHODOPHYTA (Red Algae) 252

Cell Structure	253
Class Bangiophyceae	260
Class Florideophyceae	262
Distribution and Ecology	277
Importance and Uses	277
Phylogeny	279
References	285

10 EMBRYOPHYTES 286

General Features of Embryophytes	287
Morphological Trends within Embryophytes	295
Relationships of the Embryophytes	301
References	301

11 BRYOPHYTES (Mosses, Liverworts, and Hornworts) 303

Gametophyte	303
Sporophyte	304
Life History	305
Class Hepaticae (Liverworts)	306
Class Anthocerotae (Hornworts)	315
Class Musci (Mosses)	317
References	340

12 VASCULAR PLANTS 341

Diversity	342
Distribution and Role in Vegetation	342
Interpretation of Available Information	344
Classification	344
The Plant Body and Its Growth	346
Plant Cells and Tissues	347
Growth in Plants	368
Appendages	376
Secondary Growth	378
References	382

13 RHYNIOPHYTES, TRIMEROPHYTES, AND ZOSTEROPHYLLOPHYTES 383

Subdivision Rhyniophytina (Rhyniophytes)	383
Subdivision Trimerophytina (Trimerophytes)	387
Subdivision Zosterophyllophytina (Zosterophyllophytes)	391
References	395

14 PSILOTOPHYTES (Whisk-Ferns and Their Relatives) 397

Psilotum	397
Tmesipteris	401
References	402

15 LYCOPHYTES (Lycopods) 403

General Morphology	403

	Early Lycopods	403
	Carboniferous Lycopod Forests	407
	Modern Lycopods	412
	More Fossil Lycopods	424
	References	427
16	SPHENOPHYTES (Horsetails and Their Allies)	429
	General Morphology	429
	Early Sphenophytes	429
	The Modern Horsetails	435
	Further Fossil Sphenophytes	443
	References	444
17	PTEROPHYTES	446
	Characteristics of Pterophytina	446
	References	448
18	CLADOXYLIDS AND COENOPTERIDS (Preferns)	450
	Class Cladoxylopsida	450
	Class Coenopteridopsida	456
	References	457
19	FILICOPSIDS (True Ferns)	460
	General Features	460
	Subclass Ophioglossidae	462
	Subclass Marattiidae	465
	Subclass Osmundidae	471
	Subclass Filicidae	473
	Subclass Marsileidae	487
	Subclass Salviniidae	489
	References	492
20	PROGYMNOSPERMS	493
	Class Progymnospermopsida	493
	Evolution of the Seed	500
	Evolution of the Pollen Grain	504
	References	504
21	PTERIDOSPERMS (Seed Ferns)	505
	Paleozoic Pteridosperms	505
	Mesozoic Pteridosperms	515
	References	518
22	CYCADOPHYTES	519
	Class Cycadopsida	519
	Class Cycadeoidopsida (Cycadeoids)	532
	Detached Cycadophyte Leaves	537
	Phylogeny of Cycadeoidales	538
	References	539
23	GINKGOPHYTES	540
	Ginkgo biloba	540
	Phylogeny of Ginkgophytes	549
	References	553
24	CONIFEROPHYTES	554
	Class Coniferopsida	554
	Evolution of Conifers	576
	References	580
25	GNETOPHYTES	582
	Order Ephedrales	582
	Order Welwitschiales	585
	Order Gnetales	588
	Phylogeny of Gnetophyta	591
	References	592
26	ANGIOSPERMS (Flowering Plants)	593
	Distinctive Features	593

Major Angiosperm Groups	596
Vegetative Plant Body	596
Reproductive Plant Body	606
Classification of the Angiosperms	655
References	656

27 EVOLUTION AND PHYLOGENY   658

Early Evolution	658
Evolution of Eukaryotes	660
Mitosis, Sex, and Alternation of Generations	663
Evolutionary Patterns	665
Evolution of Land Plants	672
Summary	691
References	694

GLOSSARY	697
INDEX	737

1 INTRODUCTION

The plant world exhibits a tremendous range of diversity, from unicellular bacteria and algae to immense trees. Plants occur in every conceivable environment, from aquatic (both marine and freshwater) to deserts, from tropical to alpine and arctic environments, and from hot springs to snow banks. Nutritionally, plants can be **autotrophs** or **heterotrophs,** and the latter can be **saprophytes** or **parasites.** Despite this great range of diversity, there is a natural order that exists, and the primary impetus behind much biological research has been to describe and explain this order.

Plant morphology, the theme of this book, is concerned with one aspect of biological order—the relationship between plant groups and between **tissues, structures,** and **organs** of plants exhibiting structural complexity. However, an appreciation of biological **evolution,** the mechanism that most biologists consider to be the cause of natural order, adds to the study of morphology. The purpose of this chapter is to offer a brief introduction to that mechanism and to the study of natural order.

The Process of Evolution

There are many ways to describe the process of evolution, but practically any description will include the following points: (1) evolution is a **genetic** phenomenon; (2) there are different forms of individuals in a species (variation is present); (3) variants are not equally successful in contributing progeny to the next generation; and (4) there is genetic change in species over time.

Population Genetics

Population is a general term that is used in several contexts. In evolutionary biology, a population is a group of potentially interbreeding individuals occupying a specific habitat. Because a population consists of more than one individual, its genetics is slightly different from the genetics of an individual. All traits seen in organisms are under control of units called **genes.** Genes occur on chromosomes at positions called **loci** (sing. **locus**). In diploid individuals, there are two **alleles** for each gene locus, one per chromosome. Thus, in a diploid population, the total number of alleles for any one locus is twice the number of individuals; for example, in a population of 100 plants, there will be 200 alleles for any gene locus. The two alleles at one locus in any one plant need not be the same as the two alleles for the same locus in another plant. For example, one plant can have alleles a_1 and a_2, in which case it is **heterozygous** at the a locus. Another plant can have different alleles, a_3 and a_4. For this reason many different alleles for one gene locus can exist in a population. The relative number of alleles for any gene in a population is usually called *gene frequency*, although, strictly speaking, it is allele frequency. For example, in a population of 100 plants there would be 200 alleles for the a locus. If 100 were a_1, 50 were a_2, 40 were a_3, and 10 were a_4, the gene frequencies would be 0.5 for a_1 (100/200), 0.25 for a_2, 0.20 for a_3, and 0.05 for a_4. Allele frequencies always add up to 1.00. A change in gene frequency is an integral part of evolution (see Fig. 1-1).

Variation

Variation means detectable differences that exist between members of a species. It can be due either to environmental or to genetic differences. Environmentally induced variation results from the response of plants wih similar genotypes to different environments. An example of this might be plant size, as illustrated by the brown alga *Laminaria digitata*, which grows in northern European waters. In areas where there is rapid water movement, its blades are narrow and often split longitudinally. When these plants are moved to calm waters, the blades become wider (up to 8 times as wide), and there is very little splitting. Another example is the phenomenon of **heterophylly**, the formation of different leaves. On some trees, such as bur oak *(Quercus macrocarpa)*, leaves formed in the shade are much larger, thinner, and less deeply lobed than leaves formed in the sun. Morphological responses to environmental differences such as these are often called **phenotypic plasticity**.

Genotypic differences are due to variation in the genetic makeup of plants. Such differences will usually be expressed regardless of environment, although environment may modify the degree of difference expressed. Genetic differences are most important in evolution because they are inherited.

Genetic Sources of Variation

Our concern here is with genotypic variation. There are two sources of genetic variation—**mutation** and **recombination**. Mutation is a change in base pair sequence in DNA so as to give new (or different) alleles of genes. Recombination generates variation as a result of bringing together new combinations of alleles.

There are two sources of recombination: (a) **independent assortment** of chromosomes at meiosis, with subsequent sexual reproduction, and (b) **crossing-over** between **homologous chromosomes** during meiosis. The former phenomenon results in new and unique allele combinations in the offspring. For example, if two **homozygous** individuals with the genetic constitutions $a_1a_1b_1b_1c_1c_1$ and $a_2a_2b_2b_2c_2c_2$ are crossed, the result is $a_1a_2b_1b_2c_1c_2$ in the first generation (F_1). If the F_1, which is heterozygous for these genes, is selfed (crossed with itself), there is the possibility of 27 different genotypes appearing in the second generation, the F_2. Of these, 3 will be reconstituted parental and F_1 genotypes, and 24 will be new. This is assuming the genes are on different chromosomes and thus, like chromosomes, show independent assortment. Recombination, as a result of independent assortment and subsequent sexual reproduction, can produce many genotypes. If a diploid plant that is heterozygous for ten independently assorting genes is selfed, the number of possible genotypes is 59,049 (3^n, where n = the number of independently assorting genes). Recombination within a sexually reproducing population can produce an even greater amount of variation. In a population, it is theoretically possible for a gamete from any individual to fertilize a gamete from any other individual. Thus, any allelic combination from all the alleles available in the population can be produced. The presence of **multiple alleles** in a population increases the already great potential for producing variants as a result of recombination. The following formula gives the possible number of recombinations where more than two alleles exist for a gene:

$$R = \left[\frac{r(r+1)}{2}\right]^n$$

where R = number of recombinations possible
r = number of alleles for each gene
n = number of independently assorting genes

For example, a population with ten independently assorting genes, each with four alleles can give rise to 10 billion variants as a result of recombination. A comparison of this with a number of recombinants possible when only two alleles exist for ten independently assorting genes shows that doubling the allele number increases the recombinants by about 170,000 times.

Crossing-over results in the exchange of chromosome parts between homologous chromosomes when they pair early in **meiosis**. It is significant in recombination, for it increases the number of independently assorting genes by breaking the genes linked together on one chromosome. The higher the rate of crossing-over, the greater the number of independently assorting genes.

Both mutation and genetic recombination are significant in evolution, but at different levels. Recombination is the immediate source of variation for natural selection, whereas mutation is the ultimate source for new alleles, which can be brought into new combinations by recombination.

Natural Selection

Natural populations of plants produce more progeny than can be supported by the environment. Hence, only some survive, the survivors being those better adapted to the environment. Of those that survive, not all grow equally well. Some will grow more vigorously and produce a greater proportion of progeny for the following generation (i.e., will have higher fitness). This differential reproduction by individuals in a population means that certain genotypes increase and others decrease, resulting in a change in gene frequency. Figure 1-1 presents a diagram of a hypothetical situation in which there is a change in gene frequency in a population.

Although **natural selection** results in genetic change, it does not act directly on the genotype. The genotype of an organism interacts with the environment, resulting in plants as we observe them

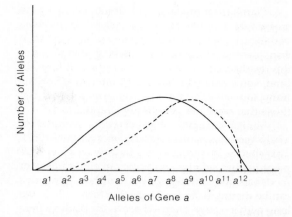

Figure 1-1 Change in gene frequency as a result of selection. Dotted line represents the change in a population as a result of selection for a_7 to a_{11}.

growing. The expression of this environmentally modified genotype is the **phenotype**. When selection acts, it does so on the phenotype.

Selection results in populations that are adapted to the habitat in which they live, and this adaptation is genetically based. Populations of one species that are genetically adapted to different environments are **ecotypes**.

There are several examples of experimental proof of adaptation resulting from natural selection. One is from a series of experiments performed on ecotypes at the Carnegie Institute at Stanford University by J. Clausen, D. D. Keck, and W. M. Hiesey (1948). These experiments involved populations of a perennial herb, yarrow *(Achillea millefolium)*, growing along an altitudinal gradient from coastal California to alpine areas in the Sierra Nevada. Plants from different altitudes could be ordered into a hierarchy and had traits that formed obvious adaptations to different habitats. Clausen, Keck, and Hiesey performed reciprocal transplants with alpine and lowland plants, transplanting alpine plants to gardens near the coast and lowland plants to gardens at timberline in the Sierra Nevada. Hence, they were able to assess the performance of plants in habitats different from their native ones. In habitats to which they had been transplanted, plants maintained certain features exhibited in their native environments. That the transplants did not "adapt" to their new environments indicates genetic control of features correlated with the native environments of the plants.

Plants from lowland and alpine habitats were crossed, and in the F_2 generation there was extreme variability. Plants of the F_2 generation were then transplanted to alpine and lowland stations, and the results were studied over a period of time. In time, some individuals were eliminated at various transplant sites as the environment selected out those that were not genetically adapted to it.

Similar adaptation has been demonstrated for a kelp species, *Laminaria groenlandica* (Druehl 1967). This species is restricted to a narrow range of temperature and salinity and occurs in either heavy or moderate surf conditions. Two morphological forms can be distinguished: a long-stipe form, 10–30 cm long with a wedge-shaped base to the blade (where it is attached to the stipe); and a short-stipe form, usually less than 10 cm long with a flat or heart-shaped blade base. The long-stipe forms occur in areas exposed to heavy surf with cool (6–13 °C), high salinity (28–32‰) waters having little seasonal variation. This habitat occurs on the outer coast of the west coast of North America (Washington and British Columbia). In contrast, the short-stipe form occurs in areas having moderate seasonal variation of temperature (6–17 °C) and salinity (24–32‰). In spring, an increase in temperature and a drop in salinity due to freshwater runoff cause death of the plants.

Other selection experiments have been performed on bacteria. Such organisms are convenient for studies on selection because many individuals can be grown in a small area, and there is very rapid generation turnover. In these experiments, bacteria were subjected to a medium with a high concentration of antibiotics. Such an environment would select for individuals tolerant to antibiotics and select against those intolerant to antibiotics. The resulting strains of bacteria were more resistant to antibiotics than normal strains. The same phenomenon is seen in many disease-causing microorganisms in which strains have evolved that are resistant to antibiotics. Among insects, strains have also evolved that are resistant to pesticides, such as DDT. Genetic studies of both bacteria and insects demonstrate that resistance to antibiotics and pesticides is inherited.

Populations of one species face different selection pressures in different environments, which result in unlike changes in their genetic makeup. This leads to an increase in the number of genetic differences between populations. With continuing selection, genetic dissimilarity between populations increases. In the early stages of the development of differences between populations, the populations will not be distinct, since intermediates between all types will exist. These intermediates may reflect intermediate environments, crossing between different populations, or the presence of some genetic similarities in the populations. Eventually, populations become morphologically and physiologically distinct, and intermediates between the populations disappear. Populations such as these are considered to be separate species. As well, different species are usually reproductively isolated (individuals of different species cannot cross and produce viable offspring). In most instances, reproductive isolation seems to be the result of an accumulation of genetic differences in physically isolated populations. There are many different kinds of isolating mechanisms. A brief outline of them is presented in Table 1-1.

Evolution is a gradual and continuing process (but see polyploidy, below). Therefore, it is possible to find populations of plants that have not diverged sufficiently to be called separate species, yet have accumulated enough genetic differences so that it is difficult to consider them as one species. Examples of this occur in the flowering plants *Stipa* (needle grass) and *Clarkia*, and in the green algae *Chara* (stonewort) and the microscopic *Gonium*. In *Stipa*, there are species that are morphologically and physiologically somewhat distinct, yet they are interfertile and, in some instances, connected by series of intermediates. In *Clarkia*, there are species that are similar morphologically and physiologically, yet they are completely intersterile. In both *Gonium* and *Chara*, there are species that are reproductively isolated in nature but can produce some viable zygotes when grown together under artificial conditions. Problematic species such as these are the focal point for many taxonomic studies.

Polyploids

The gradual nature of evolution is a constant background against which there can be periodic episodes of rapid evolution. **Polyploidy** is such an episode. Polyploidy, or chromosome doubling, is a means of evolution that is common in plants. Unlike gradual evolution, as discussed in the preceding section, evolution through polyploidy is abrupt and can result in almost instantaneous evolution of new species. There are different types of polyploids: two basic types, determined by parentage of the polyploid, and some intermediate types.

Allopolyploidy is a type of polyploidy in which two different processes are involved. The first step is the formation of a sterile hybrid; the second is doubling of chromosomes in that hybrid. Sterile hybrids are often a result of crossing between different species of plants, and hybrid sterility is due to the inability of the chromosomes to pair during meiosis (i.e., homologous chromosomes are lacking). The lack of chromosome pairing during meiosis prevents normal meiosis and results in **inviable gametes**. However, in sterile hybrids it is possible that unreduced gametes will form. Should such gametes fuse, they produce a zygote in which chromosomes are doubled as compared with the parent. Should this zygote develop into a plant, each chromosome will have an identical partner and homologous chromosomes will exist. Such a plant would be a fertile offspring from a sterile hybrid. It is also possible for somatic chromosome doubling to occur in a sterile hybrid, which would make it fertile. Either way, the newly fertile plant is capable of reproducing its "type," and a new species may be created.

Autopolyploidy, the other basic type of polyploidy, results when the chromosomes of one fertile plant double to give a plant with twice the original chromosome number. This type of polyploidy tends to produce less fertile individuals than allopolyploidy, since each chromosome will have three (instead of one) homologous partners. During meiosis, more than two chromosomes can pair, so that instead of homologous pairs on a metaphase plate, there can be groups of three or four chromosomes. Meiosis under such conditions often results in at least some genetic imbalance in the gametes—a cause of gamete inviability.

The intermediates between auto- and allopolyploids are a result of chromosome doubling in hybrid plants that have some pairing of homologous chromosomes but not enough to be completely fertile. There are two ways in which such hybrids can be created. One is crossing between two slightly different individuals of the same species, or between two very closely related species (A^1A^1AA; Fig. 1-2). Polyploids formed from such hybrids are called *segmental allopolyploids*. The other

Table 1-1 Various Mechanisms That May Prevent Hybridization between Different Plants

I. Mechanisms acting before fertilization
   A. Ecological—plants adapted to different habitats
   B. Seasonal—plants flower at different times of year
   C. Mechanical—pollination prevented by structural differences in the flowers of different species
   D. Gametic—gametes of different plants incompatible
   E. Physiological—inability of pollen to germinate on the stigma, or pollen tube to penetrate the style

II. Mechanisms acting after fertilization
   A. Hybrid inviability or weakness
   B. Hybrid sterility
   C. F_2 breakdown—F_1's viable and fertile but F_2's weak or sterile

means whereby intermediate types of polyploids arise is by crossing between an allopolyploid species and one of its parental species, followed by chromosome doubling. Interrelationships between the different kinds of polyploid species are shown in Figure 1-2. This figure is based on a study of *Bromus* by Stebbins (1950).

Hybridization and subsequent formation of polyploids is a very common phenomenon in plants. In the green alga *Nitella*, there are chromosome numbers of $n = 6, 9, 12, 18$, and 24, indicative of evolution through polyploidy. One species of *Nitella*, *N. furcata*, with $n = 12$, has been interpreted as being an autopolyploid. Polyploid series have also been demonstrated for *Chara*, with numbers of $n = 14, 28$, and 42, and for *Tolypella*, with chromosome numbers of $n = 8, 11$, and 33 reported. In the grass genus *Bromus* (brome grass) there has been extensive polyploidy. As a result of hybridization and polyploid formation, complex aggregations of species can develop (Fig. 1-3).

Our concern here has been a brief summary of the process of biological evolution. Many examples are known from vascular plants, especially in seed plants. More detailed information for many organisms is present in the works of G. L. Stebbins (1950, 1966), T. Dobzhansky (1951), and T. Dobzhansky et al. (1977).

Recent and current studies are elucidating similar processes in nonvascular plants. Evolution of species in bryophytes probably results from phenomena similar to those described; however, as very

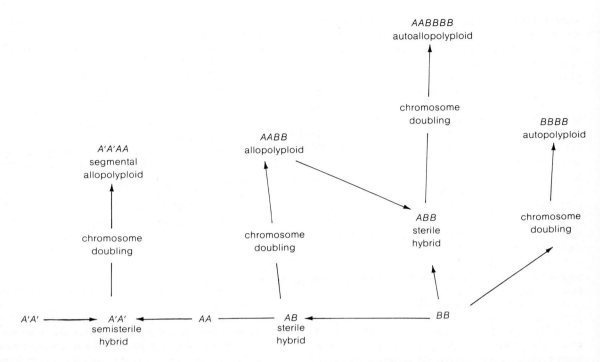

Figure 1-2 Kinds of polyploids and their origins.

little research has been done with these plants, there is a tremendous potential for investigation. Many fungi, algae, and bacteria do not undergo sexual reproduction as a regular feature. Thus, although the general mode of evolution (unequal survival of variants) is similar, details of how this occurs may be different. These are discussed in following chapters.

To this point, we have considered the process of biological evolution; we now discuss the results of evolution—natural order.

Natural Order

The biological world is not a random aggregation of individuals but consists of *groups* of individuals that are distinct from each other, are readily recognizable, and can be ordered into a hierarchy (see p. 8). The existence of these groups of individuals is called *natural order,* and an understanding of this natural order is one of the primary goals of biology. In attempting to understand it, biologists erect classifications. Thus, this section begins with a discussion of classification.

General Comments on Classification

Before discussing classification, a number of terms should be defined to avoid any ambiguity. **Class** forms the base of several words. *Class* itself denotes a group of objects that are related to each other in some manner. This relationship is usually one of similarity. **Classify,** a cognate of *class,* means to arrange objects into classes. It is also used to denote the procedure involved in identification of a plant. In this instance, a plant is being placed in a previously defined group rather than arranged with others into new groups. **Classification,** another cognate of *class,* is used in two different senses. One is the act of arranging objects into classes; the other is the end result of this act.

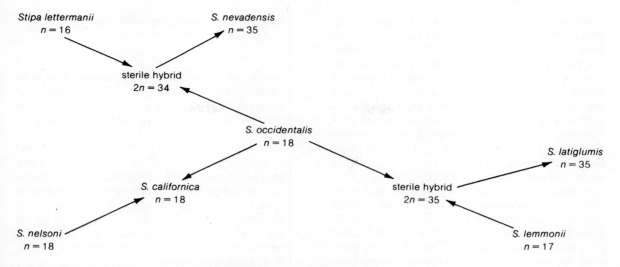

Figure 1-3 Formation of species, sterile hybrids, and polyploid species in *Stipa* (needle grass).

Although classification is commonly associated with scientific endeavors, libraries, or telephone companies, it is, in fact, one of the most general aspects of our lives. Classification is the basis for language. If we had a unique word for every individual object we perceive, language would degenerate into chaos. For example, the word *tree* applies to all objects that fit the concept of tree. If each tree in a forest had a different term, communication about that forest would be impossible.

Another term that must be understood in order to discuss classification is the general term **taxon** (pl. **taxa**), a group of related organisms. The basis for relationship can be anything; aquatic plants represent a taxon based on habitat; *Cannabis* (marijuana or hemp) represents a taxon based on many chemical, morphological, and cytological similarities.

Biological Classification

A scientist concerned with classification is called a *taxonomist* or *systematist*, and the actual practice of classifying is usually called *taxonomy* or *systematics*.

Why Classify Organisms? The primary reason for classifying is to understand natural order. However, a classification produced as an attempt to understand natural order also has many other uses. One very valuable use of classification is that names are attached to groups of plants that are recognized. These names become a means of communication about plants and about meaningful facts and theories concerning them. Thus, one can speak of "plastids of *Porphyra perforata*," a phrase that has far greater precision than the phrase "plastids of an aquatic plant." Also, when names are attached to objects of research, they add the element of repeatability. To duplicate a piece of research, the same kind of organism must be studied.

What Is Classification? In a general sense, a classification is a system of information, in that the features shared by members of a taxon convey information about the taxon. Because of the relationship between information and classification, there are different kinds of classifications depending on the amount of information they contain. A classification containing little information is very different from one containing much. For example, a classification with two taxa, one aquatic and one terrestrial, is one of very low information content. All that is known about individuals in such taxa is where they grow. There is no information on form, biochemistry, or evolutionary history. On the other hand, a classification based on many features is one of higher information content. Much is known about the individuals in such a class. Take, for example, the angiosperm family Solanaceae: many Solanaceae contain chemicals called *alkaloids*. If a biologist wants to find new plants with alkaloids, the Solanaceae would be an appropriate place to look. In studying natural order, systematists want to produce the classification with maximum information content. It has been demonstrated that classifications of maximum information content are those in which the taxa recognized are the result of evolution. Such a classification is presented as a hierarchy—a classification in which closely related classes are grouped together in higher taxa that are mutually exclusive. A classification of this kind is presented in Figure 1-4. In taxonomy, this hierarchy is referred to as a **Linnaean hierarchy** after Carolus Linnaeus, founder of biological taxonomy.

There are some important points that must be made here regarding the nature of taxa. Taxa of any one category (e.g., species or genera) will not necessarily be of the same size, exhibit the same amount of variation, or show the same degree of distinctness from other taxa of the same category. The differences in numbers of individuals in taxa are readily obvious. For example, the number of individuals of the kelp *Macrocystis angustifolia* (bladder kelp), which occurs chiefly in southern Australia and South Africa, is obviously much lower than that of *Macrocystis pyrifera* (giant kelp), which occurs mainly along the Pacific Coast between lat. 35°N–23°N and lat. 14°S–55° and is circumpolar in the Southern Hemisphere. The differences in absolute amount of variation in taxa are likewise apparent. *Macrocystis angustifolia* will express far more absolute variability than will a species of a filamentous alga. As a large kelp, it has far more cell types than a filamentous alga. The amount of relative variation also differs in taxa of the same rank. Certain species are notorious for their inherent variability. An excellent example is the green alga *Enteromorpha intestinalis*, which varies from broad tubular plants to narrow, very slender, elongate forms. Other species are not so variable, an example being the brown alga *Postelsia palmaeformis* (sea palm). The amount of relative variation can also differ in higher taxa. If one genus contains 200 species and another has only 2, the relative amount of variation in the genera will be different. This aspect of relative variation can be carried to fairly high levels. Witness the gymnosperm order Ginkgoales, with the single species, *Ginkgo biloba*.

Kinds of Biological Classification Generally speaking, there are three different kinds of biological classification: special purpose, key, and general purpose. These are considered below.

Special-Purpose Classification. Classifications based on similarities only in certain features of the organisms are known as **special-purpose classifications.** They are low-information classifications. Examples of such classifications would be those based on the physiognomy or growth habit of the plant, on the soil in which the plants grow, on leaf type, or on time of flowering. There are thousands of special-purpose classifications.

Key Classification. A classification that is used to identify objects is a **key classification.** Like the previous classification, it has low information content. This kind of classification, usually referred to as a **key**, aids in identification by erecting a series of mutually exclusive classes. In doing so, a key eliminates choices until the object in question is identified.

An example of a key classification is a key to some motile, green algae (see Fig. 7-6A):

1. Cells arranged as a slightly curved plate .... *Gonium*
1. Cells arranged in a sphere; size variable 2
    2. Cells tightly packed together, each assuming triangular shape ...... *Pandorina*
    2. Cells some distance apart, each round in form 3
3. Colony large, over 500 μ; composed of many cells; usually a few very large cells present .... *Volvox*
3. Colony less than 500 μ (average 100–200 μ); composed of (32–) 64–128 (–256) cells; cells the same size throughout .... *Eudorina*

General-Purpose Classification. The kind of classification usually associated with biology is a **general-purpose classification.** Such a classification is (1) of interest to the greatest number of biologists,

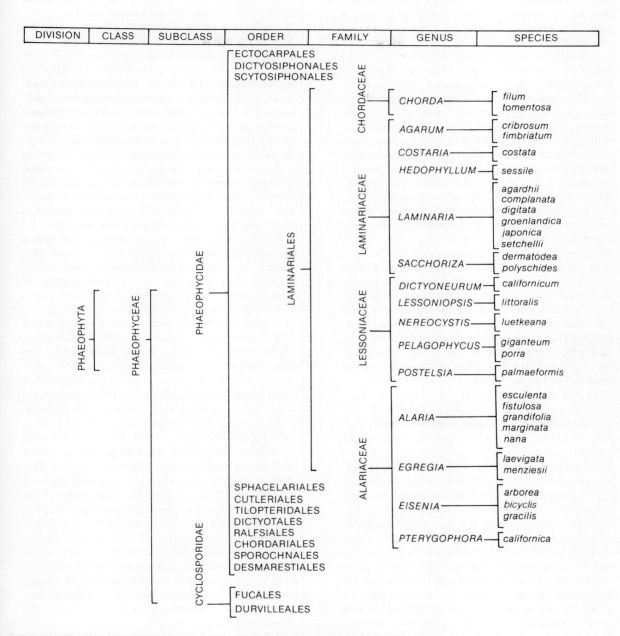

Figure 1-4 Representation of a classification of brown algae (Phaeophyta), inclusive of all orders, but including subordinal categories (families, genera, and species) only in one order, the Laminariales.

(2) concerned with names (**nomenclature**) and Linnaean hierarchy, (3) the classification that taxonomists try to produce, (4) the classification with the highest information content, and (5) consists of taxa that are evolutionary groups. We refer to such a general-purpose classification as a **natural classification**.

A natural classification is, at least theoretically, based on information derived from all aspects or characteristics of the organism: its biochemistry, parasites, genetics, physiology, cytology, life history, anatomy, development, embryology, morphology, ecology, reproductive features, and distribution. However, it must be pointed out that a natural classification is rarely, if ever, based on data derived from all these fields of study. The primary reason for this is that there is simply not enough time to gather all the information available.

Because all the available information is not actually used in erecting a classification, there is an experimental aspect that is an integral part of classifications. A taxon (group of related plants) is an hypothesis that is based on the evidence used. This hypothesis (taxon) is tested by gathering new information. For example, some taxa (hypotheses) have been erected in the Poaceae (grass family) on the basis of cell type in the leaf epidermis. Data derived from other fields of study, such as leaf anatomy, physiology, root anatomy, chromosome number, and floral morphology have also indicated the existence of the same taxa.

The Making of a Natural Classification

What we are concerned with here are the mechanics of producing a natural classification. In discussing the erection of a classification, we will consider two different methodologies: intuitive and numerically assisted. A systematist who erects a classification using an intuitive methodology will study museum specimens, specimens growing in nature, and the appropriate literature. Characteristics may or may not be recorded. Then, relying on the impressions gained from the specimens studied, a classification will be erected. This method is that most commonly used in systematic studies. Although intuitive classification may appear to be highly subjective and perhaps even suspect as a scientific methodology, it has produced remarkably stable classifications. Different systematists who use this method on a common group of plants often arrive at very similar classifications. We suspect that the efficacy of this method is a demonstration of the resolving power of the human mind.

When using intuitive methodology, a systematist often follows one of two approaches. Some try to erect a classification based on overall similarities between individuals in a taxon. Others erect a classification that stresses certain types of similarities—shared derived traits. Individuals in a taxon share the same derived traits. A derived trait is a change in a trait from the condition seen in the closest relative to the group being studied. The two approaches are often identified by name. Erecting a classification based on overall similarities is called **phenetics**; erecting a classification based on derived traits is called **phylogenetic systematics** or **cladistics**. Despite the different approaches, phenetics and cladistics can give identical classifications.

The second methodology used in erecting a classification, numerical assistance, is the mathematical formalization of an intuitive classification; it involves checking and modifying the position in each taxon by using a formula. This often leads to a more detailed understanding of relationships.

An initial step in mathematical formalization of a classification is to measure carefully and record characteristics of the organisms under study. The data derived from the organisms is called a *data matrix*, wherein the state of all characteristics has been recorded for all organisms. The data matrix is subjected to mathematical analysis. In phenetic classifications, the mathematical analysis used is called **cluster analysis**—a numerical tool designed to demonstrate the existence of groups of similar individuals amongst the organisms in the data matrix. The philosophy and mathematical methods involved in erecting a phenetic classification have been referred to as **numerical taxonomy**. In phylogenetic systematics, numerical methods, referred to here as *cladistic analysis,* have been developed by Farris (see Wiley 1981) to identify shared derived traits. The numerical methods associated with cladistic analysis are not as complicated as cluster analysis. Thus, computer assistance, although very handy, is not as necessary for cladistics as it is for cluster analysis.

Because of the mathematical precision associated with formalization of classification, it has been possible to compare phenetic and cladistic classi-

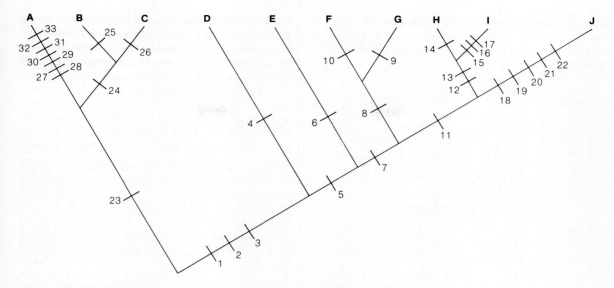

Figure 1-5 Cladogram of ten taxa. The ten taxa classified are called *terminal taxa*. Each vertical bar on the stem of the cladogram represents shared derived traits.

fications. According to these comparisons, cladistic classifications seem to be superior. Brooks (1981) has demonstrated that for any data matrix, cladistic analysis results in classifications with information content greater than or equal to those resulting from phenetic analysis. As well, the proponents of cladistics have demonstrated the compatibility between cladistics and the philosophy of science (see Wiley 1981). For these reasons, the remainder of this discussion is based on concepts taken from cladistics.

Figure 1-5 is a cladogram, which shows the results of cladistic analysis. Using this figure, we will briefly trace certain concepts of significance in classification.

Monophyly, Polyphyly, and Paraphyly The terms *monophyly, polyphyly,* and *paraphyly* are adjectives applied to taxa in a classification. A **monophyletic taxon** is one in which all members have the same derived traits and all share a single node. For example, in Figure 1-5, a taxon consisting of *H*, *I*, and *J* is monophyletic; all share the derived trait *11*. *F* and *G* likewise make up a monophyletic taxon having the unique derived trait *8*. Other examples of monophyletic taxa in Figure 1-5 are *FGHIJ*, *BC*, *ABC*, and *EFGHIJ*.

A **polyphyletic taxon** is one that includes individuals that do not share the same derived traits and a single node. For example, in Figure 1-5, a taxon consisting of *BC* and *D* would be polyphyletic. The derived trait for *BC* is *24*; for *D*, the derived traits are *1*, *2*, *3*, and *4*.

A **paraphyletic taxon** is one that does not include all individuals that share the same derived trait and a common node. For example, a taxon consisting only of *EF* and *G* is paraphyletic in that it does not include all individuals that have the derived trait *5*.

It is the goal of systematics to produce classifications that do not have paraphyletic and polyphyletic taxa, since such classifications contain less information than those with only monophyletic taxa. As a corollary of the lower information content in paraphyletic or polyphyletic taxa, knowledge of one member of such a taxon does not necessarily mean there is knowledge about other members of the same taxon.

There are other names applied to monophy-

Introduction 11

letic, paraphyletic, and polyphyletic taxa. Monophyletic taxa are sometimes called **natural,** polyphyletic and paraphyletic are called **unnatural** or **artificial.**

Homology and Analogy The terms **homology** and **analogy** refer to types of similarities between traits. Homologous similarities are those seen in shared derived traits. Analogous similarities are not shared derived traits; they are similarities between taxa that are not classed together. It is possible that a systematist may erect taxa based on analogous traits. These are a type of artificial taxon and, in the past, were rather common in plants. The old group Thallophyta, which combined together all fungi, bacteria, and algae, was one such taxon. The common analogous characteristic was a very simple plant body.

Primitive Versus Advanced The term **primitive** refers to a trait like that seen in the ancestor of a taxon; **advanced** refers to a trait different from that in the ancestor of a taxon. A derived trait would be advanced. The terms *primitive* and *advanced* are also used by some to refer to taxa; we prefer not to. It is confusing to see the same words in different contexts. As well, the different rate of evolution that different characteristics undergo means that any one extant plant can have both primitive and advanced traits. When referring to taxa, those that have fewer unique derived traits we call **generalized;** those having more unique derived traits we call **specialized.**

Sympleisiomorphy, Synapomorphy, and Homoplasy **Cladists** use special terms to refer to certain types of similarities. These terms have a distinct advantage in that they are less ambiguous than other terms.

Sympleisiomorphy refers to generalized traits in a group. **Synapomorphy** is a shared derived trait shared by all members of a taxon. **Homoplasy** is a convergent characteristic, a similarity that does not identify a natural or monophyletic taxon. Synapomorphies are those traits that confer the greatest information content on a classification; homoplasies give interesting insight into the biology of the organisms that share them, because they often represent adaptations to particular environments.

Anagenesis and Cladogenesis There are two more terms that should be understood: *anagenesis* and *cladogenesis*. **Anagenesis** refers to evolutionary change within a lineage. **Cladogenesis** refers to evolutionary splitting of a lineage to give two taxa where one existed before.

Interpretation of a Classification

We now discuss briefly the interpretation that can be put on taxa in a classification. The most common interpretation is that each monophyletic taxon in a classification is the result of evolution. As such, it has a unique history. Some biologists (see Wiley 1981) would argue that the existence of natural order that can be arranged in a hierarchy is sufficient proof of evolution.

Features Used in Classification

Characteristics are the source of information used in erecting a classification. There are certain criteria that characteristics should meet. First, they should show variation within the group being studied. Second, they should be readily measurable. The use of certain features when erecting a classification at one level does not mean that these same features will be useful when classifying organisms at a higher taxonomic level. The presence of chlorophyll, wood, tracheids, and phloem may well be worthwhile data to use when classifying pines at the level of class or division, but they are of little value when classifying pines at the species level.

In choosing characteristics, taxonomists depend on their knowledge of the organisms to know *what* features should be analyzed. Furthermore, it is an axiom of taxonomy that the quality of the characteristics is the sole determinant of the quality of the resulting classification. Neither the most brilliant intellect nor the most sophisticated computer program can do anything to compensate for poor information. Thus, the better taxonomists know the organisms, and the more they know about organisms in general, the better the classifications they produce will be.

Despite increased knowledge concerning the

biochemistry and cytology of organisms, biologists are divided in their opinions concerning the higher levels of classification, especially at the kingdom and divisional levels. The differences arise in part from the difficulty in judging the systematic weight given to the degree of similarity or difference among the different groups of organisms when assigning them to kingdom or division (and in some instances, to the level of class). These decisions are extremely arbitrary. The system adopted in this text—based on the best data available and their interpretation—recognizes five kingdoms, distinguishing all plants and plantlike organisms as follows: the Monera (the bacteria and bluegreen algae, or cyanobacteria), the Protista (the slime molds, and unicellular and colonial algal groups), the Fungi, and Plants (the predominantly multicellular algal groups, bryophytes, and vascular plants). Animals, which are not treated in this text, constitute the fifth kingdom of organisms.

What we have presented here in summary form is what seems to us the means whereby taxonomists work and think. We hasten to point out that such theoretical aspects of taxonomy have been the point of much debate, both in the literature and in various scientific meetings. Anyone who desires to follow such arguments is referred to the journal *Systematic Zoology*, or to the recent book by Wiley (1981).

References

Items marked with an asterisk (*) in the following list are of broad interest.

Anderson, E. 1949. *Introgressive hybridization.* John Wiley & Sons, New York.

Brooks, D. R. 1981. Classifications as languages of empirical comparative biology pp. 61–70, in Funk, V. A. and Brooks, D. R., *Advances in cladistics.* New York Botanical Garden, Bronx, New York.

*Clausen, J. 1951. *Stages in the evolution of plant species.* Cornell Univ. Press, Ithaca, N.Y.

Clausen, J., and Heisey, W. M. 1958. Experimental studies on the nature of species. IV. *Genetic structure of ecological races.* Carnegie Institution of Washington Publication 615.

Clausen, J.; Keck, D. D.; and Heisey, W. M. 1940. Experimental studies on the nature of species. I. *The effect of varied environment on western North American plants.* Carnegie Institution of Washington Publication 520.

——— 1945. Experimental studies on the nature of species. II. *Plant evolution through amphiploidy and autoploidy with examples from the Madiinae.* Carnegie Institution of Washington Publication 564.

——— 1948. Experimental studies on the nature of plant species. III. *Environmental responses of climatic races of Achillea.* Carnegie Institution of Washington Publication 581.

Darwin, C. 1872. *The origin of species.* 6th ed. Murray, London.

*Davis, P. H., and Heywood, V. H. 1963. *Principles of angiosperm taxonomy.* D. Van Nostrand Co., Princeton, N.J.

*Dobzhansky, T. 1951. *Genetics and the origin of species.* Columbia Univ. Press, New York. (First published in 1937.)

*Dobzhansky, T.; Ayala, F. J.; Stebbins, G. L.; and Valentine, J. W. 1977. *Evolution.* W. H. Freeman, San Francisco.

Druehl, L. D. 1967. Distribution of two species of *Laminaria* as related to some environmental factors. *J. Phycology* 3:103–8.

*Hennig, W. 1966. *Phylogenetic systematics.* Univ. of Illinois Press, Urbana, Ill.

Heslop-Harrison, J. 1953. *New concepts in flowering plant taxonomy.* Wm. Heinemann, London.

Johnson, L. A. S. 1968. Rainbow's end: The quest for an optimal taxonomy. *Proc. Linn. Soc. New South Wales* 93:8–45.

Lewis, H. 1953. The mechanism of evolution in the genus *Clarkia. Evolution* 7:1–20.

Mayr, E. 1963. *Animal species and evolution.* Harvard Univ. Press, Belknap Press, Cambridge, Mass.

*Simpson, G. G. 1961. *Principles of animal taxonomy.* Columbia Univ. Press, New York.

Sneath, P. H. A., and Sokal, R. R. 1973. *Principles and practices of numerical taxonomy.* W. H. Freeman, San Francisco.

*Stebbins, G. L. 1950. *Variation and evolution in plants.* Columbia Univ. Press, New York.

——— 1966. *Processes of organic evolution.* Prentice-Hall, Englewood Cliffs, N.J.

*Wiley, F. O. 1981. *Phylogenetics: The theory and practice of phylogenetic systematics.* John Wiley & Sons, New York.

2
MONERANS
(Monera—Bacteria and Bluegreen Algae)

The Monera comprise those organisms in which internal membranes, excluding the plasma membrane, are generally lacking. These are the **prokaryotes** (*pro* = before; *karyo* = nucleus). In the moneran cell the nuclear material is not surrounded by a nuclear envelope, and such organelles as plastids, mitochondria, golgi, and endoplasmic reticulum are absent.

The Monera can be subdivided into three groups based on such features as RNA (ribonucleic acid) composition, cell walls, cell membrane composition, and mode of nutrition. The groups are variously named and are the Schizomycophyta (or Eubacteria), the Archaebacteria, and the Cyanophyta (or Cyanobacteria).

The monerans are **autotrophic** (*auto* = self; *troph* = nourish) or **heterotrophic** (*hetero* = differ). The autotrophic monerans include some members of both the Schizomycophyta and most of the Cyanophyta. However, the majority of monerans are heterotrophic and thus dependent upon an external supply of organic compounds as their energy source. Some researchers group the Cyanophyta with the Schizomycophyta (and refer to them as the Cyanobacteria). In this text, we consider the Cyanophyta separately, primarily because their photosynthesis is like that of all photosynthetic protistans and plants (see p. 49). The Archaebacteria, with relatively few representatives, appear to be only remotely related to the other monerans and are considered by some as a separate kingdom. The three divisions are treated separately here, with comparisons among them made.

Division Schizomycophyta (Bacteria)

Bacteria are structurally simple organisms, but they are biochemically complex and display a greater variety of nutritional modes than any other group. Bacterial cells are small, and their large surface/volume ratio is accompanied by exceptionally high metabolic and growth rates. In terms of numbers of individuals, bacteria undoubtedly outnumber any other group of organisms in any particular environment. Thus, although they are inconspicuous, they have an enormous influence on the environment and its inhabitants.

Most bacteria are heterotrophs and obtain nutrients from other living or dead organisms. They can absorb small molecules directly from their surroundings and, in addition, secrete **exoenzymes** (*exo* = out) that can digest complex molecules such as proteins and polysaccharides. Bacteria are important in the preparation of certain foods (e.g., sauerkraut, yogurt) or in the destruction and spoilage of others. Bacterial diseases of man have resulted in countless deaths, yet some bacteria produce **antibiotics** (*anti* = against; *bios* = life) useful in the treatment of certain diseases caused by other bacteria and fungi. Many heterotrophic bacteria form associations with other organisms, such as **nitrogen-fixing** bacteria in root nodules and bacteria present in the digestive tract in **ruminants** (*rumina* = chew the cud). Autotrophic bacteria include both photosynthetic and chemosynthetic forms utilizing sunlight, or energy derived from oxidation-reduction reactions as their energy source. They may use only inorganic nutrients (such as CO_2, H_2S, H_2, H^+, NH_3, Fe^{2+}), or organic carbon and hydrogen donors (such as acetic acid).

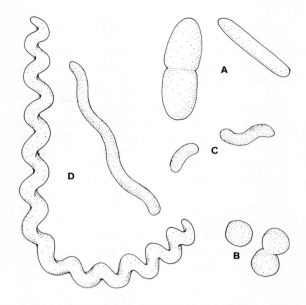

Figure 2-1 Bacterial form. **A,** bacillus; **B,** coccus; **C,** vibrio; **D,** spirillum.

Cell Form and Structure

Most bacteria are unicellular. The commonly encountered forms are spheres and straight or curved rods, although their cells can also be organized into long chains or filaments (Fig. 2-1). Cell shape is relatively constant in most species. Rod-shaped bacteria (Fig. 2-1A) are termed **bacilli** (*bacill* = little stick) and are approximately 1–5 μm long by 0.5–10 μm in diameter. The round bacteria (Fig. 2-1B), or **cocci** (*cocc* = berry), are 0.5–1.0 μm in diameter. The curved rods vary in shape: the long curved rods (Fig. 2-1D) are referred to as **spirilla** (*spirill* = curve, coil), and the very short curved rods (Fig. 2-1C) are termed **vibrios** (*vibr* = whip, vibrate).

The outermost cell layer, when sharply defined, is the **capsule** (*capsul* = little box; Fig. 2-2). The capsule material can be lost in the surrounding medium and is sometimes called a *slime layer*. Chemically, the capsule can be a single **polysaccharide** or **polypeptide,** and its presence can be influenced by environmental factors. Just under the capsule is a cell wall (Fig. 2-2) that varies in both chemistry and structure. The nature of the wall can be demonstrated by a specific stain, known as *Gram's stain*. Most bacteria either retain the stain (gram-positive) or do not (gram-negative). The process involves staining with crystal violet and then an iodine-potassium iodide solution. Cells that retain the color after being treated with a solvent such as ethanol are termed *gram-positive*. The main cell wall material is **peptidoglycan,** which is unique to monerans and is composed of sugar-amines and amino acids. Peptidoglycans constitute from as little as 5–10% of the wall to as much as 80–90%. The cell wall serves to protect the cell from osmotic damage (i.e., rupture). Differences in susceptibility to antibiotics are sometimes attributed to structural differences in the cell wall.

The plasma membrane within the cell wall (Fig. 2-2) is 6–8 nm thick and similar in structure and function to that of eukaryotic organisms. The plasma membrane can consist of a simple peripheral layer or can have extensions protruding into the cytoplasm and forming structures called **mesosomes** (*mes* = middle; *some* = body; Figs. 2-2 and 2-3A). Mesosomes are involved in the formation of new cell walls. In photosynthetic species, the pigments are associated with the plasma membrane and its infoldings, as are the enzymes functioning in cellular respiration. The bacterial cytoplasm is relatively homogeneous and is devoid of membranous organelles such as mitochondria, endoplasmic reticulum, golgi, and (generally) vacuoles. Ribosomes are present and are about 50% smaller than the cytoplasmic ribosomes in eukaryotes, but are like those in the mitochondria and chloroplasts of eukaryotes. They have a diameter of 10 nm and are called 70-*S* ribosomes. Bacterial cells each contain one or more nuclear bodies, or **nucleoids** (*nucle* = nucleus; *oid* = like); (Figs. 2-2 and 2-3A), which lack the nuclear envelope, nucleoplasm, nucleolus, and chromosomal organization of eukaryotic nuclei. The nucleoid consists of a single DNA molecule up to 1.4 μm long and in the shape of a ring. It is attached to the plasma membrane and sometimes to the mesosome. Histones, the basic proteins associated with eukaryotic chromosomes, are not present in bacteria. In addition to the nucleoid, bacterial cells often contain independent, self-replicating DNA rings termed **plasmids**.

Many bacteria exhibit a swimming motion that is dependent upon the action of one or more **flagella** (sing. **flagellum;** *flagell* = whip), which can occur singly or in tufts (Figs. 2-2 and 2-3B). A bacterial flagellum is a single strand less than 20 nm wide that consists of several protein fibers. Because of their small diameter, flagella can be seen by light microscopy only when specially stained. The electron micrographs show that the base of each flagellum is slightly bent and has a basal apparatus just beneath the plasma membrane. Many motile and nonmotile bacteria also have filamentous appendages called **pili** (sing. **pilus;** *pil* = hair; Figs. 2-2 and 2-3B). Pili are shorter and narrower than flagella and are often restricted to the poles of the cells. They are not involved in motility, but they are probably concerned with the adhesion of cells.

Reproduction

Reproduction in most bacteria is by **binary fission** (Fig. 2-4), a process in which the cell and its contents are equally divided into two. Up to 20–30 min before actual cleavage of the cell, the nucleoid divides, possibly through attachment to the mesosome (Fig. 2-4B). More than two nucleoids can be present at the time of cell division. As the meso-

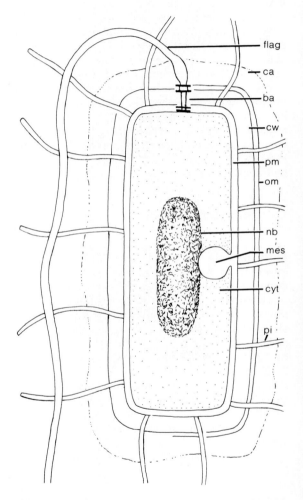

Figure 2-2 General bacterial cell structure, showing (from outside of cell inward) outer capsule (*ca*), outer membrane (*om*), cell wall (*cw*), plasma membrane (*pm*) with mesosome (*mes*) protruding into cytoplasm (*cyt*) nucleoid (*nb*): note flagellum (*flag*) with its basal apparatus (*ba*) at plasma membrane; pili (*pi*).

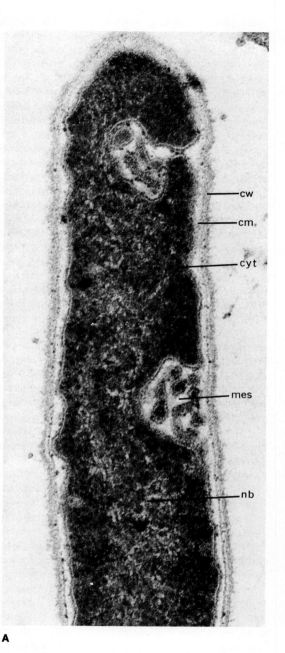
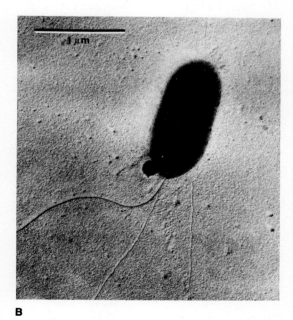

Figure 2-3 Bacterial structure, electron microscope. **A,** longitudinal section of cell, showing continuity of mesosome membrane (*mes*) and plasma membrane (*cm*) extending into cytoplasm (*cyt*): note cell wall (*cw*), nucleoid (*nb*), ×30,000. **B,** metal-shadowed cell, showing single flagellum at pole and bipolar pili of *Pseudomonas*. (**A,** courtesy of T. Bisalputra; **B,** courtesy of R. L. Weiss and H. D. Raj, with permission of *Australian Journal of Experimental Biology and Medical Science.*)

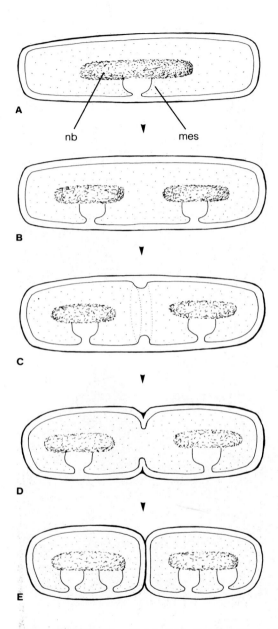

Figure 2-4 Diagram of bacterial fission. **A,** before cell division occurs, a mesosome (*mes*) attached to the nucleoid (*nb*) splits. **B,** the two parts separate with attached nucleoids. **C,** a thickened band develops on inner surface of the cell wall. **D,** cell wall grows centripetally to complete division. **E,** outer wall layer splits before transverse wall formation is completed.

some splits, two halves—each with an attached DNA strand—slowly move apart (Fig. 2-4C). Following cell division by constriction of the plasma membrane (Fig. 2-4D) and the formation of new wall material, the young cells increase in size (Fig. 2-4E). Under ideal conditions, fission occurs in many bacteria as frequently as every 20–30 min. Theoretically, this rate of reproduction could produce an immense number of bacterial cells within a few hours. Fortunately, various natural factors, such as nutrient limitations, accumulation of metabolic products, and competition prevent such increases.

Another type of vegetative reproduction involves the formation of resistant **endospores** (*endo* = within; *spor* = seed). These are thick-walled structures produced internally (Fig. 2-5) and they are extremely resistant to heat, chemicals, and desiccation. In some species, endospores can withstand boiling for more than two hours, as well as the action of chemicals that rapidly kill vegetative cells. Some can also retain their viability for more than 50 years. Endospore germination gives rise to a single cell, and thus it does not generally lead to an increase in cell number. Rather, an endospore is a mechanism for surviving unfavorable environmental conditions. **Cysts** are another type of resistant cell, but are formed when an entire vegetative cell (including wall) becomes surrounded by a thick outer wall.

Sexual reproduction, as known in eukaryotic organisms, is unknown in bacteria, since there is no fusion of gametes or nuclei, or meiosis. However, there can be genetic recombination, which involves unilateral transfer of genetic material (DNA) from one cell to another. One type of genetic recombination occurs by bacterial **conjugation** (Fig. 2-6), during which there is transfer of some genetic material through a temporary fusion bridge. After conjugation, new cells can exhibit combinations of the characteristics of both parental cells. Another type of recombination is **transformation,** in which soluble DNA from one cell is taken up by a second cell. Generally, the amount of DNA participating is no more than 5% of the total genetic material. The number of cells permanently transformed for a given genetic marker is less than 1% of the population. A third type of recombination involves the transfer of DNA fragments from one cell to another by bacterial viruses, or **bacteriophages,** also termed **phages** (*phag* = eat; Fig. 2-7). DNA introduced into cells by any of these processes can result in heritable changes in the offspring. The recipient cell is "diploid" for that portion of the DNA received, even if this is only a small fraction of the total genetic material. This partial diploid condition is ephemeral.

Classification

Morphological and nutritional differences are used to characterize the orders and families of bacteria. Since their cell morphology is relatively simple, physiological and biochemical characteristics are used for generic and specific distinctions. Generally, bacteria are stained with the Gram's stain to determine the nature of the cell wall (see p. 15).

Bacteria grown in the laboratory under controlled conditions (termed **culturing**) can be characterized by some of their nutritional and environmental requirements. Most heterotrophic species can be grown on relatively simple, chemically defined media and require only a single organic compound as an energy and carbon source, mineral salts, and water. Other heterotrophs are more exacting, requiring one or more amino acids, vitamins, or other substances. Additionally, the ability to utilize a compound and the products of its use are both of value in characterizing bacterial species. An example is glucose, of which the end products may be various acids (e.g., lactic, acetic, or succinic). Some species differ in their oxygen requirements; that is, whether they use oxygen, termed **aerobic** (*aer* = air), or not, termed **anaerobic** (*an* = without). If the requirement is absolute, it is termed **obligate** (*oblig* = obliged); if not, it is termed **facultative** (*facult* = capable).

The specificity of antigen-antibody reaction, which is often macroscopically visible, is a useful tool for classification. When certain foreign substances, called **antigens** (*anti* = against; *gen* = ori-

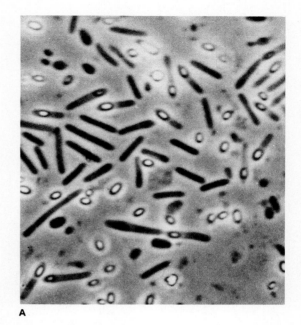

A

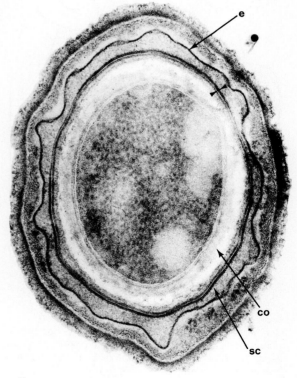

B

Figure 2-5 Bacterial endospores, *Bacillus*. **A,** a heavily sporulating culture showing numerous mature and released endospores, ×2000. **B,** electron micrograph of thin section of mature endospore showing spore walls (*e, sc, co*) and inner protoplast, ×95,000. (**B,** courtesy of P. D. Walker, with permission of *Journal of Applied Bacteriology.*)

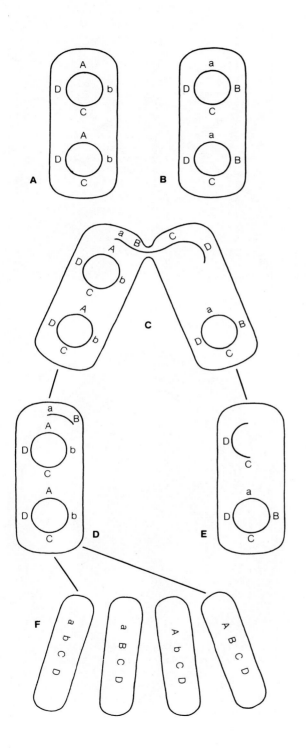

gin) are introduced into tissues of living vertebrates, the animals respond by producing protective substances called **antibodies.** Each inducing substance, which may be a polysaccharide, polypeptide, or protein, causes the formation of a specific antibody. The antibodies are blood **serum** (*serum* = watery parts) proteins that react or unite with the inducing or **homologous** (*homo* = same; *log* = study) antigens, causing an **agglutination** (*agglutin* = glued together) reaction. Serum containing antibodies is tested on unidentified bacteria. A positive reaction occurs only if the homologous antigen is present. This technique can be combined with stains incorporating an appropriate fluorescent dye. When an antibody so **labeled** is applied to a bacterial mixture, it becomes attached to cells carrying specific antigens for the labeled antibody. These cells then fluoresce when viewed by fluorescence microscopy.

Cells of some species adhere to one another, forming aggregates of relatively constant shape when grown on a solid medium. The medium contains the necessary nutrients and is solidified by 1–2% **agar,** which is a compound produced by some red algae (Chapter 9). When standard conditions are used (nutrients, light, temperature, etc.), the colony characteristics are constant.

One of the most fundamental biochemical characteristics used for classification involves the genetic material, the DNA. The analyses presently used are: (1) base-composition, or relative percentage of complementary base pairs guanine + cytosine (% G + C); and (2) degree of genetic homology between DNA strands of different species. Both analyses require extraction and purification of DNA from the organism being examined.

Figure 2-6 Bacterial conjugation. **A, B,** recipient and donor cells, each with two circular nucleoids. **C,** transfer of linear genetic material (letters a–D) from donor to recipient via temporary fusion bridge. **D, E,** donor and recipient cells after conjugation. **F,** possible types of descendants of recipient cell **D** after conjugation.

Orders of Bacteria

The bacteria are considered here as a single class, the Schizomycetes, comprised of several orders, with emphasis placed on the Eubacteriales, Pseudomonadales, and Actinomycetales because of their economic importance and the intensive studies conducted on them. Other orders are shown in Table 2-1 with some of the features that characterize them.

Order Eubacteriales (True Bacteria) Eubacteria are common in soil, water, sewage, decaying plant and animal remains, food, and the atmosphere. Many are **symbiotic** (*sym* = together; *bio* = life) in higher plants and animals. All have simple coccoid or bacilliform cells with rigid cell walls. Both gram-positive and gram-negative species are present. If the organism is motile, flagella cover the surface of the cell.

One of the most common bacteria is *Escherichia coli* (Fig. 2-8A), which normally occurs in the human intestinal tract. Usually this is a harmless bacterium, but its presence is commonly used as an indication of contaminated or polluted water. *E. coli* is commonly used as a biological "tool" because of its rapid growth and the ease with which it can be cultured. Other eubacteria growing in milk and foods can lead to spoilage or preservation of the food, depending upon the bacteria. Several species of *Streptococcus*, *Lactobacillus*, and *Leuconostoc* are regularly present in milk. They are important in the manufacture of fermented milk products, such as cottage cheese, various other cheeses, yogurt, butter, and buttermilk.

Some eubacteria are **parasitic** (*parasit* = near food), causing diseases such as anthrax (*Bacillus anthracis*), diphtheria (*Corynebacterium diphtheriae*), lobar pneumonia (*Streptococcus pneumoniae*), and typhoid fever (*Salmonella typhi*). In addition, food poisoning is caused by *Clostridium botulinum* (botulism) and *Salmonella enteritidis*. Still other examples of bacterial diseases include whooping cough, tetanus, gangrene, gonorrhea, and meningitis.

Some eubacteria (and bluegreen algae) fix atmospheric N_2 into organic nitrogen. Species of *Azotobacter* (Fig. 2-8B), *Bacillus* (Fig. 2-5), and *Clostridium* can fix nitrogen as free-living forms. In contrast, *Rhizobium* (Fig. 2-8C), lives symbiotically in the root nodules of legumes (pea family) and alders, and carries out nitrogen fixation in cooperation

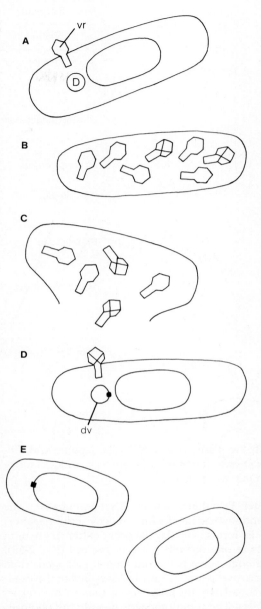

Figure 2-7 Bacterial recombination mediated by bacterial virus, or "phage." **A**, virus (*vr*) injects DNA (*D*) into bacterial cell. **B**, numerous new phage particles produced. **C**, phage particles released. **D**, defective particle (*dv*) injects DNA strand carrying small fragment of bacterial DNA into new bacterial cell. **E**, fragment transmitted by the phage is attached to bacterial chromosome (nucleoid).

Table 2-1 Characteristics of Selected Orders of Schizomycophyta, Class Schizomycetes (Bacteria)

Order	Motility	Form	Other Features
Spirochaetales	polar flagella, wound around body	spirilla	walls flexible, narrow
Chlamydobacteriales	polar flagella, when motile	bacilli, variable	ensheathed, sheaths containing Fe or Mn
Myxobacteriales	gliding, no flagella	bacilli	wall flexible
Beggiatoales	gliding, flexing, no flagella	filamentous	rigid walls; some chemoautotrophic H_2S oxidizing
Rickettsiales	nonmotile	variable (bacilli, cocci)	extremely small; obligate intracellular parasites
Mycoplasmatales	nonmotile	variable (cocci, filaments, irregular)	wall-less protoplasts

Note: See text for other orders.

with the plants (Fig. 2-8D). The legume and *Rhizobium* association fixes as much as 200 kg of nitrogen per acre per year.

Order Pseudomonadales (Pseudomonads) The Pseudomonadales include mostly gram-negative, motile forms. The flagella occur either singly or in tufts at one or both poles of the cell (Fig. 2-3B). Although some plant and animal **pathogens** (*path* = disease; *gen* = origin) are classified in the Pseudomonadales, these are fewer than in the Eubacteriales. The pseudomonads include the photoautotrophic purple sulfur bacteria, purple nonsulfur bacteria, and green sulfur bacteria. These are differentiated by the nature of their photosynthetic pigments and processes. All contain bacteriochlorophyll *a*, but purple bacteria (both sulfur and nonsulfur) have bacteriochlorophyll *b* and masking carotenoids. The green sulfur bacteria contain bacteriochlorophyll *a* and *c*, *d*, or *e*, as well as carotenoids, which do not mask the green color. The bacterial chlorophylls are similar in chemical structure to those in higher plants and algae. In addition, the pigments are associated with lamellar invaginations (*lamell* = plate) of the plasma membrane, often as flattened peripheral vesicles. In the sulfur bacteria, granules often appear as refractile granules in the cells (Fig. 2-9).

Several pseudomonads, such as *Nitrosomonas* and *Nitrobacter*, are involved in the conversion of ammonium nitrogen to nitrite or nitrate. In addition, the reduction of nitrate to nitrite, and nitrite to N_2 (known as *denitrification*) is carried out by species of *Achromobacter*, *Pseudomonas*, and *Thiobacillus*. Still other chemoautotrophic pseudomonads obtain energy through oxidation of hydrogen, carbon monoxide, methane, or sulfur. Some *Thiobacillus* species are capable of growing either auto-

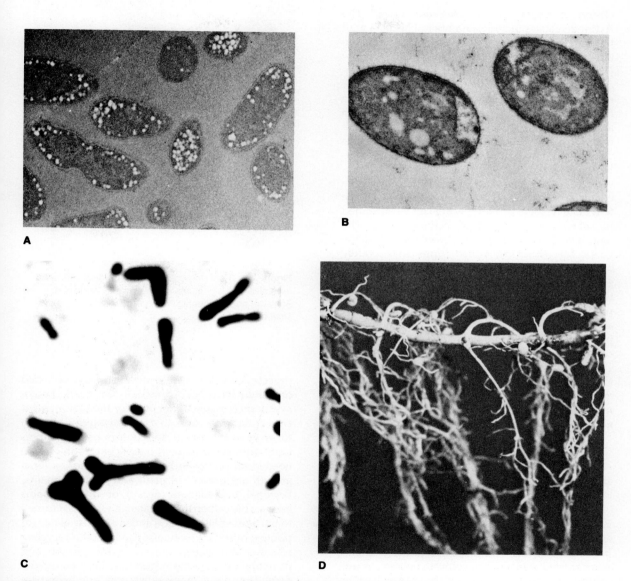

Figure 2-8 Eubacteriales. **A,** *Escherichia coli* with storage granules, ×14,500. **B,** *Azotobacter,* ×24,500. **C,** *Rhizobium* showing variation in forms, ×2000. **D,** root nodules of common clover containing *Rhizobium,* ×1.5. (**A,** courtesy of J. M. Shively, reproduced, with permission, from *Annual Review of Microbiology* Volume 28. © 1974 by Annual Reviews Inc.; **B,** courtesy of M. D. Scolofsky and O. Wyss, with permission of *Journal of Bacteriology.*)

trophically or heterotrophically and can oxidize H_2S to sulfur or sulfuric acid, thereby increasing the acidity of the surrounding environment.

Order Actinomycetales (Actinomycetes) The Actinomycetales include a number of bacteria with **hyphae** (*sing.*, **hypha**; *hyp* = web, weaving) similar to those of some fungi (Chapter 4). The narrow (1.5 μm diameter) hyphae can be short, infrequently branched units that fragment into rods, or extensive, nonfragmenting units that produce a funguslike **mycelium** (*myc* = fungus) (Figs. 2-10B and 2-11A, B). Classification within the Actinomycetales is based on wall chemistry, mycelium development, and reproductive features. All members are gram-positive and abundant in soil, freshwater, seawater, decaying material, and dung. The characteristic odor of freshly plowed soil and the foul taste of some drinking water can be attributed to substances produced by actinomycetes. The morphologically simple actinomycetes belong to the genus *Mycobacterium*, which forms short, rarely branched filaments that fragment into rod-shaped segments (Fig. 2-10A). A more extensive mycelium is produced by *Actinomyces*, whereas *Streptomyces* produces extensive, nonfragmenting hyphae that are funguslike (Fig. 2-11A, B). Special aerial branches produce chains of unicellular reproductive units, **conidia** (*sing.*, **conidium**; *conid* = dust; Fig. 2-11C). The branches can be straight or coiled, and the conidia of most species are smooth. *Streptomyces* is the major source of antibiotics such as streptomycin, nystatin, amphotericin B, chloramphenicol, aureomycin, and tetracycline.

Bacterial Ecology

Bacteria are everywhere! They occur in water, soil, air, living and dead plants, animals, fungi, and other bacteria. Many bacteria cause diseases in humans, domesticated animals, and cultivated plants, as well as destruction of stored foods. However, bacteria are indispensible in nature, and most organisms are dependent on the activities of these **microbes** (*micro* = small; *bio* = life).

Photosynthesis is sometimes called the most important single process on earth. Without this

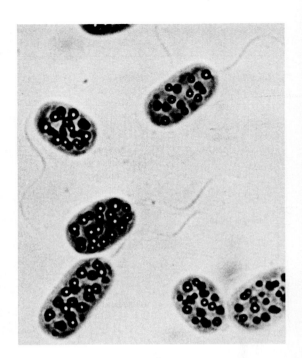

Figure 2-9 Pseudomonadales. *Chromatium* showing refractive sulfur granules and flagella. (Courtesy of H. G. Schlegel and N. Pfennig, with permission of *Archiv für Mikrobiologie*.)

process, evolution would either not have occurred or would have been drastically different. Photosynthesis is essential not only for the basic conversion of light energy into chemical energy, but also for oxygen release. Respiration is equally important, with its subsequent release of carbon dioxide needed for photosynthesis. Nutrients for bacterial growth are present in plant and animal remains. Bacterial degradation (decay) of these remains releases the carbon in the material, which is returned to the atmosphere as CO_2 and is again available for photosynthesis. Decay also leads to the recycling of other nutrients required in the growth and development of other organisms. Bacteria rapidly degrade or decay many substances, and because of their small size, they have rapid respiratory and reproductive rates. Soils, especially those used agriculturally, are teeming with bacteria, and the nutrients resulting from bacterial degradation develop and maintain soil fertility.

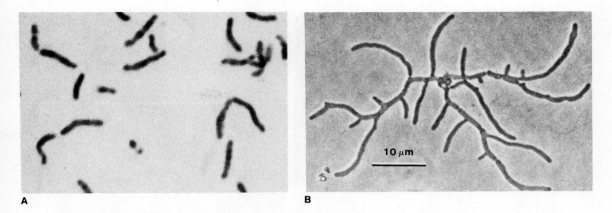

Figure 2-10 Actinomycetales. **A.** *Mycobacterium* cells growing in culture, ×3000. **B,** *Nocardia*, showing hyphae. (**B,** courtesy of B. L. Beaman and D. M. Shankel, with permission of *Journal of Bacteriology*.)

Some bacteria have **mutualistic** (*mutu* = reciprocal) relationships with land plants and animals. One of the most striking associations is with ruminant animals (Fig. 2-17B). These mammals are incapable of producing the enzymes necessary for cellulose breakdown—yet the diet of many **herbivorous** (*herb* = grass; *vor* = eat) mammals consists largely of cellulose. The ruminant digestive tract is highly modified, with the stomach consisting of four separate chambers. The first two function as fermentation chambers for the microorganisms. Newly eaten, but essentially unchewed, plant material (food) passes into the first chamber (*rumen*); it is subsequently returned to the mouth as a "cud" and is chewed. The saliva contains a dilute salt solution beneficial to growth of the microorganisms in the rumen. These include a large number of bacteria that digest cellulose and other carbohydrates, producing fatty acids, vitamins, and other substances that can be absorbed by the animals. During digestion, gases are formed, including carbon dioxide and methane, and the animal belches frequently to remove these gases. The fatty acids produced are absorbed and are the primary source of nutrients for the animal. In addition, the rumen microbial cells are digested elsewhere in the digestive tract and provide essential amino acids and other nutrients.

Division Archaebacteria

The Archaebacteria superficially resemble the Schizomycophyta. However, they differ markedly in the nature of the transfer RNA (tRNA) used in protein synthesis and the RNA enzyme, as well as in the chemistry of cell wall and cell membrane.

The Archaebacteria include the **methanogens** (*methan* = methane; *gen* = origin), **halophiles** (*halo* = salt; *phil* = love), and **thermoacidophiles** (*thermo* = temperature). The methanogens occur in a reducing environment devoid of oxygen and produce methane (marsh gas) from carbon dioxide and hydrogen. The extreme halophiles survive in saturated brine, often coloring the solution a bright red. They contain a simple photosynthetic pigment, bacterial rhodopsin, similar to one of the visual pigments in animals. The thermoacidophiles occur in hot sulfur springs (temperature about 80° C, pH less than 2).

The tRNA lacks a thymine in the key location of one of the nucleotides that is present in other prokaryotes and in eukaryotes. There are also modifications in the structure of the RNA polymerase and membrane lipids occurring in the Archaebacteria. Other differences probably segregate the Archaebacteria. Further differences are

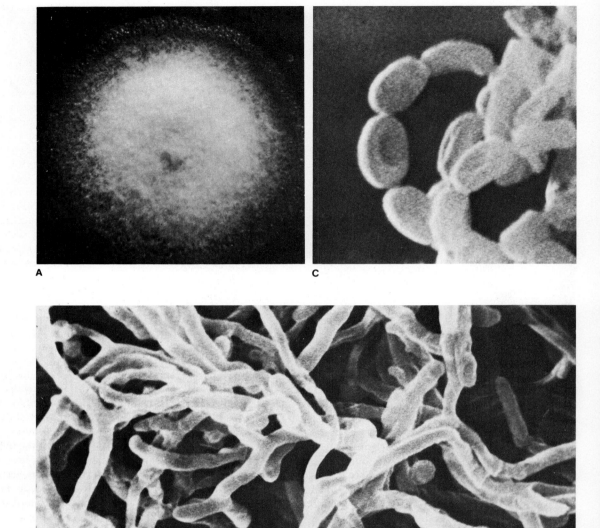

Figure 2-11 Actinomycetales. *Streptomyces.* **A,** young colony growing on agar, × 50. **B, C,** scanning electron micrographs: **B,** typical filamentous growth, × 5000; **C,** conidial chain, × 8000.

expected as more information becomes available on the known representatives and their relation to other prokaryotes.

Division Cyanophyta (Bluegreen Algae)

The division Cyanophyta (= Cyanobacteria) comprises a single class, the Cyanophyceae. The organisms are generally bluegreen but are sometimes rust red or brown to black. These photoautotrophic prokaryotes are ubiquitous, as are the bacteria. The bluegreen algae have many similarities and interrelationships with the eukaryotic algae and are considered as algae rather than Cyanobacteria in this text. Cells of bluegreen algae range from less than 1 μm to several micrometers and are rarely the size of eukaryotic algal cells.

Cell Structure and Movement

The cell is composed of a layered cell wall—generally of four layers—with the innermost one consisting of peptidoglycans (also known as **mucopeptides** or **glycoprotein**), as in bacteria. The cell wall is most similar to that of gram-negative bacteria (see p. 15). External to the wall may be a fibrous sheath (Fig. 2-12A). The sheath may be seen with the light microscope, as it is colored and/or stratified. Internal to the wall is the typical plasma membrane, which in some bluegreen algae has invaginations comparable to the mesosomes of gram-positive bacteria (see p. 15).

The photosynthetic pigments are in photosynthetic lamellae, or **thylakoids** (*thylac* = sack, pouch), that are distributed evenly throughout the cell (Fig. 2-12A). The thylakoids are closed discs containing the chlorophyll and carotenoid pigments. Associated with the thylakoids are blue and red pigments, the **phycobiliproteins** (*phyco* = alga; *bili* = bile), which are also present in some eukaryotic algae (Chapters 6 and 9).

The bluegreen algae do not have a nucleoid comparable to that in bacteria (see p. 16), and there is no condensed state of DNA or histone proteins as in eukaryotic algae. Bluegreen algal DNA, termed **nucleoplasm** (Fig. 2-12B), is like that of eukaryotic chloroplasts and mitochondria but different from that of eukaryotic nuclei. The RNA is also similar to that in chloroplasts and mitochondria of eukaryotes.

Among the cellular inclusions are polyhedral bodies as well as polyglucan granules, cyanophycin granules, and **polyphosphate granules** (Fig. 2-12A, B). The names indicate the chemical nature or the structure of the inclusion. The crystalline **polyhedral** (*poly* = many; *hedra* = seat) **bodies,** which are 0.2–0.3 μm in diameter, contain enzymes involved in carbon dioxide fixation and should be called **carboxysomes** (*some* = body). **Polyglucan** (*poly* = many; *glucan* = sweet) **granules,** 0.25 μm in diameter, are correlated with photosynthesis and are considered carbohydrate storage products. **Cyanophycin granules** (previously known as *structured granules*) contain amino acids and are prominent in resting cells (Fig. 2-12A, B). They are almost 1 μm in size. Polyphosphate granules are often present in older cells and can be seen with the light microscope (Fig. 2-12A).

One other inclusion easily observed with the light microscope as clear areas in the cell are the **gas vacuoles** (Fig. 2-14A). Structurally, these are very different from the vacuoles of eukaryotic plants. They do not contain water but are permeable to certain gases and susceptible to pressure changes. With the electron microscope, gas vacuoles are resolved into packed arrays of cylindrical vesicles (Fig. 2-12C). The vesicle membranes themselves are proteinaceous and almost completely lacking in lipids, which occur in other cell membranes.

In cell division, the plasma membrane and two inner wall layers invaginate into the cytoplasm from outside the cell to form a broad **septum** (*sept* = fence). Additional wall material is formed, and the septum continues to grow toward the center of the cell, separating the two new cells. Prior to septum initiation, there may or may not be an invagination of the thylakoids. In the multicellular forms, the outer wall layers and materials secreted by the cells hold the cells together (as in Figs. 2-13 and 2-14). Submicroscopic pores and/or larger protoplasmic connections (visible with the light microscope) sometimes connect adjoining cells, although the larger connections do not connect the protoplasts.

In contrast to some bacteria, bluegreen algae do not have any flagellated cells. However, some filamentous forms are capable of independent movements seen usually as a gliding, spiralling, or

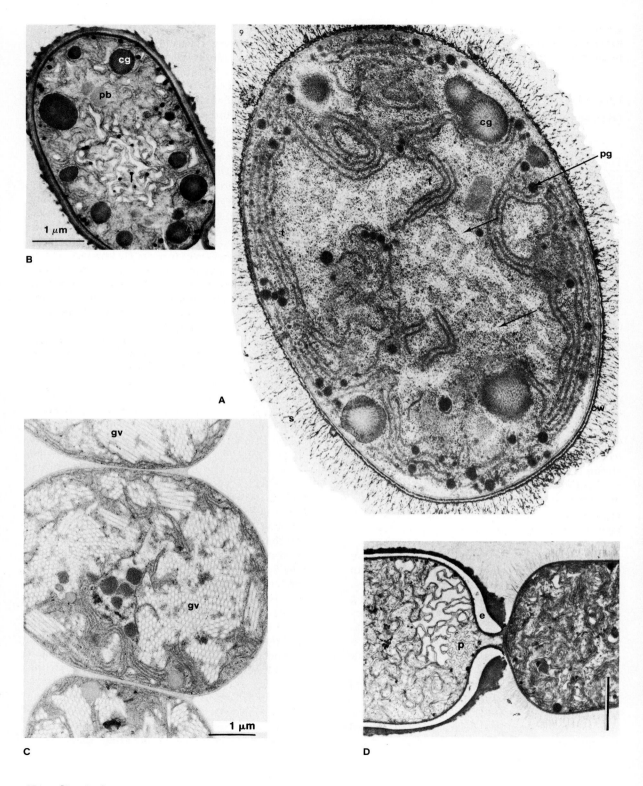

gentle waving of the loose end of the filament. At present there is no single theory to explain movement in general.

Reproduction

Reproduction in the Cyanophyceae is restricted to vegetative, or asexual, reproduction. The main method is **fragmentation,** which is a simple breaking apart of the thallus into two or more units that grow into new organisms. In filamentous forms, small fragments, termed **hormogonia** (*horm* = chain; *gon* = offspring), are released and glide away from the parent filament. Hormogonia often result from the death of a cell with a biconcave area produced from the pressure of adjacent cells.

Various types of nonflagellated spores or sporelike cells can be produced. These can serve as dormant or resting structures as well as reproductive units. True spore formation occurs in some genera, but it is not common. The most commonly produced spore is the **akinete** (*a* = without; *kine* = movement) (Figs. 2-12B and 2-16C), characteristic of many filamentous cyanophytes. The akinete is probably a metamorphosed vegetative cell that increases in size and produces a thicker wall. Akinetes serve as resting stages able to withstand unfavorable environmental conditions.

Figure 2-12 Cell structure, sections of Cyanophyceae, electron microscope. **A,** cell of *Anabaena*, showing fibrous sheath stained with ruthenium red (*s*), cell wall layers (*cw*), peripheral single thylakoids (*t*), nucleoplasm (*arrows*), cyanophycin granules (*cg*), and polyphosphate granules (*pg*), ×34,800. **B,** akinete of *Anabaenopsis*, showing large cyanophycin granules (*cg*) and smaller polyhedral bodies (*pb*), ×24,800. **C,** gas vacuoles (*gv*) of *Nostoc*, showing cylindrical shape in longitudinal section (*upper*) and packed arrays in transverse section (*right*), ×2500. **D,** heterocyst of *Anabaena* (*left*) and adjacent vegetative cell (*right*) with thick envelope (*e*) and contorted thylakoids, especially near poles (*p*), ×24,000. (Polyglucan granules not visible in figure.) (**A,** courtesy of L. V. Leak and *Journal of Ultrastructure Research*; **B,** courtesy of N. J. Lang and Blackwell Scientific Publications; **C,** courtesy of T. Bisalputra and *Journal of Ultrastructure Research*; **D,** courtesy of N. J. Lang.)

Another sporelike cell is the **heterocyst** (*heter* = different; *cyst* = bag) (Figs. 2-12D, 2-14C, and 2-16C, F), which has been shown to be involved primarily in nitrogen fixation. With the light microscope, the heterocyst appears as a translucent or shiny cell that is yellow to bluegreen. It appears to lack internal contents, and has wall thickenings at the ends where it is attached to other cells (Figs. 2-12D and 2-14C). The electron microscope shows the heterocyst to contain a complex reticular pattern of membranes within a heavy wall (Fig. 2-12D). Heterocysts can serve as a point of weakness in a filament, thus promoting fragmentation. A few heterocysts have been seen to produce new filaments; however, reproduction is not considered their prime function.

Sexual reproduction, as observed in eukaryotic algae, is not present in the cyanophytes. However, **anastomoses** (*anastomose* = coming together) of cells in filaments do occur in some strains of *Nostoc* under certain growth conditions. There is evidence for genetic transfer in bluegreen algae similar to that in bacteria, but more research is needed before we can understand the cyanophytes as well as we do the bacteria.

Morphological Variation and Classification

Morphologically some of the Cyanophyceae are as simple as the bacteria; however, most are somewhat more complex, although not as much as most eukaryotic algae. The bluegreen algae are unicellular or colonial, free-floating or attached, filamentous or nonfilamentous (Figs. 2-13 and 2-14). The cell shape is somewhat variable and has been used (with varied success) to classify nonfilamentous, or coccoid, forms. Because coccoid forms are small and morphologically simple, they can be classified using physiological and biochemical characteristics similar to those employed for bacteria. Eventually a combination of morphological and biochemical attributes will probably be used to establish a more workable system. In this text, a morphological system, based primarily on external appearance, is used to emphasize the algal features of the Cyanophyceae.

Two distinctive morphological lines of development (and possibly evolution) are evident (Fig. 2-15): the nonfilamentous and filamentous types. Morphologically, the simplest member is a unicellular, spherical, free-floating form from which the two lines can be derived. The nonfilamentous line includes unicellular, attached forms as well as unicellular or colonial forms (Fig. 2-15A). The filamentous line (Fig. 2-15B) can be derived directly from unicellular forms. If cell division occurs only in one plane, unbranched forms result. A cyanophyte filament generally consists of a row of cells, the **trichome** (*trich* = hair), and a sheath (as shown in *Lyngbya,* Fig. 2-15B). The distinction between trichome and filament is used only for bluegreen algae because of the nature of the sheath, which can be lacking altogether. Other genera have several trichomes within a single sheath (as shown in *Schizothrix,* Fig. 2-15B).

The classification system used here is a conservative one. Whereas five orders are generally recognized, representatives of only three are mentioned in detail (see Table 2-2). Classification of the bluegreens as Cyanobacteria has been officially made and is somewhat different from this more "classical" system, although incorporating some of the same characteristics.

Order Chroococcales   Most nonfilamentous forms are in the order Chroococcales. These include unicellular and colonial genera (Fig. 2-13). Because of their small size and morphological simplicity, some are difficult to distinguish from bacteria. The colonial forms result from cells remaining fastened to each other after division and/or being embedded in a common sheath. In *Merismopedia* (Fig. 2-13A, referred to as *Agmenellum* by some) cell divisions occur regularly in only two planes and form a single-layered sheet of cells. When there are no regular planes of cell division, irregular colonies like those of *Microcystis* and *Gloeothece* are formed (Fig. 2-13B, C).

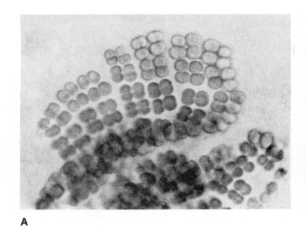

A

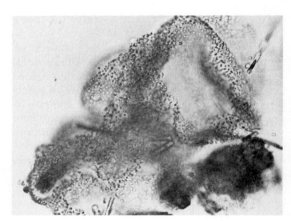

B

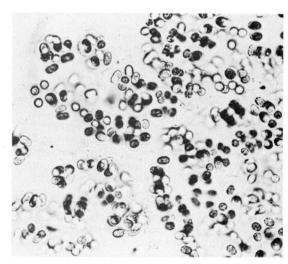

C

Figure 2-13   Cell structure and morphological diversity of colonial, nonfilamentous Cyanophyceae, light microscope. Note for each, the position of the cells within the colorless sheath material. **A,** *Merismopedia,* ×750. **B,** *Microcystis,* ×200. **C,** *Gloeothece,* ×2000.

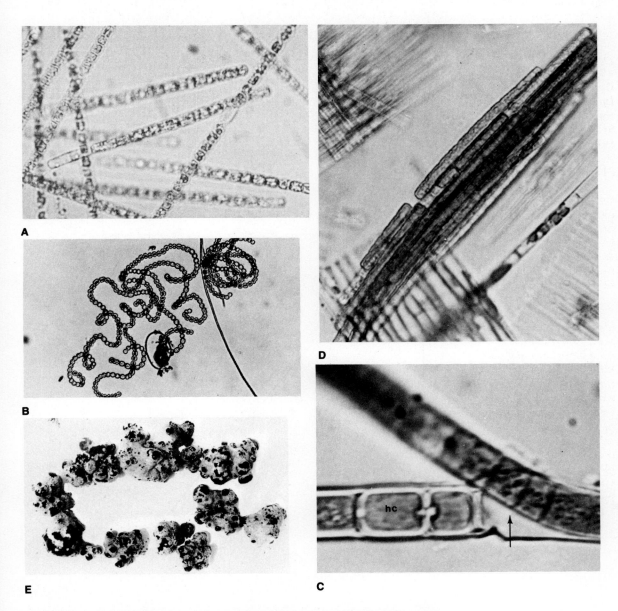

Figure 2-14 Cell structure and morphological diversity of filamentous Cyanophyceae, light microscope. **A, B,** *Anabaena;* **A,** a species with straight trichomes showing gas vacuoles, ×350; **B,** a species with coiled trichomes, ×150. **C,** *Tolypothrix,* with false branching (*arrow*) and heterocyst (*hc*), ×700. **D,** *Aphanizomenon,* with parallel trichomes without sheath, ×300. **E,** *Nostoc,* macroscopic colonies, partially encrusted with soil, ×0.4.

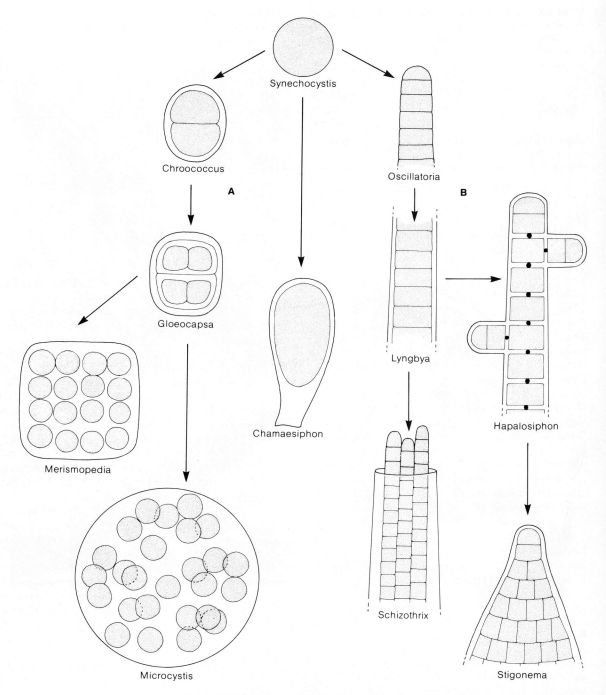

Figure 2-15 Developmental lines in Cyanophyceae resulting from plane of cell division. **A,** nonfilamentous lines, including attached forms (*left, center*). **B,** filamentous lines (*right*).

Table 2-2 Characteristics of Selected Orders of Cyanophyceae (Bluegreen Algae)

Order	Cell Arrangement	Branching	Heterocyst/Akinete
Chroococcales	unicellular or colonial; free-floating	none	none
Nostocales	filamentous	false, if present	none, either, or both
Stigonematales	filamentous or parenchymatous	true	both

Order Nostocales The largest order, Nostocales, includes most well-known cyanophyte genera. The basic structure is the trichome, which may have very little sheath, if any, as in *Oscillatoria* and *Aphanizomenon* (Figs. 2-16A and 2-14D). The cells of the trichome are in a single, linear series, forming a **uniseriate** (*uni* = one; *seri* = row) structure. The filaments can also be aggregated into various macroscopic, mucilaginous masses of regular or irregular form, as in *Nostoc* (Fig. 2-14E). In some filamentous genera, there is a break in the trichome, and one end (or both) breaks through the sheath at the point of fragmentation, as in *Tolypothrix* (Fig. 2-14C). This gives the appearance of a "branch," but since it is not attached to the rest of the trichome, and does not result from a change in plane of cell divisions, it is termed **false branching.** The break may also result from the death of an intercalary cell or extensive growth of a trichome within its sheath.

Order Stigonematales The Stigonematales are characterized by **true branching,** which is caused by a change in the plane of cell division. The order is considered the most specialized of the Cyanophyceae. Some members have several linear rows of cells (**multiseriate;** *multi* = many), as in *Hapalosiphon* (Fig. 2-16E), and some have random cell divisions, as in *Stigonema* (Fig. 2-16F). Some Stigonematales morphologically resemble certain red algae (genera of Bangiophyceae, Chapter 9), thus providing good morphological examples of the evolutionary relationships between the red and bluegreen algae.

Physiology and Biochemistry

The photoautotrophic nature of the bluegreen algae separates them from bacteria. Cyanophyte photosynthesis is an aerobic process in which water is the hydrogen donor, and molecular oxygen is released. This is identical to photosynthesis in eukaryotic, photoautotrophic plants and involves cyclic and noncyclic **photophosphorylation,** the presence of two photosystems in photosynthesis, and chlorophyll *a*. In contrast, bacterial photosynthesis is anaerobic, using other compounds as hydrogen donors, other chlorophyll pigments (bacteriochlorophyll), and so on, as discussed briefly with the Pseudomonadales (see p. 22). Some bluegreen algae can photosynthesize in anaerobic conditions, using H_2S as a source of electrons. Evidently this is not a "way of life" for most.

In addition to chlorophyll *a*, carotenoids and phycobiliproteins are present, giving the bluegreen algae their characteristic colors. These accessory pigments, having different absorption spectra, pass light energy of different wavelengths to chlorophyll *a*, which is the photoreactive pigment. Phycobiliproteins and light of specific wavelengths are also involved in the morphogenesis of some *Nostoc* species. Evidently light of red or green wavelengths is the controlling factor, and the reaction is photoreversible.

Some bluegreen algae share with bacteria the ability to fix atmospheric nitrogen (N_2). As far as is known, all nitrogen-fixing bluegreen algae are free-living, although some have symbiotic relationships with eukaryotic plants. Less is known about nitrogen fixation in cyanophytes than in bacteria; however, the processes are very similar. The heterocyst (Figs. 2-12D and 2-14C) is the main (but not the exclusive) site of fixation. Energy from photosynthesis is required, yet the main enzyme (nitrogenase) is oxygen-sensitive. Thus, the process must occur in a low-oxygen environment or under reducing conditions. In the heterocyst, this is achieved by a high respiratory rate, reduction in oxygen, and virtual absence of photosynthesis. Forms lacking heterocysts fix nitrogen in vegetative cells under anaerobic or almost anaerobic conditions. A significant amount of the nitrogen fixed is secreted into the surrounding environment, as is true in rice fields, where the bluegreen algae are the main source of usable nitrogen.

Some Cyanophyceae produce toxins, as do some bacteria. The toxins often cause the death of cattle or birds that drink water containing the algae (Fig. 2-17B). The nature of the toxins varies, but they must be released from the cell to be effective. The genera involved include *Microcystis*, *Anabaena*, and *Aphanizomenon* (Figs. 2-13B and 2-14A, B, D).

Ecology

The Cyanophyceae are probably the most ubiquitous of photosynthetic organisms. They can be free-floating, sessile, or attached in the aquatic habitat; they can be macroscopic cushions, or layers on rocks, soil, or trees. Their ability to occupy habitats with extreme environmental conditions, such as heat, cold, and moisture, is unsurpassed by any other algae or even by land plants (bryophytes, tracheophytes). This success at such extremes probably results from the prokaryotic nature of the cell. In fact, only bacteria are able to live in the more extreme and diversified environments. The lack of membrane-bound organelles and small cell size have been cited as the factors that allow survival in extreme conditions.

The ability of bluegreen algae to colonize bare surfaces has been recognized as important in the establishment of vegetation in some areas, such as the volcanic island of Surtsey, which first emerged in 1963 off Iceland. Bluegreen algae will form bluish-black patches on lawns if growth of the grass is poor or if bare areas are present. Bluegreen algae

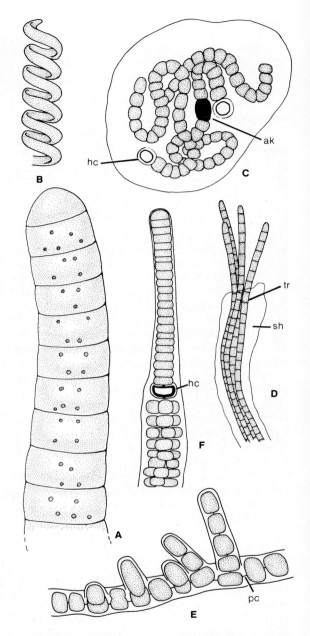

Figure 2-16 Morphological diversity of Cyanophyceae, filamentous forms. **A,** *Oscillatoria*, ×1500. **B,** *Spirulina*, ×1800. **C,** *Nostoc*, showing heterocyst (*hc*) and akinete (*ak*), ×1000. **D,** *Schizothrix*, with several trichomes (*tr*) within a sheath (*sh*), ×1000. **E,** *Hapalosiphon*, with true branching, showing cell connections (*pc*), ×1180. **F,** *Stigonema*, showing a heterocyst, nature of trichome, and random cell arrangement, ×500.

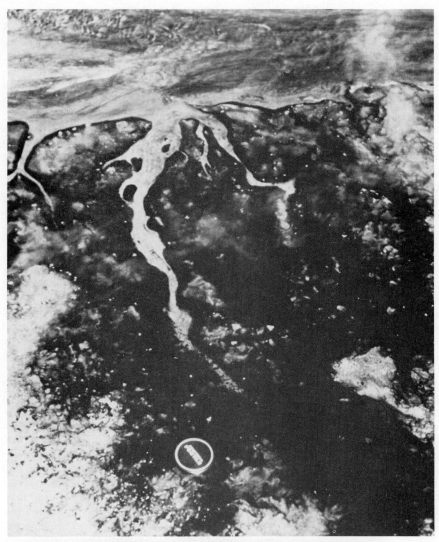

Figure 2-17 Ecology of Cyanophyceae. **A,** presence in hot springs; flow of water is left to right (camera lens cover in lower part *ca.* 50 mm). **B,** cattle drinking from pond with bluegreen bloom on surface of water, especially lower left. (**A,** courtesy of R. Castenholz; **B,** courtesy of W. Carmichael.)

are also known in hot springs and grow at temperatures as high as 73 °C (Fig. 2-17A). At higher temperatures, gliding bacteria occur, and it is often difficult to distinguish between these two types of prokaryotes.

Since cyanophytes generally have higher temperature tolerances than eukaryotes, one problem with thermal pollution can be the promotion of bluegreen algal growths. In any habitat, massive amounts of algal material are termed **blooms** because masses suddenly appear—almost overnight in the surface water (Fig. 2-17B). Actually, the sudden increase in numbers is caused by the algae from deeper waters rising to the surface. The bluegreen algae grow at depths where light intensity and oxygen are low, and available dissolved nitrogen can be low, although other nutrients are plentiful. During growth, gas vacuoles (Figs. 2-12C and 2-14A) develop, and the algae rise in the water column. Thus the "sudden" appearance of large amounts of bluegreen algae results from the surfacing of these deep-water organisms.

Another factor with regard to bluegreen algal blooms is that abundant growth can occur when inorganic nitrogen is low. If other nutrients are available, the cyanophytes can utilize atmospheric nitrogen. Bluegreen algae secrete large amounts of fixed nitrogen and carbon. In so doing, they can produce substances (including toxins) that both inhibit the growth of eukaryotic algae and provide necessary nutrients for their own growth or for that of eukaryotes. Bluegreen algae seem particularly resistant to grazing by zooplankton; thus, the standing crop is not diminished and does not enter into the normal food chain of the aquatic ecosystem.

Relationships of the Monera

The oldest fossils known are Precambrian monerans, or prokaryotes, from rocks over 3 billion years old (Chapter 27, Fig. 27-1). These are in the geologic series termed *Swaziland Supergroup*, in northeastern South Africa. Morphological and biochemical evidence indicates that anaerobic, heterotrophic, and photosynthetic prokaryotes were present at least 3.1 billion years ago. These include minute, rod-shaped bacteria and coccoid bluegreen algae. Some filamentous bluegreen algae in finely layered, calcareous formations, called **stromatolites** (*stroma* = bed; *lite* = stone), are known in rocks some 2.5 billion years old. Present-day stromatolites are known in tropical and subtropical environments, such as off the coast of Western Australia (Fig. 27-3). More diverse prokaryotic fossils are present in Precambrian sediments of the massive Canadian Shield of eastern North America (north of the Laurentian Great Lakes). The formation, which is between 2.0 and 1.6 billion years old, contains a variety of **microfossils** (*micro* = small) resembling filamentous bluegreen algae and bacteria. The geochemistry of this formation supports the occurrence of aerobic photosynthetic organisms. Younger sediments (1.4–1.2 billion years old) in southeastern California and the Northern Territory of Australia, indicate an even greater diversity of prokaryotic fossils.

The greatest evolutionary discontinuity among living organisms is that separating the prokaryotes from the eukaryotes. Although it is generally believed that eukaryotes arose from a prokaryotic ancestor, neither fossil nor living intermediates are known. A current theory that incorporates the Archaebacteria places them as a separate line from a "universal ancestor" or **progenitor** (*pro* = before; *gen* = race). Another theory, widely accepted, is that the eukaryotic cell, as we know it, evolved from the incorporation of **endosymbiotic** (*endo* = within) prokaryotes into other prokaryote cells. These endosymbiotic prokaryotes then evolved into the mitochondria and plastids present in eukaryotic cells. The mitochondria could have originated from the incorporation of purple photosynthetic bacteria, whereas chloroplasts were of cyanobacterial origin. Such contemporary endosymbiotic symbioses are known. In eukaryotic organisms, mitochondria and plastids form from existing mitochondria and plastids; that is, they are semiautonomous in their growth and division. Both organelles can self-replicate and have DNA more similar to that of prokaryotes than to that in the nuclei of eukaryotes. A good example of such a similarity is that of the cyanophyte cell and the chloroplasts of rhodophytes (red algae, Chapter 9). Both contain the phycobiliprotein pigments in special granules adjacent to the thylakoids. The pigments are very similar and in some instances are identical. Conceivably, red algae may have developed through the incorporation of a bluegreen endosymbiont in a primitive eukaryotic cell. Further discussion of the endosymbiont theory appears in Chapter 27.

References

Monera

Echlin, P., and Morris, I. 1965. The relationship between bluegreen algae and bacteria. *Biol. Rev.* 40:143–87.

Fox, G. E. 1980. The phylogeny of eukaryotes. *Science* 209:457–63.

Hall, G. 1971. Evolution of the prokaryotes. *J. Theoret. Biol.* 30:429–54.

Starr, M. P.; Stolp, H.; Trüper, H. G.; Balows, A.; and Schlegel, H. G., eds. 1981. *The prokaryotes.* 2 vols. Springer-Verlag, Berlin.

Woese, C. R. 1981. Archaebacteria. *Sci. Amer.* 244:98–122.

Schizomycophyta

Atlas, R. M., and Bartha, R. 1981. *Microbial ecology: fundamentals and applications.* Addison-Wesley Publishing Co., Reading, Mass.

Brock, T. D. 1979. *Biology of microorganisms.* 3rd ed. Prentice-Hall, Englewood Cliffs, N.J.

Buchanan, R. E., and Gibbons, N. E., eds. 1974. *Bergey's manual of determinative bacteriology.* 8th ed. Williams & Wilkins Co., Baltimore.

Bull, A. T., and Meadow, P., eds. 1978. *Companion to microbiology.* Longman, New York.

Burrows, W. 1979. *Textbook of microbiology.* 21st ed. W. B. Saunders, Philadelphia.

Campbell, R. 1977. *Microbial ecology.* John Wiley & Sons, New York.

Davis, B. D.; Dulbecco, R.; Eisen, H. N.; and Ginsburg, H. S. 1980. *Microbiology.* 3rd ed. Harper & Row, New York.

Frobisher, M.; Hinsdill, R. D.; Crabtree, K. T.; and Goodheart, C. R. 1974. *Fundamentals of microbiology.* 9th ed. W. B. Saunders, Philadelphia.

Hawker, L. E., and Linton, A. H. 1979. *Micro-organisms: function, form, and environment.* 2nd ed. Edward Arnold, London.

Nester, E. W.; Roberts, C. E.; Pearsall, N. N.; and McCarthy, B. J. 1978. *Microbiology.* 2nd ed. Holt, Rinehart & Winston, New York.

Peberdy, J. F. 1980. *Developmental microbiology.* Halsted Press, New York.

Pelczar, M. J., Jr.; Reid, R. D.; and Chan, E. C. S. 1977. *Microbiology.* 4th ed. McGraw-Hill, New York.

Stanier, R Y.; Adelberg, E. A.; and Ingraham, J. L. 1976. *The microbial world.* 4th ed. Prentice-Hall, Englewood Cliffs, N.J.

Stanier, R. Y.; Adelberg, E. A.; Ingraham, J. L.; and Wheelis, M. L. 1979. *Introduction to the microbial world.* Prentice-Hall, Englewood Cliffs, N.J.

Cyanophyta

Bourrelly, P. 1983. *Les algues d´eau douce.* Vol. 3, *Les algues bleues et rouges, eugléniens, peridiniens et cryptomonadines.* 2nd ed. N. Boubée & Cie, Paris.

Carr, N. G., and Whitton, B. A., eds., 1973. *The biology of blue-green algae.* Blackwell Scientific Publications, Ltd., Oxford.

———. 1982. *The biology of Cyanobacteria.* Blackwell Scientific Publications, Oxford.

Fogg, G. E.; Stewart, W. D. P.; Fay, P.; and Walsby, A. E. 1973. *The blue-green algae.* Academic Press, New York.

Humm, H. J., and Wicks, S. R. 1978. *Introduction and guide to the marine bluegreen algae.* Wiley-Interscience, New York.

3
NONVASCULAR EUKARYOTES

Prior to the recognition of five kingdoms, eukaryotic organisms (containing membrane-bound organelles, such as the nucleus, mitochondrion, etc.) that were attached to a substrate and produced cell walls were placed in the Kingdom Plantae. Within this kingdom, there were two major subdivisions based on the presence or absence of conducting tissue (xylem and phloem; Chapter 12). With the advent of sophisticated analytic equipment and techniques (such as the electron microscope and spectrophotometer), the diversity observed among the organisms lacking conducting tissue indicated that the concept of the plant kingdom should be restricted. Thus, two new kingdoms are now recognized. One, the Kingdom Protista, includes those organisms that are primarily unicellular (or, if multicellular, each cell is generally capable of acting as a single cell). The other is the Kingdom Fungi, which contains the nonphotosynthetic forms with morphologically simple reproductive cells.

Some members of these kingdoms—Protista, Fungi, Plantae—can be considered as the *nonvascular eukaryotes* (see Table 3-1). The Protista comprise some organisms previously classified as *thallophytes* (thall = young shoot, twig), which included the algae and the fungi. (In addition, the Protista also include the protozoa, which have been previously classified as animals.) The thallophytes were those morphologically simple plants without complex tissues and with unprotected reproductive structures. The terms *algae* and *fungi* have been used for many years; however, the term *algae* is not now used as an official taxonomic category, whereas the term *fungi* is. These terms were used to designate the photosynthetic (algae) and the nonphotosynthetic (fungi) thallophytes. To some botanists, the Cyanophyta, or Cyanobacteria (bluegreen algae), discussed in Chapter 2 as monerans, are still studied as algae.

Some of the eukaryotic algae are now classified in Protista, with the others in Plantae. The slime molds, previously included by some in the Fungi, are now also placed in Protista. The Fungi are restricted to the nonphotosynthetic organisms with firm cell walls that do not ingest particulate foods. Another group of nonvascular eukaryotes is the bryophytes. They are still classified in Plantae and "look" more plantlike than most algae and fungi. The bryophytes are easily distinguished from algae by several morphological and anatomical features, as they are structurally and reproductively more

Table 3-1 General Groups of Nonvascular Eukaryotes

Common Name (Scientific Name)	Motile Cells* Assimilative	Motile Cells* Reproductive	Form	Habitat*	Energy Source	Average Size [range]
Nonphotosynthetic Forms						
Fungi (Hyphochytridiomycota) (Chytridiomycota) (Oomycota) (Zygomycota) (Eumycota)	none	(flagella)	unicell, multicell	aquatic terrestrial	heterotrophic	10 μm–1 m [1 μm–∞]
Protists—Slime molds (Myxomycota)	amoeboid	(flagella)	acellular	terrestrial	heterotrophic; ingest particles	5 μm–10 mm
Photosynthetic Forms						
Protists—Algae (Chrysophyta) (Pyrrhophyta) (Cryptophyta) (Euglenophyta)	flagella	flagella	unicell, multicell	aquatic (terrestrial)	autotrophic	5–100 μm [1 μm–1 mm]
Plants (Chlorophyta) (Phaeophyta) (Rhodophyta)	(flagella)	(flagella)	(unicell) multicell	(aquatic) terrestrial	autotrophic	5 μm–1 m [1 μm–75 m]
Bryophytes (Bryophyta)	none	flagella (male gamete only)	multicell	(aquatic) terrestrial	autotrophic	10–30 mm [0.1–60 mm]

*Parentheses indicate some representatives, but not the majority.

complex. Although there are basic differences among the nonvascular eukaryotes, there are similarities in cell structure, reproduction, and life history that serve to unify them (see Table 3-1).

Habitat

Nonvascular eukaryotes are ubiquitous although they are limited by temperature extremes. Whereas most occur in moist situations, there is no restriction to the aquatic habitat (see Table 3-1). However, water in liquid form, often in very minute amounts, is required for sexual reproduction for most algae and for bryophytes. Fungi, with their different lifestyle, also have diverse sexual mechanisms. Fungi are limited to organic substrates for their source of energy, but otherwise they are almost as adaptable as bacteria (Chapter 2). Fungi can also grow within a temperature range of from 60 °C to less than 0 °C. Algae and bryophytes are limited generally to aerobic environments and utilize light as their energy source. Most eukaryotic algae grow in a limited temperature range as compared with some of the (prokaryotic) bluegreen algae (Chapter 2). Bryophytes generally are more restricted by temperature extremes, probably as a result of their more complex morphology and type of reproduction.

However, they occur in the tropics, the Arctic, and the Antarctic, as well as in rain forests, deserts, and at high elevations.

Cell Structure

The eukaryotic cell is easily recognized by the presence of discrete, membrane-enclosed organelles. DNA is in chromosomes within a well-defined nuclear envelope, and some of the respiratory enzymes (the pyruvic acid cycle) are within the mitochondria. The photosynthetic pigments and enzymes are in chloroplasts (Fig. 3-1A). Fungi, which are nonphotosynthetic, lack chloroplasts (Fig. 3-1B).

Some nonvascular eukaryotes are motile in the vegetative, or assimilative, stage (especially some myxomycetes and algae; see Table 3-1). Many have motile reproductive cells with one or more flagella for cell propulsion. The **flagellum** of a eukaryote is composed of one or two central **microtubules** surrounded by nine additional pairs of microtubules, which are further surrounded by the plasma membrane (Fig. 3-2A, B). If the flagellar membrane is smooth, the flagellum is referred to as a **whiplash flagellum** (Fig. 3-2C, *lower*). However, if the membrane has thin tubular hairs along its length, it is a **tinsel flagellum** (Fig. 3-2C, *upper*). In addition, some organisms have flagella bearing very delicate hairs and/or very fine scales (Fig. 3-2D). As the form, location, and number of flagella on a cell are constant for a given species, they are used to characterize some nonvascular groups (Fig. 3-3 shows this variation). Flagella will continue to beat even if removed from the cell if a chemical energy supply is available.

The nature of the outer cell covering (generally outside the plasma membrane) is almost as diverse as the organisms to be considered. Some nonvascular eukaryotes lack a cell wall but have an outer, firm, proteinaceous membrane. Others have a wall, and still others have a wall for only a part of their life history. **Polysaccharides** (*poly* = many; *sacchar* = sugar), sometimes containing additional nitrogenous compounds, are the usual components of cell coverings. In addition, or in place of the organic compounds, such inorganic materials as iron, silica, magnesium, or calcium carbonate may be present. The nature of the cell covering, like the flagellation, is specific for a given species or stage in the life history of a species. Discussion of the variety of materials included is presented later in this chapter.

Food reserves, or storage products, of nonvascular eukaryotes are also diverse. These reserves are carbohydrates, proteins, or fats and oils, and can vary depending on environmental conditions. There is more information available concerning carbohydrate storage products than others, and the polysaccharides are often composed of the same sugar as the cell coverings.

Reproduction

In many nonvascular eukaryotes, reproduction is vegetative, or asexual. This involves simple **budding** (Fig. 3-4A), **fragmentation**, development of specialized **deciduous** (*decid* = falling off) parts of the plant, or production of special reproductive structures termed **spores** (*spor* = seed; Fig. 3-4 B–H). In many forms (excluding the bryophytes), spores are produced by either **mitosis** (equational division) or **meiosis** (reduction division). In bryophytes (and in vascular plants), spores are produced only by meiosis, and may be termed **meiospores**. Spores are generally produced in a **sporangium** (*spor* = seed; *ang* = box, case; Fig. 3-4B–D, H), which is usually unicellular (Fig. 3-4B–D), but not always (Fig. 3-4H). Depending upon the organism, a cell wall can be present. Spores that move by means of flagella are **planospores** (*plano* = wandering) or **zoospores** (*zoo* = animal; Fig. 3-4D, E). Nonmotile spores are **aplanospores** (*a* = without) or **sporangiospores** (Fig. 3-4B, C, F). In some organisms a heavy wall surrounds the protoplast, and such spores serve as a resting stage (Fig. 3-4G).

Figure 3-1 Ultrastructure of eukaryote cells. **A,** cell (brown alga, *Sphacelaria*) showing nucleus (*n*), chloroplasts (*c*), mitochondria (*m*), cell wall (*cw*). **B,** cell (fungal, *Achyla*) lacking chloroplasts. (**A,** courtesy of A. Burns; **B,** courtesy of N. Ricker.)

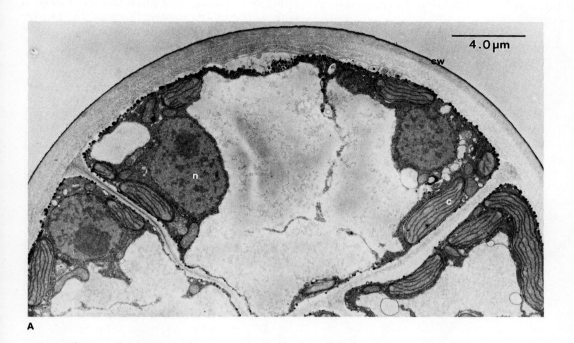

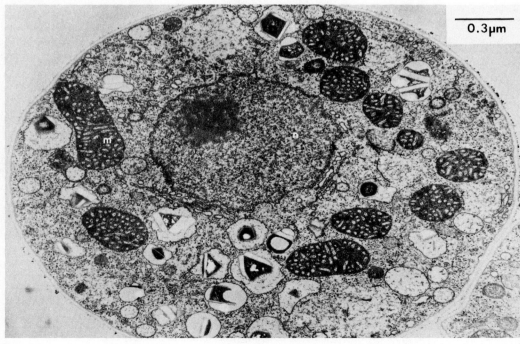

Nonvascular Eukaryotes

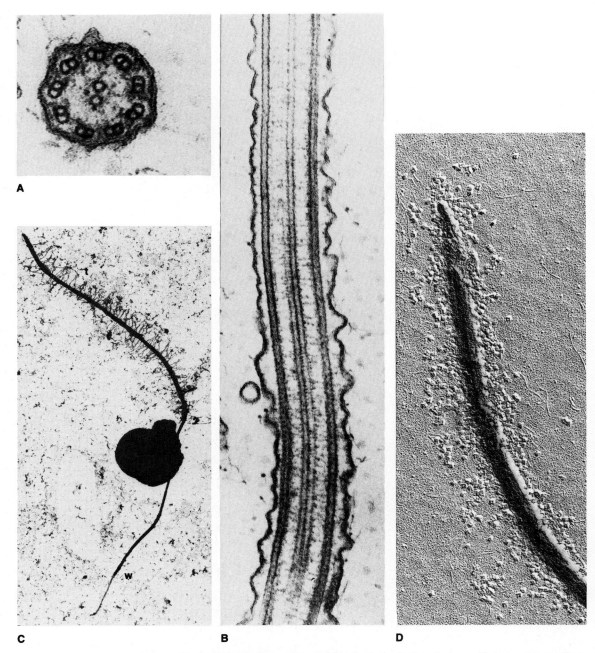

Figure 3-2 Electron micrographs of eukaryotic flagella types. **A,** transverse section (*Volvox*) showing nine peripheral pairs of microtubules surrounding two central microtubules, ×90,000. **B,** longitudinal section (*Volvox*), showing continuous membrane of flagellum and protoplast, ×90,000. **C,** tinsel (forward, *t*) and whiplash (posterior, *w*) flagella (*Chorda* zoospore), ×4,100. **D,** scaly flagellum, shadowed, ×11,000. (**A, B,** courtesy of T. Bisalputra; **C,** courtesy of R. Toth; **D,** courtesy of Ø. Moestrup).

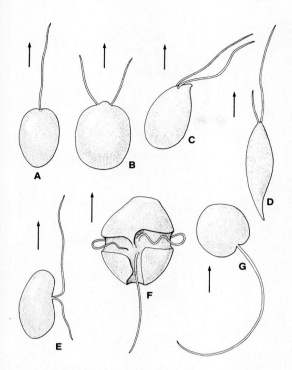

Figure 3-3 Types of flagellar arrangement in nonvascular eukaryotes (arrows indicate direction of movement). **A–D,** anterior arrangement (apical or subapical). **E, F,** lateral attachment. **G,** posterior attachment.

Sexual reproduction is present in many organisms (although certainly not universal) and involves fusion of sex cells. The morphology of the sex cells, or **gametes** (*gam* = marriage; Fig. 3-5) is diverse. Gametes are either identical in appearance (**isogamous:** *iso* = alike), as in Figure 3-5A, D, or dissimilar (**anisogamous:** *an* = not), as in Figure 3-5B, E. Both gametes can be flagellated (Fig. 3-5A, B), only one (Fig. 3-5C), or neither (Fig. 3-5D–F). If one gamete is large and nonmotile and the other markedly smaller (motile or not), sexual reproduction is **oogamous** (*oo* = egg; Fig. 3-5C, E, F). The larger gamete is regarded as female (egg) and the smaller as male (sperm). When gametes are almost identical in appearance (isogamous), they are not referred to as male and female; rather they are termed *plus* and *minus*, and are known as **mating types.** By definition, a plus gamete fuses only with a minus gamete.

Gametes are produced in structures called **gametangia** (*gam* = marriage; *ang* = box; Fig. 3-6), which in fungi and algae are typically unicellular (Fig. 3-6B, C). However, in the bryophytes, gametangia are more specialized and are multicellular (Fig. 3-6A). In some nonvascular eukaryotes, gamete cells as distinct units are not present, but nuclei function as gametes. In some of these, the nuclei move through a fertilization tube that develops after contact of the protoplasts, still referred to as *gametangia* (Fig. 3-6C). Another modification of sexual reproduction occurs through fusion of whole protoplasts, termed **conjugation** (*conjug* = joined together; Fig. 3-6D, E). In this instance, male and female gametes are usually not distinguished; however, if one protoplast moves, it is considered the male gamete.

Another feature of nonvascular eukaryotic plants is the variation in relative time and location in the life history when gamete fusion, known as **syngamy** (*syn* = with, together), and **reduction division** (meiosis) occur (Figs. 3-7 and 3-8). The time of syngamy relative to meiosis determines the type of life history, as well as whether the predominant, or obvious, phase is **haploid** (*hap* = single; *oid* = form; Fig. 3-7A) or **diploid** (*di* = double; Fig. 3-7B), or whether both phases are distinguishable (Fig. 3-8A). These types of life histories have been given different names by researchers.

Some organisms are haploid (n) in the vegetative stage, with the zygote as the only diploid ($2n$) stage (Fig. 3-7A). This life history is restricted to some algae (primarily freshwater) and many fungi. It is known as a *zygotic* or *haplontic* life history. In a second type of life history, the organism is diploid ($2n$), with gametes the only haploid (n) stage (Fig. 3-7B). This is restricted to only a few algae (mainly marine) and some fungi; however, it is typical of animals. It is known as a *gametic*, or *diplontic* life history. The third life history in which both haploid and diploid phases are present is considered the "typical plant life history" (Fig. 3-8A). In these organisms, the haploid phase produces gametes by mitosis and is termed the **gametophyte** (*phyte* = plant); the diploid plant producing meiospores is the **sporophyte.** The two plants can be identical in appearance (**isomorphic:** *iso* = alike; *morph* = form), or markedly different (**heteromorphic:** *heter* = different). This last type of life history is *sporic*, or *haplodiplontic*. It occurs in the bryophytes, as well as in all embryo-producing

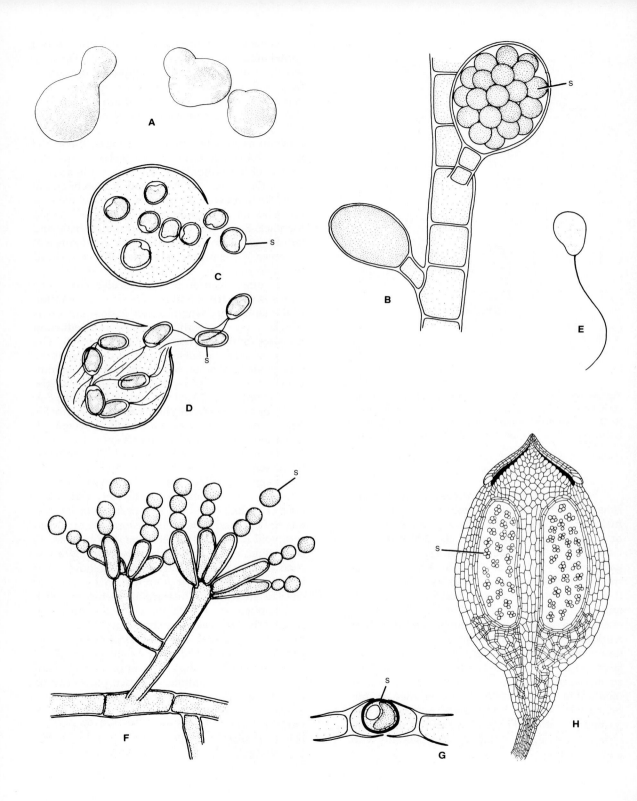

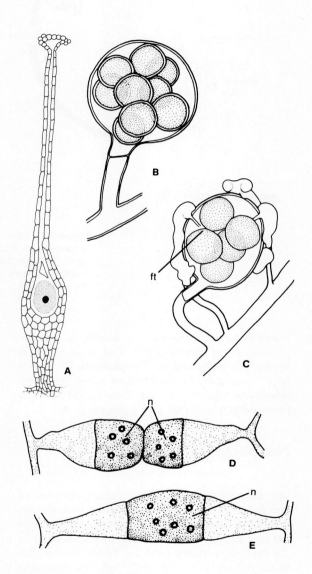

Figure 3-5 Variation in types of sexual reproduction in nonvascular eukaryotes. **A–C,** motile gametes; **D–F,** nonmotile gametes. **A–D,** isogamy; **B, E,** anisogamy (in a generalized sense, may also include **C, F**); **C, F,** oogamy (in a generalized sense, may include **E**).

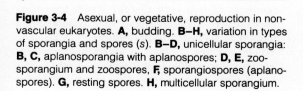

Figure 3-4 Asexual, or vegetative, reproduction in nonvascular eukaryotes. **A,** budding. **B–H,** variation in types of sporangia and spores (s). **B–D,** unicellular sporangia: **B, C,** aplanosporangia with aplanospores; **D, E,** zoosporangium and zoospores, **F,** sporangiospores (aplanospores). **G,** resting spores. **H,** multicellular sporangium.

Figure 3-6 Variation in gametangia in nonvascular eukaryotes. **A,** multicellular. **B,** unicellular. **C,** fertilization tubes. **D, E,** early and late stages in gametangial conjugation.

Nonvascular Eukaryotes 45

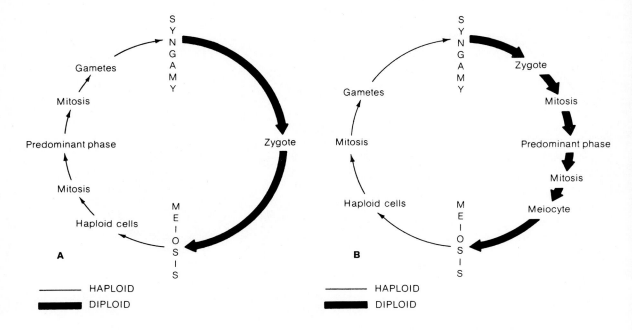

Figure 3-7 Types of alternation of life history, or phases (generations), restricted to nonvascular eukaryotes. **A,** haploid stage predominant (zygotic), with meiosis in zygote (only 2n stage). **B,** diploid stage predominant (gametic), with gametes produced by meiosis (only n stage); also typical of animals. (thin line = n stage; thick line = 2n stage)

plants (which include vascular plants, or tracheophytes), many marine algae, and some fungi. A fourth type of life history, the **dikaryotic** (*di* = two; *kary* = nucleus), is restricted to some fungi and involves a delay in time of nuclear fusion after the protoplasts have fused (Fig. 3-8B; see p. 47).

Nonphotosynthetic Forms

The nonphotosynthetic forms include slime molds and fungi, which are treated by some biologists in a single division, Mycota. However, as noted earlier (see p. 38), the slime molds are considered in the Protista, separate from the Fungi. This is due to differences in morphology, cell structure, and method of obtaining food. All slime molds are characterized by wall-less assimilative stages, and most are capable of feeding by ingesting particulate food. Fungi are primarily filamentous, although some are unicellular, and they generally have a rigid cell wall during most of their life history. Fungi cannot ingest food; instead they secrete **exoenzymes** capable of digesting materials outside the fungal cell. Subsequently, small organic and inorganic molecules produced by the enzymatic activity are taken into the cells. Food reserves of fungi are generally the polysaccharide **glycogen,** sugar alcohols, and lipids. The wall is composed of the polysaccharides, mainly **glucans** other than **cellulose,** although **chitin** and cellulose may be present in small amounts.

Slime molds Slime molds (Myxomycota) include organisms of two very distinct types: true slime molds (Class Myxomycetes), and cellular slime molds (Class Plasmodiophoromycetes, Class Dictyosteliomycetes, Class Labyrinthulomycetes). Cellular slime molds do not appear closely related to each other or to the true slime molds. All Myxo-

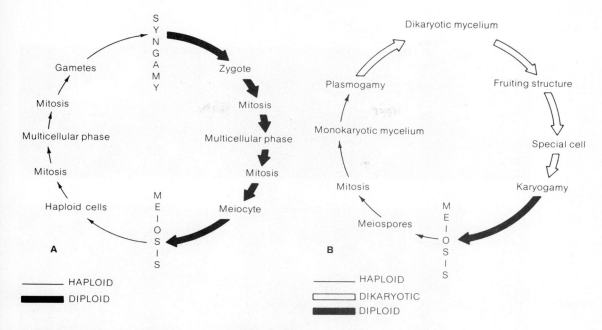

Figure 3-8 Types of alternation of life history, or phases (generations). **A,** alternation of haploid and diploid phases (sporic), with meiosis occurring in diploid plant (typical of plants). **B,** delayed nuclear fusion (dikaryon) occurring in many fungi. (thin line = *n* stage; open thick line = *n + n* stage; closed thick line = 2*n* stage)

mycota have naked assimilative stages and reproduce by spores. The assimilative body, which is multinucleate in true slime molds, is uninucleate in cellular slime molds. The Dictyosteliomycetes, Myxomycetes, and to a limited extent, Plasmodiophoromycetes, ingest solid food.

Fungi The fungi were originally classified in the plant kingdom because the conspicuous reproductive structures are anchored in the soil, and there are rigid, plantlike cell walls. Fungi comprise a large heterogeneous assemblage that presents formidable problems in classification. No system is completely acceptable to all **mycologists** (*myco* = fungus; *ology* = study). In this text several fungal divisions are recognized on the basis of morphology and reproduction. Some mycologists treat these divisions as separate classes in a single, large division. However, differences in cell wall composition, reproduction, and morphological characteristics separate the divisions. Fungi range in size from discrete units of a few micrometers to masses of material of indeterminate size (Fig. 3-9A). Many fungi are microscopic but produce easily observed reproductive structures (sometimes termed *fruiting structures,* or *fruiting bodies;* Fig. 3-9B, C). Some fungi appear closely related to flagellated algal groups; others differ markedly and are biochemically more similar to animals (including cell covering and storage reserves).

In many fungi, sexual reproduction occurs regularly, and the gametes are easily recognized. In those fungi with the modified life history mentioned earlier (Fig. 3-8B), the gametes are less obvious. In these, the time span for syngamy is divided into two distinct stages: **plasmogamy** (*plasmo* = form; *gam* = marriage), with only fusion of the cytoplasm and **karyogamy** (*kary* = nucleus), with fusion of the nuclei. Plasmogamy is followed by a period of time in which the paired gamete nuclei

Nonvascular Eukaryotes **47**

A

B

C

48 Chapter 3

divide synchronously, producing binucleate (bi = two) cells. Further divisions of the paired gamete nuclei produce more binucleate cells, forming the **dikaryon,** or $n + n$ stage. Eventually many of the nuclear pairs fuse (karyogamy), with meiosis usually following immediately to restore the haploid (n) phase.

Included with fungi are lichens (Fig. 3-10), which in the past have been treated as a separate division. The lichen association involves algal (**phycobiont:** $phyco$ = alga; bio = life) and fungal (**mycobiont:** $myco$ = fungus) partners and is one of the intriguing relationships in biology. Lichen fungi differ little from other parasitic fungi, except that the fungal filaments and algal cells associate to produce a single, characteristic "organism." As lichen reproductive structures are fungal in form, classification is essentially that of fungi.

Photosynthetic Forms

All photosynthetic eukaryotes contain chlorophyll a in their chloroplasts and release oxygen as a byproduct of photosynthesis. Other pigments are also present which absorb light energy that is passed to chlorophyll a for photosynthesis. Included in this group are eukaryotic algae, bryophytes and the vascular embryo-producing plants. Nonvascular, photosynthetic eukaryotes include only algae and bryophytes (see Table 3-2). Vascular, photosynthetic eukaryotes comprise the Tracheophyta (tracheophytes) (Chapters 12–26).

Algae The diversity of forms considered as algae has already been mentioned (see p. 38). The diversity, morphology, and physiology of the algal divisions is shown in both Table 3-2 and Figure 3-11.

Often chlorophyll a is masked by other pigments contained on or within chloroplast thylakoids. In this text the general term **chloroplast** is used for all chlorophyll-bearing organelles with membrane-limited lamellar structure.* Other pigments in algae that are of importance in photosynthesis and are used for classification are (1) other **chlorophylls** (chlorophyll b, c_1, c_2, d); (2) **carotenoids** ($carot$ = carrot), which include **carotenes** and **xanthophylls** ($xantho$ = yellow; $phyll$ = leaf); and (3) **phycobiliproteins**, composed of **phycocyanin** ($cyan$ = dark blue) and **phycoerythrin** ($erythr$ = red).

Potential chemical energy is retained in organic compounds, termed **photosynthate** ($photo$ = light; syn = together), or storage (reserve) products. Generally these are polysaccharide carbohydrates composed mainly of the **monosaccharide** glucose. They are classed either as starchlike or laminarinlike compounds. Information on fats, oils, proteins, and other compounds, such as sterols, present in the cells is less detailed, although they may be as important.

The cell covering is varied among algal groups. Often a definite wall is present at some stage in the life history. This can be cellulose, as in bryophytes and vascular plants, or some other polysaccharide. Red and brown algae (Fig. 3-11A) also have wall materials termed **phycocolloids** ($phyco$ = alga; $coll$ = glue). Additionally, inorganic substances, such as iron, calcium, and silica, are deposited in the wall (Fig. 3-11B, C). Some algae have a covering of inorganic or organic scales that are similar to the cell wall materials.

Separation of algal groups is based primarily on biochemical features such as pigments, storage products, and cell coverings, as well as on the nature and location of flagella (see Table 3-2).

Figure 3-9 Fungi. **A,** *Saprolegnia*, a water mold forming large white masses on the head and back of fish (such as this coho salmon). **B,** macroscopic "fruiting body" of the inky cap fungus (*Coprinus*). **C,** "fruiting body" of the wood-rotting polypore (*Polyporus*). (**A,** courtesy of G. Neish; **B, C,** courtesy of M. Higham.)

*Botanists generally use the term *chromatophore* for nongreen, photosynthetic organelles; however, in current usage by biologists in general, *chromatophore* applies to both nonphotosynthetic and/or nonlamellar structures. Strictly speaking, photosynthetic organelles should be named by their color. However, this can be confusing, so in this text, *chloroplast* is used for all photosynthetic organelles, regardless of color.

A

B

50 Chapter 3

Table 3-2 Major Distinguishing Characteristics of Photosynthetic Nonvascular Eukaryotes (Algae and Bryophytes)§

Division (common name)	Major Pigments*	Major Storage Products	Major Cell Covering*	Flagellation†	Habitat (approximate %)
Chrysophyta ‡ (chrysophytes)	chlorophyll a, c_1, c_2 carotenoids	laminarinlike	(cellulose) organic, $CaCO_3$, silicon	apical; 1, 2; unequal (equal); tinsel and/or whiplash	freshwater—60 marine—40
Pyrrhophyta (dinoflagellates)	chlorophyll a, c_2 carotenoids	starch	cellulose	lateral; 2; whiplash‖	freshwater—40 marine—60
Cryptophyta (cryptophytes)	chlorophyll a, c_2 (carotenoids) phycobiliproteins	starch	protein	subapical; 2; subequal; tinsel	freshwater—90 marine—10
Euglenophyta (euglenoids)	chlorophyll a, b (carotenoids)	laminarinlike	protein	apical; 1, 2; unequal; tinsel‖	freshwater—90 marine—10
Chlorophyta (green algae)	chlorophyll a, b (carotenoids)	starch	cellulose (xylan, mannan)	apical; 2, 4; equal; whiplash‖ (scales)	freshwater—60 marine—40
Phaeophyta (brown algae)	chlorophyll a, c_1, c_2 carotenoids	laminarin	cellulose phycocolloids	lateral; 2; unequal; tinsel and whiplash	freshwater—0.1 marine—99.9
Rhodophyta (red algae)	chlorophyll a, (d) (carotenoids) phycobiliproteins	starchlike	cellulose phycocolloids	none	freshwater—410 marine—90
Bryophyta (bryophytes)	chlorophyll a, b (carotenoids)	starch	cellulose	apical; 2; equal; whiplash (scales); male gamete only	aquatic—5 terrestrial—95

*Parentheses indicate a minor component.
†Flagellation: location; number; relative length; morphology.
‡Constituting a diverse assemblage in this text (Chapter 6).
§See text for details.
‖Delicate, nontubular hairs present in some representatives.

Figure 3-10 Lichens. **A,** *Peltigera*, which is fleshy and unrolled when water is available (branches of moss *Isothecium* present on left). **B,** *Cladonia*, which is morphologically unaffected by presence of water. (Courtesy of L. Goff)

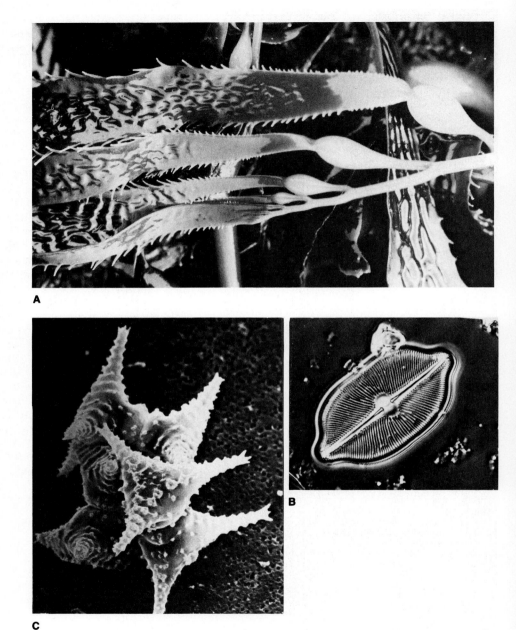

Figure 3-11 Algae. **A,** *Macrocystis,* a macroscopic brown alga, or seaweed, easily recognized in tidal areas. **B, C,** microscopic algae with inorganic materials deposited in organic cell wall: **B,** the diatom *Amphora* (ca. 30 μm long) with silica deposited to produce cell wall design (phase-contrast microscope); **C,** the desmid *Staurastrum* (ca. 40 μm long), with calcium carbonate deposited (scanning electron micrograph). (**A,** courtesy of T. Bisalputra; **C,** courtesy of H. Contant).

Bryophytes The last group of nonvascular eukaryotes, the Bryophyta, encompasses mosses and moss allies (liverworts, hornworts). Pigments, cell walls, and storage products (reserves) are the same as for most of the green algae (Division Chlorophyta) and vascular plants. However, the Bryophyta are an isolated group apparently not directly related to other green plants and probably not ancestral to any other group. Many botanists consider bryophytes as part of the large group of green terrestrial plants because of their habitat, the multicellular complexity of their reproductive structures, and the dependence of the young diploid embryo on the parent plant. Similarly, because of their life history (Fig. 3-8A) and gross morphology, bryophytes are placed near vascular plants. In both morphology and anatomy, the Bryophyta are structurally simpler than vascular plants; both gamete-producing (gametophyte) and meiospore-producing (sporophyte) phases are relatively conspicuous, each with distinctive morphology (Fig. 3-12). In a morphological and physiological sense, bryophytes are a link between those photosynthetic organisms occupying the aquatic and the terrestrial habitats, although they probably are not a link in an evolutionary sense.

The smallest bryophytes are a few millimeters, or nearly microscopic, and the largest erect forms are as much as 60 cm high; some of the creeping forms are several meters in length. Most bryophytes are strictly terrestrial and grow in humid environments for at least part of the year; however, some grow in arid sites with little humidity, and a few are aquatic.

Figure 3-12 Mosses are generally recognized by their leafy stems and projecting capsules. **A,** *Hygrohypnum*, showing the leafy gametophyte stage. **B,** *Polytrichum*, showing leafy gametophyte and apical sporophyte (Courtesy of L. Goff).

References

For references on fungi and slime molds, see Chapters 4 and 5; for specific algal divisions, see Chapters 6–9; for bryophytes, see Chapter 11.

Bold, H. C. 1977. *The plant kingdom.* 4th ed. Prentice-Hall, Englewood Cliffs, N.J.

Bold, H. C.; Alexopoulos, C. J.; and Delevoryas, T. 1980. *Morphology of plants and fungi.* 4th ed. Harper & Row, New York.

Bold, H. C., and Wynne, M. J. 1978. *Introduction to the algae.* Prentice-Hall, Englewood Cliffs, N.J.

Christensen, T. 1980. *Algae—A taxonomic survey.* vol. 1. AiO Tryk as, Odense, Norway.

Cronquist, A. 1981. *Basic botany.* 2nd ed. Harper & Row, New York.

Lee, R. E. 1980. *Phycology.* Cambridge Univ. Press, Cambridge.

Norstog, K., and Long, R. W. 1976. *Plant biology.* W. B. Saunders Co., Philadelphia.

Raven, P. H.; Evert, R. F.; and Curtis, H. 1981. *Biology of plants.* 3rd ed. Worth Publishers, New York.

Rushforth, S. R. 1976. *The plant kingdom: Evolution and form.* Prentice-Hall, Englewood Cliffs, N.J.

Scagel, R. F.; Bandoni, R. J.; Maze, J. R.; Rouse, G. E.; Schofield, W. B.; and Stein, J. R. 1982. *Nonvascular plants: An evolutionary survey.* Wadsworth Publishing Co., Belmont, Calif.

Trainor, F. R. 1978. *Introductory phycology.* John Wiley & Sons, New York.

4

FUNGI

As noted in Chapter 3, the fungi are an extremely large and heterogeneous group, placed in their own kingdom. There are approximately 100,000 species described, and estimates of the total number range to 250,000 or more. Although some species produce conspicuous reproductive structures (e.g., mushrooms, bracket fungi, and puffballs), most fungi are inconspicuous or microscopic in all phases of their life histories. Fungi are eukaryotic heterotrophs dependent upon other organisms for certain organic nutrients. Some are parasites of plants or animals; others gain nutrients from the **mutualistic symbionts** with which they are associated. For most fungi, however, dead plant or animal remains are the nutrient sources. Unlike slime molds (see Chapter 5), most fungi have rigid cell walls around their assimilative structures and do not ingest solid food particles. Instead, the fungi absorb nutrients as do heterotrophic bacteria: they produce **extracellular enzymes** that chemically break down complex substances (e.g., cellulose and starch) into smaller molecules that can be absorbed.

Both sexual and asexual reproduction occur in many species of fungi; in others, only one type of reproduction is known, which may be either sexual or asexual. The numerous types of sexual and asexual reproductive units are generally designated as **spores.** However, spores differ from one another in method of formation, general structure, and method of germination. The classification system of fungi is based largely upon the kinds of spores that are produced by each major group. The fungi are often informally classed as the *lower fungi*, which are considered primitive, and the *higher fungi*, as shown in Table 4-1.

General Characteristics

The Thallus

The assimilative structure, or **thallus,** of a fungus can be uninucleate or multinucleate. Many uninucleate thalli are spherical cells and can have rootlike absorptive structures, the **rhizoids** (*rhiz* = root; *oid* = like) (Fig. 4-1A). Multinucleate thalli are characteristic of most fungi, and these typically consist of **hyphae** (sing. **hypha;** *hyph* = web) (Fig. 4-1B). The growth of uninucleate thalli is limited; hyphae,

Table 4-1 Differences in characteristics of the divisions and subdivisions of fungi (parentheses indicate occurrences in a few representatives)

Group	Reproduction		Assimilative Features	Cell Wall Composition
	Asexual*	Sexual		
LOWER FUNGI				
Division Oomycota	zoospores; anterior, lateral; 2; whiplash, tinsel	gametangial contact; meiosis precedes gamete formation	$2n$; unicellular or nonseptate hyphae	cellulose-glucan (chitin-glucan)
Division Chytridiomycota	zoospores; posterior; 1; whiplash	motile isogametes (entire thalli); meiosis in zygote germination	n, $(2n/n)$; unicellular or nonseptate hyphae	chitin-glucan
Division Hyphochytridiomycota	zoospores; anterior; 1; tinsel	generally unknown (entire thalli?)	n?; unicellular	cellulose-chitin
Division Zygomycota	sporangiospores (nonmotile); (conidia)	gametangial fusion; meiosis in zygote formation	n?; nonseptate hyphae	chitin-chitosan chitin-glucan
HIGHER FUNGI				
Division Eumycota Subdivision Ascomycotina	budding or conidia	gametangial fusion or nuclear transfer	n, $n + n$; (unicellular), septate hyphae	chitin-glucan (mannan-glucan)
Subdivision Basidiomycotina	(conidia)	gametangial fusion or nuclear transfer	(n), $n + n$; (unicellular); septate hyphae	mannan-glucan mannan-chitin chitin-glucan
Form Class Fungi Imperfecti	conidia	unknown	?; septate hyphae	chitin-glucan

*Includes insertion, number, and type of flagella, when present

on the other hand, usually have unlimited growth potential within limits imposed by the substrate. Increase in hyphal length is restricted to approximately the first micrometer at the hyphal tip. As the hypha grows and continuously contacts new material, a pathway is digested by enzymes located at the hyphal surface or secreted into the area surrounding the hyphal tip. The products of digestion and other small molecules present are then absorbed for use within the hyphae.

Hyphae of a single thallus produce a multi-branched network referred to as a **mycelium**. Hyphae of primitive fungi are **coenocytic** (*coeno* = common; *cyte* = cell); they do not have transverse walls that divide the structure into compartments. However, transverse walls, or **septa** (*sept* = fence),

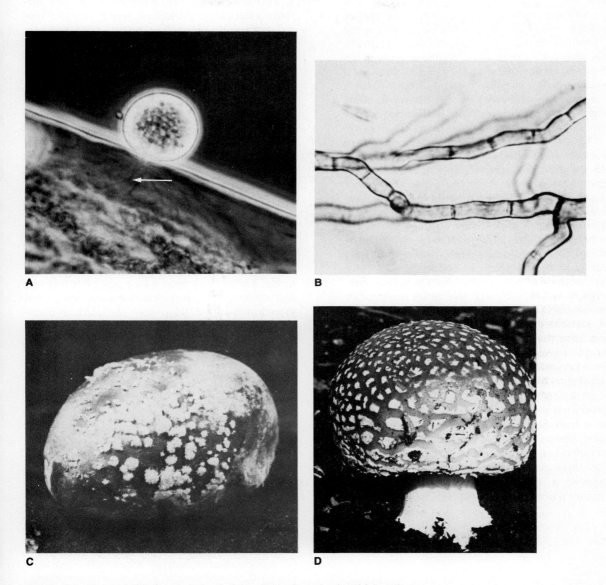

Figure 4-1 Fungal thalli. **A,** unicellular thallus, with nucleated spheroidal portion external, and rhizoids (*arrow*) penetrating host (algal cell) (*Rhizophydium;* Chytridiomycota), ×1280. **B,** hyphae, showing septa, ×650. **C,** plum covered with conidia and conidiophores produced by internal mycelium (*Monilinia,* Ascomycotina), ×1. **D,** sporocarp, or fruiting structure, of higher fungus (*Amanita;* Basidiomycotina), ×0.5. (**A,** courtesy of D. J. S. Barr; **B,** courtesy of M. Dunn.)

Fungi 57

are characteristic of most fungal hyphae (Fig. 4-1B). Most septa do not completely separate the contents of adjacent "cells" as they often have a central pore in the wall so that the cytoplasm is continuous from cell to cell.

Although most species produce either hyphae or single cells, some **dimorphic** (*di* = two; *morph* = form) fungi can produce both types of thalli. This phenomenon is most common among the primitive ascomycetes and basidiomycetes and usually entails a yeastlike haploid phase and the *n* + *n* nuclear (dikaryotic) hyphal phase. In some fungi, environmental conditions determine whether unicellular or filamentous growth occurs.

Fungi with single cells lacking absorptive structures commonly occupy habitats in which their cells are bathed in nutrient solutions. For example, some of the parasitic fungi live within the cells of their hosts, absorbing nutrients from the surrounding fluid. Others, such as yeasts, grow in water or plant exudates, such as flower nectar. If the fungal thallus is a single cell with rhizoids, the cell usually develops outside the host or substrate, with only the rhizoids embedded (Fig. 4-1A). Mycelial thalli often consist of both aerial and submerged hyphae, although the entire assimilative portion can be embedded, and only the reproductive structures exposed (Fig. 4-1C). Many parasitic fungi have the hyphae equipped with specialized absorptive branches called **haustoria** (*haus* = suck), which penetrate host cells but not the host protoplast (Fig. 4-2A).

In the higher fungi, the hyphae composing the reproductive structures, the **sporocarps** (*spor* = seed; *carp* = fruit; Fig. 4-1D) can be thick-walled or thin-walled. The cells often become inflated to the extent that individual hyphae are no longer recognizable. When these individual hyphal cells become greatly expanded, a "tissue" resembling the parenchyma of higher plants results (Chapter 10).

Cells

Initial wall material is produced at the hyphal tip where extension is occurring; secondary thickening and septa development occur a short distance back. Approximately 80–90% of the wall is polysaccharide; the remainder consists of proteins, lipids, and sometimes pigments. The polysaccharides include **chitin, chitosan, cellulose, mannan, glucan,** or **galactan.** Differences in cell wall chemistry correlate well with morphological and reproductive differences (see Table 4-1).

Within the cell wall, a plasma membrane surrounds the fungal protoplast, which contains typical eukaryotic organelles: nuclei, mitochondria, vacuoles, golgi apparatus, and endoplasmic reticulum (Fig. 4-2B, C). Vacuoles are often not a large part of the cell volume as they are in some algae and most higher plants, although very old fungal cells can have large vacuoles. Also, the amount of endoplasmic reticulum is not as great in fungi as in cells of other organisms. Food reserves are present as glycogen (a compound similar to starch) or lipids. Lipids are usually visible in hyphae and spores as conspicuous globules. Glycogen is present as small granules.

The nuclei are small and have a typical envelope consisting of a double membrane with pores, a nucleolus, and chromosomes. During mitosis, the nuclear membrane generally remains intact (Fig. 4-3). The chromosomes are difficult to observe because of their small size and the persistent nuclear membrane. Rarely is a typical metaphase plate formed, and separation of chromosomes at anaphase is often **asynchronous.** At telophase, the nuclear membrane constricts between the two poles, forming the two new nuclei. Centrioles are present in the nuclei of fungi that produce motile cells; a possibly

Figure 4-2 Fungal structure, electron microscope (Oomycota). **A,** haustorium (*h*) from fungal cell (*hmc*) in host cell: *hv* = host vacuole; *cl* = host chloroplast; *is* = intercellular space, *Albugo*, ×4240. **B, C,** *Pythium*; **B,** thin section of hypha, showing nucleus (*n*), dictyosome (*d*), endoplasmic reticulum (*er*), mitochondrion (*m*), vesicles (*v*). **C,** growing apex of hypha, showing characteristic vesicles, wall (*w*), mitochondria. (**A,** courtesy of M. D. Coffey, reproduced by permission of the National Research Council of Canada from the *Canadian Journal of Botany* 53:1285–99, 1975; **B, C,** courtesy of S. N. Grove and C. E. Bracker, with permission of *Journal of Bacteriology.*)

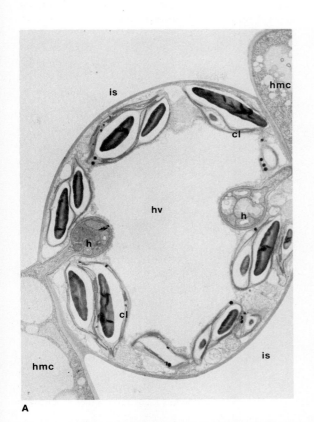

A

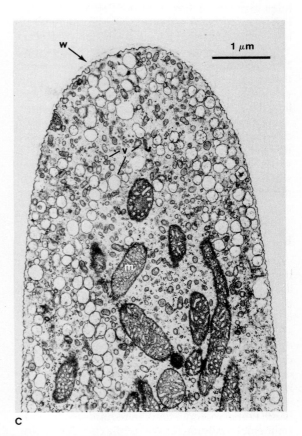

C

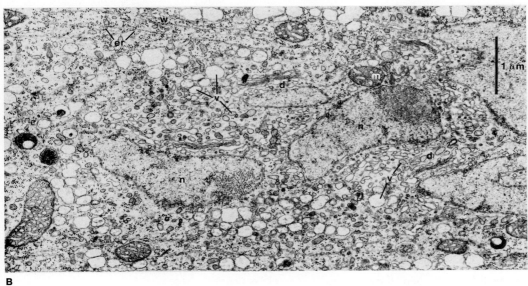

B

homologous structure is present in higher fungi. In hyphae with uni- or binucleate cells, mitosis may be accompanied by septum formation; however, the processes are independent of one another. Meiosis, too, is characterized by the persistence of the nuclear membrane. The chromosomes become more conspicuous during meiosis, which differs little from the meiotic division in other eukaryotes.

Ribosomes are abundant and sometimes associated with the endoplasmic reticulum. The molecular weight of one type of ribosomal RNA (25-S) varies among the different groups of fungi. Some scientists consider such differences great enough to indicate that fungi evolved from several different ancestral groups.

Reproduction

In morphologically simple fungi, the entire assimilative protoplast can become reproductive, with the differentiation being internal. Most fungi produce special reproductive cells (e.g., gametangia, sporangia) as distinct entities, so that reproduction does not involve the entire thallus; thus, reproductive and assimilative growth occur simultaneously. Asexual reproduction is often favored or at least permitted under the wide range of conditions in which assimilative growth occurs. On the other hand, sexual reproduction can require special environmental conditions as well as the chance encounter of gametes.

Asexual reproduction in the lower fungi generally is by the formation of sporangiospores within sporangia. The sporangiospores are produced by cleavage of the sporangial protoplast into segments containing one to several nuclei. Each segment then develops into either an unwalled zoospore (planospore) or a walled, nonmotile aplanospore (Fig. 4-7B).

The nature of the zoospore is considered basic in classification of lower fungi (see Table 4-1). Two types of flagella occur: tinsel and whiplash (Fig. 4-4A, B). The number, insertion, and type of flagella vary in the fungi (see Table 4-1; Fig. 4-4C–F). Unlike myxomycete motile cells, fungal zoospores are incapable of feeding **phagotrophically** or undergoing fission (see Chapter 5).

Walled sporangiospores, which presumably evolved from motile spores, are an adaptation to terrestrial existence. When released, these spores

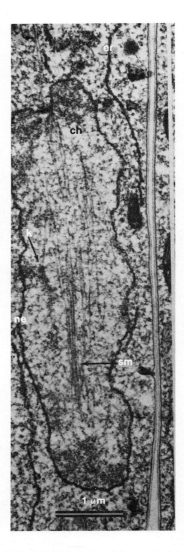

Figure 4-3 Cell division in fungi. Late anaphase of mitotic division (*Fusarium;* Ascomycotina), showing chromosomes (*ch*), spindle microtubules (*sm*), intact nuclear envelope (*ne*), kinetochore (*k*), endoplasmic reticulum (*er*). (Courtesy of J. R. Aist and P. H. Williams, reproduced from "Ultrastructure and time course of mitosis in *Fusarium oxysporum*," *Journal of Cell Biology* 55:368–89, 1972, by copyright permission of The Rockefeller University Press.)

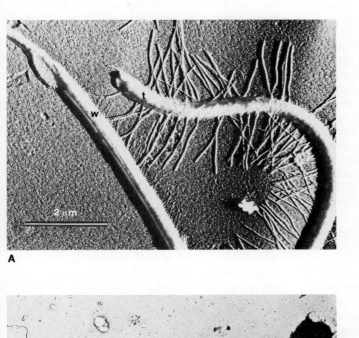
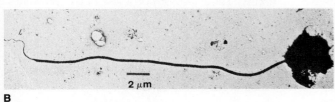
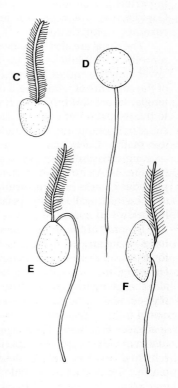

Figure 4-4 A, B, Electron micrograph of zoospore flagella. **A,** tinsel (*t*) and whiplash (*w*) flagella (Oomycota). **B,** whiplash flagellum on zoospore (Chytridiomycota). **C–F,** types and arrangement of flagella of fungal motile cells; **C,** single anterior tinsel flagellum (Hyphochytridiomycota); **D,** single posterior whiplash flagellum (Chytridiomycota); **E, F,** anterior and lateral whiplash and tinsel flagella (Oomycota), showing tinsel flagellum anteriorly directed and whiplash posteriorly directed. (**A,** courtesy of P. R. Desjardins, G. A. Zentmyer, and D. A. Reynolds, reproduced by permission of the National Research Council of Canada from the *Canadian Journal of Botany* 47:1077–79, 1969; **B,** courtesy of F. Y. Kazama, from "Ultra-structure and phototaxis of the zoospores of *Phlyctochytrium* sp., an estuarine chytrid," *Journal of General Microbiology* 71:555–66, 1972, with permission of Cambridge University Press.)

are dispersed by air currents, insects, or splashing rain. All higher fungi and some lower fungi lack sporangia. Where present, asexual reproduction is by **budding** or formation of **conidia** (sing. **conidium;** *conidi* = dust). **Budding** involves the formation of a cell outgrowth that enlarges and finally separates after receiving a nucleus and other organelles from the parent cell (Fig. 4-5A). Conidia are essentially separable portions of hyphae or of specialized hyphal branches (Fig. 4-5B). Some species produce several conidial types, and others one type or none. In one group of higher fungi (Fungi Imperfecti), reproduction is entirely by conidia. Conidia, like aplanospores, are dispersed by air currents, water, or insects.

The fungi are also varied with respect to sexual reproduction. Motile gametes occur only in some lower fungi. In other lower fungi, there is fusion of the contents of whole thalli or specialized gametangia. In most of the lower fungi, syngamy leads to the formation of either a thick-walled, resistant spore or sporangium, or to the production of a diploid thallus. Germination of the resistant spores or sporangia (basically zygotes) is preceded by meiosis (see Fig. 3-7A); however, in the Oomycota, meiosis precedes gamete nucleus formation (see Fig. 3-7B), and the thick-walled spore produces a diploid thallus or diploid zoospores.

In some of the morphologically simple higher fungi, syngamy, or the equivalent, is followed by formation of a resistant structure or diploid thallus. However, for most of the higher fungi, the equivalent of syngamy is divided into two separate phases, with protoplast fusion (**plasmogamy**) separated from nuclear fusion (**karyogamy**) in time and space (see Fig. 3-8B). Plasmogamy occurs through fusion of either hyphae or specialized reproductive cells and involves only coalescence of the protoplasm. The pair of compatible but unfused nuclei that results is called a **dikaryon,** and the cell or hypha containing the paired nuclei is said to be *dikaryotic*. Within dikaryotic hyphae, mitotic divisions of the nuclei lead to the development of many dikaryotic cells. Eventually, special cells are produced on the hyphae where karyogamy and meiosis occur.

In the Ascomycotina, the dikaryotic phase is of limited extent, whereas in the Basidiomycotina, the dikaryotic mycelium is of potentially unlimited duration. In fact, this dikaryotic phase is the dominant assimilative phase in the life history of most of the basidiomycetes (which include the mushrooms). The sexual spores (called **ascospores** or **basidiospores**) are generally produced in the sporocarps or **fruiting bodies** (Fig. 4-1D).

In many higher fungi, the sexual stage is completely absent, and in others it appears infrequently. This presents problems in identification, since the classification systems are based upon sexual structures. The problem is of special concern to applied mycologists, because the asexual stage is often the one present on diseased plants and animals. Also, many economically important fungi have no known sexual stages; thus a form class, Fungi Imperfecti, has been established for them (see p. 111).

Genetic variability in some of the higher fungi occurs by **heterokaryosis** (*heter* = differ; *kary* = nut, nucleus) or by **parasexuality** (*para* = beside). In heterokaryosis, genetically different nuclei of several types can occur in a single thallus. Such mycelia develop as a result of genetic mutation, germination of a **heterokaryotic** spore, or by fusion of hyphae from two separate mycelia. The different nuclei within a heterokaryotic mycelium exist side by side in a common cytoplasm; each nuclear type increases by mitosis, and each type exerts its influence on the growing mycelium.

Parasexuality involves the fusion of haploid nuclei within a heterokaryotic mycelium during a mitotic cycle. The diploid nuclei thus formed divide mitotically and increase in number together with haploid nuclei. Infrequently, mitotic crossing-over occurs. There also can be a loss of chromosomes. Both these phenomena lead to genetic variability (as in sexual reproduction); however, the events are not regular and do not occur in specialized cells or structures. The types of sexual and asexual reproduction are of primary importance in the classification of fungi (see Table 4-1).

Nutritional and Environmental Requirements

The nutritional requirements of most fungi are relatively simple and include organic substances for energy sources as well as water and mineral nutrients. Beyond this, the only nutritional requirement common to many species is an external supply of one or a few vitamins (often biotin or thiamine). In laboratory cultures, glucose is commonly used to provide both energy and carbon. The minerals added to culture media include relatively large amounts of nitrogen, phosphorus, potassium, and magnesium, with traces of certain other elements also necessary. Nitrogen is often provided in organic forms, such as amino acids. In nature, fungi obtain the necessary nutrients from living or dead plants or animals. The **saprobic** species are extemely versatile scavengers, utilizing many different substances. Some fungi exist as mutualistic symbionts with other organisms, both eukaryotic and prokaryotic. One of the best known associations is the lichen, which involves an algal and a fungal partner.

Many fungal species live in aquatic or semi-aquatic environments with moisture always available. Others, whose habitats are sometimes wet and sometimes dry, grow when moisture is available and become dormant during dry periods. Once established, the dry-rot fungi produce their own water as they metabolize the cellulose in wood. However, these fungi require moisture for spore germination and initial establishment of the mycelium. Many common molds are capable of slow growth even where available moisture is low, and some extensive losses or deterioration of stored wheat and other grains can result. A few fungi grow in the absence of oxygen, but most are **aerobic** (i.e., require oxygen).

Most fungi are capable of growth over a broad range of temperatures and, like bacteria, some species are adapted to either high or low temperatures. However, **thermophilic** fungi mostly thrive at temperatures of 45–60 °C. The low-temperature fungi, the **psychrophiles** (*psychr* = cold), often grow at 10 °C or less. Several psychrophilic species are economically important—they cause the snow mold

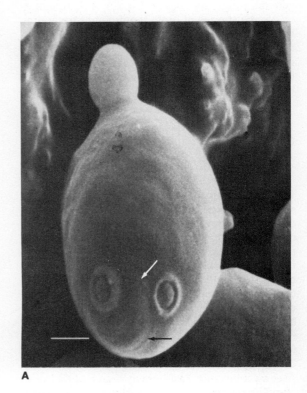

A

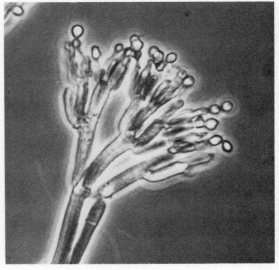

B

Figure 4-5 Asexual reproduction. **A,** budding (*Saccharomyces*, or baker's yeast; Ascomycotina) at an early stage, showing new cell at top of parental cell and scars from previous cells formed by budding (*arrow*). **B,** conidiophore bearing terminal conidia (*Penicillium*; Ascomycotina), ×1145. (**A,** courtesy of L. T. Talens, M. Miranda, and M. W. Miller, with permission of *Journal of Bacteriology*.)

diseases of wheat and other grasses that develop rapidly under an insulating blanket of snow. In general, fungi grow best in the 20–30 °C range.

Some species of fungi require light for assimilative growth or for the induction of reproductive stages. Some species require alternating light and dark periods for the production of typical sporulating structures; in some, the reproductive structures exhibit a positive or negative response to light.

Classification

The fungi are treated here as several separate divisions (see Table 4-1). The lower fungi include four divisions—Oomycota, Chytridiomycota, Hyphochytridiomycota, and Zygomycota—that are not presently thought to be related. They do share a number of common features, such as simple thalli or **nonseptate** hyphae, sporangiospores, and lack of complex fruiting bodies (sporocarps). On the basis of these features, they were once considered to belong to the single class Phycomycetes, and this name is still in use as a common name for the lower fungi. The higher fungi constitute a single division, the Eumycota, which includes two subdivisions, Ascomycotina and Basidiomycotina, and the form class Fungi Imperfecti. The fruiting bodies of Ascomycotina and Basidiomycotina are morphologically different, but life-history and other features show them to be closely related. The form class Fungi Imperfecti contains those higher fungi for which no sexual stages are known. This text considers only the more commonly occurring and better known fungi in the four divisions of the lower fungi, and the Eumycota.

Division Oomycota

The division Oomycota includes the single class Oomycetes and four orders: Saprolegniales, Leptomitales, Peronosporales, Lagenidiales. Most Oomycetes have a well-developed mycelium, with cellulose, chitin, and β-glucan present in the cell walls. Some do not produce glycogen, and their reserve carbohydrate is another β-glucan similar to that in some brown algae (Chapter 8). The multinucleate hyphae produce septa only to separate reproductive structures from the assimilative thallus. Asexual reproduction is by biflagellated zoospores (Fig. 4-4E, F) or sporangia that function as conidia. The zoospores have either anterior or lateral insertion of the flagella, although both types are present in the life histories of some forms.

Sexual reproduction in the mycelial species is by the production and contact of male and female gametangia (usually terminal antheridia and oogonia). Recent studies indicate that the assimilative thallus is diploid (Fig. 5-6). Following meiosis in the gametangia, the haploid nuclei in the antheridia are transferred to the oogonium. One or more haploid nuclei in the oogonium function as gametes, and thick-walled zygotes are formed. In the nonmycelial Lagenidiales, the gametangia are morphologically similar, and the entire protoplasts fuse. Only two orders are discussed here as representative of the Oomycota.

Order Saprolegniales The Saprolegniales inhabit fresh water or soil. They grow primarily on dead plant or animal material, although a few species parasitize algae, higher plants, or animals. The filamentous species, especially of *Saprolegnia* (Fig. 4-7) and related genera, are among the best-known aquatic fungi.

The Saprolegniales have well-developed coenocytic mycelia, with septa usually present only at the base of reproductive structures (Figs. 4-7A and 4-8). Sporangia are only slightly modified hyphal tips. The zoospores are of two types: **primary zoospores** that are pear-shaped and anteriorly biflagellated (Fig. 4-4E) and **secondary zoospores** that are bean-shaped with laterally inserted flagella (Fig. 4-4F). The sporangial protoplast is multinucleate and, following cleavage (Fig. 4-7B), produces uninucleate primary zoospores (Fig. 4-7C). They are released by internal pressure that rapidly forces them out. Once released, the zoospores swim for a short period, then withdraw their flagella and encyst (Fig 4-7D). These primary cysts germinate after a short period, giving rise to laterally flagellated secondary zoospores, which are motile for a time before encysting (Fig. 4-7E, F). Depending upon environmental conditions, this secondary cyst can germinate to produce either another secondary zoospore or a **germ tube,** which is the start of a new mycelial thallus (Fig. 4-7G).

Oogonia usually develop as terminal cells on short lateral branches, but **intercalary** (*intercal =* insert) oogonia are produced by some species. Each oogonium contains from one to twenty or more heavy-walled zygotes at maturity. The antheridium can arise from the same hypha as the oogonium, from a different hypha, or from a different thallus. The antheridia produce **fertilization tubes** that penetrate the oogonial wall, and the haploid nuclei are transferred through these tubes (Fig. 4-8). After fertilization, each zygote develops into a thick-walled resistant structure, often called the **oospore,** which can retain its viability for several years. When germination occurs, a short hypha tipped by a sporangium is usually produced.

The antheridia and oogonia of some saprolegniaceous species develop on separate thalli, and sexual reproduction occurs only if two such thalli exist side by side. For example in some species of *Achyla*, sexual reproduction occurs only when compatible thalli are brought together. The initiation and development of the gametangia occur in a series of clearly defined steps controlled by a series of hormones. Two of these hormones have been isolated.

Some species of *Achyla* and *Saprolegnia* infect living fish eggs and amphibians. These fungi start as wound parasites on adult fish; once established, however, the mycelium rapidly invades healthy tissues. Especially if crowded, fish in aquaria and in hatcheries often become infected. Salmon moving into fresh water on their spawning journey are also frequently infected (see Fig. 3-9A).

Order Peronosporales The Peronosporales are characterized by sexual reproduction similar to that of the Saprolegniales; however, only one egg is produced in the oogonium (Fig. 4-9B). Asexual reproduction of the soil- and water-inhabiting species is also similar. However, the more advanced members of the Peronosporales are obligate parasites of higher plants. Their sporangia are deciduous, functioning in dispersal, and can release zoospores, or produce hyphae directly. If zoospores are produced, they are of the secondary type (laterally biflagellated) (Fig. 4-4F). The hyphae of most obligately parasitic Peronosporales grow between the host cells and have haustoria (Fig. 4-

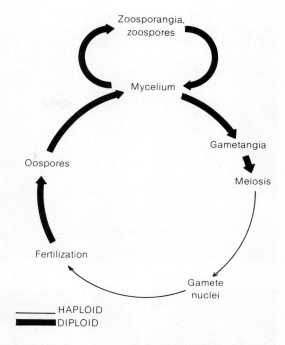

Figure 4-6 Oomycota. Life history characteristic of filamentous forms.

2A). Species of *Pythium* and *Phytophthora* occur in freshwater habitats and in soil, although some *Phytophthora* species are common in marine waters. Some soil-inhabiting species of both genera cause such plant diseases as damping off, root rots, and fruit rots.

Phytophthora infestans is a well-known parasitic species causing late blight of potato. The mycelium rapidly invades tissues of the plants, causing death of the infected parts. Lemon-shaped sporangia are produced on special aerial hyphae, which grow out through the host stomata (Fig. 4-9C). The sporangia are deciduous and airborne. At temperatures of 10–15 °C, sporangial germination is by zoospore release; at higher temperatures, germ tubes develop directly into hyphae. During periods of high humidity and relatively cool temperature, sporangia cause the rapid spread of the blight. The productivity of the potato plants is reduced by the death of leaf tissue and diminished photosynthesis. Infection and destruction of the tubers also occurs.

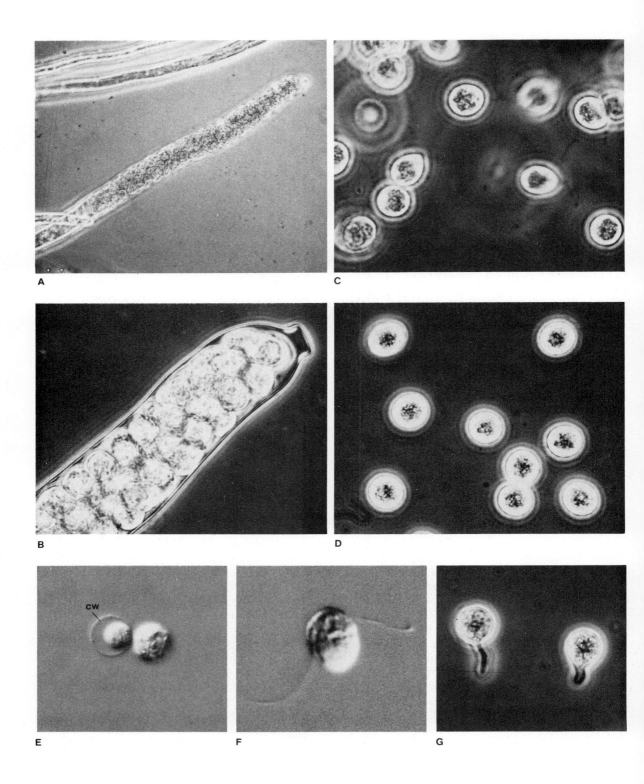

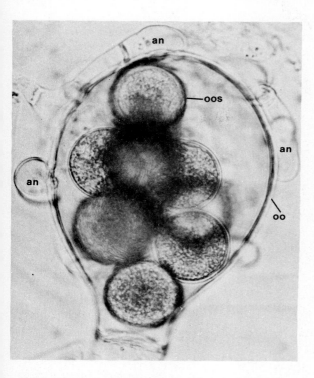

Figure 4-8 Sexual reproduction in Saprolegniales (*Saprolegnia*), showing oogonium (*oo*) with appressed antheridia (*an*) and developing oospores (*oos*); note septum at bottom of oogonium (*arrow*), × 4630.

Figure 4-7 Asexual reproduction in Saprolegniales (*Saprolegnia*). **A,** mature sporangium, with basal septum (*arrow*) and apical exit papilla (*upper right*), × 300. **B,** sporangial apex just prior to zoospore discharge, × 900. **C,** primary zoospores, × 900. **D,** primary cysts, × 900. **E,** germinating primary cyst (*cw*, cell wall), × 900. **F,** secondary zoospore, × 1400. **G,** germinating secondary cysts producing germ tubes, × 900.

P. infestans caused complete destruction of the potato crops of Ireland and parts of Europe during 1845–1847. At that time, the diet of many of the poorer people consisted largely of potatoes, and the crop loss resulted in the Irish famine. Thus, this plant disease caused much human suffering and was indirectly responsible for the emigration of many Irish to North America.

A large number of the Peronosporales are obligate parasites of terrestrial plants. They cause two types of disease: the downy mildews of grape, lettuce, onions, sugar beets, maize (corn), cabbage, and turnips; and the white rusts of cabbage and Brussels sprouts. The hyphae grow between the cells of the host, producing haustoria that penetrate the host cell walls. The sporangia of the downy mildews form on aerial hyphae; those of the white rusts develop under the host epidermis.

The fungus causing downy mildew of grapes is native to North America, but in the latter half of the nineteenth century, it was accidentally introduced into Europe. Within a few years it nearly destroyed the grape and wine industry of France. The story goes that a professor walking through the vineyards of Bordeaux observed that plants next to the trail were healthy and green; those further away had been ravaged by mildew. Upon inquiring, he learned that the vines next to the trail had been treated with a mixture of lime and copper sulfate to discourage passersby from eating the grapes. This observation led to the formulation of Bordeaux mixture, the first foliar fungicide. The mixture is also effective against a number of other fungal plant diseases.

The white rusts, caused by a species of *Albugo*, occur primarily on members of the mustard family. The sporangia are produced in compactly arranged chains below the host epidermis (Fig. 4-9D). This produces a blisterlike swelling of the epidermis, which eventually ruptures and releases the powdery white masses of sporangia.

The Peronosporales exhibit a complete range, from the strictly aquatic to well-adapted terrestrial types. In the primitive species, the sporangia remain attached to their hyphae and only produce zoospores. In the more advanced species, the deciduous sporangia are airborne, and can either produce zoospores or germinate by germ tubes. In some species, the sporangia no longer function as such and can only germinate by the production of a germ tube.

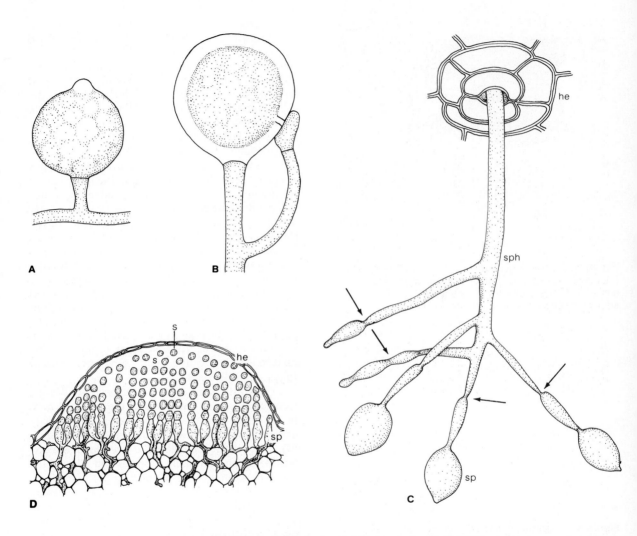

Figure 4-9 Reproduction in Peronosporales. **A, B,** *Pythium debaryanum:* **A,** hypha with young zoosporangium, ×1000; **B,** oogonium, with single oospore and appressed antheridium, ×1000. **C,** *Phytophthora infestans* sporangiophore (*sph*) from host epidermis (*he*), with sporangia (*sp*) and region of previous sporangium formation (*arrows*), ×470. **D,** *Albugo cruciferarum,* section through host tissue, showing host epidermis, host cells, and fungus with aerial hyphae, sporangia (*s*), and hyphae among host cells, ×650.

Divisions Chytridiomycota and Hyphochytridiomycota

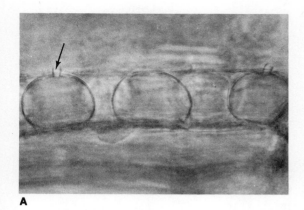

Two other divisions, previously allied with the Oomycota, are the Chytridiomycota and Hyphochytridiomycota. Extensive asexual reproduction by means of uniflagellate zoospores characterizes both divisions. In addition, both contain unicellular representatives. The division Chytridiomycota is larger and more diverse, with at least three orders recognized. Only a single class and order of Hyphochytridiomycota are known.

Chytridiomycota

The thallus in Chytridiomycota includes unicellular as well as mycelial types. The cell walls are composed of chitin and glucans but lack cellulose. The zoospores and gametes commonly have a single, posterior whiplash flagellum (Fig. 4-4D). Sexual reproduction ranges from isogamy through anisogamy to oogamy in those species with motile gametes. Some species have fusion of assimilative cells. Chytridiomycota are predominantly saprobic aquatic fungi, but many exist as parasites of other fungi, algae, aquatic animals, or terrestrial plants. Some saprobic species degrade such resistant substances as chitin and **keratin** (*kera* = horn), but few are of economic importance. Three orders are generally recognized: the Chytridiales, Blastocladiales, and Monoblepharidales.

The Chytridiales, or chytrids, are structurally simple fungi and mainly inhabit fresh water, although some occur in the marine habitat, in soil, or as parasites on higher plants. The chytrid thallus

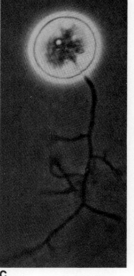
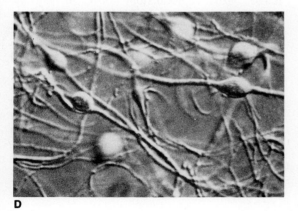

Figure 4-10 Thallus of Chytridiales. **A,** in root cells of cereal grain plants, showing exit tube (*arrow*) (*Olpidium*), ×785. **B, C,** unicellular thallus, with developing rhizoids (*Rhizophydium*). **D,** narrow filaments and swollen nucleate portions (*Nowakowskiella*), ×200. (**A,** courtesy of D. J. S. Barr; **B, C,** courtesy of D. J. S. Barr; reproduced by permission of the National Research Council of Canada from the *Canadian Journal of Botany* 48:1067–71, 1970.)

can consist of a single, uninucleate cell with or without rhizoids (Fig. 4-10A–C), or a primitive type of mycelium (Fig. 4-10D). In most chytrids, the thallus produces a single reproductive structure (gametangium or sporangium) with much or all of the protoplast being utilized in the process. When a zoospore germinates, it either casts off or withdraws its single posterior flagellum, and encysts before a germ tube develops. The tube can penetrate the host cell and discharge the protoplast to develop there (Fig. 4-11A), or the tube produces a rhizoidal system (Fig. 4-11B). Gametes, identical to zoospores, can fuse to produce a zygote, which later forms a thick-walled resistant sporangium. Meiosis probably occurs during germination of the resistant sporangium, and zoospores are produced.

The Blastocladiales produce zoospores and gametes resembling those of chytrids. Generally, a well-developed mycelium lacking transverse walls is produced, although some chytridlike thalli occur. Blastocladiales are mainly saprobic in freshwater or soil, although some can be parasitic. In a few species sexual reproduction is by motile isogametes or anisogametes. The resulting zygote ultimately produces by mitosis a diploid thallus bearing both thin-walled and resistant sporangia. The zoospores from the thin-walled sporangia are produced by mitosis and grow into more diploid thalli. In contrast, zoospores produced by the resistant sporangia result from meiosis and grow into the haploid, gamete-bearing thallus. This, then, is an alternation of gamete-bearing and spore-bearing generations, as is typical of terrestrial plants and some algae (see Fig. 3-8A). In *Allomyces,* the generations are isomorphic (Fig. 4-12); the haploid phase has orange-colored male gametangia and colorless female gametangia (Fig. 4-12**G, H**), and the diploid phase has both brown heavy-walled resistant and colorless thin-walled sporangia (Fig. 4-12**A, B, M**).

The Monoblepharidales generally occur in clean fresh water, often on waterlogged twigs. They are the only fungi in the classical sense with oogamy; that is, the female gamete is large and generally nonmotile, whereas the male gamete is small and motile. Thin-walled zoosporangia occur on the hyphae and produce zoospores that maintain the vegetative phase (Fig. 4-13A, B). Fertilization of the large egg occurs within the gametangium, and the resulting zygote forms a resistant heavy-walled oospore (Fig. 4-13C).

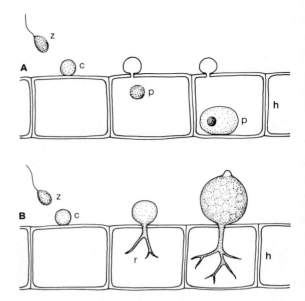

Figure 4-11 Chytridiales. Diagrams showing common zoospore (z) encystment in simple chytrids (c) in host (h): p = protoplast; r = rhizoid. **A,** *Olpidium* type; **B,** *Rhizophydium* type.

Hyphochytridiomycota

Hyphochytridiomycota are morphologically similar to the simplest Oomycetes and chytrids (of the Chytridiomycota). However, the zoospores have a single, anterior tinsel flagellum (Figs. 4-4C and 4-14A). Their cell walls contain both chitin and cellulose. Sexual reproduction, known in only one parasitic species, occurs by fusion of the whole protoplasts within the host cell. In another species, a thick-walled, resistant phase is known, but its development has not been observed. Asexual reproduction involves zoospore encystment, which produces a bulbous swelling and a rhizoidal system that grows within the substrate (Fig. 4-14B). The main part of the fungus is the **cyst,** which enlarges and functions as a zoosporangium, producing a discharge tube and a pore through which the zoosporangial protoplast is extruded. The zoospores are not discernible at discharge time, but they can develop rapidly after protoplast discharge.

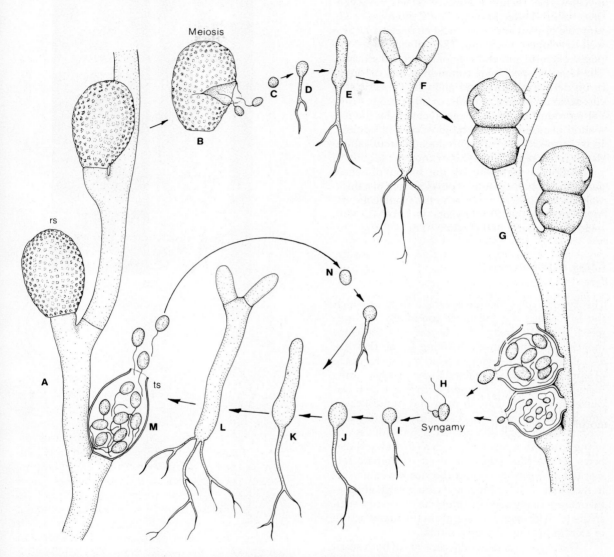

Figure 4-12 Blastocladiales. Life history and development of *Allomyces arbuscula*. **A,** diploid thallus bearing one thin-walled sporangium (*ts; lowest*) and two resistant sporangia (*rs; upper*). Zoospores (**N**) produced in thin-walled sporangia (**M**) give rise to new diploid thalli (**K, L**). Zoospores released from resistant sporangia are meiospores and develop into haploid thalli (**B–G**). The haploid thallus (**G**) bears paired gametangia; when released, gametes fuse (syngamy), and the zygote (**H**) produces a new diploid thallus (**I–M, A**). **A, G,** ×200 (approx.); other figures not drawn to scale.

Fungi 71

Division Zygomycota

Included in the Zygomycota are the classes Zygomycetes and Trichomycetes, of which the Zygomycetes are better known. The Zygomycetes are saprobic or parasitic, with most characterized by a well-developed mycelium. The cell walls are composed of chitin and chitosan or chitin and glucans. The saprobic species are common in soil, on dung, or on decaying plants and animals. Parasitic species occur on plants, animals, or other fungi. Asexual reproduction in the Zygomycota is by the formation of nonmotile sporangiospores or conidia. In most species asexual reproduction occurs abundantly and regularly. Where known, sexual reproduction is generally by the fusion of whole gametangia and subsequent development of a thick-walled structure. In contrast to the Oomycota, the mycelium of the Zygomycota is haploid, with meiosis occurring in the zygote (Fig. 4-15).

Class Zygomycetes

The Zygomycetes include several orders, of which the Mucorales are the most commonly encountered. The Mucorales include some of the commonest soil- and dung-inhabiting fungi; they are also abundant in decaying vegetation, causing decay of stored fruits and vegetables. Some parasitic forms include species of *Mucor* and *Rhizopus*, which can cause severe infections in man and domesticated animals. These infections, called **mycoses** (*myco* = fungus) often occur secondarily in individuals suffering from other diseases, such as cancer and diabetes. Some Mucorales are used to prepare fermented soybean foods and the rice beverage *sake* in the Orient. Other Mucorales are potentially useful or are actually used in industrial processes, such as the transformation of sterols in the manufacture of steroid hormonal preparations.

The mycelium generally consists of both aerial and submerged hyphae, which obtain the nutrients. The hyphae are either coenocytic or septate; the latter type occurs in those forms considered to be most advanced.

In asexual reproduction, aerial sporangiophores bear sporangia with one to many spores. The spores of some Mucorales, such as *Rhizopus*, are dry when released and are airborne. In others,

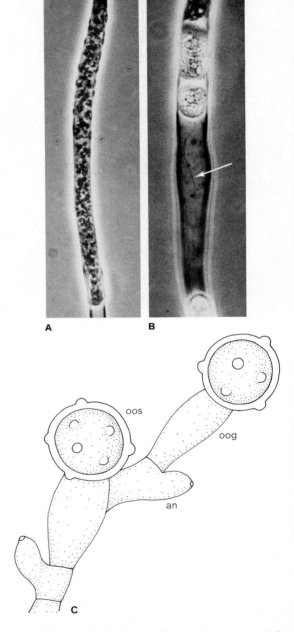

Figure 4-13 Monoblepharidales (*Monoblepharis*). **A, B**, sporangium development: **A**, immature, ×465; **B**, mature, with flagellum present on lowermost zoospore (*arrow*), ×700. **C**, diagram showing antheridium (*an*) and two oogonia (*oog*), each with oospore (*oos*) developing at mouth of oogonium, ×1000.

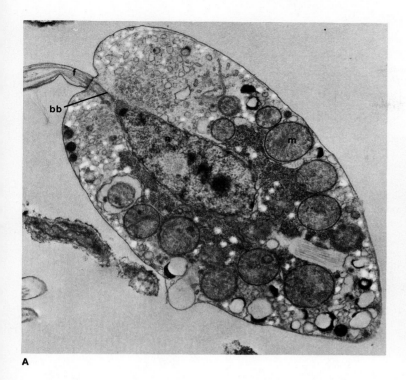
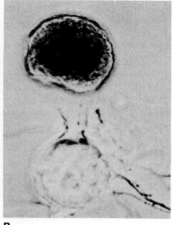

Figure 4-14 Hyphochytridiomycota (*Rhizidiomyces*). **A,** longitudinal section of zoospore, showing anterior tinsel flagellum (*f*), basal body (*bb*), nucleus (*n*), mitochondrion (*m*), ×14,515. **B,** thallus developing zoosporangium, ×335. (**A,** courtesy of M. S. Fuller and R. E. Reichle, with permission of *Mycologia;* **B,** courtesy of M. S. Fuller, with permission of *American Journal of Botany.*)

such as *Phycomyces,* they are slimy spore masses that are insect or water dispersed. Sexual reproduction involves fusion of two gametangia and the development of a thick-walled structure. The gametangia can be on the same mycelium or on morphologically indistinguishable, but genetically distinct, mycelia. Fusing gametangia on separate mycelia are termed **heterothallic** (*heter* = different). When the two fusing gametangia arise from branches of a single thallus, the species is termed **homothallic** (*homo* = same). In most species, numerous nuclei are present in the fused cell, which is commonly called a **coenozygote** (*coeno* = common). The coenozygote becomes enclosed in a thick wall. Meiosis can occur soon after the zygote has formed or after a long resting period, just prior to germination.

The common black bread mold, *Rhizopus stolonifer,* produces an extensive mycelium, which is embedded in the substrate. The sporangia arise from specialized aerial hyphae in clusters at points of contact with the substrate (Fig. 4-16A). The dry spores are exposed and then distributed by air currents at maturity (Fig. 4-16B–D). In *R. stolonifer,* as with many Mucorales, growth and asexual reproduction occur rapidly and involve airborne spores.

Both heterothallic and homothallic species of *Rhizopus* are known. In sexual reproduction, the hyphae produce special compounds that result in the formation of special aerial reproductive branches. The branches grow toward each other until the tips contact one another, and in each a septum then develops, delimiting the gametangia (Fig. 4-17A). The gametangial end walls dissolve at the contact

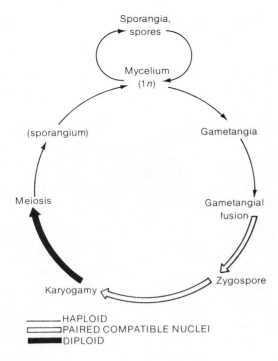

Figure 4-15 Zygomycota. Diagram of life history.

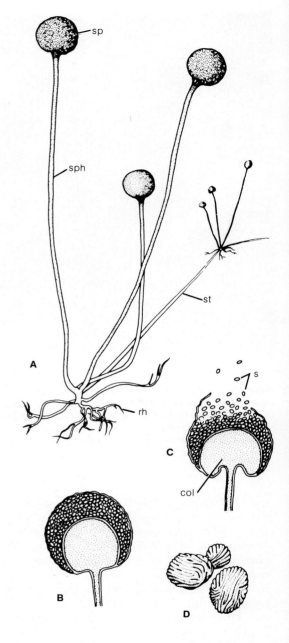

Figure 4-16 Mucorales. Asexual reproduction (*Rhizopus*, black bread mold). **A,** habit sketch, showing sporangium (*sp*), sporangiophore (*sph*), rhizoids (*rh*), stolon (*st*), ×50. **B,** sectional view of mature sporangium, ×110. **C,** ruptured sporangium, with spores (*s*) and central sterile area (*col*), ×110. **D,** sporangiospores, ×1000.

point, and a single multinucleate cell results. Within this multinucleate cell, nuclei pair and fuse. The resulting coenozygote becomes a thick-walled resistant cell by deposition of a new wall (Fig. 4-17B). Germination of these zygotes occurs typically by the formation of a short sporangial stalk and a sporangium containing haploid spores.

A few species of Mucorales exist in mutualistic associations with higher plant roots called **mycorrhizae** (*myco* = fungus; *rhiz* = root). The fungus hyphae penetrate between the cells of the root, and some branches also penetrate the cell walls. Zygomycetous mycorrhizal fungi are associated with such economically important crop plants as corn and soybeans. Evidently the fungus makes the host plant more resistant to **wilting** and assists in mineral uptake, especially in mineral-poor soils. The host plant provides nutrients and probably growth factors required by the fungus.

A second order of Zygomycetes is Entomophthorales, some species of which parasitize

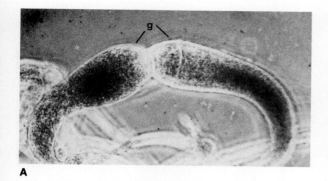

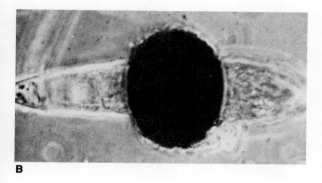

Figure 4-17 Mucorales. Sexual reproduction (*Rhizopus*). **A,** tips of gametangial hyphae (*g*) coming together. **B,** mature coenozygote with heavy wall, ×335.

and kill various insects. One species, *Entomophthora muscae*, infects the common housefly and the mycelium grows within the host's body. Infected flies are often found attached to windows in late summer or fall, with the body surrounded by a halo of whitish spores. The mycelium grows into all parts of the insect, utilizing the protein. Hyphae then grow out between the segments of the host exoskeleton and **sporulation** occurs. The spores, considered to be reduced sporangia, are borne singly and are discharged forcibly for distances up to 15 mm. Each sporangium is surrounded by a layer of adhesive material, which causes it to adhere to flies and other objects. Germination can be through production of a hypha that infects the host. Alternatively, the structure can germinate by producing a new sporangium identical to the first.

Division Eumycota

The division Eumycota includes the subdivisions Ascomycotina and Basidiomycotina, as well as the form class Fungi Imperfecti. These are known as the higher fungi, and common examples are the mushrooms, puffballs, and morels with complex fruiting bodies, or sporocarps (Fig. 4-1D). Less conspicuous but familiar higher fungi include the yeasts used in baking and brewing (Fig. 4-5A), the fungi that cause athlete's foot and ringworm, and the molds such as *Penicillium*, from which antibiotics are obtained (Fig. 4-5B).

Most Eumycota are saprobes that decay plant and animal materials. Although decay is absolutely essential in nature, it also destroys many valuable commodities (e.g., foodstuffs, timber, textiles). Many higher fungi are parasites of plants (Figs. 4-32B, 4-33A, and 4-45) and cause crop losses that can be so severe in a given area that new crops must be planted. A few species of higher fungi parasitize animals, including man; for the most part, however, they are facultative parasites normally existing as saprobes in the soil.

A substantial number of the Eumycota exist as mutualistic symbionts with plants or animals. The largest group of mutualistic fungi are those associated with algae in **lichens** (Figs. 4-19 and 4-54). A second large group, the mycorrhizal fungi, exist in mutualistic symbioses with the roots of higher plants (Fig. 4-46).

None of the Eumycota produces either sporangiospores or motile cells of any type. **Asexual reproduction,** by budding or conidia, is common in the Ascomycotina and Heterobasidiomycetes (Basidiomycotina). Sexual reproduction involves the formation of **asci** and **ascospores** for the Ascomycotina (*ascus* = sac) or **basidia** and **basidiospores** for the Basidiomycotina (*basidium* = small pedestal). Asci and basidia are specialized cells within which karyogamy and meiosis occur; ascospores develop within asci (Fig. 4-18A), and basidiospores are produced externally on basidia (Fig. 4-18B). The asci and basidia generally develop in the complex sporocarps.

More than 20,000 species of higher fungi exist as lichens (see Fig. 3-10). Lichens are sometimes treated as a separate group; such classification is unnatural, however, because the symbiotic association has originated separately among a number of different groups of fungi and with different algae. In lichens, structure and sexual reproduction differ little from those of nonlichenized fungi, and the lichen name (binomial) applies specifically to its fungal component, or mycobiont. The algal component is the phycobiont. Some fungi produce haustoria that penetrate the algal cell. However, both components have very thin walls at points of contact, and haustoria can be lacking.

Lichen thalli can mostly be categorized as **crustose** (*crust* = rind), **foliose** (*foli* = lead), or **fruticose** (*frutic* = shrub), on the basis of thallus morphology (Fig. 4-19). Crustose lichens consist of a thin layer closely adherent to the underlying substrate or sometimes submerged in it (Fig. 4-19A). The outer lichen layer, the **cortex** (*cortex* = shell), consists of tightly packed hyphae (Fig. 4-19B). Beneath the cortex is a layer of algae plus a loose hyphal network. Underlying this is the **medulla** (*medull* = narrow), composed of loosely arranged hyphae, some of which penetrate into the substratum (Fig. 4-19B). The foliose thallus is leaflike, and the margins are free from the substratum (Fig. 4-19C). In transverse section, the thallus is similar to the crustose type, but a lower cortex layer is often present (Fig. 4-19D). The fruticose lichen thallus is branched, shrublike, and can be erect or pendant (Fig. 4-19E). In transverse section, its branches are anatomically similar to those of the crustose thallus.

Lichen growth is generally slow, ranging from 0.1–2.0 mm a year in some crustose forms to 10 mm per year in foliose or fruticose thalli. Lichens generally tolerate extremes of temperature, illumination, and desiccation. During dry periods, the thallus is inactive, but water is quickly absorbed, when present. Lichens are abundant in tropical, temperate, and polar regions. In the harsh, antarctic environment, they are the dominant organisms and occur abundantly in desert and alpine habitats.

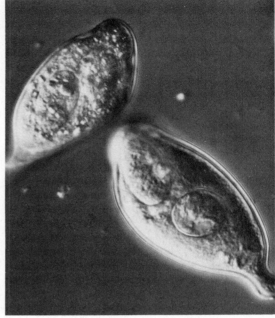

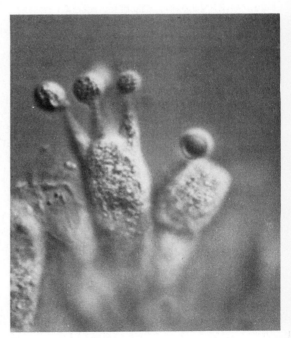

Figure 4-18 Reproductive structures in Eumycota. **A**, ascus and developing ascospores within, ×795. **B**, basidium with developing basidiospores (*center*), ×1690.

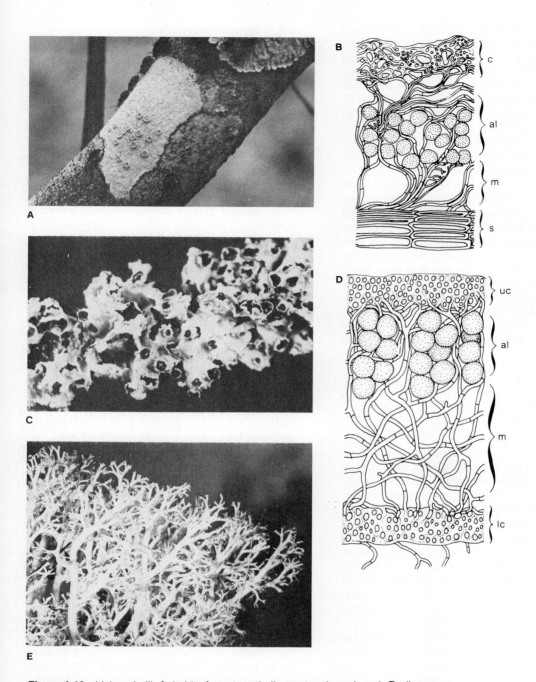

Figure 4-19 Lichen thalli. **A,** habit of crustose thallus on tree branch, ×1. **B,** diagram of section through crustose thallus, showing cortex (*c*), algal layer (*al*), medulla (*m*), substratum (*s*), ×150. **C,** habit of foliose lichen thallus, ×1.5. **D,** diagram of section through thallus, showing upper cortex (*uc*), algal layer, medulla, lower cortex (*lc*), ×150. **E,** habit of fruticose thallus (*Cladonia*), ×0.75.

Fungi **77**

Vegetative, or asexual, reproduction in lichens occurs by production of **soredia** (*sor* = heap) or **isidia** (*is* = equal; *idi* = small). Soredia, consisting of a few algal cells held together by fungal hyphae, are produced in abundant, dusty masses that can be carried by air currents (Fig. 4-20A). The isidia are more highly organized vegetative structures (Fig. 4-20B). Crustose lichens appear to lack both soredia and isidia. In sexual reproduction, the only sexual structures are those of the fungus, and they are predominantly ascomycetes.

Both lichen components can be grown separately in culture, but it is not known if all the mycobionts can grow alone in nature. Some algal species do occur as free-living forms. Since the discovery of the dual composition of lichens in the last century, there has been much debate concerning the relationship between the fungus and alga. Some consider the fungus as simply a parasite of the associated algae; others feel that the relationship is mutualistic and that both partners benefit. Probably the nature of the relationship varies, not only among species, but in time with the individual lichen.

Subdivision Ascomycotina

The sac fungi, or ascomycetes, are the largest of all fungal groups and include most of the lichen fungi. Ascomycetes are predominantly terrestrial saprobes that occur on plant or animal remains, on dung, or in soil. Some species are parasitic on plants and cause a variety of diseases, including certain mildews, leafspot diseases, root rots, and stem cankers. All ascomycetes produce asci and ascospores, but they vary widely in morphology and life history. There is no universally accepted classification, but they are treated here as constituting two classes, the Hemiascomycetes and Euascomycetes. In the Hemiascomycetes, no sporocarps are produced, and dikaryotic hyphae are characteristic of only one order (Taphrinales). The ascus develops through direct transformation of an assimilative cell or as a specialized branch of an assimilative hypha. The Euascomycetes produce sporocarps, and the asci develop within them as specialized terminal cells.

Class Hemiascomycetes

The Hemiascomycetes are all simple, microscopic forms and are generally considered the most primitive ascomycetes. Although it is a small group, it is important to man. It includes the yeasts used in baking and in manufacturing all types of alcoholic beverages. In addition, certain vitamins and other substances of value are obtained from these fungi.

The most important of the yeasts, and one of the oldest of domesticated organisms, is *Saccharomyces cerevisiae* (Fig. 4-21A), commonly known as baker's yeast or brewer's yeast. Perhaps because of its usefulness, *S. cerevisiae* is among the most intensively studied of all fungi. The cells are very small, but they contain the usual eukaryotic organelles (Fig. 4-21B). There appear to be several mitochondria present in a cell, but examination of serial sections shows that a single, large mitochondrion can be present.

The life history of *S. cerevisiae* includes both budding haploid and diploid phases. During budding (Fig. 4-21A), a small protuberance develops, gradually expands, and eventually reaches a size and form comparable to that of the parent cell. During enlargement of the bud, the nucleus divides, and one nucleus is present in the bud. A plug develops between the parent and bud cells, and after they separate, a bud scar remains on the parent cell at the site of attachment (Fig. 4-5A).

S. cerevisiae is heterothallic; the two mating strains are referred to as α and *a* (or + and −). Fusion of two haploid cells is by short, broad tubes, and karyogamy soon follows plasmogamy. The zygote produces buds, and a budding diploid phase is initiated. Under certain conditions, ascospore formation occurs after the nucleus undergoes meiosis (Fig. 4-21C). No special **dehiscence** mechanism is present, and the four ascospores eventually are released by breakdown of the ascus wall. Germination of the ascospores is by bud formation.

The second order of the Hemiascomycetes is Taphrinales. It is a small order of **obligate parasites.** A principal species, *Taphrina deformans,* causes peach leaf curl and is of widespread occurrence. The hyphae penetrate between cells of the host leaves, twigs, or other parts and induce abnormal growth. The host tissues become thickened, causing curling of the infected leaves. Some of the

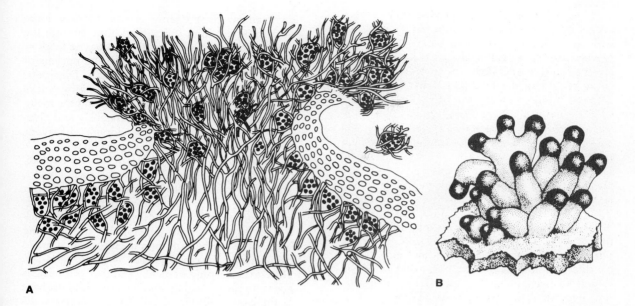

Figure 4-20 Lichen reproduction. **A,** section showing soredial production, ×140. **B,** habit sketch of isidia, ×200.

hyphae grow out between the host epidermal cells and produce asci on the surfaces of infected parts. Budding is again the typical method of ascospore germination. Interestingly, the mycelial phase in *Taphrina* apparently occurs only on or in the host and has not been observed in material grown in laboratory culture, where the fungus grows by budding.

Class Euascomycetes

The Euascomycetes are almost all mycelial fungi that produce their asci in sporocarps. In this class, the asci develop on dikaryotic hyphae that arise from the female gametangium following gametangial contact. One or more types of conidia can be produced in asexual reproduction, but budding is not common. The mycelium of most Euascomycetes is extensive. The assimilative mycelium (vegetative hyphae), excluding that of the lichenized fungi and some plant parasites, is embedded in wood or some other substrate, with only the reproductive structures exposed. Hyphae of the Euascomycetes are regularly septate, and the compartments contain one to several nuclei. Septa are incomplete, and cytoplasm is thus continuous from cell to cell. Nuclei and other organelles often move through the septal pore, probably as a result of cytoplasmic streaming.

Asexual reproduction, common among the Euascomycetes, typically occurs through the formation of conidia, although developmental details differ greatly (Fig. 4-22). Conidial stages are the repeating or reinfecting stages in many parasitic species and are formed abundantly during the growing season of the host. Some ascomycetes produce several types of conidia, whereas others produce none.

Sexual reproduction often commences later in a growing season, with maturation the following spring. Both male and female gametangia can develop on a single thallus, but the thallus may be self-sterile. The form and size of gametangia vary greatly, with the female, or **ascogonium** (*asco* = sac; *goni* = seed), typically larger than the male gametangium, or antheridium. Following contact of the

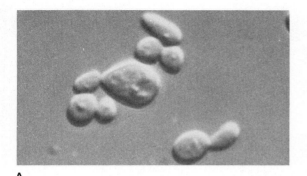

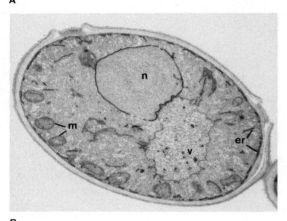

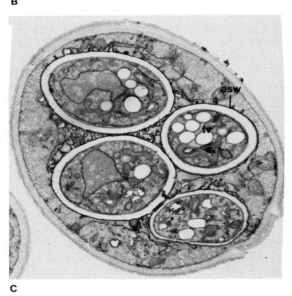

gametangia, plasmogamy occurs. One or more nuclei migrate from the antheridium through the receptive ascogonium tip, the **trichogyne** (*trich* = hair; *gyn* = female) and pair with the nucleus or nuclei in the ascogonium (Fig. 4-23). Dikaryotic hyphae then develop from the ascogonium. Nuclei within the hyphae undergo mitosis, and the members of each pair divide simultaneously and side-by-side throughout the developing dikaryotic hyphae (also termed **ascogenous hyphae**).

The dikaryotic hyphae form septate hyphae, each compartment containing a single pair of nuclei. The septa separating the compartments often differ from those of the assimilative hyphae, and formation occurs with nuclear division. In many Euascomycetes these cells develop by the growing back of the hyphal tip upon itself to form a broad crook, termed a **crozier** (*croce* = staff; Fig. 4-24). The two nuclei present divide, and then one nucleus goes into the hyphal tip and the other into the base of the crook cell (Fig. 4-24B, E). Transverse walls then develop between the nuclei and at right angles to the division plane (Fig. 4-24C). The tip cell continues to grow, contacting the basal cell and becoming confluent with it by dissolution of the intervening wall. Renewed growth occurs from what is now the terminal cell (Fig. 4-24D). Repeated branching of the tip cells produces the final branches, which develop into asci (Fig. 4-25E, F). Asci develop from terminal cells or, in some species, from both terminal and intercalary cells. Within the initial ascus cell, the two paired nuclei fuse; this is followed by cell enlargement and meiosis (Fig. 4-25A–C). After meiosis, a mitotic division generally occurs and the ascus contains eight nuclei (Fig. 4-25D). Walls develop around each nucleus together with other organelles and cytoplasm to form eight ascospores (Fig. 4-25E).

Figure 4-21 Thallus and reproduction of Hemiascomycetes (*Saccharomyces*). **A,** budding haploid cells, especially lower cell, ×1850. **B,** thin section through cell, showing nucleus (*n*), vacuole (*v*), mitochondrion (*m*), endoplasmic reticulum (*er*), ×7200. **C,** ascus containing four ascospores with thick ascospore wall (*asw*), ×9160. (**B, C,** courtesy of R. F. Illingworth, A. H. Rose, and A. Beckett, with permission of *Journal of Bacteriology*).

Ascus form and method of spore release vary greatly among the Euascomycetes. The type of ascus and certain biochemical and structural details of it are considered basic in the classification of euascomycetous fungi. In most Euascomycetes, the ascus wall appears as a single layer by light microscopy (Fig. 4-25F, G). Asci of some species open by a minute lid called an **operculum** (*opercul* = lid), and ascospores are released through the opening (Fig. 4-25G). In other species, an irregular tear or pore in the apex permits ascospore release. The asci of other species simply dissolve when the spores mature.

Haploid (or monokaryotic) hyphae around the gametangia enclose the dikaryotic ascogenous hyphae to produce the specialized sporocarp, termed an **ascocarp**. Development of the ascocarp generally commences shortly after gametangial fusion (Fig. 4-26). Several types of ascocarps occur in the Euascomycetes, depending upon the structure and location of the asci. The structure of the ascocarp is hyphal or parenchymalike (Fig. 4-26C, D), often with sterile hyphal filaments, known as **paraphyses** (*para* = beside; *phys* = bladder), among the asci (Fig. 4-33B).

Although structurally diverse, the life histories of the Euascomycetes follow a similar pattern (Fig. 4-27), with asexual and sexual reproduction occurring concomitantly on a single thallus. In general, conidia are produced on a young thallus that will later produce ascocarps. Since many species are self-sterile (also termed heterothallic), ascocarps cannot develop until fertilization has occurred.

The class Euascomycetes can be subdivided into several subclasses, each consisting of several orders. Some mycologists do not recognize the subclasses and separate Euascomycetes into "series," as this text does. The series have no official taxonomic status, but do designate general groups. The main features distinguishing each group are (1) type of ascocarp, (2) ascus arrangement, and (3) ascospore release mechanism.

Series Plectomycetes

Order Eurotiales Eurotiales include many economically important species, some of which are among the most intensively studied fungi. This is because some species produce antibiotics and

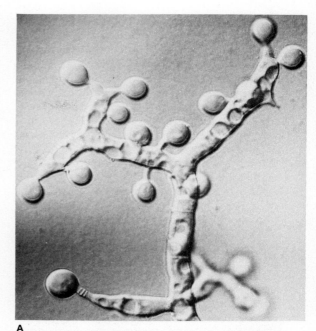

A

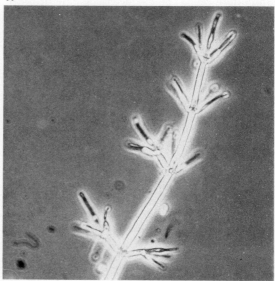

B

Figure 4-22 Coindial stages of Euascomycetes. **A,** branched conidiophore with numerous conidia (*Peziza*), ×550. **B,** conidiophore (*Verticillium*), ×525. (**A,** courtesy of K. L. O'Donnell, W. G. Fields, and G. R. Hooper, reproduced by permission of the National Research Council of Canada from the *Canadian Journal of Botany* 54:1084–93, 1976.)

organic acids, as well as being responsible for the destruction or deterioration of stored foods, leather goods, and even fine lenses. In addition, many produce toxins or carcinogenic substances. Others have the ability to break down keratin and are mainly soil saprobes found on feathers, hairs, horn, or other **keratinous** materials. However, some are **dermatophytes** (*derma* = skin; *phyte* = plant); they cause diseases of the skin or hair in man and other animals. Two of the most common of these are athlete's foot and ringworm. Only the conidial stages occur on diseased persons or animals. These stages, classified as separate genera in the Fungi Imperfecti, are known more commonly by the name given to the asexual stages.

The common green or blue molds growing on citrus fruits, jams, jellies, bread, and other foods are usually species of *Aspergillus* and *Penicillium* (Figs. 4-5B and 4-28). These names, applied to the conidial stages before correlation with the sexual stage was made, are still used for this asexual stage. The conidia are produced in long chains on special **conidiophore** cells. The characteristic blue or green color of the colonies results from the masses of colored conidia. Species of *Aspergillus* and *Penicillium* are used in the production of antibiotics, organic acids (as citric acid), and blue cheese. The organic acids are used in the preparation of foods, beverages, and medicines. The blue cheeses (e.g., Roquefort, Gorgonzola, and Stilton) owe their characteristic flavors to the action of different species of *Penicillium*; still other species are used in the manufacture of uncolored cheeses, such as Camembert and Brie.

Eurotiales are characterized by closed ascocarps, termed **cleistothecia** (*cleisto* = closed; *theca* = box), within which the asci dissolve at maturity (Fig. 4-29). Many genera have sterile appendages extending outward from the cleistothecial wall. The ascospores are freed into the center of the cleistothecium, and dispersal occurs as the cleistothecial wall disintegrates.

In the series Plectomycetes (but not the Eurotiales) there are a number of well-known representatives that cause economically important diseases of plants. These include Dutch elm disease, oak wilt, and rots of sweet potato and sugar cane. The causal agent of Dutch elm disease, *Ceratocystis ulmi*, is the most serious parasite of shade trees in North America. The fungus is carried by beetles that feed on the bark of young branches and, in so doing,

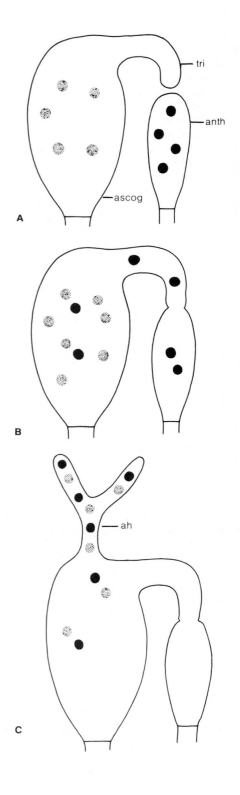

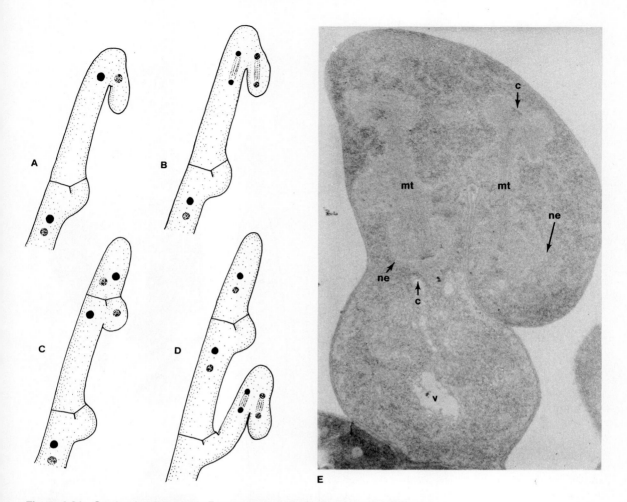

Figure 4-24 Crozier development in Euascomycetes. **A–D,** diagrammatic: **A,** hook, with dikaryon; **B,** nuclear division; **C,** fusion of tip with hypha and septum formation; **D,** two dikaryotic cells, with renewed growth of new tip cell. **E,** electron micrograph of thin section of crozier during nuclear division, showing centriolar plaques (*c*), microtubules (*mt*), nuclear envelope (*ne*), vacuole (*v*), ×9795. (**E,** courtesy of C.-Y. Hung and K. Wells, from "Light and electron microscopic studies of crozier development in *Pyronema domesticum*," *Journal of General Microbiology* 66:15–27, 1971, with permission of Cambridge University Press.)

Figure 4-23 Diagram of sexual reproduction in Euascomycetes. **A,** ascogonium (*ascog*), with trichogyne (*tri*) and antheridium (*anth*). **B,** plasmogamy. **C,** development of ascogenous hyphae (*ah*).

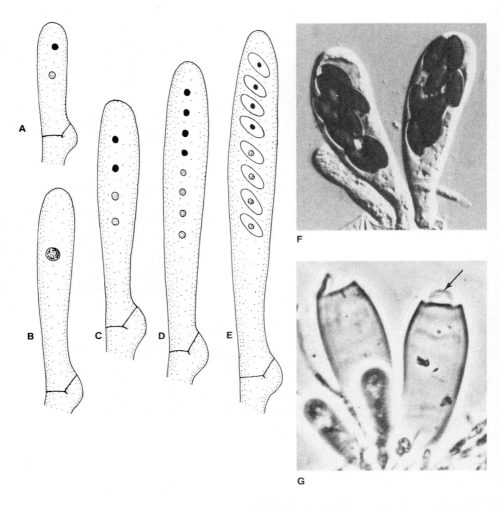

Figure 4-25 Euascomycetes asci. **A–E,** diagrams of ascus development: **A,** dikaryotic ascus cell; **B,** karyogamy; **C,** meiosis; **D,** mitosis; **E,** eight ascospore nuclei, each with a small amount of cytoplasm and becoming surrounded by ascospore walls. **F, G,** operculate asci, ×1200: **F,** before ascospore release; **G,** after release, showing operculum.

can leave fungal spores. Once established, the fungus rapidly spreads within the tree, and death soon occurs.

Series Pyrenomycetes

Order Erysiphales The order Erysiphales, which includes obligate parasites of vascular plants causing a type of powdery mildew, produces small, globose ascocarps that lack an opening for ascospore release. The hyphae grow on leaf surfaces, twigs, or fruits, producing haustoria that penetrate the host epidermal cells or those immediately below (Fig. 4-30A, B). Masses of conidia give a powdery white appearance to the infected host surface. Borne in chains, the conidia are wind-dispersed. Ascocarps develop on infected host surfaces, and appear later than conidia (Fig. 4-30C, D). A few asci are present within the ascocarp. In most genera, the ascocarp remains on fallen parts of the host until the spring. Then, as it absorbs water, the ascocarp swells and bursts. This scatters the asci, which in turn absorb water, burst, and scatter the ascospores.

Order Xylariales Members of the Xylariales produce dark, flask-shaped ascocarps, called **perithecia** (*peri* = around), each with a small opening, called an **ostiole** (Fig. 4-31A). *Neurospora* is the best known member of the Xylariales (Fig. 4-26), as it is used extensively in genetic research. Much of our knowledge of biochemical genetics was originally from studies with two species (*N. crassa, N. sitophila*). In these, the thallus is self-sterile, each producing ascogonia in **perithecial initials.** Completion of the development of perithecial initials depends upon transfer of compatible nuclei carried by conidia. The conidia germinate to form new mycelia or may function as male gametes in nuclear transfer. If a conidium of one mating type (*A* or *a*) contacts the opposite mating type, nuclear transfer occurs, and the perithecial initial completes its development by forming asci and ascospores (Fig. 4-26D).

The conidial stages are known as pink or red bread molds. If *Neurospora* becomes established in bakeries, it grows rapidly in the warm, humid environment, producing masses of colored conidia. Conidia can become established on the cooled loaves, causing rapid deterioration of the bread.

The value of *Neurospora* as a genetic tool results from the fact that it can be grown on simple, chemically defined media, and rapidly completes its life history. Additionally, the products of meiosis, the ascospores, form a linear and ordered series (similar to Fig. 4-31B), and single spores are readily isolated. Since the ascospores give rise to haploid mycelia, mutations can easily be detected.

Order Clavicipitales Another pyrenomycete is *Claviceps purpurea* (Clavicipitales), the cause of ergot of rye and other grasses (Fig. 4-32B). The ovaries of susceptible grasses become infected, developing a mycelium that bears minute conidia (Fig. 4-32A). The conidia are exuded from infected flowers in a sticky fluid, often called **honeydew,** which may attract insects that help to disperse the fungus. The hyphae continue to grow in the ovary and produce a hard, purplish **sclerotium** (*scler* = hard), which at maturity looks like a large grain produced in a healthy ovary (Fig. 4-32B). Sclerotia, or ergots (*ergot* = fungus, spur), fall to the ground and function there as resistant, overwintering structures that require exposure to low temperatures for further development. They retain their viability in the soil for several years, and upon germinating, each sclerotium produces hyphal stalks with a swollen apex, **stromata** (*stroma* = spread out, bed), in which are embedded the numerous perithecia (Fig. 4-32C, D). The asci are extremely long and narrow (Fig. 4-32E). The sclerotia contain a number of potent alkaloids including the one responsible for St. Anthony's Fire, known since the 10th century in Europe. Poisoning usually resulted from eating bread or other foods made with contaminated rye flour, and it was often fatal. Several of the alkaloids are used medicinally. The hallucinogen LSD is derived from chemicals in *Claviceps* sclerotia and may be responsible for some of the St. Anthony's Fire symptoms.

Series Discomycetes

Order Helotiales In the order Helotiales, the ascocarp, or **apothecium** (*apotheca* = storehouse), is structured so the asci are exposed at maturity (Fig. 4-33A). In this order, the asci are inoperculate (Fig. 4-33B); that is, they lack a regular opening. Generally, the small apothecia occur on wood, dead

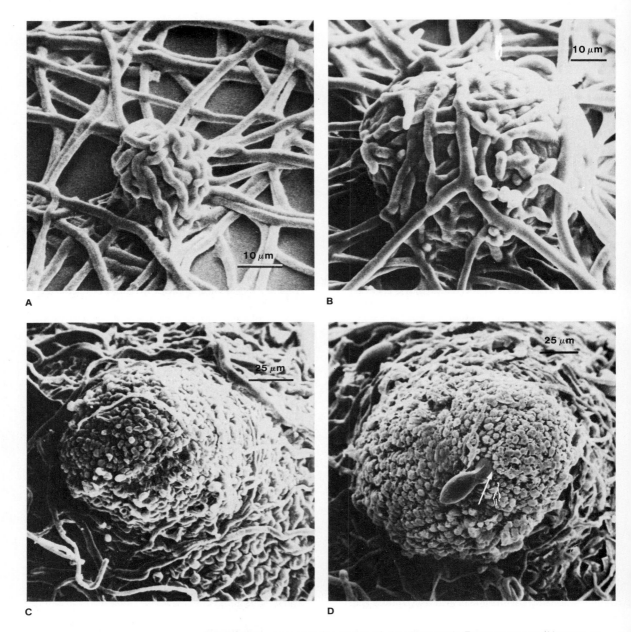

Figure 4-26 Ascocarp (perithecium) development in Euascomycetes (*Neurospora crassa*). **A,** possible ascocarp initial prior to mating. **B,** early development of ascocarp. **C,** ascocarp nearing mature form. **D,** mature ascocarp, with ascus tip and two ascospores protruding from opening (*arrow*). Note that in **C** and **D,** hyphal cells of ascocarp wall have lost their form and have become parenchymalike. (Courtesy of J. L. Harris, H. B. Howe, Jr., and I. L. Roth, with permission of *Journal of Bacteriology*.)

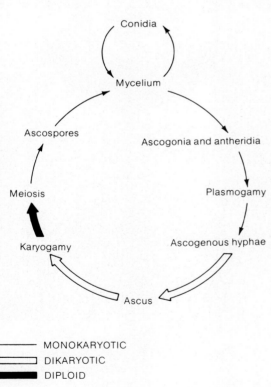

```
— MONOKARYOTIC
▭ DIKARYOTIC
■ DIPLOID
```

Figure 4-27 Euascomycetes. Diagram of typical life history.

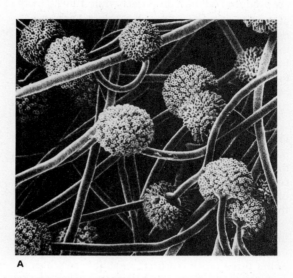

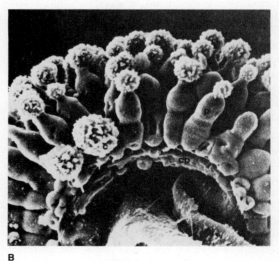

Figure 4-28 Eurotiales condiophores (*Aspergillus*). **A,** scanning electron micrograph of several conidiophores ×455. **B,** spore-producing apex, with terminal conidia, ×2190. (**A,** courtesy of J. Tokunaga, M. Tokunaga, H. Yotsumoto, and Y. Hataba, with permission of the *Japanese Journal of Medical Mycology;* **B,** courtesy of M. Tokunaga, J. Tokunaga, and K. Harada, with permission of the *Journal of Electron Microscopy.*)

leaves, or other plant materials, and some species are destructive plant parasites. Some species infect fruits or other plant parts, causing brown rot of various stone fruits (Fig. 4-33A). Abundant conidia are produced on the fruit surface while active growth of the fungus occurs (Figs. 4-33C and 4-1C). *Botrytis cinerea*, a widespread conidial fungus, has an infrequently seen apothecial stage of the type found in *Monilinia* (Fig. 4-33A). Large conidiophores are found on moribund plant materials, such as strawberries, grapes, and lettuce. Infected grapes are used to produce a most highly prized wine.

Order Lecanorales This large order contains the 8000 or more species of lichens that produce apothecia. The apothecia of lichens develop slowly and continue to produce asci and spores over a period of years. The apothecia are somewhat similar to those of the Helotiales.

Although most lichens have a typical thallus form (e.g., fruticose, foliose, crustose, see p. 76), the thalli of *Cladonia* often combine characteristics of the three types. At first, small foliose scales or a crustose structure can be present prior to the development of the more typical fruticose thallus (Fig. 4-19E). The primary scale or crustose growth often dies, leaving only the fruticose portion. One species, *C. rangiferina*, known as reindeer moss, forms extensive stands on arctic soils. This species, as well as species of *Usnea* and *Alectoria*, constitute an important part of the diet of some herbivores, especially during winter months.

Order Pezizales Pezizales are also characterized by apothecia and occur on dung or soil. The apothecia of *Peziza* often exceed 100 mm in diameter and are often bowl-shaped when young (Fig. 4-34A, B), becoming flattened with age. A familiar relative is *Aleuria*, species of which form large, open, orange-colored apothecia often on areas recently cleared by fires, road cuts, etc. In other genera, the apothecium is borne on a stemlike

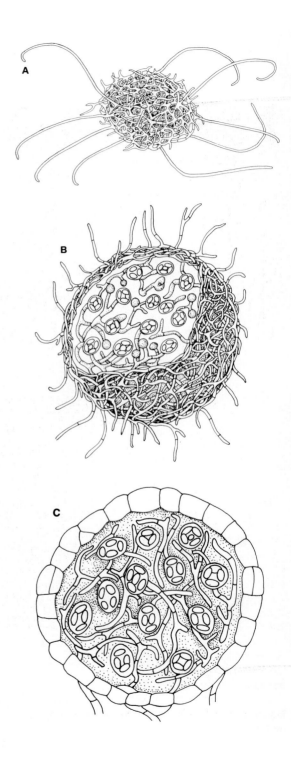

Figure 4-29 Eurotiales ascocarps; cleistothecia. **A,** with coiled appendages (*Shanorella*), ×100. **B,** section showing wall structure and scattered asci (*Talaromyces*), ×115. **C,** section showing wall structure, asci, and ascospores (*Eurotium*), ×350.

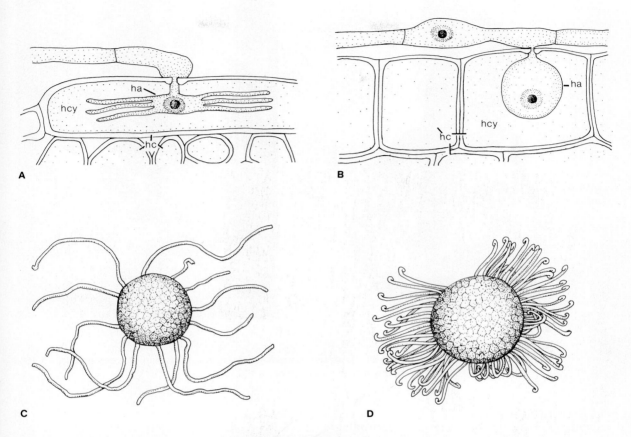

Figure 4-30 Erysiphales. **A, B,** haustoria of powdery mildews showing host cells (*hc*), host cytoplasm (*hcy*), haustorium (*ha*): **A,** *Erysiphe graminis,* ×1150; **B,** *Uncinula salicis,* ×1260. **C, D,** Ascocarps of powdery mildews with attached appendages: **C,** *Erysiphe,* ×96; **D,** *Uncinula,* ×113.

structure or can be folded and somewhat contorted. This is true of species of *Morchella*, or the morels, which are among the most highly prized of all edible fungi (Fig. 4-34C). Apothecia of the Pezizales are not as long-lived as those of lichens. In contrast to the Helotiales, the asci of Pezizales are operculate.

Possibly closely related to the Pezizales are the Tuberales, which produce ascocarps beneath the soil surface. The best known of these subterranean ascomycetes are the truffles. Their tuberlike ascocarps are hunted commercially through the assistance of trained dogs or pigs, which are attracted by the distinctive odor.

Series Loculoascomycetes

Order Pleosporales Pleosporales produce asci in the **stroma,** which is small and looks like a perithecium (Fig. 4-35A). All the asci have a double-layered wall that is apparent at the time of ascospore release. The inner wall layer of the ascus

Fungi 89

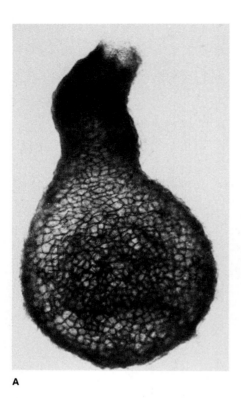 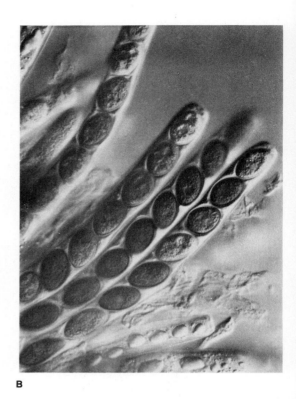

Figure 4-31 Xylariales (*Sordaria fimicola*). **A,** perithecium, with ostiole at top, ×160. **B,** mature asci and ascospores, ×490.

extends, eventually rupturing and protruding through the outer wall layer (Fig. 4-35C, D). Spores are then released through a pore at the tip of the inner wall layer. The conidial stage of *Pleospora* is often of the type classified in *Alternaria* of the Fungi Imperfecti (see p. 111). *Venturia inaequalis* is the cause of apple scab lesions on the fruit. Hyphae penetrate deep into the host tissues to produce **ascostromata** (Fig. 4-35B).

Series Laboulbeniomycetes

Order Laboulbeniales The Laboulbeniomycetes consist of the single order Laboulbeniales, which are obligate parasites of insects or, in a few instances, other arthropods. The thallus is attached to the external surface of its host, and can reach a maximum of 3 mm in height, but is generally less than 1 mm. Thalli of *Laboulbenia* attach by means of a **foot cell** embedded in the chitinous host exoskeleton. A haustorium is often present at the base of the foot cell. Cells of the thallus have relatively thick walls, and adjacent cells are interconnected by large pores. The female gametangium (ascogonium) has a receptive, hairlike extension, the trichogyne (Fig. 4-36A). Fertilization occurs through transfer of spermatia to it, and asci develop at the base of a "perithecium" (Fig. 4-36B). Two-celled ascospores are produced.

Laboulbeniomycetes are unique among parasitic fungi in the high degree of specificity shown. Species are generally able to parasitize only a single host species and, in many instances, can attack only male or female insects. Also, they can be position-

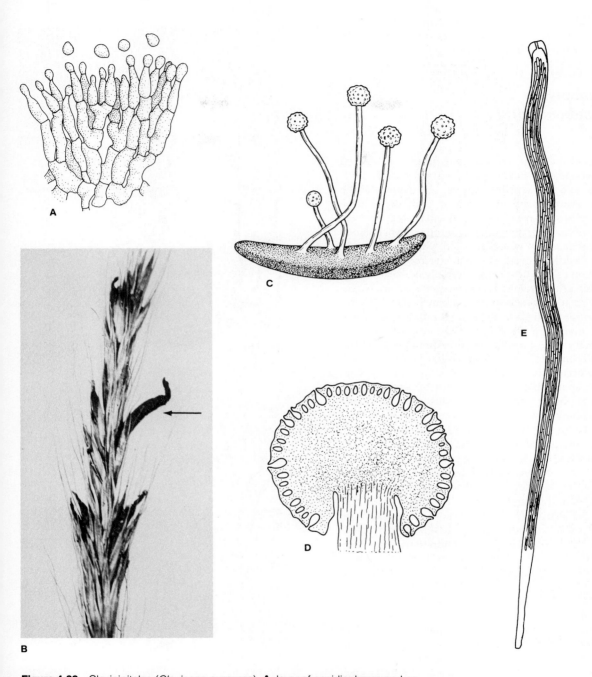

Figure 4-32 Clavicipitales (*Claviceps purpurea*). **A,** layer of conidiophores and conidia in infected grass flowers, ×805. **B,** sclerotia, or ergots (*arrow*), in grass inflorescence, ×1.5. **C,** germinated sclerotium with stalked stromata, ×3. **D,** section through a stroma, showing embedded perithecia, ×36. **E,** elongate ascus and ascospores, ×1500.

specific (always found attached to the host in the same position). The host does not appear to be greatly harmed or hampered by these parasites.

Subdivision *Basidiomycotina*

Most of the large, conspicuous fungi encountered in woods and fields, such as mushrooms, puffballs, and bracket fungi, are representatives of the subdivision Basidiomycotina. In addition to these conspicuous wood- and soil-inhabiting species, the subdivision includes numerous inconspicuous forms, many of which are plant parasites.

Basidiomycetes are characterized by the production of basidiospores externally on basidia (Fig. 4-18B). The germinating basidiospores generally give rise to a **monokaryotic** mycelium (Fig. 4-37A), although in some forms a yeast phase of a dikaryotic mycelium develops. The monokaryotic, or primary, mycelium can grow indefinitely in many basidiomycetes. However, dikaryotization typically occurs through fusion of compatible hyphae. Also spermatia, or conidia functioning as spermatia, bring about dikaryotization in some basidiomycetes.

Following hyphal fusions, nuclear migration takes place, and a dikaryotic, or secondary, mycelium with binucleate cells is produced (Fig. 4-37B). As with ascogenous hyphae, nuclear divisions of paired nuclei occur simultaneously, and septa are formed at this time. In many basidiomycetes, crozierlike structures, called **clamp connections,** are produced during divisions (Fig. 4-37C–F). In basidiomycetes, the clamp connection starts as a subterminal branch, rather than by bending back of the hyphal apex. After division of the two nuclei, septa develop as in croziers.

The secondary mycelium tends to grow radially from an initiation point. Mycelia growing in wood, dung, or similar substrates are limited in their spread by the size and shape of the substrate. However, in perennial soil-inhabiting species, the buried mycelial mass increases in diameter each year, with the older parts at the center eventually dying. The mass of fruiting bodies on the surface, called **fairy rings** (Fig. 4-38), mark the form and extent of the underground mycelium.

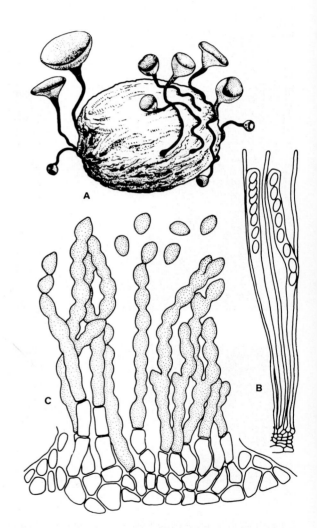

Figure 4-33 Helotiales (*Monilinia fruticola*). **A,** habit sketch of diseased peach, showing several apothecia, ×1. **B,** asci with sterile hairs (paraphyses), ×750. **C,** conidiophores with conidia, ×415.

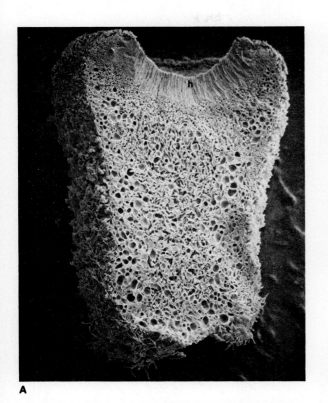

Figure 4-34 Pezizales ascocarps. **A, B,** *Peziza:* **A,** scanning electron micrograph of young apothecium, showing fertile layer (*h; upper*), ×54. **B,** habit of ascocarp, ×1. **C,** *Morchella,* with contorted fertile layer, ×0.5. (**A,** courtesy of K. L. O'Donnell, G. R. Hooper, W. G. Fields, and A. O. Ackerson, reproduced by permission of the National Research Council of Canada from the *Canadian Journal of Botany* 54:2254–67, 1967.)

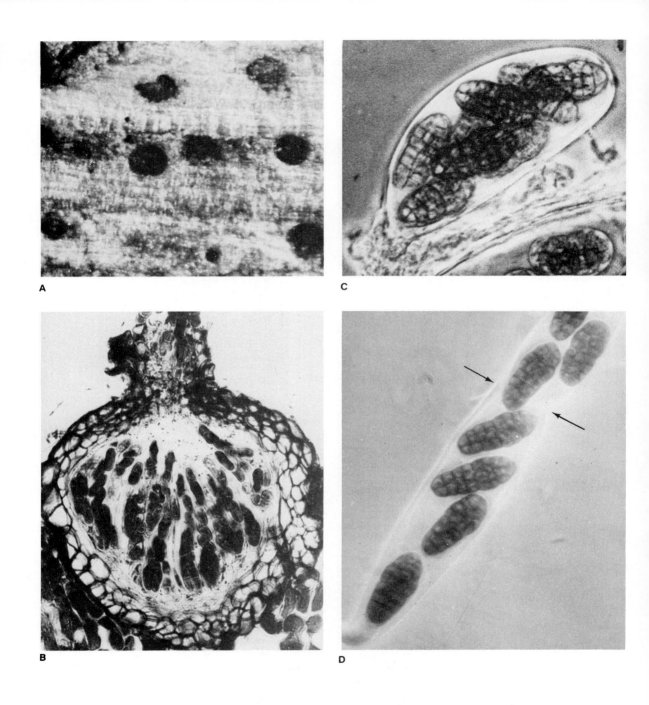

Figure 4-35 Pleosporales. **A,** habit of ascostroma on dead stem (*Pleospora*) ×22. **B,** transverse section through host leaf and ascostroma, showing asci and ascospores (*arrow*) (*Venturia*), ×540. **C, D,** *Pleospora*. **C,** mature ascus with ascospores, ×585. **D,** ascus after rupture of wall layer (*arrow* indicates edge of broken outer layer), ×665.

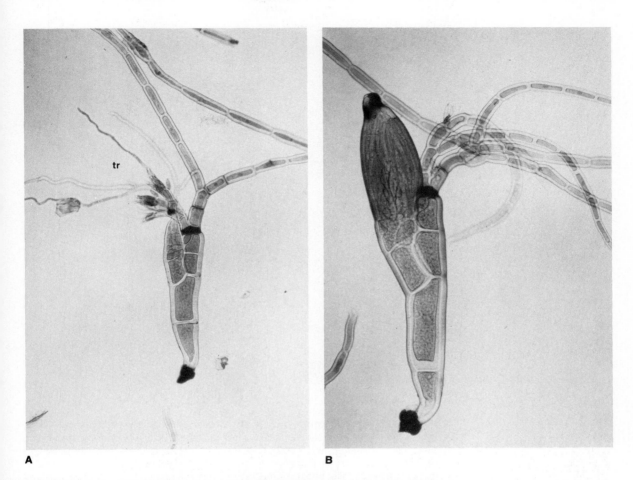

Figure 4-36 Laboulbeniales thalli (*Laboulbenia*). **A,** immature, with trichogynes (*tr*) and basal foot cell, ×200. **B,** mature, showing peritheciumlike ascocarp (*left*), ×250. (Courtesy of R. K. Benjamin.)

In general, asexual reproduction plays a less prominent role in the life history of basidiomycetous fungi than in the Ascomycotina. Conidia are produced in some basidiomycetes and their development is similar to that of ascomycetous fungi. In other basidiomycetes, conidia can be produced either on primary mycelium, secondary mycelium, or both. Budding in basidiomycetous yeasts differs in detail from budding in the Hemiascomycetes, but achieves the same end result.

The initial phase of sexual reproduction occurs when the dikaryotic mycelium is initiated. With the exception of some Uredinales, this process occurs by direct hyphal fusions. In most basidiomycetes, the secondary mycelium constitutes the main assimilative phase, and a long period of growth occurs before sexual spores are formed, generally in the **basidiocarps.**

Some of the simple basidiomycetes do not produce sporocarps; they bear their basidia directly on dikaryotic hyphae or basidia develop from thick-walled spores produced by the hyphae. In contrast

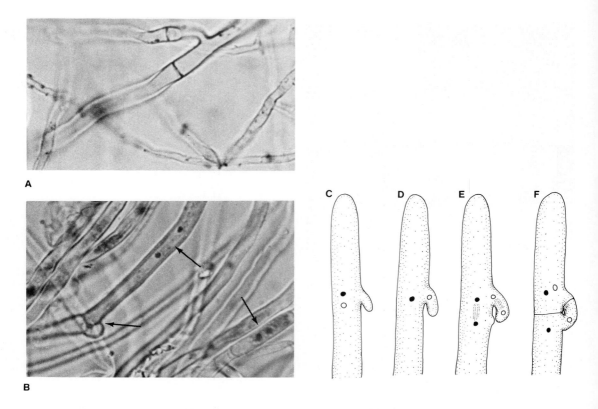

Figure 4-37 Basidiomycetes hyphae. **A,** monokaryotic, primary mycelium, ×1185. **B,** secondary mycelium of same species, showing paired nuclei (*upper arrow*) and clamp connections (*lower arrow*), ×1000. **C–F,** diagrams of clamp connection formation; **C, D,** hyphal apex with clamp connection developing; **E,** nuclear division; **F,** completed clamp connection after septum formation.

to ascocarps, the basidiocarps are entirely composed of dikaryotic hyphae, with basidia borne on or in the basidiocarp. Sterile parts of the basidiocarp are generally composed of recognizable hyphae, but parenchymalike areas can be present. In addition, numerous modifications of hyphae occur in the basidiocarp. The cytological events in the developing basidium are similar to those in the ascus (compare Fig. 4-39 with Fig. 4-25A–F). However, meiosis is generally not followed by mitosis, and (typically) four basidiospores are produced. Basidiospores develop externally, usually on narrow, tubular structures (Fig. 4-39F). One haploid nucleus and a portion of the basidial cytoplasm pass into each developing basidiospore. The mature basidiospores are often shot from the basidium, usually with a force sufficient to carry them free of the spore-producing tissue.

The subdivision Basidiomycotina is divided into two classes, Heterobasidiomycetes and Homobasidiomycetes. The basidia of most Heterobasidiomycetes are either septate or deeply divided, and in many species arise from thick-walled resistant spores. The basidia of Homobasidiomycetes are typically club-shaped, nonseptate, and do not arise from resistant structures. The basidiospores of Heterobasidiomycetes germinate by budding, conidial formation, or formation of secondary basidiospores; those of Homobasidiomycetes typically germinate by germ tubes.

Figure 4-38 Basidiomycetes. Fairy ring, with circle of basidiocarps (*Marasmuis oreades*) marking presence of a buried dikaryotic thallus, ×0.09.

Class Heterobasidiomycetes

Heterobasidiomycetes include the Tremellales, or jelly fungi, and two of the most important groups of plant parasites: the rusts (Uredinales) and smuts (Ustilaginales).

Order Tremellales The Tremellales are mainly wood-inhabiting decay fungi, but some are parasitic. Their common name, jelly fungi, is derived from the gelatinous consistency of the basidiocarps of many species (Fig. 4-40A, B). In these basidiocarps, the hyphae are embedded in a tough, gelatinlike matrix, with the exposed surface covered by basidia and/or conidiophores. Within the Tremellales there is a variety of basidial forms (Fig. 4-40C, D).

Order Uredinales The Uredinales, or rusts, are obligate parasites of vascular plants, including many economically important plants. Hyphae of rusts are intercellular and have haustoria that penetrate host

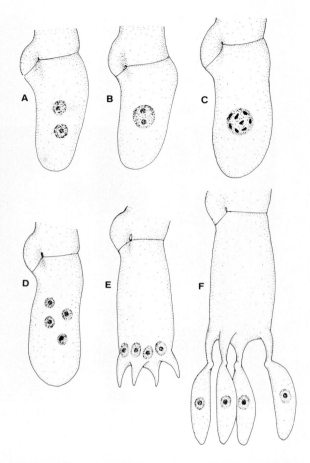

Figure 4-39 Homobasidiomycetes. Diagram of basidial development. **A,** young binucleate basidium. **B,** diploid basidium. **C, D,** meiosis. **E,** development of stalks prior to basidiospore development. **F,** basidium, with developing basidiospores.

cell walls (Fig. 4-41). As many as five types of spores are produced in the life history of some rusts, and many species require two different host species in order to complete their development (Fig. 4-42). Where two hosts are required, they are always from different plant groups: a gymnosperm and an angiosperm, a fern and a gymnosperm, or a monocot and a dicot.

The wheat rust fungus is one of the most destructive species of fungi and one that has been the subject of numerous research studies. *Puccinia*

Fungi

Figure 4-40 Tremellales. **A, B,** basidiocarps: **A,** *Tremella,* ×1; **B,** *Dacrymyces,* ×3.5. **C, D,** basidia: **C,** *Tremella,* ×1025; **D,** *Dacrymyces,* ×515.

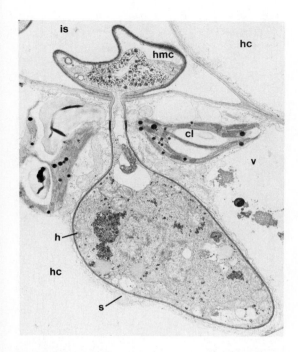

Figure 4-41 Uredinales. Electron micrograph of a rust haustorium (*h*), with originating cell (*hmc*) in host cytoplasm (*hc*); also present are host chloroplast (*cl*), intercellular space (*is*), sheath (*s*), host vacuole (*v*), ×7670. (Courtesy of M. D. Coffey, B. A. Palevitz, and P. J. Allen, reproduced by permission of the National Research Council of Canada from *Canadian Journal of Botany* 50:231–40, 1972.)

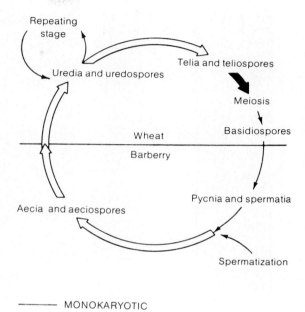

Figure 4-42 Uredinales. Outline of a rust life history (*Puccinia graminis*).

graminis is the causal agent of stem rust of wheat and other small grasses. Basidiospores released in early spring infect leaves of the common barberry, *Berberis vulgaris*. Small, flask-shaped structures, termed **pycnia** (*pycn* = dense), develop mainly on the upper surface of the leaves (Fig. 4-43A). These produce numerous minute **pycniospores** (also termed **spermatia**) and special receptive hyphae. Pycniospores are exuded in a sugary liquid, or honeydew, and are dispersed by insects. Pycniospores from one pycnium must be transferred to receptive hyphae of another to initiate the dikaryotic stage. The dikaryon is necessary for development of the second structure produced on the barberry leaves, the **aecium** (*aeci* = injury; Fig. 4-43B). Aecial initials are present before nuclear transfer, but they do not complete their development until **dikaryotization** has occurred. The aecia form mainly on the underside of the leaf and produce dikaryotic spores, called **aeciospores.** The aeciospores develop in chains and are airborne. They only infect wheat or related grasses.

Soon after the wheat plant is infected, subepidermal, blisterlike structures termed **uredia** (*ured* = blight) develop (Fig. 4-43C). These produce masses of rust-colored, dikaryotic **uredospores.** The uredospores are also dispersed by air currents, but they spread the disease to other wheat plants dur-

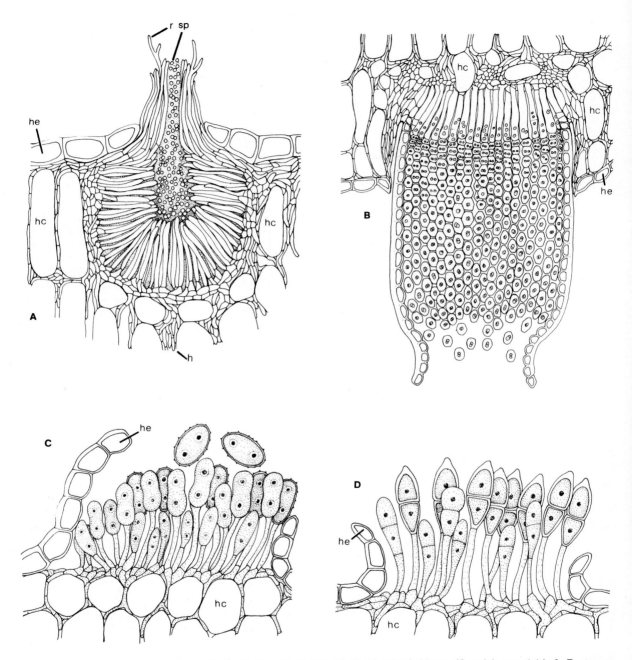

Figure 4-43 Uredinales. Stages in rust life history (*Puccinia graminis*). **A, B,** stages on barberry: **A,** pycnium with spermatia (*sp*), receptive hyphae (*r*), host epidermis (*he*), host cell (*hc*), hyphae (*h*), ×350; **B,** aecium and dikaryotic aeciospores, ×200. **C, D,** stages on wheat: **C,** uredium and dikaryotic uredospores, ×345; **D,** telium and diploid teliospores, ×345.

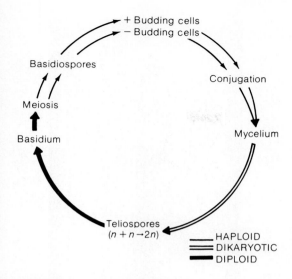

Figure 4-44 Ustilaginales. Outline of a smut life history (*Ustilago*).

ing the growing season. As the host plant approaches maturity, still another spore, the **teliospore,** is produced. These can occur in uredia, but commonly develop in separate structures called **telia** (*teli* = web; Fig. 4-43D). Teliospores remain attached to wheat culms and are the resistant, overwintering stage of the fungus. Each teliospore has two binucleate compartments in which karyogamy occurs soon after the spore forms. In spring, the teliospores germinate *in situ*, with each cell producing a basidium that produces four basidiospores. Basidiospores are airborne after being shot from their basidia.

Since it is an alternate host for *P. graminis*, eradication of barberry has been attempted in North America. However, uredospores can survive mild winters in the southernmost wheat-growing regions of North America. Thus, infected plants occur progressively northward during the growing season, with the infections spreading by windborne uredospores.

Another rust that requires two hosts is *Cronartium ribicola*, the causal agent of white pine blister rust. The pycnial and aecial stages develop on white pines, and the uredial and telial stages develop on currant or gooseberry plants. Overwintering occurs by way of mycelium in the pine host. The name blister rust is based upon the large, orange, blisterlike aecia on pines. *C. ribicola* is native to Europe but was accidentally introduced into North America on diseased nursery stock. Since its introduction around 1900, it has spread over much of the range of eastern and western white pine species and threatens the existence of all native white pines.

Order Ustilaginales Most smuts, like the rusts, are parasites of vascular plants. Haploid stages of the parasitic smut fungi can exist as saprobes, but the dikaryotic phases are obligately parasitic. All parasitic smuts complete their development on a single host plant species and form only one type of dikaryotic spore (Fig. 4-44). The resistant dikaryotic spore (teliospore), considered homologous to that of the rusts, is at first dikaryotic, then diploid. Upon germinating, the teliospore produces a cylindrical basidium.

The largest genus of smuts is *Ustilago*, species of which parasitize many economically important grains (wheat, oats, barley, maize, etc.). The basidium typically is four-celled and produces many basidiospores. These develop as successive buds from the basidial cells and continue to bud upon release. Infection of the host by a monokaryotic hypha either does not occur or is of limited duration in most species. However, if conjugation occurs on the surface of a suitable host, a dikaryotic mycelium is produced. Teliospores of some species adhere to host seeds and infect seedlings at germination. In other species, the flowers are infected, and the ovules become smutted during the same season. Still another method of infection is an infection hypha that grows down the style and into the developing embryo, where some viable hyphae can commence growth at the time of seed germination.

Within the developing host plant, the growth of smut hyphae often keeps pace with that of the host and occurs intercellularly, with haustoria. In the host, masses of teliospores are produced, so that when the host eventually disintegrates, the powdery mass of dark-colored teliospores is exposed (Fig. 4-45).

Class Homobasidiomycetes

Homobasidiomycetes are mostly saprobes that live in soil, dung, litter, and wood. Their decay of litter and wood is beneficial and vital in nature. However, in wood it can also lead to great financial loss of standing and sawn timber. Few Homobasidiomycetes parasitize green plants, but many form a symbiotic mycorrhizal association (Fig. 4-46). As with the ascomycetes, some basidiomycetes are also associated with algae in lichens (Fig. 4-54); however, the number of basidiomycetous lichens is small.

The mycorrhizal association of forest trees and orchids occurs with some basidiomycetes. In trees such as conifers, beech, oaks, and a few others, the fungus hyphae form a dense **mantle** (*mant* = cloak) around the root, with some hyphae penetrating between cells (Fig. 4-46B). In orchids, the hyphae do not form a conspicuous layer around the root but invade root cells. Normal morphology is altered by the presence of the mantle to form distinctive branching patterns (Fig. 4-46A). Root hairs are lacking, but their absence is more than compensated for in terms of surface area as the hyphae extend into the surrounding soil. These mycorrhizal fungi obtain organic compounds from the associated plants. The fungus absorbs minerals and water from the surrounding soil and helps supply these to the host.

Orchid seedlings are dependent upon the establishment of suitable mycorrhizal association. The seed contains virtually no food, and the seedling lacks chlorophyll. These mycorrhizae are capable of degrading cellulose and similar substances, as well as supplying simple sugars to the orchid. Mature orchids have varying amounts of chlorophyll and probably are less dependent upon their mycorrhizal associates. Saprophytic vascular plants, such as the Indian pipe (*Monotropa*), are completely dependent upon mycorrhizal fungi.

The life histories of the Homobasidiomycetes generally are simpler and more uniform than in the Heterobasidiomycetes (Fig. 4-47). Generally, two monokaryotic mycelia are needed for dikaryotization to occur. The dikaryotic mycelium is the main assimilative phase. Conidia can develop on the monokaryotic hyphae, dikaryotic hyphae, or both. However, many Homobasidiomycetes produce no conidia.

Homobasidiomycetes can be divided into two

A

B

Figure 4-45 Ustilaginales (*Ustilago*). Diseased plants. **A,** portion of healthy and smutted barley (left) inflorescences (*U. hordei*), × 1. **B,** corn smut, showing infected kernels (*top*) greatly enlarged and containing numerous teliospores (*U. maydis*), ×0.75.

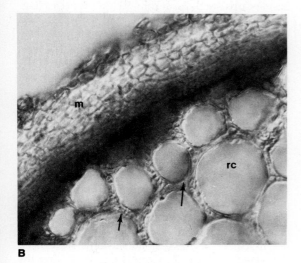

Figure 4-46 Mycorrhizae. **A,** root, showing characteristic stubby lateral branches, ×1.6. **B,** section of root, showing hyphae (*arrows*) penetrating between root cortical cells (*rc*), and thick mantle (*m*) of fungal cells on surface, ×190.

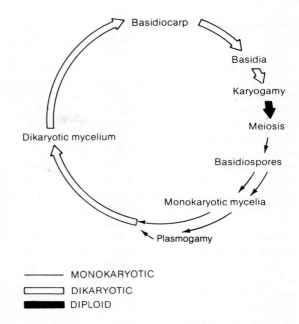

Figure 4-47 Homobasidiomycetes. Diagram of typical life cycle.

groups, or series, on the basis of general basidiocarp features. The first is Hymenomycetes, in which the basidia are arranged in a **hymenial** layer (*hymen* = membrane), which is exposed at maturity. The basidiospores are **abstricted** (*ab* = off, away; *strict* = drawn together), and the spores are then wind-dispersed. The second series is Gasteromycetes, which have basidia that are enclosed at maturity. The basidia are often scattered and not in a hymenial layer. Basidiospores of Gasteromycetes are released by enzymes that digest the basidia. Once released, these basidiospores are dispersed by a variety of mechanisms and methods. The Hymenomycetes include most mushrooms, bracket fungi, coral fungi, and tooth fungi (Figs. 4-48 to 4-54); the Gasteromycetes include the puffballs, stinkhorns, and bird's nest fungi (Figs. 4-55 to 4-57). Some Gasteromycetes are probably adapted more to arid regions than are the Hymenomycetes. However, both occur in wet and more moderate areas.

Series Hymenomycetes

Order Aphyllophorales The Aphyllophorales are predominantly wood- and soil-inhabiting fungi, and a few are important in wood decay. The basidiocarps can be fleshy, tough, corky, woody, and many can dry but then revive upon wetting.

The basidiocarp texture varies, not only with the species, but with age and moisture content, as well as with differences in the kinds of hyphae forming the basidiocarp. Corky or woody basidiocarps are largely composed of thick-walled hyphae;

fleshy basidiocarps are composed of thin-walled hyphae that are often inflated. The basidiocarp form can be simple and flattened, or **effused** (*ef* = away; *fus* = pour out) (Fig. 4-48A); it can also have varying amounts of sterile tissue, be shelflike, cap-shaped, club-shaped, or shrublike (Fig. 4-48B–E). The **hymenium** can be smooth or borne upon toothlike **spines** (Fig. 4-49A, B); more commonly they line small **tubules** (or **gills;** Fig. 4-49C, D). The pores and gills in turn increase the surface area of the hymenium, which greatly increases the total number of basidiospores produced. There is no correlation of different hymenial type to basidiocarp form. Although the hymenia are microscopically similar, there are numerous differences in types of basidia and sterile structures present. The most common sterile structures are termed **cystidia** (*cyst* = bladder, bag; *idi* = small), and are mostly of unknown function (Fig. 4-50). They have different forms and, like basidia, are used in determining relationships.

Order Agaricales Most Agaricales have soft, fleshy basidiocarps that do not revive upon drying and subsequent rewetting. All have capped, or **pileate** (*pilus* = cap) basidiocarps that are either **sessile,** as in *Pleurotus* (Fig. 4-51A), or provided with a stalk, or **stipe,** as in *Boletus* (Fig. 4-51B). The hymenium generally is borne on thin layers, or gills (Fig. 4-51A), although in some the hymenium has pores (Fig. 4-51B).

The development of many basidiocarps in the Agaricales commences with the appearance of a tuft or knot of hyphae near the substratum surface. This grows and develops into a miniature mushroom, or **button stage,** consisting of a stipe and cap (Fig. 4-52A, B). The gills of the button stage are protected by a membranous layer, termed the **partial veil,** extending from the **pileus** (or **cap**) edge to the stipe (Fig. 4-52C). As the button expands, the veil ruptures, and its remnants usually remain attached to the stipe as an **annulus,** or ring (Fig. 4-52D). Buttons of the genus *Amanita* have another membrane, the **universal veil,** which encloses most of the basidiocarp. It remains as wartlike patches on the cap (Fig. 4-1D) and forms a cuplike structure, the **volva,** near the base of the stipe (Fig. 4-53C, D). Only a universal veil is present in some genera; others have both this and the partial veil, or neither. The development of mushrooms to mature basidiocarps is often rapid.

As in Aphyllophorales, a sterile layer is present in the mushroom pileus. The basidiospores, which develop at the gill or tubule surfaces (Fig. 4-52E–G), are fired from their basidia and fall vertically. Cystidia (Fig. 4-50) similar to those in Aphyllophorales are frequently present in the hymenium, on the gill edges, or on other surfaces of the basidiocarp.

Perhaps the most complex of hymenomycete basidiocarps are those of some *Coprinus* species (Fig. 4-53A). The gills are extremely thin, numerous, and closely spaced; their separation is maintained by the presence of large cystidia. Basidia of *Coprinus* mature in groups, starting at the lower edge of the gill and moving upward. Following discharge of basidiospores by the first group, **autodigestion** (*auto* = self) of the spent gill occurs. As other groups of spores mature and are released, autodigestion again continues to dissolve the spent part of the gill (Fig. 4-53B). In this way, spore release and autodigestion occur until all parts of the gills have been digested. Autodigestion results from enzyme activity, and the dark, inky fluid produced has given rise to the common name *inky cap* (Fig. 4-53B).

The Agaricales are predominantly saprobic fungi that inhabit soil, wood, or dung, and many are mycorrhizal. Several mushrooms have been cultivated for food for many years, including the common market mushroom of Europe and North America, *Agaricus bisporus* (Fig. 4-52A), and other species in Asia (*Lentinus edodes, Volvariella volvacea*). Large quantities of wild mushrooms also are harvested and sold in markets in Europe. Many mushrooms contain toxins that can cause illness or death.

Figure 4-48 Aphyllophorales basidiocarps. **A,** flattened basidiocarp on tree branch (*Peniophora*), ×1.5. **B,** shelflike basidiocarp viewed from above (*Stereum*) ×0.75. **C,** stalked basidiocarp with hymenium lining small tubules (*Polyporus*), ×0.6. **D,** stalked basiocarp with toothed hymenium (*Hydnellum*), ×0.5; **E,** shrublike basidiocarp (*Clavulina*), ×1.6. (**C–E,** courtesy of W. McLennan.)

Fungi 105

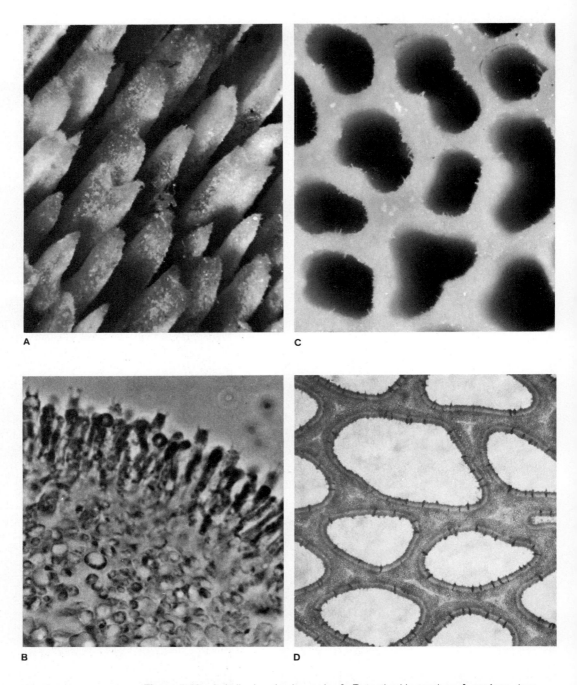

Figure 4-49 Aphyllophorales hymenia. **A, B,** toothed hymenium: **A,** surface view, ×50; **B,** section through single tooth, showing hymenial arrangement, ×800. **C, D,** poroid hymenium: **C,** surface view, ×100; **D,** transverse section through several pores, ×120.

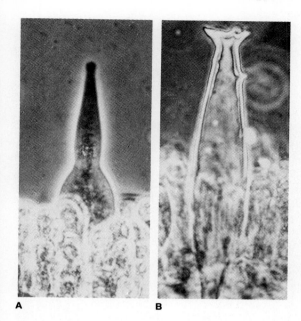

Figure 4-50 Hymenomycetes cystidia. **A,** ×1070; **B,** ×725.

Among the most poisonous are *Amanita phalloides* and the closely related *A. verna* (Figs. 4-1D and 4-53C, D), often called death angels or destroying angels.

Basidiomycetous Lichens. Basidiomycetous lichens are few and can go unnoticed. For example, *Omphalina ericetorum* (Fig. 4-54A), a common temperate species, produces basidiocarps similar in all respects to those of other small mushrooms. Algal cells of this basidiolichen are found only if hyphae leading into the base of the stipe are examined carefully. The hyphae produce a close-fitting mantle around associated algal cells (Fig. 4-54B).

Series Gasteromycetes

Order Lycoperdales Lycoperdales include many of the largest and commonest puffballs. Externally, young basidiocarps resemble the button stages of developing mushrooms (Fig. 4-55A). However, in section there is little differentiation, with only three zones visible (Fig. 4-55B). The outer layer is the **peridium** (*peridi* = little pouch), and the fleshy inner zone, the **gleba** (*gleb* = clod), is where basidia develop within minute cavities (Fig. 4-55C). Below these is the sterile basal zone. As maturity approaches, autodigestion occurs, leaving only basidiospores and a fibrous mass of special hyphae (Fig. 4-55D). The basidiocarp then dries rapidly, and the peridium is then thin and flexible. An opening develops at the top of the basidiocarp, and any force applied to the peridium, such as by large raindrops, results in a bellowslike action. Air, together with spores, is expelled through the opening, and spores are dispersed by air currents. Not all the basidiocarps of the Lycoperdales have regular openings. In those without, the peridium cracks irregularly at maturity to expose the sterile hyphae and spores.

Order Phallales In the Phallales, autodigestion of the gleba leaves a slimy, foul-smelling spore mass. The odor, and possibly the basidiocarp color, attract insects that effect spore dispersal. The odor has given rise to the common name, stinkhorns for some Phallales.

In the Phallales, basidiocarp development generally occurs at the soil surface or just below the litter layer. As the button matures, the stipe elongates and ruptures the outer peridial layer, which remains as a volvalike cup at the base of the stipe (Fig. 4-56A). The glebal mass, having undergone autodigestion, is exposed on the cap surface (Fig. 4-56B). The basidiocarps of many Phallales are brightly colored, and in some, petallike lobes surround the glebal mass. Flies are attracted to the glebal mass, some of which is eaten and some of which adheres to the legs or other appendages of the insects. All species are adapted to insect dispersal, and the spores pass unharmed through the insects' digestive tracts.

Order Nidulariales The Nidulariales, which are called bird's nest fungi, also develop on wood, dung, or soil. The basidia occur in lens-shaped cavities in the gleba, with each such cavity and its basidia surrounded by firm wall layers (Fig. 4-57A, C). The glebal material external to the lens-shaped bodies, or **peridioles** (*peridi* = little pouch; *ol* = whole), undergoes autodigestion. Thus the peridioles

Figure 4-51 Agaricales basidiocarps. **A,** cluster of sessile basidiocarps showing gills (*Pleurotus*), ×0.5. **B,** single basidiocarp, showing minute tubules (*Boletus*), ×0.47. (**B,** courtesy of W. McLennan.)

Figure 4-52 Agaricales *(Agaricus).* **A,** mushrooms commercially grown, showing button stage, ×0.55. **B–D,** diagrams of basidiocarp development, showing pileus *(pi),* stipe *(sti),* gills *(g),* partial veil *(pv),* sterile mycelium *(ms),* annulus *(an),* ×50. **E–G,** scanning electron micrographs; **E,** gill surface; **F, G,** two stages in basidiospore development on basidium. (**A,** courtesy of L. Schisler; **E–G,** courtesy of K. M. Saksena, R. Marino, M. N. Haller, and P. A. Lemke, with permission of *Journal of Bacteriology.*)

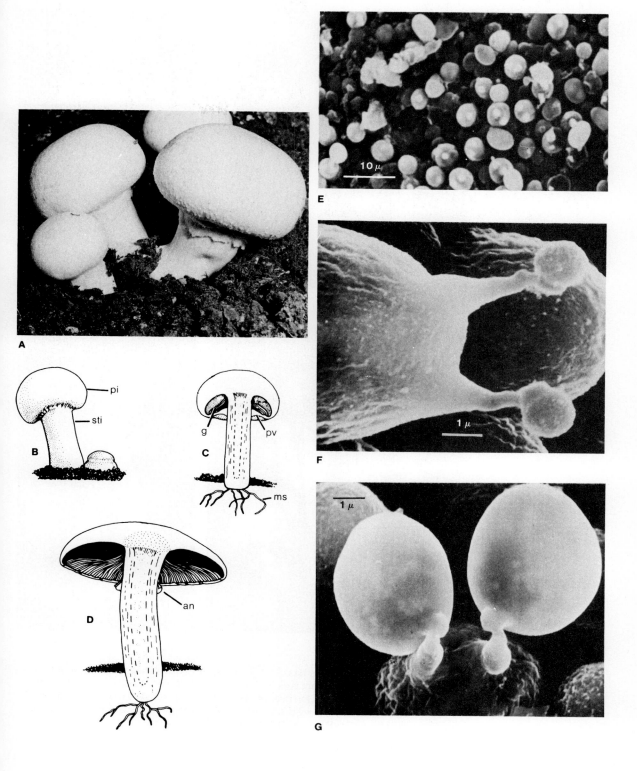

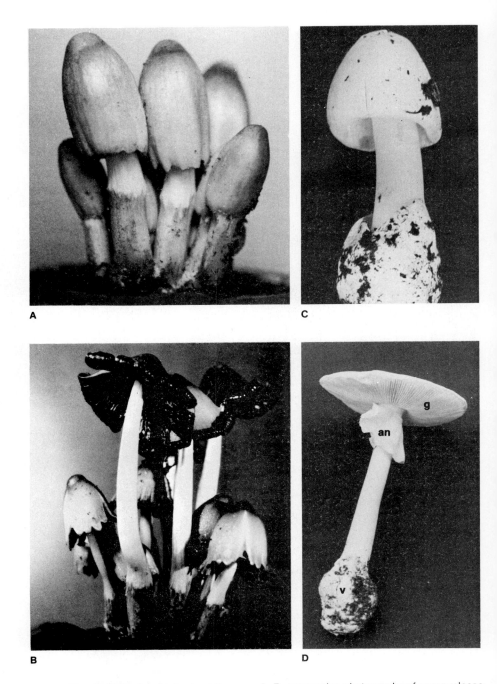

Figure 4-53 Agaricales basidiocarps. **A, B,** successive photographs of spore release in same cluster of basidiocarps (*Coprinus*), ×0.75: **A,** prior to spore release in larger basidiocarps; **B,** after autodigestion of gills. **C, D,** basidiocarps, showing splitting of universal veil with basal volva (*v*) (*Amanita verna*): **C,** young, ×0.7; **D,** mature, with skirtlike annulus (*an*) under gills (*g*), and basal volva, ×0.5.

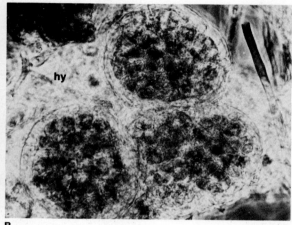

Figure 4-54 Basidiolichen (*Omphalina ericetorum*). **A,** basiodiocarps, ×0.8. **B,** modified hyphal structure (*hy*) enclosing algal cells (*al*), ×595.

become free from one another in the glebal chamber and are submerged in a liquid resulting from autodigestion of other glebal parts. In *Cyathus* and *Crucibulum*, the mature basidiocarp is an open, cup- or vase-shaped structure that contains exposed peridioles, which can be attached to the peridium by an elastic filament. When a raindrop strikes the basidiocarp opening, it tears peridioles free, and a bit of the peridium remains attached to the end of the peridiole cord (Fig. 4-57B). The entire structure (peridiole, cord, peridium remnant) resembles a small bola, and if it strikes a grass blade or similar structure, the cord wraps around it. Further dispersal presumably takes place when an animal eats the herbage and attached peridiole.

Form Class Fungi Imperfecti

Only asexual reproduction is known in many species of fungi. The relationships of some (e.g., sporangial fungi) are obvious, and they are classified with related sexually-reproducing species. However, a large number of asexual fungi cannot be so placed on the basis of asexual structures alone. These species are classified in the form class Fungi Imperfecti. Initially, it was thought that as the fungi were studied, all species in this group would eventually be placed in one of the natural classes. This has not happened, and the numbers of described Fungi Imperfecti increase annually. It is possible that some

Figure 4-55 Lycoperdales basidiocarps (*Lycoperdon*). **A,** young button stages, ×0.5. **B,** section through immature stage, showing peridium (*pe*), gleba (*gl*), sterile base (*st*), ×1.2. **C,** basidia, with elongate stalks and basidiospores, ×1235. **D,** sterile hyphae and basidiospores, ×1300.

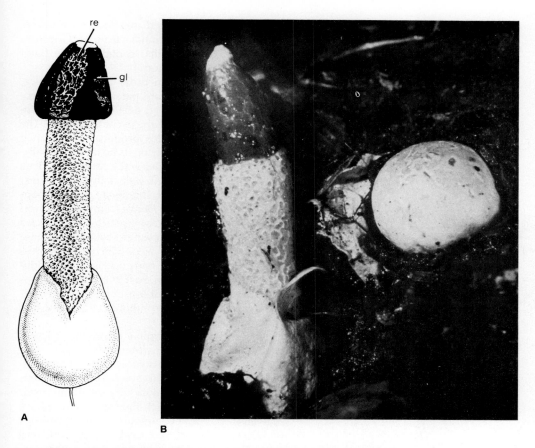

Figure 4-56 Phallales basidiocarps. **A,** habit sketch, with portion of gleba (*gl*) removed to show pitted nature of surface (*re*); note basal volvalike cap (*Phallus*), ×1. **B,** button (*right*) and mature basidiocarp (*Mutinus*), ×1.4. (**B,** courtesy of W. McLennan.)

of these fungi have lost their ability to reproduce sexually, or perhaps they never had this ability.

The conidia produced by the Fungi Imperfecti vary greatly in shape. In some instances, the form is correlated with habitat type and is probably related to dispersal within a particular habitat. Among the more striking of these correlations is found in a group of fungi that grow on wet, often submerged leaves. In fast-flowing streams, these decaying leaves support large numbers of fungi with unusual branched conidia. These conidia develop through several distinctive methods, indicating parallel evolution and a probable biological advantage of this shape. However, the exact significance is not yet known.

Since asexual and sexual stages of sexually-reproducing fungi often occur separately, it is advantageous also to include conidial stages of these fungi in keys to Fungi Imperfecti. This permits identification in the absence of the sexual material, a common condition in disease fungi and other economically important species. Asexual and sexual stages can actually be designated by different binomials, a system permitted under the International Code of Botanical Nomenclature.

The Fungi Imperfecti, also known as the Deu-

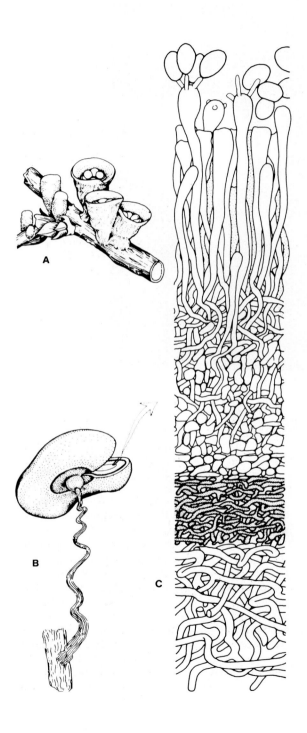

teromycetes, are classified into four form orders based primarily on their conidial features. One form order, the Mycelia Sterilia, contains those fungi lacking conidial production.

Species of the form order Sphaeropsidales, which includes many plant parasites, produce conidia within pycnidia (Fig. 4-58A). Numerous conidia-bearing hyphae line the pycnidium cavity. The form order Melanconiales produces disc-shaped structures subepidermally on its hosts (Fig. 4-58B).

The largest form order is the Moniliales. This includes yeastlike fungi as well as mycelial forms. Some budding fungi placed here are possibly haploid lines of heterothallic Hemiascomycetes or Heterobasidiomycetes. Asexual yeasts include species pathogenic to humans. The most common of these is *Candida albicans* (Fig. 4-59A), which is often present in the mouth or other body openings of healthy individuals, where it can exist as a harmless saprobe. Under certain conditions, however, the fungus invades the tissues, resulting in infections that are sometimes fatal.

Fragmentation of hyphae to produce asexual propagules occurs in many Moniliales (Fig. 4-59B). Most Moniliales produce mycelia with definite conidia-bearing hyphae and specialized conidium-producing cells. The conidia vary greatly in form, but the methods by which they are produced are relatively few. Some common conidial Moniliales are *Aspergillus* (Fig. 4-28), *Penicillium* (Fig. 4-5B), and *Verticillium* (Fig. 4-22B). Although sexual reproduction is known for some species of each genus, they are usually referred to by their form genus name.

Relationships of Fungi

Studies of fossil fungi have not been numerous, and the few that have been made tell us little about

Figure 4-57 Nidulariales basidiocarps (*Crucibulum*). **A,** young (*left*) and mature basidiocarps (*right*) on a twig, ×1.5. **B,** single peridiole and attached cord, ×20. **C,** vertical section through peridiole showing wall layers and basidia (*top*), ×950.

the evolution of fungi. Too often, fungus fossils are either fragmentary or insufficient for characterizing species. Some well-preserved Eumycota have been found, but they are of relatively recent origin. Fossil fungi reported from Precambrian sediments are considered doubtful. Both septate and nonseptate hyphae from about 300 million years ago are present in some sediments and apparently belonged to mycorrhizal fungi associated with primitive land plants. This suggests that fungi colonized terrestrial habitats at about the same time as green land plants or earlier.

Most theories on the origin of fungi and classification systems based upon them are derived from the study of extant forms. Basically, concepts of fungal origin assume either algal ancestry or protistan ancestry.

Schemes in which fungi are derived from algae are based upon the morphological similarities of certain representatives, including both vegetative and reproductive features. The Oomycota have motile cells and food reserves (**mycolaminarin**) similar to those of the brown and golden algae (Chapters 6 and 8). The schemes include comparisons of oomycetous fungi, such as *Saprolegnia*, with coenocytic filamentous algae, such as *Vaucheria* (see p. 146). The Zygomycota and the Zygnematales of the green alga may be compared (see p. 197).

Still another suggestion is that the red algae (Chapter 9) were ancestors of the ascomycetes and rusts. This is based on the lack of motile cells and on similarities in fertilization and subsequent development. The resemblances in fertilization, resulting from spermatial transfer to receptive hyphae, and in subsequent development, are striking. In both red algae and Euascomycetes, fertilization can be followed by development of complex protective structures around the meiospore-producing cells. However, there are problems; for example, the presence of clamps and croziers with accompanying dikaryotic stages, and the nutritional features (photosynthesis versus heterotrophy). Another problem with this theory is that red algae (especially the Florideophyceae), higher ascomycetes, and rusts are all highly evolved, highly specialized forms. Evolutionary pathways generally run from unspecialized to specialized groups. The ascomycetes most closely resembling the red algae are the most complex Euascomycetes—the perithecial and discomycetous types. If these are derived from red algae, the remaining simpler

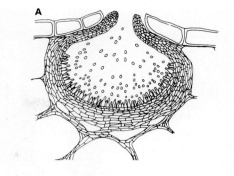

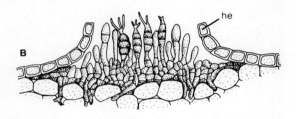

Figure 4-58 Moniliales conidial stages. **A,** vertical section through pycnidium with plant host, showing conidiophores and conidia (*Phoma*), ×300. **B,** section through sporocarp just under host epidermis (*he*), showing conidiophores and conidia (*Pestalotia*), × 315.

ascomycetes must be considered as reduced forms. This has been done in some schemes, with the yeasts being considered the end of the evolutionary line.

Others consider that fungi arose from simple protistan ancestors and attribute algal-fungal similarities to parallel evolution. To support this, they point out the ancient age of fungus groups, differences in nutritional types, and differences in cell wall chemistry and zoospore types. If slime molds are classified with true fungi, a morphological link to the protistans is apparent. The differences between certain primitive fungi, such as the Chytridiales, and some protistans are not extremely great.

The three fungal divisions with motile cells are sometimes theorized to have a relationship. The two types of uniflagellated cells (Hyphochytridiomycota, Chytridiomycota) could have been derived from the biflagellated type (Oomycota). In the

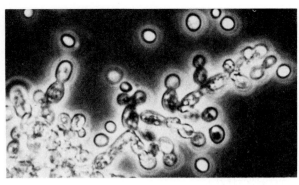

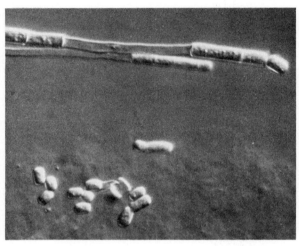

Figure 4-59 Moniliales. **A,** colony, showing limited hyphal development and numerous buds (*Candida albicans*), ×650. **B,** aerial branch fragmenting to form conidia, (*Geotrichum*), ×800.

Oomycota, the flagella are either anteriorly or laterally inserted, but in both types of insertion, the tinsel flagellum is anteriorly directed and the whiplash flagellum posteriorly directed. Through loss of either flagellum, a biflagellated ancestral form could conceivably have given rise to either the chytrid type or hyphochytrid type cell. The cell wall chemistry and the flagellar characteristics seem to suggest that Oomycota and Hyphochytridiomycota have a separate origin. The Oomycota, most of which lack chitin in the cell walls, have diploid thalli, which tends to set them apart from other fungal groups. Presently, the three groups with motile cells are considered to have evolved from three different ancestral groups.

There have been some suggestions, based on biochemical similarities, that higher fungi descended from chytridlike and/or zygomycetous ancestors. The chitin walls are similar to the walls of higher fungi. In addition, there are similar storage products, biosynthetic pathways (e.g., lysine), and production of asexual spores.

Structurally and reproductively, more advanced Zygomycota do resemble certain simple Ascomycotina. This concept is based on morphological similarities between extant Hemiascomycetes and Zygomycota: certain Hemiascomycetes closely resemble the Zygomycota. In both groups, the cell walls contain chitin, although there are other differences. Morphological similarities include lack of motile cells, production of conidia in some Zygomycota, and similarities in sexual reproduction. Initial stages in many are similar, the main difference being that a single zygospore is formed within the zygote cell of Zygomycota, whereas a number of ascospores generally develop in the ascus.

Those holding the view that Ascomycotina had zygomycetous ancestors consider the similarities between the Hemiascomycetes and the Zygomycota the result of parallel evolution. The Ascomycotina and Basidiomycotina share a number of features, primarily biochemical (e.g., storage of glycogen and lipids; cell walls containing chitin, glucans, or mannans). The similarity in ascus and basidium development is frequently shown, and the two structures are considered homologous by many mycologists. Clamp connections and croziers are remarkably similar, and are thought to be homologous structures. The presence of the dikaryotic phase in both groups, and its absence elsewhere, also suggests a relationship. Finally, in simple Hemiascomycetes and Heterobasidiomycetes, similarities in asexual reproduction also occur.

Theories concerning the origins of fungal groups are always speculative and are based upon inadequate factual information. Fungi are poorly known taxonomically, biochemically, and in every other way. Algal groups for which fungal affinities have been hypothesized are equally poorly known. As information is gained for all these groups, theories concerning their evolution, and classification systems based on these concepts, must be continuously modified.

References

All Fungal Groups

Ainsworth, G. C. 1971. *Ainsworth & Bisby's dictionary of the fungi.* 6th ed. (including Lichens, by P. W. James and D. L. Hawksworth). Commonwealth Mycological Institute, Kew, Surrey.

Ainsworth, G. C., and Sussman, A. S., eds. 1965–73. *The fungi: an advanced treatise.* vol. 1, 1965; vol. 2, 1966; vol. 3, 1968; vol. 4A, 1973; vol. 4B, 1973 (with F. K. Sparrow). Academic Press, New York.

Alexopoulos, C. J., and Mimms, C. W. 1979. *Introductory mycology.* 3rd ed. John Wiley & Sons, New York.

Barnett, H. L., and Hunter, B. B. 1972. *Illustrated genera of imperfect fungi.* Burgess Publishing Co., Minneapolis.

Beckett, A., and Heath, I. B. 1974. *An atlas of fungal ultrastructure.* Longman, London.

Brodie, H. J. 1978. *Fungi—delight or curiosity.* Univ. of Toronto Press, Toronto.

Burnett, J. H. 1976. *Fundamentals of mycology.* 2nd ed. Edward Arnold, London.

Christensen, C. M. 1975. *Molds, mushrooms, and mycotoxins.* Univ. of Minnesota Press, Minneapolis.

Cooke, R. C. 1978. *Fungi, man and his environment.* Longman, London.

Deverall, B. J. 1969. *Fungal parasitism.* Institute of Biology's Studies in Biology, no. 17. Edward Arnold, London.

Fuller, M. S., ed. 1978. *Lower fungi in the laboratory.* Palfrey Contributions in Botany, no. 1. Univ. of Georgia, Athens.

Griffin, David H. 1981. *Fungal physiology.* John Wiley & Sons, New York.

Hanlin, R. T., and Ulloa, M. 1979. *Atlas of introductory mycology.* Hunter Publishing Co., Winston-Salem, N.C.

Hudson, H. J. 1972. *Fungal saprophytism.* Institute of Biology's Studies in Biology, no. 32. Edward Arnold, London.

Ingold, C. T. 1971. *Fungal spores, their liberation and dispersal.* Clarendon Press, Oxford.

Müller, E., and Loeffler, W. 1976. *Mycology, an outline for science and medical students.* Translated by B. Kendrick and F. Barlocher. George Thieme Publishers, Stuttgart.

Peberdy, J. F. 1980. *Developmental microbiology.* John Wiley & Sons, New York.

Phaff, H. J.; Miller, M. W.; and Mrak, E. M. 1978. *The life of yeasts.* 2nd ed. Harvard Univ. Press, Cambridge, Mass.

Ross, I. 1979. *Biology of the fungi.* McGraw-Hill, New York.

Webster, J. 1980. *Introduction to fungi,* 2nd ed. Cambridge Univ. Press, London.

Lichen Fungi

Ahmadjian, V. 1967. *The lichen symbiosis.* Blaisdell Publishing Co., Waltham, Mass.

Ahmadjian, V., and Hale, M. E., eds. 1973. *The lichens.* Academic Press, New York.

Duncan, U. K. 1959. *A guide to the study of lichens.* Arbroath, Buncle.

———. 1963. *Lichen illustrations.* Supplement to *A guide to the study of lichens.* Arbroath, Buncle.

Hale, M. E. 1961. *Lichen handbook: A guide to the lichens of eastern North America.* Smithsonian Institution, Washington, D.C.

———. 1967. *The biology of lichens.* Edward Arnold, London.

———. 1979. *How to know the lichens.* 2nd ed. Wm. C. Brown Co., Dubuque, Iowa.

Seaward, M. R. D. 1977. *Lichen ecology.* Academic Press, London.

5

PROTISTANS—MYXOMYCOTA (SLIME MOLDS)

The heterotrophic eukaryotes in this text include the two groups commonly designated as slime molds and fungi. Six divisions of heterotrophs are recognized: Myxomycota, Oomycota, Hyphochytridiomycota, Chytridiomycota, Zygomycota, and Eumycota. The Myxomycota, or slime molds, are dealt with in detail in this chapter; the others were dealt with in Chapter 4. Slime molds predominantly have a motile, naked assimilative structure; that of the fungi is nonmotile and typically is enclosed within a rigid wall. However, the reproductive structures of some slime molds are funguslike and the division Myxomycota is consequently considered to be related to the fungi by many mycologists. Because of this animallike feature, the slime molds are designated as the Mycetozoa (= "fungus animals") by some biologists.

The division Myxomycota includes a heterogenous assemblage of four classes: Myxomycetes (true slime molds), Dictyosteliomycetes (dictyostelids), Plasmodiophoromycetes (endoparasitic slime molds), and Labyrinthulomycetes (labyrinthulids). All have naked, motile assimilative phases and all produce spores or similar structures. However, each class is distinctive, and their relationships to one another and to other organisms are unclear.

Class Myxomycetes (True Slime Molds or Acellular Slime Molds)

About 450 species of true slime molds are known, many of which are cosmopolitan in their distribution. Myxomycetes live in or on moist soil, wood, dung, or decaying vegetation, where they ingest microorganisms, mold spores, and other particulate material. The spore-containing structures are commonly formed in exposed positions and, in many species, are conspicuous.

Life-History Outline

Myxomycetes produce resistant spores, many of which can survive for 75 years or more. Under suitable conditions, spores germinate by the production of flagellated cells, called **swarm cells,** or **amoeboid myxamoebae.** Myxamoebae and swarm cells can ingest food particles, grow, and divide by fission. They can also form resistant cysts under unfavorable conditions, but they eventually function as gametes. The zygote is amoeboid and, through mitotic nuclear divisions, develops into a multinucleate, motile **plasmodium**. The plasmodium ingests food and enlarges; it can also form a resistant structure, the **sclerotium**, under unfavorable conditions. Plasmodia eventually produce **sporocarps**, within which spore formation and meiosis take place. An outline of the life history is illustrated in Figure 5-1.

Spores Myxomycete spores function in dispersal and are the most resistant phase in the life history. The spores are similar in appearance to those of true fungi, having a wall that is often pigmented and marked with ridges, warts, or spines (Fig. 5-2A, B). The spore wall consists of two layers (Fig. 5-2C), the innermost of which is reported to contain cellulose. The spore protoplast contains a nucleus, mitochondria, and other organelles.

Spores of many Myxomycetes germinate readily if placed in water or in aqueous extracts of decayed wood or leaves. Germination requires from 15 minutes to as much as 18 days, depending upon the species, age of the collection, and other factors. At germination (Fig. 5-3A), one or more protoplasts are released through a pore or a wedge-shaped split. The protoplast can become a flagellated swarm cell or a nonflagellated myxamoeba.

Swarm Cells and Myxamoebae Swarm cells have either one or two anterior whiplash flagella (Fig. 5-3B). Typically, one long flagellum is anteriorly directed and a shorter one is recurved. However, two flagella of equal length or a single long flagellum can be present. Swarm cells are elongate and pear-shaped when swimming, but they are capable of amoeboid movement, and their form then varies.

Myxamoebae (Fig. 5-3C) are similar in motility and appearance to other amoebae. They are potentially flagellated and can become swarm cells; conversely, swarm cells can become myxamoebae. Flagella are often lacking in the absence of free water (e.g., on agar media).

Both swarm cells and myxamoebae ingest food particles and absorb dissolved nutrients. Swarm cells have an adhesive posterior to which particles such as bacterial cells adhere; they are then engulfed.

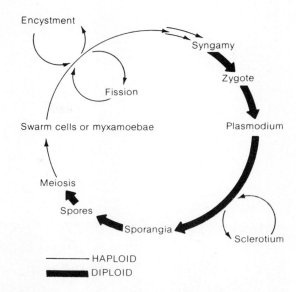

Figure 5-1 Outline of a myxomycete life history.

Myxamoebae engulf particles by "surrounding" them (Fig. 5-3D). The particles are subsequently enclosed within digestive vacuoles.

Myxamoebae can undergo fission directly, but swarm cells retract their flagella before dividing. Under unfavorable conditions, such as drying, myxamoebae and swarm cells form resistant cysts. Eventually, however, these cells act as gametes. Syngamy can be by fusion of two myxamoebae or two swarm cells; some species require distinct mating types for fusion.

Zygote If the zygote is formed by fusion of swarm cells, flagella are withdrawn, and the zygote becomes amoeboid. Fusion of gamete nuclei occurs soon after the protoplasts have fused. Within a few hours, the zygote nucleus divides mitotically, and the plasmodial stage is initiated. As the plasmodium ingests food and grows, subsequent nuclear divisions are synchronous.

Plasmodium The plasmodium (Fig. 5-4) is the main assimilative structure in the myxomycete life history. Plasmodia are all multinucleate and **acellular**,

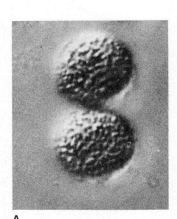

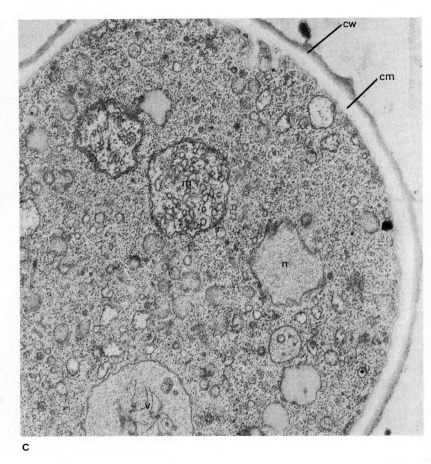

Figure 5-2 A, B, myxomycete spores showing characteristic surface patterns, ×2000. **C,** *Fuligo;* thin section through spore, ×25,000: *cm,* cytoplasmic membrane; *cw,* cell wall consisting of dark outer layer and light inner layer; *m,* mitochondrion; *n,* nucleus; *v,* vacuole. Note: only a small lobe of the nucleus is visible in this photograph. (Photograph courtesy of Nancy Ricker.)

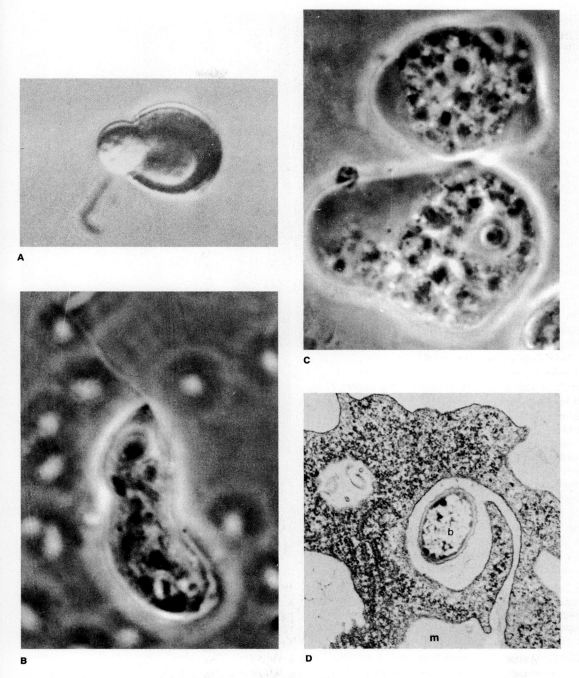

Figure 5-3 **A, B,** *Fuligo*. **A,** germinating spore, ×2700. **B,** swarm cell, ×3400. **C,** myxamoebae of *Didymium*, ×3550. **D,** electron micrograph of *Physarum* myxamoeba engulfing a bacterial cell, ×27,650: *m*, myxamoeba; *b*, bacterium. (**D,** courtesy of F. Y. Kazama and H. C. Aldrich, with permission of *Mycologia*.)

and they lack rigid walls. However, some remain microscopic and amoebalike throughout their development. Others, such as shown in Figure 5-4, consist of macroscopic networks of veinlike strands. Migrating plasmodia of the latter type are roughly fan-shaped; they have conspicuous, infrequently branched strands posteriorly and abundant branching toward the broad anterior. The advancing margin consists of a continuous sheet of protoplasm. Within the tubelike plasmodial strands, the protoplasm exhibits remarkably rapid, rhythmic streaming. The cytoplasm flows in one direction for a few seconds, slows, stops, then reverses its flow.

The plasmodium moves as an amoeba does, and the movement is an oriented response to external stimuli. During assimilative growth, movement is toward increasing moisture and dissolved nutrients. Like swarm cells and myxamoebae, the plasmodium ingests particulate food **phagotrophically** and absorbs dissolved nutrients.

The plasmodium is bounded by a **plasma membrane** (Fig. 5-5) and by a surface layer of slime. The plasma membrane is often extensively invaginated and forms minute channels that run parallel to the surface of the plasmodium. The peripheral cytoplasm of a strand is relatively free of organelles; toward the strand center, organelles and pigment granules are abundant. Minute fibrils, chemically similar to the muscle protein actomyosin, are present near the plasma membrane and in the external layer of slime. At the present time, such proteins are thought to be responsible for both streaming and locomotion.

In culture, plasmodia often fragment into two or more smaller plasmodia, which then go their separate ways. Small plasmodia can also coalesce to form a single large plasmodium. However, coalescence is genetically regulated, and not all strains of a given species can fuse in this way.

Sclerotium Under favorable conditions, plasmodia continue to assimilate food and grow. If drying occurs, or temperature becomes unfavorable, the plasmodium converts to a hardened, irregular mass called a sclerotium. Internally, the sclerotium is composed of numerous walled, cell-like compartments. A sclerotium can remain viable for up to three years, reverting to a plasmodium with the return of favorable conditions.

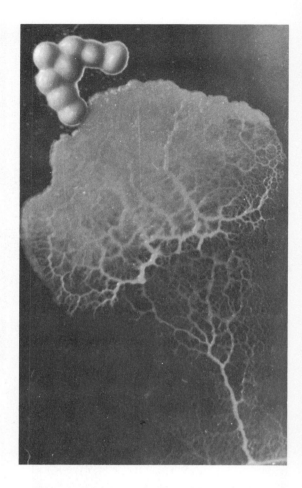

Figure 5-4 Small plasmodium of *Physarum*, shown moving toward mass of yeast cells, upper left, ×3.

Sporulation Factors governing the change from assimilative to reproductive development are not completely known. Starvation, drying, changes in pH, and temperature changes are known to stimulate this change. Exposure to light is also necessary before fruiting occurs in some species. At the initiation of the reproductive phase, the plasmodium moves to an exposed position favorable to spore dispersal. There it forms one or more fructifications or **sporocarps.**

The commonest type of sporocarp is called a **sporangium,** and a single plasmodium typically

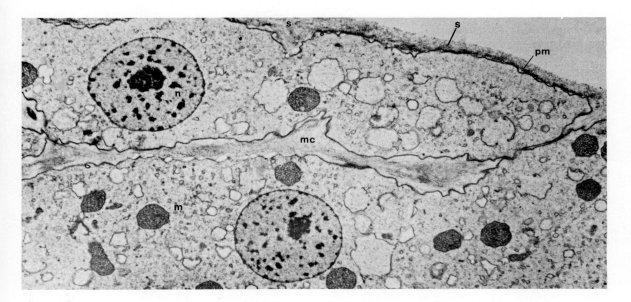

Figure 5-5 Electron micrograph of thin section of *Physarum* plasmodium, ×7300: *s*, surface slime; *pm*, plasma membrane; *mc*, microchannel; *n*, nucleus; *m*, mitochondrion. (Photograph courtesy of J. W. Daniel and U. Järlfors, with permission of *Tissue and Cell*.)

produces numerous sporangia. The developmental pattern shown for *Didymium* in Figure 5-6A–C is one of several known in the group. In this species, protoplasm accumulates at a number of points in the plasmodial network. Small amounts of protoplasm become isolated at such points, then develop into sporangia. The protoplast first assumes a columnar form, and the apex then expands to produce a shape similar to that of the mature sporangium. At this time, the sporangial initial is soft and white or brightly colored. Mitosis occurs in the sporogenous zone of the fructification, and the cytoplasm becomes cleaved into uninucleate portions. After cleavage, a thin wall layer develops outside the spore cleavage membrane, and a thicker layer is then deposited immediately within the membrane. Meiosis occurs after cleavage of the spores. Three of the four nuclei so produced degenerate in many species. The mature spore is thus haploid and commonly uninucleate.

Threadlike structures, collectively called the **capillitium,** are produced as the spores are developing. Capillitial strands are interspersed with the spores, but are not attached to them. The surface layer of the sporangium develops into a wall, or **peridium,** in most myxomycetes, but only the plasma membrane is present at maturity in others.

The parts of the sporangium are diagrammed in Figure 5-7A. The sporangia of many species are stalkless (**sessile**); in others, the stalk can continue into the spore case to form the **columella.** Capillitial strands are free in many types of sporangia; in others, they arise as branches of the columella. If present, the peridium consists of a thin, membranous layer with or without limy deposits. A shiny plasmodial remnant, the **hypothallus,** is often visible at the base of the sporocarp.

Other types of sporocarps produced include **plasmodiocarps** and **aethalia.** In the development of a plasmodiocarp (Fig 5-7B), portions of the plasmodial strands appear to develop directly into fructifications. Plasmodiocarps are sessile and can

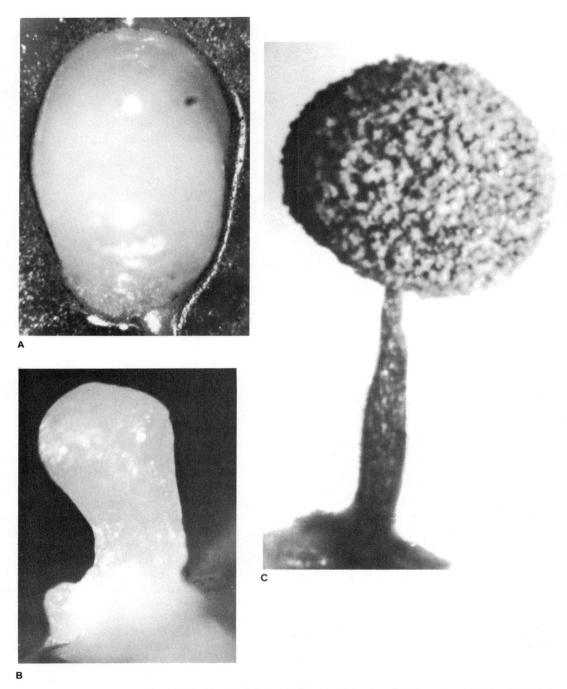

Figure 5-6 Sporangial development in *Didymium*. **A,** pillow-shaped thickening or primordium still attached to plasmodial strand, ×50. **B,** the primordium, now isolated from the plasmodium, has become columnar and erect, ×66. **C,** mature sporangium, ×115.

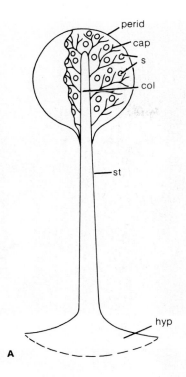

be branched networks, doughnut-shaped, or simply elongate. In the production of an aethalium (Fig. 5-7C), an entire plasmodium heaps up into one or a few pillow-shaped or rounded masses. The internal development of aethalia and plasmodiocarps is as in sporangia; these structures also have an outer peridium and can contain a capillitium in addition to the spores.

Sporangia and plasmodiocarps intergrade, and both can be produced by a single plasmodium in some species. The sporangia of other species develop into crowded masses that closely resemble, or are indistinguishable from, aethalia. When mature, all of these sporocarps are dry, and their spores exposed to air currents through dehiscence or rupture of the peridium.

Myxomycete Classification

Myxomycetes are classified on the basis of sporangial development and mature form, presence or absence of lime (calcium carbonate) in the fructifications, and spore characteristics. In the ubiquitous *Ceratiomyxa fruticulosa* (Fig. 5-8) and related species (Ceratiomyxales), the fructifications are unique, and the relationship to other slime molds is unclear. The fructification typically consists of erect or **procumbent** strands (Fig. 5-8B, C). Numerous spores, each with its own stalk (Fig. 5-8D), are borne on the surface of the strands. When such a spore germinates, it releases a protoplast that divides to form eight swarm cells, rather than the usual one to four. Consequently, the spores of *Ceratiomyxa* are considered by some to be homologous to the sporangia of other Myxomycetes.

In the remaining slime molds, spores are produced within the sporocarp, rather than on exposed stalks. Fructifications of *Lycogala* and *Dictydium* (Fig. 5-9 C, D; Liceales) lack true capillitia, although

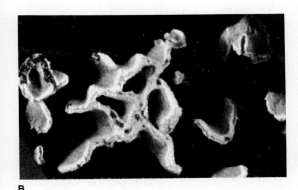

Figure 5-7 Myxomycete fructifications. **A,** diagram of sporangium showing types of structures present: *cap,* capillitium; *col,* columella; *perid,* peridium; *s,* spore; *st,* stipe or stalk; *hyp,* hypothallus. **B,** plasmodiocarps of *Physarum bivalve,* ×5. **C,** aethalium of *Fuligo septica,* ×0.5.

Protistans—Myxomycota (Slime Molds) **125**

capillitium-like vestiges (Fig. 5-9B) of the plasmodium are present in *Lycogala* fructifications. Aethalia of *L. epidendrum* (Fig. 5-9A, B), a cosmopolitan species, can reach 15 mm in diameter and resemble small puffballs. When mature, the flexible peridium ruptures apically and the aethalium functions as a small bellows. Raindrops striking the peridium cause air and spores to be expelled.

Echinostelium species (Echinosteliales) all have microscopic plasmodia (Fig. 5-10A, B), each of which produces a single sporangium. The sporangia (Fig. 5-10C) are minute and typically have a columella and capillitium. *Echinostelium* species often fruit on bark, wood, or mosses placed in moist chambers.

The peridium is membranous and often brightly colored in species of *Hemitrichia* and *Arcyria* (Trichiales; Fig. 5-11). At maturity of the sporangium, the peridium dehisces irregularly or along preformed lines. Subsequently, the capillitium can expand permanently; in some species, hygroscopic expansion and contraction can occur. The capillitial strands are marked by regular spiral patterns in *Hemitrichia* (Fig. 5-11B) and by spines or cogs in *Arcyria* (Fig. 5-1D).

In *Stemonitis* (Stemonitales; Fig. 5-12), capillitial strands arise as branches of the columella (Fig. 5-1B). Elongate sporangia of *Stemonitis* (Fig. 5-12A) are commonly produced in dense clusters and may reach 20 mm in height. The peridium is ephemeral in *Stemonitis;* only a coarse network of capillitial threads encloses the mature spores. *Diachea* species have sporangia similar to those in *Stemonitis*, but the stipe and columella are covered by a conspicuous limy layer (Fig. 5-12C). The peridium, more conspicuous and persistent than in *Stemonitis*, is membranous and beautifully iridescent.

In the largest order of true slime molds, the Physarales, lime is present in the peridium, stalk, or capillitium, or often in all three. *Physarum polycephalum*, a common species, is widely used in biol-

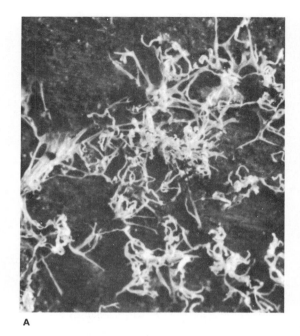

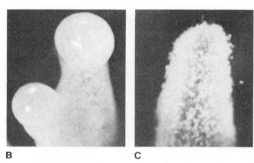

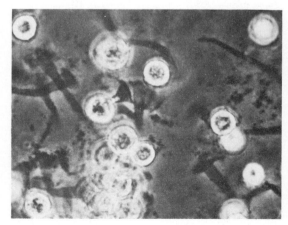

Figure 5-8 **A–D,** *Ceratiomyxa fruticulosa.* **A,** habit of mature fructification, ×12.5. **B,** single strand or pillar during early development, ×150. **C,** pillar nearing maturity, ×160. **D,** spores and individual stalks, ×1000.

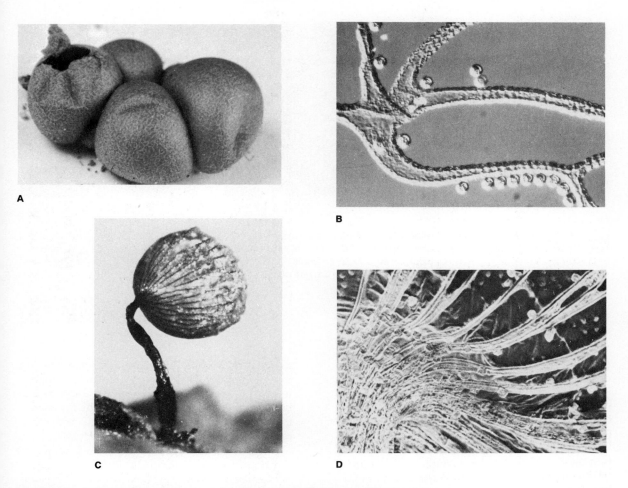

Figure 5-9 A, B, *Lycogala epidendrum.* **A,** aethalium, ×6. **B,** false capillitium and spores, ×565. **C, D,** *Dictydium.* **C,** sporangium, ×31; **D,** scanning electron micrograph at base of sporangium, × 1630.

ogy classes. The plasmodium of this species has been studied intensively, and it will grow on defined media in **axenic** culture. *Physarum* sporocarps are either sporangia or plasmodiocarps (Fig. 5-7B), the capillitia of which consist of **hyaline tubules** interconnecting knots of lime.

Fuligo septica (Fig. 5-7C) produces aethalia that can reach 20 cm in diameter and 3 cm in thickness. The thick peridial layer is limy, the color ranging from whitish through tan, yellowish, greenish, or dull reddish. Within the aethalium, a capillitium and spores similar to those of *Physarum* are present. *Didymium* species (Fig. 5-6) resemble *Physarum* in the general appearance of their sporangia, but the lime present is crystalline rather than amorphous.

Class Dictyosteliomycetes (Dictyostelids)

The Dictyosteliomycetes are predominantly inhabitants of soil, dung, or litter, where they "prey" upon bacteria. Dictyostelids are also called *cellular slime molds*, since the assimilative phase consists of uninucleate amoeboid cells. As in the slime bacteria

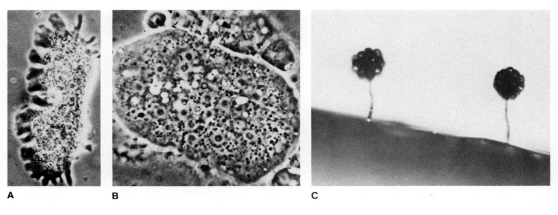

Figure 5-10 **A–C,** *Echinostelium minutum.* **A, B,** plasmodium showing amoeboid form: **A,** ×1200; **B,** ×930. Note the frilled advancing margin in **A** and the numerous nuclei in **B. C,** two sporangia, ×160. (Photographs courtesy of E. F. Haskins, with permission of *Canadian Journal of Microbiology.*)

(p. 22), the assimilative cells aggregate to form communal fruiting bodies. Most species grow readily if cultured with bacteria; some have also been grown on synthetic media in axenic culture. Although economically unimportant, the dictyostelids have become the subjects of intensive laboratory study.

The life history of *Dictyostelium discoideum*, the best-known of the species, is outlined in Figure 5-13. The following description is also based primarily upon this species.

Life-History Outline

Spores The spores of *Dictyostelium discoideum* (Fig. 5-14) have thick walls composed of cellulose, protein, and an outermost layer of mucopolysaccharide. Within the plasma membrane, a nucleus, mitochondria, and other organelles are present. During germination, a single amoeboid cell, a myxamoeba, emerges from each spore.

Myxamoebae Myxamoebae, morphologically like other amoebae, move over the substrate and engulf bacterial cells (Fig. 5-14B). They enlarge, undergo fission, and (during this phase) act independently of one another. At this time, individual amoebae tend to move away from other amoebae and toward bacterial cells. Like other amoebae, the myxamoebae have food vacuoles, a nucleus, and other organelles typical of eukaryotes (Fig. 5-14C).

The assimilative phase may go on indefinitely if moisture remains adequate, the food supply is not depleted, and the myxamoebae do not become crowded. At the end of this phase, myxamoebae stop feeding, and further development is at the expense of stored cellular reserves. The cells decrease in size, food vacuoles disappear, and granules appear in the cytoplasm; the plasma membrane also becomes more adhesive.

Aggregation In **aggregation,** many myxamoebae move toward a common central point (Fig. 5-14D). Branched "streams" develop that lead to the aggregation center. Cells contacting such a stream adhere to and become oriented with other myxamoebae in the stream. Within the streams, the cells are elongate and have a definite polar organization.

It is not known what causes a cell or group of cells to function as an aggregation center. However,

Figure 5-11 **A, B,** *Hemitrichia:* **A,** sporangia, ×45; **B,** capillitium and spore, ×550.
C, D, *Arcyria:* **C,** sporangia, ×12; **D,** capillitium and spores, ×1000.

movement toward the streams and to the aggregation center is controlled by the interaction of a hormone, an enzyme that inactivates the hormone, and an inhibitor of the enzyme. The hormone, called **acrasin,** is cyclic AMP.

Cyclic AMP acts as a chemical signal, diffusing out from the aggregation center and from the branched streams leading to it. Neither the signal nor the response to it are continuous, the hormone being released intermittently. Consequently, bands or spiral patterns of moving and stationary cells are visible in early aggregation (Fig. 5-17B). As myxamoebae respond to the hormone, they are also stimulated to produce it. Interestingly, individual amoebae can distinguish between signals from the center and signals from cells peripheral to themselves.

Migration The myxamoebae heap up at the aggregation center and a rounded, papillate mass of cells develops (Fig. 5-14E). The mass tilts, elongates, and becomes a sluglike **grex** (Fig. 5-14F). The grex, consisting of many myxamoebae, has the appearance of a single multicellular organism. It migrates away from the aggregation point at a speed of 0.25 to 2.0 mm per hour, depositing a slime trail as it glides. The slime trail is actually a sheath formed by the grex and collapsing immediately behind it. On the grex, the sheath is thin at the anterior and thicker toward the posterior; it remains stationary with respect to the substratum. The mechanism responsible for gliding movement is not known.

Cells of the grex number up to 2000 or more, and all appear morphologically alike. Nevertheless, differentiation into two types has occurred. About one-third of the cells, mainly at the anterior of the grex, are destined to become stalk cells. The remaining two-thirds, the prespore cells, will eventually form spores.

The ratio of prestalk to prespore cells is relatively constant, regardless of grex size. Differentiation normally occurs before aggregation commences. However, if a grex is cut into two portions and these are induced to continue migration, the 1:2 ratio will be restored.

Migrating grexes move toward faint light and toward warmer areas, the sensitivity to both stimuli being remarkable. Even the faint light given off by bioluminescent bacterial cultures or luminescent watch dials will elicit a response. These responses insure that movement is toward areas favorable to spore dispersal.

Culmination The migration period varies; drying, bright light, and high pH tend to decrease or eliminate the migration period. When migration ceases, the grex assumes a form similar to that at the end of aggregation and then commences **sorocarp** formation (Fig. 5-15). Directly below the papilla, prestalk cells enlarge, become vacuolate, and secrete cellulose walls. This process takes place within a vertical, noncellular stalk sheath. More cells migrate to the apex of the stalk initial, become appropriately positioned, and secrete walls. As the stalk, or **sorophore,** is produced, the mass of prespore cells migrates upward around it. At completion of the sorophore, walls form around individual prespore cells, and they become spores. There is no increase in the number of prespore cells prior to spore formation. The mature spores are embedded in a droplet of slime, the mass being called a **sorus.** In *D. discoideum,* a few cells remain at the base of the sorophore and are transformed into a basal disc.

Sexual Reproduction In the dictyostelids, sexual reproduction occurs through formation of large spherical-to-oval structures called **macrocysts.** Macrocyst development can be induced by growing myxamoebae under a thin film of water, maintaining cultures in darkness, and incubating at around 25 °C. As with sorocarp formation, macrocyst development is preceded by the aggregation of myxamoebae. Within the aggregate, clumps of cells become surrounded by thin primary walls (Fig. 5-16A). Within the primary wall, a central cell enlarges, ingesting other myxamoebae phagotrophically (Fig. 5-16B). When all of the myxamoebae have been ingested, a thick secondary wall is deposited within the primary layer (Fig. 5-16C).

The giant protoplast within the macrocyst is at first binucleate, then uninucleate. Meiosis appears to occur at the time when most myxamoebae have been ingested. The macrocyst then requires an aging period before germination will take place readily.

At germination of the macrocyst (Fig. 5-16D-F), myxamoebae again become visible within the

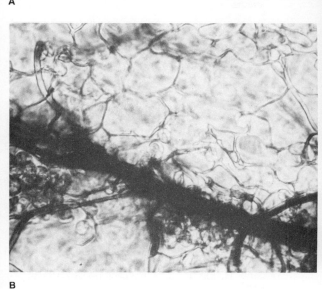

Figure 5-12 Sporangia of *Stemonitis*. **A,** × 14. **B,** photomicrograph of *Stemonitis* sporangium showing origin of capillitial threads from columella, × 350. **C,** *Diachea* sporangia, showing vestiges of the membranous peridium; a thick deposit of lime covers stipe, × 60.

cyst and are released through breakdown of the wall (Fig. 5-22). They may either initiate the assimilative phase or form a sorocarp.

In *Dictyostelium discoideum*, two compatible mating strains are required for macrocyst formation. When a cyst germinates, all of the myxamoebae are of a single mating type. However, both mating types are present in the progeny of a cross. This indicates that three of the four meiotic nuclei degenerate in each macrocyst.

Other Dictyostelids Free migration of the grex does not occur in most dictyostelids. Instead, sporocarp development commences at the point of grex formation. This type of development occurs in species of *Polysphondylium*, as shown diagrammatically in Figure 5-17A–D. Grex migration is limited to that occurring during stalk formation, the base of the stalk lying flat upon the substratum. The upper portion of the stalk is erect; as it is added to apically, portions of the grex remain behind at a number of points. These portions produce whorls of branches at right angles to the main axis, and each branch is tipped by a sorus. The main axis is tipped by a slightly larger sorus (Fig. 5-23B).

Class Plasmodiophoromycetes (Endoparasitic Slime Molds)

The Plasmodiophoromycetes are endoparasites of vascular plants or fungi and are variously classified as true fungi or as slime molds. Since they produce multinucleate plasmodia and have anteriorly biflagellate swarm cells, they are treated here with the slime molds. Two of the Plasmodiophoromycetes cause economically important diseases of plants: *Plasmodiophora brassicae* causes club root of cabbage and other crucifers, and *Spongospora subterranea,* the cause of powdery scab of potato.

Plasmodiophora brassicae is of widespread distribution and often severely damages cabbage crops. Its life history is incompletely known, and many reports on this species are contradictory. The fol-

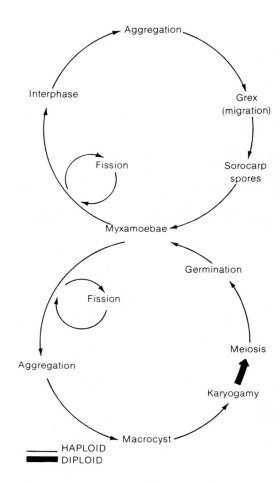

Figure 5-13 Life history of *Dictyostelium discoideum*.

Figure 5-14 A–F, *Dictyostelium discoideum*. **A**, spores, ×2250. **B**, assimilative myxamoeba, ×2000: note partially digested remains of bacteria (*arrows*) in large vacuoles. **C**, portion of an aggregating mass of myxamoebae, the branched streams leading to an aggregation center, ×38. **D**, migrating grex, ×45: note slime trail at posterior. **E**, grex at end of migrating phase, ×45. **F**, mature sorocarp, ×32.

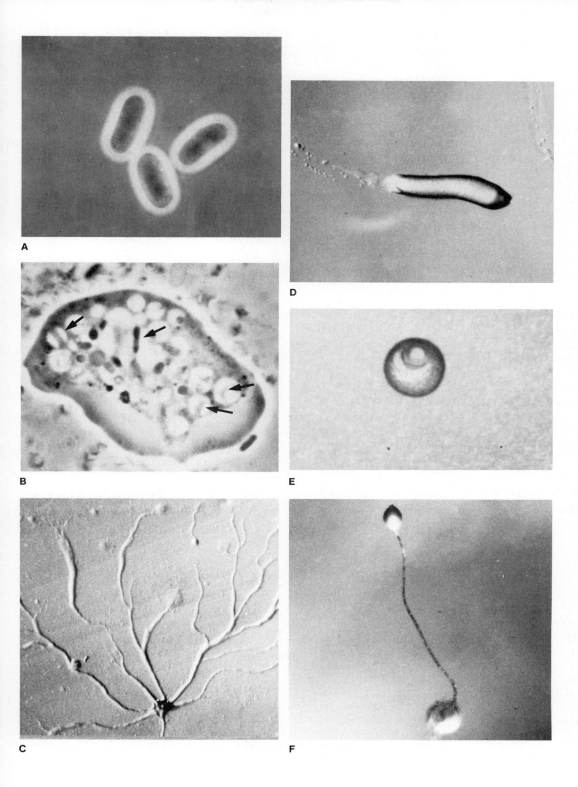

Protistans—Myxomycota (Slime Molds) 133

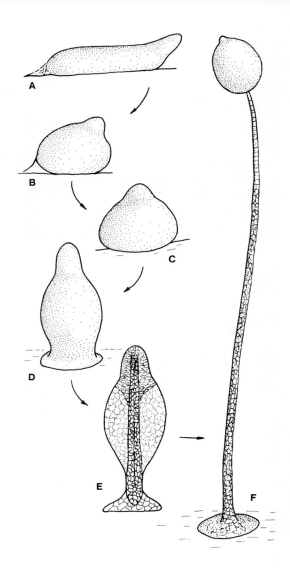

Figure 5-15 Culmination and sorocarp formation (semidiagrammatic). **A,** Migration ceases and the grex becomes short, erect, and mammiform (**B–D**). Within this structure, prestalk cells encyst at the stalk apex and develop rigid cellulose-containing walls. Other cells migrate upward on the stalk (sorophore) as it is produced. The sectional view (**E**), shows the typical form, concentration of prestalk cells in the apical region, the incomplete central sorophore, and a mass of prespore cells. Mature sorocarp (**F**) consists of the sorophore, basal disk, and apical sorus. The sorus, which lacks a limiting wall, is made up of encysted amoebae (= spores) embedded in a slime droplet.

Figure 5-16 Macrocysts in *Dictyostelium mucoroides*. **A–C,** macrocyst formation. **A,** masses of amoebae surrounded by thin primary wall, ×375. **B,** engulfment of amoebae by giant cell (*arrows*) has commenced, ×1215: *e*, engulfed cells; *n*, nucleus; *p*, peripheral amoebae not yet engulfed. **C,** mature macrocyst, ×1600: *s*, fibrillar sheath; *w*, macrocyst wall. **D–F,** macrocyst germination. **D,** dormant (*left*) and germinating macrocyst (*upper right*), ×752. The germinating macrocyst contains numerous myxamoebae. **E, F,** rupture of the outer wall and release of myxamoebae: **E,** ×800; **C,** ×575. (Photographs **A–C,** courtesy of M. F. Filosa and R. E. Dengler, "Ultrastructure of macrocyst formation in the cellular slime mold *Dictyostelium mucoroides*: Extensive phagocytosis of amoebae by a specialized cell," *Developmental Biology* 29:1–16, 1972, with permission of Academic Press. **D–F,** courtesy of A. W. Nickerson and K. B. Raper, with permission of *American Journal of Botany*.)

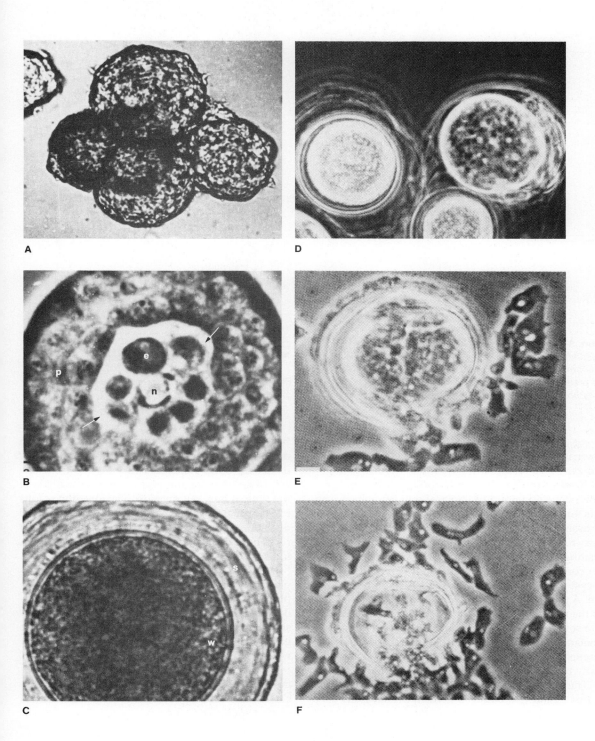

Protistans—Myxomycota (Slime Molds) 135

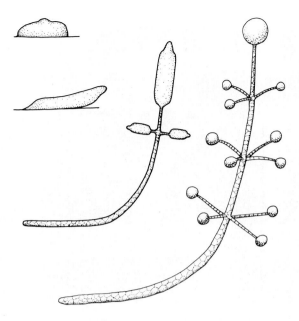

Figure 5-17 **A–D,** diagram of sorocarp formation in *Polysphondylium*. After aggregation, grex migration and stalk formation commence. Portions of the grex separate and produce lateral branches. **E,** mature sorocarp of *Polysphondylium*, ×50.

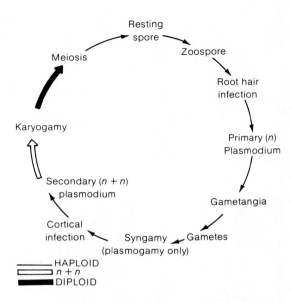

Figure 5-18 Diagram of suggested life history of *Plasmodiophora brassicae*.

lowing discussion and the life-history diagram shown in Figure 5-18 are based upon several recent studies.

Resting spores of *P. brassicae* (Figs. 5-19B and 5-20D) develop within the cortical cells of cabbage roots and are released into the soil upon decay of the tissue. They remain viable in soil for up to six years, releasing a biflagellate swarm cell upon germinating (Fig. 5-19C, D). The swarm cell encysts after contact with a host root hair (Fig. 5-19D), and the cyst protoplast enters the root hair via a minute pore. A primary plasmodium develops (Fig. 5-20A), and the root hair becomes somewhat swollen. After six to seven days, this plasmodium cleaves to form gametangia (Fig. 5-20B). Mitotic divisions occur, and gametes, similar to the swarm cells but smaller, are mostly released to the outside of the root hair.

The gametes fuse in pairs prior to reinfection, but this fusion is thought to involve **plasmogamy** only. After reinfection of the root, migration to cortical cells occurs. There the **secondary plasmodium** develops, becoming multinucleate through mitotic divisions. Such plasmodia can migrate from one cell to another, or the plasmodium can divide at the time of host cell division. The infection results in abnormal host cell enlargement, increased cell division, and altered tissue development. Infected roots have the swollen, "clubbed" appearance typical of the disease (Fig. 5-19A).

Within the host cytoplasm, the secondary plasmodium (Fig. 5-20C) is surrounded only by a plasma membrane. Host plastids and mitochondria are sometimes present in plasmodial vacuoles, showing that some phagotrophism occurs. The infected host cell remains alive, degenerating only when sporulation occurs.

Late in the development of secondary plasmodia, the haploid nuclei fuse in pairs. **Karyogamy** is soon followed by meiosis and the formation of resting spores (Fig. 5-20D).

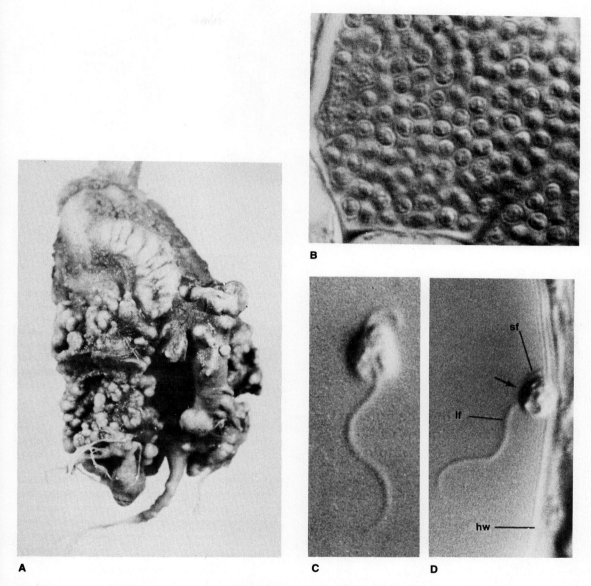

Figure 5-19 *Plasmodiophora brassicae.* **A,** infected root of cabbage plant, showing characteristic "clubbed" form, ×1. **B,** spores in cabbage root cell, ×1500. **C, D,** living zoospores, one of which (**D**) has just attached to a root hair: **C,** ×4400; **D,** ×3375. (Photographs **C, D,** courtesy of J. R. Aist and P. H. Williams, with permission of *Canadian Journal of Botany.*)

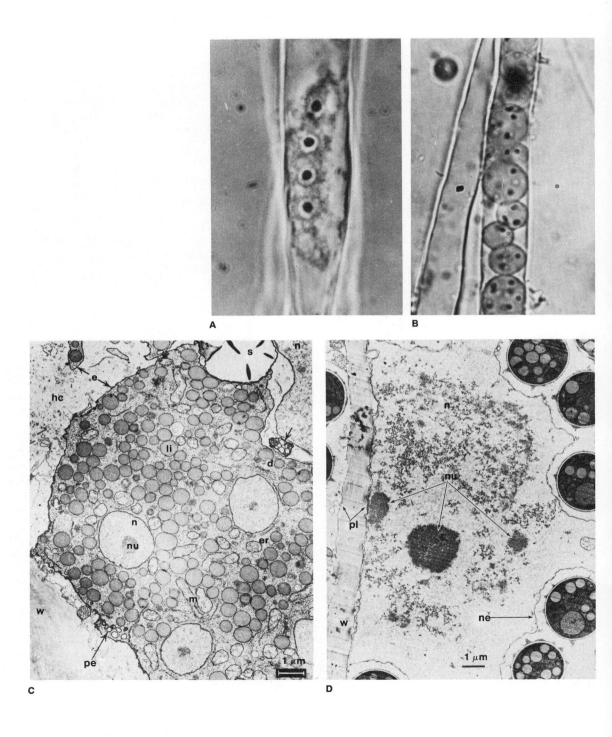

138　Chapter 5

Class Labyrinthulomycetes (Labyrinthulids)

The labyrinthulids, predominantly marine or freshwater organisms, are parasites of algae, higher plants, or animals. A serious disease of eel grass (*Zostera marina*), an ecologically important marine angiosperm, has been attributed to parasitism by *Labyrinthula macrocystis*.

The assimilative phase in *Labyrinthula* consists of uninucleate, oval- or spindle-shaped cells (Fig. 5-21A, B); the cells are motile but are neither flagellate nor amoeboid. The cells appear to be bounded only by a plasma membrane in some species, but a thin wall is present in others. Assimilative cells of *Labyrinthula* glide within a network of tubular "slimeways," the **ectoplasmic net**. The tubules of this network have membrane-limited outer and inner surfaces (Fig. 5-21C), but no organelles are present in the matrix of the tubules. Each assimilative cell has a special organelle thought to function in net formation.

The tubules are produced ahead of the advancing cells of a colony. Cells put forth filaments up to 10 times their own length, and the filaments of adjacent cells fuse to form the network. Although the network matrix lacks organelles, digestion of food organisms occurs only when the net is in contact with such organisms.

Within the ectoplasmic net, assimilative cells grow and undergo fission to form two or four daughter cells. Cells of *Labyrinthula* can also heap together to produce a membrane-bound sorus of "spores." Each such spore gives rise to an assimilative cell when it germinates.

Biflagellate motile cells are produced by some labyrinthulids; these cells have a tinsel flagellum and a whiplash inserted anteriorly. Amoebic and plasmoidal stages have also been reported for some species of labyrinthulids.

Relationships of Myxomycota

No fossil record of the slime molds exists; all concepts concerning relationships are based upon characteristics of extant forms. The Myxomycetes are generally placed either with the protozoa or with the fungi. Several of the characteristics, such as naked, amoeboid assimilative structures and phagotrophy, suggest a relationship to protozoa rather than to fungi. Their sporangia and spores are superficially similar to those of fungi, but the development of these structures and the behavior of swarm cells and myxamoebae have no parallel in any fungal cells. That is, some fungal motile cells are capable of amoeboid movement, but not of phagotrophy or fission.

The Dictyosteliomycetes are possibly related to the true slime molds (Myxomycetes). The Protostelida, a recently discovered group of organisms, shares characteristics with both the true slime molds and the dictyostelids. Protostelids are amoebae that form stalked spores; typically a single spore is produced by an amoeba, and both the spore wall and the stalk contain cellulose. Some protostelid amoebae are multinucleate, although most are uninucleate.

Plasmodiophoromycetes, because of their anteriorly biflagellate reproductive cells, phagotrophy, and plasmodia, also appear to be related to the true slime molds. Finally, the labyrinthulids seem at this time to be unrelated to other classes included in this chapter. The ectoplasmic net and gliding cells differ markedly from structures in other slime molds.

Figure 5-20 *Plasmodiophora brassicae*. **A,** primary plasmodium in cabbage root hair, ×1500. **B,** gametangia in root hair, ×1000. **C,** secondary plasmodium in root cell; *scale line* shown is 1 μm: *m*, mitochondrion; *li*, lipid; *n*, nucleus; *nu*, nucleolus; *er*, endoplasmic reticulum; *hc*, host cytoplasm; *w*, host cell wall; *s*, starch grain; *d*, dictyosome; *pe*, plasmodial envelope. **D,** resting spores in dead host cell; *scale line* shown is 1 μm: *n*, disintegrating host nucleus; *ne*, host nuclear envelope; *w*, host wall; *pl*, host plasma membrane. (Photograph **A,** courtesy of P. H. Williams, S. J. Aist, and J. R. Aist, with permission of *Canadian Journal of Botany*. **B,** courtesy of D. S. Ingram and I. C. Tommerup, with permission of the Royal Society, London. **C, D,** courtesy of P. H. Williams and S. S. McNabola, with permission of *Canadian Journal of Botany*.

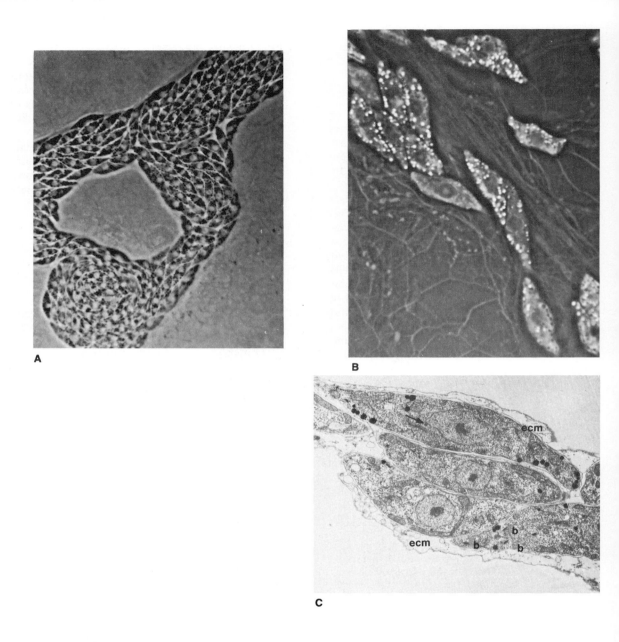

Figure 5-21 **A,** colony of gliding cells of *Labyrinthula vitellina,* the ectoplasmic nets of which are not visible, ×1300. **B,** cells and ectoplasmic net of *Labyrinthula minuta,* ×1100. **C,** electron micrograph of *Labyrinthula* cells within ectoplasmic net elements, ×5640: *ecm,* ectoplasmic matrix. (Photographs **A, B,** courtesy of S. W. Watson, from "*Labyrinthula minuta* sp. nov.," by S. W. Watson and K. B. Raper, *Journal of General Microbiology* 17:368–77, with permission of Cambridge University Press. Photograph **C** courtesy of D. Porter, with permission of *Protoplasma*.)

References

Alexopoulos, C. J., and Mimms, C. W. 1979. *Introductory mycology*, 2nd ed. John Wiley & Sons, New York.

Ashworth, J. M., and Cappuccinelli, P., eds. 1977. *Development and differentiation in the cellular slime molds.* Elsevier/North-Holland Biomedical Press, Amsterdam.

Bonner, J. T. 1967. *The cellular slime molds.* 2nd ed. Princeton Univ. Press, Princeton, N.J.

Erdos, G. W.; Raper, K. B.; and Vogen, L. K. 1975. Sexuality in the cellular slime mold *Dictyostelium giganteum. Proc. Nat. Acad. Sci.* 72:970-73.

Gray, W. D., and Alexopoulos, C. J. 1968. *Biology of the Myxomycetes.* Ronald Press Co., New York.

Ingram, D. S., and Tommerup, I. C. 1972. The life history of *Plasmodiophora brassicae* Woron. *Proc. Roy. Soc. London,* B, 180:103-12.

Martin, G. W., and Alexopoulos, C. J. 1969. *The Myxomycetes.* Univ. of Iowa Press, Iowa City.

Olive, L. S. 1975. *The mycetozoans.* Academic Press, New York.

Porter, D. 1974. Phylogenetic considerations of the Thraustochytriaceae and Labyrinthulaceae. In *Marine Mycology.* 2nd Internat. Symposium (Bremerhaven, Sept. 1972). Veröff. Inst. Meeresforsch. Bremerh. (Suppl.) 5:19-44.

6

PROTISTAN ALGAE

As noted in Chapter 3 (Table 3-3), there are several algal divisions. The largest forms in the aquatic environment belong to Chlorophyta, Phaeophyta, and Rhodophyta. Representatives of other divisions are generally less prominent and consist of many small, often motile forms (see Table 6-1), presently classified by many as plantlike Protista. These organisms are primarily unicellular. Those that are multicellular have little if any division of labor. Thus each cell of a multicellular protist is capable of metabolizing and reproducing by itself. In this text, these photosynthetic protists are divided into two groups for ease of discussion: the yellow-green to yellow-brown forms classified as Chrysophyta, and the primarily motile divisions containing the Pyrrhophyta, Euglenophyta, and Cryptophyta. This second group, often thought of as the "photosynthetic animals," will treat each division separately.

Division Chrysophyta (Chrysophytes)

The yellow-green to yellow-brown algae constitute the algal classes Tribophyceae (formerly known as Xanthophyceae; *xantho* = yellow), Raphidophyceae, Eustigmatophyceae, Chrysophyceae, Bacillariophyceae, and Prymnesiophyceae. Although they are treated here as a single division, many authors recognize at least two divisions and possibly more. The classes have several important features in common:

1. Presence of accessory chlorophyll and carotenoid pigments.
2. Chloroplast composed of many units of three associated but unfused thylakoids.
3. Chloroplast surrounded by endoplasmic reticulum that also interconnects with the outer nuclear envelope.
4. Storage product not starchlike, but laminarinlike (termed **chrysolaminarin**), as well as fats and oils and simple carbohydrates.
5. Storage product stored outside the chloroplast.
6. At least one tinsel flagellum (except in the Prymnesiophyceae).

Table 6-1 Characteristics of Chrysophyta, Pyrrhophyta, Euglenophyta, and Cryptophyta (Items in parentheses occur in some forms)

Characteristics	Division			
	Chrysophyta	Pyrrhophyta	Euglenophyta	Cryptophyta
Predominant vegetative stage	motile, nonmotile	motile	motile	motile
Flagella				
number	2 (1)	2	2 (1, 3)	2
insertion	apical (subapical)	lateral	apical groove	subapical groove
Chlorophyll a +	chlorophyll c_2, c_1	chlorophyll c_2	chlorophyll b	chlorophyll c_2
Dominant pigments	carotenoids	chlorophylls + carotenoids	chlorophylls	chlorophylls + phycobiliproteins
Nonphotosynthetic (%)	less than 1	45	66	5
Storage product	chrysolaminarin fats, oil	starch	paramylon	starch
Cell covering				
form	wall, scales	plates	strips (spiral)	strips (straight)
chemical composition	cellulose; organic silica, ($CaCO_3$)	cellulose	proteinaceous	proteinaceous

7. Cell covering, when present, outside the **plasmalemma** and often not a wall and not cellulose.

Only four classes, the Tribophyceae, Chrysophyceae, Bacillariophyceae and Prymnesiophyceae are discussed, since they are more likely to be encountered in aquatic or damp environments.

Class Tribophyceae (Yellow-Greens, or Xanthophytes)

There are approximately 120 genera with some 575 species referred to the Tribophyceae. They occur predominantly in freshwater habitats and soil. Most species are nonmotile, small, and easily overlooked. Many are ephemeral and delicate—so delicate that they die when collected in soft (soda) glass containers.

Cell Structure and Reproduction The yellow-green algal cell is superficially similar to other eukaryotic algal cells, with the usual organelles present. One to several disc-shaped, smooth-edged, green or yellow-green chloroplasts occur peripherally in the cell (Fig. 6-1A, B). The main green pigment is chlorophyll a, with chlorophyll c present in most species. Carotenoid pigments mask the chlorophylls to some extent. The chloroplasts are composed of bands of three unfused thylakoids (Fig. 6-2). When **pyrenoids** are present, they are always traversed by thylakoids and appear to be different from those in Chlorophyta. The pyrenoids may or may not be associated with photosynthate storage, which always occurs outside the chloroplast. The storage product varies, with reports of low molecular weight polysaccharides present, as well as fatty acids and sterols. Since the storage product is not in the chloroplasts, they have smooth edges.

The main cell wall materials are cellulose, **pectic** substances, and sometimes silica. In some, the

Protistan Algae

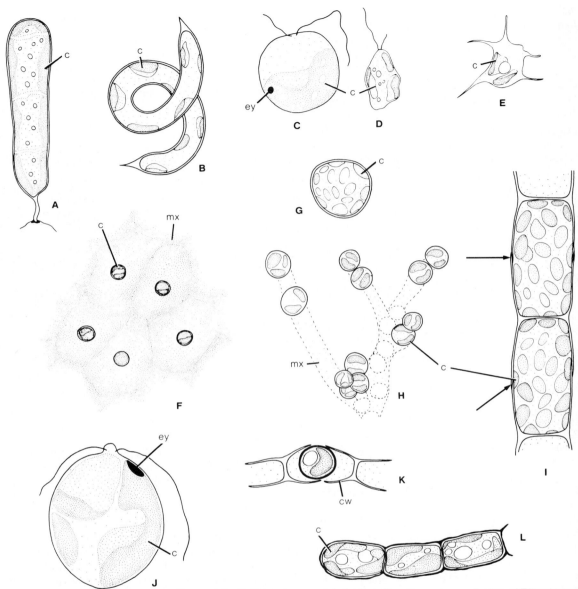

Figure 6-1 Cell structure, morphological diversity, and reproduction of Tribophyceae. **A, B,** coccoid form, *Ophiocytium* spp., showing peripheral, disc-shaped chloroplasts (*c*): **A,** ×235; **B,** ×500. **C, D,** motile, flagellated forms: **C,** *Xanthophycomonas,* with eyespot (*ey*), ×2150; **D,** *Heterochloris,* ×560. **E,** amoeboid form, *Rhizochloris,* ×830. **F,** palmelloid form, *Chlorosaccus,* portion of colony in matrix (*mx*), ×650. **G, H,** coccoid forms: **G,** *Pleurochloris,* ×940; **H,** *Mischococcus,* within matrix, ×855. **I,** filamentous form, *Tribonema,* vegetative filament showing disc-shaped chloroplasts and overlapping cell walls (*arrows*), ×915. **J–L,** reproduction: **J,** heterokont zoospore of *Tribonema,* ×8000; **K,** internal cyst of *Tribonema* in cell wall (*cw*), ×400; **L,** akinete formation in *Bumilleria,* ×450. (**D, E, K, L** after Pascher, in Rabenhorst, with permission of Akademische Verlagsgesellschaft, Geest & Portig K.-G., Leipzig.)

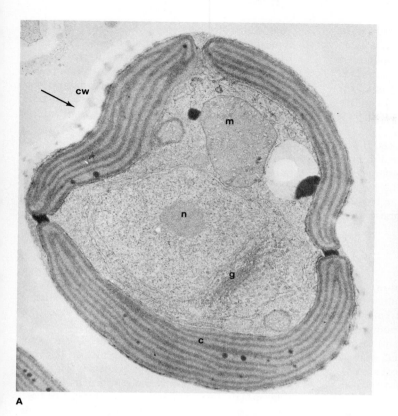
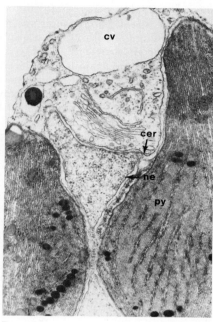

Figure 6-2 Cell structure of Tribophyceae, electron microscope. **A,** cell of *Tribonema* showing nucleus (*n*), chloroplast (*c*) with unfused thylakoids, golgi (*g*), mitochondrion (*m*), and overlapping cell wall (*cw*) at *arrow*, ×10,000. **B,** section of zoospore of *Bumilleria* at anterior, showing relationship between nucleus and chloroplasts; nuclear envelope (*ne*) continuous with endoplasmic reticulum of chloroplast (*cer*); note pyrenoid (*py*) and contractile vacuole (*cv*), ×40,000. (**A,** courtesy of T. Bisalputra; **B,** courtesy of A. Massalski, G. F. Leedale and *British Phycological Journal*.)

cell wall is composed of two overlapping, equal or unequal parts (Figs. 6-1I and 6-2A). Generally the cells are uninucleate when young, although at maturity they can become multinucleate. The endoplasmic reticulum is continuous from the nuclear envelope to the chloroplast (Fig. 6-2B). This nuclear-chloroplast-endoplasmic reticulum feature also occurs in Chrysophyceae (Fig 6-6A) and in the Phaeophyta (see Chapter 8).

Few Tribophyceae are motile in the vegetative stage. The few forms that are can be motile by flagella, amoeboid movement, or a combination of both (Fig. 6-1C–E). Thus, motile cells are generally either zoospores or gametes, usually with two unequal, anteriorly inserted flagella (Figs. 6-1C, D, J and 6-3A). The shorter flagellum, which can be trailing, is of the whiplash type; the longer one is anteriorly directed and contains a double row of tubular tinsels (Fig. 6-3A). The motile cells of the ubiquitous **coenocytic** genus *Vaucheria* differ from those of any other genus of algae (Figs. 6-3C and 6-4B, D, H; see p. 146).

The chief method of reproduction is by spore production. Zoospores and aplanospores are produced singly in the cell, or the protoplast can divide to produce several spores. The zoospores (Fig. 6-1J) are generally bilaterally symmetrical, with two dorsiventrally arranged chloroplasts. Near the base of the shorter flagellum is a swelling that is associated with the eyespot (Fig. 6-3B). In multicellular species, fragmentation also serves as a means of vegetative reproduction. Some species produce internal **cysts** (Fig. 6-1K), similar to those of Chrysophyceae (see p. 149). In some species the entire cell (including the cell wall) is enclosed in a heavy wall to form an **akinete** (Fig. 6-1L).

Sexual reproduction is not widespread in the Tribophyceae, although it is common in *Vaucheria* (see p. 146). In the freshwater forms studied, meiosis apparently occurs on germination of the zygote. Thus, the haploid phase predominates, and the life history has a predominant haploid phase (Fig. 3-7A).

Morphological Variation, Classification, and Ecology The same morphological variation present in the Chlorophyceae is present in Tribophyceae. However, most yellow-greens are nonmotile, unicellular, and small (less than 50-75 μm) forms. Some of these are of common occurrence, such as *Ophiocytium* (Fig. 6-1A, B). However, most are easily overlooked because of their size, and because they are often **epiphytes** and occur in specialized habitats, such as alpine pools, ditches, or sphagnum bogs. Typical **coccoid** forms include the unicellular *Pleurochloris* (Fig. 6-1G) and the multicellular *Mischococcus* (Fig. 6-1H). Other morphological forms include both motile vegetative cells and nonmotile **colonial** genera (Fig. 6-1C–E). The most ubiquitous and commonly encountered Tribophyceae are species belonging to the **filamentous** *Tribonema*, and the coenocytic *Vaucheria* (Figs. 6-1I and 6-4), which occur free-floating or attached in water or on moist soil.

Tribonema is an unbranched, filamentous alga generally present in cool waters (around 12 °C or less), usually in the spring, as light green or almost pale yellow floating masses. The cells are often thick in the middle and slightly constricted at the transverse walls. This is the result of the H- or I-shaped transverse walls that overlap at the center of the cell (Figs. 6-1I and 6-2A). This transverse wall is formed from the center outward, and gradual separation of the parental half-cells follows. Sometimes yellow- to brown-colored materials settle irregularly on the older H-shaped pieces. One or two zoospores (Fig. 6-1I) are formed in ordinary vegetative cells. They are **pyriform** to **ovoid**, with typical yellow-green algal flagellation. They show some plasticity when swimming; ultimately, they settle to form an elongated holdfast cell. Sexual reproduction is not commonly observed but is isogamous, with one gamete settling prior to **syngamy**. The zygote is a resting stage, and it is assumed that meiosis occurs during its germination.

Species of *Vaucheria* occur in water or on moist mud. It is unique in producing unusually large, multinucleate and multiflagellated zoospores. The flagella occur in pairs; one member of each pair is slightly longer than the other, and both are whiplash. It is thought that this zoospore is a compound one, composed of many flagellar units (Fig. 6-4B, D). Sperms of *Vaucheria* have typical Tribophyceae flagellation, except the longer flagellum is the whiplash type (Fig. 6-3C). *Vaucheria* is a **coenocyte** (Fig. 6-4A) that produces transverse walls only when injured or during formation of reproductive cells. The zoospore is produced at the end of a somewhat swollen tube and ultimately becomes separated from the rest of the plant. The zoosporangium produces a single, large (over 100 μm) multiflagellate, multinucleate, zoospore (Fig. 6-4B, E). Each nucleus is associated with a pair of flagella (Fig. 6-4D). The zoospore swims sluggishly for a few hours before settling and growing into a new coenocyte. The coenocytic tubes produced by the zoospore are narrower than the zoospore, so it is easy to recognize the origin of the new plant (Fig. 6-4E).

In sexual reproduction, the gametangia are generally adjacent to one another and are easily recognized (Fig. 6-4F, G). A single, large, nonmotile green egg is retained within the female gametangium. Many small (ca. 5μm), elongate, colorless sperms are produced in a hook-shaped male gametangium (Fig. 6-4F–H). The flagella are laterally inserted; the forward, short flagellum has tinsels, and the longer flagellum is smooth (Fig. 6-3C)—the reverse of their condition in most motile Tribophyceae cells. The sperm fertilizes the egg *in situ*, and a heavy-walled, resting zygote results. Germination occurs under favorable conditions of light, temperature, and moisture. Meiosis is believed to occur upon germination, with only one meiotic nucleus surviving.

The Tribophyceae are primarily freshwater organisms, often occurring in small ponds or pools. Because of their delicate nature and small size, their

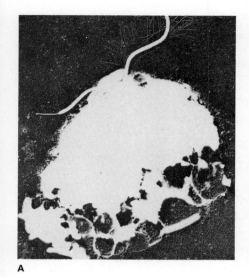
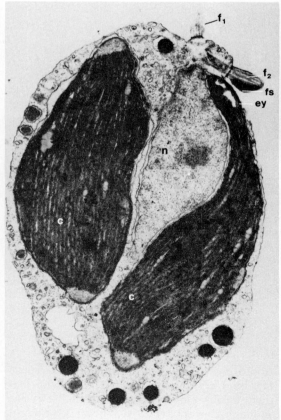
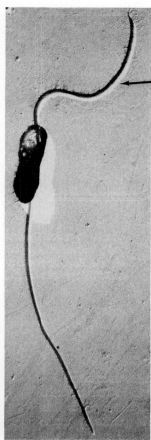

Figure 6-3 Motile cells of Tribophyceae, electron microscope. **A, B,** zoospores of *Bumilleria:* **A,** shadow-cast to show flagella; forward projecting one with tinsels, ×4000; **B,** longitudinal section showing parietal chloroplasts (*c*), with eyespot (*ey*), nucleus (*n*), flagellar insertion (f_1, f_2), and flagellar swelling (*fs*), ×10,000. **C,** sperm of *Vaucheria,* showing lateral insertion of flagella; forward flagellum with tinsels (*arrow*), ×6000. (**A, B,** courtesy of A. Massalski, G. F. Leedale and *British Phycological Journal;* **C,** courtesy of Ø. Moestrup from "On the fine structure of the spermatozoids of *Vaucheria sescuplicaria* and the later stages in spermatogenesis," *Journal of the Marine Biological Association U.K.* 50:513–23, 1970, with permission of Cambridge University Press.)

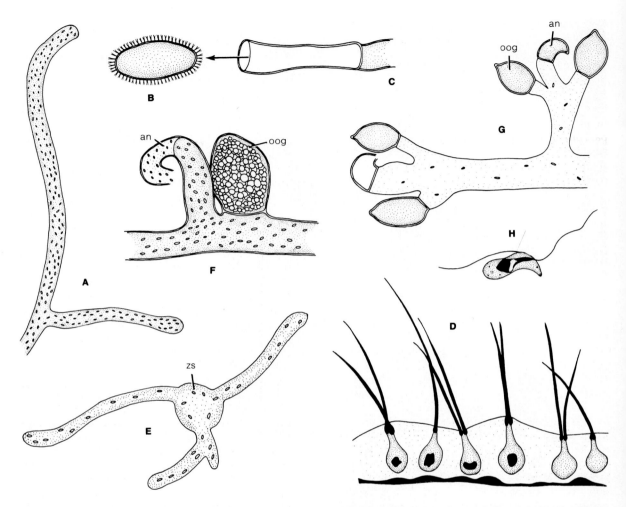

Figure 6-4 Coenocytic Tribophyceae, *Vaucheria*. **A,** vegetative coenocyte, ×60. **B,** zoospore, after release from zoosporangium (**C**), ×200. **D,** detailed structure of edge of zoospore, showing relationship of flagellar pair and single nucleus, ×1240. **E,** young coenocytic germling from zoospore (*zs*), ×400. **F, G,** gametangial arrangement (*an,* antheridium; *oog,* oogonium), ×140. **H,** sperm, ×1250. (**D, H,** courtesy of W. J. Koch and *Journal of the Elisha Mitchell Scientific Society.*)

ecology is not well known. A common inhabitant of swimming pools, known as mustard alga, is a unicellular xanthophyte. *Tribonema* occurs in cool water (around 10 °C) in the spring, and *Vaucheria* grows in muddy areas, often on the edges of drying ponds or puddles. Some species of *Vaucheria* are extensive colonizers of brackish water marshes and embayments. Most yellow-greens occur in acid pools, sphagnum bogs, or alpine and subalpine pools.

Class Chrysophyceae (Golden Algae)

Approximately 100 genera and 500 species constitute the Chrysophyceae (*chryso* = golden). Most representatives are motile and occur in fresh water, although a few species are significant in the marine **plankton** and **nannoplankton** (*nan* = dwarf).

Cell Structure and Reproduction The typical uninucleate golden algal cell contains one or two smooth-edged, golden to yellow-brown chloroplasts (Fig. 6-5A–D). The color results from the predominance of the xanthophyll **fucoxanthin,** which masks the chlorophyll *a*. Additional pigments include other carotenoids and chlorophyll *c*. In some motile forms, an eyespot occurs near the anterior part of the chloroplast. The photosynthate is chrysolaminarin, which can be a large refractive granule at the cell posterior (Fig. 6-6A) or adjacent to the pyrenoid (but outside the chloroplast).

Most species are unicellular or colonial flagellates (Fig. 6-5A–E) that generally lack a definitive cell wall but can have **scales** (Figs. 6-5C, D and 6-6B) or a loose covering known as a **lorica** (Fig. 6-5E, F). The lorica (*lorica* = armor) is essentially a shell to which the cell may or may not be attached by cytoplasmic processes. It is microfibrillar or scalar in nature and can contain cellulose or minerals, such as iron, magnesium, and carbonate. When present, the scales are inorganic, usually composed of silica. Nonmotile forms have a cellulosic wall with pectic substances. As in the Tribophyceae, the nuclear membrane and endoplasmic reticulum of the chloroplast are associated (Fig. 6-6A).

In motile cells, the flagella are anteriorly inserted, sometimes in a small depression, or pit. In the biflagellated condition, the longer flagellum is tinsel and the other whiplash. If the two flagella are markedly unequal in length, the whiplash one is reduced, and reduction may be so complete that only the tinsel flagellum protrudes. In some biflagellated species, the second flagellum is very delicate and is easily shed. Some of the loricate forms are motile, with only the flagellum projecting. Other loricate species are stationary even when functional flagella are present to move water currents around the cell (Fig. 6-5F).

Reproduction in flagellated forms can occur while they are still motile. Prior to nuclear division in these motile forms, there has been replication of the organelles. In many of the loricate forms, following cell division, one of the new cells remains within the parental lorica. In nonmotile vegetative forms, the zoospores produced resemble either the biflagellated or uniflagellated vegetative cells (Fig. 6-5A, B). Some nonmotile genera form aplanospores.

In many Chrysophyceae, a resting cyst, the statospore (*stat* = standing) is produced within a cell. The cyst wall becomes heavily silicified, except for a small opening containing a plug composed of silica and polysaccharide substances that closes the statospore (Fig. 6-5J). Under favorable conditions, this resting stage germinates, generally liberating one or two motile cells, or zoospores, which escape through the pore left after the plug dissolves (Fig. 6-5I). Some statospores result from syngamy, but sexual reproduction is uncommon—at least it has not been observed often. Syngamy is usually isogamous, with flagellated gametes.

Morphological Variation, Classification, and Ecology The Chrysophyceae are almost as diverse in form as the Chlorophyceae (Chapter 7), although most Chrysophyceae genera are motile in the vegetative stage. These include the unicellular flagellates, *Chromulina, Ochromonas,* and *Mallomonas* (Fig. 6-5A–C); and the multicellular flagellates, such as *Dinobryon* and *Synura* (Fig. 6-5D, E). Loricate forms can be multicellular, such as *Dinobryon* (Fig. 6-5E), which also occurs as a single cell, or *Epipyxis* (Fig. 6-5F), which is usually epiphytic on other algae.

Nonmotile golden algae include colonial forms, such as *Hydrurus* (Fig. 6-5G, H); coccoid types, such as *Epichrysis* (Fig. 6-5K); and filamentous forms, such as *Phaeothamnion* (Fig. 6-5L). These nonmotile forms, which are not commonly collected, comprise less than a third the genera in the class. When motile reproductive cells are produced, they resemble either *Chromulina* (uniflagellated) or *Ochromonas* (biflagellated) cells (Fig. 6-5A, B).

Freshwater flagellated species of Chrysophyceae are common in the phytoplankton of lakes, particularly during colder seasons of the year, when these algae are abundant. Many nonmotile, freshwater species occur commonly in cold (8–10 °C), fast-running streams, springs or spring pools, and shallow ponds. Acid habitats, such as those present in sphagnum bogs, support a large variety of smaller, delicate forms. Some golden algae occur in tidepools and salt marshes, often in large enough numbers to color the water a golden-brown. Motile Chrysophyceae in the nannoplankton of colder waters, in both the Northern and Southern Hemispheres, are considered important primary food producers. A few Chrysophyceae are **phago-**

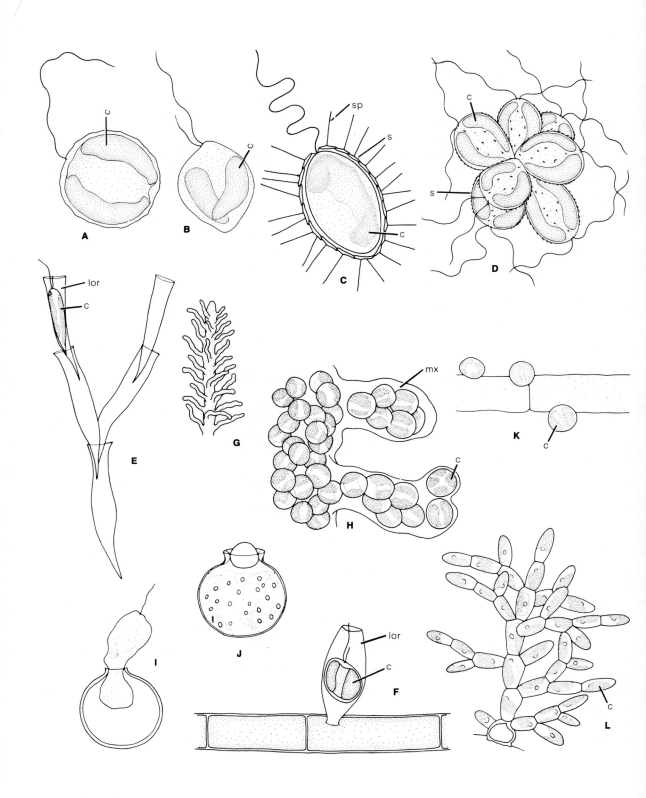

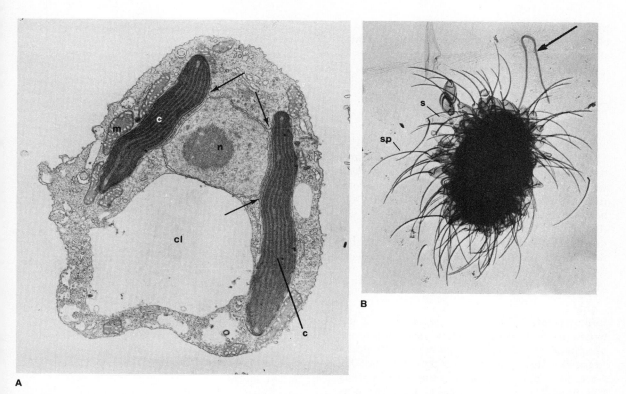

Figure 6-6 Cell structure of Chrysophyceae, electron microscope. **A,** longitudinal section of *Ochromonas,* showing two lobes of chloroplast (*c*), nucleus (*n*), nuclear-chloroplast endoplasmic reticulum (*arrows*), mitochondrion (*m*), and chrysolaminarin vacuole (*cl*), ×9000. **B,** dried cell of *Mallomonas* showing scales (*s*) with spines (*sp*) and obvious tinsel flagellum (*arrow*); smooth flagellum not shown, ×2500. (**A,** courtesy of T. Slankis and S. P. Gibbs and *Journal of Phycology;* **B,** courtesy H. Belcher and *Nova Hedwigia.*)

Figure 6-5 Cell structure and morphological diversity of Chrysophyceae. **A–E,** flagellated form, motile. **A–C,** unicellular form showing chloroplasts (*c*), scales (*s*), and spines (*sp*): **A,** *Chromulina,* ×2700; **B,** *Ochromonas,* ×3200; **C,** *Mallomonas,* optical section to show cell contents and side view of scales and spines. **D, E,** multicellular form: **D,** *Synura,* with scales, ×2700; **E,** *Dinobryon,* with lorica (*lor*) around individual cells, ×1000. **F-H, K, L,** nonmotile forms. **F,** attached, nonmotile loricate (*lor*) form, *Epipyxis,* (on an algal filament), with two flagella for creating water currents, ×1000. **G, H,** *Hydrurus,* showing habit (**G,** ×50) and detail of colony in matrix (*mx*); **H,** ×1000. **K,** coccoid form, *Epichrysis,* (epiphytic on algal filament), ×930. **L,** filamentous form, *Phaeothamnion,* ×5000. **I, J,** statospores: **I,** germination with single zoospore being released, ×1000; **J,** statospore with plug, ×1000. (**I,** after Smith.)

trophic (*phag* = eat), often ingesting bacteria or bluegreen algae. This occurs in species containing chloroplasts as well as in those without.

Class Bacillariophyceae (Diatoms)

The Bacillariophyceae (*bacill* = little stick) are a conspicuous element in both marine and freshwater habitats and are well represented in the fossil record. The outstanding feature is the nonmotile, box-like vegetative cells with silica in the cell wall (Fig. 6-7). Over 5600 living species in approximately 200 genera belong to this class. Including fossil forms, there are probably over 10,000 species described.

Cell structure The uninucleate diatom cell contains one, two, or many yellow or golden-brown peripheral chloroplasts, which are disc-shaped or ribbonlike. The ultrastructure of the chloroplast (Fig. 6-8A, B) is similar to that of other Chrysophyta. Chlorophylls (a, c_1, c_2) are masked by carotenoids, especially fucoxanthin and other xanthophylls. Food reserves are stored as fats, oils, or chrysolaminarin. Pyrenoids are often present in the chloroplasts (Fig. 6-8A, B), but the photosynthate is always external to the chloroplast envelope.

The cell walls or **frustules** (*frust* = little piece) are complex. They consist of two overlapping halves (Fig. 6-7), each termed a **theca** (*thec* = case, box). Because of these two parts, the common name for the class is *diatom*, meaning "cut in two." The outer part of the cell wall (comparable to the top of a box) is the **epitheca** (*epi* = above), and the inner part of the cell (or bottom of the box) is the **hypotheca** (*hypo* = below), as shown in Figure 6-7. When the top or bottom surface of the cell is viewed, it is the **valve view** (Figs. 6-7 and 6-9A, C). Each theca consists of a flattened or convex valve with a connecting band attached along its edge (Fig. 6-7). The overlapping walls of the cell form a **girdle** region, and when the cell is observed from the side, it is the **girdle view** (Figs. 6-7 and 6-9B). The girdle area is often wider because of the presence of **intercalary bands** (Fig. 6-9A). As much as 95% of the cell wall is silica, deposited in an organic (noncel-

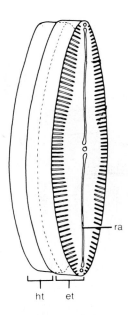

Figure 6-7 Cell structure of Bacillariophyceae (*Pinnularia*) showing girdle (*left*) and valve (*right*) and to illustrate the hypotheca (*ht*), epitheca (*et*), raphe (*ra*), and linear markings, ×500.

lulosic) framework. The pattern of silica deposition is generally species specific, although environmental factors can affect the amount of deposition.

Due to the uneven deposition of silica, the wall contains a variety of markings. Depositions that form large openings in the valve are known as **areolae** (*areol* = little space). Each areola is covered by a thin siliceous layer containing slits or holes (Fig. 6-9D). Smaller openings are **puncta** (*punct* = sting, prick), which lack the siliceous layer over the opening (Fig. 6-9A). The puncta and areolae are often in close linear order, forming a **stria** (*stria* = furrow, streak; Fig. 6-9A), which appears as a thin line with the light microscope. A linear thickening, or ridge, on the valves is the **costa** (*cost* = rib). The valve sometimes consists of chambers with openings both inward and outward, or several thickened layers. In addition, wall projections, spines, or granules of various types can be present.

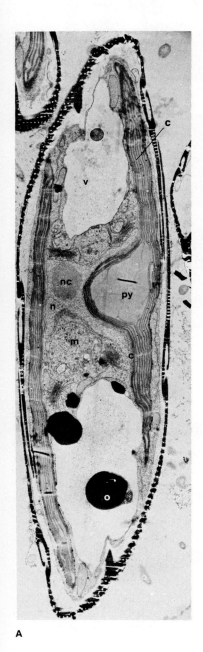
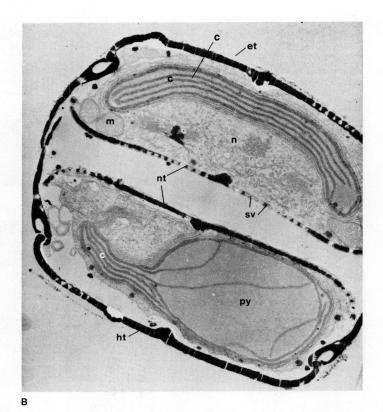

Figure 6-8 Cell structure and division of Bacillariophyceae (*Amphipleura*), transmission electron microscope. **A,** longitudinal section, slightly oblique to valve, showing chloroplast (*c*) with pyrenoid (*py*), central cytoplasmic bridge with nucleus (*n*) and nucleolus (*nc*), mitochondria (*m*), vacuole (*v*), and oil bodies (*o*), ×6750. **B,** transverse section of recently divided cell, showing chloroplasts with pyrenoid, nucleus, mitochondria, silicalemma deposition vesicles (*sv*), new hypothecae (*nt*), and parental thecae (hypotheca, *ht*; epitheca, *et*), which become epithecae of new cells. (Courtesy of E. F. Stoermer and *American Journal of Botany*.)

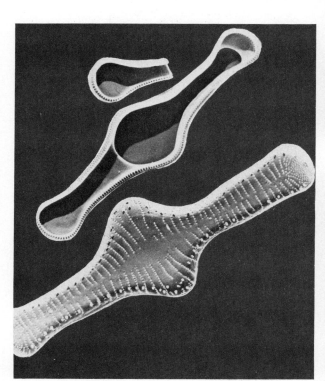
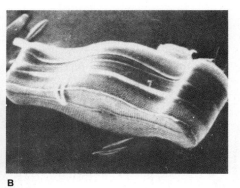

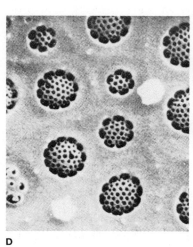
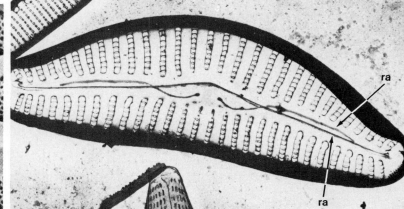

Cell shape and valve markings are radial (including triangular), rectangular, or asymmetrical. The radial form delimits **centric** diatoms (Fig. 6-10A–D); rectangular and asymmetrical forms delimit the **pennate** diatoms (Fig. 6-10E–H).

Many pennate diatoms have an unsilicified groove or channel on the valve, termed the **raphe** (*raph* = seam, suture; Figs. 6-7 and 6-9E). The raphe can be straight, undulate, or S-shaped and extends through the frustule (Fig. 6-9E). Part of the raphe or the raphe channel can contain cytoplasm, which is enclosed by the plasma membrane.

Only those pennate diatoms with a raphe can move either by a smooth or somewhat jerky motion. Rates of movement have been calculated at 0.2–25 µm/second. The mechanism is believed to involve cytoplasmic streaming within the raphe system and may involve oscillations of protoplasmic protein fibrils within the cells.

Reproduction The chief method of reproduction in diatoms is vegetative cell division, which involves the formation of two new valves after nuclear and cytoplasmic division. Silica, which is an essential element for diatom walls, is deposited in the **silicalemma** forming the valve mold typical of the species. The new valves (both of which are the hypotheca) are completed before the girdle or intercalary bands are formed (Fig. 6-8B).

Since the parent cell valves are the epitheca for the new cells, there can be a gradual size diminution in some populations. Other populations do not show this decrease in size, because weakly silicified cells resume original size and shape after cell division. Those populations with a gradual size diminution may produce a large cell, termed an **auxospore** (*aux* = grow, enlarge) as described below.

Sexual reproduction is variable, and in contrast to many algae, the vegetative diatom cell is diploid (Fig. 3-7B gives the diploid life history). Meiosis occurs during gamete formation, and after syngamy the resulting zygote may differ morphologically from the vegetative cells.

Details of sexual reproduction have been studied in only a few diatoms. In pennate forms, the gametes are usually isogamous and lack flagella. Parental cells pair along the girdle, and meiosis results in one or two viable gametes. In contrast, centric diatoms have oogamy; the tiny male gametes have at least one tinsel flagellum. After syngamy an auxospore is formed that differs from the vegetative cell (Fig. 6-11A, B). When cell division occurs, the first cells are "atypical" and the "typical" mold for the new valves is lacking. This probably is indicative of a certain amount of genetic plasticity in thecal structure.

Cysts or resting cells are also produced by diatoms. These are smaller than the cell producing them and have a thickened wall. The shape is unlike that of the parental cell (Fig. 6-11C). Cyst production probably does not involve sexual reproduction.

Figure 6-9 Frustule structure of Bacillariophyceae. **A, B,** scanning electron microscope. **A,** theca of *Tabellaria*, showing valve (*lower*) and intercalary bands (*upper*): note rows of punctate striae on valve, ×3300. **B,** external features of two cells of *Didymosphaenia*, showing valve (*lower*) and girdle views, ×650. **C,** phase-contrast microscope of centric diatom (*Coscinodiscus*) in valve view, showing areolae, ×1200. **D, E,** transmission electron microscope: **D,** areolae with perforate siliceous plate with puncta of *Triceratium*, ×5000; **E,** carbon replica of frustule of *Cymbella*, showing raphe (*ra*) from both outside (*straight*) and inside (*curved*) of the valve, ×3600. (**A,** courtesy of J. D. Koppen and *Journal of Phycology*; **B,** courtesy of P. A. Dawson and the *British Phycological Journal*; **C,** courtesy of D. Walker; **D,** courtesy of R. Ross and the *British Phycological Journal*; **E,** courtesy of R. W. Drum, from "Electron microscope observations of diatoms," *Oesterrich Botanishe Zeitschrift* 116:321–30, 1969, with permission of Springer-Verlag.)

Morphological Variation and Classification
Several orders of diatoms are recognized, but all are the same as the nonmotile, nonfilamentous orders of other algal classes (e.g., Chlorococcales of Chlorophyceae). The main morphological separation into centric and pennate diatoms is adequate to show the diversity within the Bacillariophyceae. In addition to wall markings and symmetry, centric and pennate diatoms are separated on the basis of shape, chloroplast number, and motility, with developmental lines possibly analogous to these groups. Centric diatoms are radially symmetrical, usually with circular valves,

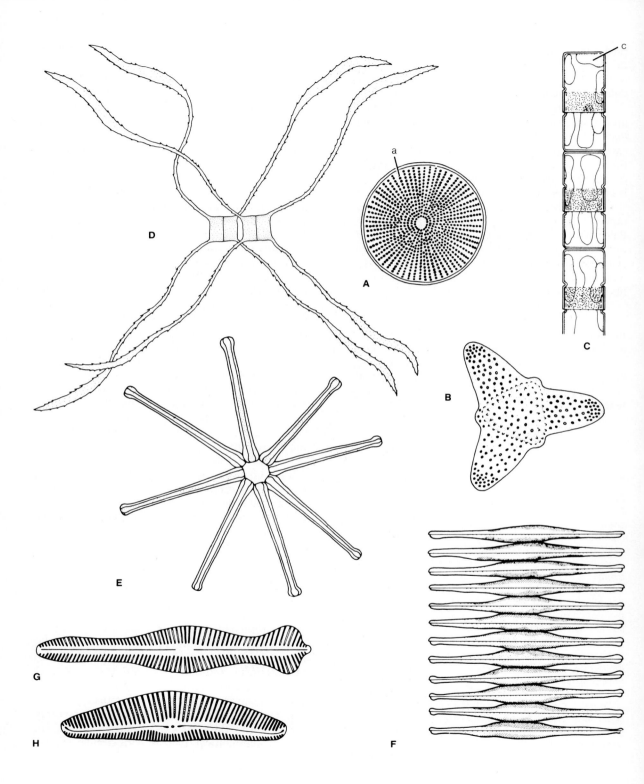

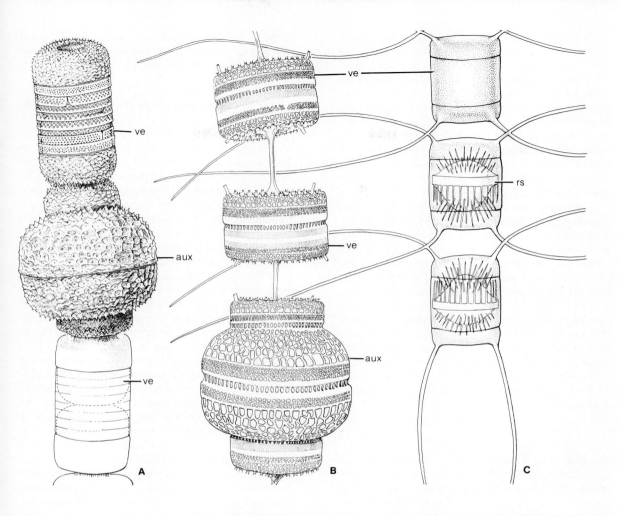

Figure 6-11 Reproductive cells and morphological forms of Bacillariophyceae (centric forms). **A, B,** auxospores (*aux*) in filamentous colonies with vegetative cells (*ve*): **A,** *Melosira;* **B,** *Thalassiosira,* ×500. **C,** resting spore (*rs*), or cysts, in filamentous colony of *Chaetoceros,* with vegetative cells, ×500. (After Hendey.)

Figure 6-10 Morphological diversity and wall markings of Bacillariophyceae. **A–D,** centric diatoms: **A,** *Coscinodiscus,* valve view, showing rows of areolae (*a*), ×125; **B,** *Triceratium,* valve view, ×545; **C,** *Melosira* cells, girdle view, showing chloroplasts (*c*), ×640; **D,** *Chaetoceros* cells, girdle view with long spines, ×500. **E–H,** pennate diatoms: **E,** *Asterionella* colony, girdle view, ×1000; **F,** *Fragilaria* colony, girdle view, ×750; **G,** *Gomphonema,* valve view (note transverse asymmetry), ×600; **H,** *Cymbella,* valve view (note longitudinal asymmetry), ×545.

Protistan Algae 157

as in *Coscinodiscus* (Figs. 6-9C and 6-10A); sometimes, however, they are triangular, as in *Triceratium* (Fig. 6-10B), or elongate, as in *Melosira* (Fig. 6-10C). Pennate diatoms have bilateral symmetry with elongate valves, as in *Navicula* and *Pinnularia* (Fig. 6-7). Sometimes they are asymmetrical in girdle view (*Gomphonema, Cymbella;* Figs. 6-9E and 6-10G, H). Valve markings occur in two longitudinal series on either side of a median line (Fig. 6-7), and the valve and/or girdle is longitudinally or transversely symmetrical or asymmetrical. The raphe is generally present in the valve (Fig. 6-9E).

Most diatoms occur as single cells. Others are loosely aggregated in irregular filamentous chains, with adjacent cells interconnected by mucilaginous pads that join the cells into a zig-zag colony. Species of *Gomphonema* form stalked, treelike colonies, often with large amounts of stalk materials. *Cymbella* also produces stalks, and some species live within extensive mucilaginous tubes. The stalk and mucilaginous tubes are carbohydrate or sulfate polysaccharides similar to those present in brown and red algal cell walls (Chapters 8 and 9). Still other species form colonies by adhesion of the valve faces (*Melosira, Fragilaria;* Figs. 6-10C, F and 6-11A) or long, spinelike processes (*Chaetoceros;* Figs. 6-10D and 6-11C).

Ecology The Bacillariophyceae are widely distributed throughout freshwater and marine environments. Centric species are primarily marine, and pennate diatoms occur more often in fresh water. Diatoms occur free-floating in plankton, as well as on the surfaces of solid substrates, on mudflat surfaces, in salt marshes, and attached as epiphytes to other algae and aquatic plants. Diatoms also occur in soil, on old brickwork or rock walls, and in almost any moist place, including within human livers and kidneys! Diatoms occur in vast numbers in phytoplankton, often forming extensive blooms.

Some diatoms can tolerate only a narrow range of environmental conditions and are used as indicators of the physical-chemical characteristics of some waters. The composition of diatom communities is also used to indicate varied ecological conditions. For example, in rivers free from pollution, the population is composed of many different species, each present in relatively small numbers. In contrast, polluted waters have few diatom species, often in very large numbers. Lake sediment cores containing diatom frustules (essentially fossils) help establish the history of the lake.

Class Prymnesiophyceae (Prymnesiophytes)

Prymnesiophyceae are important with regard to productivity in marine waters of some areas, especially the tropics and subtropics, where they can be the predominant phytoplankton. Only 60 genera with nearly 200 species are known. They are separated from Chrysophyceae primarily by cellular differences as well as life-history differences. Many Prymnesiophyceae have been shown to have alternate nonmotile and motile stages.

The motile cells generally lack cell walls but have scales of organic materials and/or calcium carbonate (Figs. 6-12B and 6-13A, B). The calcium carbonate scales are large enough to be seen with the light microscope and are termed **coccoliths** (Fig. 6-13B, C). Cell structure is similar to that of Chrysophyceae (Fig. 6-6A). However, one outstanding difference is the apical **haptonema** (Figs. 6-12A, 6-13D, and 6-14C). The term means "fastening thread," which was originally thought to be its function. As with the flagellum, the cell plasma membrane is continuous around the haptonema. However, in contrast to the flagellum with its typical 9 + 2 arrangement of microtubules (see Fig. 3-2A), the haptonema contains extensions of the endoplasmic reticulum surrounding six or seven longitudinally oriented microtubules (Fig. 6-13E). In some instances, the haptonema is long and coiled (Fig. 6-12A); in others it is short (Figs. 6-13D and 6-14C).

The motile cells of prymnesiophytes differ from those of other Chrysophyta by having only smooth, whiplash flagella (Fig. 6-13D). The two flagella are equal in length and are apically or subapically inserted. The swimming motion is a rotating one, typical of other flagellated forms; however, the flagella can bend back on the cell and trail, or push the cell.

Diversity in form is similar to that in Chrysophyceae. However, in Prymnesiophyceae, many of the diverse morphological forms are apparently stages in the life history of a species. For example, the life history of one prymnesiophyte can produce (1) coccolith-bearing motile cells (Fig. 6-14A), (2) coccolith-free motile cells with haptonema (Fig. 6-

14C), (3) coccolith-free motile cells without haptonema, (4) coccolith-bearing nonmotile cells, (5) coccolith-free nonmotile cells, and (6) filamentous plants (Fig. 6-14B).

Division Pyrrhophyta (Class Dinophyceae) (Pyrrhophytes)

The Pyrrhophyta (*pyrrh* = red, burning) include over 3000 species in at least 125 genera, generally placed in a single class, Dinophyceae, the dinoflagellates (*dino* = whirling). Most organisms are flagellated, although there are a few nonmotile genera (Fig. 6-15). In addition to the expected photoautotrophic forms, some are colorless. Some species may be phagotrophic, whether chlorophyll is present or not.

Cell Structure and Reproduction

The photosynthetic dinoflagellates have chloroplasts similar in structure to those of most Chrysophyta. The pigments include chlorophyll *a* and c_2, along with some dominant carotenoids specific to dinoflagellates. The thylakoids are in threes; pyrenoids, when present, are varied in position. The photosynthate is starch, but unlike the starch in Chlorophyta, it is stored outside the chloroplast. Sometimes fats and oils accumulate in the cell in large quantities and are red or orange. Eyespots are varied and may or may not be associated with the chloroplasts.

The cells are generally uninucleate with a unique nucleus, as the chromosomes remain condensed and visible throughout cell division and interphase (Figs. 6-16A and 6-17A). During mitosis the nuclear membrane and nucleolus remain present. Cytoplasmic microtubules separate the new nuclei in division but are not involved in actual separation of the chromosomes, which appear attached to the nuclear envelope.

The nonmotile forms (filamentous or nonfilamentous) have a cellulosic wall. However, most dinoflagellates are motile, biflagellated cells (Figs. 6-15A–E and 6-17A, B) that lack a cellulosic wall but usually contain cellulosic plates of varied thickness

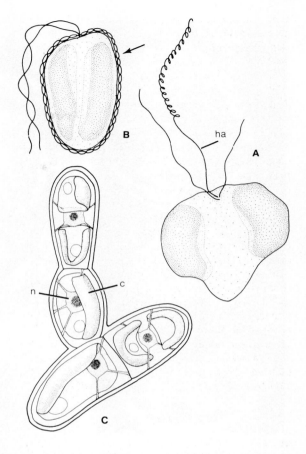

Figure 6-12 Cell structure and morphological diversity of Prymnesiophyceae. **A, B,** motile form: **A,** *Chrysochromulina* with coiled haptonema (*ha*) between apical flagella, ×700; **B,** *Hymenomonas* with surface coccoliths (arrow), ×1000. **C,** nonmotile form, *Pleurochrysis* (*Apistonema*), with chloroplast (*c*) and nucleus (*n*), ×1200.

internal to the plasma membrane (Fig. 6-16B, C). Those forms in which the plates are obvious with the light microscope (Fig. 6-17C) are known as **armored,** or **thecate** (*thec* = box, case), dinoflagellates. Many of the supposedly **unarmored** (or naked) species have been shown to have very delicate plates. The outer cell covering involves a system of **vesicles** (*vesicul* = little bladder) that can contain cellulosic plates adjacent to the outer cell membrane (Fig. 6-16B, C). The number of plates in a species is generally fixed, and there is a correlation between

Protistan Algae 159

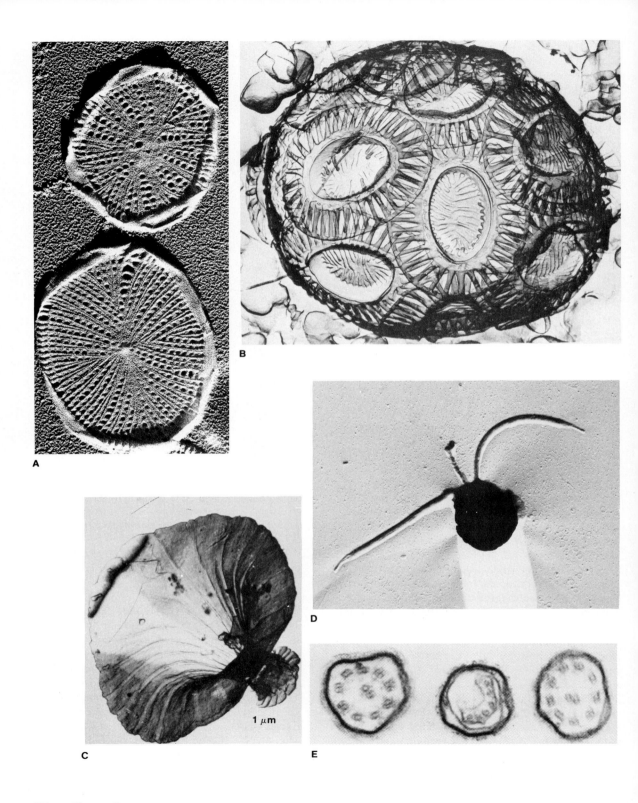

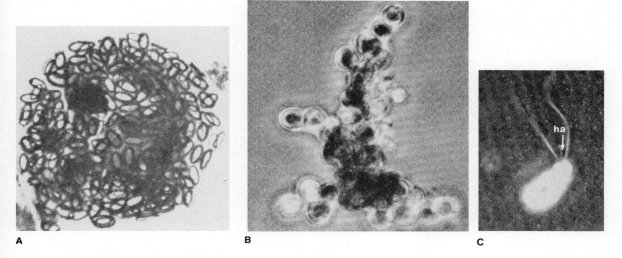

Figure 6-14 Life history of a Prymnesiophyceae. **A,** coccolith-bearing stage (*Hymenomonas*-stage), ×3000. **B,** nonmotile thallus (*Apistonema*-stage), ×750. **C,** coccolith-free swarmer from nonmotile thallus, with short haptonema (*ha*), ×2000. (Courtesy of B. S. C. Leadbeater and the *British Phycological Journal*.)

Figure 6-13 Cell structure of Prymnesiophyceae, electron microscope. **A–C,** scales: **A,** organic scales of *Chrysochromulina,* ×29,000; **B, C,** coccoliths made of calcium carbonate; **B,** whole cell of *Coccolithus,* ×15,000; **C,** single scale, ×13,000. **D, E,** flagella and haptonema: **D,** zoospore showing central haptonema with bulbous tip and two flagella, ×6000; **E,** transverse section of two flagella and central haptonema (*Phaeocystis*), ×60,000. (**A,** courtesy I. Manton and M. Parke from I. Manton and M. Parke, "Preliminary observations on scales and their mode of origin in *Chrysochromulina polylepis* sp. nov.," *Journal of the Marine Biological Association, U. K.* 42:565–78, 1962, with permission of Cambridge University Press; **B, C,** courtesy of A. McIntrye, from A. McIntrye, and A. W. H. Bé, "Modern Coccolithophoridae of the Atlantic Ocean, I. Placoliths and cryoliths," *Deep-Sea Res.* 14:561–97, 1967, with permission of Pergamon Press Ltd.; **D, E,** courtesy of J. C. Green and I. Manton, from M. Parke, J. C. Green and I. Manton, "Observations on the fine structure of zoids of the genus *Phaeocystis* (Haptophyceae)," *Journal of the Marine Biological Association, U. K.* 51:927–41, 1971, with permission of Cambridge University Press.)

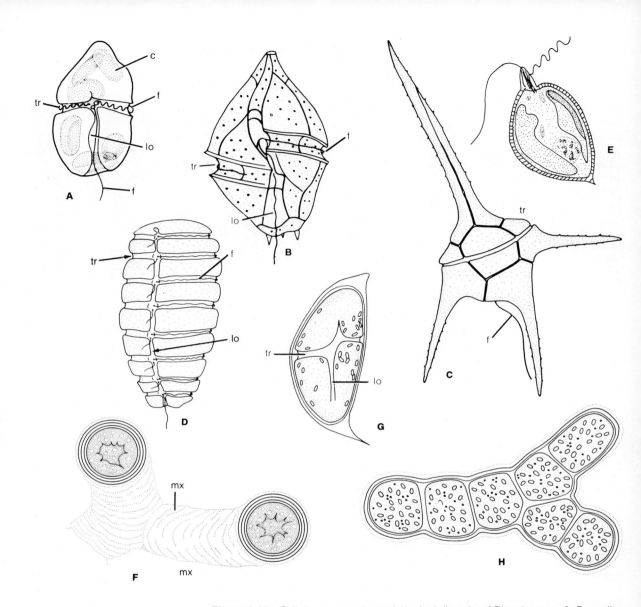

Figure 6-15 Cell structure and morphological diversity of Dinophyceae. **A–D,** motile form, viewed from ventral side (**A, B, D**) to show orientation of flagella (*f*) and grooves (transverse, *tr;* longitudinal, *lo*): **A,** *Gymnodinium,* unarmored form, showing chloroplast (*c*), ×250; **B,** *Gonyaulax,* armored form, ×750; **C,** *Ceratium,* armored form with horns (dorsal view), ×550; **D,** *Polykrikos,* a coenocytic form, ×75. **E,** apical flagellation of *Prorocentrum,* ×400. **F–H,** nonmotile forms: **F,** *Rufusiella,* nonfilamentous type with extensive mucilaginous matrix (*mx*), ×245; **G,** coccoid form, containing characteristic dinoflagellate swarmer, *Cystodinium,* ×650; **H,** filamentous form of *Dinothrix,* ×720. (**B, D, H,** after Schiller in Rabenhorst, with permission of Akademische Verlagsgesellschaft, Geest & Portig K.-G., Leipzig; **F,** from R. H. Thompson, "Algae" in W. T. Edmonson ed., *Freshwater Biology,* 2d ed., copyright ©1959, by John Wiley & Sons. Reprinted by permission of John Wiley & Sons, Inc.)

number and thickness—the thicker the plates, the fewer their number. The arrangement of plates is used as a generic and specific characteristic.

The typical dinoflagellate cell has the two flagella laterally inserted (Fig. 6-15A–D). One flagellum encircles in a transverse groove, whereas the other trails posteriorly in a second groove (Fig. 6-15A, B). The flagella can emerge from a common pore or two separate ones, and there can be a series of small plates in the grooves. The transverse flagellum is ribbonlike due to the presence of a **striated** band of material within the flagellar membrane. This striated band is shorter than the flagellar microtubules, causing a "buckling" of the flagellum similar to that of a hoop within a skirt hem (Fig. 6-16E). Very fine hairs can be present on this flagellum. The trailing flagellum is whiplash and is responsible for most of the forward and rotating movement of the cell. This motion of the cell explains the derivation of the name *dinoflagellates*, or "whirling" flagellates. The location of the transverse groove with regard to the overall cell appearance is used as a taxonomic character. In one order of Dinophyceae, the flagella are apically inserted, with one encircling the other (Fig. 6-15E).

Vegetative reproduction in Pyrrhophyta is varied. Following nuclear and organelle division in motile forms, separation of new cells may involve the parental cell. Generally there is an invagination, or pinching in half, of the parental outer layers (plates or vesicles and associated membranes). In some forms with thick plates, there is diagonal splitting of the parental plates, with each new cell receiving some parental plates. A third type of cell division occurs completely within the parental cell, with the new cells being released as two or more spores and forming new plates, vesicles, and external membranes.

Nonmotile species with a cellulose wall produce motile cells that may or may not have plates. These swarmers are often similar to described motile species and have the lateral flagellation typical of dinoflagellates. Some dinoflagellates produce cysts within the cell, which may resemble the parental cell (Fig. 6-17D) or differ radically (Fig. 6-17E).

Sexual reproduction is not well known in Pyrrhophyta. In some genera, such as *Peridinium, Gymnodinium, Woloszynskia* (Figs. 6-15A and 6-17A), the isogametes resemble the parental cell. The gametes are produced by mitosis, and after fusion there is a swimming period of several days before the **planozygote** (*plano* = wandering) forms a heavy-walled resting cell. In germination the zygote forms one, two, or four swarmers by a prolonged meiosis that may not be completed for several days. In the armored, photosynthetic *Ceratium* (Figs. 6-15C and 6-17B), anisogametes are produced by haploid cells. The larger (female) gamete evidently absorbs the smaller gamete, which is considered the male. In these, the zygote is the only diploid stage.

Morphological Variation, Classification, and Ecology

The most commonly occurring type of dinoflagellate is a photosynthetic, motile unicell with lateral flagellar insertion (Figs. 6-15A–C and 6-17A–C). There are also motile cells, such as *Prorocentrum* (Fig. 6-15E) that have apical flagellation, with the encircling flagellum having the striated rod. There is a coenocytic genus, *Polykrikos* (Fig. 6-15D), as well as colorless genera. Not all motile forms are obviously flagellated. *Noctiluca*, for example, has a motile tentacle. However, it does produce dinoflagellatelike swarmers. Nonmotile forms include nonfilamentous genera, such as *Rufusiella, Cystodinium,* and *Pyrocystis* (Figs. 6-15F, G and 6-17F, G), as well as the filamentous *Dinothrix* (Fig. 6-15H). In all, the motile cells are similar to the "typical" dinoflagellate cell (Figs. 6-15A, B and 6-17A). Note that in the dinoflagellates, as in many of the other algal classes (Chlorophyceae, Tribophyceae, Chrysophyceae), the range of morphological variation includes motile and nonmotile as well as filamentous and nonfilamentous forms.

Dinoflagellates occur in both freshwater and marine habitats. In abundance, they are second only to the diatoms (see p. 158) in marine phytoplankton, and in tropical oceanic waters they can be the dominant phytoplankters. They can occur in numbers sufficient to color the water a reddish-brown, often termed a **red tide**. This red color explains the name, Pyrrhophyta, which is derived from the term *pyrrhos*, meaning "flaming" or "burning." Under optimum growth conditions, rapid asexual reproduction occurs, producing a bloom similar to that occurring in Cyanophyceae (Chapter 2).

Some dinoflagellates (*Gonyaulax, Gymnodinium*; Figs. 6-15A, B and 6-17A) produce substances

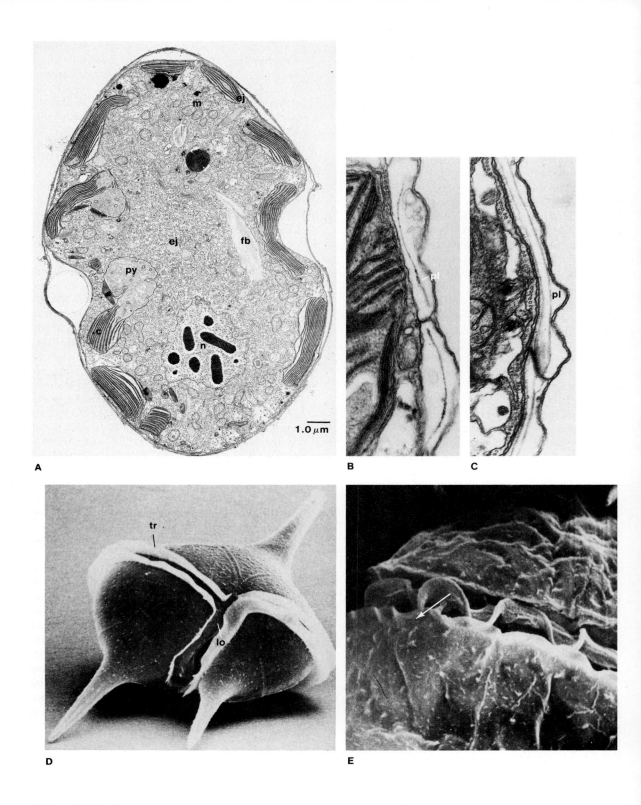

extremely poisonous to man and many animals. Poisoning occurs indirectly when people eat shellfish that have concentrated the dinoflagellate toxin, which does not usually affect the shellfish. In some vertebrates, especially mammals, the toxin inhibits the diaphragm, causing respiratory failure. Dinoflagellate blooms produce the red tides that cause concern in coastal waters. Not all red tides are caused by dinoflagellates, nor are they all toxic.

Division Euglenophyta (Class Euglenophyceae) (Euglenophytes)

There are approximately 40 genera and 450 species of algae or algalike organisms in Euglenophyta. All euglenoid genera are unicellular and motile, except for one sessile taxon (Fig. 6-18). This group has both plantlike and animallike features, and the representatives can be considered equally well as algae or protozoa. The euglenoids, as well as the dinoflagellates, are considered excellent examples of Protista, since they have both photosynthetic and nonphotosynthetic representatives that are unicellular and motile.

Cell Structure and Reproduction

Approximately one-third of the euglenophytes are photosynthetic. The bright-green chloroplasts contain both chlorophyll *a* and *b* and additional carotenoids. One, two, or many disc-shaped, elongate, platelike, or ribbon-shaped chloroplasts are present. The thylakoids are arranged generally in groups of 3 to 12. Pyrenoids are present in some euglenoids, and their occurrence and position are regular features of certain chloroplast types. The photosynthate is like laminarin in the brown algae and is known as **paramylon** (*para* = near; *amyl* = starch). The photosynthate is always outside the chloroplast and sometimes associated with the pyrenoid (Fig. 6-19C). With the light microscope, paramylon appears as refractive granules of various sizes and shapes and is deposited in a helical fashion (Fig. 6-19A).

Most euglenoids are colorless and thus **obligate heterotrophs.** Many secure food saprobically, as do slime molds and fungi. However, a few euglenoids are phagotrophic and ingest particulate material, sometimes with a special organelle that captures and helps ingest the food (Fig. 6-18E).

The euglenophyte cell contains a single, large, centrally located nucleus. The chromosomes appear condensed throughout interphase (Figs. 6-18A and 6-19B), and the nucleolus and nuclear envelope are present during division. Superficially, the nucleus is similar to that in the Dinophyceae (see Figs. 6-16A and 6-17A).

The cell lacks a firm wall; however, the outer part of the protoplast consists of a **pellicle** (*pell* = skin) covered by the plasma membrane. The pellicle consists of articulating, ridged, proteinaceous strips that spiral from the posterior of the cell to the apical end and can be conspicuous (Fig. 6-19A). Associated with these strips are mucilage-producing bodies and microtubules that are involved in the characteristic plasticity of the euglenoid cell. In addition to the pellicle, some forms have a lorica (Figs. 6-18C and 6-19C).

At the anterior end of the cell is an invagination or groove from which the flagella emerge (Figs. 6-18A and 6-19A). The invagination consists of a

Figure 6-16 Cell structure of Dinophyceae, electron microscope. **A–C,** transmission EM. **A,** entire cell of *Cachonina* with transverse flagellum groove in middle, showing many chloroplasts (*c*) with pyrenoid (*py*), mitochondria (*m*), ejecting organelles (ejectosomes, *ej*), fibrillar band (*fb*), and distinctive interphase dinoflagellate nucleus (*n*) with interphase chromosomes, ×6190. **B, C,** section through periphery of cell showing plates (*pl*) ×4900; **B,** thin plates of *Katodinium;* **C,** thicker plates and junction of *Glenodinium*. **D, E,** scanning EM of *Peridinium:* **D,** entire cell, ventral view, showing plates, pores, and furrows (transverse, *tr*, and longitudinal, *lo*), ×14,000; **E,** portion of girdle flagellum showing straight, striated strand (*arrow*) within flagellar membrane, ×13,000. (**A,** courtesy E. M. Herman, B. Sweeney and *Journal of Phycology;* **B, C,** courtesy J. D. Dodge from J. D. Dodge and R. M. Crawford, with permission from the *Botanical Journal of the Linnaean Society* Vol. 63, 1970, Copyright The Linnaean Society of London; **D, E,** courtesy of F. J. R. Taylor.)

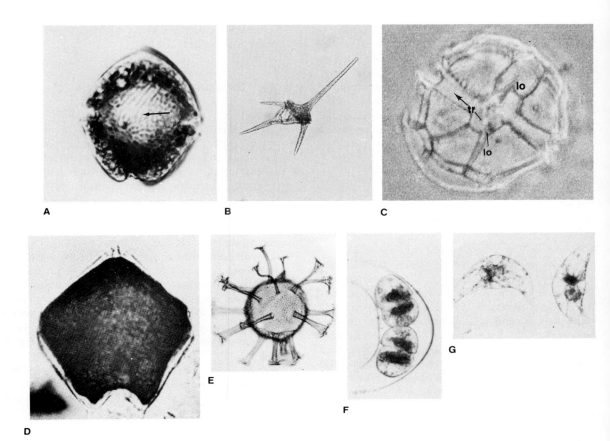

Figure 6-17 Cell structure, morphological diversity, and reproduction of Dinophyceae, light microscope. **A–C,** motile unicellular form: **A,** *Gymnodinium,* showing characteristic dinoflagellate nucleus (*arrow*), ×1000; **B,** *Ceratium,* with plates and horns, ×500; **C,** *Peridinium* theca, showing plate arrangement and flagellar grooves (*tr,* transverse; *lo,* longitudinal), ×800. **D, E,** resting cysts; **D,** living cyst resembling parent cell, ×450; **E,** fossil cyst differing from parent cell, ×360. **F, G,** nonmotile form, *Pyrocystis,* in vegetative stage (**G,** ×400) and producing motile cells (**F,** ×500). (**A,** courtesy H. A. von Stosch and *British Phycological Journal;* **D,** courtesy of W. R. Evitt and *Stanford University Publications, Geological Sciences;* **E,** courtesy of D. Wall and *Journal of Phycology;* **F, G,** courtesy of E. Swift, E. G. Durbin, and *Journal of Phycology.*)

narrow canal that opens into a larger reservoir. Pellicular strips occur only in the canal. Contractile vacuoles empty into the reservoir and flagella are inserted on the reservoir surface (Fig. 6-18A). Adjacent to the reservoir is a large eyespot (Figs. 6-18A and 6-19B) that is independent of the chloroplasts (contrary to the situation in many algae) and probably acts as a light screen to a photoreceptive flagellar swelling at the base of one flagellum in photosynthetic forms (Fig. 6-18A). The basic flagellar arrangement is two flagella, with unilateral hairs present on at least one. In most euglenoids, the second flagellum is very short and may not extend beyond the canal (Fig. 6-18A). The

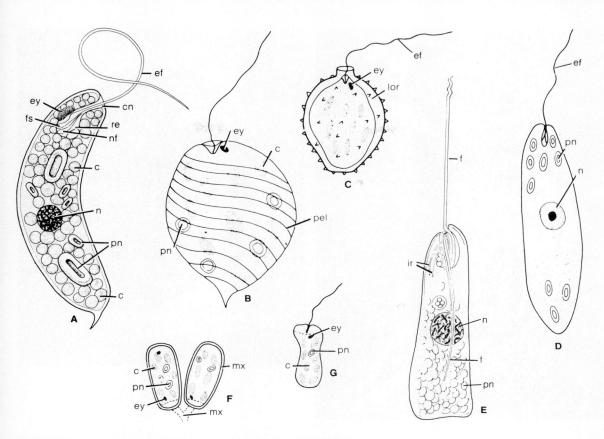

Figure 6-18 Cell structure and morphological diversity of Euglenophyceae. **A**, *Euglena*, showing disc-shaped chloroplasts (*c*), paramylon (*pn*), nucleus (*n*), reservoir (*re*), eyespot (*ey*), canal (*cn*), flagellar swelling (*fs*), emergent flagellum (*ef*), nonemergent flagellum (*nf*), ×1400. **B**, *Phacus*, with pellicle striations (*pel*), ×1000. **C**, *Trachelomonas*, with lorica (*lor*), ×155. **D**, *Astasia*, a colorless form, ×1915. **E**, *Peranema*, with two flagella (*f*) and showing ingestion rods (*ir*), ×1000. **F, G**, *Colacium*, showing attached stage with matrix (*mx*) (**F**, ×800) and motile stage (**G**, ×530). (**A, E**, after Leedale.)

emergent flagellum is often easily seen with the light microscope, as it has additional thickening material within the membrane.

Only asexual reproduction is known in Euglenophyceae, and it generally occurs in the motile condition. In cell division, the nucleus migrates to the anterior of the cell. At this time the eyespot, chloroplast, and other cytoplasmic organelles, including the reservoir and canal, can divide. The flagellar apparatus either completely replicates itself or only produces flagella and basal bodies for the new cell, while retaining the parental apparatus. Upon completion of organelle replication, cell cleavage occurs by an invagination of the pellicle at the anterior end of the cell between the openings of the two new canals. This cleavage line progresses posteriorly, following the helix of the pellicle, until the new cells are connected by a narrow, posterior, protoplasmic bridge. New cells can remain attached for varied periods of time.

In unfavorable conditions a cell can encyst, becoming resistant to extreme environmental con-

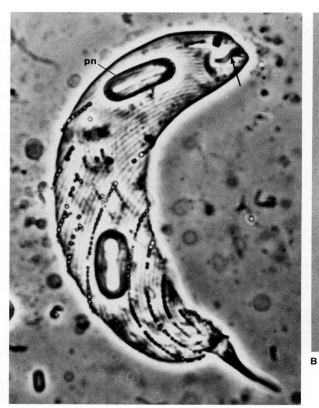
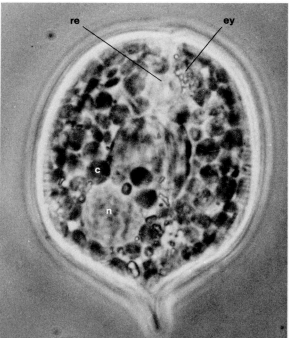
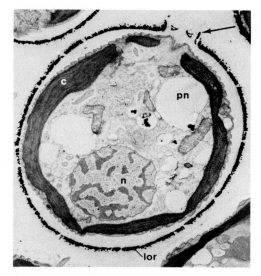

Figure 6-19 Euglenophyceae. **A, B,** light microscope: **A,** *Euglena* showing pellicle striations with warts, canal (*arrow*), and two paramylon (*pn*) granules (anoptral phase contrast), ×1500; **B,** cell contents of *Phacus,* showing chloroplasts (*c*), nucleus (*n*), and eyespot (*ey*) adjacent to reservoir (*re*) (phase-contrast), ×2000. **C,** electron microscope; oblique section of *Trachelomonas,* showing chloroplast, paramylon, nucleus, and lorica (*lor*) with pore and collar (*arrow*), ×4000. (**A, B,** courtesy of G. F. Leedale, *Euglenoid Flagellates,* © 1967, p. 97, 172. Reprinted by permission of Prentice-Hall, Inc., Englewood Cliffs, N.J. **B, C,** courtesy of G. F. Leedale and *British Phycological Journal.*)

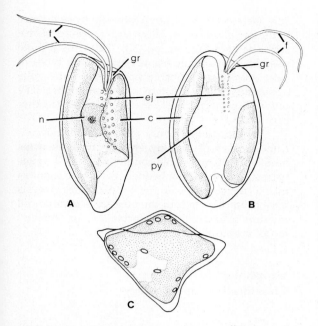

Figure 6-20 Cell structure and morphological diversity of Cryptophyceae, ×2850. **A, B,** motile forms: **A,** *Cryptomonas*, showing chloroplast (*c*), groove (*gr*), flagella (*f*), and groove ejectosomes (*ej*); **B,** *Chroomonas*, also showing pyrenoid (*py*); compare ejectosome distribution and size with *Cryptomonas*. **C,** nonmotile form, *Tetragonidium*, ×2000. (**A, B,** after Butcher; **C,** from R. H. Thompson, "Algae," in W. T. Edmonson, ed., *Freshwater Biology* 2d ed., copyright © 1959 by John Wiley & Sons. Reprinted by permission of John Wiley & Sons, Inc.)

ditions. In the encysted form, some cells appear red because they accumulate great amounts of carotenoids, termed **hematochrome** (*hema* = blood; *chrome* = color) but shown to be the xanthophyll *astaxanthin*. These cysts usually float on the water surface. Germination of thick-walled cysts results in only one motile cell; in thin-walled cysts, the contents often divide into a number of cells.

Sexual reproduction seems to be absent. There are a few records reporting union of gametes, but these reports are unsubstantiated and some are apparently due to the presence of other organisms or to misinterpretation.

Morphological Variation, Classification, and Ecology

Classification of the Euglenophyceae is based on flagellar organization. All genera but one are motile in the vegetative stage. This exception is the photosynthetic *Colacium*, which retains two flagella within the reservoir but attaches at its anterior end. It can undergo cell division in this attached, nonmotile state (Fig. 6-18F, G). As noted previously, only a third of the Euglenophyceae contain chlorophyll. These include such common genera as *Euglena*, *Phacus*, and the loricate *Trachelomonas* (Fig. 6-18A–C). Some more common colorless saprobic forms are *Astasia* (Fig. 6-18D) and *Peranema* (Fig. 6-18E). *Peranema* also contains ingestion rods and thus has a phagotrophic mode of nutrition.

Most euglenoids are freshwater organisms, although a few occur in brackish or marine habitats—especially on mudflats and areas where organic material collects. Anywhere decaying organic matter is present—including barnyards and pasture ponds and ditches—there are likely to be some euglenoids. Since two-thirds of the genera are colorless, it is interesting to note that even photosynthetic forms require at least one vitamin, generally cyanocobalamin (B_{12}). A few euglenoids occur endozoically in invertebrates such as rotifers, nematodes, flatworms, oligochaetes, and copepods. One faintly green taxon even occurs only in the lower intestinal tract of tadpoles.

Division Cryptophyta (Class Cryptophyceae) (Cryptophytes)

The division Cryptophyta contains approximately 100 species in less than 25 genera, which occur in both fresh and marine waters. Most cryptophytes are flagellated, and a single class, the Cryptophyceae, is recognized (Fig. 6-20).

Cell Structure and Reproduction

The typical cryptophyte is a dorsiventrally flattened, oblong or ovoid motile unicell. It is biflagellated, generally with an anterior groove from which the flagella arise, and one or two elongate chloroplasts are present (Figs. 6-20A, B and 6-21B). The color of the cells varies and can be green, blue-green, golden, brown, or red, depending in part on environmental conditions. The main pigments include chlorophyll a and c_2, and specific carotenoids unique to the division. However, the red and blue masking pigments, the phycobiliproteins, are also present. These phycobiliproteins are chemically similar to those of the Cyanophyta (Chapter 2) and Rhodophyta (Chapter 9). The chloroplast is composed of pairs of unfused thylakoids (Fig. 6-21B), with phycobiliproteins in the space within a thylakoid, not in the form of discrete units as in Cyanophyta and Rhodophyta. Pyrenoids can be present, with the storage starch produced outside the chloroplast (Fig. 6-21B). This photosynthate is enclosed within the nuclear-chloroplast endoplasmic reticulum. Eyespots are uncommon; when present, however, they are in the chloroplast, usually near the nucleus and adjacent to the flagellar base.

The two flagella are subequal in length, and both have tinsels along their length (Fig. 6-21A). The shorter flagellum generally has tinsels arranged unilaterally; the longer flagellum has bilateral tinsels. The flagella are attached near the base of the groove and emerge subapically or almost laterally.

The nucleus is generally posteriorly located and irregular in outline, often with cytoplasmic invaginations. The interphase nucleus has large chromatin areas and a prominent nucleolus. The nuclear envelope, unlike that of Pyrrhophyta and Euglenophyta, breaks down during division, and microtubules extend into the chromosome mass.

As in Dinophyceae and Euglenophyceae, a cell wall is lacking. The outer layer of the cell, the **periplast** (*peri* = around; *plast* = form), differs from the pellicle of Euglenophyceae and plates of Dinophyceae. The cryptophyte periplast is composed of a series of regularly aligned, longitudinal, plate-like areas delineated by a number of shallow ridges extending the length of the cell. In addition, lateral furrows are present, resulting in an alternating pattern (Fig. 6-21A). The longitudinal plate areas do not extend into the flagellar groove. At the junction of the ridges and lateral furrows are located small ejecting organelles (**ejectosomes**), which appear numerous in the cell posterior (Fig. 6-21B). In the flagellar groove are located larger, possibly more complex ejectosomes, approximately twice the size of the peripheral ones. These larger ones are visible with the light microscope (Fig. 6-20A, B) and are used as a taxonomic characteristic. Both types of ejectosomes are probably produced by the golgi apparatus. Their function is uncertain but may involve protection of the cell against other organisms.

Reproduction is only asexual, with no confirmed reports of sexuality. The cell can divide while motile or in a nonmotile condition where it comes to rest, becoming embedded in a mucilaginous matrix. In some genera, thick-walled internal cysts have been reported. In the nonmotile genera, cell division results in packet-formation, or "typical" motile cryptophyte cells.

Morphological Variation, Classification, and Ecology

Cryptophyceae are classified on the basis of vegetative morphology—motile versus nonmotile. Only two, rarely collected genera are nonmotile. Motile genera are separated on the basis of cell shape, appearance and location of the groove, number and location of the large ejectosomes, and chloroplast number. Previous systems incorporating color are unsatisfactory, since this is an environmental response. Motile genera include *Cryptomonas* and

Figure 6-21 Cell structure of Cryptophyceae, electron microscope, *Chroomonas*. **A,** platinum-carbon replica of cell, showing tinsel flagella emerging from groove region (*top*), and surface pattern formed by plate areas; small bulges are the ejecting organelles (*arrows*), ×18,400. **B,** longitudinal section, showing chloroplast (*c*) with pyrenoid (*py*) and thylakoids, starch (*sg*), nucleus (*n*) golgi (*go*) mitochondrion (*m*), small peripheral ejecting organelles (ejectosomes, *ej*), and groove (*arrow*); note transverse section of the two flagella (*f*) above groove (*top*), ×17,850. (Courtesy of E. Gantt and *Journal of Phycology*.)

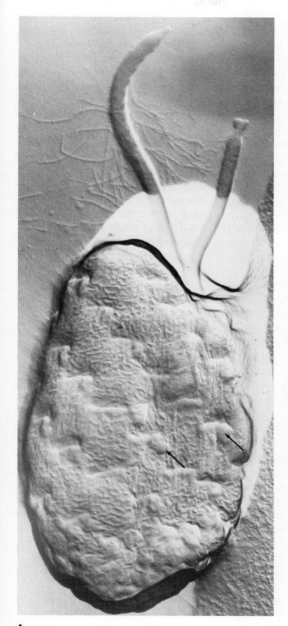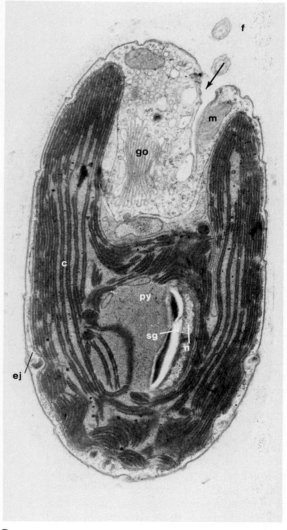

Chroomonas (Fig. 6-20A, B), which are pigmented. *Tetragonidium* (Fig. 6-20C) is one of the two pigmented, nonmotile genera. *Chilomonas* is a nonpigmented genus studied by many biochemists.

Cryptophyta occur more commonly in fresh water than in marine waters, but species of some genera, such as *Cryptomonas* and *Chroomonas*, occur regularly in saltmarsh pools with salinities varying from 5 to 30 parts per thousand at all times during the year. In fresh water, cryptophytes are often common when some organic material is present. They are sometimes the dominant phytoflagellates in early spring blooms of temperate lakes. The few species studied in culture are obligate photoautotrophs and require the vitamin cyanocobalamin (B_{12}). A red-colored cryptophyte occurs in a few marine ciliate hosts. These ciliates can be present in such large numbers as to create red tides similar to those caused by Dinophyceae (see p. 163). How the ciliates acquire the cryptophyte is unknown, but the chloroplasts are photosynthetically active.

Relationships of the Protistan Algae

As mentioned at the beginning of this chapter, there are few similarities among the divisions discussed. The interrelationships of the classes constituting Chrysophyta are complex, with pigmentation, cell covering, and storage products generally serving to unify the classes. However, in the broad sense used in this text, there are some deviations, such as pigment differences (lack of fucoxanthin in Tribophyceae), and the flagellation of Prymnesiophyceae (whiplash flagella), which may be important in showing relationships of the organisms among themselves as well as to other classes.

The nuclear morphology of the Pyrrhophyta, Euglenophyta, and Cryptophyta seems superficially similar—all show persistence of condensed chromatin in interphase and persistence of some nuclear envelope and nucleolus during division. However, the paucity of certain key proteins and the lack of intranuclear microtubules during division separate pyrrhophytes from euglenophytes. Nuclear division in cryptophytes is similar to that in Chrysophyta and not to that in either euglenophytes, dinoflagellates, or the phycobiliprotein-containing rhodophytes (Chapter 9).

The fossil record for Chrysophyta (in the broad sense) includes primarily diatoms (Bacillariophyceae) and coccolithophorids (Prymnesiophyceae), with a few cysts allied to Chrysophyceae and Tribophyceae reported. Some diatoms occur in Mesozoic strata about 150 million years old. However, they are present in much larger numbers since the beginning of the Cenozoic era (65 million years ago). Traces of Chrysophyceae are in the Cenozoic, but have not been studied much. Many Prymnesiophyceae are little known, as organic scales do not generally preserve well. Presently, it is impossible to derive chrysophyte classes from each other using the fossil record. However, the fossil record is helpful in indicating trends within some classes. Of the other three divisions considered here, only the Pyrrhophyta have any fossil record. This record goes back possibly to late Precambrian (650 million years ago) and is known primarily from cysts and cells that produced them. Neither Euglenophyta nor Cryptophyta are known much before 200,000 years ago (late Mesozoic).

All divisions resemble brown algal lines (Phaeophyta, Chapter 8) more closely than green algal lines (Chlorophyta, Chapter 7). Chrysophyta—especially Chrysophyceae—are the most similar to Phaeophyta; however, specific differences set the other divisions slightly apart. In Pyrrhophyta these are the nuclear condition, variation in carotenoids, and storage of starch. In Euglenophyta, the presence of chlorophyll *b* and laminarinlike storage products are outstanding features. In Cryptophyta, there is the presence of phycobiliproteins and starch. The possibility that cryptophytes are closely related to either Cyanophyta (Chapter 2) or Rhodophyta (Chapter 9) based on pigmentation is untenable because of more basic differences, such as the lack of flagellated cells in either bluegreen or red algae. At this time it is not (and probably never will be) possible to indicate positively the relationships among these divisions.

References

Aaronson, S. 1980. Descriptive biochemistry and physiology of the Chrysophyceae (with some comparison to Prymnesiophyceae). In Levandowsky,

M. and Hutner, S. H., eds. *Biochemistry and physiology of protozoa*. Vol. 3. 2nd ed. Academic Press, New York and London.

Boney, A. D. 1975. *Phytoplankton*. Edward Arnold, London.

Bourrelly, P. 1981. *Les algues d'eau douce*. Vol. 2, *Les algues jaunes et brunes, Chrysophycées, Phaeophycées, Xanthophycées et Diatomées*. 2nd ed. N. Boubée & Cie, Paris.

Bourrelly, P. 1983. *Les algues d'eau douce*. Vol. 3, *Les algues bleues et rouges, End Chleniens, Peridiniens et Cryptomonadines*. 2nd ed. N. Boubée & Cie, Paris.

Cox, E. R., ed. 1980. *Phytoflagellates*. Elsevier/North Holland, New York. (Chapters on xanthophytes, chrysophytes, prymnesiophytes, free-living dinoflagellates, euglenoid flagellates, cryptophytes.)

Dodge, J. D. 1979. The phytoflagellates: fine structure and phylogeny. In Levandowsky, M. and Hutner, S. H., eds., *Biochemistry and physiology of protozoa*. Vol. 1. 2nd ed. Academic Press, New York and London.

Hibberd, D. J. 1976. The ultrastructure and taxonomy of the Chrysophyceae and Prymnesiophyceae (Haptophyceae): a survey with some new observations on the ultrastructure of Chrysophyceae. *Bot. J. Linn. Soc.* 72:55–80.

Leedale, G. F. 1967. *Euglenoid flagellates*. Prentice-Hall, Inc., Englewood Cliffs, N. J.

Sargeant, W. A. S. 1974. *Fossil and living dinoflagellates*. Academic Press, New York.

Werner, D., ed. 1976. *The biology of the diatoms*. Univ. of California Press, Berkeley and Los Angeles.

7
PLANTS—CHLOROPHYTA (GREEN ALGAE)

In an evolutionary survey of photosynthetic plants, green algae (Division Chlorophyta) should be considered just prior to any discussion of embryo-producing plants. Logically, algal groups containing **accessory pigments** that mask chlorophyll *a* are studied first, since they have few characteristics in common with land plants (which are the majority of plants). The Chlorophyta is a large division, with morphological and reproductive diversity.

Chlorophyta occur throughout the aquatic environment and are known by a variety of names, such as slime, frog spit, green seaweed, and other more descriptive (and unpublished) ones. Many occur in fresh water and a significant number are in the marine habitat. Generally marine forms are larger than the freshwater ones. There are over 7000 species in more than 450 genera.

Vegetatively the Chlorophyta are relatively simple. The range in morphological diversity includes unicellular, multicellular, motile, nonmotile, filamentous, nonfilamentous, parenchymatous, and siphonous forms, and various combinations of these. They exhibit a greater variety of life histories and methods of reproduction than any other division of plants.

Several classification systems are used for Chlorophyta. In this text, four classes are recognized (see Table 7-1). Some workers believe that some of these should be separate divisions; however, this does not necessarily facilitate an overall understanding (especially by the neophyte) of the relationships of these algae. Morphologically, the largest and most diverse class is Chlorophyceae, which in this text is considered in restricted terms. An outstanding feature of Chlorophyceae is that their cells are uninucleate. Recognition of a second class, Bryopsidophyceae, is controversial, but its members differ from Chlorophyceae in several ways, including the occurrence of multinucleate cells. The third class, Charophyceae, is delimited in a narrow sense in this text by its very characteristic morphology and reproduction. Since 1975 the class has been expanded by some botanists to include genera previously included in the Chlorophyceae and Prasinophyceae (pp. 214). Still other botanists interpret the Charophyceae in a narrow sense and as a separate division. Finally, recognition of a fourth class, Prasinophyceae, is in a state of flux. As defined here, the Prasinophyceae are a small group of

Table 7-1 Characteristics of the Classes of Chlorophyta (parentheses indicate occurrence in some members)

Class and Habitat	Size	Nuclei per Cell	Carotenoids*	Cell Wall	Dominant Life History†
Chlorophyceae freshwater marine	microscopic (few μm) to several meters	1	usual	cellulose pectin	haploid haploid/diploid
Bryopsidophyceae marine (freshwater)	macroscopic to several meters	many	usual; additional may be present	xylan, mannan cellulose ($CaCO_3$)	diploid haploid/diploid
Prasinophyceae marine (freshwater)	microscopic (few μm)	1	usual	often lacking (cellulose?, pectin)	?
Charophyceae freshwater (marine)	macroscopic	1 to many	usual	cellulose (pectin) $CaCO_3$	haploid

*Usual for division; includes those occurring in land plants: additional carotenoid pigments are present only in some Bryopsidophyceae.
†Haploid = zygote only $2n$ stage (see Fig. 3-7A); diploid = gametes $1n$ stage (see Fig. 3-7B); haploid/diploid = both multicellular phases present (see Fig. 3-8A).

uninucleate green algae with characteristic motile cells that differ from those typical of Chlorophyceae.

As the Chlorophyceae constitute the largest class of Chlorophyta and are encountered in most aquatic habitats, they are discussed first and in greatest detail. The features separating the other three classes are considered with regard to those characteristics that set them apart from Chlorophyceae.

Class Chlorophyceae

Cell Structure

Observations of chlorophycean cells confirm that they are eukaryotic. The most obvious part of the cell is the chloroplast, and in larger cells the single nucleus can be readily observed (Fig. 7-1A, B). The protoplast is generally surrounded by a firm **fibrillar** wall composed of cellulose with an outer **amorphous** pectic layer. Within the cell wall and plasma membrane are chloroplast, nucleus, mitochondria, golgi, endoplasmic reticulum, and vacuoles (Fig. 7-1 C, D).

Chloroplasts contain the photosynthetic and accessory photosynthetic pigments in **thylakoids.** Chlorophyll *a* and *b* and carotenoids are present to give the characteristic grass-green color to cells. Sometimes, when there is an abundance of carotenoids (often outside the chloroplast), the cells are orange or brick-red. Thylakoids are arranged in bands, with two to six fused thylakoids per band (Fig. 7-1C, D); they can also be in stacks, similar to the **grana** occurring in the chloroplasts of land plants. The number of chloroplasts in a cell varies with a given species or genus. When numerous chloroplasts are present, they are small, disc-shaped, and sometimes interconnected (Fig. 7-2A). More often only one or two chloroplasts occur, shaped like a half-filled cup, flat band, open ring, net, spiral, or star (Fig. 7-2B–G). The chloroplast position in the cell is either peripheral (Fig. 7-2D, F) or central (Fig. 7-2C, G). Within the chloroplast there can be a dense region, the **pyrenoid** (*pyren* = fruit, stone; *oid* = like), which is proteinaceous. With the light microscope, the pyrenoid appears in the chlo-

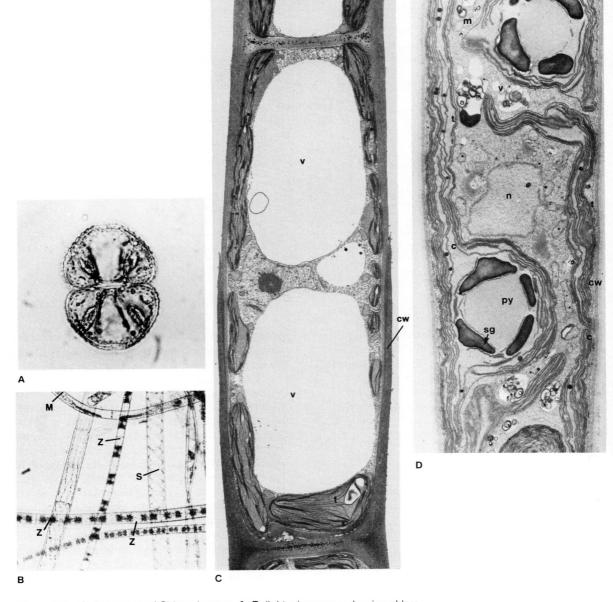

Figure 7-1 Cell structure of Chlorophyceae. **A, B,** light microscope showing chloroplasts (*c*): **A,** *Cosmarium,* ×1000; **B,** *Spirogyra* (*S*), *Zygnema* (*Z*), *Mougeotia* (*M*), ×1000. **C, D,** electron microscope, cells showing chloroplasts (*c*) with pyrenoid (*py*), starch (*sg*), and thylakoid (*t*); nucleus (*n*); vacuoles (*v*); mitochondria (*m*); and cell wall (*cw*): **C,** *Microspora,* ×5000; **D,** *Stigeoclonium,* ×9200. (**C,** courtesy of J. D. Pickett-Heaps and *New Phytologist;* **D,** courtesy of G. L. Floyd, K. D. Stewart, K. R. Mattox, and *Journal of Phycology.*)

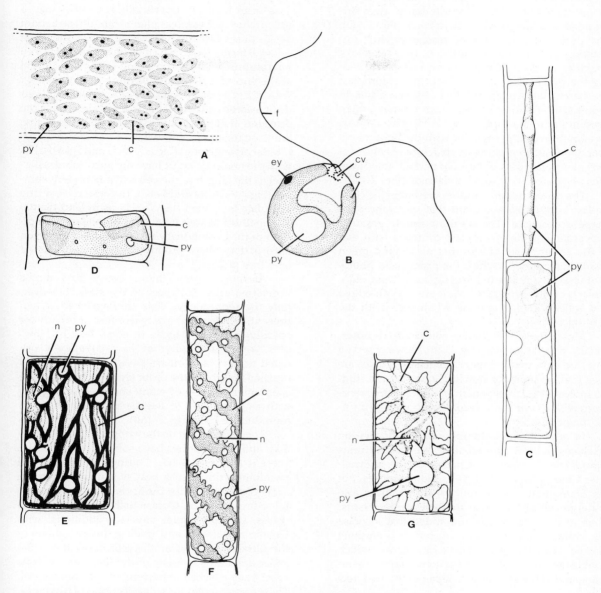

Figure 7-2 Chloroplast arrangement in Chlorophyceae: chloroplast (*c*); pyrenoid (*py*); contractile vacuole (*cv*); eyespot (*ey*); nucleus (*n*); **A, B, D–F,** parietal (peripheral) position in cell. **A,** disc-shaped (*Trentepohlia*); **B,** cup-shaped (*Chlamydomonas*) with apical eyespot and contractile vacuoles; **D,** open ring-shaped (*Ulothrix*); **E,** netlike (*Oedogonium*); **F,** spiral (*Spirogyra*). **G, G,** axile (central) position in cell; **C,** band-shaped (*Mougeotia*); **G,** star-shaped (*Zygnema*).

Plants—Chlorophyta (Green Algae) 177

roplast as a central, clear area, often hexagonal or round in form (Fig. 7-2B–G). The electron microscope shows it as an electron-dense area surrounded and traversed by thylakoids (Fig. 7-1D). The pyrenoid is an area of storage-product formation, and probably contains the necessary enzymes for such. Electron-clear areas surrounding the pyrenoid and throughout the chloroplast are starch, the storage product of green algae. In motile cells of some green algae, an **eyespot** is often present in the chloroplast (Fig. 7-2B). This appears as a bright red or orange dot with the light microscope, but sections show its structure to consist of two to several rows of densely packed lipid granules associated with some thylakoids (Fig. 7-15B).

The starch storage product present in the chloroplast is often so abundant that the chloroplast shape is obscured. The resulting rough, granular appearance of the chloroplast contrasts with that of the chloroplasts of all other algae (and green plants generally), in which the products of photosynthesis accumulate in the cytoplasm. Thus, chloroplasts of all *but* the green algae are smooth-edged and well-defined in outline when seen with the light microscope.

Most motile cells contain **contractile vacuoles,** usually at the apical end (Fig. 7-2B). When observed with the light microscope, contractile vacuoles are seen to fill and empty slowly. Ultrastructural studies indicate the involvement of endoplasmic reticulum with a contractile vacuole delimited by a single membrane.

In unicellular Chlorophyceae, cell division produces several reproductive, unicellular units (spores) as discussed in Chapter 3. The multicellular Chlorophyceae can also produce spores, but the filamentous and parenchymatous forms also undergo cell division that produces larger plants. Thus in these, new cells are added that increase the overall size of the plant. In cell division there may be formation of a cell plate or a furrow, which establishes the location of the new plasma membrane and cell wall that will separate the two new cells.

The method of cell wall formation varies in Chlorophyceae and can involve the ability of the cell to expand following nuclear division to allow for the process of **cytokinesis** (*cyto* = cell; *kinesi* = movement). In some filamentous forms, the cell wall is formed by a furrowing inward from the outside of the cell (Fig. 7-3A, B). In others, the cell wall begins from the center of the cell and proceeds to form toward the outside (Figs. 7-3C, D and 7-4B). This latter method superficially resembles that of bryophytes and vascular plants, but ultrastructural details differ. Electron microscope studies show that cell wall formation involves the mitotic spindle, cell **microtubules,** and the position of the new nuclei at the end of mitosis and during cytokinesis (Fig. 7-4A, C). The position of the new nuclei and their distance from one another evidently determine the direction of fibrils forming the cell wall. Nuclei that are close together tend to have a transverse microtubular arrangement (Figs. 7-3B, C and 7-4A). Cells with the nuclei far apart have the microtubules parallel to the length of the cell and a persistent spindle (Figs. 7-3A, D and 7-4C). In these cells, a true **cell plate** is formed (Fig. 7-3D), as in land plants. The method of wall formation is used as an indicator of the evolutionary relationships of the green algae to the embryo-producing plants, as discussed later (see p. 217).

As noted, some Chlorophyceae have motile vegetative cells. For most of the class, however, only the reproductive cells are flagellated. When present, motile cells have two or four anteriorly (or apically) oriented flagella of equal length. However, some Chlorophyceae have subapically oriented flagella, and others have flagella of slightly unequal length. In the order Oeodogoniales, there is an anterior circle of many short, equal flagella, with as many as 128 flagella reported on some reproductive cells (Fig. 7-16B).

The flagella are of the whiplash type (Fig. 7-5A), and some species have submicroscopic scales present (see Fig. 3-2D). During normal forward swimming, the flagella beat synchronously in the same plane, much like the arms of a person swimming breast stroke. Occasionally cells swim backwards, possibly under adverse conditions, with flagella undulating and trailing. Investigations of the ultrastructure of motile cells show that a flagellar apparatus consists of the flagellum with its **basal body** attached to microtubules within the cell. These microtubules form flagellar roots of two types. In the more commonly occurring type there are four **microtubular roots** arranged as a cross (Fig. 7-5A, B). In the second type, there is only one broad band or root (Fig. 7-5C). Both types of flagellar roots may have a striated appearance. The single broad band is typical of that in the motile male gametes of bryophytes and vascular plants. Motile cells with

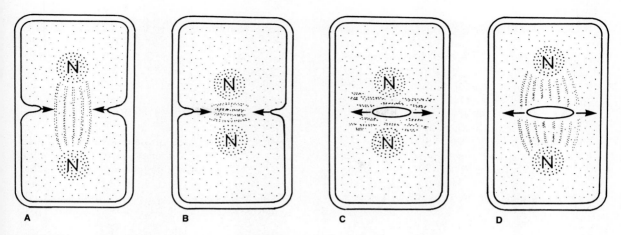

Figure 7-3 Diagrammatic representation of cytokinesis following nuclear division, showing cell plate and wall formation (*arrows* indicate direction). **A, B,** furrowing from outside of cell toward center. **C, D,** forming from inside of cell outward. **A, D,** presence of longitudinal microtubules with nuclei widely separated. **B, C,** parallel microtubules, nuclei close together.

the cross arrangement are symmetrical, whereas the other motile cells are asymmetrical (compare Fig. 7-5A and C).

Reproduction

Both vegetative (asexual) and sexual reproduction occur commonly. Many multicellular forms, especially filamentous ones, increase vegetatively by fragmentation. For the filamentous members of the ubiquitous freshwater order Zygnematales (see p. 197) this is the only means of asexual reproduction. Most green algae produce spores that are either motile (zoospore or planospore) or nonmotile (aplanospore). These spores are usually formed by cell divisions that do not utilize the parental wall in what appears to have been an ordinary vegetative cell. Upon release, spores are dispersed and ultimately settle, producing new plants. Many spores are released at one time, but only a few grow into mature plants. In a few green algae, the spores are retained within the parental wall, forming a multicellular **colony** with a characteristic shape.

Some multicellular Chlorophyceae have a specific number of divisions, and new cells remain united when released from the parental cell. This colony, consisting of a definite number of cells when released from the parental cell, is known as a **coenobium** (*coen* = common). Any increase in plant size results from expansion of the cells, not from an increase in their number. Most **colonial** Volvocales (see p. 181 and Fig. 7-7B–F) and some colonial Chlorococcales (see p. 188 and Fig. 7-10D–F) are coenobial.

Many Chlorophyceae, especially in freshwater habitats, undergo extensive and rapid asexual reproduction. In fact, for some this is the only means of reproduction and is important in maintaining and increasing the population. In contrast, some Chlorophyceae (especially marine species) have minimal asexual reproduction. Rather, they alternate a haploid, gamete-producing stage with a diploid meiospore-producing stage. Both types of reproductive cells are usually motile, and the alternation is often correlated with the tides. Massive release of the reproductive cells can be observed in tidepools or similar areas of slack water at low tide. The entire water mass can be green with millions

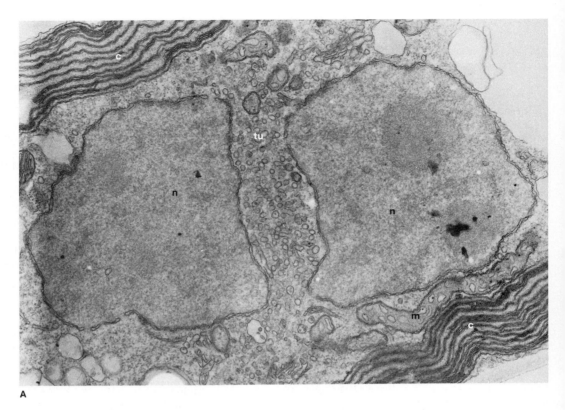

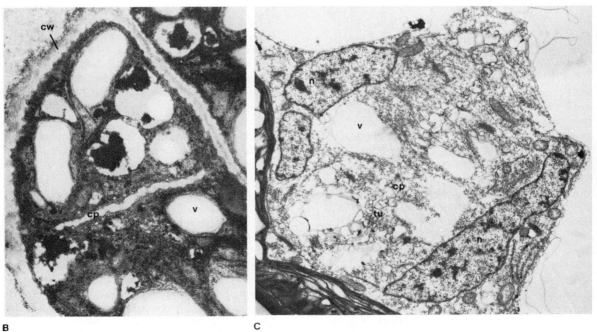

of cells about 10 μm in size. Most of these cells (termed *swarmers,* if it is not known whether they are gametes or meiospores) probably do not produce new plants. Many serve as food for small animals, especially filter feeders, and many die.

The relationship between gamete fusion, or syngamy, and the time of meiosis is more varied in Chlorophyceae than in any other class of photosynthetic plants. Three of the four plant life history types (Chapter 3, Figs. 3-7 and 3-8) occur, with only the dikaryotic type lacking. The two most common life-history types in Chlorophyceae are the predominantly haploid (see Fig. 3-7A) and the haploid alternating with diploid (see Fig. 3-8A).

Morphological Variation and Classification

In the restricted sense of Chlorophyceae of this text, there are forms that are unicellular (motile or nonmotile), multicellular (motile or nonmotile), simple or branched filaments, and large parenchymatous plants. This diversity is shown by the representative orders in Table 7-2.

At least three distinct groups (or lines) can be delimited in Chlorophyceae. Orders can be arranged in a morphological and possibly evolutionary series within these groups (Fig. 7-6). These three groups are (1) the motile group; (2) the nonmotile, nonfilamentous group; and (3) the nonmotile, filamentous, group. The filamentous group has the greatest morphological diversity and has been studied extensively since 1970. There is excellent evidence for accepting at least two (possibly three) main lines in it, and probably two side branches.

Motile Chlorophyceae

The motile line (Fig. 7-6A) comprises primarily a single, well-defined order, Volvocales. All representatives are flagellated in the vegetative stage (Figs. 7-7 and 7-9A).

Vegetatively, one of the simplest is the unicellular *Chlamydomonas* (Fig. 7-7A). It can be considered representative of primitive stock from which the majority of Chlorophyceae may have evolved. *Chlamydomonas* is a biflagellated, motile alga, which under certain environmental conditions reverts to a nonmotile state and becomes embedded in a gelatinous matrix. In this state, it divides vegetatively, forming a dense, unorganized, amorphous mass of separate cells, superficially resembling some genera in the nonmotile, nonfilamentous group.

The majority of motile Chlorophyceae have each cell similar to that of a vegetative *Chlamydomonas* cell (Fig. 7-7A). The multicellular forms remain fastened together by a common matrix after cell division, and are sometimes connected by cytoplasmic strands or attached cell walls. Most colonial Volvocales contain a constant number of cells. In these forms, such as *Gonium, Pandorina,* and *Eudorina* (Fig. 7-7B–E), each cell can produce progeny colonies identical to the parental colony. In *Pleodorina* and *Volvox* only a few large cells are reproductive (Figs. 7-7F and 7-9A). Hence, in this motile series, there is a trend toward an increase in the number of cells in the colony and toward specialization of cells. In *Gonium, Pandorina,* and *Eudorina* (Fig. 7-7B–E), all cells are reproductive. *Pleodorina* and *Volvox* (Figs. 7-7F and 7-9A) have small vegetative and large reproductive cells. *Volvox* contains 500 to 50,000 cells, with only a few scattered reproductive cells (Fig. 7-9A). Sexual reproduction in the Volvocales shows another series involving differentiation of gametes: *Gonium* and *Pandorina* are isogamous, whereas *Eudorina, Pleodorina,* and *Volvox* are oogamous.

The two genera at the morphological extremes of the motile series (unicellular, colonial) are *Chlamydomonas* and *Volvox*. Both occur in freshwater and are ubiquitous. As it is possible to maintain them easily in the laboratory, they are both well studied.

Figure 7-4 Transverse wall formation in Chlorophyceae, transmission electron microscope. **A,** *Ulothrix,* showing transverse microtubules (*tu*) and nuclei (*n*) close together, × 11,500. **B,** *Fritschiella,* showing cell plate (*cp*) formation starting in center (but note microtubules lacking) × 13,500. **C,** *Coleochaete,* showing cell plate, longitudinal microtubules, and nuclei well separated, × 8200. *Note* also vacuole (*v*), mitochondria (*m*), nucleus (*n*), and chloroplast (*c*). (**A,** courtesy of G. L. Floyd, K. D. Stewart, K. R. Mattox, and *Journal of Phycology;* **B,** courtesy of G. E. McBride and *Archiv für Protistenkunde;* **C,** courtesy of J. D. Pickett-Heaps, H. J. Marchant, and the international journal *Cytobios.*)

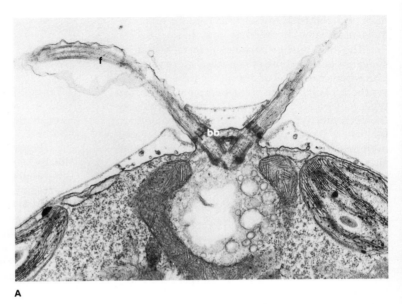

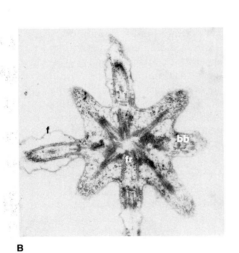

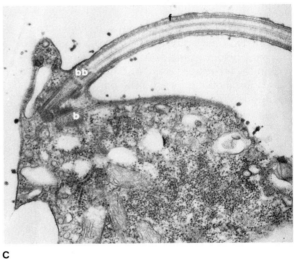

Figure 7-5 Flagellar arrangement and structure in Chlorophyceae in motile cells, transmission electron microscope. **A, B,** symmetrical cells with cross-shaped flagellar roots: **A,** longitudinal section at apex, showing flagella (*f*), flagellar insertion, basal bodies (*bb*), *Tetraspora,* ×1700; **B,** tangential section just below flagellar insertion near basal bodies, showing flagellar roots (*fr*), *Fritschiella,* ×22,500. **C,** asymmetrical cells sectioned at flagellar insertion, showing single broad band (*b*), *Klebsormidium;* ×34,500. (**A,** courtesy of J. D. Pickett-Heaps, copyright The Linnean Society of London; **B,** courtesy of M. Melkonian and *Protoplasma;* **C,** courtesy of H. J. Marchant, J. D. Pickett-Heaps, K. Jacobs, and the international journal *Cytobios*.)

Table 7-2 Representative Orders of Chlorophyceae, Including Selected Genera (items in parentheses refer to some representatives)

Order and Genera	Cell Arrangement	Dominant Life History*	Habitat	Distinctive Features
VOLVOCALES *Chlamydomonas, Eudorina, Gonium, Haematococcus, Pandorina, Pleodorina, Volvox*	unicellular, colonies	haploid	freshwater	motile, nonfilamentous
TETRASPORALES *Tetraspora*	colonies (unicellular)	haploid	freshwater	nonmotile; cells within matrix
CHLOROCOCCALES *Chlorella, Chlorococcum, Hydrodictyon, Kentrosphaera, Pediastrum, Prototheca, Scenedesmus, Trebouxia*	unicellular, colonies	haploid	freshwater (marine)	nonmotile, no vegetative divisions
ULOTRICHALES *Klebsormidium, Ulothrix*	filaments	haploid	freshwater (marine)	vegetative divisions (never unicellular)
ULVALES *Enteromorpha, Monostroma, Percursaria, Ulva, Ulvaria*	parenchymatous	haploid haploid/diploid	marine (freshwater)	vegetative divisions (never unicellular)
CHAETOPHORALES *Coleochaete, Draparnaldia, Fritschiella, Trentepohlia*	filaments (parenchyma)	haploid haploid/diploid	freshwater	heterotrichy often present; vegetative divisions (never unicellular)
OEDOGONIALES *Oedogonium*	filaments	haploid	freshwater	swarmer with ring of flagella (never unicellular)
ZYGNEMATALES *Closterium, Cosmarium, Micrasterias, Mougeotia, Spirogyra, Staurastrum, Zygnema*	unicellular, filaments	haploid	freshwater	lack flagellated cells

*Haploid = zygote only $2n$ stage (see Fig. 3-7A); diploid = gametes only $1n$ stage (see Fig. 3-7B); haploid/diploid = both multicellular phases present (see Fig. 3-8A).

In *Chlamydomonas*, vegetative reproduction takes place by cell division, usually in the motile condition, forming 4, 8, or 16 biflagellated cells. These are morphologically similar to the parent cell, but are smaller when released and grow to adult size. In this large genus, the gametes vary from isogamous to oogamous, depending on the species.

A generalized life history for the commonly occurring isogamous species is shown in Figure 7-8. Vegetative cells (Fig. 7-8A, B) function as gametangia by dividing to produce 4–32 biflagellated gametes (Fig. 7-8D, E), termed *plus* and *minus* **mating types**. After gametes are released (Fig. 7-8F, G) they swim about before fusing to form a zygote (Fig. 7-8H, I). The zygote rounds up and secretes a sometimes sculptured, heavy cellulose wall (Fig. 7-8J). This resting zygote (also termed a *zygospore*) is dormant in most freshwater species, permitting

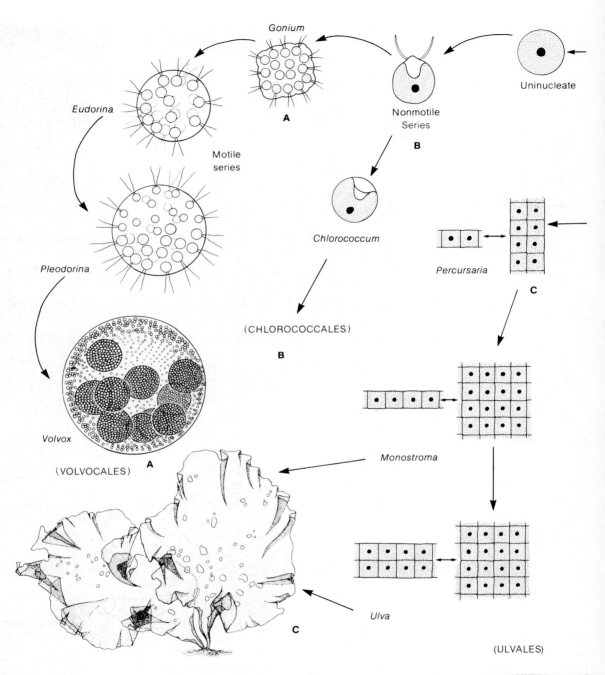

Figure 7-6 Groups in Chlorophyceae based on developmental series resulting from planes of nuclear and cell divisions (order names in *parentheses*). **A,** motile series (Volvocales), flagella not shown for *Volvox*. **B–G,** nonmotile series culminating in different plant types. **B,** nonfilamentous series (Chlorococcales); **C, D,** filamentous-parenchymatous series with symmetrical zoospores, showing surface and transverse views for parenchymatous algae (**C,** Ulvales; **D,** Ulotrichales and Chaetophorales, in part); **E,**

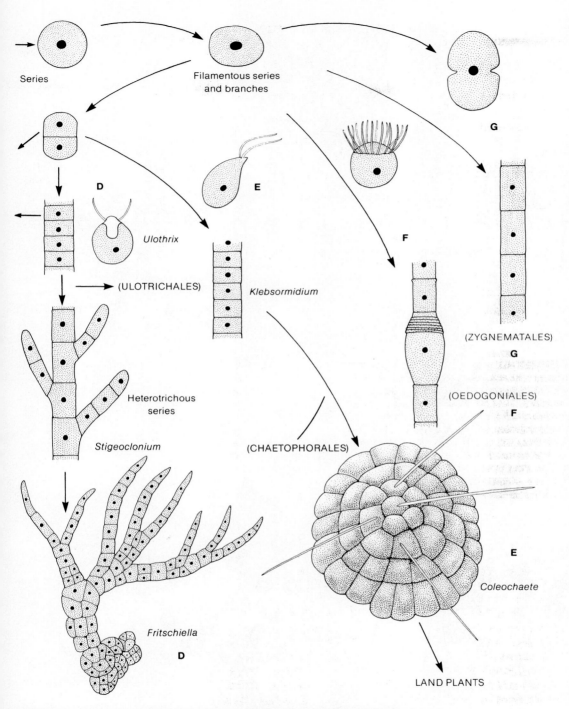

filamentous series with asymmetrical zoospores (Ulotrichales and Chaetophorales, in part), possibly leading to land plants; F, G, filamentous side-branches (F, Oedogoniales; G, Zygnematales) (genus names for reference only).

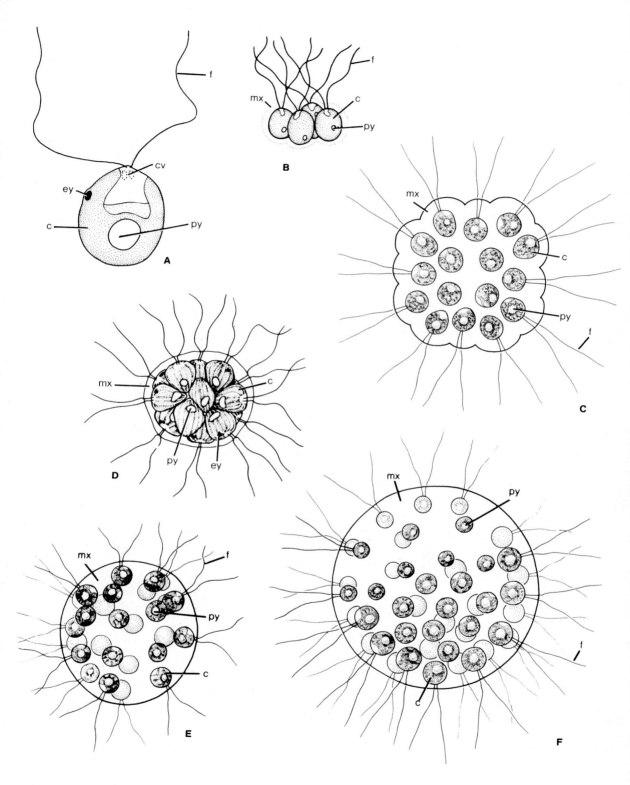

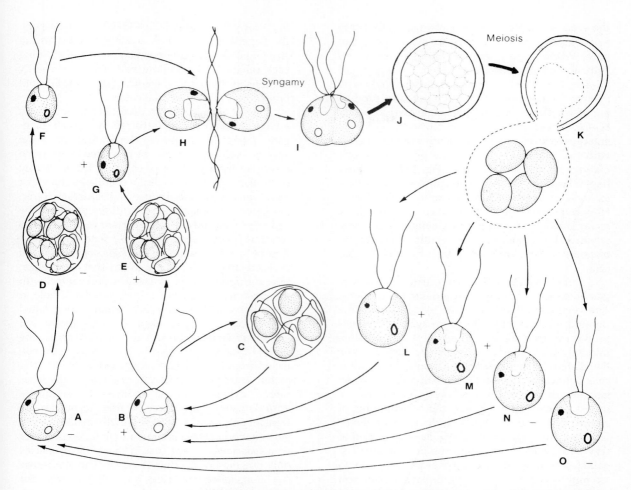

Figure 7-8 Predominantly haploid life history, *Chlamydomonas* (Volvocales). **A, B,** vegetative condition. **C,** asexual (or vegetative) reproduction. **D, E,** gamete production (by mitosis). **F, G,** gametes. **H,** fusion of isogametes (syngamy). **I,** motile zygote (or planozygote). **J,** resting zygote (zygospore) with heavy wall. **K,** germinating zygote (after meiosis). **L-O,** motile meiospores, + and − indicate mating types. *Thin lines* indicate haploid phase, *thick lines* indicate diploid phase.

Figure 7-7 Morphological diversity in the motile series (Volvocales). **A,** unicellular, *Chlamydomonas,* showing eyespot (*ey*), contractile vacuoles (*cv*), pyrenoid (*py*) in chloroplast (*c*), and flagella (*f*), ×2800. **B–F,** coenobial: **B, C,** *Gonium* in side view (**B**) and surface view (**C**), showing colony matrix (*mx*), ×800; **D,** *Pandorina,* ×800; **E,** *Eudorina,* ×400; **F,** *Pleodorina,* ×450.

Plants—Chlorophyta (Green Algae)

the species to tolerate extreme changes in moisture and temperature that can occur over a period of time. When appropriate conditions of moisture, light, and temperature are provided, the zygote germinates (Fig. 7-8K). The zygote nucleus divides by meiosis (sometimes followed by mitosis), producing at least four biflagellated meiospores (Fig. 7-8L–O). These zoospores are released, and each enlarges to form a typical vegetative cell. In species with two mating types, half the progeny are one mating type and half the other (Fig. 7-8L–O). These haploid cells are similar in morphology and behavior to zoospores produced by mitosis (Fig. 7-8C). However, they differ genetically as a result of recombination of genes during syngamy and segregation during meiosis. The results of any recombination are immediately apparent in the haploid cells.

Some species can be manipulated to produce gametes by regulating environmental conditions. The vegetative (haploid) stage is maintained in a mineral medium in liquid or on agar. When cells grown on agar are suspended in water or dilute mineral solution and illuminated, they become flagellated. Within two to four hours, these motile cells behave as gametes, and mating occurs when gametes of opposite types (plus and minus) are mixed. In mating, the initial step is a clumping of compatible gametes, which results from a cell-surface enzyme produced by the flagellar tips that serves for cellular recognition and/or adhesion. The gametes of opposite mating types pair, and five to ten minutes later a tubelike protoplasmic extension produced by one or both gametes holds them together. The paired gametes swim freely for a period of time by the activity of the flagella of only one gamete. During swimming, fusion proceeds to produce a single-celled zygote. After cytoplasmic fusion, chloroplast and nuclear fusion follow. A heavy zygote wall forms, and the cell becomes a resting stage. When transferred to fresh mineral medium in the light (after maturing in the dark), zygotes germinate within 24 to 48 hours. The heavy wall expands, and the diploid zygote nucleus undergoes meiosis followed by mitosis.

The colonial *Volvox* is 0.5 to 2.0 mm in diameter and only a few large cells divide vegetatively (Fig. 7-9A). When cell division is completed, the cells of the new colony are oriented with the apical or flagellar end pointing toward the inside of the new colony. Thus, the developing colony must literally turn inside out so the flagella are on the outside of the colony (**inversion;** Fig. 7-9B–G). As this occurs, the flagella are elongated and begin to expand in size (Fig. 7-9G, H). Inversion is essentially a folding back on itself of a hollow ball. The new colony remains in the parental colony (Fig. 7-9A, H) until breakdown of the parental colony.

Sexual reproduction in *Volvox* is oogamous, with a large, nonmotile egg fertilized by a small, elongate, biflagellated sperm. Depending upon the species, both gametes may be produced by the same colony or by two different colonies. Studies using controlled cultures and culture conditions show that some species produce inducer substances that trigger gamete production and/or differentiation. The inducers vary and can cause certain vegetative cells to act as female gametes (eggs), cause the production of sperm-containing colonies, or initiate cell divisions to produce a new colony that forms eggs. Sperm colonies swim as a unit. Upon contact with an egg-containing colony, they dissolve a hole in the colony matrix, and individual, biflagellated sperms burrow to the eggs. A single sperm and egg effect fertilization. As in *Chlamydomonas*, the diploid zygote develops a heavy wall. Zygotes collected from a pond have remained viable as long as ten years when stored in dry soil. In the laboratory, zygotes can be induced to germinate approximately three weeks after formation by transferring them to fresh growth medium. The germination process is similar to that described for *Chlamydomonas* except that in most *Volvox* species only one meiotic product is viable. The meiospore is a biflagellated, spherical, pale cell that swims for a short period before dividing to produce a small colony of 8–128 cells with a few reproductive cells.

Nonmotile, Nonfilamentous Chlorophyceae

The morphology of the nonmotile, nonfilamentous Chlorophyceae is more variable than in the motile series. Organisms range from simple, unicellular plants to more complex, multicellular plants. The cells and plants are of varied shapes, sizes, and forms, but never filamentous or parenchymatous as in the filamentous Chlorophyceae. When present, motile cells are only a reproductive stage, and many are similar to the **chlamydomonad** form.

Chlorococcales is the main order illustrating this unicellular to multicellular line (Fig. 7-6B). Basic forms are the unicellular *Chlorococcum*, which pro-

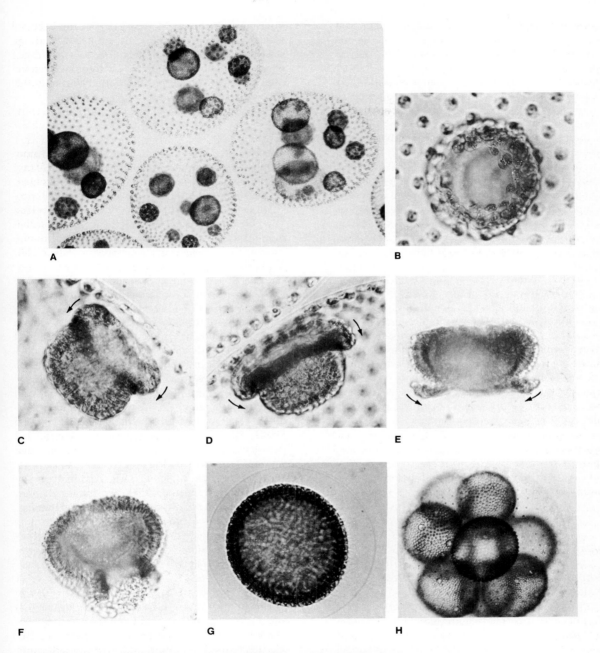

Figure 7-9 *Volvox* (Volvocales). **A,** vegetative cells, showing new colonies (coenobial), ×80. **B–H,** new colony development, showing a type of inversion. **B,** end view of young colony to show opening through which colony will invert. **C,** beginning of inversion, with new colony starting to fold back; *arrows* indicate movement of cells during inversion. **D–F,** further stages in inversion. **G,** inversion complete, with flagella extended. **H,** several young colonies within parental colony (**B–D,** ×500; **E–G,** ×560; **H,** ×110). (**A,** courtesy of R. Starr; **B–H,** courtesy of W. H. Darden and *Journal of Protozoology*.)

duces motile reproductive cells, and *Chlorella*, which produces only nonmotile spores, or aplanospores (Fig. 7-10A–C). In some genera, such as *Scenedesmus*, *Pediastrum*, and *Hydrodictyon* (Fig. 7-10D–F), zoospores are retained within parental cells until a new colony is produced having the parental form. In these algae, individual zoospores are never released from the parental cell. The number of cells in the new colony is equal to the number of zoospores produced originally.

Members of the order Chlorococcales are generally small and often overlooked, but they are ubiquitous in standing water and are an important component of some soils. Those living **symbiotically** with animals, or with fungi to form lichens (Chapter 4), provide a ready energy source for the heterotrophic partner. Other Chlorococcales occur in **bloom** conditions (as do some bluegreen algae, Chapter 2), including *Scenedesmus* (Fig. 7-10D).

Some Chlorococcales are effectively exploited for laboratory studies of photosynthesis and other biochemical research. These include *Scenedesmus* and *Chlorella* (Fig. 7-10C, D), which were used in early investigations on the splitting of water and fixation of carbon in photosynthesis. Because of their small size and simple nutrient requirements (only minerals, no vitamins or growth factors), these algae are easy to grow and maintain in the laboratory with minimal space and equipment.

Chlorococcales are probably an evolutionary sideline in green algae, much as Volvocales are. Their great diversity and widespread distribution indicate that they probably form a heterogeneous order of artificially grouped genera.

Filamentous Chlorophyceae

Morphological variation in the large, diverse group of filamentous Chlorophyceae is great. This group is composed primarily (although not exclusively) of multicellular forms that are filamentous or parenchymatous (Fig. 7-6C–E). Motile cells are present only as reproductive cells and can be symmetrical or asymmetrical (Fig. 7-5A, C). However, not all representatives produce motile reproductive cells. The main orders recognized in this text are Ulotrichales, Ulvales, and Chaetophorales (sometimes placed in the single order Ulotrichales). These algae have vegetative divisions in a regular manner (incorporating the parental cell wall), and they produce reproductive cells (spores). On the basis of cytological features (primarily ultrastructural), it is apparent that each of the orders is heterogeneous. There are also two other orders of filamentous Chlorophyceae, the Oedogoniales and the Zygnematales.

Ulotrichales, Ulvales, Chaetophorales If a nonmotile *Chlorococcum*like cell divides vegetatively (not to produce spores), and the cell wall is involved, its derivatives dividing in the same plane will produce a simple **uniseriate** (*uni* = one; *seri* = series, row) filament of cells. Essentially, this type of division occurs in a linear fashion anywhere in the filament, as in *Ulothrix* or *Klebsormidium* (Fig. 7-11A, B). If every cell in a primary, unbranched, uniseriate axis divides once in a second plane, a **biseriate** (*bi* = two) plant comparable to the marine *Percursaria* results (Fig. 7-11C). And if, in addition, division occurs repeatedly in a second plane, then a single-layered plant typical of *Monostroma* is derived (Fig. 7-11D–F). It is also possible for early cell divisions to result in **foliose** (*foli* = leaf) plants in the form of a **monostromatic** (*mono* = single; *stroma* = layer) tube, as in *Enteromorpha*, or a **distromatic** (*di* = two) blade, in which the two cell layers are usually in close contact, as in *Ulva* (Figs. 7-11G, H and 7-13D). Both these foliose types can be considered parenchymatous, as cell divisions occur somewhat randomly in the growing plant.

It is possible to achieve other plant types by additional cell divisions of the uniseriate filament with occasional cell divisions in a second plane (Fig. 7-6D, E). Thus a branch initial is established. A uniseriate, branched filament, such as the freshwater *Stigeoclonium* (Figs. 7-12A and 7-13A), can achieve growth in two directions: upright and horizontal (Fig. 7-6D). This growth habit with erect and prostrate parts is known as **heterotrichy** (*hetero* = different; *trich* = thread, filament; Figs. 7-12B and 7-13A, B) and is considered a necessary precursor for the development in land plants of any erect **assimilative system** and prostrate attaching system. Some species of *Stigeoclonium* have both prostrate and erect systems almost equally well represented under natural conditions. In *Fritschiella* (Figs. 7-12B and 7-13B), both systems develop. The prostrate portion further differentiates into the multicellular, colorless **rhizoids** (*rhiz* = root; *oid* = like), which anchor the plant in mud,

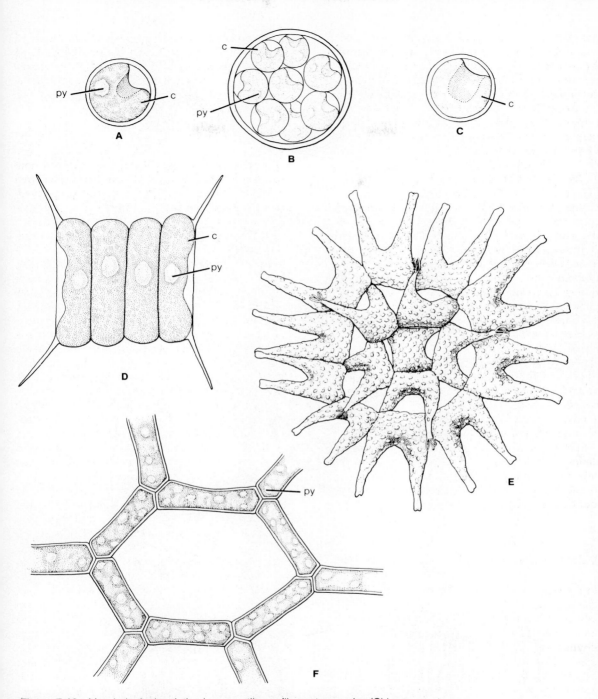

Figure 7-10 Morphological variation in nonmotile nonfilamentous series (Chlorococcales). **A, B,** *Chlorococcum,* ×1000: **A,** vegetative cell, showing pyrenoid (*py*) in chloroplast (*c*); **B,** zoospore production (flagella not shown). **C,** *Chlorella,* ×1000. **D–F,** coenobial forms: **D,** *Scenedesmus,* ×1000; **E,** *Pediastrum,* ×400; **F,** *Hydrodictyon,* portion of plants, ×200.

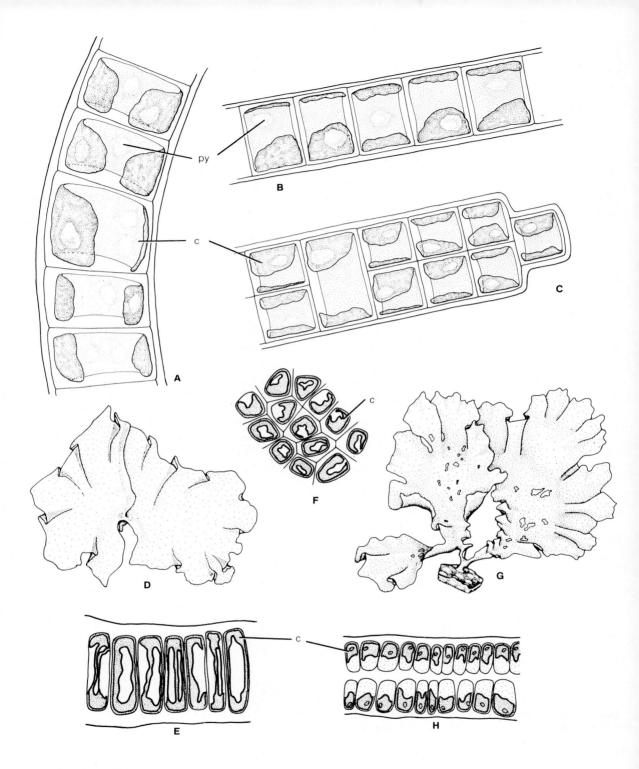

and the parenchymatous, photosynthetic portions on the surface. In other genera, one or the other of these systems has been reduced or even eliminated. The prostrate system is completely suppressed in *Draparnaldia* (Fig. 7-12C), whereas in *Coleochaete* (Figs. 7-12D and 7-13C) the prostrate system is usually better developed, and the upright system is reduced, sometimes to only a few colorless hairs.

Representatives of the basic group described here are interesting, in that one series, *Ulothrix* to *Ulva*, occurs primarily in the marine environment (Fig. 7-6C), and another, *Ulothrix* to *Fritschiella*, occurs primarily in the freshwater habitat (Fig. 7-6D). Some species of *Ulothrix* grow in fresh water and others in seawater, indicating that it is either a primitive algal form or has secondarily returned to the marine habitat. A third series (Fig. 7-6E), *Klebsormidium* to *Coleochaete*, which also occurs in fresh water, appears more important with regard to the origin of land plants. The significance of this last line is considered in the section on the relationships of Chlorophyta (see p. 217). At this point, the important feature is the occurrence of cell divisions in three planes with growth in two more directions. These are basic for true parenchymatous tissue, which makes possible the development of complex tissues in land plants. This does not indicate that any developmental series or line led directly to land plants; rather, it shows that the genetic potential for development is present in ulotrichalean algae. Therefore, they contain the potential for the basic form of land plants.

Of the many representatives of the Ulotrichales, Ulvales, and Chaetophorales, relatively few are known in any detail. These three orders are distinguished on the basis of their growth type: they are parenchymatous (= Ulvales), uniseriate filaments with branches (= Chaetophorales), or uniseriate unbranched filaments (= Ulotrichales). Cells of *Ulothrix* (Ulotrichales) are identical, except for the basal **holdfast cell.** Each cell is capable of producing isogametes or zoospores. Zoospores are biflagellated or quadriflagellated, and gametes are biflagellated. Freshwater species form a resting zygote following syngamy, as is typical of freshwater algae (*Chlamydomonas*; Fig. 7-8). In contrast, marine species, which are usually larger, have two generations (isomorphic or heteromorphic), with the zygote developing immediately into the diploid phase. Similar in appearance to *Ulothrix*, except lacking a holdfast cell, is *Klebsormidium* (previously known as *Hormidium* or *Chlorohormidium*; Fig. 7-11B). It is known only from the freshwater habitat and grows readily on moist soil or in cool, fast-moving water. Both *Ulothrix* and *Klebsormidium* have a single, open, ring- or bracelet-shaped chloroplast per cell. At least one pyrenoid is present, and the chloroplast is peripheral in position (Fig. 7-2D). *Ulothrix* readily produces symmetrical motile cells with apical flagella and cruciately arranged flagellar roots (Fig. 7-5B). *Klebsormidium* does not produce motile cells readily; when it does, they are asymmetrical. Their flagella are subapically attached and have a single, broad band root (Fig. 7-5C). Ultrastructural differences between the two genera involve the manner of cytokinesis and wall deposition (cell wall formation).

The freshwater *Fritschiella* (Figs. 7-12B and 7-13B), which grows naturally on mud banks beside streams or ponds, will produce an extensive basal portion, complete with colorless rhizoids and parenchymatous patches, when grown on a solid substrate with a thin liquid layer on top (as in the laboratory on mineral agar). *Stigeoclonium* (Figs. 7-12A and 7-13A) has a similar plasticity in form, depending upon available water. Growth in liquid culture enhances the upright system at the expense of the basal system. *Fritschiella* and *Stigeoclonium*, like *Ulothrix*, are isogamous and produce symmetrical motile reproductive cells. There are reports of alternation of isomorphic phases in *Fritschiella*, but not in *Stigeoclonium*.

Coleochaete consists of branched filaments, and is reduced primarily to the prostrate system (Figs. 7-12D and 7-13C). These filaments are closely

Figure 7-11 Morphological variation in filamentous, parenchymatous series. **A, B,** Ulotrichales. **A,** *Ulothrix*, showing unbranched, uniseriate filament; note single chloroplast (*c*) with several pyrenoids (*py*), ×400. **B,** *Klebsormidium*, morphologically similar to *Ulothrix*, but with only one pyrenoid per chloroplast, ×500. **C–H,** Ulvales. **C,** *Percursaria*, showing biseriate form, ×400. **D–F,** *Monostroma*. **D,** habit, ×0.5. **E,** transverse section, showing single layer of cells, ×530. **F,** surface view, ×750. **G, H,** *Ulva*. **G,** habit, ×0.5. **H,** transverse section showing two layers of cells, ×500; surface view of *Ulva* similar to that of *Monostroma* (**F**).

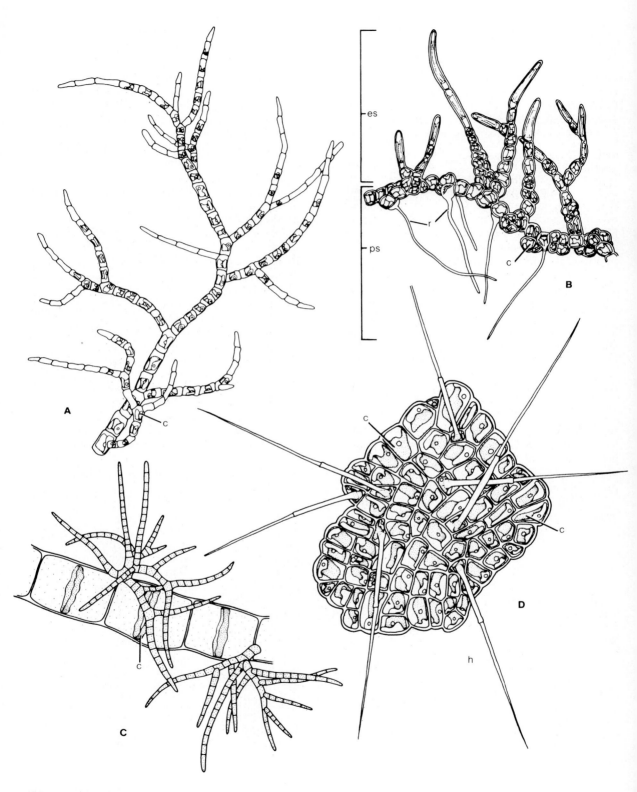

appressed to form a flat disc that appears superficially parenchymatous. In vegetative cell division, persistent longitudinal microtubules remain following mitotic spindle formation (Fig. 7-4C), as in land plants. However, the new wall of *Coleochaete* forms first from the outside of the cell, as is typical of Chlorophyceae. Asexual reproduction is by production of biflagellated, asymmetric zoospores with scaly flagella and a single, broad, striated flagellar root. *Coleochaete* differs further from *Ulothrix, Stigeoclonium,* and *Klebsormidium* by being oogamous. The resting zygote remains *in situ* on the parent plant and is surrounded by haploid vegetative cells. The sperm is similar to the zoospore.

Ulva grows in the marine habitat (Fig. 7-13D) and produces a juvenile vegetative plant that starts as a uniseriate filament but quickly becomes biseriate. Divisions occur in both planes, eventually resulting in a flat, foliose plant consisting of two layers, usually in close contact (Fig. 7-11H). Basal cells are rhizoidal, and they aggregate to form a compact holdfast. *Ulva* has an alternation of isomorphic phases with no apparent morphological difference between haploid gametophyte (male and female) plants and diploid sporophyte plants. In some species, the mature male plant appears more orange or yellow due to an abundance of carotenoid pigments in the smaller male gametes. Except for basal cells, which usually remain sterile, all cells are potentially capable of producing gametes, starting at the plant margin.

Ulva undergoes sexual reproduction readily. The uninucleate cells of haploid gametophytes (Fig. 7-14A, B) divide mitotically to produce biflagellated, symmetrical gametes (Fig. 7-14C–F). Gamete fusion occurs in water after a period of **swarming**. As in *Chlamydomonas,* gametes of opposite mating types clump before pairing. Cytoplasmic contact near the anterior end is established within ten seconds after pairing (Fig. 7-15A), and fusing cells swim away from light. Fusion is completed in three minutes (Fig. 7-15B), and quadriflagellated zygotes (Fig. 7-15C) start to settle after about five minutes (Fig. 7-14H, I). The zygote attaches to the substrate by the anterior end, and the flagella (which lie along the cell) are absorbed in 15 minutes; nuclear fusion is complete 30 minutes from the start of mating.

The zygote secretes a cellulose wall before germinating, then produces a multicellular filament by repeated mitotic divisions (Fig. 7-14J, K). Subsequently, the typical, flat, foliose plant is formed (Fig. 7-14L). This plant is identical to the gametophyte, except that the cells are diploid. At maturity, all cells (except those near the base) are potentially capable of producing meiospores (Fig. 7-14M), although usually only marginal cells do. Meiosis is followed by several mitotic divisions that produce quadriflagellated, symmetrical zoospores (Fig. 7-14N, O). The zoospores have a period of swimming and then settle down, losing their flagella and secreting a cellulose wall (Fig. 7-14P, Q). Cell divisions form a filament (Fig. 7-14R–U), and then the typical foliose plant, which is the new gametophytic phase (Fig. 7-14A, B). Both the gametes and zoospores are similar to those of *Ulothrix, Stigeoclonium,* and *Fritschiella.*

Other Filamentous Chlorophyceae

The Oedogoniales and Zygnematales are restricted to fresh water and have somewhat specialized morphological and reproductive features. Both are ubiquitous and have the predominantly haploid life history characteristic of freshwater algae. The Oedogoniales include only three filamentous genera, whereas the Zygnematales include several unicellular and multicellular genera. Some workers consider both orders distinct from other filamentous Chlorophyceae, and in some instances they recognize these orders as separate classes, subclasses, or even divisions (see Fig. 7-6F, G).

Oedogoniales The order Oedogoniales is set apart on the basis of cell division and reproductive cell morphology. The vegetative cells contain a single, peripheral, netlike chloroplast (Figs. 7-2E and 7-16A). The motile reproductive cells characteristically have a ring, or crown, of flagella at the ante-

Figure 7-12 Morphological variation in heterotrichous, filamentous series (Chaetophorales). **A,** *Stigeoclonium,* showing branched growth, ×150. **B,** *Fritschiella,* showing heterotrichy with prostrate (*ps*) and erect (*es*) systems, and rhizoids (*r*), ×375. **C,** *Draparnaldia,* showing main axis, erect form, and smaller branches; note chloroplast (*c*). **D,** *Coleochaete,* prostrate species with erect hairs (*h*), ×435.

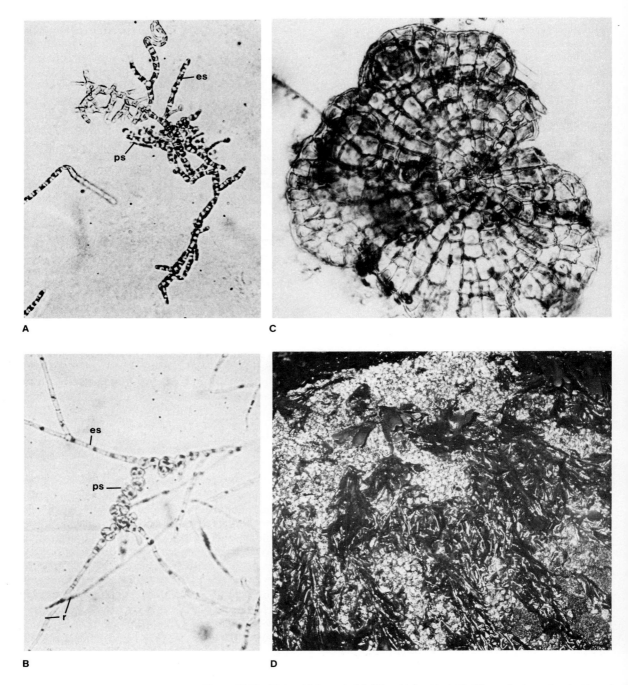

Figure 7-13 Heterotrichous habit (Chaetophorales). **A,** *Stigeoclonium,* showing basal prostrate system (*ps*) and start of erect filaments (*es*), ×150. **B,** *Fritschiella,* young plant, showing rhizoids (*r*) and start of prostrate and erect system, ×150. **C,** *Coleochaete,* prostrate plant, top view, ×1000. **D,** *Ulva,* habitat on intertidal rocks, ×0.5. (**D,** courtesy of M. Higham.)

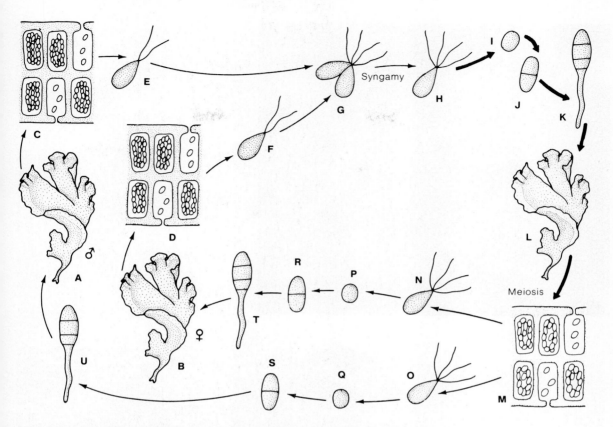

Figure 7-14 Life history of *Ulva* (Ulvales), with alternation of isomorphic haploid (*thin lines*) and diploid phases (*thick lines*). **A, B,** mature gametophytes (pale margin indicates region of gamete discharge). **C, D,** gamete production by mitosis. **E, F,** gametes (biflagellated). **G,** fusion of isogametes (syngamy). **H,** motile zygote (or planozygote). **I,** zygote. **J, K,** filamentous juvenile sporophyte. **L,** mature sporophyte (pale margin indicates region of spore discharge). **M,** zoospore production (by meiosis followed by mitosis). **N, O,** motile meiospores (quadriflagellate). **P–U,** filamentous juvenile gametophytes.

rior end of the cell (Fig. 7-16B), with zoospores having more flagella (approximately 128) than male gametes (about 30). Cell division is unique for this order, in that there are cellulose rings produced at the anterior end of vegetative cells (Fig. 7-16A). These rings, remnants of broken parental cell walls, can be very obvious and are used in recognizing genera, especially the ubiquitous *Oedogonium* (Fig. 7-16A).

All three genera of Oedogoniales are oogamous, and both gametes are produced in distinctive gametangia (Fig. 7-16C, D). In syngamy, the sperm is attracted to the egg, which is retained in the female gametangium on the haploid filament (Fig. 7-16D, E). Fusion of gametes produces a heavy-walled zygote. Germination of the zygote is similar to that in *Chlamydomonas* and *Volvox*. The diploid nucleus undergoes meiosis, producing one or two motile, haploid meiospores with the typical ring of flagella. Upon settling with its flagellar end on the substrate, the meiospore forms a filamentous, haploid plant with a holdfast cell.

Zygnematales The order Zygnematales includes forms such as *Mougeotia*, *Zygnema*, and *Spirogyra* (Figs. 7-1B and 7-17A–C), and the desmids, such as *Staurastrum*, *Micrasterias*, *Cosmarium*, and *Clos-*

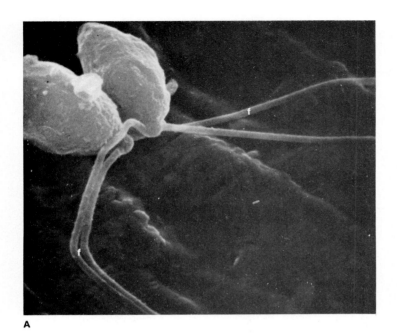
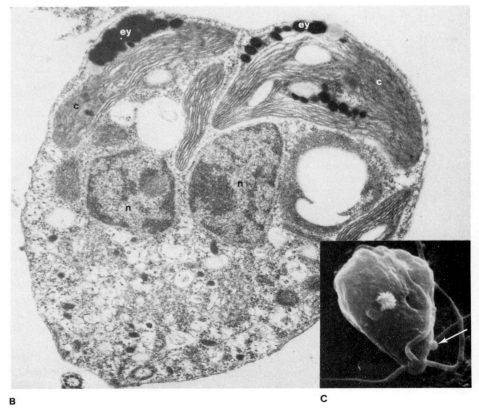

198 Chapter 7

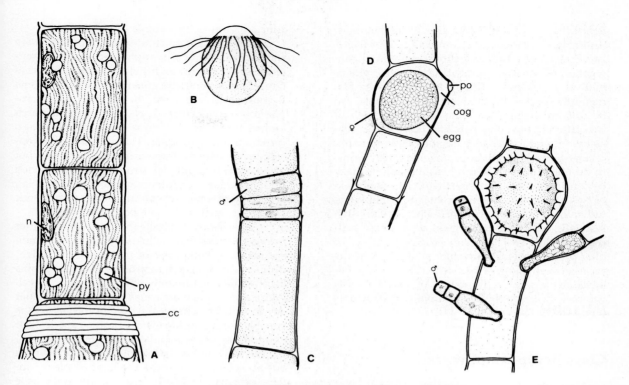

Figure 7-16 Morphology of *Oedogonium* (Oedogoniales). **A,** part of vegetative filament, showing netlike chloroplast with pyrenoids (*py*), cell wall rings comprising cellulose cap (*cc*), and nucleus (*n*), ×260. **B,** multiflagellate sperm, with anterior crown of flagella, ×500. **C,** filament with narrow, immature male gamete-producing cells (♂), ×700. **D,** filament with mature female (♀) gametangium (*oog*) containing single egg (*egg*) and subapical fertilization pore (*po*), ×700. **E,** filament with spiny resting zygote, showing three epiphytic dwarf male plants; two plants immature, each with two male gametangia (*left*); third plant (*right*) mature with discharged male gametangia, ×825.

Figure 7-15 Sexual reproduction in *Ulva* (Ulvales). **A, C,** scanning electron microscope. **A,** mating gametes, fixed prior to cytoplasmic contact while held together by flagella (*f*), ×10,000. **B,** transmission electron microscope of young zygote five minutes after start of syngamy, showing two nuclei (*n*), chloroplasts (*c*), and eyespots (*ey*) visible, ×24,000. **C,** young zygote, showing four flagella and extruding vacuole (*arrow*), ×8000. (Photographs by T. Braten, courtesy of *Journal of Cell Science*.)

terium (Figs. 7-1A and 7-17D–G). One of their most distinctive features is the complete lack of flagellated cells. Sexual reproduction results from conjugation of somewhat amoeboid gametes, one or both of which move (Fig. 7-18A, B). In the filamentous species, the conjugating gametangia can be adjacent cells of the same filament, or cells of two different filaments parallel to each other. Vegetative reproduction is by fragmentation in multicellular forms and cell division in unicellular forms. In desmids (most of which are unicellular or may be considered so), the vegetative cell is composed of two halves that are mirror images if split longitudinally or horizontally (Fig. 7-18C). Cell division results in each **half-cell** regenerating a new half-cell to produce two new, identical cells (Fig. 7-18D). This process begins with nuclear division in the area between the two half-cells (Fig. 7-18C), followed by a gradual formation of new half-cells complete with duplicated cytoplasmic structures (Fig. 7-18D).

Class Bryopsidophyceae

The recognition that multinucleate Chlorophyta should be considered as a separate class results from cytological, biochemical, and life-history differences that set them apart. Essentially, this class encompasses the forms resulting in a **siphonous** type of development. Bryopsidophyceae are poorly represented in fresh water; most occur in warmer marine waters, and many are macroscopic.

Cell Structure

Cells of the Bryopsidophyceae are multinucleate at maturity, and many genera form transverse walls only when producing reproductive structures. Thus, each protoplasmic unit contains many nuclei, chloroplasts, and other organelles. A striking feature of siphonous forms is the large, central vacuole occupying the main part of the cell (Fig. 7-19A, C), with cytoplasm only a thin layer appressed to the cell wall. In some cells there are two zones in the cytoplasm, with numerous chloroplasts in the inner zone adjacent to the vacuole (Fig. 7-19C).

Generally, the small, disclike chloroplasts are similar to those of Chlorophyceae, except pyrenoids can be absent, and thylakoids are present in larger bands, similar to grana in land plants. Some Bryopsidophyceae have, in addition to chloroplasts, a second, colorless plastid containing starch but no thylakoids. In these algae, starch is also present in the cytoplasm. The pigments of Bryopsidophyceae are the same as in Chlorophyceae, but some contain additional carotenoids not reported for other algae. In addition to mitochondria, golgi, and endoplasmic reticulum, other cytoplasmic inclusions, such as protein bodies, are present.

Cell walls are generally thick and contain other polysaccharides than cellulose; the most common are **mannan**, composed of **mannose**, and **xylan**, composed of **xylose**. Interestingly, xylan maintains its crystalline nature only when wet, whereas cellulose and mannan are equally crystalline whether dry or wet. The chemical nature of the cell wall is variable during the life history of some species. For example, in *Bryopsis* and *Derbesia* (Fig. 7-22A–D), the cell walls of the diploid plant contain mannan but no cellulose, whereas the haploid plant contains primarily cellulose or xylan. The walls have alternating layers of parallel-oriented microfibrils, which spiral around the length of the plant and form helices. This probably imparts strength to the wall, which is perhaps necessary for a large, multinucleate plant. Many tropical species have calcium carbonate in the wall and are important in coral reef formation.

Motile reproductive cells are uninucleate, generally resembling the symmetrical reproductive cells of the Chlorophyceae. Contractile vacuoles (a feature of many freshwater motile cells) are usually not present, and the occurrence of eyespots is variable. Motile cells can have two equal, anteriorly attached, whiplash flagella similar to those of *Chlamydomonas* (Fig. 7-7A) or a ring of anterior, whiplash flagella, as in Oedogoniales (Fig. 7-16B). The arrangement of flagellar roots is a cross, as in many Chlorophyceae (Fig. 7-5B).

Division of organelles in the multinucleate cells occurs regularly. Some genera in the Cladophorales and Siphonocladales produce transverse walls, creating multinucleate protoplasmic units. Genera in the Codiales and Dasycladales form incomplete transverse walls, and the cytoplasmic continuity is maintained (see Fig. 7-19B).

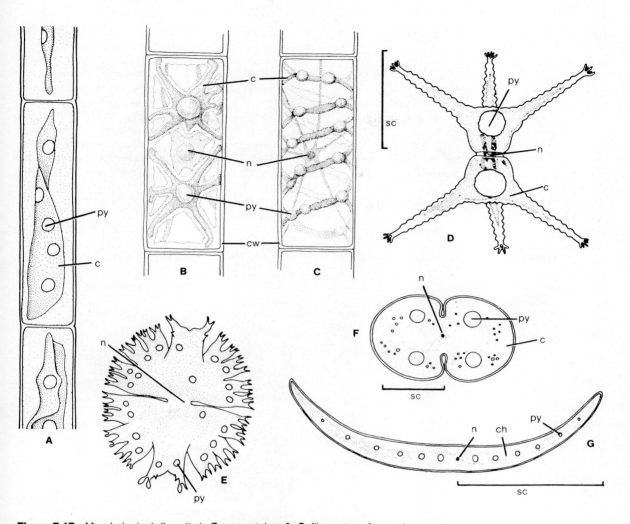

Figure 7-17 Morphological diversity in Zygnematales. **A–C,** filamentous forms showing cell wall (*cw*), nucleus (*n*), and chloroplast (*c*) with pyrenoid (*py*): **A,** *Mougeotia*, ×510; **B,** *Zygnema*, ×700; **C,** *Spirogyra*, ×700. **D–G,** unicellular desmids, showing the half-cells (*sc*): **D,** *Staurastrum*, ×270; **E,** *Micrasterias*, ×270; **F,** *Cosmarium*, ×220; **G,** *Closterium*, ×525.

Reproduction

Many Bryopsidophyceae are predominantly diploid, with gametes the only haploid cells (see Fig. 3-7B). Other members have two different morphological phases (see Fig. 3-8A). The predominantly haploid life history (see Fig 3-7A) common in Chlorophyceae is rare. Asexual reproduction is usually by fragmentation of vegetative parts. Asexual spores can be produced but are irregular, and their production is possibly restricted to conditions of environmental stress.

Many Bryopsidophyceae are anisogamous, with small, almost colorless, biflagellated male gametes. In contrast, the female gamete contains a great deal of cytoplasm and is often motile for only a short time. After syngamy, the zygote produces a new plant without any resting period.

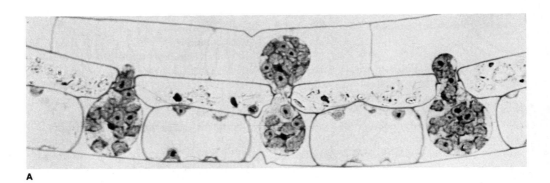

A

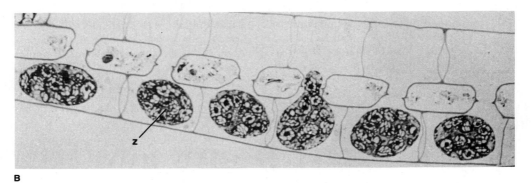

B

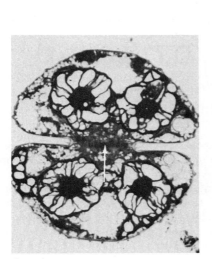

C

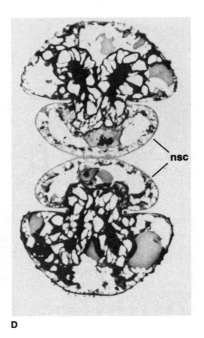

D

202   Chapter 7

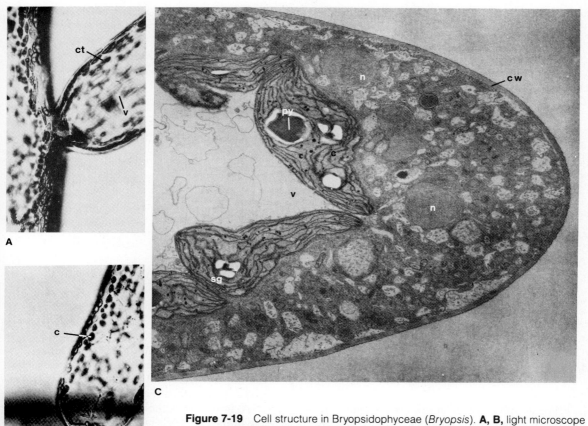

Figure 7-19 Cell structure in Bryopsidophyceae (*Bryopsis*). **A, B,** light microscope view, showing central vacuole (*v*), peripheral cytoplasm (*ct*), and chloroplasts (*c*) at point of lateral branch attachment, ×300. **C,** electron microscope view of growing tip, showing cell wall (*cw*), several nuclei (*n*), central vacuole, chloroplasts with pyrenoids (*py*) and starch (*sg*); note chloroplasts are in a zone adjacent to the vacuole, ×4000. (**A, B,** courtesy of F. A. Burr, J. A. West and *Journal of Ultrastructure Research,* from F. A. Burr and J. A. West, 1971, "Protein bodies in *Bryopsis hypnoides:* their relationship to wound-healing and branch septum development," *Journal of Ultrastructure Research* 35:476–98. **C,** courtesy of F. A. Burr, J. A. West, and *Phycologia.*)

Figure 7-18 **A, B,** sexual reproduction (conjugation) in Zygnematales. **A, B,** *Spirogyra,* ×400: **A,** migrating gametes; **B,** final stages in migration of male gamete (*arrow*); note also young zygotes (*z*). **C, D,** cell division in the desmid *Cosmarium:* **C,** division with metaphase chromosomes at center (*arrow*), ×1300; **D,** expansion of new half-cells (*nsc*) after transverse wall formation, ×1100. (**A, B,** courtesy of L. C. Fowke, J. D. Pickett-Heaps, and *Journal of Phycology.* **C, D,** courtesy of J. D. Pickett-Heaps and *Journal of Phycology.*)

Table 7-3 Orders of Bryopsidophyceae, Including Selected Genera (parentheses indicate presence in some representatives)

Order, *Genera*	Cell Walls Transverse	Cell Walls Constituent	Pigments*	Starch Plastids	Dominant Life History†
CLADOPHORALES					
Cladophora	regular	cellulose	regular	absent	haploid/diploid (haploid)
SIPHONOCLADALES					
Siphonocladus, Valonia	irregular	cellulose	regular	absent	diploid
CODIALES					
Bryopsis, Codium, Derbesia, (Halicystis), Halimeda	rare	(cellulose) xylan mannan ($CaCO_3$)	extra carotenoids	present	diploid haploid/diploid
DASYCLADALES					
Acetabularia, Dasycladus, Neomeris	rare	mannan cellulose ($CaCO_3$)	regular	absent	diploid

*Regular refers to those pigments characteristic of the division.
†Haploid = zygote only 2*n* stage (see Fig. 3-7A); diploid = gametes only 1*n* stage (see Fig. 3-7B); haploid/diploid = both multicellular phases present (see Fig. 3-8A).

Morphological Variation and Classification

The Bryopsidophyceae include at least four orders separated on the basis of location and the time in development when transverse walls (or septa) are formed (Table 7-3). Whether the morphological variation present could be derived from a single ancestor is speculative. A possibility is an alga similar to a present-day chlorophyceaen in the nonfilamentous group (such as the Chlorococcales; Fig. 7-6B). At maturity, many forms become multinucleate prior to spore formation. If the alga remains multinucleate but does not produce spores, then a subsequent increase in size could produce the bryopsidophyceaen forms (Fig. 7-20). For example, in the freshwater chlorococcalean *Hydrodictyon* (Fig. 7-10F), nuclear division occurs while the cells are increasing in size, so that each cell of the large net is multinucleate for a period of time before motile cells are formed.

The Cladophorales show a series in which there is regular (and complete) transverse wall formation of multinucleate units (Fig. 7-20A). The simplest form is the unbranched filament, as in *Urospora* (Fig. 7-21A, B). Similar development occurs in Siphonocladales (Fig. 7-20B); however, transverse wall formation is irregular, and the plant, such as *Valonia*, is never filamentous (Fig. 7-21D, E).

A second morphological series with incomplete transverse walls is possible, as shown by the Codiales (Fig. 7-20C). In these, further elongation produces a branched, multinucleate tube, or **coenocyte** (*coeno* = common; *cyte* = cell; Fig 7-22A, C, D). There can also be an aggregation of dense, intertwined coenocytes (as in *Codium* or *Halimeda*; Fig. 7-22E–G), so that the superficial appearance is of parenchymatous tissue.

Members of the fourth order, the Dasycladales, are multinucleate only late in development, after extensive growth and differentiation while uninucleate (Fig. 7-20D). When transverse walls are formed, however, they are complete.

Cladophorales, Siphonocladales Algae in the orders Cladophorales and Siphonocladales contain many multinucleate cells. Cladophorales are present in both freshwater and marine habitats; some freshwater species are nuisance algae in lakes or rivers. In contrast, Siphonocladales are restricted to tropical and subtropical marine waters. The life

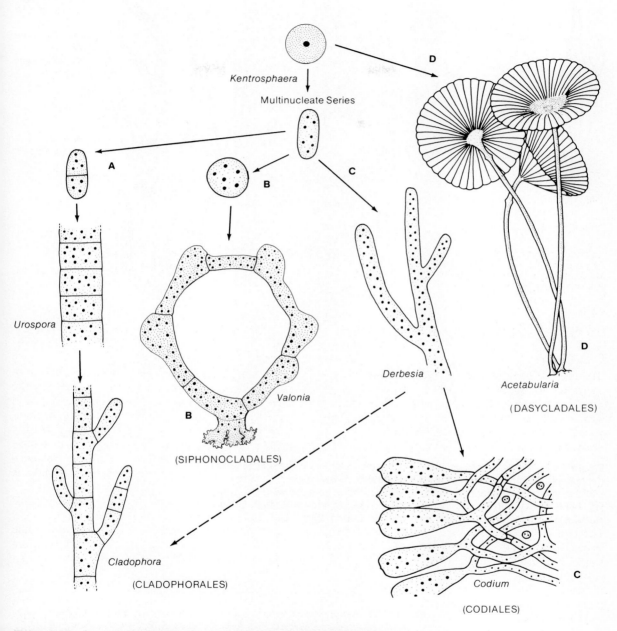

Figure 7-20 Groups of Bryopsidophyceae forming developmental series based on cell number (order names in *parentheses*). **A, B,** multinucleate, multicellular with transverse walls (**A,** Cladophorales; **B,** Siphonocladales). **C,** multinucleate (primarily one cell), lacking transverse walls (Codiales). **D,** multinucleate only at maturity, uninucleate during development (Dasycladales) (genus names for reference only).

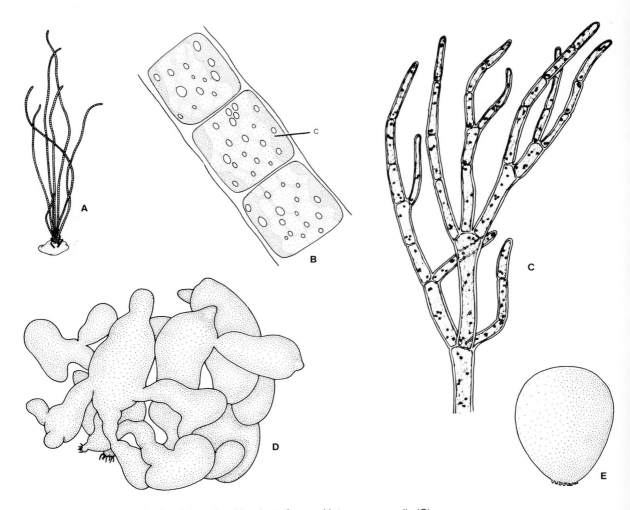

Figure 7-21 Morphological variation of multinucleate forms with transverse walls (Cladophorales, Siphonocladales). **A–C,** Cladophorales. **A, B,** *Urospora,* showing unbranched, uniseriate form: **A,** habit, ×9; **B,** vegetative cells, showing chloroplast (*c*), ×450. **C,** *Cladophora,* showing branched, uniseriate form, ×75. **D, E,** Siphonocladales, habit of *Valonia* spp. (cell wall detail not shown): **D,** ×1; **E,** ×0.5.

Figure 7-22 Morphological variation of siphonous, multinucleate forms, lacking transverse walls (Codiales, Dasycladales). **A–G,** Codiales. **A, B,** *Derbesia:* **A,** coenocytic stage, ×110; **B,** globose stage, showing differentiated gametangial region (*g*), ×4.5. **C, D,** *Bryopsis:* **C,** habit, ×1.5; **D,** coenocytic branch, ×110. **E, F,** *Codium:* **E,** habit, ×1; **F,** coenocytic branch tip (*arrow*) with gametangia (*g*), ×1000. **G,** habit of *Halimeda* (cellular detail not shown), ×0.5. **H,** Dasycladales, *Acetabularia,* showing apical cap (*cap*), ×1.5.

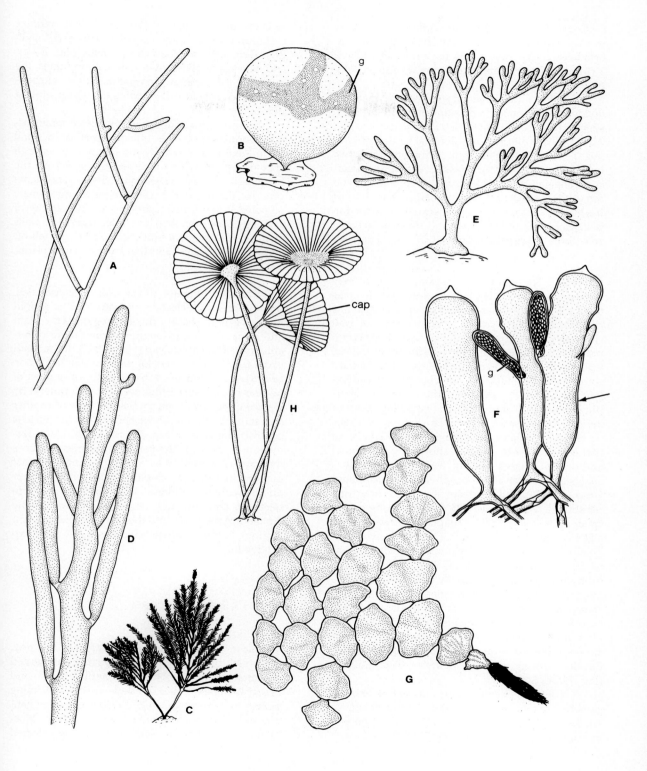

Plants—Chlorophyta (Green Algae)

histories of the two orders are different, but those in Cladophorales are so variable that the Siphonocladales could conceivably have developed from a closely related form, if not from a simple cladophoralean alga.

Variability in life history is used by many to divide Cladophorales into a number of orders. Some species (especially marine *Cladophora* species; Fig. 7-21C) have a life history identical to that of *Ulva* (Fig. 7-14), with an alternation of isomorphic phases. In others, there are two phases of different appearance: one generation is large and predominant, and the other is small and almost microscopic. In most freshwater *Cladophora* species, sexual reproduction is lacking, but biflagellated and quadriflagellated zoospores are readily produced.

Codiales The Codiales are almost exclusively restricted to the marine habitat, and in some areas they are the dominant vegetation. The order is characterized by specific differences in cell wall chemistry and pigmentation (see Table 7-3). When present, cellulose is a minor constituent in the cell wall, with mannan and xylan the predominant components. In addition, the chloroplasts contain two xanthophylls specific to the order. The morphological and reproductive variability present indicate that the order may not be a natural grouping.

Many Codiales have a life history similar to that of land plants, with an alternation of haploid and diploid phases (see Fig. 3-8A). An example is *Derbesia*, in which the sporophytic (diploid) phase is a coenocytic, multinucleate, irregularly branched plant (Figs. 7-22A and 7-23I, J). Depending on the species, the gametophyte (haploid) phase is either a globose, coenocytic plant (e.g., *D. tenuissima*; Figs. 7-22B and 7-23A, B), or a regularly branched, coenocytic plant (e.g., *D. neglecta*). Both of these gametophyte stages were originally described as different genera (i.e., the globose plant was a species of *Halicystis*, whereas the branched plant was a species of *Bryopsis*). The gametes are biflagellated and often anisogamous (Fig. 7-23C, D), but the meiospores (zoospores) are multiflagellated, with the flagella forming a crown as in the Oedogoniales (Fig. 7-23L, M).

In contrast, *Codium* plants are diploid, and the gametes are the only haploid cells. The distinct male and female gametes (Fig. 7-24E, F) are released into the water. After syngamy (Fig. 7-24G), the zygote remains motile (Fig. 7-24H) for several minutes before rounding up and becoming attached (Fig. 7-24I). Germination occurs immediately (Fig. 7-24J), and depending upon environmental conditions, the germling can form an unorganized mass of coenocytic threads before developing the typical diploid plant (Figs. 7-22E, F and 7-24A, B).

Sexual reproduction is unknown for some *Codium* populations. The effectiveness of asexual reproduction is shown by the massive increase in area occupied by one species (*C. fragile*) on the northeastern coast of the United States. The plants produce large motile cells, resembling female gametes, but no male gametes or indications of syngamy have been seen. In winter, transverse walls form on the terminal branches. These tips fall off and act as vegetative units that grow into new plants.

Dasycladales Finally, in the order Dasycladales, the cells are uninucleate until maturity. The diploid nucleus is in the basal, rhizoidal region of the plant and remains there throughout growth and differentiation of the apical cap (Fig. 7-22H). If the plant is damaged during development, it will heal. *Acetabularia* has been used extensively for studies of nuclear-cytoplasmic influence on development. By replacing the cap of one species with that of another (cap transplant), it has been shown that the nucleus controls the type of cap differentiated. Upon formation of the cap, the primary nucleus produces derivatives, probably as a result of meiosis, that migrate to the cap. Haploid anisogametes, produced within individual units in the cap, are released, and syngamy occurs in the water. The diploid zygote grows into the new *Acetabularia* plant without further nuclear divisions until after cap production.

Class Charophyceae

The class Charophyceae, as delimited in this text, is a small, distinctive group occurring submerged and attached to the bottom, primarily in fresh and brackish waters. Although there are only six living genera with approximately 250 species, many genera and species are known only as fossils. Most members of Charophyceae have heavily calcified

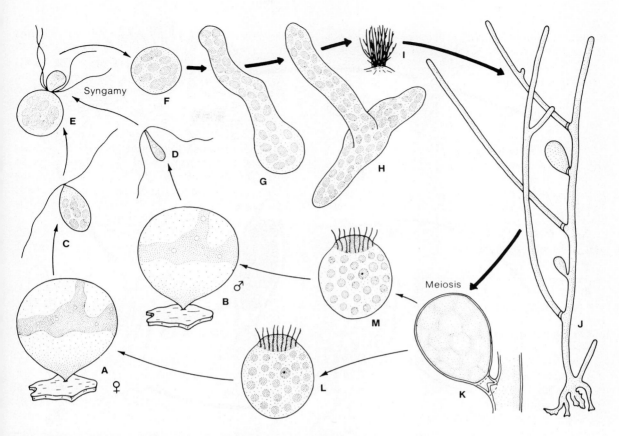

Figure 7-23 Life history of *Derbesia* sp. (Codiales) with alternation of heteromorphic phases (*thin lines* indicate haploid phase, *thick lines* indicate diploid phase). **A, B,** globose, coenocytic stage: **A,** mature female gametophyte; **B,** mature male gametophyte. **C, D,** biflagellated gametes (produced by mitosis). **E,** fusion of anisogametes (syngamy). **F,** zygote. **G, H,** coenocytic juvenile sporophyte. **I,** habit of mature sporophyte. **J,** mature coenocytic sporophyte with two meiosporangia. **K,** mature meiosporangium. **L, M,** multiflagellated zoospores (produced by meiosis).

(calcium carbonate) walls—thus the common name, stoneworts.

The members of this class are so distinctive morphologically (Fig. 7-25A) that some botanists place them in a separate division. The charophytes in the restricted sense are considered taxonomically remote from other groups of green algae, but their biochemical, cytological, and some morphological characteristics indicate that they may have a common evolutionary origin with some Chlorophyceae. Because some features and the overall morphology of reproduction are more like those of bryophytes (Chapter 11), it is interesting to note that some botanists have excluded charophytes from the algae and consider them closer to bryophytes.

Structure and Reproduction

Cells of Charophyceae are similar to those of land plants; they have a large central vacuole and

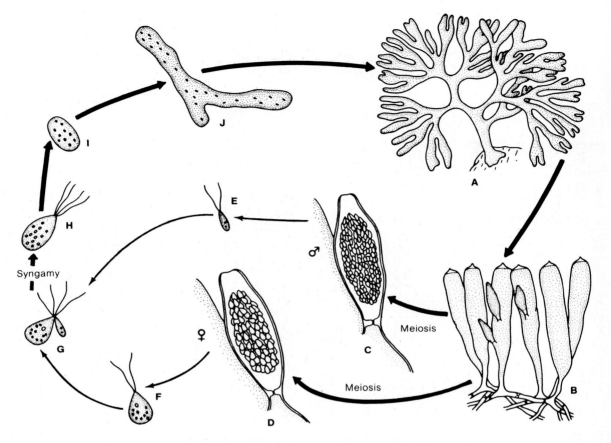

Figure 7-24 Predominantly diploid life history of *Codium* (Codiales) (*thin lines* indicate haploid phase, *thick lines* indicate diploid phase). **A,** mature diploid plant. **B,** coenocytic tips of gametangia. **C,** male gametangium after meiosis. **D,** female gametangium after meiosis. **E,** biflagellated male gamete. **F,** biflagellated female gamete. **G,** fusion of anisogametes (syngamy). **H,** planozygote. **I,** settled zygote. **J,** coenocytic juvenile stage.

Figure 7-25 Charophyceae (Charales), *Chara.* **A,** habit view, showing characteristic whorled branches, ×5. **B,** cell and plant structure; longitudinal section of apical region, showing apical cell (*ac*), internode cells (*i*) with large central vacuole (*v*), and node cells (*nd*); enclosing internodal filaments (*cf*) seen at right of internodal cell (chloroplasts not obvious), ×200. **C,** *Chara* sperm (dried), showing flagellar insertion (*fi*) away from apex, ×3000. **D,** fossil female reproductive structure (gyrogonite), ×50. (**B,** courtesy of J. D. Pickett-Heaps. **C,** courtesy of Ø. Moestrup and *Planta*. **D,** courtesy of R. E. Peck, G. A. Morales, and *Micropaleontology*.)

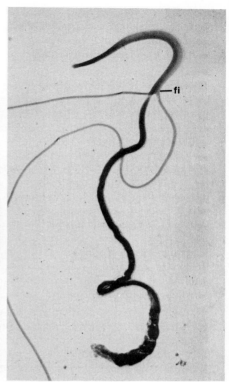
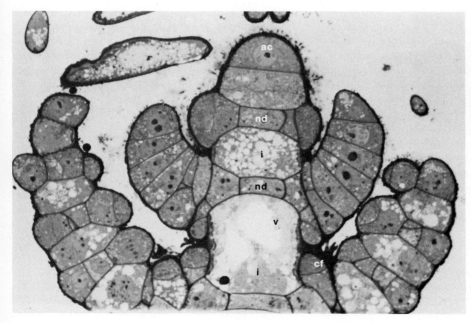

Plants—Chlorophyta (Green Algae)

numerous disc-shaped chloroplasts in the peripheral cytoplasm (Fig. 7-25B). The cell organelles are like those of Chlorophyceae; starch is present in the chloroplast, although pyrenoids are absent. Cytoplasmic streaming is especially evident in large cells, and chloroplasts remain stationary.

Charophyceae are distinguished from most Chlorophyceae by producing plants differentiated into short **node** (*nod* = knot, swelling) and long **internode** (*inter* = between) regions (Fig. 7-26B). Longitudinal growth is primarily apical and involves a single, large, apical cell for each branch (Fig. 7-25B). From the nodes, whorls of branches of limited growth arise. In *Chara*, the long internodal region is covered by enclosing filaments of cells, which arise at the nodes and extend up and down over the internodal cells (Fig. 7-26A). In *Nitella* (Fig. 7-26B, C) these covering filaments are absent, and the long internodal cells, often several centimeters long, are obvious. Cells of the nodal region are uninucleate, but the large internodal cells become multinucleate. At the base of the plant there is a colorless, branched rhizoidal system.

Vegetative reproduction is uncommon but occurs by fragmentation of specialized groups of cells. No zoospores are produced. Sexual reproduction occurs readily, with definite male and female gametes produced. The zygote is the only diploid stage.

The female gametangium is borne at a node and contains a single egg (Fig. 7-26A, D). Surrounding the nonmotile egg is a sterile vegetative sheath, which originates beneath the gametangium and grows up spirally to form a protective layer of cells. At maturity, this sheath covers the tip of the gametangium with one or two rows of crown cells (Fig. 7-26D). A sperm must penetrate this crown to contact the egg. The female structure, termed **nucule** (*nucl* = small nut), appears to be multicellular but is actually unicellular, with secondarily produced protecting cells (as happens in the chaetophoralean *Coleochaete*). The immature female gametangium is white, but after syngamy it becomes black.

In contrast, the male reproductive structure, or **globule** (*glob* = ball), is the most complex one in all the algae. It also occurs at a node and is generally a spherical, orange structure attached by a short, stalklike cell (Fig. 7-26A, F). Sterile cells surround the fertile, sperm-producing cells (Fig. 7-26E, F). Formation of sterile and fertile cells occurs simultaneously, so that the male reproductive unit is multicellular and is interpreted as a complex branch. In the globule, fertile cells are in a linear series, and each produces a single, coiled, asymmetric, biflagellated sperm. The coiled flagella extend backward from the point of attachment, which is some distance from the anterior tip of the sperm (Fig. 7-25C). The sperm body and flagella are both covered with scales, and the microtubular flagellar roots form a flat band. The sperm is comparable to that of some land plants and to the motile cells of *Klebsormidium* (Fig. 7-5C) and *Coleochaete*.

The female gametangium containing the nonmotile egg is retained on the plant after fertilization, and the zygote becomes a resting stage within a thick, protective covering of vegetative cells. After a resting period, the zygote nucleus divides by meiosis, producing four haploid nuclei. Three of these nuclei degenerate, and the functional nucleus divides to produce a filamentous stage (Fig. 7-26G). This filamentous stage, which develops into the typical plant with apical growth (Fig. 7-25A), is similar to the **protonema** of mosses (Chapter 11).

Morphological Variation and Diversity

There is little diversity within this class compared to other Chlorophyta. All living genera are placed in a single order, Charales, but extinct representatives are placed in four orders, including Charales. Living genera are separated on the basis of

Figure 7-26 Morphology and reproduction of the Charophyceae (Charales). **A,** *Chara*, showing enclosing filaments (*cf*) and reproductive structures (gametangia) at nodes (*nd*), ×30. **B, C,** *Nitella:* **B,** habit, showing characteristic nodes and internodes (*in*), ×0.5; **C,** axis, ×30. **D–G,** *Chara:* **D,** female structure with twisted cover cells (*co*) and one row of crown cells (*cr*) protecting single gamete (*oog*), ×750; **E, F,** developmental stages of male reproductive structure, showing sterile cells (*st*) and reproductive filaments (*af*), ×750; **G,** juvenile, filamentous plant from germinating zygote (*z*), showing basal system (*bs*), rhizoids (*r*), and erect systems (*es*), ×225. (**E–G,** after Smith, with permission of McGraw-Hill Book Co.)

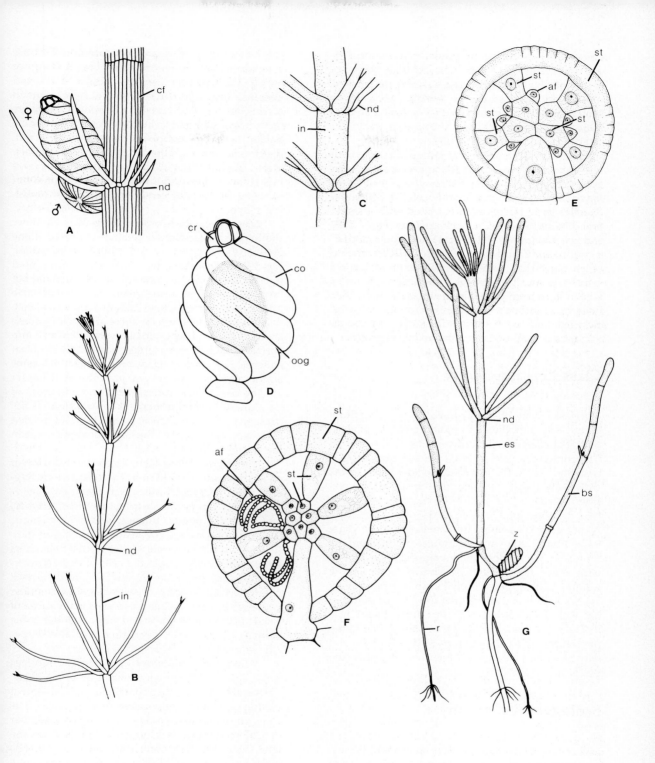

Plants—Chlorophyta (Green Algae)

gametangial features and gametophyte morphology, including the presence of enclosing filaments at the internodes. Extinct genera are distinguished mostly on features of the female reproductive structure that have survived as calcified fossils, known as **gyrogonites** (*gyro* = round, turning; *gon* = seed; Figs. 7-25D and 27-14).

In 1975, the concept of the class Charophyceae was expanded to include those green algae with persistent spindle and microtubules during cytokinesis, and asymmetric motile cells in which the flagellar basal bodies are associated with a single broad band (Fig. 7-5C). *Klebsormidium* (Fig. 7-11B) and *Coleochaete* (Figs. 7-12D and 7-13C) are included, as well as the Zygnematales (which lack any motile cells). Since then, with the completion of further cytological and physiological studies, a few other genera have been placed in the class. Until further research elucidates which genera should also be included, this text considers the Charophyceae as a restricted, well-defined morphological group.

Class Prasinophyceae

The class Prasinophyceae presently consists of those uninucleate, unicellular, or multicellular green algae with flagellated cells different from the typical Chlorophyceae (see p. 178; compare Fig. 7-5A, B with Fig. 7-27A, B). The class includes those forms with scaly flagella arising from an apical depression or groove (Fig. 7-27B). Motile cells have one, two, or more equal or unequal flagella (Fig. 7-27A). Although most genera are flagellated in the vegetative phase, there are nonmotile, nonfilamentous representatives. Most genera are less than 30 μm, with more representatives known from marine waters. As most prasinophytes are small and easily overlooked, it is possible that they are present but not well-documented in freshwater habitats. It is thought that some prasinophytes are similar to the progenitors of various groups of Chlorophyta and may be in an evolutionary line to land plants (see p. 217 and Fig. 7-28B).

Ecology of Chlorophyta

Few other groups of algae have such a wide distribution and variety of habitat as Chlorophyta. They can occur wherever some moisture and light are available. They are commonly free-floating or swimming in plankton of lakes, ponds, rivers, and streams. To a lesser extent, green algae occur in oceans but can be abundant in tidepools. There are many larger, attached forms, especially in the marine habitat, where Chlorophyta have a broad, vertical distribution through the intertidal zone into deep water. In cold waters, they are more conspicuous in shallow regions and in the intertidal zone, but in warm waters they can extend to depths exceeding 100 meters.

Most genera (even orders) are restricted to either the marine or freshwater habitat. However, some genera (e.g., *Ulothrix*, Fig. 7-11A; *Chlamydomonas*, Fig. 7-7A; *Cladophora*, Fig. 7-21C) are represented in both. Rarely is the same species an inhabitant of both, but there are some green algae that can live in estuarine waters, where salinity fluctuates widely.

Although most green algae are aquatic, some are terrestrial and occur almost everywhere on damp surfaces, such as tree trunks, moist walls, and leaf surfaces. Some green algae occur in soil or in lichens as the algal component, or **phycobiont** (Chapter 4). The commonest cause of red snow in areas of permanent or semipermanent snow and ice is the presence of *Chlamydomonas* species and related genera containing masking carotenoid pigments in the cells.

Other green algae occur epiphytically on larger algae or aquatic seed plants. The filamentous Zygnematales (*Zygnema*, *Mougeotia*, *Spirogyra*, Fig. 7-1B) occur in large, free-floating or submerged mats in freshwater ponds, ditches, and slow streams. Some Chlorophyta grow inside the shells of molluscs, or within tissues of higher plants. Still others live symbiotically within the cells of certain invertebrates and protozoa. A few Chlorophyta lack chloroplasts and occur as saprobes. One colorless genus (*Prototheca* in Chlorococcales) is the cause of a skin disease (*protothecosis*) in humans and in some animals (dogs, cattle, deer, sheep), it affects specific organs.

Almost any container of water that receives some light will become colonized by green algae. Urns and bird baths often contain the reddish-brown resting cells of the volvocalean *Haematococcus*. The water in fish tanks can have any one of a number of Chlorococcales as phytoplankton, and on the tank sides, the filamentous *Oedogonium* (Fig. 7-16A) often occurs. Some Chlorophyta are present in large

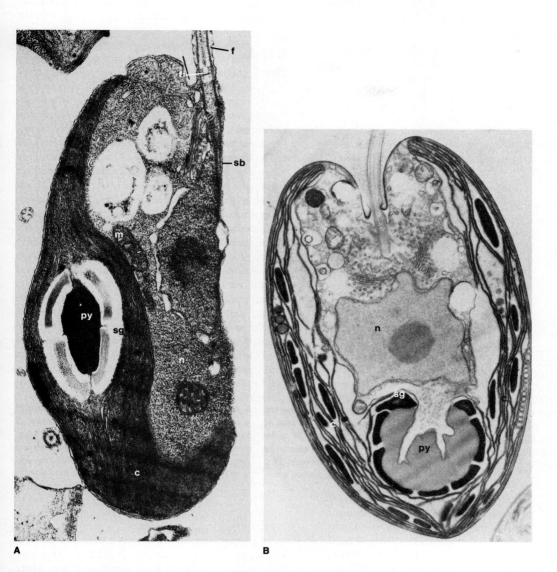

Figure 7-27 Prasinophyceae. Cell structure of motile cells, transmission electron microscope. **A,** *Pedinomonas,* longitudinal section of vegetative cell, showing chloroplast (*c*) with pyrenoid (*py*) and starch (*sg*); note nucleus (*n*), flagellar insertion (*fi*), and attachment of flagellum (*f*) to striated band (*sb*), and mitochondrion (*m*), ×2125. **B,** *Platymonas,* longitudinal section of vegetative cell showing off-center pyrenoid; note nucleus and apical depression, ×9000. (**A,** courtesy of J. D. Pickett-Heaps, D. W. Ott, and the international journal *Cytobios*. **B,** courtesy of I. Manton and M. Parke, 1965, from "Observations on the fine structure of two species of *Platymonas* with special reference to flagellar scales and the mode of origin of the theca," *Journal of the Marine Biological Association of the U. K.* 45:743-54, with permission of Cambridge University Press.)

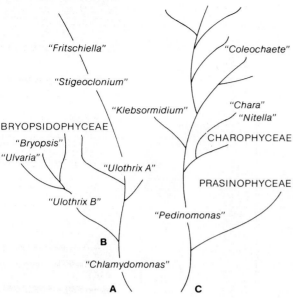

Figure 7-28 Possible derivation of Chlorophyta and land plants from flagellated, unicellular progenitors. **A,** primarily Chlorophyceae, starting with symmetrical *Chlamydomonas*like motile cell. **B,** branch from line A containing those Chlorophyceae and Bryopsidophyceae lacking microtubules at cytokinesis. **C,** selected Chlorophyceae starting from asymmetrical *Pedinomonas*like motile cell and leading toward land plants (embryophytes) with a sideline to Prasinophyceae and Charophyceae. Generic names (in quotation marks) indicate genera similar to present-day forms.

numbers in domestic water supplies, giving a disagreeable flavor to the water or clogging filters. Other chlorophytes grow in great masses, as does *Cladophora* (Fig. 7-21C), which grows in some nutrient-rich waters. One example is *C. glomerata* in the Laurentian Great Lakes between Canada and the United States (Lakes Erie, Ontario, Michigan). Evidently growth is favored by high light intensity, high water turbulence, and large amounts of calcium and carbonate, as well as nitrogen, phosphorus, silica, and sometimes thiamin. The algal masses interfere with boating, swimming, and fishing, and can be responsible for fish kills by depleting oxygen lev-

els. When it washes ashore, the odor and appearance of the rotting alga make it even more unpopular.

Relationships of the Chlorophyta

The division Chlorophyta contains representatives that are more like higher plants than those of any other algal division. Their fossil history goes back to the Precambrian, with the oldest supposed green algae being microfossils from sediments in eastern California (1.3 billion years old). The interpretation of these, however, is subject to question. It has been shown in some instances that their internal contents, rather than being remnants of eukaryotic organelles, may be degraded prokaryotic cells in which the cell wall and sheath are intact, but the protoplasm is a globular mass. More recent data on the ultrastructural features and cell size of similar microfossils support their recognition as eukaryotes, possibly green algae.

In general, the fossil record of most Chlorophyta is poor, as they are morphologically simple and do not contain material that withstands decomposition. The best known fossil Chlorophyta are representatives of Bryopsidophyceae (Dasycladales, Codiales; see pp. 200 and Fig. 27-8) and Charophyceae (see p. 208 and Fig. 27-14). These fossils are calcified remnants resulting from calcium carbonate deposition in the cell walls. There are indications of filamentous forms in the Cambrian and younger rocks, as well as geochemical evidence of biological compounds attributable to green algae. At present, the main speculation concerning the phylogeny of Chlorophyta is based primarily on living organisms.

It is thought that possibly the four classes of Chlorophyta had a common origin, but that at least two or three separate lines developed very early. Because of physiological and biochemical similarities, Chlorophyta are accepted as obvious progenitors of land plants. Exactly how this has occurred is unknown and probably will remain so. However, it is possible to construct a series based on present-day algae. This is a **phenetic** (*pheno* = appear) series, since it does not consider ancestral forms or actual **phylogeny** (*phyl* = tribe; *gen* = origin) because of the poor fossil record. The starting point of the series is a unicellular, flagellated alga. The series continues through a filamentous, heterotrichous form, ultimately producing a prototype of a land plant adapted to the harsher terrestrial environment.

Such a series includes not only biochemical features but also morphological and cytological features and the genetic potential implicit in sexual reproduction. The obvious biochemical features are pigmentation (chlorophyll *a*, *b*, carotenoids), cell walls (cellulose as a substance that maintains its structure when out of water), food manufacture (typical carbon fixation into simple carbohydrates), and storage products (starch). The morphological features include specialized tissues for anchoring, assimilation, storage, and reproduction, as well as cytological features. These cytological features are divided into those of vegetative and motile cells. Vegetative cell features include (1) type of cell division (origin and orientation of microtubules in cell wall formation; Fig. 7-4A, C); (2) direction of cytokinesis (from center out or toward center; Fig. 7-3); and (3) presence of cytoplasmic connections (**plasmodesmata:** *plasmo* = mold, form; *desma* = bond) between cells. The features of the motile cells are (1) asymmetry of cell; (2) presence of a single, broad flagellar root; (3) presence of flagellar and/or body scales; and (4) lack of a firm cell wall (Fig. 7-5C). The importance of sexual reproduction involves not only the type of gametes but also an understanding of the alternation of phases (generations), which necessitates multicellular haploid and diploid stages.

The occurrence of multiple and somewhat parallel series in Chlorophyceae may explain the diversity of forms in green algae and in land plants (Fig. 7-28). One series, with a *Chlamydomonas*like form as the flagellated unicell (Fig. 7-7A), could lead through a simple filament (*Ulothrix*; Fig. 7-11A) to heterotrichous taxa (*Stigeoclonium*, *Fritschiella*; Figs. 7-12A, B and 7-13A, B). This series did not evolve into land plants, although certain forms (e.g., *Fritschiella*) are somewhat adapted to fluctuating moisture conditions (Fig. 7-28A). A second series culminating in land plants (Fig. 7-28C) may have a *Pedinomonas*like flagellated cell with a simple, filamentous *Klebsormidium*like form (Fig. 7-11B). Intermediate stages are similar to *Coleochaete* and *Chara* (Figs. 7-12D, 7-13C, and 7-25A), although neither should be considered in a direct line to land plants. This second series is the expanded concept of the Charophyceae proposed in 1975. Current research shows there may be even a third series (Fig. 7-28B). This last series involves at least one species each of *Ulothrix*, *Monostroma*, and *Ulva*, as well as several Bryopsidophyceae (including species of *Bryopsis* and *Derbesia*). In these algae, microtubules at cytokinesis are lacking, and the motile cells have a cruciate root system and body scales. As more species and genera are examined (primarily with regard to ultrastructure), the validity of these phenetic schemes may be clarified. At the same time, there will probably be taxonomic changes of the species involved at the generic level.

In establishing such phenetic series, it is important to remember that present-day green algae are not as morphologically complex as representatives of some other algal divisions (Phaeophyta, Rhodophyta; Chapters 8, 9). However, the most highly evolved green algae may actually be land plants, such as mosses or ferns! The series in Figure 7-28 are based primarily on cytological features, with some consideration of the habitat. This is especially true for *Klebsormidium*, which often grows in moist, temporary habitats. Further studies of other heterotrichous and parenchymatous green algae are needed to fill the blank spaces in the land plant phenetic series. It is apparent that this series could lead to a *Chara*like organism as an offshoot terminating in rather complex forms. These branches, though, are actually dead ends in an evolutionary series to land plants. The similarities to land plants of the Charophyceae and their long fossil history indicate early divergence (probably 500 million years ago) in the early Cambrian period.

Termination of such a line as charophytes is somewhat self-imposed, since the complexities produced are the ultimate that a group of organisms can achieve from a basic stock. This, then, is natural selection in one form. Similar dead ends occur in various groups of Chlorophyceae (e.g., Volvocales) and Bryopsidophyceae.

References

Bold, H. C., and Wynne, M. J. 1978. *Introduction to the algae*. Prentice-Hall, Englewood Cliffs, N. J.

Bourrelly, P. 1972. *Les algues d'eau douce*. Vol. 1, *Les algues vertes*. 2nd ed. N. Boubée & Cie, Paris.

Cox, E. R., ed. 1980. *Phytoflagellates*. Elsevier/North-Holland: New York.

Dodge, J. D. 1979. *The phytoflagellates: fine structure and phylogeny.* In Levandowsky, M., and Hutner, S. H., eds., *Biochemistry and physiology of protozoa.* Vol. 1. 2nd ed. Academic Press, New York.

Pickett-Heaps, J. D. 1975. *Green algae: structure, reproduction, and evolution of selected genera.* Sinauer Associates, Sunderland, Mass.

———. 1979. Electron microscopy and the phylogeny of green algae and land plants. *Amer. Zool.* 19:545–54.

Stewart, K. D., and Mattox, K. R. 1975. Comparative cytology, evolution and classification of the green algae, with some consideration of the origin of other organisms with chlorophyll A and B. *Bot. Rev.* 41:104–35.

———. 1978. Structural evolution in the flagellated cells of green algae and land plants. *Biosystems* 10:142–52.

Wood, R. D., and Imahori, K. 1965. *Revision of the Characeae.* 2 vols. J. Cramer, Weinheim.

8

PLANTS—PHAEOPHYTA (BROWN ALGAE)

The brown algae (division Phaeophyta) form a large, conspicuous group that contains about 260 genera and about 2000 species. They are restricted to the sea, with the exception of four freshwater genera that are rare and inconspicuous. Phaeophyta are generally classified (Table 8-1) in 11 to 14 orders, which can be grouped in one class or in three separate classes. Their separation into classes based on life-history characteristics has been traditional; however, recent culture studies suggest that this is a highly artificial arrangement. Their uniformity with respect to many biochemical characteristics supports the concept of a single class, Phaeophyceae, which is the system adopted in this text. At the same time, increasing knowledge concerning the ultrastructural features of flagella suggests that there are two main evolutionary series within brown algae. These series are placed in this text in two distinct subclasses, Phaeophycidae (with 11 orders), comprising the ectocarpalean evolutionary line, and Cyclosporidae (with one order), comprising the fucalean evolutionary line. Hence, in this text, 12 orders of brown algae are recognized and placed in one class, Phaeophyceae, with two subclasses.

Cell Structure

Cell Wall

Except for reproductive cells (gametes and spores), which have naked protoplasts, the cells of brown algae are bounded by a well-developed cell wall. In general, the cell wall is composed of an inner, firm, **fibrillar,** cellulose layer and an outer, amorphous, mucilaginous, pectic or pectinlike layer. Various carboxylated (**alginic acid**) and sulphated (**fucoidan**) polysaccharides, termed **phycocolloids,** also occur in the matrix of the cell walls or in the intercellular spaces of brown algae. Alginic acid and fucoidan occur in some larger genera of **kelps** (Laminariales) and **rockweeds** (Fucales) in sufficient quantity to be commercially important. Alginic acid can represent 10–40% of the dry weight of the plant.

Alginic acid is a polyuronide, a linear polymer of β-1, 4-linked *D*-mannuronic and *L*-guluronic acids, although the proportion of the two types of resi-

Table 8-1 Orders of Division Phaeophyta and Representative Genera

Class	Subclass	Order	Representative Genera
Phaeophyceae	Phaeophycidae	Ectocarpales	*Ectocarpus, Pilayella (Pylaiella), Ralfsia,*[1] *Streblonema*
		Chordariales	*Analipus (Heterochordaria),*[1] *Haplogloia, Leathesia, Myrionema*
		Sporochnales	*Carpomitra, Perithalia, Sporochnus*
		Desmarestiales	*Desmarestia, Himantothallus*
		Sphacelariales	*Cladostephus, Halopteris, Sphacelaria*
		Cutleriales	*Cutleria* (= *Aglaozonia*), *Zanardinia*
		Tilopteridales	*Tilopteris*
		Dictyotales	*Dictyota, Glossophora, Padina, Syringoderma,*[2] *Zonaria*
		Dictyosiphonales	*Phaeostrophion, Punctaria, Soranthera*
		Scytosiphonales	*Hydroclathrus, Petalonia, Scytosiphon*
		Laminariales	*Agarum, Alaria, Cymathere, Egregia, Eklonia, Laminaria, Lessonia, Macrocystis, Nereocystis, Pelagophycus, Postelsia, Pterygophora, Thalassiophyllum*
	Cyclosporidae	Fuscales	*Ascophyllum, Ascoseira,*[3] *Bifurcariopsis, Carpophyllum, Cystophora, Cystoseira, Durvillaea,*[4] *Fucus, Halidrys, Himanthalia, Hormosira, Pelvetiopsis, Sargassum*

[1] The crustose *Ralfsia*, parenchymatous *Analipus*, and certain other genera are sometimes placed in a separate order, Ralfsiales.
[2] The genus *Syringoderma* should probably be placed elsewhere, possibly in a separate order.
[3] The genus *Ascoseira* is sometimes placed in a separate order, Ascoseirales.
[4] The large genus *Durvillaea* is sometimes placed in a separate order, Durvillaeales.

dues present varies. In structure it appears to be somewhat similar to cellulose and pectic acid. Fucoidan is typified by the presence mainly of L-fucose (a methyl pentose sugar) residues esterified by sulphuric acid. Lesser amounts of D-xylose, galactose, and D-glucuronic acid are also present.

Calcification ($CaCO_3$) of the cell wall occurs in one genus, *Padina* (Fig. 8-1), and small amounts of silica occur in some brown algae.

Chloroplast

Pigments are present in one or more usually peripheral chloroplasts, which vary to some extent in size and shape (Fig. 8-2). Where they are numerous in the cell, chloroplasts are peripheral and usually small and discoid, but they can be elongate, ribbon-shaped, netlike, **laminate**, or irregular. In a few instances they are **stellate** and **axile**. Electron micrographs show that the chloroplasts are bound by two membranes and contain photosynthetic **lamellae** surrounded by a granular matrix, the **stroma**. The photosynthetic lamellae are formed by associated but unfused thylakoids in stacks of three (Fig. 8-3), forming parallel bands running the full length of the **plastid**. The bands are separated by a uniform space, but there can be some exchange of thylakoids between them. It appears that the peripheral, or girdle, band encircles the other bands, following the contour of the chloroplasts. Small, membrane-free DNA regions occur between peripheral and parallel bands, forming a ring at the ends of the chloroplast. Longitudinal division of the chloroplast is preceded by division and separation of these DNA regions. The chloroplast lies within a double-membrane sac, the chloroplast endoplasmic reticulum, which is continuous with the outer membrane of the nuclear envelope.

The characteristic brownish color of these algae results from the masking of chlorophyll pigments by an excess of accessory carotenoid pigments, especially the xanthophyll pigments **fucoxanthin** and **β-carotene**. Chlorophylls *a* and *c* are present. In addition to the dominant fucoxanthin and β-carotene, additional xanthophyll pigments are usually present in smaller amounts. Pigments occur

Figure 8-1 View of population of *Padina*, a tropical member of the Dictyotales with calcified cell walls; concentric zones on the fanlike branches are due to rows of microscopic filamentous "hairs," ×0.25.

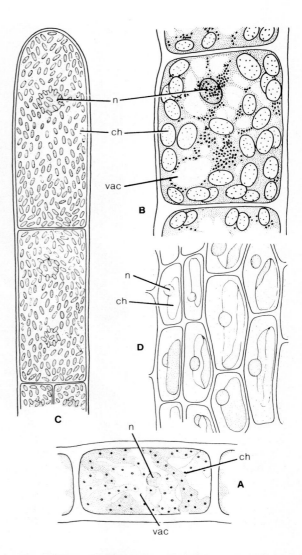

Figure 8-2 Chloroplast (*ch*) types in Phaeophyta. **A,** reticulate (*Ectocarpus*), ×1500. **B, C,** discoid: **B,** *Pilayella*, ×690; **C,** *Sphacelaria*, ×320. **D,** laminate (*Scytosiphon*), ×300. *n,* nucleus; *vac,* vacuole.

in the photosynthetic lamellae (thylakoids). According to the conditions in which they occur, brown algae range in color from light yellow-brown or olive-brown through a rich, golden brown to practically black. This variation is a reflection of the relative proportions of chlorophylls to other pigments present.

Pyrenoids are not universally present in the brown algae. When present, the pyrenoid is excentric, stalked, inside the chloroplast, projecting from either an end or a side (Fig. 8-3), capped by and separated from the cytoplasm by a dilated sac. It is not penetrated by the thylakoids.

Other Cell Components and Storage Products

A central vacuole can be present in the cell, but usually there are numerous small vacuoles scattered throughout the cytoplasm (Fig. 8-2B). Typical **dictyosomes, ribosomes, endoplasmic reticulum,** and **mitochondria** are present in the cytoplasm. The principal product of photosynthesis that is stored as food reserves is the polysaccharide **laminarin,** which is a polymer of D-glucose, characterized by β, 1-3, and 1-6 linkages. The sugar alcohol, D-mannitol, is a terminal residue of the polymer. Fat globules and oil droplets can also be stored. In electron micrographs, laminarin appears as opaque granules scattered throughout the cytoplasm,

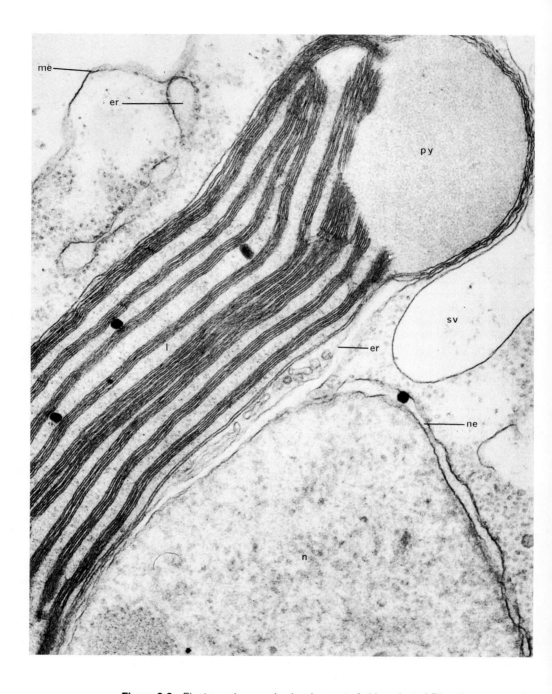

Figure 8-3 Electron micrograph, showing part of chloroplast of *Pilayella* near pyrenoid (*py*), with characteristic compound lamellations, each composed of fused thylakoids in stacks of three (*l*), and dense lipid droplets lying between them: *n*, nucleus; *ne*, nuclear envelope; *er*, layer of endoplasmic reticulum surrounding chloroplast; *me*, membrane; *sv*, part of storage vesicle, ×54,600. (Photograph courtesy of L. V. Evans.)

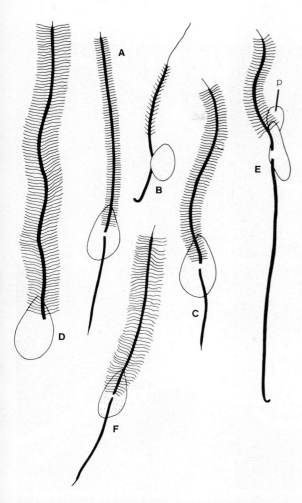

Figure 8-4 Arrangement of flagella of various Phaeophyta (anterior end toward top of each diagram). **A–C,** zoospores: **A,** *Pilayella*; **B,** *Chordaria*; **C,** *Laminaria*. **D–F,** sperms: **D,** *Dictyota*; **E,** *Fucus* (*p*, proboscis); **F,** *Alaria*.[1]

---

[1]See footnote on p. 245.

activity, such as dividing cells. Sterols, such as fucosterol and sargasterol, are also reported in brown algae. Unsaturated hydrocarbons have been found in a number of brown algae, and several have already been shown to function as **gamones,** including ectocarpen (in *Ectocarpus* and *Cutleria*), multifiden and aucanten (in *Cutleria*), and fucoserraten (in *Fucus*).

Nucleus

The cells are generally uninucleate (Fig. 8-2); however, in certain regions of the thallus (as in certain holdfast cells of some kelps), they can be multinucleate. The nucleus, with a well-developed nuclear envelope, contains one or more nucleoli. Interphase chromatin is granular, and chromosomes contain DNA and histones similar to those in higher plants. During mitotic and meiotic divisions there are polar gaps in the nuclear membrane until metaphase, then the nuclear membrane breaks down completely. Divisions complete with centrosomes, centromeres, and spindles have been reported. The spindle is reported to be **intranuclear** or **extranuclear,** whereas the centrosomal apparatus is extranuclear. In cell division, a fusion of small vesicles produces a membrane that extends **centrifugally,** and becomes thickened with a fibrillar layer to form the cell plate. Although chromosomes are usually small and morphologically similar, in some Laminariales an XY chromosomal sex determination has been reported; the larger X chromosome appears in the female gametophyte and the smaller Y chromosome in the male gametophyte.

Flagellated Cells

No brown alga has a motile vegetative stage, although motile cells (gametes and zoospores) are produced (Figs. 8-4 and 8-5). Motile cells often have a reddish eyespot. They are almost always biflagellate and somewhat **reniform** or **pyriform,** and have flagella of unequal length. In the ectocarpalean line, the longer flagellum is of the tinsel type, with two rows of **mastigonemes,** and is directed anteriorly; the shorter flagellum is of the whiplash type and is directed posteriorly. Usually mastigo-

external to the chloroplast and frequently adjacent to the pyrenoid, when present.

Numerous small, highly refractive vesicles, commonly referred to as **fucosan vesicles,** or **physodes,** originate within the chloroplast and are usually aggregated about the nucleus. They contain a tanninlike substance, and are sometimes referred to as **phaeophyte tannin.** These vesicles appear to be most abundant in regions of high metabolic

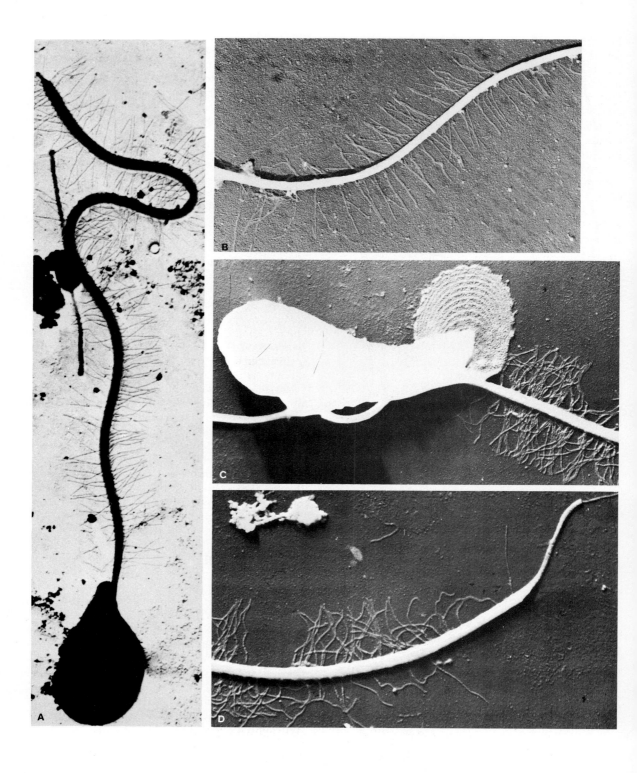

nemes extend almost the full length of the flagellum. However, in some instances, as in *Chordaria* (Fig. 8-4B), the anterior flagellum does not have mastigonemes along its full length; the distal end terminates in a fairly extensive smooth point. In Dictyotales, the posterior flagellum is absent (Figs. 8-4D and 8-5A). In the fucalean line, the longer flagellum is usually directed anteriorly, and the shorter one trails posteriorly (Fig. 8-4E). In addition, the motile cell (sperm) is described as being somewhat bell-shaped and usually has a striking, transparent, striated, beaklike **proboscis** (Figs. 8-4E and 8-5C). The proboscis is less well developed in some fucalean genera than it is in *Fucus,* and it is absent in most. Other modifications occur in other fucalean genera.

Basal bodies of the flagella are joined laterally, and both are in close contact with the nucleus. The anterior flagellum is attached to the nucleus by a fibrous connection (**rhizoplast**). The posterior flagellum is attached to the cell surface immediately above the eyespot, and the flagellar membrane is dilated at this point. In Dictyotales, despite the absence of a posterior flagellum, there are two basal bodies present in the sperm. The posterior part of the one without a flagellum projects as a fibrillar root to the surface of the nucleus. It would appear that the sperm is a reduced type of motile cell derived from a biflagellate prototype through loss of the emergent portion of a posterior whiplash flagellum.

Figure 8-5 Electron micrographs of flagellar structure of Phaeophyta. **A,** *Dictyota* sperm, showing tinsel flagellum, ×7500. **B,** anterior portion of tinsel flagellum of *Chordaria* zoospore, ×20,000. **C, D,** *Fucus* sperm, showing (**C**) body of sperm, proboscis, and insertion of posterior whiplash and anterior tinsel flagella, and (**D**) distal end of tinsel flagellum; **C,** ×26,000, **D,** ×28,000. (**A,** courtesy of I. Manton, B. Clarke, and A. D. Greenwood, with permission of *Journal of Experimental Botany* and Oxford University Press. **B,** photograph courtesy of J. B. Petersen, J. B. Hansen, and B. Caram, with permission of *Botanisk Tidsskrift.* **C, D,** photographs courtesy of I. Manton and B. Clarke, with permission of *Journal of Experimental Botany* and Oxford University Press.)

Classification and Morphological Diversity

The 12 orders of Phaeophyta in the class Phaeophyceae can be placed in two evolutionary series, based on their biochemical and flagellar characteristics. One of these, the fucalean line (subclass Cyclosporidae), comprising the single, oogamous order Fucales, also has a distinctive type of life history (gametic) common to all its genera.[2] The other orders, placed in the ectocarpalean line (subclass Phaeophycidae), can be arranged along several lines of development, in which the main features emphasized relate to life history and alternation of generations, vegetative construction (filamentous or parenchymatous), and reproductive characteristics (isogamy, anisogamy, and oogamy) (see Fig. 27-13).

The Ectocarpalean Line

Orders in the ectocarpalean line (subclass Phaeophycidae) have an alternation of isomorphic or heteromorphic generations, sexual reproduction ranging from isogamy through anisogamy to oogamy, and morphological diversity ranging from simple, filamentous types to complex, parenchymatous forms. Only representative genera of the better-known orders are dealt with in this text.

Since there is no motile or nonmotile unicellular or colonial brown alga, the simplest organization is an unbranched or sparingly branched, **uniseriate** type. In more specialized members, well-developed tissues are present, and there can be considerable differentiation of cells in the thallus and in the method of growth from one phase to another in the life history. No other group of algae has attained the diversity of form, complexity of vegetative construction, or size of Phaeophyta. They range from microscopic, filamentous forms (less than 1 mm) to massive plants 50 m or more in length. Their diversity and size are achieved largely as a result of the great variety of growth types occurring in brown algae.

---

[2]See footnote on p. 239.

Filamentous and Pseudoparenchymatous Organization In the ectocarpalean line, the simplest type of growth, which results from **intercalary cell divisions** occurring at any point in the thallus in one plane, produces an unbranched, uniseriate filament (Fig. 8-6A, C). An occasional division in a second plane, on the same basic theme, results in a sparingly branched, uniseriate filament, as in *Pilayella* (Fig. 8-7G). With more frequent but still intermittent division in the second plane, a profusely branched, loosely filamentous plant is derived, as in *Ectocarpus* (Figs. 8-6B and 8-7C). Erect, unbranched filaments arising from a prostrate basal system of branched filaments are characteristic of some genera, such as *Myrionema* (Fig. 8-7K). This type of growth habit, referred to as **heterotrichy**, is common in brown algae. In some brown algae, there can be an aggregation of only slightly branched, erect filaments, which adhere laterally to form a compact, encrusting layer, as in *Ralfsia* (Fig. 8-7I, J). In other instances, a **pseudoparenchymatous** thallus can be formed (Fig. 8-8) by an aggregation or intertwining of branched filaments that are loosely organized (*Leathesia*; Fig. 8-8C, D) or densely compacted (*Analipus*; Fig. 8-8A, B). In the mature thallus of *Leathesia*, the central portion becomes hollow.

Parenchymatous Growth In some brown algae, the cells of a primary, uniseriate axis can divide in more than one plane (Fig. 8-9) to produce a **corticated** thallus, as in *Sphacelaria* (Figs. 8-9C–E and 8-10). In other instances, diffuse parenchymatous growth occurs as a result of divisions only in surface cells. **Periclinal** divisions in these surface cells increase the thallus thickness, and occasional **anticlinal** divisions permit surface growth to keep pace with the increase in girth (Fig. 8-12D, F). The resulting thallus can be a solid, foliose one, as in *Phaeostrophion* (Fig. 8-11E, F), or it can become hollow, as in *Scytosiphon* (Fig. 8-11C, D). In some instances, one phase in the life history can be an erect, parenchymatous thallus and can alternate with another free-living phase, which may be a prostrate, pseudoparenchymatous disc or a branched filament.

Apical Growth Growth is usually intercalary, although in Sphacelariales (Figs. 8-9C and 8-10C), Dictyotales (Figs. 8-9F–H and 8-12B, C), and Fucales (Fig. 8-14) there is a marked apical development of the thallus. There can be a single apical cell, as in *Sphacelaria* (Figs. 8-9C and 8-10C). Most genera of Fucales, such as *Fucus*, also have a single apical cell (Fig. 8-16K), but in some there can be a group of several apical cells at a branch apex. Some of the Dictyotales also have a single apical cell, as in *Dictyota* (Figs. 8-9F–H and 8-12B, C), but in others there is a margin of apical cells, as in *Padina* (Fig. 8-1), *Zonaria*, and *Syringoderma*[3] (Figs. 8-9K, L and 8-11A, B).

Intercalary Meristems The ultimate in size and complexity in brown algae is achieved through **meristematic activity** in specific, localized regions. Some meristems have potentialities somewhat similar to those in vascular plants. One type of meristem, unique to brown algae, results in **trichothallic growth** (Figs. 8-6D, E and 8-13C–E), which is the result of a special type of cell division at the base of a hairlike, uniseriate filament. This region can be recognized as a subapical band of narrow, actively dividing cells in the filament. Trichothallic growth reaches an advanced stage of development in some pseudoparenchymatous forms and is especially characteristic of Desmarestiales (Fig. 8-13).

In *Desmarestia*, cell divisions are confined to these localized, subapical portions of the thallus apex, and cells are added both **distally** and **proximally** to these apical filaments (Fig. 8-13C, D). Those added distally increase the length of the terminal filament only to a limited extent, as older cells are gradually eroded away. However, those added proximally increase the length of the thallus, and from this axial row of cells arise lateral orders of branching (Fig. 8-13G). These lateral branches also establish similar subapical meristems, thus increasing the girth of the thallus by the same type of development. This combined system of branched, aggregated, and compacted filaments leads to the development of a mature pseudoparenchymatous organization (Fig. 8-13F, G) that permits a variety of morphological types and the attainment of great size. Some straplike species of *Desmarestia* (Fig. 8-

---

[3] Recent studies suggest that *Syringoderma* should be removed from the Dictyotales and possibly referred to a new order; the plant should probably be interpreted as a pseudoparenchymatous precursor.

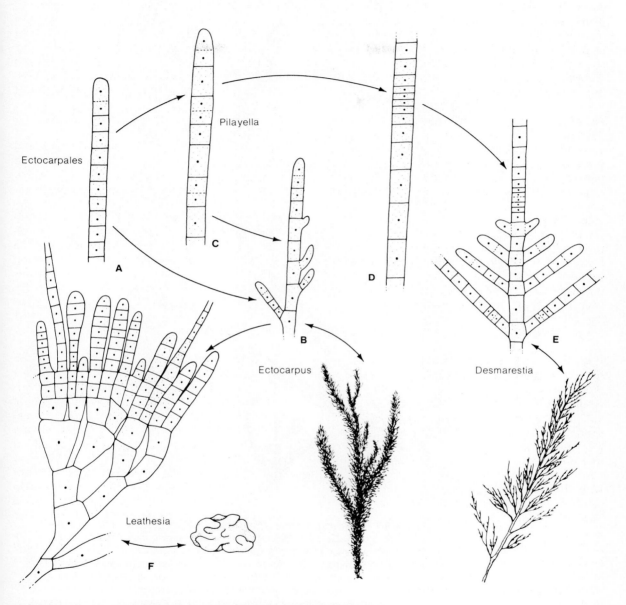

Figure 8-6 Growth types of filamentous and pseudoparenchymatous genera of Phaeophyta with possible evolutionary lines. **A–C,** unbranched and branched uniseriate filamentous forms: **A, B,** with apical growth; **C,** with intercalary cell division in addition to apical growth. **D, E,** trichothallic growth. **E, F,** pseudoparenchymatous forms: **E,** trichothallic growth; **F,** apical growth.

Plants—Phaeophyta (Brown Algae)

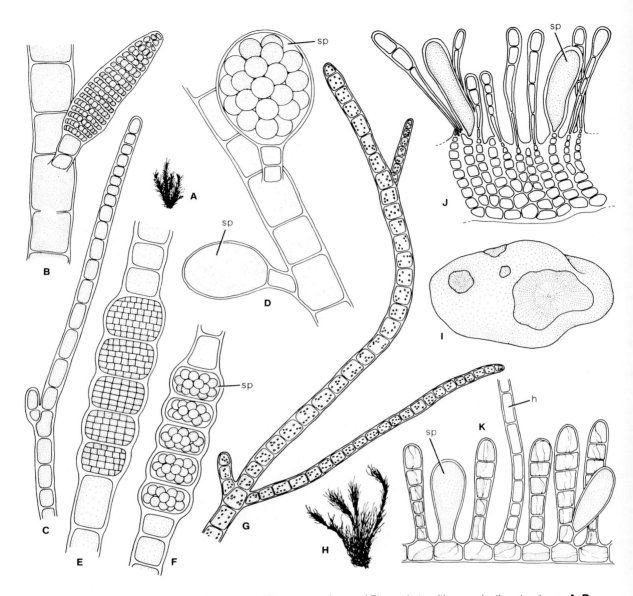

Figure 8-7 Filamentous forms of Phaeophyta with reproductive structures. **A–D,** *Ectocarpus:* **A,** habit, ×0.5; **B,** portion of filament bearing plurilocular structure, ×250; **C,** vegetative portion of thallus, showing initiation of a branch, ×235; **D,** portion of filament, showing immature (*below*) and mature (*above*) unilocular meiosporangia (*sp*), ×250. **E–H,** *Pilayella:* **E,** fertile portion of a filament, showing intercalary plurilocular structure, ×110; **F,** fertile portion of filament showing intercalary unilocular meiosporangia, ×110; **G,** vegetative portion of thallus, ×110; **H,** habit, ×0.5. **I–J,** *Ralfsia:* **I,** habit showing several encrusting thalli growing on a rock, ×0.5. **J,** transverse section through portion of encrusting thallus, showing erect compacted filaments and two unilocular meiosporangia (*sp*), ×225. **K,** *Myrionema,* showing prostrate basal system, hair (*h*), erect vegetative filaments of cells, and two unilocular meiosporangia (*sp*), ×800.

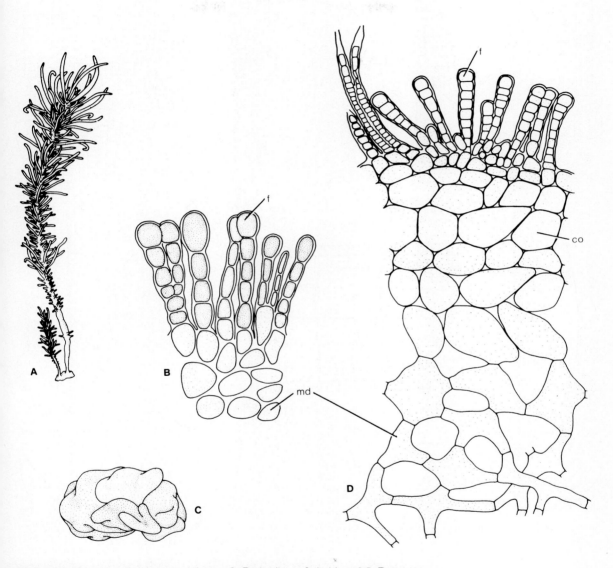

Figure 8-8 Pseudoparenchymatous forms. **A, B,** *Analipus:* **A,** habit, ×0.5; **B,** transverse section through outer portion of thallus, showing pseudoparenchymatous medulla (*md*) and free surface filaments (*f*), ×300. **C, D,** *Leathesia:* **C,** habit, ×0.5; **D,** transverse section through outer portion of thallus, showing loose filamentous medulla, pseudoparenchymatous cortical region, and free surface filaments, ×155.

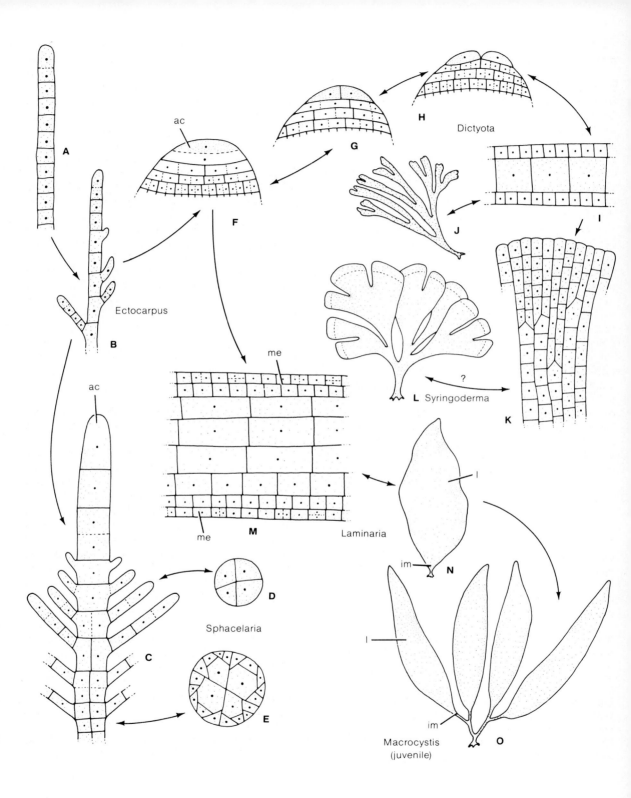

230 Chapter 8

13A) are up to a meter in width and four or more meters in length.

The most highly developed type of meristematic activity in brown algae occurs in the parenchymatous kelps, or Laminariales (*Laminaria*, Figs. 8-9M, N; *Nereocystis*, Figs. 8-14 and 8-19; *Macrocystis*, Fig. 8-9O). A primary meristem called the **transition zone** (Fig. 8-9N–O) is an intercalary meristem, located between the bladelike distal portion and the narrow, stemlike proximal region of the plant. This meristem is the main growth region: it contributes cells distally, increasing the length and development of the blade; proximally, increasing the length and development of the basal portion of the plant; and in the second and third planes, increasing the girth of the plant. In addition, surface meristematic activity (Figs. 8-9M and 8-14B, D) occurs in the epidermal region, which is referred to as the **meristoderm.**

Most large, conspicuous brown algae have these well-developed intercalary meristems, which account for their great size. However, brown algae can attain macroscopic dimensions in two additional ways: (1) forms with **diffuse** intercalary meristematic regions form an extensive, multilayered, or parenchymatous, thallus by undergoing cell divisions in three planes, as in *Phaeostrophion* (Fig. 8-11E, F) and *Scytosiphon* (Fig. 8-11C, D); (2) simple intercalary cell divisions, trichothallic growth, strict apical growth, or a combination of these can result in a basically filamentous thallus. However, an aggregation, lateral fusion, or intertwining of these branched filaments results in a large pseudoparenchymatous thallus, as in *Leathesia* (Figs. 8-6F and 8-8C, D) and *Desmarestia* (Figs. 8-6E and 8-13). This second type is fundamentally simpler than the first, but neither generally results in forms as large as those with intercalary meristems, as in Laminariales.

In *Fucus*, as well as other Fucales, there is also a highly developed type of meristematic activity, and a variety of different types of growth occur during **ontogeny** of the conspicuous diploid phase (Fig. 8-15). The embryo develops a definite polarity at an early stage of development (Fig. 8-15B–E). After the developing embryo elongates to produce a short filament, filamentous rhizoids develop at one end, and a parenchymatous apex is soon established at the other (Fig. 8-15G). Although surface cells continue to divide, following the development of a prominent apical depression (Fig. 8-15K), trichothallic growth becomes established (Fig. 8-15I). However, this method of growth soon ceases. The establishment of the first apical cell (Fig. 8-15K) in the apical depression initiates apical growth, which continues throughout the life of the mature plant.

Cellular and Structural Differentiation Differentiation in more complex brown algae, such as Laminariales (Figs. 8-14 and 8-16) and Fucales (Fig. 8-15), is such that structures somewhat comparable to higher plant organs can be distinguished. Superficially, these parts can resemble leaves, stems, and roots, but they do not function in exactly the same way as these organs do in higher plants, because vascular tissues are lacking. The leaflike part of a brown alga is referred to as a **lamina,** or blade; the stemlike part as a **stipe;** and the attachment part as a **holdfast,** often with rootlike structures, the **haptera.** Despite the absence of vascular tissues in these structures, they are well adapted for some of the functions of leaves, stems, and roots: blades have expanded surfaces that increase their photosynthetic efficiency, stipes are frequently stiffened or almost woody and provide excellent support, and holdfasts are effective anchors.

Definite anatomical regions can be differentiated (Figs. 8-14B–E and 8-16A–E) as meristoderm (epidermis), **cortex,** and innermost **medulla.** In the cortex, pitted elements (Fig. 8-16A) and **mucilage canals** with adjacent **secretory cells** (Figs. 8-14C and 8-16C) can occur, whereas phloemlike ele-

Figure 8-9 Growth types of parenchymatous genera of Phaeophyta, with possible evolutionary lines from filamentous prototypes. **A, B,** filamentous forms: **A,** unbranched filament; **B,** branched filament. **C–J,** apical growth followed by intercalary and parenchymatous cell divisions. **C–E,** with a single apical cell (*ac*) at the tip of axis. **D, E,** transverse views of axis at points indicated in **C** by arrows. **F–J,** also with a single apical cell, but with dichotomous branching as shown in **G, H,** and **J. I,** transverse section of thallus, showing three cell layers. **K, L,** marginal row of apical cells (can be interpreted as a pseudoparenchymatous precursor). **K,** detail of portion of **L. M–O,** growth by meristoderm (*me*) and intercalary meristem (*im*). **M,** detail of transverse section of lamina (*l*) of **N** or **O,** showing intercalary meristem (transition zone) (*im*) and meristoderm (*me*).

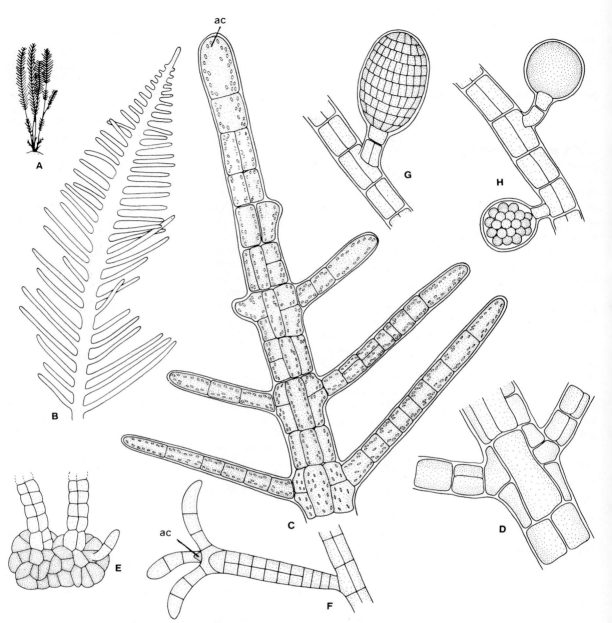

Figure 8-10 Morphology and reproduction of parenchymatous form with apical growth (*Sphacelaria,* Sphacelariales). **A,** habit, ×1. **B,** apical portion of thallus, showing pinnate branching, ×30. **C,** apical portion of thallus, showing apical cell (*ac*), branches, and parenchymatous development of main axis, ×150. **D,** more mature portion of axis, showing cells in surface view, ×160. **E,** basal attachment region of a young thallus, showing prostrate basal system and erect axes, ×120. **F,** portion of axis, showing asexual reproductive unit, the propagule, ×120. **G, H,** branches with reproductive structures: **G,** plurilocular structure, ×180; **H,** immature (*above*) and mature (*below*) unilocular meiosporangia, ×180.

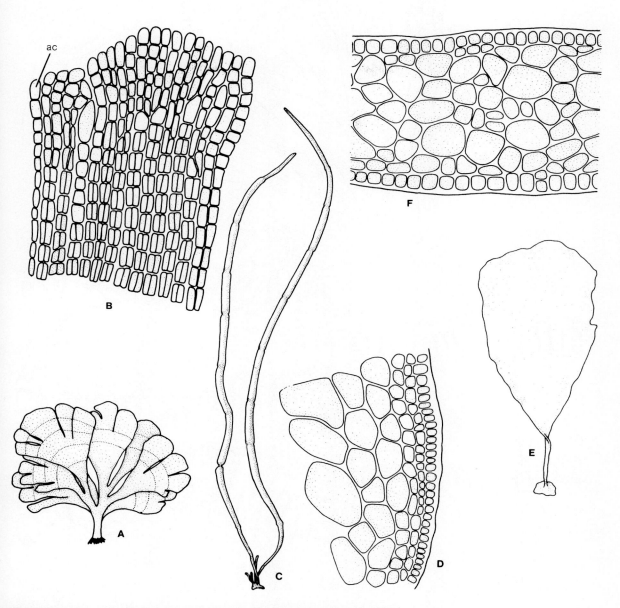

Figure 8-11 Parenchymatous forms of Phaeophyta. **A, B,** *Syringoderma:* **A,** habit of fan-shaped thallus, ×2.5; **B,** outer region of monostromatic thallus, showing marginal row of apical cells (*ac*), ×125. **C, D,** *Scytosiphon:* **C,** habit of hollow thallus, × 0.5; **D,** transverse section of outer portion of thallus, showing parenchymatous organization of cells, ×500. **E, F,** *Phaeostrophion:* **E,** habit, ×1; **F,** transverse section through portion of thallus, showing parenchymatous organization, ×300 (Note: *Syringoderma* can be interpreted as a pseudoparenchymatous precursor.)

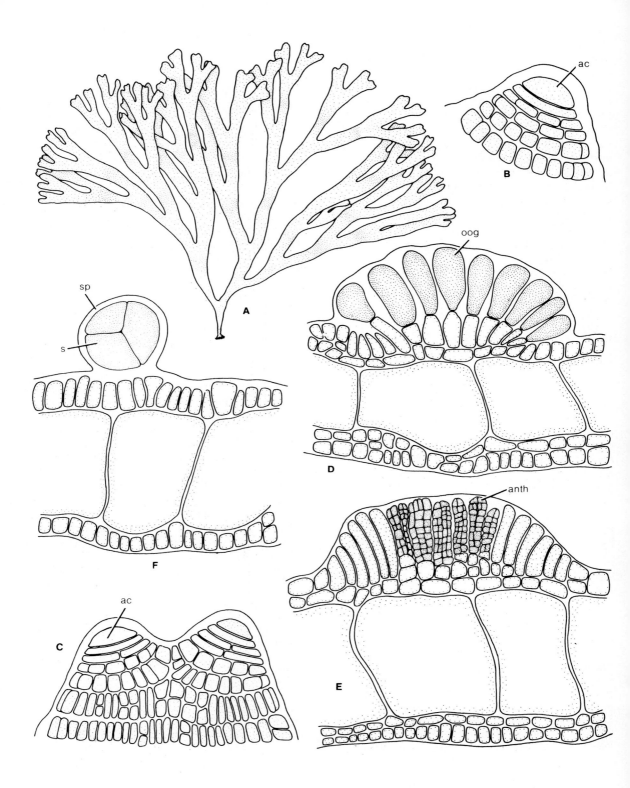

ments (Figs. 8-14E and 8-16D, E) can occur in the medulla. Pitted walls with **plasmodesmata** appear to be widespread in brown algae. In Laminariales they are common between adjacent cells in the outer cortex, between adjacent secretory cells, and between secretory cells and adjacent cortical cells. Morphologically and chemically, the phloemlike elements are very similar to sieve tube elements of vascular plants. Movement of photosynthate formed in the laminae can take place through these conducting elements to the basal regions of the plants. Kelp **callose** (Fig. 8-16D) similar to that present in sieve tube elements of higher plants has been demonstrated.

Some regions of the stipe can become hollow and develop into floats, or **pneumatocysts** (Figs. 8-14A and 8-17P), which hold the plant more or less erect in the water. These pneumatocysts contain a mixture of gases, including nitrogen, oxygen, and carbon dioxide and/or carbon monoxide.

Reproduction and Life Histories

Although many brown algae are complex vegetatively, they all have relatively simple reproductive structures. Sexual reproduction varies greatly, especially in the ectocarpalean line, where isogamy, anisogamy, and oogamy are represented. In this group of orders, there is typically an alternation of two multicellular phases or generations. The fucalean line is entirely oogamous, with a gametic type of life history[4] in which the diploid plant is the conspicuous and only multicellular phase in the life history.

---

[4]See footnote on p. 239.

Figure 8-12 Morphology and reproduction of parenchymatous form with apical growth (*Dictyota*, Dictyotales). **A**, habit, ×0.5. **B, C**, apical region of branch with apical cells (*ac*), showing dichotomous division in **C: B**, ×300; **C**, ×215. **D–F**, transverse sections through portions of different thalli, showing reproductive structures at surface: **D**, mature oogonia (*oog*), ×140; **E**, mature antheridia (*anth*), ×145; **F**, mature meiosporangium (*sp*) with three of four meiospores (*s*) visible × 145.

Vegetative and Asexual Reproduction

Vegetative reproduction can occur by simple thallus fragmentation in simpler filamentous types, or in somewhat specialized multicellular portions such as **propagules** of *Sphacelaria* (Fig. 8-10F). Thallus fragmentation is the only method of reproduction known in some free-floating species of *Sargassum*. **Mitospores** (also termed **neutral spores**) from many-chambered (**plurilocular**) mitosporangia can also occur as an accessory method of reproduction. Meiosis occurs in an unpartitioned (**unilocular**) meiosporangium (Figs. 8-7D, F, 8-10H, and 8-12F). Meiospores (zoospores or aplanospores) are produced by **free-nuclear division**.

Sexual Reproduction

In sexual reproduction, plants with isogamy, anisogamy, and oogamy occur, and apparently these various types have evolved repeatedly along diverging evolutionary lines. Motile gametes (sperms) (Fig. 8-4D–F) are morphologically similar to the zoospores (Fig. 8-4A–C) of brown algae, although often smaller. Isogamous forms produce gametes that are morphologically alike and of the same size. Anisogamous gametes are also morphologically alike, but one gamete (female) is consistently larger than the other (male). In oogamous forms, a nonmotile egg is fertilized by a small, motile sperm. Generally syngamy occurs in the water, although an egg can remain fastened on the haploid female plant. In oogamous species, such as those of Dictyotales, Laminariales, and Fucales, one to several eggs or sperms form in a gametangium. However, in some isogamous and anisogamous species (e.g., of Ectocarpales and Sphacelariales), gametangia are subdivided into many compartments, each of which produces a single, motile gamete. These gametangia are referred to as being **plurilocular** (Figs. 8-7B, E and 8-10G).

Life Histories

Sexual plants of brown algae generally produce either male or female gametangia. In some instances

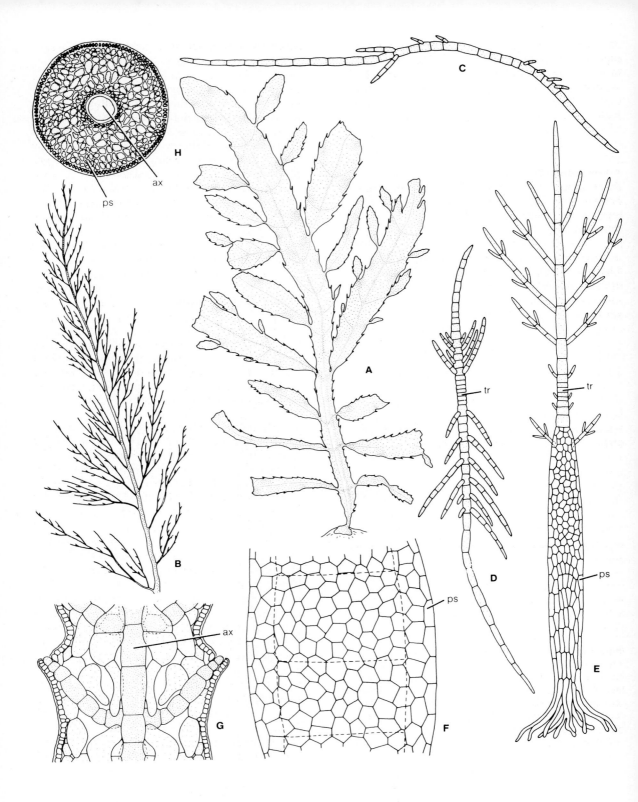

236   Chapter 8

male and female gametophytes are morphologically distinguishable, as in *Nereocystis* (Fig. 8-17). In other brown algae, gametophytes are indistinguishable, as in some species of *Ectocarpus*. Less commonly, other species of brown algae produce both kinds of gametes on the same plant.

Alternation of Isomorphic Generations Except in Fucales, brown algae usually have an alternation of free-living **multicellular,** gametophytic and sporophytic generations. In simpler filamentous brown algae, such as Ectocarpales, there is usually an alternation of isomorphic generations (Fig. 8-18), in which haploid and diploid phases are morphologically identical and physiologically independent in the vegetative condition. However, even in the same genus, as in *Ectocarpus* and *Sphacelaria*, there are some species that have an alternation of isomorphic generations, and others in which there are slightly **heteromorphic** generations.

Ectocarpus. In *Ectocarpus* (Ectocarpales), motile isogametes (Fig. 8-18E, F) are produced in many-chambered (plurilocular) gametangia (Fig. 8-18C, D). It has been shown in one species that a highly volatile sexual attractant, or gamone, called **ectocarpen** (S-1-but-1'-enyl-2,5-cycloheptadiene) is produced by female gametes and is responsible for the clumping of male gametes. The female gamete settles to the substratum shortly after it is released from the gametangium, withdraws its flagella, and at this point apparently releases the highly volatile gamone, which attracts male gametes to it. The male gamete has a longer period of motility (about 24 hours) and, when attracted to the female gamete, attaches to it by the smooth tip of the tinsel flagellum. The flagella are then retracted, and the male gamete fuses with the female gamete to form the diploid zygote (Fig. 8-18H), which immediately divides to produce the diploid sporophytic generation.

The sporophytic generation produces plurilocular mitosporangia (Fig. 8-18T), which in turn produce diploid mitospores capable of maintaining the diploid generation by asexual means. Meiosis occurs in unilocular meiosporangia (Fig. 8-18K–M). The resulting meiospores grow into two types of haploid gametophytes (Fig. 8-18A, B), each producing one type of gamete. In recent studies it has been shown that haploid gametophytes can also produce many-chambered mitosporangia that in turn produce mitospores capable of maintaining the haploid generation by asexual means.

Dictyota. The life history of *Dictyota* (Dictyotales) is very similar to that of *Ectocarpus*. However, in this oogamous genus, only four nonmotile meiospores (Fig. 8-12F) are produced in the meiosporangium. Two develop into male and the other two into female gametophytic plants.

Alternation of Heteromorphic Generations In more specialized orders, such as Laminariales (Fig. 8-19) and Desmarestiales, there is typically an alternation of heteromorphic generations (Fig. 8-19), in which the haploid phase is a free-living, physiologically independent, diminutive (although multicellular) generation, and the diploid phase is the conspicuous, relatively large, generation. Some of these large, more complex, sporophytic generations are annual, as in *Nereocystis*, but they still reach a length of 35 meters or more in less than four months. Others are perennial, as in *Pterygophora*, and can reach a length of about four meters or more in 10–15 years.

Figure 8-13 Morphology and anatomy of pseudoparenchymatous form (*Desmarestia*, Desmarestiales). **A, B,** species diversity; except for symmetry of thallus, anatomical detail is similar in all species. **A,** flattened, strap-shaped species. **B,** terete species (distal portion only shown), ×0.3. **C–E,** juvenile sporophyte plants, showing development of trichothallic meristem. **C, D,** early filamentous phases, ×150; **D,** showing initiation of trichothallic growth (*tr*), ×150. **E,** more mature phase, showing trichothallic growth and pseudoparenchymatous habit (*ps*) of cortical filaments, ×110. **F,** enlarged surface view of basal portion of **E,** showing pseudoparenchymatous cells surrounding central axis (*broken lines*), ×200. **G,** enlarged longitudinal section of more mature portion of the thallus, showing axial row of cells (*ax*), ×250. **H,** transverse section of mature portion of plant, ×150.

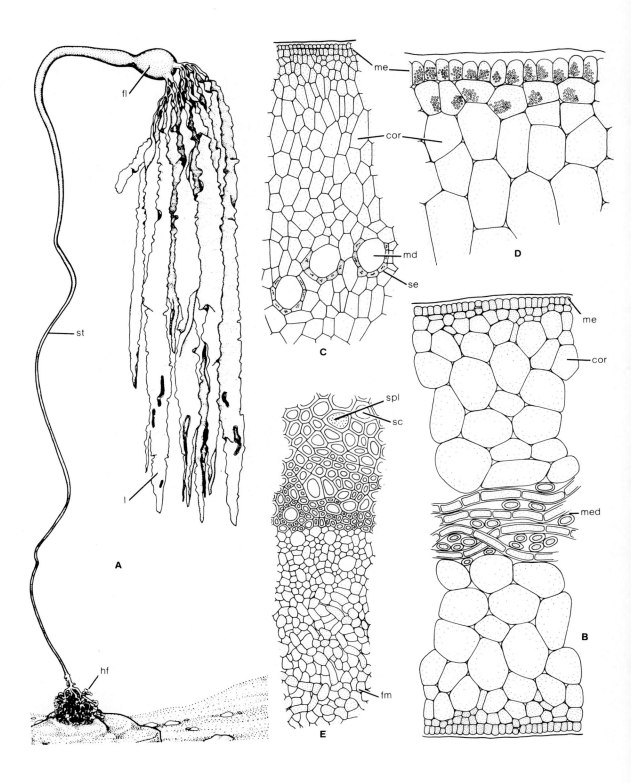

Nereocystis. In the typical kelp (Laminariales), such as *Nereocystis*, the haploid gametophytes are microscopic (Figs. 8-17F, G and 8-19A, B). The filamentous female gametophyte (Figs. 8-17F and 8-19B) is made up of somewhat larger cells than the male, and the filaments are slightly less branched (Figs. 8-17G and 8-19A). The female gametophyte produces one egg (per oogonium), which is extruded from but held at the open end of the oogonium (Figs. 8-17F and 8-19D). The male gametophyte is made up of smaller cells, and the filaments are profusely branched (Figs. 8-17G and 8-19A). It produces many motile sperms (Fig. 8-17H), usually one per antheridium, which swim to the egg (Fig. 8-19E). On fertilization, the resulting zygote (Figs. 8-17I and 8-19F), as in Ectocarpales, undergoes mitosis to form the embryonic sporophyte (Figs. 8-17L–M and 8-19G–I). The sporophyte continues growing into the typical kelp plant (Figs. 8-17O, P and 8-19J). In specialized areas on the sporophyte blade, meiosis occurs in unilocular meiosporangia (Figs. 8-16F and 8-19K), each of which produces 64 meiospores (Fig. 8-16F). Motile meiospores (Figs. 8-17A and 8-19M, N) swim for a brief period, then settle to the substratum, retract their flagella, and secrete a cell wall. The mucilaginous cell wall cements the settled zoospores to the substratum, and they develop into the two types of haploid gametophytes.

Fucus. In *Fucus* (Fucales), the diploid generation is the only multicellular, conspicuous, and free-living phase (Fig. 8-20). The diploid phase is **perennial,** lasting in some species up to four years. The haploid generation has been reduced essentially to single cells that become the gametes themselves. In other words, meiosis occurs at gamete formation in the diploid plant. Thus, gametangia are comparable to the unicellular unilocular sporangia of other brown algae. By comparison with other brown algae, the spores function essentially as gametes, fusing immediately instead of developing into free-living gametophytes.[5]

The embryo undergoes a complex development (Fig. 8-15A–J) before it becomes the typical, dichotomously branched (Fig. 8-20G), diploid thallus. The zygote is spherical when first formed, but soon develops a definite **polarity** (Fig. 8-15E). **Rhizoids** (Fig. 8-15G) always develop at the end of the zygote directed away from the light, and the apex develops toward the light.

At maturity, swollen tips (**receptacles,** Fig. 8-20H) on the thallus produce numerous small cavities, called **conceptacles** (Figs. 8-20I and 8-21A), in which gametangia are formed (Fig. 8-21A). Meiosis occurs (Figs. 8-20J and 8-21B–E) in the one-chambered female gametangium (oogonium), resulting in eight functional eggs (Figs. 8-20K and 8-21F–H). Meiosis in the one-chambered male gametangium (antheridium), results in 64 sperms (Figs. 8-20L, M and 8-21M–O). Eggs (Figs. 8-20B and 8-21I–K) and sperms (Figs. 8-20A and 8-21P) are released through the **ostiole** of the conceptacle into the water, where

---

Figure 8-14 Morphology and anatomy of parenchymatous form (*Nereocystis*, Laminariales). **A,** habit view, showing branched holdfast (*hf*) of rootlike structures, erect stalklike stipe (*s*), enlarged spherical float (*fl*), and dense cluster of terminal, bladelike laminae (*l*), ×0.1. **B–E,** anatomical details of portions of **A. B,** transverse section through sterile portion of lamina, showing meristoderm (*me*), cortex (*cor*), and medulla (*med*), ×350. **C–E,** transverse sections through portions of stipe. **C, D,** outer portion of stipe, showing meristoderm (*me*), compact arrangement of cells in outer cortex (*cor*), mucilage ducts (*md*), and secretory cells (*se*): **C,** ×150; **D,** ×600. **E,** inner portion of stipe, showing compact cells in transition region between inner cortex region and outer medulla region with sieve cells (*sc*), sieve plates (*spl*), and central filamentous medulla (*fm*), ×150.

---

[5] An early interpretation by Strasburger (1906) of the life history of *Fucus* as being sporic, which has been adopted by other authors (Smith 1938), has been reinforced by a recent study (Jensen 1974) of *Bifurcariopsis*, another member of the Fucales. However, until the order has been more thoroughly investigated, we prefer to follow the interpretation of Kylin (1933), Fritsch (1945), and Papenfuss (1951) and consider the typical life history as being gametic. In Strasburger's hypothesis regarding *Fucus*, now elaborated by Jensen in *Bifurcariopsis*, the stage immediately following meiotic division of the diploid nucleus in the "sporangium" (= gametangium of the interpretation followed in this text) is regarded as a much-reduced endosporic "gametophyte" retained within the "unilocular sporangium."

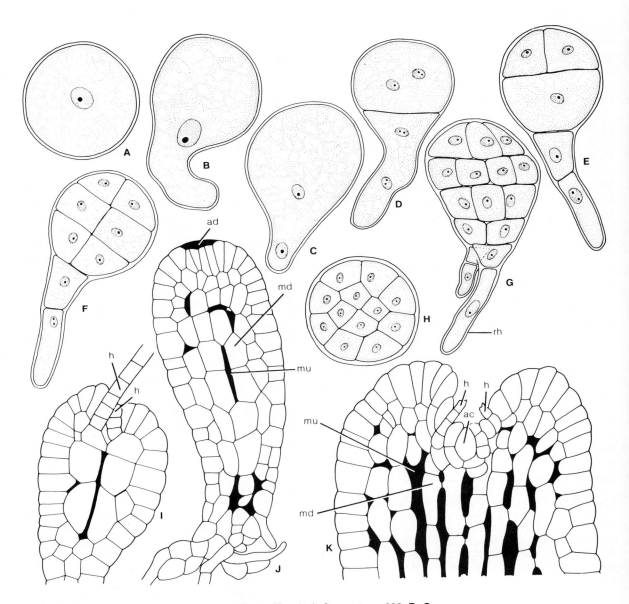

Figure 8-15 Early developmental stages of *Fucus* (Fucales). **A,** zygote, ×600. **B–G,** stages in development of juvenile thallus, showing nuclei, ×600. **H,** transverse sectional view of terminal portion of **G,** ×600. **I,** upper portion of more mature juvenile, showing two terminal trichothallic hairs *(h)*, ×325. **J,** juvenile, showing early development of apical depression *(ad)*, filamentous medulla *(fm)*, and mucilage *(mu)*, ×250. **K,** upper portion of more mature juvenile thallus, showing apical depression, remnants of terminal hairs *(h)*, and initiation of apical cell *(ac)*, ×325. (**I,** after Nienburg; **J, K,** after Oltmanns)

fertilization occurs (Fig. 8-20C, D). Motile sperms are attracted to eggs by the gamone, **fucoserraten**. The zygote sinks to the bottom and develops a firm mucilaginous wall, which adheres to the substratum. It germinates soon after fertilization and grows into the typical plant.

Other Life-History Complexities In some brown algae, isogametes or anisogametes develop directly into haploid plants, thus behaving like mitospores. This suggests that vegetative reproduction is more primitive, that fusion of cells has occurred secondarily, and that sexuality is perhaps relative. For example, in some isogamous species of algae, both gametes are produced by the same plant; in other isogamous forms, gametes fuse only with those from different plants. The next step in the development of sexual differentiation was probably anisogamy. The most advanced type of sexual reproduction is oogamy. One of the advantages of oogamous sexual reproduction is that the large, nonmotile female gamete, or egg, can survive longer before being fertilized without exhausting its stored food reserves. Also, because of the egg's relatively large contribution to it, the zygote too can survive longer before developing.

The detailed examples used in the foregoing discussion are well known in respect to their life histories and reproduction. However, the number of brown algae that are equally well known represents only a small fraction of this large and heterogeneous group of algae. Recent culture studies continue to yield surprises among genera whose life histories are completely unknown or incompletely known. In one instance, a *Ralfsia*like stage has been reported in cultures initiated with *Scytosiphon*; in another, a *Scytosiphon*like stage has been reported in cultures initiated with *Streblonema*. Many genera whose life histories are still incompletely known will probably be found to be linked to other entities presently considered as separate genera.

Lines of Evolution

Repeatedly, in the highly evolved groups of algae, we find a trend from isogamy through anisogamy to oogamy. In the brown algae this trend appears to have occurred along several distinct lines of evolution, especially within orders having an alternation of heteromorphic generations. Three of these evolutionary lines are particularly well distinguished.

The order Ectocarpales contains the simplest morphological forms of brown algae. A number of genera in other orders, especially in Dictyosiphonales, Scytosiphonales, and Chordariales, merge with this basic group, both in respect to their vegetative structure and life histories. This is the chief reason why an arbitrary main division into several classes based primarily on life-history types does not result in a tenable phylogenetic or phenetic arrangement. *Ectocarpus* illustrates well the generalized features of brown algae and serves to illustrate a possible prototype from which all other groups of brown algae (with the possible exception of Fucales) may be derived (see Figure 27-13).

From the *Ectocarpus* type of life history, forms with anisogamy and an alternation of isomorphic generations can readily be derived, as in *Sphacelaria*. Similarly, a more advanced type of sexual reproduction, oogamy, still with an alternation of isomorphic generations, as in *Dictyota,* is in a direct line of evolution from this basic group. Although vegetatively *Nereocystis* has a reduced, *Ectocarpus*-like, filamentous gametophytic stage, the conspicuous generation has become a massive, complex, and highly developed sporophytic stage. *Fucus*, on the other hand, has the ultimate in reduction in the haploid (n) phase, namely the gametes, but also has a complex, highly developed diploid ($2n$) phase.

Relationships of Brown Algae

There are no fossils that can be referred to brown algae with any degree of certainty from periods earlier than the Triassic, although some plants resembling *Fucus* apparently occurred in the early Paleozoic Period. Some **phylogeneticists** have suggested a possible Precambrian origin for brown algae in the Proterozoic or late Archeozoic periods. Except for the **extant** genus *Padina* (Fig. 8-1), which is only lightly calcified, all extant brown algae are noncalcareous. Hence, the fossil record for the Phaeophyta, in contrast to that for Chlorophyta and Rhodophyta, in which certain groups are heavily calcified, is relatively poor.

From Paleozoic rocks in the Silurian and Devonian, certain fossil plants have been reported that bear a similarity to modern members of Fucales, Dictyotales, Chordariales, and Sphacelariales. These observations suggest that Phaeophyta had evolved before late Devonian time.

However, a number of less dubious Cenozoic records from Miocene diatomite deposits in southern California provide some of the best fossil evidence known. Certain of these, such as *Julescraneia* (Fig. 8-22C), have been assigned to Laminariales. Most have been referred to Cystoseiraceae (Fucales), and because of their similarity in appearance to the extant genera *Halidrys*, *Cystoseira*, and *Cystophora*, they bear the names *Paleohalidrys* (Fig. 8-22A), *Cystoseirites*, and *Paleocystophora* (Fig. 8-22B). *Julescraneia*, the largest brown algal fossil known, bears some remarkable similarities to the extant kelps *Pelagophycus* and *Nereocystis* (Fig. 8-14A).

Brown algae are obviously not closely related to any other group of living algae. One can assume that nuclear condition, flagellation, and biochemical characteristics (pigments and nature of food reserves) followed one another in evolutionary sequence. Most likely, brown algae are derived from some laterally biflagellated ancestral stock comparable to the characteristic motile cells of brown algae. Because of similarities in flagellation, a common ancestry has been suggested for brown algae and Chrysophyta. This suggestion is further supported by the presence of chlorophyll *c* and fucoxanthin in Chrysophyta and their storage of food reserves as a laminarinlike polysaccharide. A less likely relationship to Pyrrhophyta has been suggested because of the presence of chlorophyll *c* in that group. If these three divisions had a common origin, it is likely that brown algae had some prechrysophyte ancestor.

The most primitive order of brown algae is thought to be Ectocarpales. From this assemblage, or an *Ectocarpus*like ancestry, all other orders of brown algae, except for Fucales, may have evolved (see Fig. 27-13). Biochemical characteristics, including pigment complex, support a common origin for the Fucales within the same division (Phaeophyta) and class (Phaeophyceae), possibly at some early stage of evolution. However, it would appear that the Fucales comprise an evolutionary series (fucalean line) distinct from the ectocarpalean evolutionary line.

Within the ectocarpalean line, there have evolved two distinct evolutionary series, based on vegetative construction, from a simple uniseriate prototype. One of these, comprising filamentous forms with various degrees of aggregation of filaments, culminates in complex pseudoparenchymatous types. The other, comprising parenchymatous forms with highly organized meristems and differentiated tissues, culminates in the largest and most complex aquatic plants (Laminariales). The fucalean line is entirely oogamous, but within the ectocarpalean line there is an advance from the primitive isogamous condition through anisogamy to oogamy in both filamentous and parenchymatous series.

Distribution and Ecology

Brown algae are almost entirely marine in distribution, including brackish water and salt marshes. Some species of *Sargassum* can occur free-floating far from shore. Floating populations of *Fucus* and *Macrocystis* have also been reported. However, in the unattached condition these algae generally reproduce only by vegetative propagation. A number of taxa, such as *Nereocystis* and *Macrocystis*, have large thalli which grow attached in deeper water (to 35 m or more) in the North Pacific, form dense subtidal forests with distal branches floating in dense mats at the sea surface.

The most conspicuous brown algae belong to Desmarestiales, Laminariales and Fucales, which

---

Figure 8-16 Anatomy and reproduction of parenchymatous form (*Nereocystis*, Laminariales). **A, B,** longitudinal sections of portions of stipe, showing (**A**) arrangement of cortical cells, with pits between some of the adjacent cells, and (**B**) filamentous organization of medulla region: **A,** ×300; **B,** ×450. **C,** transverse section through mucilage duct, showing secretary cells *(se)* surrounding duct *(md)*, ×400. **D,** longitudinal section through sieve tube in region of sieve plate *(spl)*, showing adjacent sieve tube members (note heavy deposit of callose, *cal*, in sieve tube members), ×650. **E,** enlarged transverse section of sieve plate, ×1200. **F,** transverse section through outer portion of fertile region of lamina, showing almost mature unilocular meiosporangia *(mei)* with meiospores *(msp)* and paraphyses *(par)*, ×720.

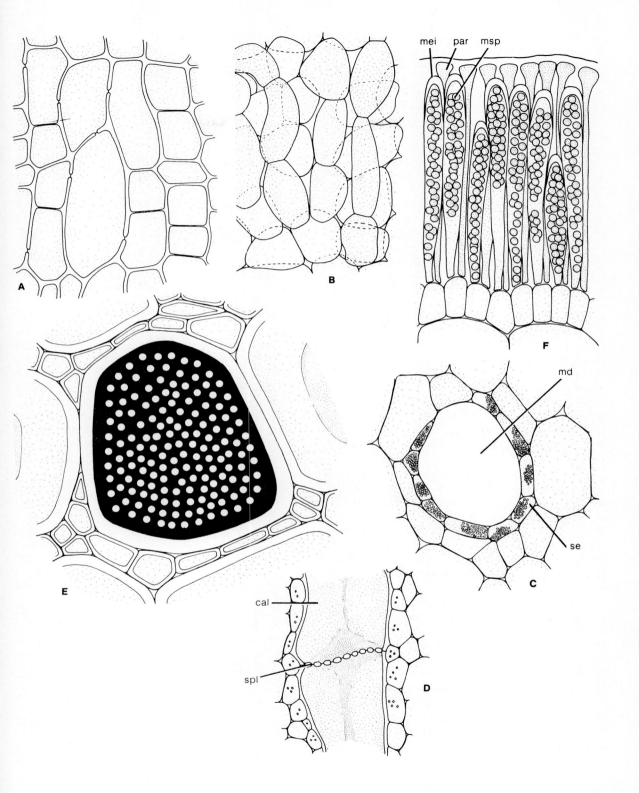

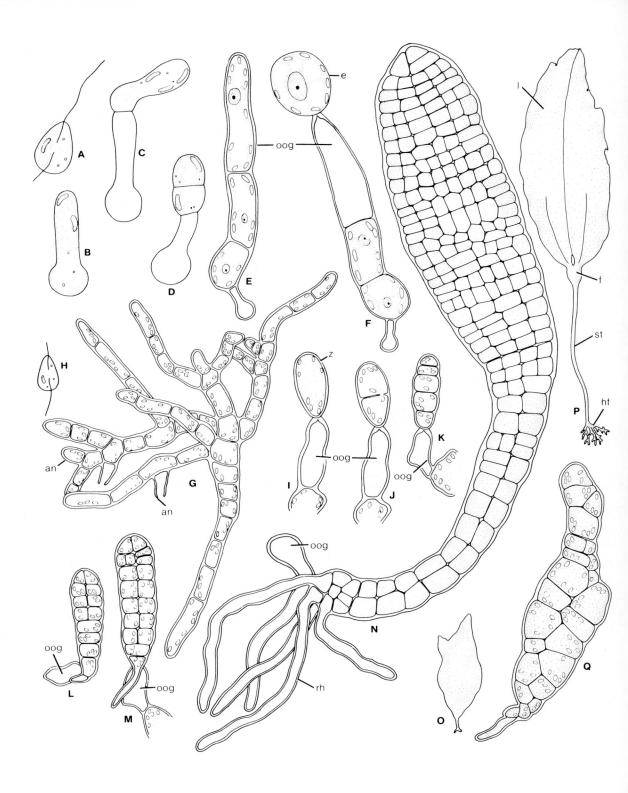

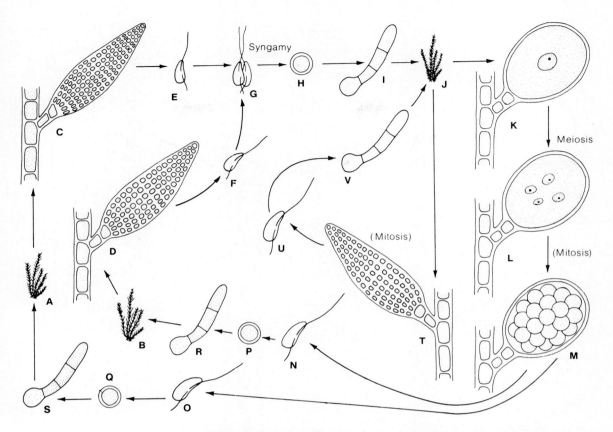

Figure 8-18 Life history of *Ectocarpus* (Ectocarpales). **A, B,** habit of gametophyte plants. **C, D,** plurilocular gametangia. **E, F,** gametes. **G,** syngamy. **H,** zygote. **I,** developing sporophyte. **J,** habit of mature sporophyte. **K–M,** maturation of meiosporangium. **N, O,** meiospores. **P–S,** developing gametophytes. **T,** plurilocular mitosporangium on sporophyte. **U,** mitospore. **V,** developing sporophyte. (For structural details and relative sizes of structures refer to Figs. 8-2A and 8-7B–D.)

Figure 8-17 Stages in life history of *Nereocystis* (Laminariales). **A,** zoospore (meiospore), ×1000. **B,** germinating zoospore, ×1000. **C–E,** stages in development of female gametophyte, ×1000. **F,** mature female gametophyte with empty oogonium (*oog*) and discharged, unfertilized egg (*e*), ×1000. **G,** mature male gametophyte with immature and mature (*empty*) antheridia (*an*), ×1000. **H,** sperm,[6] ×1000. **I–K,** zygote (*z*) and early divisions in young sporophyte attached to empty oogonium, ×1000. **L–N,** stages in development of young sporophyte (**L, M,** ×800; **N,** ×400). **O, P,** later stages (macroscopic) of young sporophyte (**O,** ×0.5; **P,** ×0.3): *f*, float; *hf*, holdfast; *l*, lamina; *rh*, rhizoid; *st*, stipe.

[6]Recent critical ultrastructural studies (Henry, E. C. and Cole, K. M., 1982, *J. Phycology,* Vol. 18, pp. 550–579) indicate that the distal whiplash portion of the anterior tinsel flagellum of the sperm and the zoospore in the Laminariales is much longer than previously believed; and that in the sperm, the posterior flagellum is, in fact, *longer* than the anterior tinsel flagellum. These features should be noted in comparing Figures 8-4F and 8-19C.

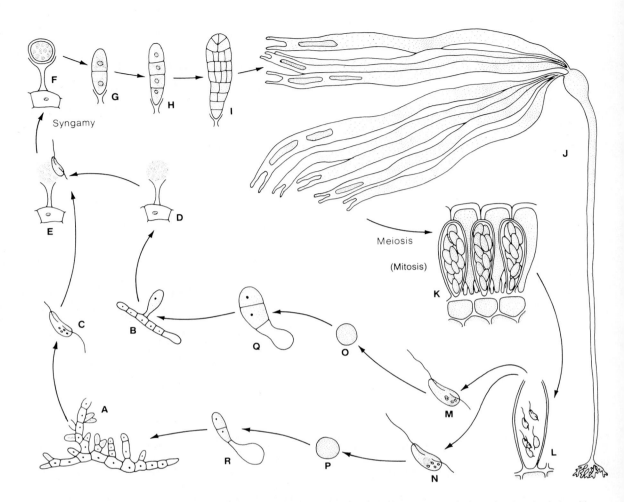

Figure 8-19 Life history of *Nereocystis* (Laminariales). **A**, male gametophyte with antheridia. **B**, female gametophyte with oogonium. **C**, sperm. **D**, unfertilized egg. **E**, fusion of egg and sperm (syngamy). **F**, zygote. **G–I**, development of young sporophyte on female plant. **J**, mature sporophyte with dark, fertile sori on distal portion of lamina. **K**, meiosporangia. **L**, release of meiospores. **M, N**, meiospores. **O–R**, development of gametophytes. (For details and relative sizes of structures refer to Figs. 8-14A and 8-19A–P.)

have representatives in both the Northern and Southern Hemispheres. However, Laminariales are richest in genera and species in the North Pacific. In contrast, Fucales are most abundant in genera and species in the Southern Hemisphere. Certain genera in each of these groups occur in only one or the other region. In Fucales, for example, the large *Durvillaea* does not occur in the Northern Hemisphere; it is found only in colder waters of Australia, New Zealand, Chile, and other subantarctic regions. *Fucus*, on the other hand, occurs only in the Northern Hemisphere, where it is present in both the North Atlantic and North Pacific. *Ascophyllum*, however, is restricted in the Northern

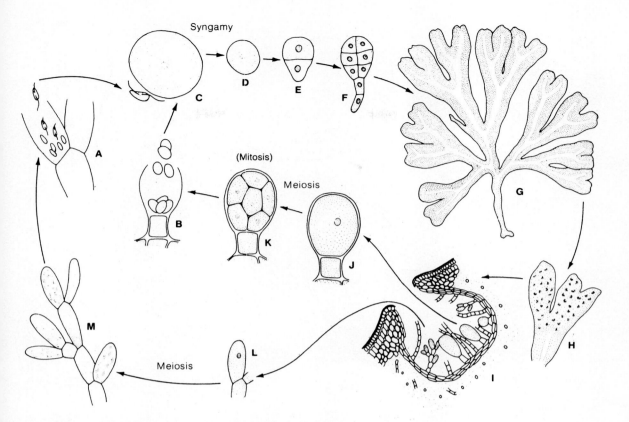

Figure 8-20 Life history of *Fucus* (Fucales). **A,** mature antheridium liberating sperm. **B,** mature oogonium liberating eggs. **C,** mature sperm and egg prior to fusion. **D,** zygote. **E, F,** early development of thallus. **G,** mature thallus. **H,** fertile receptacle with numerous conceptacles. **I,** conceptacle with male and female structures. **J, K,** development within conceptacle of oogonium and eggs. **L, M,** development within the conceptacle of branch with several antheridia. (For structural details and relative sizes of structures refer to Figs. 8-16 and 8-23.)

Hemisphere to the North Atlantic. The ecological niche occupied by *Fucus* in the Northern Hemisphere is occupied in Australia and New Zealand by another member of Fucales, *Hormosira*. In Laminariales, *Macrocystis* (Fig. 8-23) is found in colder waters in both Northern and Southern Hemispheres; it does not occur in the western Pacific, or in the North Atlantic Ocean.

Apart from *Sargassum,* and a few other members of Fucales that are widespread especially in tropical oceans, brown algae tend to be small and inconspicuous in the tropics.

Importance and Uses

The direct importance to humans of the brown algae in coastal regions is related to their potential as **primary producers.** The plankton algae are prob-

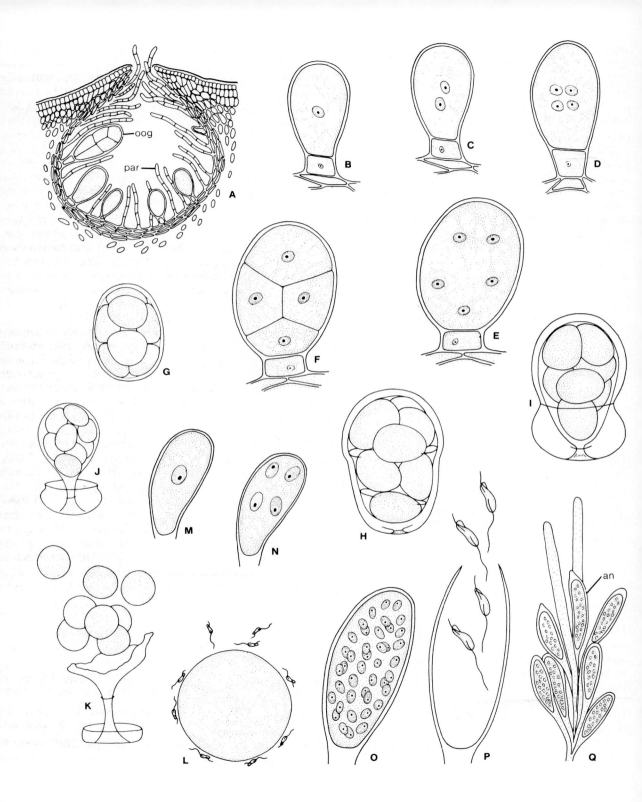

ably the most important primary producers in the sea as a whole; a much greater area of the globe and volume of water are available to support these free-floating or free-swimming microscopic plants than for **benthic** algae. Benthic, or attached, marine algae (the seaweeds), cover a small area—a relatively limited area of **continental shelf,** or water shallow enough to allow the penetration of sufficient light for plant growth. However, in coastal areas, seaweeds such as kelps are important contributors to the economy of the sea, and can be more significant than phytoplankton in areas where dense growth of attached algae occurs. Their large plant bodies provide a substrate for small, sessile animals; among their rootlike branches, the holdfasts can house many diverse small animals. Dense subtidal forests can provide protection and **spawning** areas for larger, free-swimming animals, including fish. Some marine animals, including some fish, browse directly on the tissues of larger seaweeds, and others make indirect use of these large algae by feeding upon **detritus** that originates from their decaying tissues. Specimens of *Nereocystis* (Fig. 8-14A) individually reach a length of over 35 m and can weigh over 125 kg (fresh weight). Such plants provide an abundant source of detritus in coastal regions for filter-feeding animals. Larger algae, especially floating kelps such as *Macrocystis* and *Nereocystis*, also contribute significantly to phytoplankton through the motile stages in their life history (zoospores and gametes). The total primary production in such an area also includes the reproductive cells liberated by other, smaller kelps and other algae growing around the holdfasts of the larger algae.

Figure 8-21 Reproduction in *Fucus* (Fucales). **A,** transverse section of fertile portion of thallus, showing conceptacle containing immature and mature oogonia (*oog*), and paraphyses (*par*), ×125. **B–K,** stages in development and release of eggs in oogonium: **B–F,** ×500; **G,** ×400; **H, I, K,** ×500; **J,** ×350. **L,** sperms swimming about unfertilized egg, ×1500. **M–P,** stages in development of antheridia and sperm release, ×1800. **Q,** branch bearing several antheridia (*an*), ×800. (**H–L,** after Thuret.)

The larger brown algae are also directly important to humans. Species of Laminariales (especially *Laminaria*) have been used as food (called *kombu*) in the Orient since at least 1730. Most kombu is produced in China and Japan; although much of the domestic production in Japan is consumed by the Japanese, a large amount is exported elsewhere.

Algin, or alginic acid, is produced exclusively from seaweeds. Valuable algin industries have developed in Britain and California, and to a limited extent in Japan, Australia, Canada, and other countries. The greatest production of algin is from *Macrocystis* in California. Dense stands are harvested by mowers on floating barges (Fig. 8-23).

Although alginic acid as such is limited in use, its salts—the **alginates**—both in soluble and insoluble form, have a great variety of uses. Alginic acid forms a wide variety of alginates having interesting properties, as well as complex compounds of the plastic type. For example, ammonium alginate is used in fireproofing fabrics, calcium alginate in plastics and as a laundry starch substitute, and sodium alginate as a stabilizer in ice cream and other dairy products.

There are many other uses of alginates and their derivatives in industrial processes. They are used as binders in printer's ink, in soaps and shampoos, in molding material for artificial limbs, and in the manufacture of buttons. They are used in photography for film coatings, in paints and varnishes, in dental impression materials, in leather finishes, and in insecticides. Pharmaceutical preparations such as toothpaste, shaving cream, lipsticks, medicines, and tablets are manufactured from alginates. They are widely used in the food industry as stabilizers, and are also used as a clarifier in the manufacture of beer.

Brown algae have been used at times for fertilizer. Because the numerous elements present in sea water are also found in marine algae, deficiency diseases in land plants can often be remedied with seaweed fertilizer. Brown seaweeds contain more potassium, less phosphorous and about the same proportion of nitrogen as barnyard manure.

Iodine was reclaimed from seaweeds on a large scale in the last century until cheaper sources of supply were discovered. This industry has largely disappeared, although it still exists to a small extent in Japan and northern Europe.

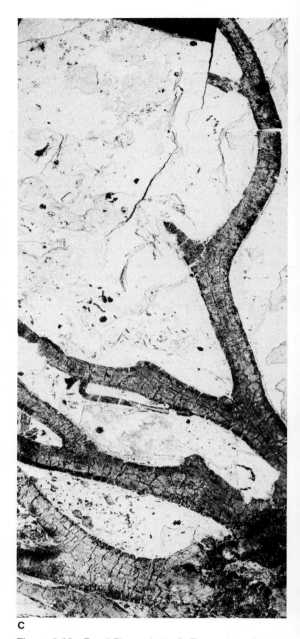

Figure 8-22 Fossil Phaeophyta. **A, B,** species referred to the family Cystoseiraceae (Fucales): **A,** *Paleohalidrys lompocensis*, ×1.0; **B,** *Paleocystophora delicatula*, ×1.3. **C,** *Julescraneia grandicornis*, a species referred to the family Lessoniaceae (Laminariales) and believed to be closely related to *Pelagophycus*, ×0.38. (Photographs courtesy of B. C. Parker, with permission of *Nova Hedwigia*.)

Figure 8-23 Harvesting barge at work in dense bed of *Macrocystis pyrifera* in California. (Photograph courtesy of Kelco Company.)

References

Abbott, I. A., and Hollenberg, G. J. 1976. *Marine algae of California*. Stanford Univ. Press, Stanford, Calif.

Bold, H. C., and Wynne, M. J. 1978. *Introduction to the algae*. Prentice-Hall, Englewood Cliffs, N.J.

Boney, A. D. 1965. Aspects of the biology of the seaweeds of economic importance. *Advances Mar. Biol.* 3:105–253.

Henry, E. C., and Cole, K. M., 1982. Ultrastructure of swarmers in the Laminariales (Phaeophyceae). I. Zoospores. *J. Phycology*, 18:550–569.

Henry, E. C., and Cole, K. M., 1982. Ultrastructure of swarmers in the Laminariales (Phaeophyceae). II. Sperm. *J. Phycology*, 18:570–579.

Jensen, J. A. 1974. Morphological studies in Cystoseiraceae and Sargassaceae (Phaeophyceae) with special reference to apical organization. *Univ. Calif. Publ. Bot.* 68:1–63.

Lobban, C. S., and Wynne, M. J. (eds.) 1981. *The biology of seaweeds*. Bot. Monogr. Vol. 17. Blackwell Scientific Publs., Oxford.

Russell, G. 1973. The Phaeophyta: a synopsis of some recent developments. *Oceanogr. Mar. Biol. Ann. Rev.* 11:45–88.

Scagel, R. F. 1966. The Phaeophyceae in perspective. *Oceanogr. Mar. Biol. Ann. Rev.* 4:123–94.

Scagel, R. F.; Bandoni, R. J.; Maze, J. R.; Rouse, G. E.; Schofield, W. B.; and Stein, J. R. 1981. *Nonvascular plants: an evolutionary survey*. Wadsworth Publishing Co., Belmont, Calif.

Setchell, W. A., and Gardner, N. L. 1925. The marine algae of the Pacific Coast of North America. III. Melanophyceae. *Univ. Calif. Publ. Bot.* 8:383–898.

Taylor, W. R. 1960. *Marine algae of the eastern tropical and subtropical coasts of the Americas*. Univ. of Michigan Press, Ann Arbor.

9
PLANTS—RHODOPHYTA (RED ALGAE)

The red algae (division Rhodophyta) are an extremely large and diverse group of algae with over 800 genera and over 5200 species. The group is predominantly marine, with less than 4% of the species (in about 35 genera) occurring in fresh water. With few exceptions, genera are either strictly marine or occur exclusively in fresh water. In this text, the division is considered in two classes: Bangiophyceae and Florideophyceae (see Table 9-1). Each class has several orders, some of which are still poorly known.

Flagellated cells are absent in red algae. Although the vegetative plant is relatively simple, most red algae have a complex reproductive system. This is especially true of the female reproductive system, postfertilization processes, and **embryogeny.** Characteristics associated with the female reproductive system provide the primary basis for distinguishing orders of Florideophyceae. Red algae as a whole are distinguished by having oogamous sexual reproduction, usually with a regular alternation of **syngamy** and meiosis. A nonmotile female gamete, retained on the haploid gametophyte, is fertilized *in situ* by a nonmotile male gamete carried by water currents. Various types of spores are also produced, some as a result of mitosis, and others as a result of meiosis. Male and female reproductive structures can occur on the same thallus or on separate thalli, with male and female plants vegetatively indistinguishable. In some instances, there are **unisexual** and **bisexual** species in the same genus. Most genera have an alternation (or sequence) of free-living generations; in addition, there is generally a morphologically and cytologically distinct diploid phase or generation remaining attached to the haploid female gametophyte.

In red algae, the **antheridium** (or male gametangium) is referred to as a **spermatangium,** and nonmotile male gametes are called **spermatia.** Each spermatangium usually produces a single spermatium. The **oogonium,** or female gametangium, is called a **carpogonium** and usually has an elongate, emergent, threadlike receptive portion called the **trichogyne.** The uninucleate protoplast of the carpogonium functions as the female gamete. The carpogonium can be borne at the end of a special filament, the carpogonial branch.

The spermatium fuses with the **trichogyne,** and the male nucleus enters it and passes down into the enlarged base of the carpogonium, where it

Table 9-1 Orders of Division Rhodophyta and Representative Genera

Class	Order	Representative Genera
Bangiophyceae	Porphyridiales	*Asterocystis, Goniotrichum, Porphyridium*
	Bangiales	*Bangia, Porphyra (Conchocelis)*
	Rhodochaetales	*Rhodochaete*
	Erythropeltidales	*Compsopogon, Erythrocladia, Erythrotrichia, Smithora*
Florideophyceae	Acrochaetiales	*Audouinella*
	Nemaliales	*Batrachospermum, Bonnemaisonia (Trailliella), Cumagloia, Gelidium, Pseudogloiophloea (Acrochaetium), Nemalion (Acrochaetium)*
	Cryptonemiales*	*Bossiella, Callophyllis, Choreocolax, Corallina, Cryptosiphonia, Dudresnaya, Farlowia (Cruoriopsis), Gloiopeltis, Gloiosiphonia (Cruoriopsis), Harveyella, Lithothamnium, Melobesia, Mesophyllum, Thuretellopsis (Erythrodermis), Weeksia*
	Gigartinales	*Neoagardhiella (Sarcodiotheca), Ahnfeltia (Porphyrodiscus), Chondrus, Eucheuma, Gigartina (Petrocelis), Gracilaria, Gymnogongrus, Iridaea, Halarachnion (Cruoria), Turnerella (Cruoria)*
	Rhodymeniales	*Rhodymenia, Champia, Fauchea, Cenacrum*
	Palmariales	*Halosaccion, Palmaria*
	Ceramiales	*Antithamnion, Ceramium, Nitophyllum, Odonthalia, Polyneura, Polysiphonia, Rhodomela, Spermothamnion*

*On the basis of recent studies, the Corallinaceae (including among other genera *Bossiella, Lithothamnium, Melobesia, Mesophyllum*) are considered to warrant being placed in a separate order.

fuses with the female nucleus, thus completing the process of fertilization. The cytoplasm in the carpogonium does not become a discrete female gamete, nor does the cytoplasm of the fertilized carpogonium form a discrete zygote. Following fertilization, the zygote nucleus undergoes further divisions *in situ,* or it (or a derivative) is transferred to another cell where further nuclear divisions occur.

Division in the zygote nucleus is mitotic and results directly or indirectly in the production of **carposporangia,** which produce $2n$ **carpospores.** Carposporangia can occur in filaments called **gonimoblast filaments.** Gonimoblast filaments, including the carposporangia, make up the **carposporophyte,** which is also referred to as a **gonimoblast.** The carposporophyte is not a free-living phase, but remains attached to the female gametophyte.

In simpler members of Florideophyceae, gonimoblast filaments arise directly from the fertilized carpogonium or from a cell very closely associated with it into which a diploid nucleus has been transferred. In more specialized members, a diploid nucleus is generally transferred by a **connecting filament** to one or more specialized cells, called **auxiliary cells,** which are often remote from the carpogonium. Gonimoblast filaments then emanate from the "diploidized" auxiliary cell.

Carpospores generally give rise directly to a free-living sporophyte generation—the **tetrasporophyte.** The tetrasporophyte is generally morphologically similar, but can be distinctly dissimilar, to the free-living gametophytes in the life history. It is in the free-living diploid phase that meiosis occurs. The meiosporangium of Florideophyceae is referred to as a **tetrasporangium,** and each produces four **tetraspores.** Tetraspores then germinate into free-living gametophytes.

Cell Structure

Cell Wall Composition

The red algal cell wall is typically differentiated into an inner, cellulosic layer and an outer, pectic layer. Cellulosic fibrils are generally randomly scattered throughout a granular matrix. A xylan (containing both β-1,3 and 1,4 linkages) forms a significant portion of the matrix. A number of complex, colloidal, mucilaginous substances (**galactans**) such as **agar** and **carrageenan,** can also occur in the cell wall and intercellular spaces of many red algae. These complex compounds are mixtures of sulfated polysaccharides (mainly galactan sulfate, containing 1,3-linked galactose together with certain proportions of 3,6-anhydrogalactose residues and sulfate). Other mucilages or gelans can also occur. In certain genera, such as coralline algae, there is a heavy calcification of the outer part of the cell wall, which is composed chiefly of calcium carbonate. In some coralline genera the whole thallus is encrusted with lime (encrusting forms, such as *Lithothamnium*), and the plant is rigid. In other genera, such as *Bossiella*, certain regions of the plant remain uncalcified (articulated forms). In many calcified Nemaliales, the deposition is light, and the plants retain a good deal of flexibility.

Nucleus, Chromosomes, and Cell Division

Cells in simpler forms are usually uninucleate. In more advanced forms, they are frequently multinucleate, except for very young or reproductive cells. In some instances, large cells of red algae can contain hundreds, even thousands, of nuclei. The nucleus in red algae has a well-defined nuclear envelope and a nucleolus. Mitosis is essentially similar to that in higher plants, with spindle fibers, distinct chromosomes, and **kinetochores.** However, the nuclear envelope can still be intact at metaphase, except for fenestrations (gaps) at the poles in some. In some instances, a structure referred to as a **polar ring** occurs during division at the poles. However, no true centrioles have been confirmed. Not only the nuclei, but the chromosomes also are very small in red algae. As chromosomes separate, the nucleus elongates and constricts in the middle, resulting in two daughter nuclei. Division of the nucleus is usually followed by the annular ingrowth of a **septum** from the lateral walls, resulting at first in a rimmed aperture through which the cytoplasm is continuous from cell to cell. This is followed by the development of a highly structured plug (Fig. 9-1A), which usually completely blocks the aperture and apparently eliminates all vestiges of cytoplasmic continuity. This unique red algal structure, which is characteristic of all Florideophyceae and is present in certain stages of some Bangiophyceae, is referred to as a **primary pit connection.**

Chloroplasts and Other Cell Organelles

Chloroplasts can be central (axile) or peripheral (parietal) in position. In some simple forms, there may be only one chloroplast per cell. Generally, more complex forms have numerous, small, discoid chloroplasts in each cell, although irregular or bandlike ones also occur. Pyrenoids frequently occur within the chloroplast in simpler forms (Bangiales, Nemaliales) but are absent in more advanced forms (Ceramiales). The pyrenoid has a dense matrix and is generally intimately associated with penetrating thylakoids of the chloroplast (Fig. 9-2). The chloroplast is limited by a double membrane and has a homogeneous stroma traversed by a number of widely spaced thylakoids lying separately (never stacked) (Fig. 9-3A). DNA has been demonstrated in the matrix; small dense granules and larger, lipidlike granules have been reported. Thylakoids usually occur as single, unassociated, unstacked, photosynthetic lamellae, roughly parallel to each other (Fig. 9-3A). Just within the chloroplast envelope a single thylakoid sometimes passes all the way around the chloroplast (Fig. 9-4A), and others seem to arise from it by invagination. In other instances, this outer, encircling thylakoid is absent (Fig. 9-4B). Dictyosomes, endoplasmic reticulum, mitochondria, ribosomes, and starch granules occur in the cytoplasm, and one or more vacuoles can be present.

Pigments

Red algae owe their characteristic color to the presence of accessory, water-soluble, proteinaceous **phycobilin** pigments in the chloroplasts. These bilin pigments (phycoerythrobilin and phycocyanobilin) are present as prosthetic groups of proteins (biliproteins), to which they are firmly bound. The **phycoerythrins** that occur in red algae include *R*-phycoerythrin and *B*-phycoerythrin; the **phycocyanins** include *C*-phycocyanin, *R*-phycocyanin, and allophycocyanin. Accessory carotenoid pigments (carotenes and xanthophylls) are also present in the chloroplast. These include two carotenes, α-carotene and β-carotene, and the xanthophylls, lutein and zeaxanthin. β-carotene is the principal carotene, although in some instances α-carotene predominates. Green pigments include the universal chlorophyll *a* and, in some instances, chlorophyll *d*. The green pigments are generally masked by accessory pigments. However, depending on environmental conditions, red algae exhibit a variety of shades from green through red-brown, bright red, blue, bluegreen, purple-red, and even black. These shades depend on the relative amounts of pigments present. Plants living in subdued light or shaded locations tend to have a higher proportion of phycoerythrin present.

Biliproteins are aggregated in small, regularly arranged granules (**phycobilisomes**), which are attached to the outer surfaces of the thylakoid membrane.

Polysaccharides and Other Storage Products

Food reserves are stored chiefly as a polysaccharide, generally referred to as **floridean starch** (α, 1,4- and 1,6-linked D-glucose). The floridean starch occurs as granules outside the chloroplast (Fig. 9-3A), often associated with the outer surface of the nucleus or, when present, the pyrenoid. This starch is identical to the branched, or amylopectin, fraction of starch that occurs in green algae and other plants; however, a slightly different staining reaction—a red-violet color—is obtained when testing for the presence of floridean starch with iodine-potassium-iodide solution. Certain sugar alcohols, such as mannitol, are abundant in some forms. Phytosterols stored include cholesterol, β-sitosterol, and fucosterol.

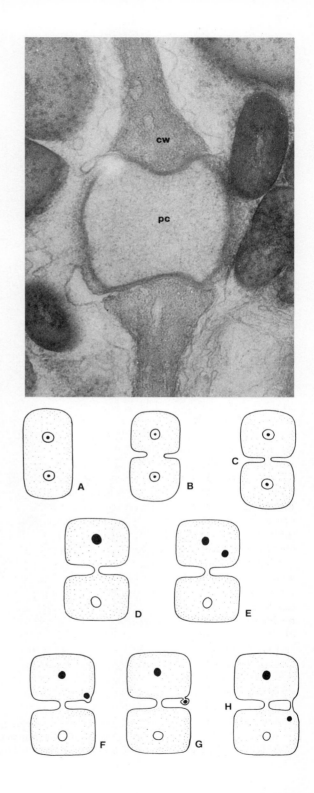

Figure 9-1 Pit connections in red algae. **A,** section through plugged pit connections (*pc*) in cell wall (*cw*) between cells of conchocelis stage of *Porphyra* (Bangiophyceae), ×65,000. **B,** diagrammatic representation of formation of pit connections in Rhodophyta: *a–c*, formation of primary pit connection; *d–h*, formation of secondary pit connection. (**A,** courtesy of K. Cole, with permission of *Phycologia*.)

Plants—Rhodophyta (Red Algae)   255

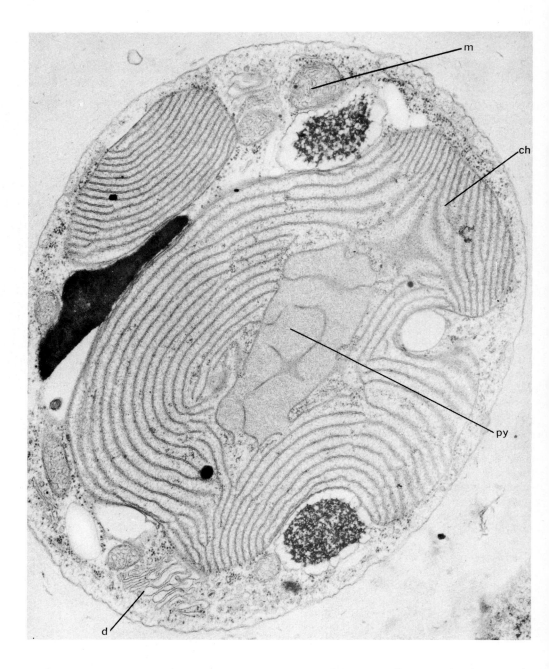

Figure 9-2 Ultrastructure of whole cell of *Porphyridium* (Bangiophyceae), ×25,000. Note close association of pyrenoid and penetrating thylakoids: *ch*, chloroplast; *d*, dictyosome; *m*, mitochondrion; *py*, pyrenoid. (Photograph courtesy of A. D. Greenwood.)

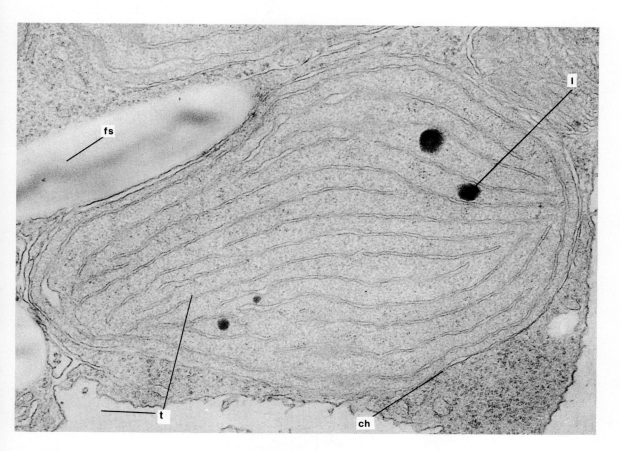

Figure 9-3 Ultrastructure of chloroplast of cortical cell in *Ceramium* (Florideophyceae), ×53,000. Note thylakoids *(t)*, consisting of single membrane pair, are not elaborately layered as is typical of other algal chloroplasts (see Figs. **6-2D, 7-1C, D,** and **8-3**); also compare this chloroplast ultrastructure with photosynthetic lamellae in Cyanophyta (see Fig. **2-12D**). Large translucent areas are floridean starch granules *(fs)* outside the chloroplast *(ch)*: *l*, lipid globule. (Courtesy of A. D. Greenwood.)

Mode of Nutrition

Most red algae are autotrophic. A number occur as **epiphytes** or **endophytes**. Some are very pale, or completely white, and it is presumed that they are in part, if not wholly, dependent on the host for their nutrition. In *Harveyella*, a diminutive, almost colorless red algal parasite that occurs on *Odonthalia* and *Rhodomela*, a physiological dependence of the parasite on host tissues has been demonstrated.

The main distinctions that separate Bangiophyceae from Florideophyceae relate to general morphology, reproduction, and life histories.

Figure 9-4 Red algal chloroplast ultrastructure. **A,** portion of chloroplast of vegetative cell of conchocelis stage, with characteristic girdle thylakoid (*gt*) inside chloroplast envelope (*ce*) and encircling inner parallel thylakoids (*pt*), ×80,000. **B,** portion of chloroplast of vegetative cell of foliose *Porphyra* thallus with parallel thylakoids (*pt*) terminating at chloroplast envelope (*ce*); note absence of encircling thylakoid, ×80,000. (**A,** courtesy of K. Cole, with permission of *Phycologia*.)

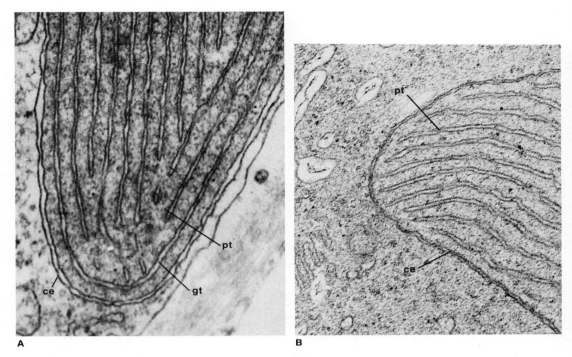

Figure 9-5 Morphological diversity in Bangiophyceae. **A, B,** unicellular forms: **A,** *Porphyridium,* ×2500; **B,** *Asterocystis,* after division, before separating into two new cells, ×1200. **C, D,** unbranched and branched filamentous forms: **C,** *Erythrotrichia* with rhizoidal holdfast, ×825; **D,** *Goniotrichum* with undifferentiated holdfast cell, ×800. **E–I,** parenchymatous types. **E–G,** *Bangia:* **E,** basal region of thallus with rhizoidal cells, ×500; **F,** distal vegetative portion of parenchymatous region of thallus, ×500; **G,** fertile portion of thallus, ×500. **H, I,** *Erythrocladia:* **H,** unistratose, prostrate, discoid species, viewed from above, ×1000; **I,** multistratose species, in transverse section, ×1000.

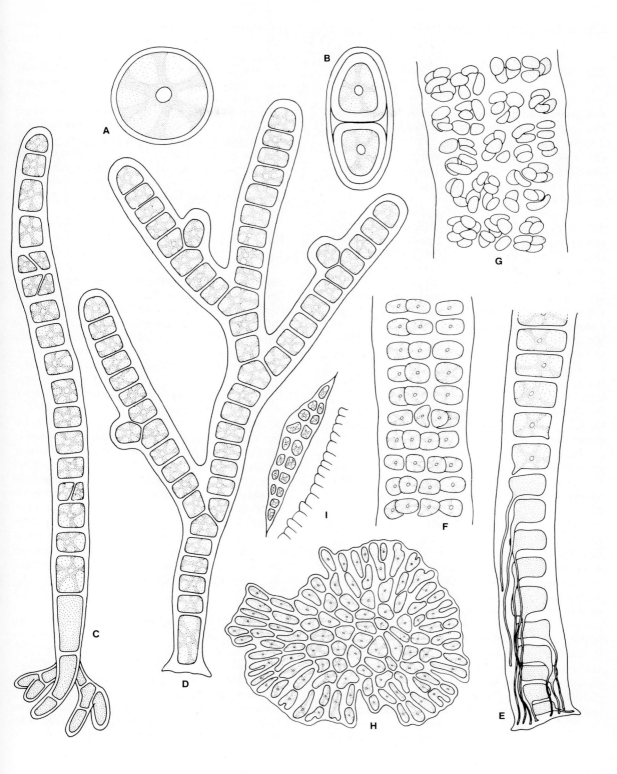

Plants—Rhodophyta (Red Algae)

Class Bangiophyceae

Classification and Morphological Diversity

Bangiophyceae is a small class, with about 125 species in 29 genera. The group is chiefly marine in distribution, although almost 25% of the species (in 14 genera) occur in fresh water. They are the simplest of all red algae, and they exhibit a wide range in morphology. Most occur on rocks or as epiphytes.

The simplest type of thallus is unicellular (Fig. 9-5A). A number of genera are filamentous. These can be erect, uniseriate or multiseriate, branched (Fig. 9-5D) or unbranched (Fig. 9-5C), or prostrate and **monostromatic** (Fig. 9-5H) or **polystromatic** (Fig. 9-5I). In others, there can be a few regular divisions in the erect simple branches, indicative of parenchyma (Fig. 9-6F). The most advanced parenchymatous growth attained in the group is in the diffuse **foliose** type; here cells divide in two planes to form a diffuse monostromatic (Fig. 9-6A, C) plant body or in three planes to form a distromatic (Fig. 9-6D, E) thallus. The lines along which these various types may have evolved can be suggested: it is possible to derive the upright parenchymatous thallus from a unicellular (Fig. 9-5A) form through either the upright filamentous form (Fig. 9-5C–D) or the prostrate discoid type (Fig. 9-5H). Apical growth is known only in the uniseriate branched *Rhodochaete*, which is also unusual in apparently having unplugged primary pit connections between its cells.

The erect thallus can be attached by a differentiated holdfast cell (Fig. 9-5D), by a few basal, **rhizoidal**, multicellular filaments (Fig. 9-5C), or by a number of rhizoidal filaments emanating from cells at the thallus base (Fig. 9-5E). These filaments anchor the thallus to the substrate. There can be a very extensive production of nonseptate, rhizoidal processes (Fig. 9-6F) aggregated to form a massive, cushionlike holdfast structure.

Although many genera occurring as epiphytes are not selective as to a substrate alga, in some instances they are highly specific, being confined almost entirely to a single algal species (as *Porphyra nereocystis* on *Nereocystis*) or marine seagrasses (as *Smithora* on *Zostera* and *Phyllospadix*).

The class Bangiophyceae is generally divided into three to five orders (see Table 9-1). Representatives of only a few of these are considered in detail in this text. Disposition of the taxa in these orders is based primarily on chloroplast features and vegetative organization of the thallus. The number and shape of chloroplasts, and the presence or absence of pyrenoids are significant. Distinctive features of vegetative organization include whether the form is unicellular or multicellular, has a branched or unbranched uniseriate or multiseriate filament, has a prostrate disc or cushion of cells, is **saccate**, or has an erect parenchymatous (diffuse) monostromatic or distromatic, foliose thallus.

Life Histories and Reproduction

Vegetative and Asexual Reproduction Vegetative reproduction by **fragmentation** is common. Various types of spores are also formed. A single spore, produced by simple metamorphosis of a vegetative cell, is called a **monospore,** and the cell a **monosporangium.** In other instances, metamorphosis and division of the protoplast of a vegetative cell result in several spores, called **aplanospores,** and the cell is an **aplanosporangium.** The occurrence of sexual reproduction has been reported in certain genera. In some instances, it would appear that some cells previously described as gametes in Bangiophyceae behave more like spores. Perhaps sexual reproduction has been lost in some members of the group, resulting in **parthenogenetic development** of gametes. However, despite the somewhat incomplete nuclear and chromosomal evidence of a sexual cycle in Bangiophyceae, there is a morphological alternation or sequence of stages in the life history of certain genera (*Porphyra* and *Bangia*), with very different morphological phases and several types of spores.

Types of Life History in Porphyra In *Porphyra*, three different types of life histories (Figs. 9-7–9-9) have been established. These seem to vary with the species and, to some extent, with environmental conditions. In the simplest type (Fig. 9-7), the foliose stage produces monospores (Fig. 9-7B, C). These germinate and develop directly into a similar foliose thallus. In another simple type (Fig. 9-8), the foliose thallus produces aplanospores (8 or 16

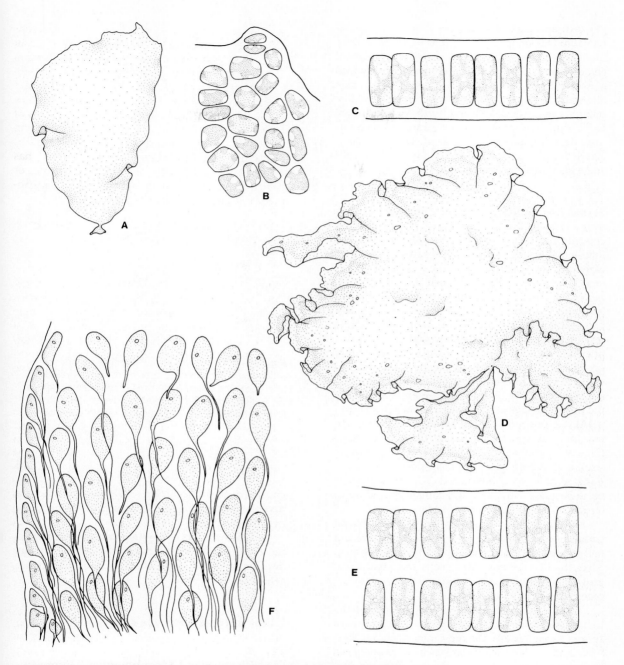

Figure 9-6 Morphological diversity in parenchymatous Bangiophyceae (*Porphyra*). **A–C,** monostromatic species: **A,** habit, ×1.5; **B,** surface view of cells, ×500; **C,** transverse section through portion of thallus, ×500. **D, E,** distromatic species: **D,** habit ×0.5; **E,** transverse section through portion of thallus, ×500. **F,** portion of basal attachment region of thallus as seen in longitudinal section, showing rhizoidal cells, ×500.

per cell) and each aplanospore (Fig. 9-8E) germinates and develops directly into a similar foliose thallus. A more complex type of life history is most common and occurs in still other instances, in which a filamentous, shellinhabiting, conchocelis stage alternates in the life history (Fig. 9-9). This filamentous stage (Figs. 9-9F and 9-10A–E) was known for a long time as a separate genus, *Conchocelis*, until its affinities were recognized. The same characteristics of reproduction involving a conchocelis stage also occur in *Bangia*.

Sexual Reproduction and the Conchocelis Stage
In the type of *Porphyra* life history in which a conchocelis stage (Figs. 9-9F and 9-10A–E) occurs, the foliose thallus (Fig. 9-9A) produces two types (sizes) of reproductive cells that are borne on separate plants or on the same plant. The larger of these (4 to 64 per cell) are highly pigmented carpospores (Fig. 9-9C), and they develop following fertilization of a carpogonial cell (carpogonium), which is a slightly modified vegetative cell usually with a very indistinct trichogyne. Mature carpospores frequently occur as highly pigmented patches. The smaller ones (16 to 128 per cell), which are almost colorless, are spermatia (Fig. 9-9B) and frequently appear on the mature thallus as marginal whitish patches (Fig. 9-9A). Carpospores germinate to produce the filamentous, uniseriate, branched, conchocelis stage (Figs. 9-9F and 9-10A–E). Foliose plants can be bisexual, with both types of patches occurring intermingled or in clearly separated areas; they can also be unisexual, with male and female reproductive structures borne on separate individuals.

Although the foliose *Porphyra* thallus lacks pit connections, primary pit connections are present between cells in the conchocelis stage of this alga. Also, the filamentous conchocelis stage has apical growth, a feature that is consistent in Florideophyceae but rare in Bangiophyceae.

Conchocelis filaments (Figs. 9-9F and 9-10A–E) become branched and can produce two types of spores. One of these, the typical monospore (Figs. 9-9G and 9-10D), which arises in a swollen cell (monosporangium) on a conchocelis filament, germinates and develops into a similar conchocelis filament (Fig. 9-9F). The other is produced on a specialized branch (conchosporangial branch), each cell of which (**conchosporangium**) gives rise to 1-4 spores, known as **conchospores** (Fig. 9-9H–J). The

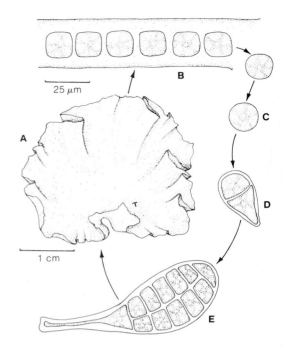

Figure 9-7 Direct type of life history in a monospore-producing monostromatic species of *Porphyra* (Bangiophyceae). The foliose thallus (**A**) produces monospores (**B, C**) that germinate and develop directly (**D**) into a juvenile thallus (**E**) that produces a new foliose plant.

chospore germinates (Fig. 9-9J) and produces a juvenile thallus (Fig. 9-9K), which develops into a new foliose plant.

In all instances in Bangiophyceae where a sexual cycle has been indicated, the foliose thallus appears to be the haploid stage, and the conchocelis stage is diploid.

Class Florideophyceae

Classification and Morphological Diversity

Florideophyceae is a large class with nearly 5100 species in about 790 genera. The group is predominantly marine, with less than 3% of the species

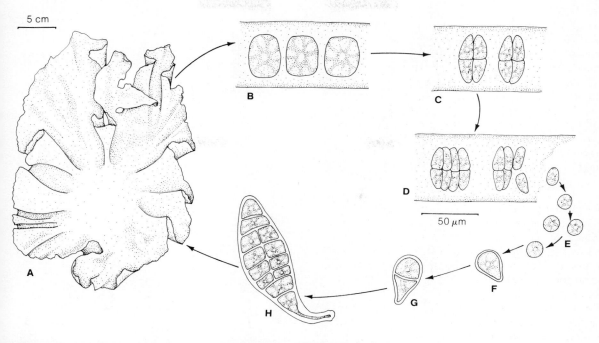

Figure 9-8 Direct type of life history in an aplanospore-producing monostromatic species of *Porphyra* (Bangiophyceae). The foliose thallus (**A**) produces aplanospores (**B–E**) that germinate and develop directly (**F–G**) into a juvenile thallus (**H**), which produces a new foliose plant.

(about 145 species in 22 genera) occurring in freshwater habitats. They grow on rocks or other usually solid substrata. Florideophyceae are also vegetatively simple, although somewhat more complex than Bangiophyceae.

Florideophyceae consistently have primary pit connections (Fig. 9-1B) between adjacent cells in contrast to Bangiophyceae, in which primary pit connections are uncommon. In more complex thalli, secondary pit connections (Fig. 9-1B) can also occur between adjacent cells in the same filament and between cells in adjacent filaments; in parasitic taxa (e.g., *Harveyella*), they can occur between the cells of parasite and host. The formation of secondary pit connections is more complex (Fig. 9-1B). When a nucleus in a cell (uninucleate or multinucleate) divides, one daughter nucleus can migrate to the periphery of the cell. A protuberance in the cell wall is formed, the daughter nucleus migrates into it, and, in some instances, is separated as a very small cell. As this protuberance or small cell elongates, it comes in contact with an adjacent cell, and the daughter nucleus passes into it. The resulting connection, known as a *secondary pit connection,* is a permanent one. As in formation of a primary pit connection, there may initially be protoplasmic continuity between adjacent cells, but this aperture soon disappears. The secondary type of pit connection is a unique feature in Florideophyceae. At certain stages (especially in early **postfertilization** development), these plugs apparently break down and thereby permit subsequent transfer of materials, including nuclei, from one cell to the next. Although primary and secondary pit connections differ in their methods of formation, they are structurally similar when mature. In parasitic red algae, such as *Harveyella*, secondary pit connections between host and parasite are reported to be initiated by the host.

There are no unicellular members in the Flor-

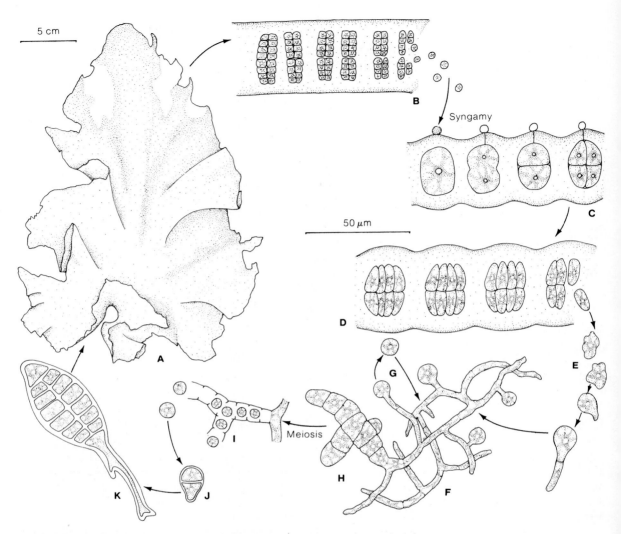

Figure 9-9 Conchocelis-producing type of life history of monostromatic species of *Porphyra* (Bangiophyceae). **A,** foliose macroscopic thallus with fertile margin. **B–D,** transverse sections of foliose plant: **B,** showing spermatium formation and release; **C,** showing carpogonia, fertilization, and early stages in carpospore formation; **D,** showing carpospore formation and release. **E,** germination of carpospores. **F,** conchocelis stage. **G,** formation and release of monospore from monosporangium. **H,** formation of conchosporangial branch. **I,** release of conchospores. **J,** germination of conchospore into juvenile thallus (**K**) that develops into a new foliose plant.

Figure 9-10 Vegetative and reproductive stages of conchocelis development in Bangiophyceae. **A–C,** vegetative stages in *Porphyra*. **A, B,** conchocelis filaments of *P. nereocystis* in shell: **A,** ×800; **B,** ×175. **C,** conchocelis filaments of *P. nereocystis* in culture, ×400. **D,** monosporangium on conchocelis stage of *Bangia*, ×2000. **E,** conchosporangial branch on conchocelis stage of *Porphyra*, ×1200. (**A, E,** courtesy of T. F. Mumford. **B, C,** courtesy of M. W. Hawkes. **D,** courtesy of K. Cole.)

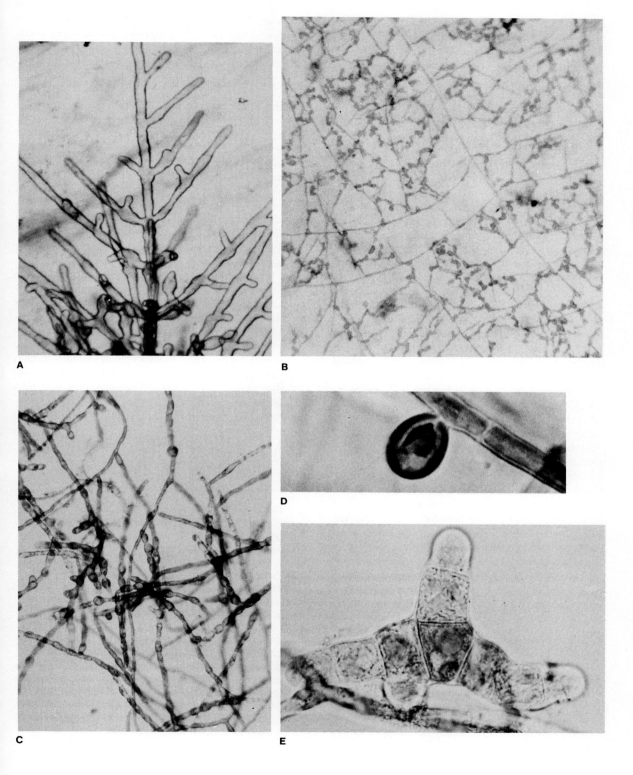

Plants—Rhodophyta (Red Algae) 265

ideophyceae. The simplest are uniseriate, branched forms (Fig. 9-11). The more complex forms are basically filamentous, but branches are laterally fused to varying degrees, so that a **pseudoparenchymatous** thallus results (Figs. 9-12–9-14). There is little tissue differentiation, although the aggregation of filaments often results in a compact, small-celled **cortex** and a **medulla** that can be loose and filamentous (Fig. 9-13B, C) or compacted large cells (Fig. 9-14B). The growth of each filamentous branch is by means of a simple, apical cell. In rare instances, intercalary cell divisions occur, resulting in a type of marginal growth.

Pseudoparenchymatous and loosely branched forms can have either **uniaxial** or **multiaxial** types of growth. In the former, the main axis consists of a single row of large cells (Fig. 9-11E). In multiaxial types, the main axis is composed of a number of parallel or almost parallel filaments (Figs. 9-12B and 9-13B, C). In each instance there are principal, main axes of **unlimited growth** accompanied by secondary axes of **limited growth,** with the latter often surrounding the former. In some cases, the juvenile thallus has a uniaxial type of growth at first and becomes multiaxial as it matures. The lines along which the uniaxial and multiaxial types may have evolved are diagrammed in Figures 9-15 and 9-16.

Considering the relatively simple vegetative growth that occurs in Florideophyceae, the great variety in gross morphology is remarkable. They vary from filamentous microscopic plants to large, profusely branched, or foliose plants. They are usually less than a meter in length, but some (e.g., *Gracilaria*, Fig. 9-20) are as long as four meters.

The class Florideophyceae is generally divided into five to eight orders (see Table 9-1). The disposition of genera in these orders is based primarily on the female reproductive system, especially the postfertilization stages and embryogeny. Some orders appear to be especially clear-cut and relatively uniform in their reproductive features, whereas others are much less uniform.

Life Histories and Reproduction

Vegetative and Asexual Reproduction With some more complex thalli, vegetative reproduction is common from a perennial base. Erect branches die back to the base, and new branches arise later from a basal, undifferentiated region. A variety of spores is produced, especially in some simpler forms. As in the Bangiophyceae, monospores can be formed, or highly differentiated sporangia can divide to produce two or more nonmotile spores.

Sexual Reproduction On a morphological basis, sexual reproduction appears to be well established in Florideophyceae; there is usually an alternation or sequence of free-living generations. When a free-living tetrasporophyte occurs, it is often vegetatively indistinguishable from the haploid gametophytes, although many examples are now known where gametophyte and free-living tetrasporophyte phases are distinctly heteromorphic. Each spermatangium gives rise to a single spermatium. Similarly, each carposporangium gives rise to only one carpospore in Florideophyceae.

The female sexual apparatus and fertilization processes are usually extremely complex and variable in Florideophyceae. It is impossible in this text to summarize in this respect the characteristic variation that occurs in all orders of this class. The cytological evidence for alternating phases in red algae is still very limited. **Genotypic** sex determination has been demonstrated in a few Florideophyceae; that is, from a single tetrasporangium that produces four tetraspores by meiosis, two spores produce male gametophytes and two produce female

Figure 9-11 Morphology and reproductiion in uniaxial Florideophyceae. **A, B,** *Antithamnion:* **A,** habit epiphytic on kelp stipe, ×0.5; **B,** terminal portion of thallus, showing uniaxial, loose, filamentous construction, and immature tetrasporangia (*tspn*), ×240. **C–K,** *Batrachospermum.* **C,** habit, ×1. **D,** apical region of thallus, showing apical cell (*ac*) and filamentous branches, ×475. **E,** portion of main axis, showing uniaxial construction and whorls of filamentous branches, ×165. **F,** more mature portion, showing development of corticating filaments that cover the axial row of cells, ×165. **G,** unfertilized carpogonium (*cp*), ×915. **H,** postfertilization stage, showing initiation of carposporophyte (gonimoblast) filaments (*gb*), ×915. **I,** branch bearing several mature spermatangia (*sp*), ×960. **J,** later stage of carposporophyte development, showing chains of carposporangia (*cspn*), ×790. **K,** portion of thallus, showing mature and empty monosporangia (*mspn*), ×790.

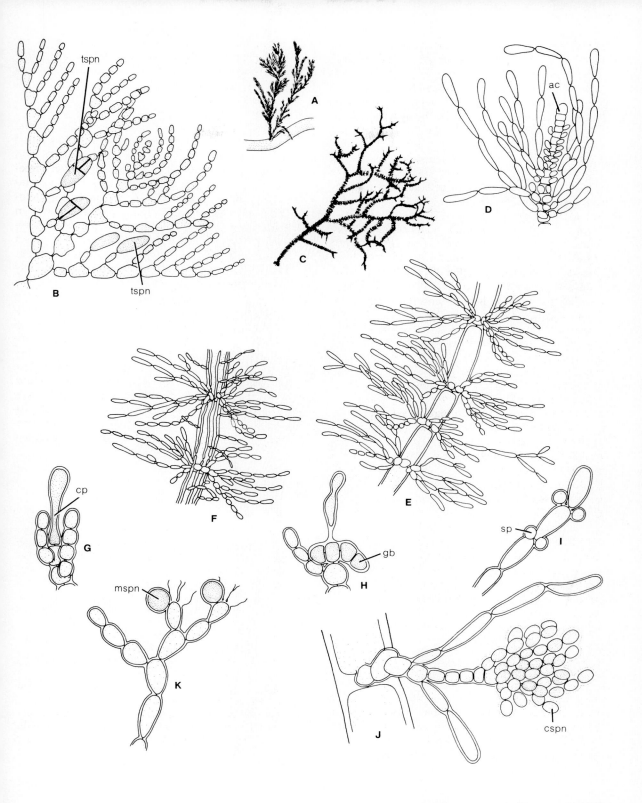

Plants—Rhodophyta (Red Algae)

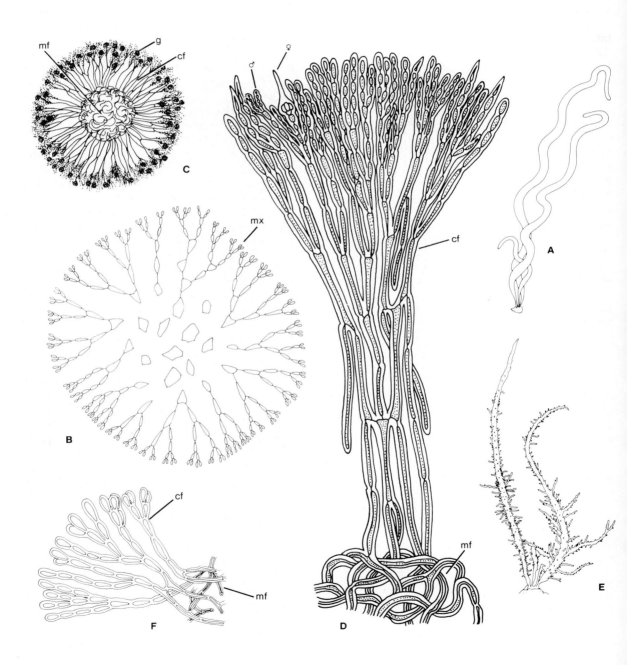

ones. However, genetic studies of red algae in general are very few.

Life Histories The most commonly encountered and best-known type of life history in the Florideophyceae is that occurring in the more complex orders (e.g., Ceramiales). It comes close to an alternation of isomorphic generations, except for the introduction of an additional diploid generation, the carposporophyte, or gonimoblast. The free-living, gamete-producing generations (male and female) "alternate" with a free-living, tetraspore-producing generation, and an intermediate, diminutive, diploid carposporophyte develops and remains *in situ* on the female gametophyte.

In some instances, within the same genus, species can have unisexual or bisexual gametophytes. The additional diploid carposporophyte generation evidently occurs as a result of a delay in the time and place of meiosis. In a few species, one or another of these phases appears to have been lost (or perhaps is not yet known and may even be so morphologically different as not to have been associated in the life history).

The characteristics of the female reproductive system and the events following fertilization vary greatly in the Florideophyceae. *Polysiphonia* (Ceramiales) is the only example of this type of life history that is discussed in detail.

Figure 9-12 Morphology of multiaxial Florideophyceae. **A–D,** *Nemalion.* **A,** habit of several plants, ×0.7. **B,** transverse section through young, vegetative thallus, showing simple multiaxial construction and organization of branched filaments embedded in mucilaginous matrix, ×200. **C,** transverse section of mature, fertile portion of thallus (diagrammatic), showing densely compacted, intertwined, medullary filaments (*mf*), looser cortical filaments (*cf*), and mature gonimoblasts (*g*), ×50. **D,** transverse sectional view of portion of mature thallus, showing densely compacted, intertwined, medullary filaments (*mf*), looser, cortical filaments (*cf*), with male (♂) and female (♀) reproductive structures at the surface of the thallus, ×500. **E, F,** *Cumagloia:* **E,** habit of a cluster of plants, ×0.3. **F,** transverse section through portion of thallus, showing more complex multiaxial construction and filamentous organization, ×350.

The Polysiphonia-type Life History. In *Polysiphonia* (Fig. 9-17) there is an alternation of free-living, vegetatively similar, haploid male and female gametophytes with a morphologically identical diploid tetrasporophyte. A diminutive diploid carposporophyte, which remains attached to the female gametophyte (enclosed in a **pericarp,** Fig. 9-17L) and produces diploid carpospores, precedes initiation of the free-living tetrasporophyte. Tetraspores (meiospores) produced on the free-living tetrasporophyte germinate and develop into the free-living gametophytes. Genotypic sex determination occurs, and two of the tetraspores from each tetrasporangium give rise to male gametophytes; the other two give rise to female gametophytes. The cytological alternation, as outlined above, accompanies the morphological sequence of generations.

The mature plant of *Polysiphonia* is a branched thallus with cells in regular tiers (Fig. 9-18B). Each mature tier or segment consists of a **central cell** surrounded by a number of **pericentral cells** of equal length. This is often referred to as a **polysiphonous** type of construction. The number of pericentral cells in vegetative axes is usually constant in any one species, although it varies in number from species to species; 4 is a common number, but 5, 6 or even as many as 24 can occur. Even where the number of pericentral cells in vegetative axes is 4 or more than 5, reproductive axes usually revert to the basic 5.

Growth in *Polysiphonia,* as in all members of the Rhodomelaceae (Ceramiales) (Fig. 9-19), is strictly apical, and the apical cell cuts off cells from its posterior face. If the apical cell segment is to produce only a central cell and pericentral cells, daughter nuclei in the dividing apical cell line up along the longitudinal axis (Fig. 9-19B–D). The cell wall that follows this nuclear division develops at right angles to the longitudinal axis. The primary cell undergoes division in a regularly alternating sequence within a few segments of the apical cell— that is, from left to right of the first cell cut off (Fig. 9-19J–O). This results in the typical arrangement of a central cell surrounded by four or more pericentral cells.

If the apical cell segment is to produce a branch, daughter nuclei in the dividing apical cell line up diagonally, and the cell wall following is oblique to the longitudinal axis (Fig. 9-19E–I). From the high side of the resulting subapical, wedge-shaped cell, a protuberance arises that initiates the branch (Fig.

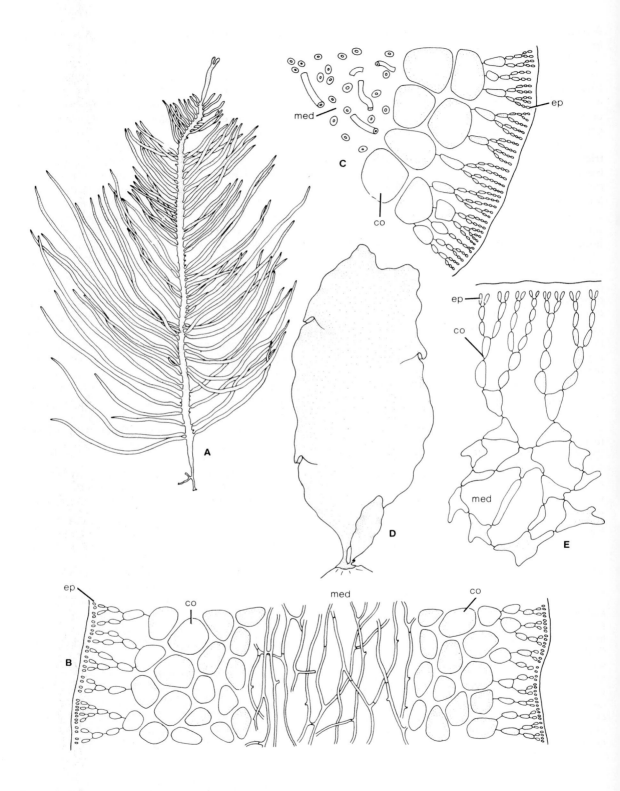

9-19H, I). The primary cell thus first forms a branch initial. Subsequently the primary cell divides in typical fashion, producing a central cell surrounded by four or more pericentral cells. Branch initials can be **indeterminate** and repeat the development of the main axis; they can also be **determinate** and of limited development, remaining partly or wholly **monosiphonous**. These monosiphonous determinate branches are referred to as **trichoblasts** and can be further branched. Trichoblasts can be sterile or fertile. Sexual structures are formed on this determinate type of branch, resulting in distinctive female and male trichoblasts.

On the female trichoblast (Fig. 9-17C), usually only the basal three segments next to the supporting axis become polysiphonous (Fig. 9-17D); the distal portion, which can be simple or branched, remains monosiphonous and is generally rather **evanescent**. It is the first segment above the basal one that becomes fertile; the last-formed pericentral cell in the fertile segment produces the **carpogonial branch**.

Before considering the ontogeny of the female reproductive system, we must consider the vegetative development of some adjacent cells (Figs. 9-17E–H and 9-18E). The last-formed pericentral cells of both the basal segment and the segment above the fertile one cut off a series of vegetative cells. Similarly, the pericentral cells on each side of the fertile one also cut off a series of vegetative cells. The filamentous series of cells resulting from these four vegetative pericentral cells fuse laterally to form an urn-shaped pericarp (Figs. 9-17L and 9-18E, F), which encloses the developing female reproductive system. These filaments do not fuse completely; at the apex, a small pore, or **ostiole** (Fig. 9-18F), remains. The trichogyne extends through it into the water. Development of the sterile pericarp starts before fertilization but is accelerated after fertilization, keeping pace with the developing carposporophyte within it. Intimately associated cells of the gonimoblast ($2n$) and surrounding pericarp ($1n$) are frequently referred to collectively as the **cystocarp**.

Prior to fertilization, the fertile pericentral cell enclosed within the pericarp undergoes three primary divisions (Fig. 9-17E–G). The first cuts off a cell to one side and results in a lateral sterile cell (Fig. 9-17E); the second cuts off toward the top of the carpogonial branch initial (Fig. 9-17F); and the third cuts off (Fig. 9-17G) a basal sterile cell toward the base. The remaining portion of the fertile pericentral cell, which is attached to the central cell of the fertile segment, is referred to as the **supporting cell** (Fig. 9-17G). Thus, there are three initials attached to the supporting cell at this stage. The lateral and basal sterile cells each divide a few times to form short filaments of sterile cells. These may later serve as a source of nutriment for the developing carposporophyte. The carpogonial branch initial divides to produce a four-celled carpogonial branch (Figs. 9-17H and 9-18E), the terminal cell of which develops into the carpogonium. It has a long, narrow trichogyne that emerges through the pericarp ostiole and protrudes into the water, forming a receptive structure for the spermatium (Fig. 9-17H).

The male reproductive system matures at about the same time as the female. In contrast to the female trichoblast, the male trichoblast (Fig. 9-17N) forms a large number of spermatangia. The basal segment of the fertile trichoblast generally remains monosiphonous, as does the often-branched distal portion (Figs. 9-17Q and 9-18C). However, between these two monosiphonous regions, several segments divide to form a central cell and five pericentral cells. All five pericentral cells in each of these fertile, polysiphonous segments become fertile.

Each fertile pericentral cell undergoes a number of transverse divisions to produce three or more spermatangial parent cells (Fig. 9-18D). In turn, each of these divides obliquely, cutting off two or more spermatangia (Fig. 9-18D), which (at maturity) liberate a single, colorless, naked spermatium (Fig. 9-17R). Due to the number of divisions resulting in successively smaller cells, the mature male trichoblast appears as a stalked, club-shaped structure

Figure 9-13 Morphology of multiaxial Florideophyceae. **A–C,** *Neoagardhiella* (*Sarcodiotheca*). **A,** habit ×0.5. **B,** longitudinal section through portion of thallus, showing filamentous medulla (*med*) and pseudoparenchymatous cortex (*co*), ×60. **C,** transverse section through outer portion of thallus, showing filamentous medulla and pseudoparenchymatous cortex, ×70. **D, E,** *Iridaea*. **D,** habit of plant, ×0.5. **E,** transverse section of outer part of thallus, showing loose filamentous inner, medulla region (*med*), and more compact, outer, small-celled cortex (*co*), ×700.

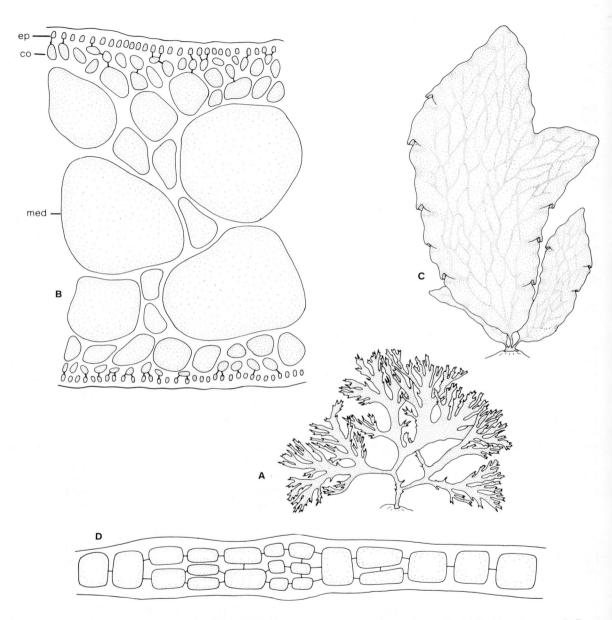

Figure 9-14 Morphological diversity of complex multiaxial Florideophyceae. **A, B,** *Callophyllis:* **A,** habit, ×0.5; **B,** transverse section through portion of thallus, showing pseudoparenchymatous organization (obscuring multiaxial nature of thallus) with compact, large-celled medulla (*med*), small-celled cortex (*co*), and epidermis (*ep*), ×150. **C, D,** *Polyneura,* a membranous form: **C,** habit of veined thallus, ×0.5; **D,** transverse section through portion of thallus in vicinity of vein: *ep,* epidermis.

with a large number of small, almost colorless, refractive cells (spermatangia) on its surface (Fig. 9-18C).

When a spermatium, which is passively carried by water currents, comes in contact with the trichogyne, fusion occurs, and the male nucleus passes down the trichogyne to the base of the carpogonium, where it fuses with the female nucleus. As soon as fertilization occurs, the trichogyne degenerates, and a cell is cut off from the top of the supporting cell. This is the auxiliary cell (Fig. 9-17I). A protuberance very much like that produced during the formation of a secondary pit connection arises from near the base of the fertilized carpogonium. The diploid zygote nucleus migrates into it and is eventually transferred to the auxiliary cell. The haploid nucleus within the auxiliary cell apparently degenerates, and the diploid (zygote) nucleus now present divides mitotically. The first true diploid cell of the carposporophyte generation is formed when a new diploid nucleus is cut off in the gonimoblast initial (Fig. 9-17K). This initial cell then divides repeatedly by mitosis to produce a multicellular gonimoblast, or carposporophyte.

As the gonimoblast develops, pit connections between cells of the adjacent female gametophyte (auxiliary cell, carpogonial branch, sterile cells, supporting cell) tend to break down and open up, resulting in extensive cytoplasmic continuity between cells in this area. This probably facilitates the transfer of food reserves from the female gametophyte to the developing gonimoblast. Terminal cells of the filamentous gonimoblast become much enlarged and develop into carposporangia (Fig. 9-17M). At maturity, each carposporangium liberates a single, elongate, densely pigmented, naked carpospore, which escapes through the ostiole. The carpospore is then carried passively in the water and, on contacting a suitable substrate, germinates and develops into a free-living tetrasporophyte generation (Fig. 9-17S) that is morphologically identical to the gametophyte generations (Fig. 9-17A, B).

Reproduction in the free-living tetrasporophyte is less complicated than that of the gametophyte phase. Tetrasporangia (meiosporangia) are produced in ordinary, vegetative, polysiphonous axes of the plant (Figs. 9-17T and 9-18H). However, when the plant becomes fertile, a large number of consecutive segments toward the apices of the branches become fertile. Each segment generally produces one tetrasporangium, which arises indirectly from one of the pericentral cells. The fertile pericentral cell undergoes a number of divisions (Fig. 9-18G, H), somewhat comparable to the primary divisions that occur in the development of the female reproductive system. Two cells are cut off from the outer face of the fertile pericentral cell. Each of these is a sterile **cover cell** (Fig. 9-18G, H). Then, from the upper portion of the fertile pericentral cell, a single cell, the tetrasporangium, is cut off and protected by external cover cells. The small cell, to which the tetrasporangium and two cover cells remain attached, is the **stalk cell** (Fig. 9-18G). The tetrasporangium enlarges considerably, and its diploid nucleus divides meiotically to form four haploid nuclei. The cytoplasm then divides into four uninucleate portions. At maturity, the wall of the tetrasporangium breaks down, the intercellular material between the cover cells breaks down, and naked tetraspores (meiospores) are liberated from between the cover cells. These meiospores (Fig. 9-17U) are carried passively in the water, and they germinate to develop into free-living gametophyte generations.

Antithamnion (Ceramiales), which also has a *Polysiphonia*-type life history, is simpler in its vegetative organization (Fig. 9-11A, B). Male branches are simpler in organization, although freely exposed as in *Polysiphonia*. However, the female reproductive system, mature carposporangia, and tetrasporangia are also all freely exposed. A number of genera having a *Polysiphonia*-type life history produce a variety of additional types of spores. Some of these are apparently meiospores, homologous to tetraspores, whereas others are thought to be **mitospores,** and apparently function in the same way as the monospores of Bangiophyceae in producing the same type of plant on which they were formed.

Many Florideophyceae having the *Polysiphonia*-type life history have some of their reproductive structures, especially the female system, deeply immersed in the cortical tissues of a pseudoparenchymatous thallus (Fig. 9-20). In these pseudoparenchymatous forms, the spermatangia are usually modified epidermal cells (usually smaller and colorless), although in some instances they are produced in conceptaclelike depressions, as in *Gracilaria* (Fig. 9-20B). The tetrasporangia are also usually modified epidermal cells (usually larger and more densely pigmented), sometimes scattered (Fig.

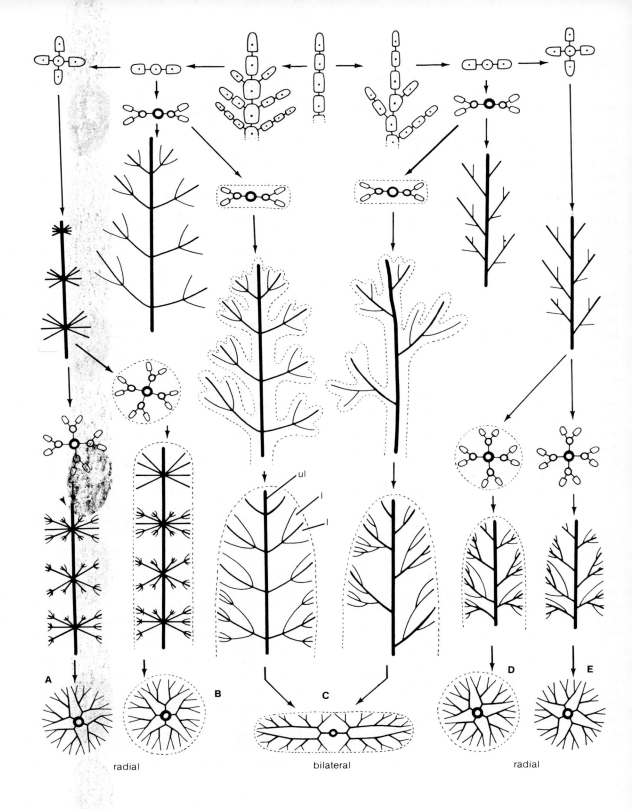

radial   bilateral   radial

9-20J) and sometimes densely aggregated. In some instances, however, tetrasporangia are located deep within the cortical tissue, sometimes scattered and sometimes densely aggregated. In these pseudoparenchymatous types, although the trichogyne emerges through the surface cells and is in a receptive position to receive the spermatia, the carpogonium is usually deeply immersed in the cortical tissue of the female gametophyte, where fertilization and subsequent development of the carposporophyte occur. In such instances, the deeply embedded carposporophyte can be completely obscured or apparent only as a slight swelling. The eventual release of carpospores can occur at the surface through a definite ostiole or by a breakdown of the outer tissues of the thallus.

The Nemalion-type Life History. It is in the simpler forms, especially in Nemaliales, that some of the greatest variation in life history occurs. The long-standing interpretation that *Nemalion* (Fig. 9-21) and other members of Nemaliales have meiosis immediately following syngamy, resulting in haploid carpospores, is no longer valid. Culture studies of Florideophyceae have now shown that some genera produce filamentous or crustose diminutive tetrasporophytes. For example, in Nemaliales, *Nemalion* (Fig. 9-21) has an **acrochaetioid** tetrasporophyte phase; and *Bonnemaisonia* has a *Trailiella*like tetrasporophyte stage (Fig. 9-22). A number of red algae in other orders have been shown to have life histories differing from the *Polysiphonia* type (Ceramiales). For example, some species of *Gigartina* (Gigartinales) have a *Petrocelis*like crustose tetrasporophyte stage. On the other hand, *Gracilaria*, also in the Gigartinales (Fig. 9-20), has the *Polysiphonia*-type life history (Fig. 9-17).

It is now clear that many Florideophyceae have a sequence of heteromorphic phases or generations, in which there is a dominant or conspicuous, free-living gametophyte state (or stages, in which the plant is bisexual) "alternating" with a diminutive, free-living, filamentous or crustose tetrasporophyte. And again, as in the *Polysiphonia*-type life history, a diminutive, intervening, diploid carposporophyte develops *in situ* on the gametophyte and produces diploid carpospores. *Nemalion* is the only example of this heteromorphic type of life history that will be discussed in detail.

In *Nemalion*, the gametophyte is the conspicuous generation. The plant is an erect, branched or unbranched multiaxial thallus (Figs. 9-12A and 9-21A, B) of densely compacted filaments, embedded in a mucilaginous matrix (Fig. 9-12B). The medullary region has an axial core of intertwined longitudinal filaments, and lateral branches from this axis form a progressively looser filamentous cortex toward the surface. Colorless spermatia are produced superficially in spermatangia borne on the tips of filamentous branches (Fig. 9-21C). Carpogonia, also freely exposed at the tips of filamentous branches, have a well-defined, diminutive trichogyne (Fig. 9-21D). The spermatium rests on the trichogyne and fuses with it; the nucleus passes to the enlarged basal part of the carpogonium (Fig. 9-21G), where it fuses with the female nucleus. The zygote nucleus divides (Fig. 9-21I) mitotically; daughter nuclei are cut off in cells, which then divide repeatedly, forming chains of diploid cells in a tight cluster. Collectively, these diploid cells make up the gonimoblast filaments, or carposporophyte (Fig. 9-21K). The carposporophyte is freely exposed, and terminal cells mature directly into carposporangia, each of which produces a single, diploid carpospore (Fig. 9-21L). The carpospore germinates and develops into a free-living, acrochaetioid filament (Fig. 9-21M–P). In other members of the Nemaliales, the zygote nucleus is transferred to a cell beneath the carpogonium, and from this lower (**hypogynous**) cell, the gonimoblast filaments emanate. The filamentous acrochaetioid stage produces tetrasporangia, and meiosis occurs during tetraspore production (Fig. 9-21P, Q). The liberated tetraspores germinate and develop first into a pros-

Figure 9-15 Growth types in uniaxial Florideophyceae, showing possible evolutionary lines leading to freely branched (**A, E**) and pseudoparenchymatous (**B–D**) types, with varied amounts of fusion of filaments of unlimited and limited growth, and with radial (**A, B, D, E**) and bilateral (**C**) symmetry. Diagrammatic cellular details of filaments are shown across top of figure (and above transverse section of each type). Throughout diagrams, filaments of axes are of unlimited (*ul*) growth, shown by heaviest lines, or limited (*l*) growth, shown by lighter lines; in transverse sectional views, filaments of unlimited growth are shown as heavy circles. Dotted lines indicate fusion of axes and laterals in a common mucilaginous matrix.

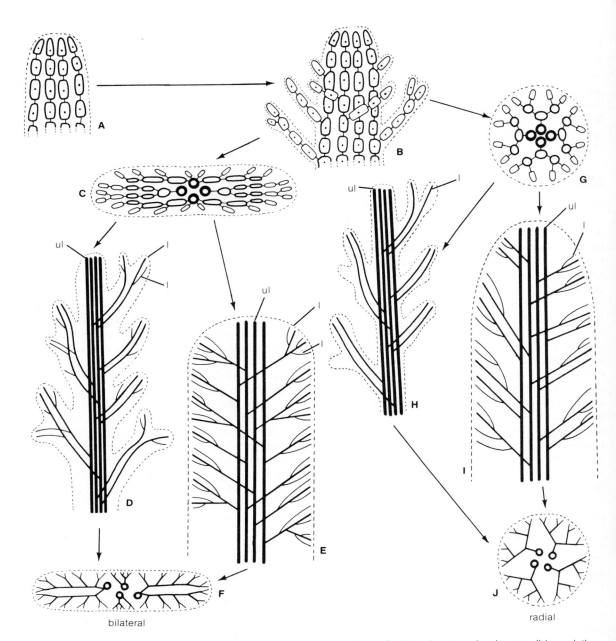

Figure 9-16 Growth types of multiaxial Florideophyceae, showing possible evolutionary lines leading to pseudoparenchymatous forms, with varied amounts of fusion of filaments of unlimited and limited growth, and with bilateral (*left*) and radial (*right*) symmetry. Filaments of axes are of unlimited (*ul*) growth, shown by heaviest lines, and limited (*l*) growth, shown by lighter lines. **A, B,** diagrammatic cellular detail of filaments. **C, F, G, J,** diagrammatic transverse sectional views. **D, E, H, I,** diagrammatic longitudinal sectional views. In transverse sections (**C, F, G, J**), positions of filaments of unlimited growth are shown as heavy circles. Dotted lines indicate fusion of axes and laterals in a common mucilaginous matrix.

trate filamentous stage, from which the typical *Nemalion* gametophyte plants eventually arise (Fig. 9-21A, B).

The florideophycean examples discussed in detail in this chapter have a type of postfertilization development that is relatively easy to follow. Postfertilization development in representatives of other orders is exceedingly complex and beyond the scope of this text.

Distribution and Ecology

The red algae are predominantly marine and are more widely distributed than brown algae. Some, such as *Porphyridium* (Fig. 9-5A), can be terrestrial, forming reddish scums on damp surfaces in greenhouses or around the margins of drying garden pools. Many freshwater species are confined to cold, fast-running streams.

Marine Rhodophyta are the most abundant and have attained great diversity in form and color. They are also the most widely distributed geographically and vertically of all the larger algae. Some, such as certain species of *Porphyra* and *Gloiopeltis*, occur high in the intertidal region. Others occur at great depths, especially in the tropics, where they have been recorded below 120 meters. Although more species of red algae occur in tropical waters, they are well represented in colder waters. However, they are seldom as conspicuous as the larger brown algae.

In color, freshwater species characteristically are a dull green, blue, or sometimes brown to black. The extreme range in color in marine red algae follows vertical variations in habitat with respect to light conditions. Intertidal forms achieve the greatest diversity in color. Depending on the relative proportion of the various pigments present, they range from a dull green or black high in the intertidal zone to purplish-red, brown, or rosy-red lower in the intertidal zone. In the intertidal area, the phycoerythrin pigment is masked by the chlorophylls and other accessory pigments. Those from greater depths are generally a bright rosy-red, with phycoerythrin masking the chlorophyll.

Although coralline algae of both the encrusting and articulated types are widespread in cold and warm waters, they are particularly significant in coral reef areas, where the cementing action of the encrusting coralline algae is thought to be primarily responsible for maintaining the reef structure.

Importance and Uses

Red algae provide, directly and indirectly, a source of detritus and food for marine animals. Reproductive cells of the red algae, although nonmotile, are liberated in profusion and form part of the phytoplankton.

Red algae are used directly by humans and are important for industrial and domestic uses, including food. **Dulse,** prepared in several countries from *Palmaria palmata* (= *Rhodymenia palmata*), is used in many ways. It is eaten like candy, used as a relish with potatoes, or cooked in soups. *Porphyra,* known as **purple laver,** or **nori** in Japan, is also used widely as food. The Japanese have artificially cultivated this red alga for many years. Nori is prepared as a dry, flavored product. It is also used in the preparation of Japanese macaroni and in soups and sauces.

Another important use of red algae is as a source of certain valuable phycocolloids, such as **funoran** (funori), carrageenan, and agar. Funoran, which is obtained from *Gloiopeltis*, is used as a water-soluble sizing. It is also used in the preparation of certain water-base paints, as an adhesive in hair dressings, and as a starch substitute in laundering.

Carrageenan is a mucilaginous extract obtained chiefly from *Chondrus* (**Irish Moss**) of the North American Atlantic Coast and from *Eucheuma* from southeast Asia. It is used in the food industry as a stabilizing or thickening agent in chocolate milk, cheese, ice cream, puddings, and jellied foods; it is also used in cosmetics, insect sprays, and water-base paints.

Agar is probably the most widely used and most valuable product obtained from red algae. It has many of the same uses as carrageenan, and many additional uses. It is widely used in microbiology in the preparation of media for culture work. It is a useful therapeutic agent in intestinal disorders, since it is a nonirritant bulk producer that can absorb and hold water and at the same time act as a mild laxative. It is used also to make capsules to enclose antibiotics, sulfa compounds, vitamins, and other substances where a slow release of the medicant is desired at a point beyond the stomach. It is used as a dental impression material. In certain special breads for diabetics, agar replaces starch; however, agar and carrageenan are not good sources of nutrition for human consumption because they are relatively indigestible.

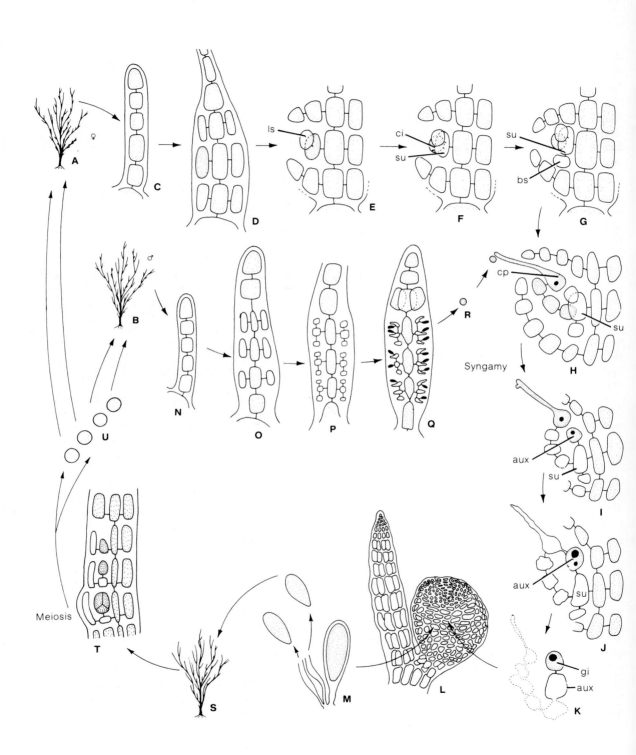

278 Chapter 9

Phylogeny

Although some red algae have left excellent fossil records, the origin of the group as a whole is obscure. Most fossil records are of the lime-encrusting forms, such as coralline algae. It is apparent that encrusting coralline algae, such as *Lithothamnium*, were as important and active in reef building in past periods (especially during the Jurassic) as they are now. Some believe that red algae probably had their origin during the Archeozoic Era from some unknown, probably uncalcified Precambrian ancestor. During the Paleozoic Era, calcareous red algae were represented by the Solenoporaceae and closely related forms, which appear in the fossil record during the Late Cambrian but were rare until late in the Ordovician. By the mid-Silurian, two structurally distinct types had differentiated in the Solenoporaceae: *Solenopora* and *Parachaetetes*. During the Pennsylvanian period, *Archaeolithophyllum*, *Cuneiphycus*, and several closely related forms appeared. Before the end of the Paleozoic, forms that clearly forecast the appearance of true coralline algae, including Recent encrusting and articulated types, are recognizable. However, many recent genera were well established by the end of the Mesozoic Era; from the Cretaceous onward, and especially in the Tertiary, genera such as *Lithothamnium* and *Lithophyllum* are clearly recognizable.

On the basis of recent morphological and culture studies, especially concerning patterns of spore germination, it is suggested that a systematic arrangement of modern coralline algae into articulated and nonarticulated groups is unnatural. It now appears that some of the articulated taxa are more closely related to certain nonarticulated genera than to other articulated genera. These phenetic data suggest that articulated genera may have evolved from two distinct groups of early, nonarticulated ancestors.

It has been suggested that red algae and bluegreen algae are closely related, because of a number of similarities in their biochemical and certain ultrastructural characteristics. The presence of phycobilin pigments, forms of starch and oil stored as reserves, and unstacked thylakoids have been given greatest attention, but the absence of motile cells in each group and the presence of primary pit connections (in some of each group) have also been emphasized. The major obstacle in this suggested relationship is the difference in organization of the genetic material. In the eukaryotic red algae, the genetic material is clearly organized in a nucleus, whereas in the prokaryotic bluegreen algae, it is present as a nucleoid. The presence of other organelles in the red algae, that are not present in the bluegreen algae, present another obstacle to this suggested relationship.

However, since the fossil history of recent Rhodophyta is known with certainty only as far back as the Cretaceous, there is little phylogenetic evidence for an origin of red algae from bluegreen algae, which are well represented in the Precambrian. Perhaps red algae arose from a nonflagellated but nucleated organism after bluegreen algae had evolved. The absence of centrioles (as far as is presently known) in any stage of the life history of a red alga would argue against a derivation from a flagellated ancestor. Rhodophyta are obviously taxonomically remote from other groups of algae. The similarity of the sexual reproductive processes of some Florideophyceae to those of certain ascomycetous fungi has led some botanists to suggest that fungi may have descended from a red algal prototype, but this speculation has few adherents.

The relationship between the two classes of red algae is also largely speculative. Similarity in their biochemical and ultrastructural characteris-

Figure 9-17 Life history of *Polysiphonia* (Ceramiales, Florideophyceae). **A, B,** female and male gametophytes. **C–H,** stages in development of female branch (trichoblast), showing formation of lateral sterile cell (*ls*), carpogonial branch initial (*ci*), and basal sterile cell (*bs*). Note: details of surrounding pericarp are omitted in **H–K.** The four-celled carpogonial branch with carpogonium (*cp*) in **H** is fully developed and attached to supporting cell (*su*). **I,** fertilization is complete, and auxiliary cell (*aux*) is formed. **J,** movement of zygote nucleus into auxiliary cell. **K,** formation of carposporophyte initial (*gi*). **L,** mature carposporophyte enclosed in pericarp. **M,** carposporangia, liberating carpospores. **N–R,** stages in development of male branch (trichoblast), showing formation of spermatangia and release of spermatia. **S,** tetrasporophyte. **T,** fertile branch of tetrasporophyte with tetrasporangia forming (lowest tetrasporangium mature). **U,** tetraspores (meiospores). For structural details and representative sizes of structures refer to Fig. 9-18.

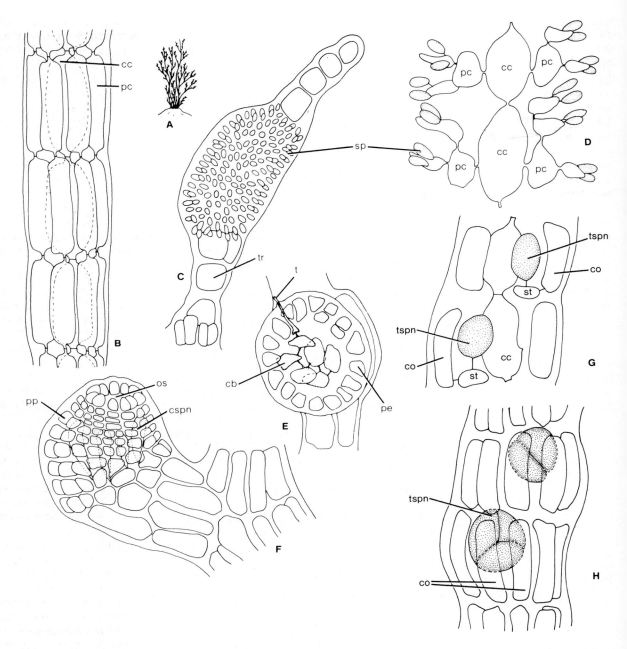

Figure 9-18 Structure and reproduction in *Polysiphonia* (Ceramiales, Florideophyceae). **A**, habit, ×1. **B**, polysiphonous sterile portion of thallus, showing central series of cells (*cc*) surrounded by pericentral cells (*pc*), ×250. **C**, mature male branch, or trichoblast, showing dense mass of spermatangia (*sp*), ×200. **D**, longitudinal section through two fertile segments of male trichoblast, showing derivation of spermatangia from pericentral cells and relationships to central cells, ×500. **E**, carpogonial branch (*cb*) with terminal carpogonium and emergent trichogyne (*t*) surrounded by pericarp

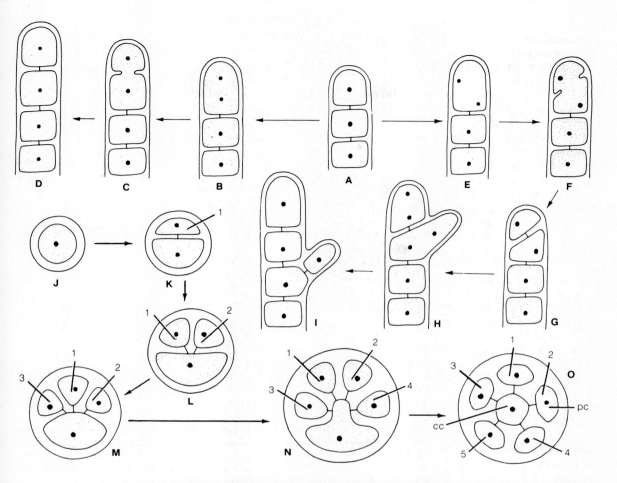

Figure 9-19 Apical growth and polysiphonous structure in *Rhodomelaceae* (Ceramiales, Florideophyceae). **A–D,** stages in division of apical cell to produce nonbranching segments. **E–I,** stages in division of apical cell (note position of nuclei in apical cell) to produce a branch-forming segment. **J–O,** divisions in segments (as seen in transverse section) cut off from apical cell, resulting in polysiphonous structure of axis with central (axial) cell and several (in this example, five) pericentral cells. Sequence in division is indicated by number: *cc,* central cell; *pc,* pericentral cell.

(sectional view), ×200. **F,** cystocarp, showing ostiole (*os*) and outer sterile pericarp wall (*pp*), through which some of fertile terminal carposporangia (*cspn*) of carposporophyte are shown, ×150. **G,** longitudinal section through two fertile segments of tetrasporophyte, showing attachment of immature tetrasporangia (*tspn*) with stalk (*st*) and cover cells (*cov*), ×250. **H,** surface view of fertile portion of thallus, showing mature tetrasporangia as seen through protecting cover cells (*cov*), ×250.

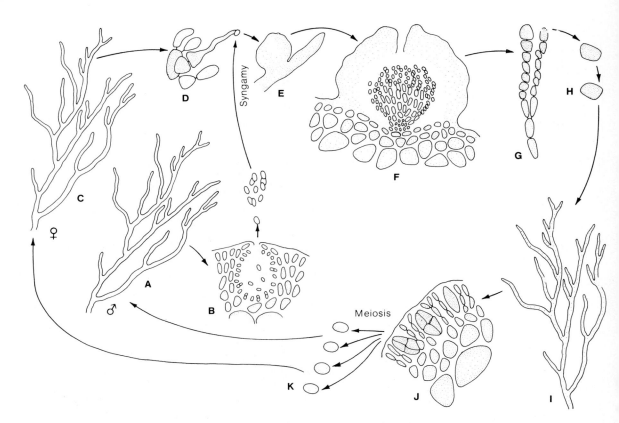

Figure 9-20 Life history of *Gracilaria* (Gigartinales, Florideophyceae). **A,** male gametophyte. **B,** conceptacle of spermatangia liberating spermatia. **C,** female gametophyte. **D,** spermatium attached to trichogyne of carpogonium before fertilization. **E,** immature cystocarp. **F,** transverse section through mature cystocarp, showing carposporophyte enclosed within ostiolate sterile pericarp wall of gametophytic tissue. **G,** portion of carposporophyte, showing chains of carposporangia. **H,** carpospore. **I,** tetrasporophyte. **J,** tetrasporangia (meiosporangia) in surface region of mature thallus. **K,** tetraspores released from tetrasporangium; these germinate to produce new male and female gametophytes.

tics, both with respect to pigments and storage products, provides a strong basis for keeping these otherwise morphologically different groups together. The absence of a sexual process in many Bangiophyceae provides a striking difference in the two classes. However, it may well be that sexual reproduction has been lost in most Bangiophyceae. Perhaps the filamentous conchocelis stage of *Porphyra* represents a prototype in the red algal line from which Florideophyceae have arisen, and the Bangiophyceae may represent a dead end culminating in the parenchymatous type of plant exemplified by *Porphyra*. The presence of a girdle thylakoid in the chloroplast of the conchocelis stage (and its absence in the foliose *Porphyra* phase) suggests that a conchocelislike alga may be considered as a prototype of Florideophyceae.

The filamentous line is obviously the one that has been most highly developed in red algae. Although one can suggest relationships among

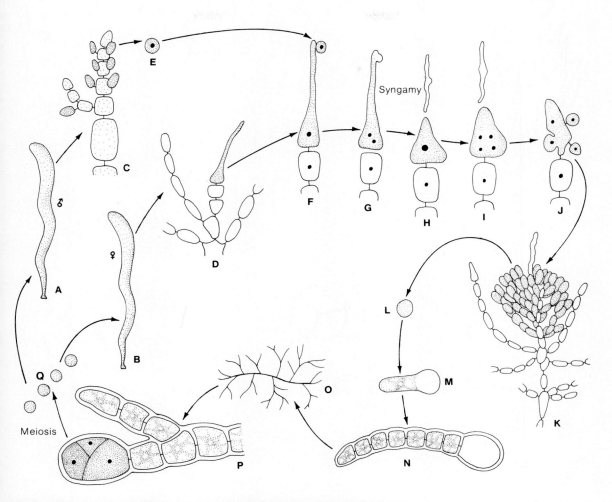

Figure 9-21 Life history of *Nemalion* (Nemaliales, Florideophyceae). **A, B,** gametophytes. **C,** branch of male gametophyte with several spermatangia and one released spermatium (**E**). **D,** branch of female gametophyte with carpogonium. **F,** spermatium adjacent to trichogyne on carpogonium. **G,** spermatium nucleus in base of carpogonium. **H,** syngamy completed. **I,** after meiosis, trichogyne degenerates. **J,** initiation of gonimoblast filaments from fertilized carpogonium. **K,** mature carposporophyte with numerous carposporangia. **L,** carpospore. **M, N,** germinating carpospore. **O,** filamentous tetrasporophyte (acrochaetioid stage). **P,** portion of branch of tetrasporophyte, showing tetrasporangium with three of four tetraspores visible. **Q,** tetraspores (meiospores). For details and representative sizes of comparable structures, as in *Batrachospermum*, refer to Fig. 9-19C–K.

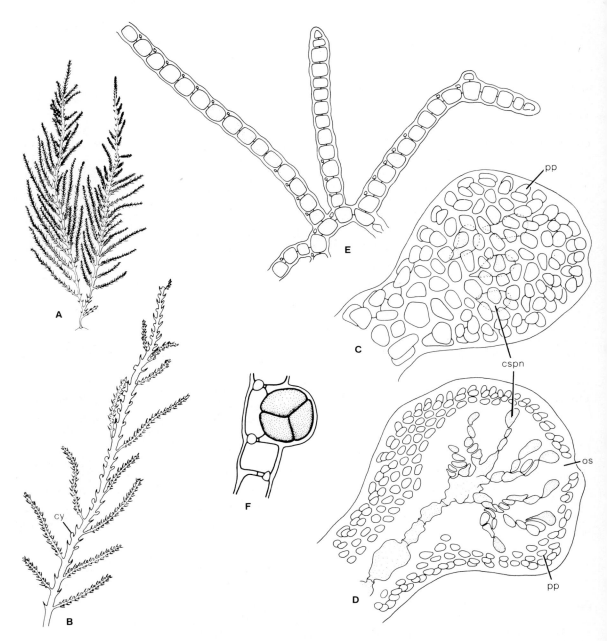

Figure 9-22 *Bonnemaisonia* (Nemaliales, Florideophyceae). **A–D,** female gametophyte. **A,** habit, ×0.3. **B,** fertile portion of branch, showing position of cystocarps (*cy*), ×1. **C,** surface view of cystocarp, showing sterile pericarp wall (*pp*) through which some terminal carposporangia (*cspn*) of carposporophyte are seen, ×300. **D,** transverse section of cystocarp, showing outer sterile pericarp wall, ostiole (*os*), and enclosed carposporophyte, ×250. **E, F,** filamentous tetrasporophyte (*Trailiella*-stage): **F,** portion of filamentous thallus, showing tetrasporangium with three of four tetraspores (meiospores) visible, ×400.

Florideophyceae, in the present state of uncertainty concerning orders, this too is highly speculative (see Fig. 27-12).

In respect to red algal life history, it would appear on the basis of present information that the basic type of life history is a modification of an alternation of isomorphic generations. One can assume that a delay in meiosis, with retention of the zygote on the female gametophyte, has led to development of the diploid "parasitic" carposporophyte in the life history of Florideophyceae and that a further delay has resulted in the production of diploid carpospores. This is the *Polysiphonia*-type life history. A modification of this basic type, in which an alternation of heteromorphic generations occurs, would be a second step. This is the *Nemalion*-type life history. This would suggest that the branched filament (which in *Porphyra* is believed to be diploid) and the acrochaetioid stage (which has been shown to be diploid) in some Florideophyceae are homologous.

Certainly among Florideophyceae, if one accepts the significance placed on the nature of the female reproductive structure, postfertilization stages, and embryogeny, it appears that Nemaliales is the most primitive order, and that all other orders are progressively advanced.

References

Abbott, I. A., and Hollenberg, G. J. 1976. *Marine algae of California.* Stanford Univ. Press, Stanford, Calif.

Bold, H. C., and Wynne, M. J. 1978. *Introduction to the algae.* Prentice-Hall, Englewood Cliffs, N.J.

Bourrelly, P. 1983. *Les algues d'eau douce.* Vol. 3, *Les bleues et rouges.* N. Boubée et Cie, Paris.

Chiang, Y.-M. 1970. Morphological studies of red algae of the family Cryptonemiaceae. *Univ. Calif. Publ. Bot.* 48:vi + 1–98.

Dixon, P. S. 1973. *Biology of the Rhodophyta.* Oliver & Boyd, Edinburgh.

Dixon, P. S., and Irvine, L. M. 1977. *Seaweeds of the British Isles.* Vol. 1, *Rhodophyta.* Part 1, *Introduction, Nemaliales, Gigartinales.* British Museum (Natural History); The George Press, Kettering, Northhamptonshire.

Fan, K. 1961. Morphological studies of the Gelidiales. *Univ. Calif. Publ. Bot.* 32:315–68.

Garbary, D. J.; Hansen, G. I.; and Scagel, R. F. 1980. A revised classification of Bangiophyceae (Rhodophyta). *Nova Hedwigia* 33:145–66.

Hawkes, M. W. 1978. Sexual reproduction in *Porphyra gardneri* (Smith *et* Hollenberg) Hawkes (Bangiales, Rhodophyta). *Phycologia* 17:329–53.

Hommersand, M. H. 1963. The morphology and classification of some Ceramiaceae and Rhodomelaceae. *Univ. Calif. Publ. Bot.* 35:165–366.

Irvine, D. E. G., and Price, J. H. eds. 1978. *Modern approaches to the taxonomy of red and brown algae.* Academic Press, London.

Littler, M. M. 1972. The crustose Corallinaceae. *Oceanogr. Mar. Biol. Ann. Rev.* 10:311–47.

Lobban, C. S., and Wynne, M. J., eds. 1981. *The biology of seaweeds.* Bot. Monogr. Vol. 17. Blackwell Scientific Publs., Oxford.

Norris, R. E. 1957. Morphological studies on the Kallymeniaceae. *Univ. Calif. Publ. Bot.* 28:251–334.

Scagel, R. F. 1953. A morphological study of some dorsiventral Rhodomelaceae. *Univ. Calif. Publ. Bot.* 27:1–108.

Scagel, R. F.; Bandoni, R. J.; Maze, J. R.; Rouse, G. E.; Schofield, W. B.; and Stein, J. R. 1982. *Nonvascular plants: an evolutionary survey.* Wadsworth Publishing Co., Belmont, Calif.

Searles, R. B. 1968. Morphological studies of red algae of the order Gigartinales. *Univ. Calif. Publ. Bot.* 43:viii + 1–100.

Sparling, S. R. 1957. The structure and reproduction of some members of the Rhodymeniaceae. *Univ. Calif. Publ. Bot.* 29:319–96.

Taylor, W. R. 1957. *Marine algae of the northeastern coast of North America.* 2nd ed. Univ. of Michigan Press, Ann Arbor.

Wagner, F. S. 1954. Contributions to the morphology of the Delesseriaceae. *Univ. Calif. Publ. Bot.* 27:279–346.

10

EMBRYOPHYTES

Embryophytes are plants that exhibit many features that sharply differentiate them from algae, fungi, and bacteria. Since they have a great variety of cell types, tissues, and organs, embryophytes are structurally far more complex than these other plant groups. Higher embryophytes also have physiologically and structurally complex meristematic tissues that are involved with growth in length (apical meristems) and growth in width (vascular and cork **cambia**). Embryophyte sex organs have a multicellular jacket of sterile protective cells enclosing the gametes. Additionally, as the name implies, their zygotes develop into multicellular embryos, which, like the gametes, are contained in protective multicellular jackets. The embryo represents an early stage of the **sporophyte** generation.

Heterospory, the formation of different types of spores (called **megaspores** and **microspores**), is found in many embryophytes, primarily in more specialized classes of the division Tracheophyta. As the prefixes *mega-* (large) and *micro-* (small) indicate, the two types of spores are commonly differentiated by size, but they also differ physiologically and develop in different **sporangia**. Megaspores develop into female **gametophytes (megagametophytes),** and microspores into male gametophytes **(microgametophytes).** This physiological difference is far more significant than the size difference. In fact, many so-called megaspores are smaller than their microspore counterparts. This apparent contradiction in spore terminology has prompted some botanists to coin new terms for megaspores and microspores. However, the original terms are so well entrenched in botanical literature and terminology that we have retained them. They are used here, not necessarily to denote size differences in spores, but to indicate their function, namely, whether they will ultimately develop into male or female gametophytes.

Another feature unique to embryophytes, and generally associated with heterospory, is **endosporic development** of gametophytes. In many **homosporous plants,** the spore germinates and grows into a gametophyte completely outside the spore wall. This is called **exosporic development.** On the other hand, in most extant **heterosporous** plants, the spore germinates and grows into a gametophyte within the spore wall. Such endosporic development is a prerequisite for the evolution of the **seed.**

Among algae, the closest relatives of the

embryophytes are generally conceded to be members of the division Chlorophyta (green algae). Features common to embryophytes and green algae are:

1. Presence of chlorophyll *a* and *b*
2. Presence of carotenoids, and absence of other pigments
3. Nature of cell wall formation (phragmoplast formation [between one alga *(Fritschiella)* and embryophytes])
4. Plastid ultrastructure (fused thylakoids)
5. Nature of starch (it stains with IKI in embryophytes and green algae, but not in other algal groups)
6. Presence of cellulose in cell walls
7. Specific nature of the whiplash flagellum.

It is the general consensus of modern botanists that embryophytes comprise two divisions: Bryophyta (mosses and their relatives), in which the gametophyte is the dominant generation and has the greater diversity of form, and Tracheophyta (vascular plants), in which the sporophyte is the dominant generation and shows much more diversity of form than does the gametophyte.

General Features of Embryophytes

Although the two groups within embryophytes are distinct in many ways, they also show some similarities. Some of these features have been alluded to previously; others are discussed below.

Biochemical

The striking biochemical differences seen among the various divisions of algae do not exist among embryophyte groups. Also, lignin, which is unknown in algae and fungi, is found in many different embryophyte groups. Additionally, the phenolic compounds present in embryophytes are apparently biosynthetically distinct from the phenolics of algae, although little is known about the latter.

Plant Cells

There are certain features common to many living cells in embryophytes. What follows is an introduction to some of these common features and a very brief review of plant cytology.

Figure 10-1 presents a classification of the cellular components of embryophyte cells. A plant cell in general consists of a **cell wall** which surrounds the cell contents, the **protoplast.** The protoplast can be divided into nonliving and living portions, the latter being called protoplasm. The nonliving portion is further divided into **vacuoles** and **ergastic substances.** Each of these components is discussed below. Electron micrographs of two "typical" plant cells are presented in Figure 10-2.

Protoplast

Protoplasm **Protoplasm** consists of **cytoplasm** and the nucleus, the control center of the cell (Fig. 10-2).

Cytoplasm Cytoplasm has several different components. The ground substance in which other structures are embedded is referred to as **hyaloplasm.** Surrounding the cytoplasm is a membrane, the **plasma membrane** or **ectoplast** (Fig. 10-2A, B). This membrane controls the entrance of ions, and once they have entered, it prevents them from readily leaving the cytoplasm. Another bounding membrane in the cytoplasm, the **tonoplast,** surrounds vacuoles (Figs. 10-2B and 10-3A). There is an internal membrane system in the hyaloplasm, the **endoplasmic reticulum,** which courses throughout the hyaloplasm (Figs. 10-2A and 10-3D). Also present are **dictyosomes** (Figs. 10-2A, B, 10-3B, and 10-3C)—series of stacked membranes called **cisternae** with their associated **vesicles** (Fig. 10-3B). They have been implicated in synthesis and in the release of synthesized materials. There is evidence that they are derived from the endoplasmic reticulum and hence can be interpreted as part of the membrane system.

There are three major organelles in the cytoplasm: **mitochondria, ribosomes,** and **plastids.** Mitochondria are double-membrane-bounded bodies that are important in respiration. The inner membrane is invaginated as a series of **cristae** (Fig. 10-3C). The assumed function of the cristae is to provide sites for respiratory enzymes. Also present in mitochondria are DNA, RNA, and ribosomes. Mitochondrial DNA may be involved in cytoplasmic inheritance.

Ribosomes are small bodies that can exist free within the hyaloplasm or in association with the endoplasmic reticulum (Fig. 10-3D). In the latter instance, the endoplasmic reticulum is referred to as rough endoplasmic reticulum, since it appears "rough" in micrographs (Fig. 10-3D). These small organelles contain much RNA and are believed to be sites of protein synthesis.

There are three different kinds of plastids: **chloroplasts, chromoplasts,** and **leucoplasts.** The most common are the disc-shaped chloroplasts containing chlorophyll and associated pigments. As seen using the electron microscope (Fig. 10-3A, E), the chloroplast is a double-membrane-bounded structure with an internal membrane system. The internal membranes are organized into stacks of cisternae, called **grana,** and an inter-connecting series of membranes among the grana—the fretwork or intergranal lamellae (Fig. 10-3E). These membranes are embedded in a matrix referred to as the **stroma.** Chloroplasts have been found to contain DNA and, like mitochondria, they may be involved in cytoplasmic inheritance.

Chromoplasts contain yellow or orange pigments and have a reduced internal membrane system. Some are **ontogenetically** related to chloroplasts, in that they develop from chloroplasts. This commonly occurs in fruits, such as the orange, which changes from green to orange as it ripens.

Leucoplasts are colorless plastids that serve as storage locations for starch, oil, or protein. They too are ontogenetically related to chloroplasts. Some leucoplasts, like those of potatoes, will revert to chloroplasts when exposed to the sun.

Nucleus The nucleus is a spherical body, bounded by a double membrane that has pores (Fig.

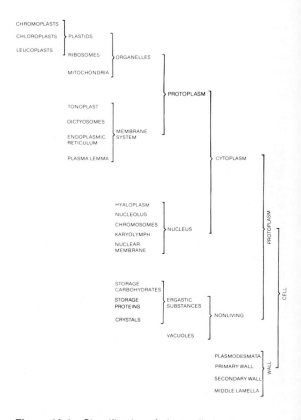

Figure 10-1 Classification of plant cell structures.

10-2A). Sometimes the outer layer of the nuclear membrane is seen to be continuous with the endoplasmic reticulum (Fig. 10-2A). Therefore, the endoplasmic reticulum may be considered an extension of the nuclear membrane. Inside the nucleus is the nuclear sap, or karyolymph—the substance in which the other nuclear structures are embedded. One of these is the nucleolus, a nonmembranous body that contains much RNA. Also in the nucleus are the **chromosomes,** which contain most of the DNA in the cell.

Nonliving Part of the Protoplast

The cell parts mentioned above can be demonstrated to be actively involved in processes necessary for life: synthesis, storage, control, and the

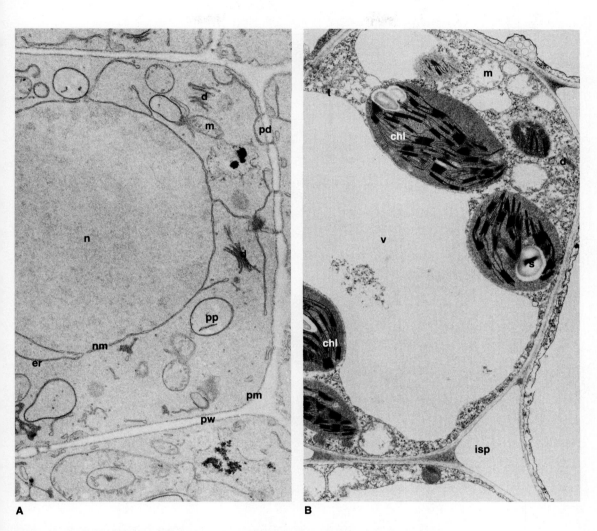

Figure 10-2 Two "typical" plant cells. **A,** meristematic cell with a nucleus surrounded by a nuclear membrane, endoplasmic reticulum, dictyosomes, plasmodesmata, mitochondria, plasma membrane, and cell wall, ×11,000. **B,** photosynthetic cell, showing chloroplasts and some assimilatory starch, vacuole with its surrounding tonoplast, mitochondria, dictyosome, and intercellular space, ×8,000: *n,* nucleus; *nm,* nuclear membrane; *er,* endoplasmic reticulum; *d,* dictyosome; *pd,* plasmodesmata; *m,* mitochondria; *pm,* plasma membrane; *pw,* cell wall; *chl,* chloroplast; *s,* assimilatory starch; *v,* vacuole; *t,* tonoplast; *isp,* intercellular space; protoplastid. (From M. C. Ledbetter and K. R. Porter, 1970.)

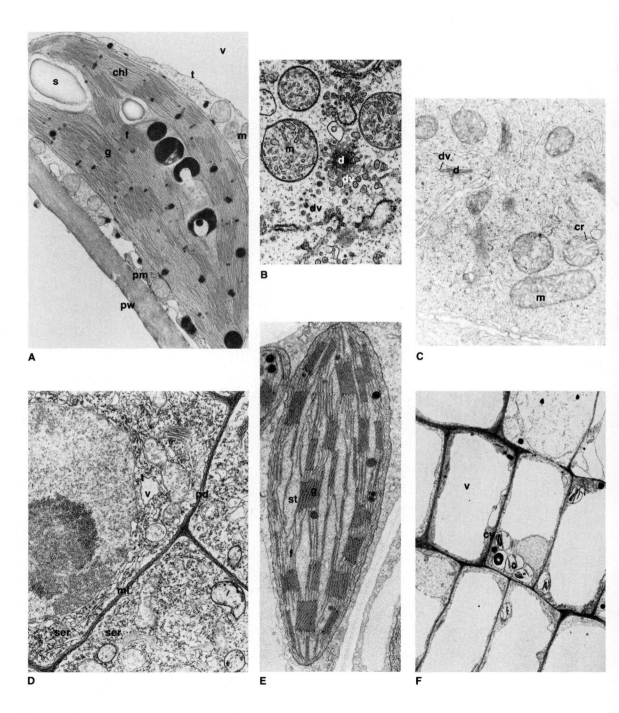

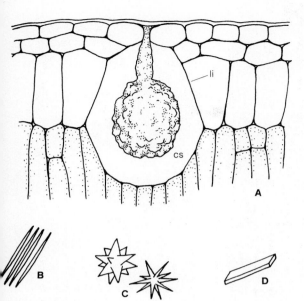

Figure 10-4 Crystals. **A,** cystolith in lithocyst, ×300. **B,** raphides, ×1000. **C,** druses, ×1000. **D,** prismatic crystals, ×1000: *cs*, cystolith; *li*, lithocyst.

Figure 10-3 Electron micrographs of various cellular components. **A,** chloroplast and assimilatory starch, vacuole and its surrounding tonoplast, mitochondria, plasma membrane, and cell wall, ×10,000. **B,** face view of a dictyosome, showing vesicles (to appreciate the three-dimensional form of a dictyosome, compare with the dictyosome shown in Fig. 10-3C), ×15,000. **C,** mitochondria with cristae, ×15,000. **D,** ribosomes, some associated with endoplasmic reticulum to give rough endoplasmic reticulum, and some free; also smooth endoplasmic reticulum, small vacuole, and tonoplast, ×10,000. **E,** chloroplast, showing detail of grana, fretwork (intergranal lamellae), and stroma, ×9000. **F,** series of plant cells with large vacuoles, showing most of the cytoplasm being addressed to the cell wall, ×5000: *chl*, chloroplast; *cr*, crista; *cy*, cytoplasm; *d*, dictyosome; *dv*, dictyosome vesicles; *f*, fretwork; *g*, grana; *m*, mitochondria; *ml*, middle lamella; *pd*, plasmodesmata; *pm*, plasma membrane; *pw*, cell wall; *r*, ribosomes; *rer*, rough endoplasmic reticulum; *s*, assimilatory starch; *ser*, smooth endoplasmic reticulum; *st*, stroma; *t*, tonoplast; *v*, vacuole, (**A, B, D, F,** from M. C. Ledbetter and K. R. Porter, 1970. **C,** courtesy W. H. Jensen; **E,** from W. H. Jensen and F. B. Salisbury, 1972.)

trapping and release of energy. There are also parts of the protoplast, called nonliving components, that are not obviously involved in such processes. Even though they are nonliving, they are built up by living parts of the protoplast that can also break them down and utilize them. Hence, these so-called nonliving parts of the protoplast are in a state of dynamic equilibrium with the living components.

Vacuoles One of the more obvious nonliving bodies is the vacuole (Figs. 10-2B and 10-3F). This is a membrane-bounded aqueous solution of sugars, amino acids, pigments, andother solutes and colloidal particles. The size of the vacuole depends on the degree of matruation of the cell. Immature cells have many small vacuoles. As the cell grows and differentiates, most increase in bulk occurs in the vacuoles. In mature cells, the bulk of the protoplast consists of vacuoles.

Ergastic Substances There are three types of ergastic substances: crystals, starch, and certain kinds of proteins.

Crystals. Crystals are by-products of cellular metabolism and, with rare exceptions, are composed of calcium oxalate. The exceptions are in *Ficus* (fig) (Fig. 10-4A) and *Cannabis* (hemp), where crystals of calcium carbonate, called **cystoliths,** occur. Crystals can take on many forms, including needlelike structures (**raphides,** Fig. 10-4B), spherical bodies with many projections (**druses,** Fig. 10-4C), and prismatic bodies (**prismatic crystals,** Fig. 10-4D). Various functions have been attributed to the crystals. One is to remove poisonous oxalic acid from cells by precipitating it out as a calcium oxalate.

Starch. Starch is the most common ergastic substance found in plants. It occurs in grains (Fig. 10-5) and forms the stored energy of plants. Starch grains develop from leucoplasts that store starch and, in doing so, become so distended that they lose their identity as plastids. Starch is also sometimes formed in chloroplasts during active photosynthesis; this is called assimilatory starch (Figs. 10-2B and 10-3A), and it is not a storage product in embryophytes.

Protein. Some proteins, other than those found in the protoplasm, are classed as ergastic substances.

Embryophytes **291**

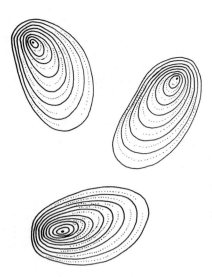

Figure 10-5 Starch grains, ×300.

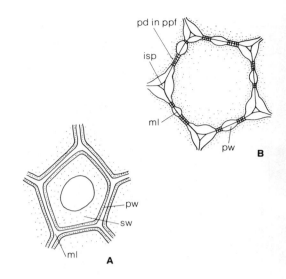

Figure 10-6 Structure of plant cell walls. **A,** cell with secondary wall, primary wall, and middle lamella, ×300. **B,** cell wall with plasmodesmata in primary pit fields, middle lamella, and intercellular space, ×300: *sw,* secondary wall; *pw,* primary wall; *ml,* middle lamella; *pd,* plasmodesmata; *ppf,* primary pit field; *isp,* intercellular space.

Like starch, these proteins are storage products. The gluten found in wheat flour and the **aleurone** grains of some grasses are examples of such proteins.

Cell Wall

The rigid cell walls that are characteristic of plant cells are of great importance. For one thing, they provide support for the plant body. Secondly, they have significant functions in tissues with which they are associated: the cell walls of tissues involved in support will be thick and hard, whereas those of storage tissues will usually be thin. It is the cell walls that determine the texture of a tissue, a matter of prime consideration in edibility. The main component of cell walls is **cellulose,** a polymer composed of many glucose molecules. Other major substances contained in cell walls are **hemicellulose** and **pectins.**

In many cells there are two types of cell walls: the **primary wall** and the **secondary wall** (Fig. 10-6A). The primary cell wall, deposited during cell enlargement, undergoes stretching as it develops. Even though it stretches, the primary wall retains a standard thickness and can even increase in thickness. This indicates that new primary wall material is being deposited throughout the growth of the cell. A primary cell wall is composed of cellulose, hemicellulose, and pectic substances, with cellulose making up 10–20% of the wall. The cellulose and hemicellulose are mostly structural materials. Pectic substances can occur in the wall proper, but they also occur between cell walls as substances that "cement" cells together. This cementing material, calcium pectate, occurs in a structure called the **middle lamella** (Fig. 10-6A). The middle lamella forms from the **cell plate,** which forms during **cytokinesis.**

Plasmodesmata are another structural feature associated with the primary wall. These are protoplasmic strands passing through the wall that connect the protoplasts of adjacent cells (Fig. 10-2A). It is believed that the endoplasmic reticulum traverses plasmodesmata from one cell to another (Fig. 10-2A). Because of the connection between endoplasmic reticulum and nuclear membrane, and reported intercellular connection by endoplasmic reticulum through plasmodesmata, there has been

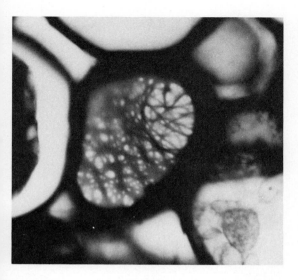

Figure 10-7 Primary pit fields as seen in face view in *Abies* root cortex, × 1000. (Photograph courtesy of K. Esau, from K. Esau, *Plant anatomy*, 2nd ed., John Wiley & Sons.)

some speculation that this system might be a means whereby information, originally genetic in nature, is passed from one cell to another. Theoretically, plasmodesmata can occur anywhere along the cell wall. However, they are often aggregated in certain thinner areas of the primary wall, referred to as **primary pit fields** (Figs. 10-6B and 10-7).

The secondary wall is deposited after the cell has ceased to enlarge and is composed mostly of cellulose, having little hemicellulose and no pectin. Secondary walls are usually supportive. This support can be for an individual cell (annular or helical secondary walls, see p. 363) or, along with the secondary walls of many other cells, for a tissue or an organ. Many cells that have secondary walls are dead (have no protoplast) when performing the supportive function for which they appear to be designed. In such instances, the empty middle of the cell, where the protoplast once occurred, is called the **lumen.**

In the secondary cell walls of many cells, there are **pits** or breaks in the wall. These pits may, or may not, occur over a primary pit field. The actual break in the wall is called the **pit cavity** or **pit chamber;** the primary wall at the bottom of the pit is called the **pit membrane** (Fig. 10-8A, C, G, H, J, K). Two basic types of pits—simple and bordered—are recognized. **Simple pits** are those with straight sides (Fig. 10-8A, B); **bordered pits** are those in which the secondary wall overarches the pit chamber. The portion of secondary wall overarching the pit chamber is called the border (Fig. 10-8C, G, H, M, R).

There is much variation seen in bordered pits. As seen in surface view, they can vary from circular to oval (Fig. 10-8E, F). Variations are also seen in pit structure. In many **angiosperms,** the border is thick, and a **pit canal** connects the pit chamber to the lumen of the cell (Fig. 10-8G, H). The opening of the pit canal into the lumen of the cell is called the **inner aperture,** and the opening into the pit chamber is called the **outer aperture.** Sometimes the inner aperture is slitlike and the outer is circular (Fig. 10-8G–I). Some simple pits in angiosperms look as if they are circular bordered pits that have lost their pit chambers. These consist of a slitlike opening to the lumen of the cell and a circular configuration at the pit membrane (Fig. 10-8J–L). The pits of **gymnosperms** are commonly bordered, and the inner aperture varies from round to oval (Fig. 10-8N). In some gymnosperms, the middle lamella is swollen to form the **torus** (Fig. 10-8M). In such pits, that portion of the pit membrane between the torus and the pit wall is called the **margo** (Fig. 10-8M, R).

A pit in one cell is usually associated with a pit in an adjacent cell. The two pits are referred to collectively as a **pit pair.** The members of a pit pair may or may not be the same type of pit. It is possible to have a **simple pit pair,** in which both pits are simple (Fig. 10-8O). Conversely, both pits can be bordered, and there is then a **bordered pit pair** (Fig. 10-8Q, R). The third alternative, one simple pit and one bordered pit in a pit pair, is known as a **half-bordered pit pair** (Fig. 10-8P).

Aside from cellulose, hemicellulose, and pectin, cell walls contain several other compounds, including **cutin, suberin,** and **lignin.** These substances are commonly embedded in the cellulose. Cutin and suberin are both long-chain fatty compounds found mostly in cell walls at the external surface of the plant. Cutin occurs in the epidermis and is impermeable to water, thus functioning in the prevention of dehydration. This is accomplished in two ways. As a result of its being impermeable to water, cutin prevents water loss. When cutin occurs in thick layers, it causes the leaf to be lighter in color and hence reflect light. The

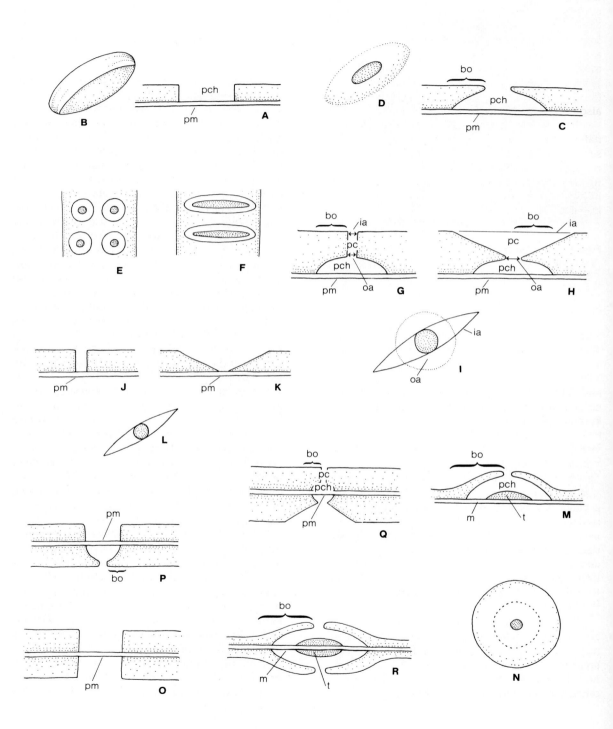

reflectance of light results in a cooler leaf which, in turn, also results in less water loss. It has also been implicated in the prevention of pathogen invasion. Cutin has also been reported in the endodermis of some roots. Its presence there seems to be associated with the physiology of the plant. Suberin is most common in cork cells. Like cutin, it is impervious to water and obviously serves a similar function in preventing dehydration. Like cutin, it has also been reported in the root endodermis.

The most common compound found impregnated in plant cell walls is lignin, the substance that serves as a hardening agent for cells in wood, seed coats, and fruits. Lignin is a complex polymer made up of many **phenylpropanoid units**. It is deposited in plant cell walls after the secondary wall has been initiated. The first site of deposition is in the primary wall, and from there it spreads into the secondary wall.

It must be emphasized that the information contained in this chapter is only introductory. For more detail on plant cells, the student is referred to the works cited in the list of references at the end of the chapter.

Figure 10-8 Pits in secondary walls, × 500. **A, B,** simple pit with pit membrane: **A,** transverse sectional view; **B,** oblique view from above pit chamber. **C, D,** bordered pit with pit chamber, border, and pit membrane: **C,** transverse sectional view; **D,** oblique view from above pit chamber. **E, F,** two different shapes of bordered pits as seen from above pit chamber: **E,** circular bordered pit; **F,** oblong bordered pit. **G, H, I,** bordered pit with slitlike inner aperture, circular outer aperture, and pit canal: **G, H,** transverse sectional views of the same bordered pit, from two different angles (90° removed); **I,** view from above pit chamber. **J, K, L,** reduced bordered pit: **J, K,** transverse sectional views of the same reduced bordered pit from two different angles (90° removed); **L,** view from above pit chamber. **M, N,** coniferous bordered pit with margo and torus: **M,** transverse sectional view; **N,** view from above pit chamber. **O,** simple pit pair, as seen in transverse section. **P,** half-bordered pit pair, as seen in transverse section. **Q, R,** bordered pit pairs of angiosperm (**Q**) and conifer (**R**), transverse sectional view. (Labeled features for **O, P, Q,** and **R** are same as for **A, C, G, H,** and **M,** respectively): *pm,* pit membrane; *bo,* border; *pc,* pit canal; *pch,* pit chamber; *ia,* inner aperture; *oa,* outer aperture; *t,* torus; *m,* margo.

Morphological Trends Within Embryophytes

Vegetative Trends

One of the most obvious, and interesting, trends seen in embryophytes is the change in relative size of gametophytes and sporophytes (Fig. 10-9A, B, E, H, I, O, Q). In Bryophyta, the gametophyte (haploid generation) is structurally and physiologically dominant, independent, and usually **perennial.** The sporophyte (diploid generation), on the other hand, is **ephemeral** (at least in part), nutritionally dependent on the gametophyte, and bears a single sporangium (Fig. 10-9A, B). In contrast, the sporophyte of Tracheophyta is structurally and physiologically dominant (Fig. 10-9E, I), in many instances perennial, and bears many sporangia (Fig. 10-9E, N), whereas the gametophyte usually shows varied degrees of dependence upon it. The gametophyte of the more generalized Tracheophyta is small but nutritionally independent of the sporophyte, being either saprophytic or photosynthetic (Fig. 10-9H). With a few exceptions, this more generalized group of Tracheophyta is homosporous and, like Bryophyta, has well-developed gametangia and exosporic development of gametophytes. In other Tracheophyta (some ferns and some lycopods, gymnosperms, and angiosperms) the gametophytes are short-lived. These groups are heterosporous, with reduced gametangia and endosporic gametophyte development. In these heterosporous plants, there are differences in the degree of dependence of the megagametophyte (female) and microgametophyte (male) on the sporophyte. Both types of gametophytes in the heterosporous lycopods and ferns develop more or less independently of the sporophyte, at least in extant genera. Megagametophytes in the gymnosperms and angiosperms are completely dependent on the sporophyte (Fig. 10-9P, Q) and are retained on the sporophyte until shed as seeds (Fig. 10-9R). The microgametophyte of angiosperms and gymnosperms is shed from the sporophyte and is physically independent of it and is called a pollen grain (Fig. 10-9O).

Other significant trends seen in embryophytes involve changes in anatomical features, many resulting from adaptations to terrestrial existence.

One such adaptation is the development of absorbing structures. Their significance is obvious. Without these adaptations, plants would have no means of obtaining water and mineral nutrients from the substrate. Different groups of embryophytes have different types of absorbing organs. Bryophyta—structurally the simplest of embryophytes—can absorb water and nutrients through any cell, although in some instances simple cells or filaments of cells called **rhizoids** assist in water absorption, and in others rootlike structures are present (Fig. 10-9A). The most generalized of extant vascular plants, *Psilotum* and its relatives (see Chapter 14), also have rhizoids, but most vascular plants have more complex structures for absorption, the **roots**. In almost all instances, rhizoids and roots also serve to anchor the plant to its substrate.

Another anatomical adaptation to the terrestrial environment is the presence of conducting or **vascular tissues**. Their function is to conduct water and mineral nutrients from the point of absorption to sites of photosynthesis and to conduct photosynthetic products from sites of formation to sites of storage (usually underground organs) or nonphotosynthesizing areas of utilization. In Bryophyta, conducting tissue, when present, is a central column (Fig. 10-9C) of differentiated cells: the **hydrome** conducts water and mineral nutrients, and the **leptome** conducts photosynthetic products. Both also function in support. Tracheophyta have highly complex and specialized vascular tissues serving both for support and conduction. In this group, conducting tissues are of two types: **xylem** conducts water and mineral nutrients and is the major support tissue, and **phloem** conducts photosynthetic products and provides some support (Fig. 10-9F, J, L).

Still another anatomical feature associated with adaptation to a terrestrial habitat is an outer covering for protection against undue water loss. Unlike most aquatic plants, terrestrial plants have much of their bodies exposed to air. Without some sort of protection against dehydration, it would be impossible for living tissue to survive on land. Bryophytes generally have a very poorly developed protective layer, and this is reflected by the fact that they are physiologically active only under humid conditions. Tracheophytes have a well-developed epidermal layer that is covered by a waxy cuticle (Fig. 10-9K). Woody gymnosperms and angiosperms have an additional suberin-containing protective tissue called **cork** (Fig. 10-9M).

There are physiological as well as anatomical adaptations to the terrestrial environment. These include various features to promote **dormancy,** allowing plants temporarily to tolerate unfavorable conditions, such as aridity and temperature extremes. Physiological adaptations are often manifested in reproductive cycles. For example, many embryophytes survive unfavorable growth periods as seeds or spores. Another means of surviving harsh conditions is through vegetative dormancy. In this instance, the whole plant "turns off"; normal physiological processes are halted or slowed to a very low level, cell cytoplasm decreases in viscosity, and the water potential of cell sap decreases. These changes allow cells to survive conditions that otherwise would kill them. A final physiological adaptation for extreme environmental conditions is the ability of cells to withstand low water concentrations due to dehydration. Such a feature is seen in *Selaginella* and in the gametophytes of many bryophytes.

Reproductive Trends

Male Structures In homosporous embryophytes—most bryophytes and ferns and their allies—male gametes are produced in gametangia called **antheridia.** They are borne on the gametophyte. Bryophyte antheridia are rather large, discrete structures (Fig. 10-10A), whereas those of lower Tracheophyta become progressively reduced (Fig. 10-10B) until, in some ferns, the antheridial wall consists of only three cells (Fig. 10-10C).

In the heterosporous vascular plant, microspores develop into microgametophytes, which in turn produce male gametes. There is no specialized structure that bears gametes; instead, most of the microgametophyte is composed of gametes. As in the antheridia of homosporous forms, there is a reduction series in the microgametophyte (Fig. 10-10D–F). It has several vegetative (nonreproductive) cells in heterosporous lycopods (Fig. 10-10D), and there is a gradual reduction, with some exceptions, in the number of vegetative cells through gymnosperms (Fig. 10-10E) to angiosperms (Fig. 10-10F). In the latter group there is, characteristically, only

Figure 10-9 Vegetative trends in embryophytes. **A–D,** moss: **A,** gametophyte with rhizoids; **B,** sporophyte; **C,** transverse section of gametophyte stem; **D,** spores. **E–H,** fern: **E,** sporophyte with roots; **F,** transverse section of stem (rhizome); **G,** spores; **H,** gametophyte. **I–R,** angiosperm: **I,** sporophyte with roots; **J,** transverse section of young stem; **K,** outer portion of young stem; **L,** transverse section of older stem; **M,** outer portion of older stem; **N,** longitudinal section of flower containing sporangium-bearing structures; **O,** microgametophyte; **P,** ovule (megasporangia and megaspore plus integuments); **Q,** ovule with mature megagametophyte in megasporangium; **R,** seed: *cc,* central column; *ck,* cork; *cu,* cuticle; *e,* epidermis; *ph,* phloem; *sp,* sporangia; *spb,* sporangium-bearing structures; *vt,* vascular tissue; *x,* xylem. Note: Figs. A–U not drawn to scale. (**A, B,** × 120; **C, D, F-H, J, L, O-R,** × 250; **E,** × 0.5; **K, M,** × 400; **I,** × 0.25; **N,** × 1)

one vegetative cell and a total of only three cells in the entire microgametophyte (Fig. 10-10F).

Female Structures In homosporous forms, the female gamete, or egg, is borne in a flask-shaped structure, the **archegonium.** Within these homosporous plants, there is a progressive reduction in the complexity and discreteness of the archegonium (Fig. 10-11A–D). In Bryophytes (Fig. 10-11A), it is generally large and stalked, whereas in homosporous vascular plants, it is small and partially-embedded in gametophytic tissues (Fig. 10-11B).

There are two aspects to consider in the female reproductive structures of heterosporous vascular plants: the **megagametophyte** and the actual gamete bearer. We will consider the gamete bearer first. In some plants (Fig. 10-11C, D), the egg is borne in an archegonium basically like that of homosporous embryophytes. As before, archegonia become progressively reduced in size and complexity from heterosporous lycopods (Fig. 10-11C) to gymnosperms (Fig. 10-11D). In fact, were it not for the connecting series between the archegonia of gymnosperms and those of bryophytes, it would be difficult indeed to detect the existence of archegonia in gymnosperms. Angiosperms have no archegonium; the functional equivalent of the archegonium is the egg apparatus, which consists of the egg and two **synergids** (Fig. 10-11E).

The megagametophyte in heterosporous embryophytes is larger and exhibits a greater reduction series than is seen in the microgametophytes of heterosporous forms. The megagametophyte goes through three growth phases: an initial free-nuclear period, formation of cell walls, and differentiation of gametangia (or gametes). The reduction series in the megagametophyte involves both size (Fig. 10-11F–H) and amount of **free-nuclear division.** In heterosporous lycopods and gymnosperms, there is an extended period of free-nuclear division. These plants have larger gametophytes (Fig. 10-11F, G). In most angiosperms, there are only three free-nuclear divisions (to produce eight nuclei), after which cell walls (or at least membranes) form. The megagametophyte in angiosperms is often called an embryo sac and gives an impression of greater organization than the gametophyte of other heterosporous embryophytes (Fig. 10-11H).

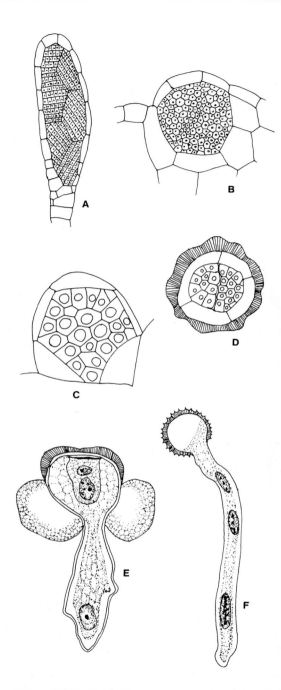

Figure 10-10 Trends in male reproductive structures of embryophytes. **A–C,** trends in antheridia: **A,** moss, ×100; **B,** *Lycopodium*, ×200; **C,** fern, ×400. **D–F,** trends in microgametophytes: **D,** *Selaginella*, ×800; **E,** pine, ×1000; **F,** angiosperm, ×1000.

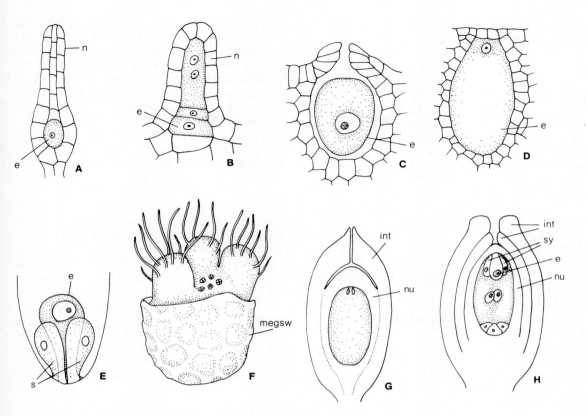

Figure 10-11 Trends in female reproductive structures in embryophytes. **A–D,** archegonia: **A,** moss, ×50; **B,** fern, ×500; **C,** *Selaginella*, ×400; **D,** conifer, ×50. **E,** egg apparatus of angiosperm, ×250. **F–H,** megagametophytes: **F,** *Selaginella*, ×500; **G,** conifer, ×5; **H,** angiosperm, ×5: *n*, neck; *e*, egg; *sy*, synergids; *megsw*, megaspore wall; *int*, integument; *nu*, nucellus. Note: **G** and **H** include structures associated with megagametophyte, namely nucellus (megasporangium) and integuments.

Another trend seen in heterosporous forms is increased retention of the megagametophyte within the sporangium. Truly heterosporous embryophytes have endosporic development of the gametophyte. In gymnosperms and angiosperms, the megagametophyte develops within the spore and within the sporangium (Fig. 10-11G, H). When this occurs, it is often difficult to discern the exact limits of the sporangium. Therefore, the sporangium is usually considered to be composed of the tissue immediately surrounding the female gametophyte. This tissue has a specific name, the **nucellus,** and it is surrounded by one or two **integuments.** The sporangium with its megagametophyte and surrounding integuments is referred to as the **ovule.** After fertilization and development of the zygote, the ovule matures into a **seed,** which contains the **embryo** and nutritive tissue.

The seed is of great evolutionary significance and occurs in two of the dominant plant groups of the world today, gymnosperms and angiosperms. The seed offers two distinct advantages: protection of the enclosed embryo and increased potential for dispersal. The seed is a mass of sterile tissue surrounding an embryo; the sterile tissue can be modified as a result of evolution to promote dispersal

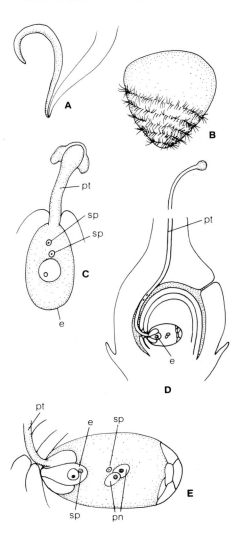

Figure 10-12 Trends in modes of fertilization as reflected in male gametes and their bearers. **A**, moss sperm, ×1000. **B**, cycad sperm, ×750. **C**, microgametophyte and sperm in conifer. **D**, microgametophyte and sperm in angiosperm, ×200. **E**, details of fertilization in angiosperm, showing double fertilization, ×400: *pt*, pollen tube; *e*, egg; *sp*, sperm nuclei; *pn* polar nuclei.

of the **propagule** with no detrimental effect on the embryo itself. Conifer seeds, for example, can be winged, as in *Pinus* (pine), or fleshy, as in *Taxus* (yew). Still another significant aspect of the seed is that it allows the evolution of various structural and physiological mechanisms that enforce seed dormancy under certain conditions, thereby preventing the embryo from beginning its growth under adverse conditions. These statements should not be taken to mean that evolution of the seed automatically means evolutionary success. Many groups of seed plants (e.g., seed ferns, Chapter 21) are extinct.

Fertilization Embryophytes also exhibit trends in their manner of fertilization. In lower embryophytes (bryophytes, and ferns and their allies) an external source of water is necessary for fertilization. In these plants, sperms (male gametes) swim through water from antheridium to archegonium. This mode of fertilization is reflected in the flagellated structure of the sperm (Fig. 10-12A). Fertilization does not require vast quantities of water; the sperm can swim in a thin surface film.

In the seed plants (gymnosperms and angiosperms), the microgametophyte is deposited on a portion of the plant removed some distance from the megagametophyte. This phenomenon, called **pollination,** is a necessary prerequisite for fertilization. In gymnosperms, pollen grains are carried by gravity or wind and deposited close to the ovule. In cycads and *Ginkgo*, flagellated sperms (Fig. 10-12B) are released from the pollen grain into a chamber, and swim to the archegonium. In conifers, a pollen tube is produced by the pollen grain and grows down to the archegonium. The nonflagellated sperm passes down this tube to the egg (Fig. 10-12C). In both instances, fertilization takes place, and a seed is produced.

Fertilization in angiosperms is remarkably different. The pollen grain is deposited on a specialized structure, the **stigma,** that is some distance from the ovule but forms part of the structure in which the ovule is contained. After pollination, a pollen tube containing two nonflagellated male gametes, or sperm cells, grows down to the ovule, and through an opening in the integuments, the **micropyle,** to the megagametophyte (Fig. 10-12D). Both male gametes are involved in the process of fertilization. One fuses with the egg to form the zygote, which develops into an embryo. The other fuses with two (rarely one or more than two) nuclei **(polar nuclei)** in the female gametophyte to give rise to the **primary endosperm nucleus,** which ultimately develops in the seed into nutritive tissue, the **endosperm** (Fig. 10-12E). Because fertilization in angiosperms involves fusion of two male gametes with nuclei of the female gametophyte, it is referred to as **double fertilization.** It can be viewed as an efficient procedure, since it ensures that mobilization of resources to form seed storage products will occur only after fertilization and not before, as can happen in gymnosperms.

Relationships of the Embryophytes

We follow current botanical convention in considering embryophytes in two distinct divisions, Bryophyta and Tracheophyta. Bryophytes and vascular plants are markedly distinct, as already mentioned. Despite this, it is highly probable that bryophytes and vascular plants evolved from one evolutionary line of structurally complex but extinct terrestrial green plants. The great number of biochemical and cytological similarities between Bryophyta and Tracheophyta support this view, as do similarities between reproductive structures in bryophytes and lower tracheophytes. However, we do not wish to imply that bryophytes are ancestral to tracheophytes. In all probability, these two groups are specialized for the habitats they occupy, and both could very well be equally advanced groups.

Distinctive Features

These were discussed in the preceding section on the distinctive features of morphological trends. They are presented again here in a slightly different context to emphasize the differences between bryophytes and tracheophytes.

Bryophyta In bryophytes, gametophytes are the morphologically and physiologically dominant part of the life history and are usually perennial. The gametophyte reaches its highest degree of structural complexity in these plants. The sporophyte bears a single sporangium and is partially dependent on the gametophyte (Fig. 10-9A, B); it is always annual and ephemeral. There is little structural complexity in the bryophyte sporophyte.

Tracheophyta Tracheophyta have a dominant, multisporangiate sporophyte that is usually **autotrophic** and an ephemeral, dependent gametophyte. The gametophytes are structurally simple. This group possesses the most complex and specialized sporophytes in plants; hence, it is usually interpreted as being the most highly evolved. Many features serve to distinguish vascular plants from all other plants. The most obvious is the presence of specialized conducting (vascular) tissues (xylem and phloem) in the sporophyte. In fact, it is the xylem that inspired the name Tracheophyta: because some xylem cells resemble the tracheae of insects, they are called **tracheary elements,** and give the division its name. Other features that serve to distinguish Tracheophyta from other groups of plants are (1) supporting cells with thick cell walls impregnated with lignin, a substance that serves to harden cell walls; (2) an epidermis with specialized cells for gaseous exchange; (3) true roots and leaves in most groups; (4) cork tissues in many tracheophytes; and (5) an ability for growth to great size in both weight and girth without external support for the organisms. Practically all of these distinguishing features are obvious adaptations to growth in a terrestrial environment.

For more information on the Bryophyta and Tracheophyta, the reader is referred to the specific chapters treating these groups in detail (Bryophyta, Chapter 11; Tracheophyta, Chapters 13–27).

References

Bold, H.; Alexopoulos, C.J.; and Delevoryas, T. 1980. *Morphology of plants.* 4th ed. Harper & Row, New York.

Bower, F. O. 1935. *Primitive land plants.* Macmillan & Co., London.

Cutter, E. G. 1969. *Plant anatomy: experiment and interpretation.* Part 1, *Cells and tissues.* Edward Arnold, London.

De Robertis, E. D. P.; Nowinski, W. W.; and Saez, F. A. 1970. *Cell biology.* 5th ed. W. B. Saunders Co., Philadelphia.

Eames, A. J. 1961. *Morphology of angiosperms.* McGraw-Hill Book Co., New York.

Esau, K. 1965. *Plant anatomy.* 2nd ed. John Wiley & Sons, New York.

———. 1976. *Anatomy of seed plants.* 2nd ed. John Wiley & Sons, New York.

Foster, A. S., and Gifford, E. M. 1974. *Comparative morphology of vascular plants.* W. H. Freeman, San Francisco, Calif.

Gunning, B. E. S., and Steer, M. W. 1975. *Ultrastructure and the biology of plant cells,* Edward Arnold, London.

Jensen, W. H., and Park, R. B. 1967. *Cell ultrastructure.* Wadsworth Publishing Co., Belmont, Calif.

Ledbetter, M. C., and Porter, K. R. 1970. *Introduction to the fine structure of plant cells.* Springer-Verlag, Berlin and New York.

Lewy, A. C., and Siekevitz, P. 1969. *Cell structure and function.* 2nd ed. Holt, Rinehart & Winston, New York.

O'Brien, J. P., and McCully, M. E. 1969. *Plant structure and development: a pictorial and physiological approach.* Macmillan & Co., London.

Robards, A. W., ed. 1974. *Dynamic aspects of plant ultrastructure.* McGraw-Hill Book Co., New York.

Troughton, J., and Sampson, F. B. 1974. *Plants: a scanning electron microscope survey.* John Wiley & Sons, New York.

Scagel, R. F.; Bandoni, R. J.; Rouse, G. E.; Schofield, W. B.; Stein, J. R.; and Taylor, T. M. C. 1965. *Plant diversity.* Wadsworth Publishing Co., Belmont, Calif.

Scagel, R. F.; Bandoni, R. J.; Rouse, G. E.; Schofield, W. B.; Stein, J. R.; and Taylor, T. M. C. 1965. *An evolutionary survey of the plant kingdom.* Wadsworth Publishing Co., Belmont, Calif.

11

BRYOPHYTES (MOSSES, LIVERWORTS, AND HORNWORTS)

Mosses and moss allies belong to the division Bryophyta. Both **sporophyte** and **gametophyte** are generally conspicuous, with the sporophyte structurally different from the gametophyte and physiologically dependent upon it. The sporophyte is annual and penetrates the gametophyte. It is structurally simple, unbranched, and bears a single terminal sporangium. The gametophyte, generally the conspicuous part of the life cycle, is often "leafy," autotrophic, and usually perennial. Growth is usually by an apical cell.

In this text, the division Bryophyta includes liverworts (class Hepaticae), hornworts (class Anthocerotae), and mosses (class Musci). Bryophytes show no close relationships to any living plant group, nor do they appear to have been ancestral to any other plant group. Of all plant groups, the subdivision Rhyniophytina, division Tracheophyta (see p. 383), shows the greatest similarities to bryophytes and is suggested to have shared a common ancestor with them. Another theory is that bryophytes and vascular plants arose independently from green algal ancestors.

Bryophytes are relatively small, the smallest being almost microscopic. The largest erect forms are up to 80 cm tall; some aquatic species are more than a meter long, and some epiphytic mosses reach lengths of more than 60 cm.

Most bryophytes absorb moisture and dissolved nutrients directly through the walls of the cells in which they are used. Conduction of water in most is over the surface of the plant, often aided by capillary spaces among leaves or rhizoids. Mineral nutrients come from atmospheric moisture or from water near the soil surface. Most bryophytes absorb and lose water rapidly.

Gametophyte

In bryophytes, the gametophyte reaches its greatest diversification in the plant kingdom. The most structurally complex part of it is the **gametophore,** which produces the sex organs. The spore of a bryophyte usually germinates to produce a structurally simple **protonema.** The structurally more complex gametophore is produced from the protonema. The gametophore is a flattened thallus in many liverworts and most hornworts, but in mosses and

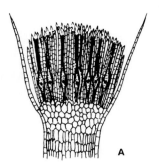

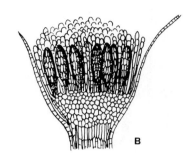

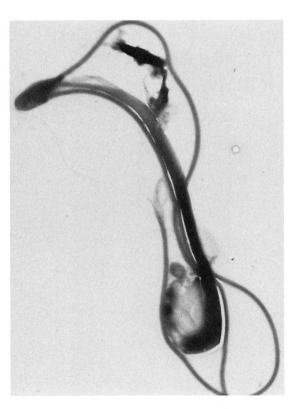

most liverworts it is leafy. Thus the gametophyte consists of two distinctive phases, protonema and gametophore.

Bryophytes lack true leaves and stems, but since they bear structures that are similar in structure and function to true leaves and stems, the traditional terms leaf and stem are used here.

The flask-shaped, multicellular **archegonium** is barely visible to the unaided eye (Figs. 11-1A, 11-2E, F, and 11-3G, H). The **unistratose** neck surrounds a single row of neck-canal cells within. The lower swollen portion of the archegonium is the **venter.** Its walls are often more than one cell thick **(multistratose)** and enclose a single egg. The archegonium is usually attached to the gametophore by a short, multicellular stalk. At maturity, a cap cell at the archegonium apex folds back, exposing the tip of the canal. Meanwhile the neck-canal cells have dissolved, leaving semifluid material that flows into the surrounding water that initiated the rupture of the apex of the archegonium.

The multicellular male sex organ, the **antheridium,** is elongate or spherical, and is generally attached to the gametophore by a stalk (Figs. 11-1B, 11-2E, and 11-3F). Its outer, sterile jacket is unistratose, but there are numerous inner cells, each of which produces a motile sperm that has two anterior whiplash flagella (Fig. 11-1C).

Sporophyte

Although a few bryophytes are not known to produce sporophytes, most develop one by mitotic division of the zygote. Generally only a single egg is fertilized on each sexual branch of the gametophore. The sporophyte produces a single sporangium at the apex of an unbranched stalk termed

Figure 11-1 Sex organs of bryophytes. **A,** archegonia of a moss with intermixed paraphyses, ×10. **B,** antheridia of a moss with intermixed paraphyses, ×10. **C,** sperm of the liverwort *Blasia pusilla*, showing flagella, ×9500. (**C,** after Carothers, with permission of the Council of Linnaean Society of London.)

the **seta.** The base of the sporophyte, the **foot,** penetrates the apex of the sexual branch into the tissue of the gametophore. As the sporophyte develops, the cells of the venter and stalk of the archegonium undergo many further divisions, enlarging to become a protective cap **(calyptra),** which encloses the young, chlorophyll-rich embryo.

Life History

The life history of a bryophyte begins with germination of a spore. When the spore coat is ruptured, a germ tube usually emerges to begin the structurally simple protonema (Figs. 11-2 and 11-3) that precedes production of the more complex gametophore. Ultimately a bud of undifferentiated cells is formed, one cell of which achieves apical dominance and cuts off cells that initiate the gametophore.

The gametophore bears the sex organs. Expelled in a mass from the mature antheridium, sperms float to the surface of the water film that initiated antheridial rupture. They spread quickly and rotate randomly near the water film surface, moving horizontally largely by movement of the water film. As sperms enter the vicinity of the fluid that has diffused from the neck-canal cell, their movement is guided into the archegonium, which has the greatest concentration of this fluid.

The zygote begins to divide immediately after fertilization, producing an embryonic sporophyte. As the embryo develops, cell division and enlargement are also initiated in the venter and stalk of the archegonium, producing a sheathing calyptra that protects the zygote.

The sporangium usually has a multistratose,

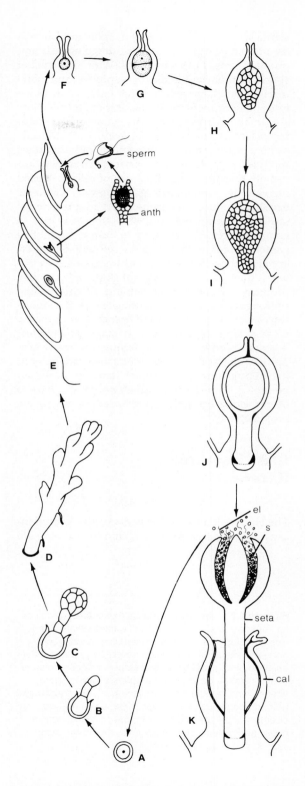

Figure 11-2 Life cycle of typical leafy hepatic. **A–D,** germination of spore and development of gametophyte. **E,** detail of apex of fertile shoot, showing antheridia, archegonium, and fertilization. **F–J,** development of zygote and differentiation of sporophyte, **K,** mature sporangium bursting to shed spores: *an*, antheridium; *cal*, calyptra; *el*, elater; *seta*, seta; *s*, spore; *sperm*, sperm. (After Schuster.)

sterile jacket and, enclosed by it, spore mother cells are ultimately differentiated, undergo meiosis and produce tetrads of meiospores. Wind dispersal of spores is important in most bryophytes, and many mechanisms have evolved that enhance it. Moisture and light are prerequisites of spore germination.

In most bryophytes, the gametophytes also reproduce asexually (Figs. 11-4 and 11-5). Indeed, some mosses and liverworts known only as gametophytes reproduce only asexually. **Gemmae** are usually undifferentiated masses of cells produced on various parts of the gametophyte. Gemmae usually produce the same type of protonemal structures as spores, and ultimately give rise to gametophores. Sometimes they form as diminutive gametophores with an apical cell already differentiated; these occur on either stems or leaves and when they fall from the parent gametophore, they produce a gametophore without an intervening protonemal phase. Fragmenting leaves, deciduous leaves, and fragmenting short branches are all means of vegetative reproduction in many bryophytes. Asexual reproduction is so frequent in many bryophytes that it doubtless diminishes variability in the population. On the other hand, if a random mutant does produce a gametophore, its chances of persisting are greatly enhanced by vegetative propagation.

Class Hepaticae (Liverworts)

The Hepaticae form a highly distinctive evolutionary line in the bryophytes. Within the hepatics are

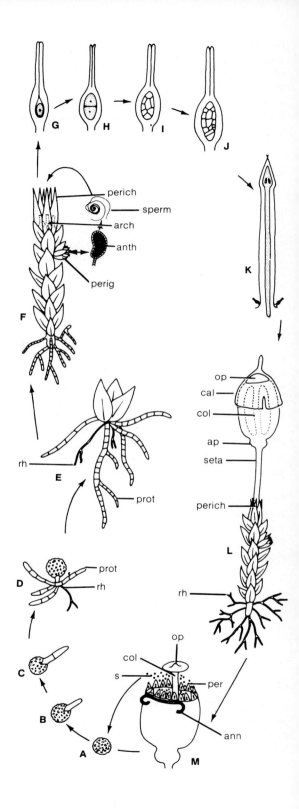

Figure 11-3 Life cycle of typical moss. **A–E,** germination of spore and development of gametophyte. **F,** mature gametophore, showing antheridium, archegonia, and fertilization. **G–K,** development of zygote and differentiation of sporophyte. **L,** gametophore bearing mature sporophyte. **M,** detail of sporangium, showing spore release: *ann*, annulus; *anth*, antheridium; *ap*, apophysis; *arch*, archegonium; *cal*, calyptra; *col*, columella; *op*, operculum; *per*, peristome tooth; *perich*, perichaetium; *perig*, perigonium; *prot*, protonema; *rh*, rhizoid; *seta*, seta; *s*, spore; *sperm*, sperm.

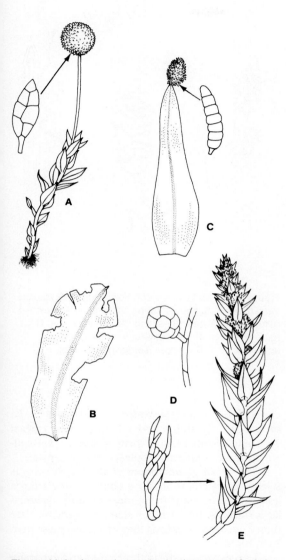

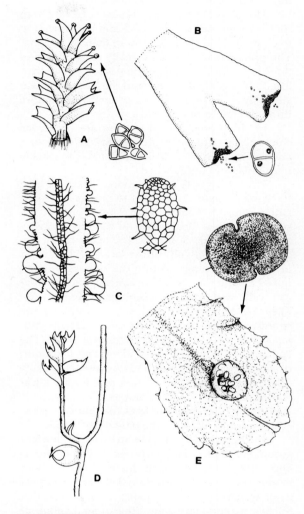

Figure 11-4 Asexual reproduction in mosses. **A,** *Aulacomnium androgynum,* with gemmiferous shoot, ×3. **B,** *Tortula fragilifolia,* showing gaps, fragments from which serve as propagants, ×5. **C,** *Ulota phyllantha* with gemmae at costa apex, ×5. **D,** rhizoidal gemma of *Bryum violaceum,* ×25. **E,** gemma clusters of *Pohlia annotina,* ×4.

Figure 11-5 Asexual reproduction in hepatics. **A,** *Lophozia ascendens,* with gemmae on apical leaf lobes, ×15. **B,** fragment of gemmiferous section of *Riccardia palmata* thallus, showing terminally produced gemmae, ×15. **C,** *Metzgeria* sp., with marginal gemmae on thallus, ×15. **D,** *Plagiochila tridenticulata* shoot showing barren stem from which propagating caducous leaves have fallen, ×14. **E,** *Marchantia polymorpha* thallus with gemma cup containing gemmae, ×2. (**A–D,** after Schuster.)

several divergent evolutionary lines, recognized as separate orders. Although hepatics reach their greatest diversity in the tropics, they are most luxuriant in humid, temperate climates, where they flourish chiefly in shaded sites. Some are highly tolerant of desiccation, and some are aquatic.

Most hepatics are relatively small and vary from tiny, leafy strands less than 0.5 mm in diameter to thalli up to 5 cm broad and more than 20 cm long. The protonemal stage is normally reduced and of brief duration. The gametophore is sometimes thallose, but is generally leafy, and the leaves are usually in three ranks. Leaves are generally unistratose throughout and never have a midrib (**costa**).

In chlorophyllose cells, there are numerous chloroplasts. Characteristic compound oil bodies are also frequent in hepatics. Rhizoids are usually unicellular and unbranched.

The sporophyte is even more distinctive than the gametophyte. In contrast to mosses, an **operculum** is rare and a **peristome** is never present; the seta is of thin-walled cells held rigid by turgidity and usually persists briefly. The sporangium frequently has outer cells of the jacket with annular or spiral thickenings. Within the sporangium and mixed among the spores are **elaters**; they are elongate cells with spiral thickenings on their walls. They are **hygroscopic** and aid in scattering the spores, which are carried away by air currents. **Stomata** are never present in sporangial walls. The sporangium generally opens along four longitudinal lines. The sporangium and its contents are fully differentiated before elongation of the seta, a feature in marked contrast to mosses. The calyptra is ruptured by the sporophyte as the seta elongates, and is left as a sleeve at the base of the seta.

The hepatics appear to be an extremely ancient group. Fossil material interpreted to be hepatics is known from the Devonian. Material obviously of modern genera is known from Baltic amber deposits of Tertiary age. The hepatics show several distinctive evolutionary lines (the orders), none of which can be convincingly considered as ancestral to any others.

Subclass Jungermanniae

Order Calobryales This order contains two families (Fig. 11-6): Takakiaceae (*Takakia*, two species)

Figure 11-6 Morphology of Calobryales. **A,B,** *Haplomitrium hookeri*. **A,** habit showing sporophyte with elongated stem calyptra sheathing base of seta, ×4. **B,** habit of gametophore, showing terminal clusters of antheridia, ×4. **C,** *Takakia lepidozioides*, habit, ×4.

and Calobryaceae (*Haplomitrium*, ten species). *Takakia* is mainly alpine and subalpine in southeast Asia and North Pacific North America. *Haplomitrium* is predominantly tropical and subtropical.

The small gametophore of the Calobryales is a leafy erect chlorophyllose shoot arising from a generally subterranean colorless rootlike system; it lacks rhizoids. *Haplomitrium* stems are generally less than 2 cm high; in *Takakia* they are usually less than 3 cm high.

Leaves are usually in three ranks and generally all are structurally alike. *Takakia* has **mucilage hairs** of two types on erect shoots and on roots. In *Haplomitrium*, roots lack such mucilage hairs. Where roots occur, they appear to be absorptive in function; no other bryophytes appear to have evolved such a rootlike system.

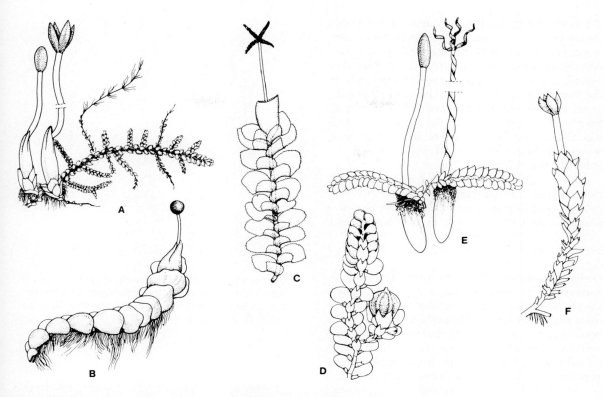

Figure 11-7 Morphology of Jungermanniales, showing various sporophytes and gametophytes. **A,** *Lepidozia reptans* shoot with ripe sporangia, ×5. **B,** *Solenostoma hyalinum* shoot with sporophyte, ×8. **C,** *Scapania nemorosa* shoot with open sporangium, ×5. **D,** *Lejeunea flava* shoot, showing terminal perigonial branch and lateral perianth-bearing branch (note also notched amphigastria and lobules of lateral leaves), ×15. **E,** *Calypogeja fissa* shoot, showing fleshy subterranean perigynium and emergent sporophytes, ×4. **F,** *Marsupella sprucei* shoot with mature sporophyte, ×12. (**A, C, E, F** after Müller; **D,** after Schuster.)

Haplomitrium has flattened, somewhat fleshy, unistratose, cutinized leaves that are roughly elliptical in outline. Within the thin-walled leaf cells are oil bodies and chloroplasts. Cortical cells of the stem are also cutinized, and the innermost stem cells form a central strand, which conducts fluid from the absorptive rootlike system to the leafy gametophore.

Takakia leaves are generally less than 0.5 mm long and consist of one to four tapered segments, each an attenuated cone. Each cell is thin-walled and has several chloroplasts and oil bodies.

Haplomitrium gametophores are unisexual (**dioicous**). When a hepatic gametophore produces the archegonia laterally, rather than utilizing the apical cell of the main shoot, it is termed **anacrogynous**. This is the case for *Haplomitrium*. Archegonia in *Haplomitrium* are few, near the apex, and lateral on the shoot. Each archegonium has a rather massive stalk. In *Haplomitrium*, antheridia are often in a terminal cluster surrounded by several enlarged leaves. Antheridia are numerous, short-stalked, and spherical to ovoid. Archegonia in *Takakia* are similarly located and few. Antheridia have never been observed in *Takakia*.

The sporophyte of *Haplomitrium* terminates the

shoot. In early development of the sporophyte, both the calyptra and the shoot below it grow and enlarge to form a massive, protective **stem-calyptra.** As the cylindrical sporangium matures, the seta elongates rapidly, and the sporangium ruptures the apex of the stem-calyptra and pushes upward on the pale green seta. The sporangium opens by one or four longitudinal lines. Elaters are numerous, and admixed among the spores. The sporangium wall is one cell thick, and each cell has vertically oriented thickened bands.

In *Takakia,* the brittle leaves are readily deciduous and probably serve as vegetative propagules. The rhizomatous extensions from gametophores extend the colonies of all Calobryales.

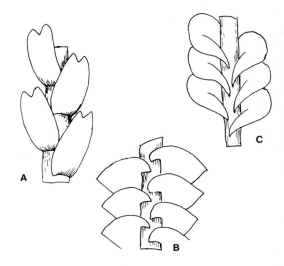

Figure 11-8 Leaf arrangement in Jungermanniales (diagrammatic), showing leaf attachment, as seen from the *upper* surface of the shoot. **A,** transverse. **B,** incubous. **C,** succubous.

Order Jungermanniales The order Jungermanniales (scale mosses) contains over two-thirds of all known hepatics. Approximately 230 genera and more than 7000 species in nearly 40 families have been described. These are distributed from the Arctic and Antarctic to the tropics, and reach their greatest diversity in humid forests in the tropics. They grow as epiphytes, or on rocks and soil, or in water.

The gametophore is generally dorsiventrally flattened with two lateral rows of leaves. On the under surface of the (normally) reclining stem is usually a row of ventral leaves, termed **amphigastria** (Fig. 11-7D). Leaf arrangement follows three basic patterns, illustrated in Figure 11-8. Elaboration of the leaf outline shows great variety in this order. Leaves are generally unistratose, and their cells are usually **isodiametric** and of the same general shape throughout the leaf. Cell walls are of uniform thickness or markedly thickened at the corners, exhibiting conspicuous **trigones** (Fig. 11-9A–C, F). Most genera have numerous complex, colorless oil bodies (Fig. 11-9) in each leaf cell. These tend to be larger than the numerous chloroplasts, and are of a consistent morphology in each species.

The stem is structurally simple. Outer cortical cells are generally chlorophyllose and have oil bodies. Their outermost wall is generally smooth and somewhat thickened. Inner cells generally lack chloroplasts and are thin-walled. The stem is circular or flattened in transverse section. Rhizoids are unicellular in most genera. They arise from the surficial cells of the stem cortex and sometimes from leaf bases.

Branching follows several regular patterns and has been used as a feature to suggest interrelationships among genera. The modes of branch origin differ entirely from those in other bryophytes.

Most species of Jungermanniales are unisexual, but there are various manifestations of the bisexual **(monoicous)** condition. Sex organs are always **discrete,** stalked, and on the surface rather than immersed. Antheridia are spherical or ovoid, and each is borne on a slender, multicellular stalk. They are ensheathed in the axil of a lateral leaf, termed a **perigonium.**

Archegonia terminate the main shoot, utilizing the apical cell in their production, and inhibit further growth of that shoot. This condition is termed **acrogynous,** and is restricted to the Jungermanniales. Several archegonia are on each shoot and are frequently surrounded by the **perianth,** a protective sleeve formed of fused leaves (Fig. 11-7A–D).

The maturing seta of the sporophyte usually elongates rapidly and pushes the sporangium well above the gametophore and its protective envelopes. The seta varies from slender (less than 0.1 mm in diameter) to massive (nearly 1 mm in diam-

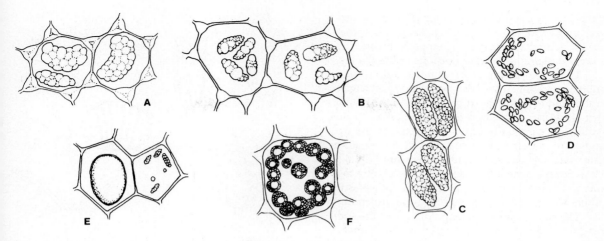

Figure 11-9 Oil bodies of Hepatics (magnification approx. ×600). **A,** median cells of *Cheilolejeunea clausa*. **B,** *Cryptocolea imbricata*, also showing trigones. **C,** basal cells of *Frullania kunzei*, also showing trigones. **D,** *Lejeunea minutilobula*. **E,** *Diplasiolejeunea rudolphiana*, showing differing oil bodies in each cell. **F,** *Lophozia silvicola*. (After Schuster.)

eter). The sporangium is either cylindrical or spherical. The jacket is two to eight cells thick, and its outer cells are generally thickened by brown to black nodular to semiannular thickenings.

Spores are derived from a four-lobed spore mother cell. Among the spores are numerous unicellular, cylindrical elaters which vary in length and in the number of spiral thickenings. The sporangium generally opens by four longitudinal lines that run the full length of the sporangium. Upon drying, the jacket divisions fold back and expose the sporangium contents (Fig. 11-10A–C). Elaters immediately dry out and leap into the air, loosening the spores and discharging them to be carried away by air currents. Spores are generally small, 6–30 μm in diameter, numerous, and are usually unicellular when dispersed.

Spore germination patterns are highly diverse. The spore can produce a uniseriate or multiseriate filament, which soon initiates an apical cell that produces the leafy gametophore. In some instances, it produces an **endosporic** or **exosporic** cell mass that differentiates an apical cell.

Asexual reproduction is frequent in the Jungermanniales. Many species have brittle leaves; leaf fragments or even entire leaves can produce new gametophores (Fig. 11-5D). **Gemmae** are often produced on leaf margins or at shoot apices (Fig. 11-5A).

The Jungermanniales do not appear in the fossil record until the Tertiary. Some beautiful preserved material in Baltic amber deposits can be referred to extant genera, implying that the order must have originated before the Tertiary.

Order Metzgeriales The order Metzgeriales contains 8 families with approximately 25 genera. The order is worldwide but reaches its greatest diversity in the humid tropics. Most species grow in damp, shaded habitats, but some tolerate exposed sites.

The gametophore is usually thallose but is sometimes leafy (Fig. 11-11). Thalli are highly diverse and are generally prostrate on the substrate; branching is usually irregular.

Except for cells with perforated walls, found in the thalli of several genera, cells are mainly thin-walled. Upper cells are chlorophyllose, but those deeper in the thallus lack chlorophyll. In most genera, many chloroplasts and numerous oil bodies are in each cell. The generally smooth rhizoids are confined mainly to thicker portions of the thallus.

Thalli are always anacrogynous and are generally unisexual. Antheridia are variously arranged: embedded in pits in the upper surface of the thallus (Fig. 11-11B), exposed on the upper surface of the thallus, or surrounded by a flap of tissue. Archegonia are generally in a chamber made up of an extended flap of tissue, the **perigynium** (Fig. 11-11A). Several genera have an extensive calyptra that surrounds the developing sporophyte, whereas in others the perigynium serves the same function.

The seta is generally long, and raises the sporangium well above the surface of the gametophore. The sporangium wall is two to five cells thick, and outer cells have **semiannular** thickenings. The sporangium opens somewhat irregularly or by one, two, four, or even six longitudinal lines. Elaters are scattered among the spores. Spores are derived from a four-lobed spore mother cell. When shed, spores are generally unicellular. The germinating spore usually produces a short filament, but an apical cell is differentiated early, and initiates production of the more complex gametophore. In the genus *Fossombronia*, the gametophore is so uniform that species are distinguished on the basis of spore sculpturing.

Hepaticites devonicus of Devonian age is the earliest material attributed to the Hepaticae (Fig. 11-12A, B). It is structurally similar to the Metzgeriales. Carboniferous fossils presumed to be hepatics also resemble several metzgerialean genera. All are placed conservatively in the form genus *Hepaticites*.

Subclass Marchantiae

Order Sphaerocarpales The order Sphaerocarpales contains three genera and approximately 30 species in two extant families. The fossil family Naiaditaceae contains only *Naiadita*, with one species. The extant genera are found primarily in warm temperate to subtropical countries. Both *Sphaerocarpos* and *Geothallus* grow on somewhat clayey soil, whereas *Riella* and the fossil *Naiadita* are submerged aquatics, growing on mud and in quiet pools. Some species of *Riella* are **halophytes**.

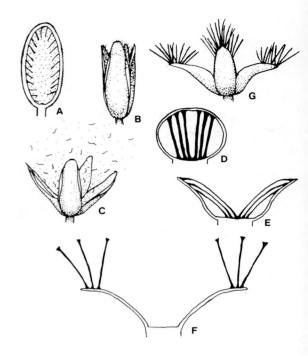

Figure 11-10 Spore discharge in Jungermanniales (diagrammatic). **A–C**, *Cephalozia bicuspidata*. **A**, longitudinal section through sporangium, showing elaters and spores (diagrammatic). **B**, sporangium, showing longitudinal splitting of sporangium wall, exposing the elaters to air. **C**, sporangium, showing rapidly uncoiling elaters leaping from sporangium, dislodging the spores, and throwing them into the air. **D–G**, *Frullania dilatata*. **D**, longitudinal section through sporangium, showing elaters attached to walls and base of the sporangium (diagrammatic). **E**, sporangium opening, showing elaters strongly stretched by the opened sporangium. **F**, sporangium, showing that the elaters have become rapidly detached from base of sporangium, thus flicking the sporangial walls outward quickly, and throwing the spores into the air. **G**, view of ruptured sporangium, showing diagrammatically in **F**. (Modified after Ingold, with permission of Oxford University Press.)

A distinctive, unistratose "bottle" of tissue surrounds each sex organ and thus the sporophyte (Fig. 11-13C–E). The sporophyte is also distinctive. A seta is essentially absent, and the sporangium jacket is unistratose and lacks any ornamentation on the cell walls. Elaters are absent (but **nurse cells**

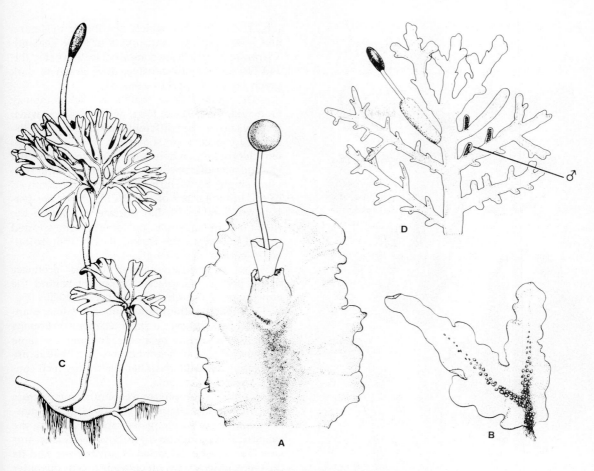

Figure 11-11 Metzgeriales; habits of a variety of morphological types. **A, B,** *Pellia neesiana;* **A,** sporophyte-bearing thallus lobe, ×6. **B,** *P. neesiana* antheridial thallus, ×6. **C,** *Hymenophytum flabellatum* thallus with sporophyte, ×3. **D,** *Riccardia multifida*, portion of thallus with sporophyte and antheridial lobes, ×5.

are present), and spores are dispersed after decay of the sporangial jacket, rather than by a specialized dehiscence.

In *Sphaerocarpos*, the thallus is unisexual and has a thickened, stemlike central portion with leaflike lobes (Fig. 11-13C, E). Mature thalli are usually 0.2–1.5 cm in diameter and are pallid green. The "bottles" cover much of the upper surface of the thallus. Simple, smooth rhizoids attach the thallus to the substrate. Spores usually remain united in tetrads when shed. In *Geothallus*, **tubers** are formed within the thallus behind the growing points. These serve as vegetative propagules.

The genetics of *Sphaerocarpos* is better understood than that of any other bryophyte. In species of this genus, sex-correlated chromosomes were first discovered in plants. In the archegonium-bearing plants, one chromosome is much larger than the seven others; in the antheridium-bearing plant, one chromosome is much smaller than the others. In

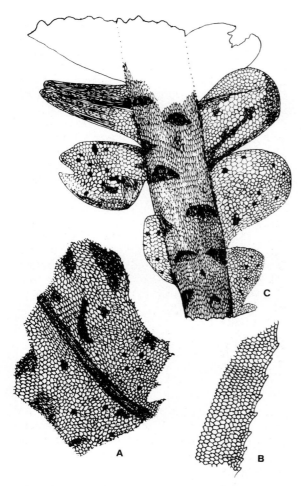

Figure 11-12 Fossil Metzgeriales. **A,** fragment of the Devonian hepatic *Hepaticites devonicus*, showing similarity to the modern genus *Pallavicinia*, ×20. **B,** margin of thallus of *H. devonicus*, showing teeth, ×20. **C,** the Carboniferous hepatic *Hepancites kidstoni*, ×40. (**A, B,** modified from Hueber; **C,** modified from Walton.)

each tetrad of spores, two spores produce male gametophores, and the others produce female gametophores.

It is impossible to suggest any close relationships of the Sphaerocarpales to any other hepatics.

Order Monocleales The order Monocleales has one genus, *Monoclea*, and two very similar species (Fig. 11-14), both are widely distributed in Central and South America and one is in New Zealand. The unisexual thalli are somewhat leathery (Fig. 11-14A, B) and form extensive, olive green to dark green mats in humid to wet sites.

The multistratose thallus is dichotomously branched, up to 20 cm long, and up to 5 cm wide; it shows no marked differentiation into discrete tissues. Upper cells are somewhat smaller and contain chloroplasts, whereas deeper in the thallus the chloroplasts disappear. Some scattered cells contain a single, brown oil body. Rhizoids are confined to the undersurface of the thallus. Most are perpendicular to the thallus surface and attach the thallus to the substrate, but others form a tufted central line along the thallus length and are oriented parallel to the length.

The antheridium-bearing thallus produces small, padlike receptacles, generally behind the growing point of each dichotomy of the thallus (Fig. 11-14A). Each antheridium embedded in the antheridial pad occupies a small chamber that opens to the upper surface by a pore. In much the same position on an archegonium-bearing thallus are elongate, sleevelike chambers, each of which contains several archegonia.

Multiple sporophytes frequently emerge from each archegonial chamber (Fig. 11-14B). The massive, colorless seta elongates rapidly, pushing the cylindrical sporangium high above the thallus surface. The sporangial jacket is unistratose, and its cells have thickenings on all walls except the outer one (Fig. 11-14D). The sporangium opens by a single, longitudinal line, exposing elongate elaters that spring from the sporangial cavity and throw the spores into the air.

The order Monocleales appears to represent an independent evolutionary line, closely allied to the Marchantiales.

Order Marchantiales The most specialized of the hepatics are in the order Marchantiales (Fig. 11-15), which contains 11 families with approximately 30 genera and 450 species. The order is worldwide in all climatic regions. Most grow on moist earth or mud, a few are commonly found in water, and most are perennial.

The prostrate gametophore is thallose, generally dichotomously branched, and varies from a few millimeters to more than 2 cm in diameter. It is generally internally differentiated into several distinct tissues. The upper surface forms a unistratose epidermis, the outer walls of which are often cutinized. The epidermis is often punctured by special pores that open to intercellular air chambers beneath. The pores are usually continually open and are bounded by several cells.

The air chambers enclose the photosynthetic tissue. Chloroplasts are in cells that form the chamber walls or in filaments that often arise from the floor of the chamber. Scattered cells of the thallus contain, instead of chloroplasts, a single, large, complex oil body. Cells below the chlorophyllose region are predominantly parenchymatous; some cells are filled with mucilage, and occasional ones have pits but do not form a continuous conducting system. This parenchymatous tissue is often invaded by fungi. There is a ventral epidermis.

Some of the lower epidermal cells, particularly those of a thickened band of tissue of the middle of the thallus, produce colorless rhizoids of several types: some that are smooth-walled (either broad or slender), some with peglike thickenings on the inner walls, and some with alternate constrictions in the walls, giving them a corkscrewlike appearance (Fig. 11-16E). Rhizoids are either perpendicular to the thallus surface and attach it to the substrate, or they diverge backward away from the growing point and are longitudinal to the thallus. These latter rhizoids, plus the scales on the undersurface of the thallus, serve to conduct water along the length of the thallus. There are various other types of tissue organization represented in the order; Figure 11-16 illustrates some of these.

Thalli are bisexual or unisexual. Each antheridium is enclosed within a chamber that opens to the upper surface of the thallus. In many genera, antheridia are embedded in a **peltate** receptacle that terminates a specialized extension of the gametophore termed the **antheridiophore** (Fig. 11-15E). Sperms sometimes are dispersed widely by splashing raindrops.

Archegonia are sometimes embedded in the thallus, as in *Riccia*, or are on the thallus surface and surrounded by a protective envelope, as in most genera. Archegonia are also often borne on a specialized receptacle, the **archegoniophore**.

The sporangium is from spherical to elongate; its jacket is unistratose and the walls may be spirally thickened or lack ornamentation. Within the sporangium, spores are produced from an unlobed spore mother cell and are generally mixed with elaters. The sporangium opens in different ways: by four longitudinal lines, by an operculum, by irregular rupturing, or even by decomposition of the sporangial jacket. Spore ornamentation is variable and often elaborate. Spores are usually unicellular, and their size varies from 10 μm to about 40 μm.

Asexual reproduction by gemmae is uncommon in the order. When present, the gemmae are discoid; they occur on slender stalks and are formed in gemmae cups on the surface of the thallus (Fig. 11-5E). Asexual reproduction also results from fragmentation of the thallus.

The fossil record of the Marchantiales is rather scanty. The oldest fossil material suggestive of this order is of Carboniferous age. The order Marchantiales represents the peak of tissue organization in the hepatics. It also represents an independent evolutionary line within the hepatics, set apart from the others in both morphology and details of gametophyte development.

Class Anthocerotae (Hornworts)

Hornworts superficially resemble some thallose liverworts. They show sporophytic and gametophytic features that are absent from other bryophytes and separate them as an independent evolutionary line:

1. Generally a single chloroplast is in each cell.
2. A **pyrenoid** is often present in the chloroplast.
3. The archegonium is not a discrete organ.
4. Sex organs are derived from a **superficial** cell, but one that never becomes **papillate**.
5. Occasionally, stomatelike structures occur in the thallus.
6. The sporophyte is of indeterminate growth.
7. An **intercalary** meristem is present at the base of the sporophyte.
8. Elaters are generally multicellular.

The class includes a single order, the Anthocerotales, and a single family, Anthocerotaceae.

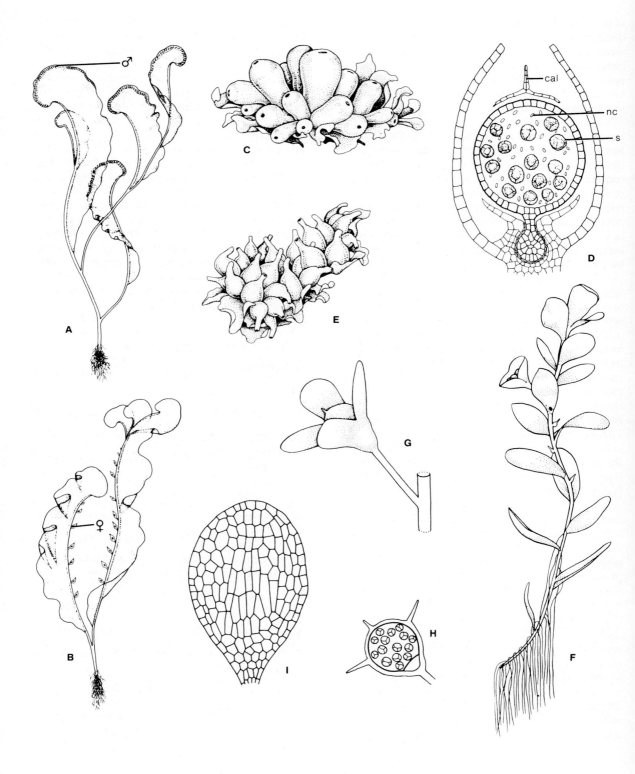

316 Chapter 11

There are five genera: *Notothylas, Anthoceros,* and *Phaeoceros* are rather widely distributed, particularly in milder climates; *Dendroceros* and *Megaceros* are essentially tropical and subtropical. Most species grow on shaded, humid soil.

The gametophore is flattened, thallose, and normally composed of radiating, overlapping lobes. The thallus varies from 1–4 cm in diameter and is occasionally larger. A midriblike, thickened central band is sometimes present, which generally thins gradually toward the margins. The thallus is either annual or perennial.

Most cells of the thallus are thin-walled; the upper cells are chlorophyllose, but the inner cells lack chloroplasts. In most genera, each cell contains a single, large, lens-shaped chloroplast associated with the pyrenoid. The thallus often contains mucilage-filled intercellular cavities, some of which open to the ventral surface by stomatelike pores (Fig. 11-17B). These cavities are often invaded by colonies of the bluegreen alga, *Nostoc*. The thallus is attached to the substrate by thin-walled, unicellular rhizoids.

Most species are bisexual, with the sex organs embedded in the upper surface of the thallus. Their mode of origin is unlike that in other bryophytes. There are usually several antheridia within a chamber. Cells surrounding the contents of the archegonium do not differ conspicuously from others in the thallus; thus the archegonium is not a discrete organ.

The sporophyte is always cylindrical and tapers to the apex (Fig. 11-17A). As it elongates, the sporophyte pierces the upper surface of the thallus and is generally perpendicular to it. The enlarged foot is embedded in the thallus (Fig. 11-17E). The sporophyte varies from a few millimeters up to 16 cm tall and is green until it begins to shed spores. Early in sporophyte development, an intercalary meristem is formed near the base, above the foot. This meristem continues to differentiate new sporophytic tissue throughout the life of the sporophyte, which gives it indeterminate growth. Most of the sporophyte is sporangium. The jacket is four to five cells thick, and the outer cells have chlorophyll. The epidermis of most species has characteristic stomata (Fig. 11-17C). The sporogenous layer generally forms a dome-shaped region that overtops a central **columella**. Among the spores of Anthocerotae are pseudoelaters (Fig. 11-17F) that are often multicellular with irregularly thickened or spirally thickened walls.

The sporangium usually opens by two opposite, longitudinal lines that meet at the apex of the sporangium. The dehiscence lines extend downward toward the sporangium base as the sporogenous tissue matures; thus spores are shed over a relatively extended period. Spores are sometimes viable for only a brief period, but they may be viable after more than ten years.

Some features in rhyniophyte (see p. 383) fossils suggest that the Anthocerotae may have been derived from an ancestor common to this group. The rhyniophyte *Horneophyton* has a columella; the spore ornamentation resembles that of some species of *Anthoceros*; and the sporophyte is a naked axis, albeit one that is dichotomously branched. Hornworts, hepatics, and mosses have so many features in common that it is reasonable to assume a relationship.

Figure 11-13 Sphaerocarpales. **A,** habit of *Riella americana* thallus with antheridia on margin, ×3. **B,** *R. americana* thallus with sporophytes on stem, ×3. **C,** *Sphaerocarpos texanus* archegoniate plants, showing bottles containing sporangia, ×15. **D,** longitudinal section through *Sphaerocarpos* sporangium and bottle, ×35. **E,** *S. texanus* antheridial plant, showing antheridium-containing bottles, ×50. **F,** reconstruction of the Triassic hepatic *Naiadita lanceolata*, ×6. **G,** sporangium of *N. lanceolata* with subtending bracts, ×10. **H,** longitudinal section through sporangium of *N. lanceolata*, showing tetrads of spores, ×8. **I,** leaf of *N. lanceolata*, ×50: *col.,* columella; *nc,* nurse cells; *s,* spores. (**A, B,** after Studhalter; **F–I,** after Harris.)

Class Musci (Mosses)

Musci represent the most diverse assemblage of bryophytes, with more than 680 genera and somewhat fewer than 15,000 species. These can be separated readily into six distinctive evolutionary lines, corresponding to the subclasses.

In many ways the mosses resemble the "leafy" liverworts (Order Jungermanniales). Mosses, however, show several fundamental differences:

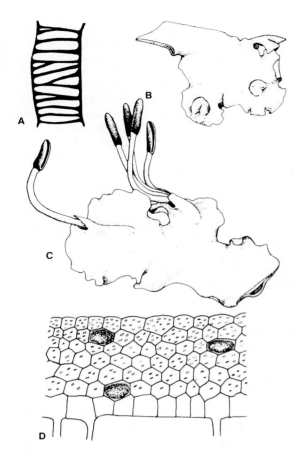
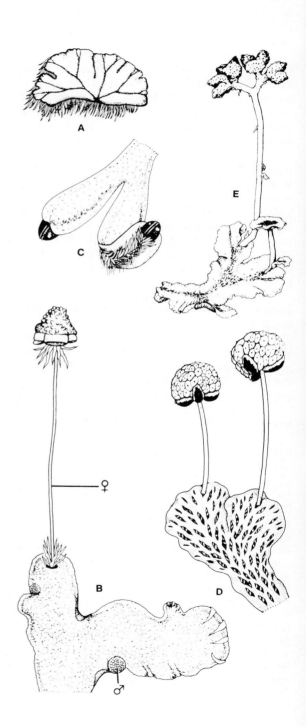

Figure 11-14 Monocleales: *Monoclea forsteri*. **A,** portion of thallus with antheridial pads, ×0.75. **B,** portion of thallus with sporophytes, showing longitudinal dehiscence of the sporangia, ×0.75. **C,** transverse section of thallus, showing oil bodies and ventral rhizoids, ×120. **D,** cell of jacket from longitudinal section of sporangium, showing thickenings of walls, ×100.

Figure 11-15 Marchantiales, showing variation in morphology. **A,** *Ricciocarpus natans*, habit of aquatic thallus, ×6. **B,** *Mannia sibirica*, unisexual thallus with single carpocephalum (note operculum on sporangium), ×5. **C,** *Targionia hypophylla*, portion of thallus with ventral sporangia, ×3. **D,** *Mannia rupestris*, portion of thallus with carpocephala (note slits forming openings into air chambers of thallus), ×6. **E,** *Neohodgsonia mirabilis*, unisexual thallus with carpocephalum (*left*) and antheridiophore (*right*), ×4. (**B, D,** after Schuster.)

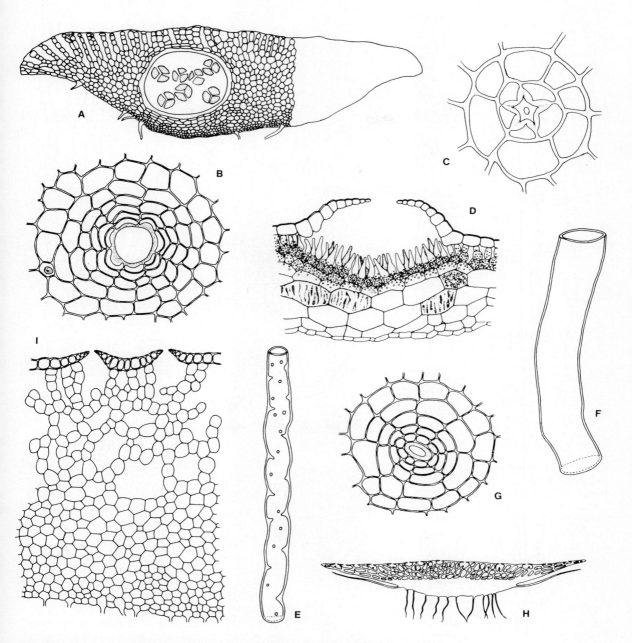

Figure 11-16 Anatomy of Marchantiales (transverse sections through thalli). **A,** *Riccia beyrichiana*, showing sporangium containing spores, ×20. **B,** surface view of air-chamber pore of *Conocephalum conicum*, ×180. **C,** surface view of compound pore of *Athalamia hyalina*, ×200. **D,** air chamber of *C. conicum* (transverse section of thallus), showing chlorophyllose filaments, ×180. **E, F,** rhizoids of *C. conicum*, ×360. **G,** surface view of air-chamber pore of *Reboulia hemisphaerica*, ×180. **H,** transverse section of thallus of *R. hemisphaerica*, ×7.5. **I,** detail of transverse section of thallus and air chambers of *R. hemisphaerica*, ×180. (**A, B, G–I,** after Schuster.)

Bryophytes (Mosses, Liverworts, and Hornworts)

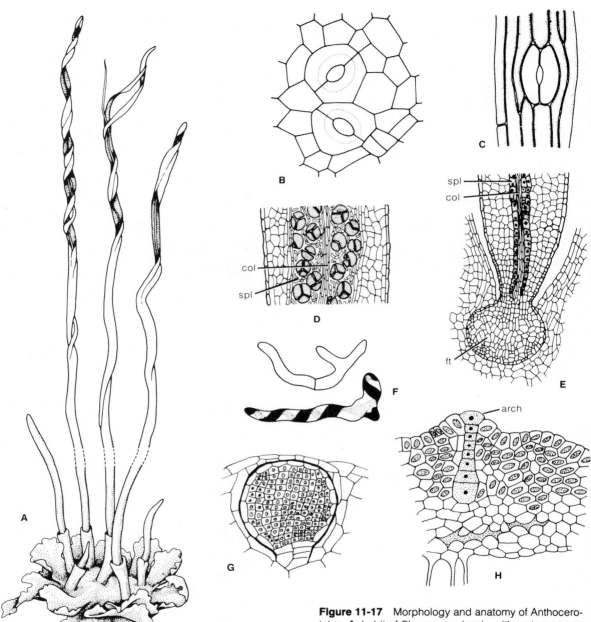

Figure 11-17 Morphology and anatomy of Anthocerotales. **A,** habit of *Phaeoceros laevis,* with mature sporangia, ×5. **B,** pores of gametophytic thallus of *Anthoceros,* ×225. **C,** stoma with guard cells of *Anthoceros* sporangium wall, ×150. **D,** longitudinal section through mature sporangium (note pseudoelaters among spore tetrads), ×100. **E,** base of *Anthoceros* sporangium, ×160. **F,** pseudoelaters of Anthocerotales (*upper, A. punctatus; lower, Megaceros endivaefolius*), ×500. **G,** transverse section through fertile thallus, showing mucilage chambers (*black*) ×80. **H,** young antheridium filling antheridial chamber (note single lenticular chloroplast in each cell), ×200: *arch,* archegonium; *col,* columella; *ft,* foot; *spl,* sporogenous layer. (**B,** after Goebel.)

320   Chapter 11

1. The protonema is normally extensive and usually forms a uniseriate, much-branched system that precedes the appearance of the gametophore.
2. The gametophore is leafy, and leaves are usually in more than three ranks.
3. Complex oil bodies are absent.
4. Rhizoids are multicellular.
5. Leaves commonly have a multistratose midrib (costa).
6. Leaf cells are commonly elongate and rarely have trigonous corners.
7. Leaves are rarely lobed.
8. The sporangial wall frequently has stomata and lacks spiral or nodular ornamentation of the outer cells.
9. The sporangium usually opens by an apical lid, beneath which hygroscopic **peristome teeth** are usually present.
10. A columella is in the interior of the sporangium.
11. The seta is usually long and wiry, and the sporophyte may persist for a long period so that the spores can be shed over an extended interval.
12. The seta elongates *before* the sporangium is fully differentiated, and the calyptra is torn from the gametophore to form a protective cap over the apex of the elongating sporophyte.

Gametophyte

Spore germination is essentially exosporic. The protonema is chlorophyllose and is attached to the substrate by means of multicellular, uniseriate, nonchlorophyllose rhizoids that have oblique end walls to their cells. The structure of the gametophore varies greatly among mosses, from unbranched shoots to a complex system of branches.

The position of the archegonia takes two main patterns: (1) they may terminate the main shoot, in which case the apical cell of that shoot is used to produce archegonia and their associated structures; in these mosses the sporophyte also terminates the main shoot, and is described as **acrocarpous;** (2) they may occur on reduced lateral branches, so that the apical cell of the main shoot is not used up in producing archegonia, but continues to differentiate new growth from year to year; in these mosses the sporophytes are on budlike lateral branches and are termed **pleurocarpous.**

Leaves are generally spirally arranged on the stem and generally unistratose except for the multistratose costa. Leaf cells are variously ornamented by thickenings or bulging of the exposed surfaces. Leaf shapes vary considerably (Fig. 11-29). Besides bearing leaves, the stem sometimes has small, unistratose, branched, chlorophyllose structures termed **paraphyllia** (Fig. 11-26). These may be important in external water conduction along the stem.

Moss gametophores are frequently unisexual. In bisexual mosses, archegonia are generally on different branches than antheridia. Archegonia are generally surrounded by specialized leaves termed **perichaetial leaves,** and among the archegonia are scattered filaments termed **paraphyses**. Antheridia also usually have paraphyses intermixed among them, and these are enclosed by **perigonial leaves** (Fig. 11-1A, B).

Moss stems are usually structurally more complex than those of hepatics. The epidermal cells, usually termed the outer cortical cells, often differ structurally from those of the center of the stem. Often the center of the stem consists of a central strand of smaller cells. In some mosses, especially in the central strand, some of the cells resemble sieve-tube elements (see p. 363) in both structure and function and are termed **leptoids.** In other mosses, especially in the subclass Polytrichidae, there are also clusters of tracheidlike cells, the **hydroids** (Fig. 11-18). The stem is sometimes densely clothed in rhizoids.

Vegetative propagation of the gametophore is by gemmae and fragmentation. Most expansion of local gametophore populations appears to be by vegetative propagation (Fig. 11-4).

Sporophyte

The sporophyte generally consists of a sporangium at the apex of a usually rigid seta that is composed of longitudinally elongate cells. In some mosses, the seta has both hydroids and leptoids; leptoids alone may be present, or both may be lacking.

While the sporophyte is growing, it is rich in chlorophyll. Elongation of the seta is generally very slow. The sporangium wall is always several cells thick and often has stomata. The sporangium usually opens by a line of weakness that circles it near the apex; the lid cut off is the operculum. The calyptra usually caps the sporangium until the operculum is shed.

Most mosses have teeth around the mouth of the sporangium. These are the peristome teeth and are exposed when the operculum is shed. Various devices have evolved that assist in releasing spores from the sporangium. In most instances, moving air is the main mechanism for spore dispersal. The sporangium usually has a central mass of sterile tissue, the columella, surrounded by a cylinder of spores.

Subclass Sphagnidae

The genus *Sphagnum* (peat moss) is the sole representative of the subclass Sphagnidae. About 150 species can be distinguished. *Sphagnum* is worldwide in distribution, but forms extensive peatlands only in the Northern Hemisphere. All species grow in wet sites.

The gametophore is often large, occasionally exceeding 50 cm in length, but the living portion rarely exceeds 10 cm. The main axis bears inconspicuous, nonchlorophyllose, widely spaced, spirally arranged leaves. At every fourth leaf of the spiral, a **fascicle** of branches generally emerges. Three to eight branches emerge from each point on the stem, two or three of which generally diverge; the remainder droop downward against the stem. The branch leaves are **imbricate**.

The unistratose leaves lack costae, and are made up of a network of elongate chlorophyllose cells that encircle larger, swollen, dead **hyaline** cells (Fig. 11-19D). The outer walls of the hyaline cells are generally perforated by several pores, and the walls are reinforced by fibrillar thickenings (Fig. 11-20B).

The stem apex generally bears a condensed mass of branches (Fig. 11-19A). The mature gametophore is perennial and lacks rhizoids. The main stem of the gametophore is somewhat brittle, and aggregations of gametophores support each other or are supported by their aquatic medium. The stem is differentiated into a cortex, in which cells are often enlarged, hyaline, porose, and sometimes have fibrillar wall thickenings; and a central cylinder made up of a layer of thick-walled cells that enclose thin-walled cells.

Gametophores are bisexual or unisexual. Spherical antheridia are on specialized divergent branches, and each is borne on a slender stalk—one at the base of each leaf of a **perigonial branch**. **Perichaetial branches** are reduced to a few enlarged leaves that enclose three or four archegonia.

After fertilization of one egg, the apex of the perichaetial shoot, just beneath the archegonia, begins to elongate somewhat. As the sporophyte matures, the gametophytic tissue beneath it elongates rapidly and extends the sporophyte on a leafless **pseudopodium** (Fig. 11-19L) well above the **perichaetium**. The sporophyte is made up of a subspherical sporangium and an enlarged foot.

As it enlarges, the sporangium is completely enclosed within a calyptra, but this is ruptured early in sporangium development. The walls of the chlorophyllose sporophyte become brownish upon maturity. The sporangium opens by means of an operculum. Within the sporangium, the sporogenous layer forms an inverted cup over the central, dome-shaped columella. Rupturing of the operculum is remarkably specialized. As the sporangium matures, cells of the columella collapse and are replaced by gaseous material. As the sporangium dries, it shrinks. Internal pressure can build up to five atmospheres, and becomes so great that the operculum is thrown off explosively. The spores are ejected into the air and carried away (Fig. 11-21).

The spore germinates to produce a short filament terminated by a tiny, fragile, multicellular, unistratose, chlorophyllose thallus with short rhizoids extending to the substrate (Fig. 11-19J). This thallus ultimately initiates the leafy gametophore. Expansion of colonies of *Sphagnum* is mainly by fragmentation. When broken off, young branches can ultimately produce an entire gametophore.

Because of the abundance of dead, porose hyaline cells, gametophores of *Sphagnum* absorb great quantities of fluid, sometimes up to 20 times their own weight. This characteristic is significant both in their commercial use and in their role in vegetation dynamics.

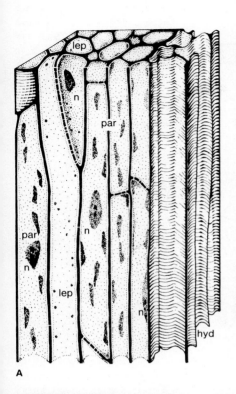
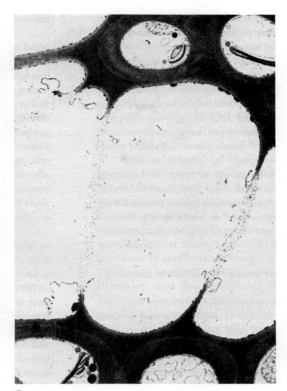
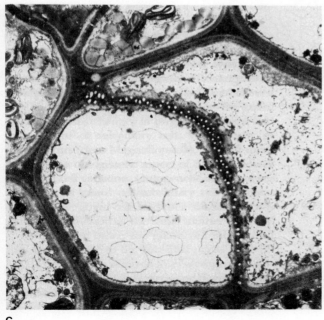

Figure 11-18 Conducting systems in mosses. **A,** diagrammatic representation of the organization of conductive tissue in a leafy stem of *Polytrichum commune*, showing hydroids (*hyd*) and leptoids (*lep*), arranged much as are xylem and phloem tissues in vascular plants: *n*, nucleus; *par*, parenchyma. **B,** transmission electron micrograph of hydroids of *P. commune*, ×2000: *st*, stereid. **C,** transmission electron micrograph of leptoids of *P. commune*. (Courtesy of C. Hébant.)

Bryophytes (Mosses, Liverworts, and Hornworts) 323

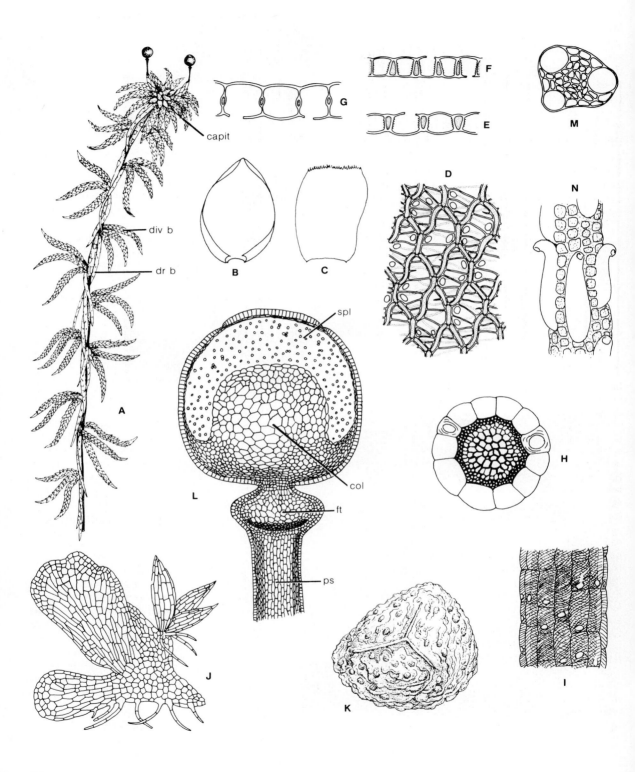

Floating mat species of *Sphagnum* growing on lake and pool margins add to the acidity of these water bodies; the water becomes turbid, highly acidic, and low in oxygen, and it inhibits the decomposition of organic material. In time, the floating mat, first restricted to the perimeter of the lake, expands toward the center. Accumulation of organic material in the marginal shallows makes these areas solid enough for other plants to invade them. The ultimate result of such succession and persistent organic sedimentation is the filling in of the water body with organic peat. However, most peatland is not a result of lake-deposited peat but of the establishment of *Sphagnum* and other plants tolerant to waterlogged habitats.

Sphagnum peat has been used for many years as insulating material for homes. It has also been used as fuel for fires. Its replacement of cotton as a surgical dressing was of particular importance during much of the first World War when cotton was unavailable.

Reliably identified *Sphagnum* spores and leaves are known from as early as the lower Jurassic, whereas modern species can be determined from Quaternary material.

Subclass Andreaeidae

The Andreaeidae include a single order, Andreaeales; a single family, Andreaeaceae; and two genera, *Andreaea*, with possibly fewer than 50 distinct species, and *Andreaeobryum*, with one. The order is widespread in temperate to frigid climates throughout the world.

Andreaea grows on exposed siliceous rock, where it usually forms blackish to red-brown cushions. *Andreaeobryum* grows on limestone in northwestern North America.

Andreaea is similar to *Sphagnum* in that the sporangium is raised above the perichaetium on a pseudopodium and the sporogenous tissue that overarches the columella is in the same position (Fig. 11-22D). In sporangium dehiscence, it is unique among mosses in having longitudinal lines of opening. Thus, when mature, the sporangium wall gapes open, often by four elongate slits (Fig. 11-22A, B).

The protonema is generally multiseriate and has some biseriate as well as uniseriate rhizoids. Although very simple in morphology and anatomy, the gametophore most resembles that of some Bryidae. The brownish to black gametophore is irregularly branched and usually tightly affixed to the substrate by rhizoids at the base of the shoot (Fig. 11-22A). Leaves are spirally arranged, crowded, and generally ovate to lanceolate. Except for those around the sex organs, the leaves are all similar. They are either costate or lack a costa. All leaf cells are small and chlorophyllose, although this is obscured by the darkly pigmented cell walls. The stem is of thick-walled cells throughout, and cell walls are pigmented. The mature gametophore varies from 1–20 cm in length, with leaves 0.5–3 mm long. Gametophores are bisexual or unisexual. Enlarged perichaetial leaves terminate the main shoot; thus the gametophore is acrocarpous. Several archegonia are in each perichaetium, intermixed with a few short paraphyses. Antheridia are on reduced, bulbiform lateral branches, and peri-

Figure 11-19 Morphology and anatomy of subclass Sphagnidae. **A,** habit of sporophyte-bearing shoot of *Sphagnum palustre*, ×1. **B,** branch leaf of *S. palustre*, ×12. **C,** stem-leaf of *S. palustre*, ×12. **D,** leaf cells of *S. palustre* (note network of **E,** *Sphagnum palustre*. **A,** chlorophyllose cells surrounding porose hyaline cells; also the fibril thickenings of walls of hyaline cells), ×165. **E,** transverse section of *S. palustre* leaf, showing relationships of hyaline and chlorophyllose cells, ×165. **F,** same for *S. squarrosum*, ×215. **G,** same for *S. magellanicum*, ×235. **H,** transverse section of branch-stem of *S. palustre* (note hyaline outer cells and dark-walled cells of central axis), ×295. **I,** external view of stem of *S. palustre*, showing hyaline porose outer cells with their fibril thickenings, ×225. **J,** prothallial protonema of *Sphagnum* bearing young leafy shoot, ×75. **K,** spore of *S. papillosum*, showing spore-wall sculpturing and trilete face, ×1000. **L,** longitudinal section through *Sphagnum* sporangium showing anatomy, ×95. **M,** transverse section of branch-stem of *S. tenellum*, ×120. **N,** external view of branch-stem of *S. tenellum*, showing the hyaline retort cells, ×120: *col*, columella; *ft*, foot; *ps*, pseudopodium; *spl*, sporogenous layer. (**H, I,** after Breen, with permission of University Presses of Florida. **L,** after Flowers, with permission of Brigham Young University Press, from Flowers, Seville, *Mosses: Utah and the West*. Provo, Utah: Brigham Young University Press, 1973.)

A

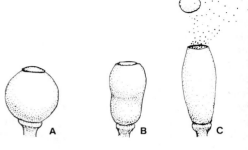

Figure 11-21 Spore dispersal in the genus *Sphagnum* (diagrammatic), showing the "air-gun" mechanism. **A,** mature sporangium on pseudopodium. **B,** as the sporangium dries, the diameter decreases, placing the sporangium contents under high pressure. **C,** when the pressure is sufficient, the operculum is thrown off, and the spores are jetted into the air (see text). (After Ingold, with permission of Oxford University Press.)

B

Figure 11-20 Scanning electron micrographs of *Sphagnum* gametophore. **A,** divergent and pendent branch, showing leaf orientation, ×70. **B,** portion of the abaxial surface of the leaf, showing pores (*black*), fibril thickenings, and outlines of the hyaline cells of the leaf, ×260. (After Mozingo et al., with permission of *The Bryologist*.)

gonial leaves envelop the several elongate antheridia with a few **filiform** paraphyses.

In *Andreaea*, the sporophyte consists of a small, elliptical sporangium that, when mature, is raised above the perichaetium by a pseudopodium. In *Andreaeobryum*, the sporangium is top-shaped and has a short seta. The sporangium in *Andreaea* opens along four to six longitudinal lines of weakness. As the sporangium dries, it shortens so that the openings gape apart, and spores fall out. In *Andreaea*, the calyptra forms a small apical cap on the sporangium and is ephemeral; in *Andreaeobryum*, it is large and persistent (Fig. 11-22E), enveloping the sporangium until it opens.

Spores of *Andreaea* are usually multicellular when shed. When the spore coat is ruptured, the spore contents produce a multiseriate filament, or thallus, and this branched structure produces rhizoids that affix it to the hard rock substrate. The upper surface of this thallus produces clusters of biseriate, determinate flaps of tissue (Fig. 11-22C) that serve as vegetative propagules or to extend the photosynthetic surface. Ultimately, buds are produced on the thallus, from which leafy stems arise.

The thallus is microscopic and disappears as soon as the leafy gametophores are produced.

Despite some superficial features similar to those of *Sphagnum*, *Andreaea* is probably not closely related to it, nor can it be considered as derived from or very closely related to any other subclass of mosses.

Subclass Tetraphidae

The Tetraphidae consist of the single order Tetraphidales, with one family, Tetraphidaceae, containing the genera *Tetraphis* (two species), and *Tetrodontium* (two species). *Tetraphis* is restricted to the Northern Hemisphere, largely in temperate forests, where it is especially common on shaded, rotting logs and stumps. *Tetrodontium* is confined to humid, shaded crevices and overhangs of siliceous rock; it is widely scattered, but local, in the Northern Hemisphere and very rare in New Zealand.

The gametophores of *Tetraphis* form extensive, short turfs of erect, unbranched shoots (Fig. 11-23A). The shoots that first arise from the protonema produce apical clusters of gemmae (Fig. 11-23A) that are surrounded by a cuplike group of blunt leaves. Each disclike gemma surmounts a slender stalk, which is easily broken (Fig. 11-23B). Upon germination, the gemma produces a primary protonema of uniseriate filaments, from which emerges a secondary protonema of perpendicular, unistratose flaps (Fig. 11-23D). From the bases of these flaps, both gemmiferous shoots and archegonium-bearing shoots arise, on which the leaves are in three to eight ranks. At the apex of the shoot is the perichaetium, surrounding eight to ten archegonia. If fertilization does not occur, during the next year this shoot gives rise to one or two lateral branches that emerge just beneath the perichaetium. Each of these branches is terminated by a cluster of larger and blunter leaves that surround the antheridia and their intermixed filamentous paraphyses.

Leaves of the gametophores are costate, but are otherwise unistratose. The cells are hexagonal and filled with numerous chloroplasts. Stems are made up of thicker-walled outer cells and a slender central cylinder of hydroids. The entire gametophore is rarely taller than 4 cm.

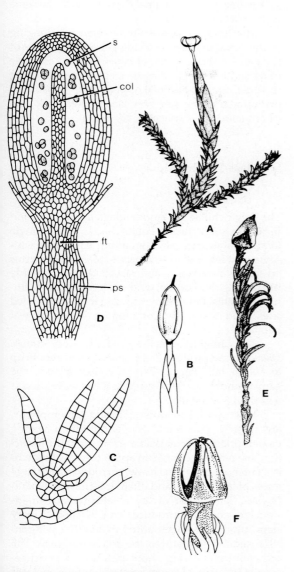

Figure 11-22 Morphology and anatomy of subclass Andreaeidae. **A,** habit sketch of *Andreaea rupestris* bearing dehiscing sporangium, ×8. **B,** sporangium of *A. rupestris* when moist, ×15. **C,** portion of straplike protonema of *Andreaea* with frondiform flaps, ×130. **D,** longitudinal section through sporangium of *Andreaea*, ×130. **E,** *Andreaeobryum macrosporum* habit, showing sporophytes with seta, ×3. **F,** *A. macrosporum*, detail of opening sporangium, ×5: *col*, columella; *ft*, foot; *ps*, pseudopodium; *s*, spore. (**F, G,** after Ruhland.)

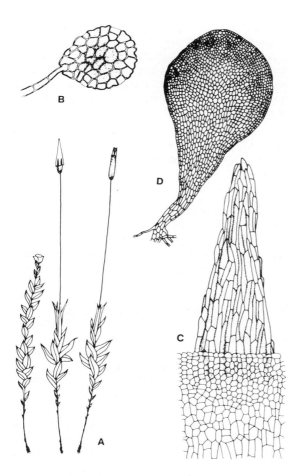

Figure 11-23 Morphology and anatomy of subclass Tetraphidae: *Tetraphis pellucida*. **A,** habit (gemma-bearing plant at *far left*), ×4. **B,** lenticular gemma, ×225. **C,** multicellular peristome tooth, ×150. **D,** frondiform protonemal flap, ×35. (**D,** after Ruhland.)

The sporophyte always consists of an elongate seta surmounted by a cylindrical sporangium. The sporangium is almost completely enclosed by a pleated, conical calyptra (Fig. 11-23A). Beneath a conical operculum are four multicellular peristome teeth (Fig. 11-23C). These are formed from four equal fractions of all cells enclosed by the operculum. Within the sporangium, the sporogenous layer forms a cylinder around the central cylinder of the columella. Spores are small and are shed when the operculum falls off. They germinate to produce protonema and protonemal flaps, in much the same fashion as do gemmae.

The Tetraphidae are often treated as belonging to the Bryidae because of the general similarity of the gametophore to that of many Bryidae. Because of its multicellular peristome teeth, as well as other features, the Tetraphidae appear to be an isolated evolutionary line, not very closely related to any extant subclass of mosses.

Subclass Polytrichidae

The Polytrichidae are worldwide in distribution and include a single order, Polytrichales, with a single family, Polytrichaceae, containing 20 genera with somewhat fewer than 400 species. Most species grow on somewhat acidic to neutral soil. Some are highly tolerant of desiccation, and grow in brightly illuminated sites. Others are restricted to humid shaded sites, and some are common in boggy or wet habitats.

The gametophores are generally conspicuous. *Dawsonia* can produce erect, unbranched shoots up to 80 cm tall (Fig. 11-24I). In other genera, the gametophore can be reduced to a few leaves and is less than 3 mm tall.

Most species are unisexual, and gametophores are usually perennial. The antheridium-bearing gametophore is remarkable in its annual production, in some species, of an apical rosette of brownish, enlarged perigonial leaves surrounding the large mass of antheridia and their intermingled paraphyses. The center of the rosette remains meristematic, and thus produces an extension of leafy stem, ultimately surmounted by another perigonial rosette that matures the following year. The archegonium-bearing shoot produces a perichaetium at the apex of the main shoot, which encloses the many archegonia and their associated paraphyses. Further growth of the archegonium-bearing shoot follows intercalary production of a new apical cell from beneath the perichaetium and from the stem cortex.

The erect shoot is usually unbranched, but occasionally it is multibranched near the apex of the main shoot. The stem is often flexible, and aggregated gametophores frequently form tall turfs.

The leaves are spirally arranged (Fig. 11-24A, C, I). The leaf base is often colorless and sheathes the stem. The lamina diverges outward to expose the upper surface, which is often covered by numerous, perpendicular, longitudinal, unistratose plates of chlorophyllose tissue, the **lamellae** (Fig. 11-24F). Sometimes the edge of the lamina curves over the lamellae, partially enclosing them.

In some Polytrichidae, the stem is differentiated into conspicuous tissues comparable to those in the shoots of lignified plants (Figs. 11-18 and 11-24G). The outer layer of the stem constitutes the epidermis, generally composed of thickened, darkly pigmented cells. The cortex forms a broad multistratose layer, the outer cells of which are thick-walled, whereas the inner cells are parenchymatous and predominantly chlorophyllose. Occasionally leaf traces are in this cortical region, and sometimes they are continuous with the axial cylinder. The axial cylinder is composed of a central cylinder of conducting cells, the hydroids, comparable to the tracheids of vascular plants. Surrounding the hydroids is an interrupted cylinder of leptoids, comparable to sieve-tube elements. The hydroids are water-conducting, whereas the leptoids conduct metabolites. Leaf traces also contain hydroids and leptoids; thus there is a vascular system in the gametophore of the Polytrichidae comparable to that in the sporophyte of Tracheophyta. Rhizoids that emerge from the epidermal cells are uniseriate and much branched.

The sporophyte in the Polytrichidae has a wiry seta that bears an erect or suberect, cylindrical sporangium that is sometimes distinctly four-angled (Fig. 11-24H). Stomata are sometimes present in the epidermis of the sporangium.

The axial strand of the seta is of hydroids, outside of which there are sometimes leptoids. These are surrounded by several rows of thick-walled cells and bounded on the outside by the epidermis. The foot pierces the upper end of the gametophore, thus forming a conducting system leading from the gametophore upward to the sporangium. When young, the seta is chlorophyllose, but as it ages, the cells usually lose their chlorophyll. The mature seta is usually brownish.

The sporangium is sometimes of complex anatomy (Fig. 11-24D) and opens by an operculum, beneath which is usually a membrane composed of the expanded apex of the columella, the **epiphragm**. Overarching the edge of the epiphragm are the blunt, multicellular peristome teeth (Fig. 11-24H, J). There are usually 16, 32, or 64 of these, depending on the genus, and they are not hygroscopic. The spores are minute and must pass between the small openings between the peristome teeth and the epiphragm. Any slight movement of the mature sporophyte causes some spores to be released; thus they are released gradually rather than all at once. In most Polytrichidae, each tooth is relatively short and stout. They are derived in most instances from four concentric layers of cells (Fig. 11-24E, K).

The calyptra sheathes the immature sporangium and is often completely invested with a mass of tawny hairs. Generally less than 10 μm in diameter, the spores germinate to produce an extensive, multibranched, heterotrichous protonema. The protonema generally disappears with the appearance of the leafy gametophores.

Relationships of the Polytrichidae are far from clear. Fossils attributed to this subclass are reported from the Eocene.

Subclass Buxbaumiidae

The subclass Buxbaumiidae combines four genera in two families, the Buxabaumiaceae with *Buxbaumia* (10 species) and Diphysciaceae with *Diphyscium* (19 species), *Theriotia* (probably 1 species) and *Muscoflorschuetzia* (1 species). All are restricted to acidic or neutral sites; some grow on soil or rotten wood and others on rock. They are rather widespread in north temperate to subtropical climates and are rare in the tropics and Southern Hemisphere.

Buxbaumia often grows on decaying wood. The gametophore is extremely reduced and consists of a multibranched, uniseriate protonema that gives rise to reduced perichaetia composed of six to ten nonchlorophyllose leaves surrounding one to five archegonia (Fig. 11-25B, C). A stem is barely evident. The perigonium is borne directly on the same protonema and is even more reduced: it consists of a simple bud made up of a unistratose flap of tissue folded around a single, spherical antheridium (Fig. 11-25E).

In the Diphysciaceae, the gametophore is more apparent (Fig. 11-25F). The stem is short and bears several ranks of spirally arranged costate leaves. In

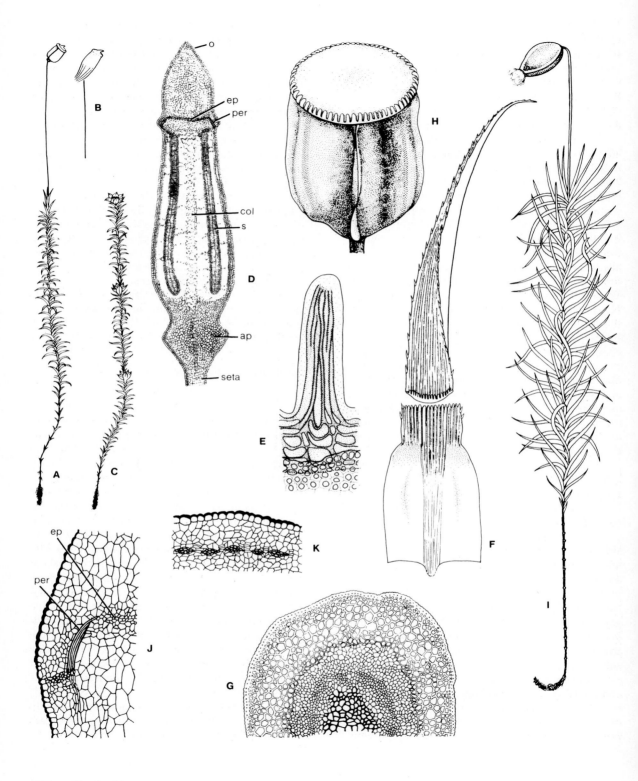

Diphyscium, most of the leaf cells are chlorophyllose. As in *Buxbaumia*, the gametophores are unisexual, but with several archegonia intermixed with reduced paraphyses. The perichaetial leaves are often brownish with ciliate margins. The antheridial plants arise as dwarf shoots beneath the perichaetial shoot. Filiform paraphyses are intermixed among the antheridia.

In the Diphysciaceae, the uniseriate, multibranched protonema produces protonemal flaps that are shaped rather like a golf tee (Fig. 11-25G). These flaps disappear as the leafy gametophores are formed.

In *Buxbaumia*, the sporophyte always has a conspicuous, rough seta that is terminated by an obliquely oriented sporangium (Fig. 11-25A). When immature, the sporophyte is green.

As in many Bryidae, the peristome of *Buxbaumia* consists of an **endostome** (inner peristome) and an **exostome** (outer peristome) (Fig. 11-25D). The endostome is a colorless, twisted, truncate, pleated cone with 32 pleats. Any external pressure, even a gentle touch on the jacket of the sporangium, causes spores to be puffed out through the tip of the endostome into the air. Up to four concentric rows of peristome teeth alternate with the grooves of the pleats; they are not hygroscopic and appear to serve no function in spore dispersal. Outside the exostomial teeth there is often a ring of cells termed a **false annulus.** Thus, in *Buxbaumia* the peristome is the most elaborate known in mosses.

In *Diphyscium*, each sporangium is partially sheathed by perichaetial leaves and has a very short seta, evident only after careful dissection. The shape and anatomy of the sporangium are identical to that of *Buxbaumia*, but the endostome is of 16 pleats rather than 32, and exostomial teeth are usually absent. The subclass seems most closely related to the Bryidae.

Subclass Bryidae

The subclass Bryidae contains most mosses. There are at least 75 families, more than 650 genera, and approximately 14,000 species. Species of Bryidae are found nearly everywhere a plant can grow.

Features essentially restricted to the Bryidae include, in the gametophore, the pleurocarpous condition, paraphyllia (Fig. 11-26) on the stem, leaf gemmae, and strong differentiation of the **alar cells** (Fig. 11-27F) in leaves. Features that unite the Bryidae are largely sporophytic, particularly the details of peristomial structure and development. The peristome teeth are never derived from more than three concentric rows of cells. This is in strong contrast to all other subclasses of peristomate mosses, in which more than three concentric rows are involved in tooth formation. The peristome teeth are always **articulate** and consist, in large part, of only cell wall fragments, not of complete cells. Elaboration of cell wall ornamentation and of the gross morphology of the peristome teeth reaches its peak of complexity in this subclass. The peristome teeth are generally hygroscopic and are usually important in spore dissemination.

The gametophore of the Bryidae is leafy, and its height varies from less than 2 mm to more than 5 cm. In reclining forms or in species that festoon branches or clothe tree trunks, lengths can exceed 40 cm, and in aquatic forms the gametophore can approach a meter in length. In those with a more elaborate gametophore, leaf arrangement is generally spiral and in more than three ranks. Leaves are strongly divergent or imbricate, and frequently

Figure 11-24 Morphology and anatomy of subclass Polytrichidae. **A–H,** *Polytrichum commune.* **A,** sporophyte-bearing plant after calyptra has fallen, ×1. **B,** sporangium with calyptra in place, ×1. **C,** antheridial plant, showing three successive years' crops of perigonia, ×1. **D,** longitudinal section through sporangium (idealized), ×25. **E,** multicellular peristome tooth, ×75. **F,** leaf, showing lamellae on upper surface, ×12. **G,** transverse section of stem, showing differentiation into tissuelike areas (hydroids in center), ×100. **H,** sporangium, showing peristome teeth and epiphragm, ×8. **I,** habit of *Dawsonia superba*, ×1. **J,** longitudinal section of *Polytrichum commune* sporangium near apex, ×50. **K,** transverse section of sporangium near apex, showing multicellular teeth (*darkened bands*), ×50: *ap*, apophysis; *col*, columella; *ep*, epiphragm; *o*, operculum; *per*, peristome; *seta*, seta; *s*, sporogenous layer. (**D,** after Gibbs; **G, J, K,** after Flowers, with permission of Brigham Young University Press, from Flowers and Seville, *Mosses: Utah and the West.* Provo, Utah: Brigham Young University Press, 1973.)

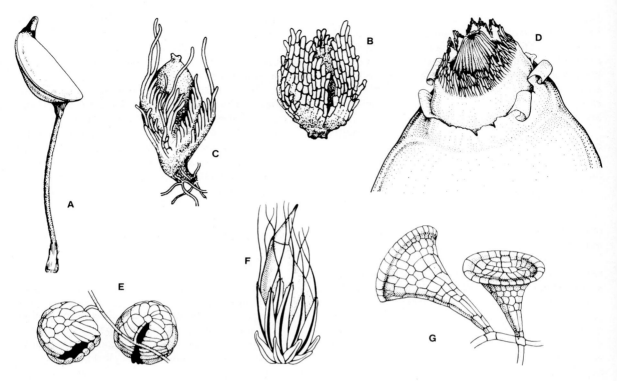

Figure 11-25 Morphology and anatomy of subclass Buxbaumiidae. **A–E,** *Buxbaumia aphylla.* **A,** habit, ×7. **B,** gametophore with single archegonium, ×120. **C,** young gametophore with developing sporophyte, ×120. **D,** apex of sporangium, showing peristome (note *outer* multicellular peristome and *inner* pleated cone), ×13.5; **E,** antheridial "branches," external view, ×225. **F, G,** *Diphyscium foliosum.* **F,** habit of sporangium-bearing gametophyte, ×8. **G,** protonemal flaps, ×25. (**E,** after Goebel; **G,** after Ruhland.)

alter their orientation in response to changes in moisture.

The Bryidae show great diversity in growth forms. Most acrocarpous genera form turfs or cushions (Fig. 11-28B, D, G), whereas pleurocarpous genera also form mats, wefts, or loose colonies of **dendroid** gametophores (Fig. 11-28A, C, E). The growth form is strongly affected by the microclimate, particularly by relative humidity and shade.

Leaf shape is more variable in Bryidae than in other mosses (Fig. 11-29) and may differ from the main stem to the branches. The leaf apex is generally acute, but can be blunt or obtuse; sometimes it extends as a long **awn** (Fig. 11-27G). The margins are toothed or entire, and marginal cells are sometimes markedly different in shape from those of most of the leaf (Fig. 11-27A–D).

The leaf is generally unistratose, but the margins can be thickened and multistratose, and the costa is generally multistratose (Fig. 11-27H, J). The leaf is costate in most acrocarpous mosses as well as in many pleurocarpous genera. Sometimes there is more than one costa, or a costa is absent. The anatomy of the costa is often a distinctive arrangement of stereid cells and cells of much greater diameter (Fig. 11-27H, J).

The arrangement of cells within the leaf is termed its **aerolation** (Fig. 11-27). The basal mar-

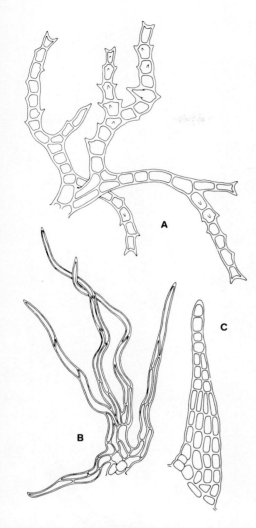

Figure 11-26 Variety in paraphyllia of subclass Bryidae. **A,** *Thuidium delicatulum,* ×340. **B,** *Hylocomium splendens,* ×135. **C,** *Lescuraea incurvata,* ×340.

gins of the leaf are often made up of cells distinctly different in shape and color from those of the rest of the leaf. These are the alar cells, and are a diagnostic feature of several genera and species (Fig. 11-27F). Leaf cells are smooth, or their exposed walls are variously thickened with complex **papillae** (Fig. 11-27H–J) or swollen outward **(mammillose)**. Walls shared by adjacent cells are often perforated by pits. Irregular thickening of the cell walls often gives the leaves a striking aerolation. Leaf cell shape is highly diverse (Fig. 11-29).

Some pleurocarpous Bryidae also have paraphyllia on the stems. Paraphyllia are chlorophyllose, unistratose, determinate outgrowths from the stem epidermis and are mixed among the leaves. (Fig. 11-26). Most Bryidae also have rhizoids on the stems. Rhizoids of Bryidae are uniseriate, and their cell surfaces are usually smooth. Besides attaching the gametophore to the substrate, rhizoids also assist in capillary conduction of water over the surface of the stem. Sometimes they produce rhizoidal gemmae, or "tubers," that are important in asexual reproduction (Fig. 11-4D).

The stem structure in the Bryidae appears to be less complex than in either Polytrichidae or Sphagnidae. Epidermal cells are either extremely thick-walled or swollen and thin-walled. These cells generally have smooth outer walls. Cortical cells tend to be thick-walled stereids. The central cylinder of the stem is generally of parenchymatous cells, but the very center of the stem sometimes has a strand of smaller cells, usually hydroids. In most Bryidae, water conduction is largely external, with lesser conduction in the central cylinder.

The gametophores are often unisexual in Bryidae. In acrocarpous Bryidae, the perigonia are generally terminal on a main branch. The perichaetia are always terminal in acrocarpous Bryidae. In pleurocarpous Bryidae, both perigonia and perichaetia are on reduced lateral branches, and many are produced by each gametophore.

Archegonia are always surrounded by a well-differentiated perichaetium of closely sheathing leaves. Generally the archegonia have intermixed, chlorophyllose, uniseriate paraphyses.

In most Bryidae, the sporophyte consists of an elongate seta terminated by a sporangium (Fig. 11-30). If present, stomata are often restricted to the base of the sporangium. The stomata are usually exposed on the surface, but in some instances they are deeply immersed (Fig. 11-31).

The sporangium is generally operculate. The operculum is released as the sporangium dries out and shrinks in diameter. In many Bryidae, a ring of specialized cells separates the operculum from the rest of the sporangium. These cells are somewhat thicker-walled than the other sporangial cells and are elastic. As they dry, the cells of this ring (the **annulus**) curl back, thus breaking the operculum loose.

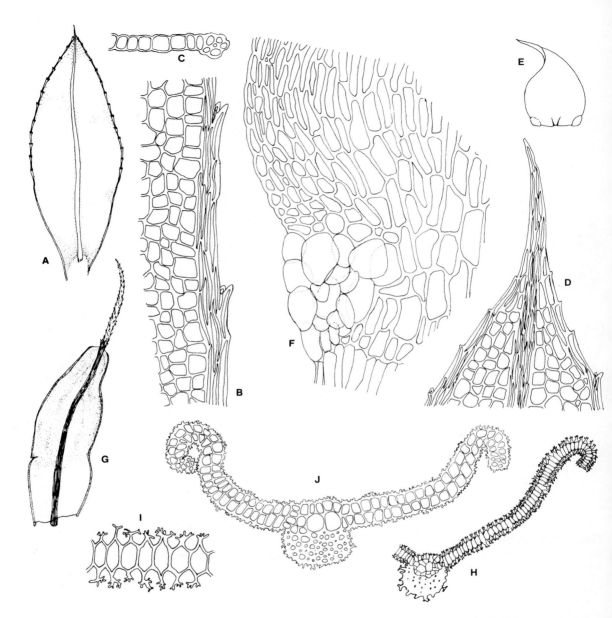

Figure 11-27 Morphology of leaves of subclass Bryidae. **A–D,** *Mnium marginatum.* **A,** leaf outline, showing costa and differentiated margin, ×8. **B,** differentiated leaf margin, showing cellular detail, ×120. **C,** transverse section of leaf margin, ×120. **D,** leaf apex, ×120. **E, F,** *Hypnum lindbergii.* **E,** leaf outline, showing alar region and double costa, ×20. **F,** swollen alar cells (detail), ×385. **G–I,** *Tortula papillosissima.* **G,** leaf, ×8. **H,** transverse section of leaf, showing remarkable papillae and stereid cells making up much of costa, ×24. **I,** detail of leaf cells in transverse section, ×120. **J,** *Tortula bistratosa,* transverse section of bistratose papillose leaf, ×80. (**A–D, G–J,** after Flowers, with permission of Brigham Young University Press, from Flowers, Seville, *Mosses: Utah and the West.* Provo, Utah: Brigham Young University Press, 1973).

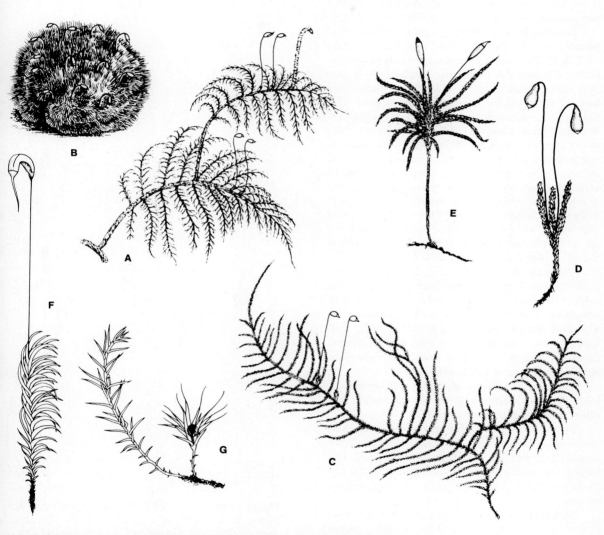

Figure 11-28 Variety in gametophytes and sporophytes of subclass Bryidae. **A,** *Hylocomium splendens*, ×1. **B,** *Grimmia pulvinata*, ×1. **C,** *Kindbergia oregana*, ×2. **D,** *Bryum* sp., ×1. **E,** *Climacium dendroides*, ×1. **F,** *Archidium alternifolium*, ×6. **G,** *Dicranum scoparium*, ×1.

In many acrocarpous Bryidae, the peristome consists of a single row of 16 peristome teeth (Fig. 11-32B, E). Each of these teeth can be forked, and is opaque with brownish or yellowish pigment and varied ornamentation. In such mosses, the teeth are articulate and formed of cell fragments from two concentric rows of adjacent cells. Only the inner wall of the outer concentric row and the outer wall of the inner concentric row remain to form the teeth; the remainder of the cell row disappears and does not participate directly in the peristome structure.

In other Bryidae, the peristome is of two concentric rows of peristome teeth, 16 in each row (Fig. 11-32A, C, D). The outer row is generally of opaque teeth, whereas the inner row tends to be fragile and transparent. Such "double" peristomes are formed from three concentric rows of cells. The teeth of the exostome, or of the single row of peristome teeth in many Bryidae, are hygroscopic. In many species, the teeth curl inward when humid, catch spores on their jagged inner faces, and curl outward when dry, exposing spores to the air. In other Bryidae, the teeth curve inward when dry and outward when humid. Spores are shed gradually from the sporangium, improving the chances that some will be released when conditions are favorable for germination. A cylindric columella is usually present in the capsule.

The seta is generally smooth and wiry, and both the epidermis and the cortex are of stereid cells. The central portion of the seta is often of somewhat thinner-walled cells and occasionally has a central strand. This strand is of hydroids, and leptoids are sometimes associated with it. Sometimes the seta is very short, and the sporangium is immersed among perichaetial leaves.

The usually smooth calyptra is well developed in all Bryidae. It always covers the operculum and often sheathes the entire capsule.

Spores of Bryidae vary in number, size, and shape (Fig. 11-33). They are generally unicellular when shed, but are sometimes multicellular. The largest unicellular spores are occasionally 200 μm in diameter (Fig. 11-33A), but are usually less than 20 μm.

Moisture and light are required for germination of spores in nature. Most germinating spores produce a protonema of a uniseriate multibranched filament, similar to that in many Polytrichidae. Erect branches of this protonema are strongly chlorophyllose and sometimes rather brittle; they serve

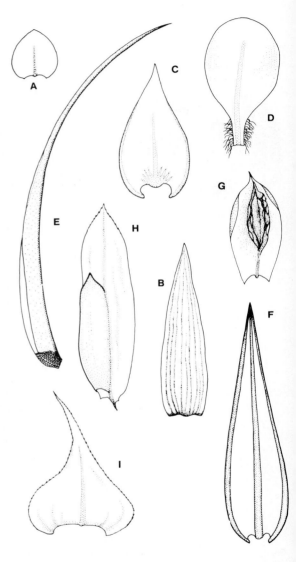

Figure 11-29 Variety in leaves of subclass Bryidae. **A,** *Hygrohypnum smithii*, ×15. **B,** *Orthothecium rufescens*, ×25. **C,** *Antitrichia curtipendula*, ×20. **D,** *Oedipodium griffithianum*, ×15. **E,** *Dicranum scoparium*, ×15. **F,** *Sciaromium tricostatum*, ×20. **G,** *Pterigoneurum ovatum*, ×30. **J.** *Fissidens adianthoides*, ×15. **I,** *Kindbergia oregana*, ×15.

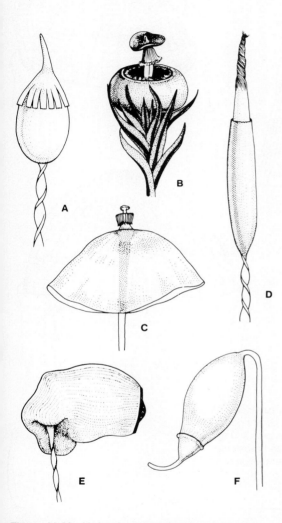

Figure 11-30 Variety in sporangia and calyptrae of subclass Bryidae. **A,** *Rhacomitrium lanuginosum,* with calyptra, ×20. **B,** *Scouleria aquatica,* dehiscing, ×10. **C,** *Splachnum luteum,* showing extensive hypophysis (sporangium dehiscing), ×6. **D,** *Tortula princeps,* dehiscing, ×7. **E,** *Philonotis fontana,* with operculum in place, ×8. **F,** *Stokesiella oregana,* with operculum in place, ×10.

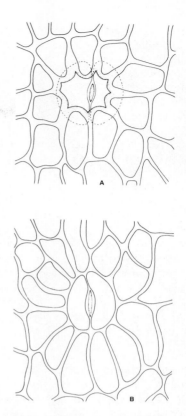

Figure 11-31 Stomata of sporangia in subclass Bryidae. **A,** immersed stoma of *Orthotrichum ohioense,* ×15. **B,** superficial stoma of *Funaria hygrometrica,* ×35.

as vegetative propagants. The protonema ultimately produces gametophores in response to a substance that inhibits further growth of the protonema. This results in a "bud" of undifferentiated cells, in which an apical cell is differentiated. Many buds can be produced on a single protonema; thus a population of gametophores is produced. The protonema usually disappears as the leafy gametophores grow larger.

The relationships of the Bryidae appear closest to the Buxbaumiidae, which shares both sporophytic and gametophytic similarities.

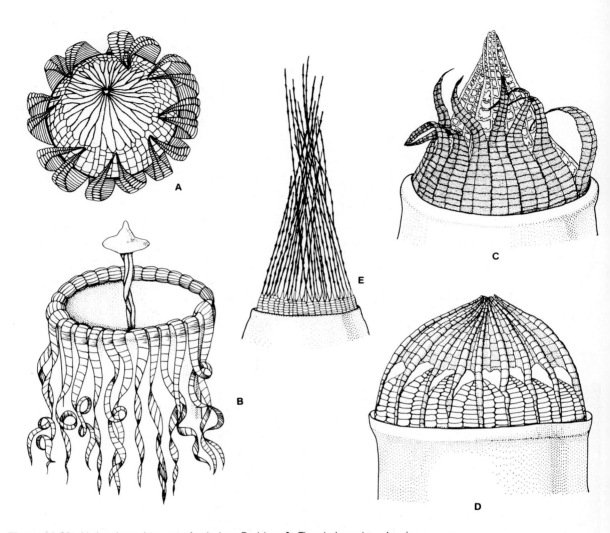

Figure 11-32 Variety in peristomes of subclass Bryidae. **A,** *Timmia bavarica,* showing endostome and exostome, ×35. **B,** *Tayloria splachnoides,* showing conic endostome and exostome, ×50. **C,** *Fontinalis antipyretica,* showing conic endostome and exostome, ×30. **D,** *Cinclidium stygium,* showing domelike endostome and exostome, ×50. **E,** *Rhacomitrium canescens,* ×35. (**A,** after Lazarenko with permission of *The Bryologist;* **B, D,** after Schimper.)

Figure 11-33 Variation in size and morphology of spores of subclass Bryidae, ×350. **A,** *Archidium alternifolium,* showing outline of spore to indicate size and wedge to show sculpturing. **B,** *Ulota megalospora.* **C,** *Macromitrium comatum,* showing different spore sizes. **D,** *Ephemerum minutissimum.* **E,** *Encalypta ciliata.* **F,** *Encalypta rhabdocarpa.* **G,** *Octoblepharum albidum.* **H,** *Bruchia brevifolia.* **I,** *Dicnemon calycinum,* multicellular *spore.* **J,** *Bruchia drummondii.* **K–N,** scanning electron micrographs of spores of *Encalypta.* **K,** *E. alpina,* ×1520. **L,** *E. ciliata,* ×1850. **M,** *E. rhaptocarpa,* ×1320. **N,** *E. affinis,* ×1700. (**K–N,** courtesy of D. H. Vitt.)

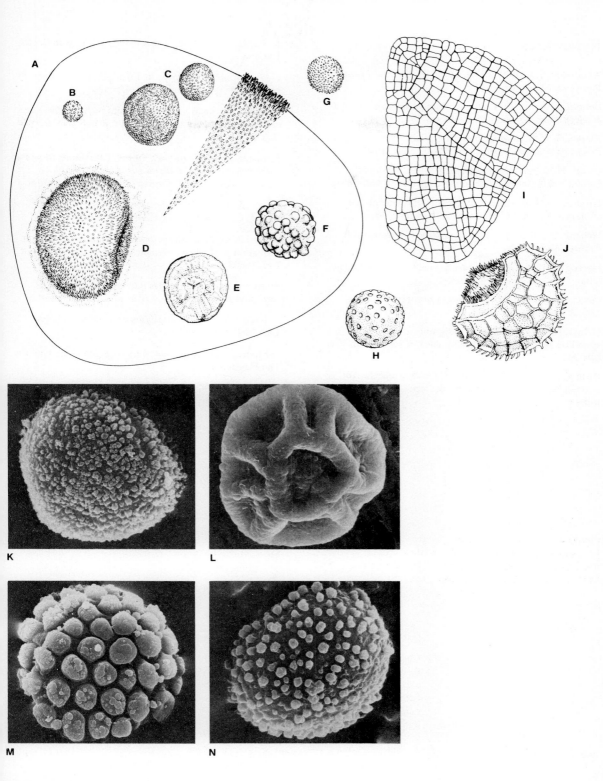

Bryophytes (Mosses, Liverworts, and Hornworts) 339

References

Anderson, L. E. 1974. Bryology: 1947–72. *Ann. Missouri Bot. Gard.* 61:56–85.

Bonnot, E. J., ed. 1974. *Les problèmes modernes de la bryologie.* Soc. Bot. France. Colloque, 1972, Paris.

Clarke, G. C. S., and Duckett, J. G., eds. 1979. *Bryophyte systematics.* Academic Press, New York.

Crum, H. A., and Anderson, L. E. 1981. *Mosses of Eastern North America.* Columbia Univ. Press, New York.

Grout, A. J. 1928–40. *Moss flora of North America, north of Mexico.* 3 vols. Publ. by author, Newfane, Vermont.

Hébant, C. 1978. *The conducting tissues of bryophytes.* J. Cramer, Vaduz.

Herzog, T. 1926. *Geographie der Moose.* G. Fischer Verlag, Jena.

Lewis, K. R. 1961. The genetics of bryophytes. *Trans. Brit. Bryol. Soc.* 4:111–30.

Nehira, K. 1974. Phylogenetic significance of the sporeling pattern in Jungermanniales. *J. Hattori Bot. Lab.* 38:151–60.

Parihar, N. S. 1965. *An introduction to embryophyta.* Vol. 1. *Bryophyta.* 5th ed. Central Book Depot, Allahabad, India.

Richards, P. W. 1950a. *A book of mosses.* Penguin Books, London.

——— 1950b. Bryophyta. In Turrill, W. B., ed. *Vistas in botany.* vol. 1. Pergamon Press, London.

Savicz-Ljubitzkaja, L. I., and Abramov, I. I. 1959. The geological annals of the Bryophyta. *Rev. Bryol. Lichenol.* 28:330–42.

Schuster, R. M. 1966, 1969, 1974, 1980. *The hepaticae and anthocerotae of North America east of the hundredth meridian.* vols. 1, 4. Columbia Univ. Press, New York.

Smith, A. J. E. 1978. *The mosses of Great Britain and Ireland.* Cambridge Univ. Press., Cambridge.

Stotler, B. C. 1972. Morphogenetic patterns of branch formation in the leafy Hepaticae. A resumé. *Bryologist* 75:381–403.

Suire, C., ed. 1978. *Congrès international de bryologie.* Bryophytorum Bibliotheca. Vol. 13. J. Cramer, Vaduz.

Thieret, J. W. 1955. Bryophytes as economic plants. *Econ. Bot.* 10:75–91.

Verdoorn, F. 1932. *Manual of bryology.* Martinus Nijhoff, The Hague.

Watson, E. V. 1971. *The structure and life of bryophytes.* Hutchinson Univ. Library, London.

12

VASCULAR PLANTS

Vascular plants are structurally the most complex members of the plant kingdom, and much of this complexity is associated with their adaptation to a terrestrial environment. Because of a terrestrial existence, vascular plants face special problems. Since they stand free in a nonsupportive medium, they must have support tissue. Furthermore, most vascular plants are not bathed in water as are most algae; thus they obtain water only from that portion of the plant embedded in the substrate. Since the source of water is localized, transport tissues are necessary. Much of a terrestrial plant exists in a very dry environment; consequently, tissues are necessary for protection against desiccation. As well as serving specific functions, the common occurrence of special cells and tissues is used by many botanists to indicate that vascular plants are a monophyletic group, comprising the division Tracheophyta.

All vascular plants, or tracheophytes, are included in the division Tracheophyta. They comprise the major component of terrestrial vegetation of the world. It is a remarkably distinct evolutionary line, with a fossil record extending back to the Silurian, and possibly earlier. The presence of vascular plants on land has determined the presence of all other terrestrial organisms and has been a basic influence in their evolution. Vascular plants supply the major source of food and shelter for terrestrial organisms. Without vascular plants, it is unlikely that most of the major classes and orders of terrestrial animals would have evolved. The continued persistence of vascular plants determines both the well-being and survival of mankind. Tracheophytes control the appearance and character of the landscape; their beauty has inspired artists and craftsmen to create works that are the stamp of civilization.

Tracheophytes have several fundamental characteristics:

1. Both physically and morphologically, the sporophyte is the dominant part of the life history, and it is normally autotrophic.

2. The sporophyte has lignified cells and a vascular system made of water-conducting cells (xylem) and metabolite-conducting cells (phloem).

3. In the sporophyte, most tracheophytes have a

true stem, true leaves, and roots, all of which show complex tissue differentiation.

4. In the sporophyte, most tracheophytes have a definite apical shoot meristem, root meristem, and vascular cambium; many exhibit secondary growth.

5. The sporophyte is generally perennial and bears numerous sporangia.

6. The gametophyte is generally small and inconspicuous in comparison to the sporophyte; it is never leafy, and it is frequently dependent on the sporophyte.

7. The gametophyte is generally annual, and in most instances of very brief duration.

Diversity

Tracheophytes are represented by more than 220,000 extant species in approximately 500 families. The flowering plant family Asteraceae appears to be the largest, with approximately 1,000 genera and nearly 20,000 species. Many families contain a single genus and a single species.

The sporophyte of vascular plants shows considerable diversity—from gigantic trees to nearly microscopic duckweeds in the seed plants (Fig. 12-1) and from tree ferns to miniature filmy ferns (Fig. 12-2) the size of hepatics. This array of diversity in extant plants creates numerous problems in determining interrelationships, and evidence is used from as many sources as possible: morphology, cytology, anatomy, physiology, and biochemistry. Evidence from paleobotany is so fragmentary that it must be used cautiously in interpreting interrelationships or as a source of convincing evidence concerning the origins of modern groups. Utilizing cytological characteristics (cellulose cell walls, biochemical and ultrastructural nature of chloroplasts, and nature of storage products), green algae are assumed to have been ancestral to all green plants.

Tracheophytes have evolved three generalized plant forms: those that are leafless (**aphyllous**), those with small, needlelike leaves that generally have a simple vascular strand (**microphyllous**), and those with elaborate, large leaves with a complex of branched vascular strands (**megaphyllous**). Within microphyllous and megaphyllous types, there are examples of plants that have been reduced to an aphyllous condition, but other details of morphology and anatomy indicate that, technically, these plants are microphyllous or megaphyllous despite the nature of the leaves. Such trends pose problems for both evolutionists and taxonomists but shed much light on phenomena that influence the process of evolution.

Distribution and Role in Vegetation

Tracheophytes grow in most environments on the earth's surface, including the sea, where the angiosperm *Zostera* (eelgrass) is widely distributed. A species of *Najas* has been found in fresh water at 60 °C. Vascular plants are found on land nearest the North Pole, even on soil that has moved away from land surface and is carried into the Arctic Sea on an ice island. In high elevations of the Himalayan Mountains, at 6000 m, a few small flowering plants can grow. Only two species of flowering plants now persist in a few sites on the Antarctic continent. Many vascular plants can tolerate arid climates, including the frigid deserts of arctic regions and the cold and hot deserts of the middle latitudes of both hemispheres. Aquatic tracheophytes in bodies of fresh water are confined to depths at which light will penetrate effectively for photosynthesis, resulting in stratification of kinds of plants that can survive with increase in water depth. Rooted plants can grow as deep as 10 m in lakes where the water is very clear.

Such stratification of plants is even more conspicuous in the tropical rain forest, where the canopy affects light penetration to the forest floor and thus controls the many layers of understory species. In such forests, there are many epiphytic tracheophytes on both the trunks and branches of trees.

Some tracheophytes are parasitic on other vascular plants (Fig. 12-3A). Many are partially or wholly dependent on fungi for their food requirements: many gymnospermous trees and most orchids have an obligate mycorrhizal association with fungi, and several flowering plant families are saprophytic (completely dependent on other plants,

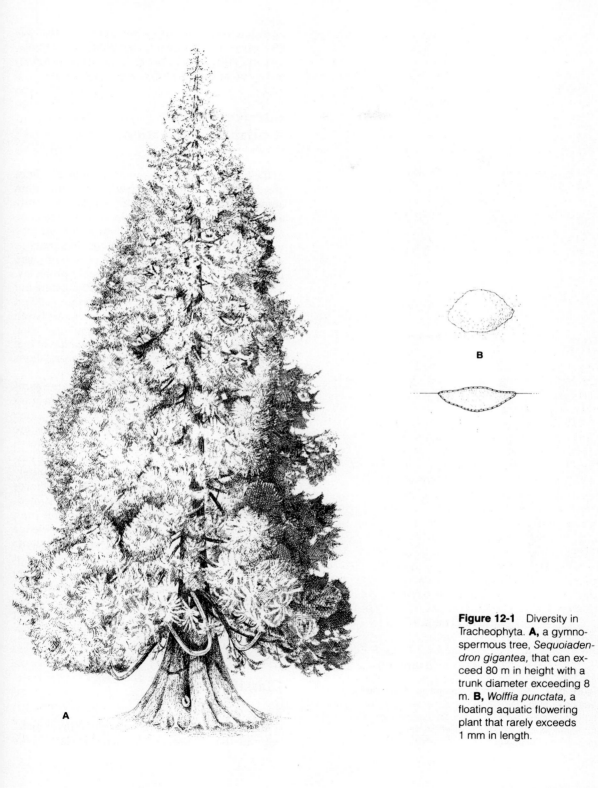

Figure 12-1 Diversity in Tracheophyta. **A,** a gymnospermous tree, *Sequoiadendron gigantea,* that can exceed 80 m in height with a trunk diameter exceeding 8 m. **B,** *Wolffia punctata,* a floating aquatic flowering plant that rarely exceeds 1 mm in length.

including fungi, to supply their nutrition) (Fig. 12-3B). Few flowering plants complete their life history underground, but a small, subterranean Australian orchid, *Rhizanthella*, appears to do so.

Interpretation of Available Information

In the following chapters, various evolutionary lines are discussed. Interrelationships are uncertain among all major groups and are based on currently available evidence. Further evidence, as well as more careful assessment of available information, may well alter many current assumptions. In studies as complex as evolution, the solution of each problem uncovers an array of further intriguing problems. This makes the study intellectually stimulating but also frustrating to a scientist who seeks unqualified answers. All evolutionary interpretations are based on currently available evidence and involve what is assumed to be a reasonable interpretation of that evidence. In consequence, various interpretations of the evidence lead to different opinions.

The most primitive vascular plants are vascular archegoniates—a group in which the female sex organ is similar to that of bryophytes. Plants somewhat similar to living vascular archegoniates are also the first vascular plants to appear in the fossil record. The assumption that many of the fossil vascular plants are indeed archegoniates is based purely on circumstantial evidence, since in most instances only sporophytic material has been demonstrated. Reconstruction of the form of vascular plants from fossil evidence always relies heavily on knowledge of living vascular plants. When plant fragments show no organic connection in the fossil record, it is often assumed that particular associated fragments may have belonged to the same organism.

The first fossils generally accepted as vascular plants are those of *Cooksonia*, of late Silurian age. It is considered to belong to the subdivision Rhyniophytina, as discussed later (p. 384).

Classification

Considering the age of vascular plant fossil material and the subsequent erosional cycles on the

Figure 12-2 Diversity in Tracheophyta. **A,** the tree fern *Cyathea cunninghamii*, that can reach heights over 20 m and bear leaves up to 3 m long. **B,** the epiphytic, filmy fern *Microgonium tahitiense*, in which the leaf is sometimes 6mm in diameter.

earth's surface, it is remarkable that any fossil record persists. The coincidence of favorable conditions for fossilization involves a highly random sequence of events, further decreasing the chances of fossilization. Thus the discovery of ancient fossil materials is an event of much importance, and the meticulous study of such materials yields further insight into the nature and origin of land plants.

In this text we follow a classification that places the vascular plants that are structurally the most simple first and the more complex ones last. With some exceptions, this follows the time of appearance in the fossil record, but many major subdivisions of vascular plants were contemporaneous and appear suddenly with no earlier evidence to reveal their predecessors.

The system of classification followed here for the division Tracheophyta is essentially that proposed by Banks in 1969; it reflects the great progress made in understanding vascular plants in recent years.

Subdivisions of Division Tracheophyta

Subdivision Rhyniophytina

Subdivision Trimerophytina

Subdivision Zosterophyllophytina

Subdivision Psilotophytina

Subdivision Lycophytina

Subdivision Sphenophytina

Subdivision Pterophytina

The position of the Psilotophytina is still highly controversial, as will be discussed later. The earlier used name, (subdivision) Psilophytina, has been omitted from this text, since it has been so diversely interpreted.

The subdivision Pterophytina includes all megaphyllous plants. Since their diversity and evolutionary lines are so complex, the various classes are treated as representatives of these lines of diversity.

Classes of Subdivision Pterophytina

Class Cladoxylopsida

Class Coenopteridopsida

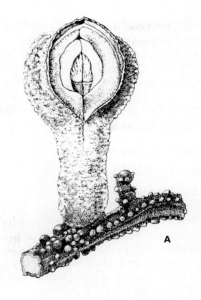

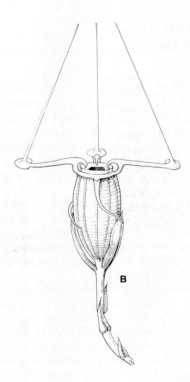

Figure 12-3 Parasitic and saprophytic flowering plants. **A,** *Hydnora africana*, a parasite on roots of *Euphorbia* in Africa. **B,** *Thismia neptunis*, a saprophyte in tropical Asia.

Class Filicopsida

Class Progymnospermopsida

Class Pteridospermopsida

Class Cycadopsida

Class Cycadoidopsida

Class Ginkgopsida

Class Coniferopsida

Class Gnetopsida

Class Magnoliopsida

There still remain many puzzling fossil plants of uncertain relationships. In the future, these problematic fossil specimens may lead us to a more comprehensive understanding of plants, but at present their status is so speculative that they are omitted from consideration in this text.

The Plant Body and Its Growth

The plant body consists of the **root system,** which is usually subterranean, and the **shoot system,** which is usually aerial. The root system is made up entirely of roots; the shoot system is made up of a series of axes (stems) and appendages (leaves and reproductive parts) (Fig. 12-4). Sites of leaf attachment are called **nodes,** and the areas between are called **internodes**.

Although there is continuity between root and shoot systems, they mainly perform different functions. The root system is involved primarily in water and ion uptake, anchorage, and storage. The shoot system functions mainly in photosynthesis and sexual reproduction.

To present a clearer understanding of the plant body, the growing tip of a stem of a **dicotyledon** is described (Fig. 12-5A). At the very tip, there is a dome-shaped structure with little tissue differentiation, the shoot **apical meristem.** A meristem consists of tissue that is involved in growth; an apical meristem is responsible for growth in length. By continual division, cells of the shoot apical meristem produce tissues that ultimately mature into the plant body below the tip of the plant. During maturation, cells of the tissues derived from the apical meristem undergo considerable elongation, a phenomenon that results in an increase in length. Because the apical meristem is the source of new cells in the plant body, cells nearest the apical meristem are immature, whereas those farther away become progressively more mature. Immediately below the apical meristem, some tissue differentiation can be discerned, the primary meristems. At the surface is a layer one cell thick, the **protoderm** (Fig. 12-5A, B). The inner portion of the stem has discrete strands in it, the **procambium** (Fig. 12-5A, B). The tissue in which the procambium is embedded is the **ground meristem** (Fig. 12-5A, B). The protoderm matures into the epidermis (Fig. 12-5B, D), a covering tissue that protects the plant from external injury and desiccation. Procambium develops into two tissues, **xylem** and **phloem** (Fig. 12-5B, D). These tissues are involved primarily in conduction and support and are classified together as **vascular tissues.** Vascular tissues, like the procambium from which they mature, occur as discrete strands, referred to as **vascular bundles.** That portion of the plant into which the ground meristem matures is defined on the basis of both the tissues present and location. Tissues derived from ground meristem are (1) **parenchyma,** a generalized tissue; (2) **collenchyma,** a specialized type of support tissue; and (3) **sclerenchyma,** a supportive and protective tissue. Regions that develop from the ground meristem are (1) **pith,** in the center of the stem; (2) **cortex,** between the epidermis and the vascular bundles; and (3) **pith rays,** between the vascular bundles (Fig. 12-5B, D). These tissue types and regions are referred to collectively as *ground tissue.* Parenchyma, collenchyma, or sclerenchyma can occur in any one of the regions, although certain tissues are more common in particular regions. For example, tissues commonly in the cortex are parenchyma, collenchyma, and sclerenchyma. The most common tissue of the pith and pith rays is parenchyma.

These tissues and regions that mature from the apical meristem are the **primary tissues,** and growth derived from the activity of the apical meristem is **primary growth.** The same primary tissues are in roots and leaves, although their patterns of distribution differ in roots, stems, and leaves (Fig. 12-5B, F, G). There are basic structural differences between roots, stems, and leaves.

At some distance behind the apical meristem, another meristematic tissue exists in many plants.

This meristem, the vascular cambium, completely encircles the stem and produces xylem inwardly and phloem outwardly (Fig. 12-5I). Growth from the activity of the vascular cambium is **secondary growth,** and the tissues produced by it are **secondary tissues.** Secondary growth is the main cause of increase in the girth of plants. In plants with secondary growth, some increase in girth results from primary growth, but this is a result of cell enlargement and random cell divisions and not of the functioning of a specific meristem some distance back from the apical meristem. The massive plant bodies of many trees, such as *Eucalyptus* (gum), *Pseudotsuga* (Douglas fir), and *Sequoia* (redwood) are the result of secondary growth and consist mostly of **secondary xylem** or **wood.** However, even in the largest trees, primary growth still occurs at the tips of the branches.

Some plants that lack secondary growth exhibit much growth in girth. This results from the activity of the **primary thickening meristem,** which occurs just behind the apical meristem (Fig. 12-5J). In some plants, another meristem appears behind the apical meristem. This is the **cork cambium** or **phellogen,** and it produces cork, a protective tissue (Fig. 12-5I). These various meristems are considered in great detail later.

Plant Cells and Tissues

Vascular plants contain several different kinds of cells. These cells are often aggregated into **tissues** (a group of cells that share some feature). This feature may be function, structure, or ontogenetic origin. The distinction between different tissues is often very subtle.

Parenchyma Tissue

Parenchyma tissue is a simple tissue composed entirely of parenchyma cells. Parenchyma cells are unspecialized, commonly **isodiametric,** and usually have thin primary walls (Fig. 12-6A). They remain living throughout their functional lifetime. Parenchyma tissue is the most common type of ground tissue in vascular plants, and parenchyma

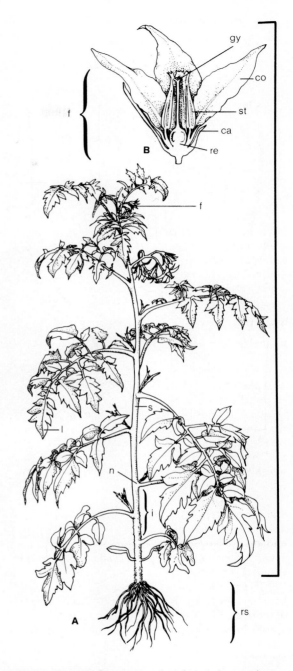

Figure 12-4 A, B, vascular plant (an angiosperm), showing its general morphology. **A,** complete plant, × 0.1. **B,** longitudinal section of a flower, × 10: *ca,* calyx; *co,* corolla; *f,* flower; *gy,* gynoecium; *i,* internode; *l,* leaf; *n,* node; *r,* receptacle; *rs,* root system; *s,* stem; *ss,* shoot system; *st,* stamen.

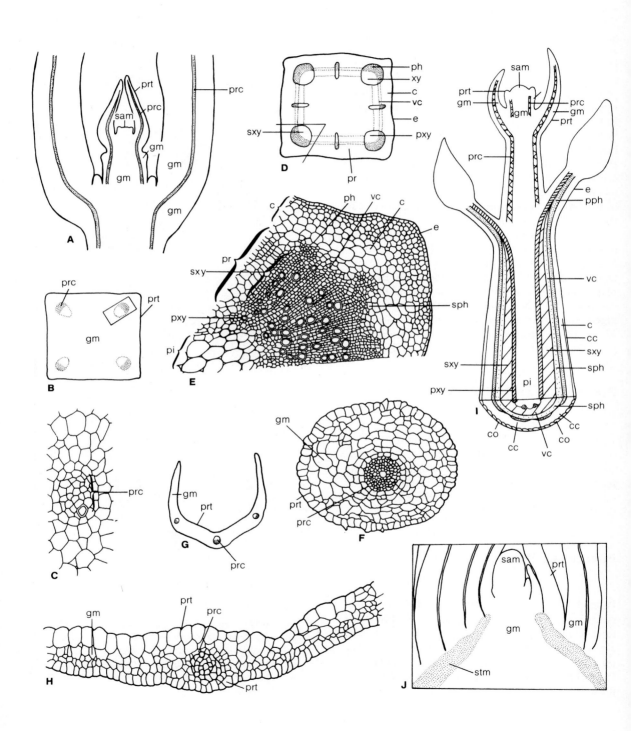

348   Chapter 12

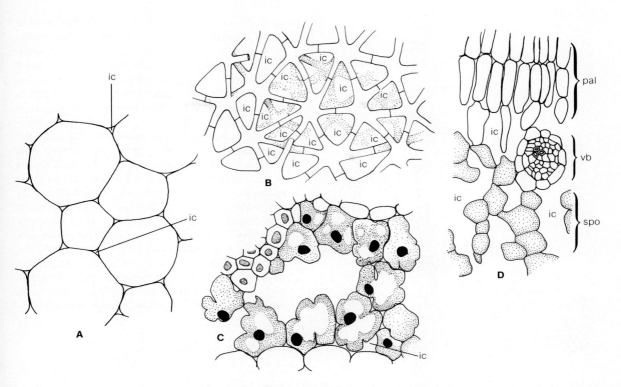

Figure 12-6 Types of parenchyma cells, × 500. **A,** cortical parenchyma of *Coleus*. **B,** aerenchyma of *Juncus*. **C,** lobed mesophyll of *Pinus*. **D,** palisade and spongy mesophyll of *Coleus: ic,* intercellular space; *pal,* palisade mesophyll; *spo,* spongy mesophyll; *vb,* vascular bundle.

Figure 12-5 Developing plant. **A,** longitudinal section of shoot tip, showing apical meristem and primary meristems. **B,** transverse section of a shoot tip, showing distribution of primary meristems, × 200. **C,** procambial strand and adjacent ground meristem, showing detail from rectangle in **B,** × 200. **D,** transverse section of *Coleus* stem, showing distribution of primary tissues and start of secondary growth, × 500. **E,** detail of vascular bundle and surrounding tissue from pie-shaped region in **D,** × 200. **F,** transverse section of developing root tip, showing distribution of primary meristems, × 500. **G,** transverse section of young leaf, showing distribution of primary meristems, × 500. **H,** detail of transverse section of young leaf, × 200. **I,** diagram summarizing plant development in a dicotyledon, × 500. **J,** shoot tip of a monocotyledon, showing primary thickening meristem, × 500; *c,* cortex; *cc,* cork cambium; *co,* cork; *e,* epidermis; *gm,* ground meristem; *ph,* phloem; *pi,* pith; *pph,* primary phloem; *pr,* pith ray; *prc,* procambium; *prt,* protoderm; *pxy,* primary xylem; *sam,* shoot apical meristem; *sph,* secondary phloem; *stm,* primary thickening meristem; *sxy,* secondary xylem; *vc,* vascular cambium.

anaerobic environment. Such parenchyma tissue is called **aerenchyma**, a tissue found in other plants of similar habitat. Photosynthetic tissue is composed of parenchyma cells that contain a great number of chloroplasts and is called *chlorenchyma*.

Most chlorenchyma is found in leaves and is called **mesophyll**. In some conifers, mesophyll cells are lobed (Fig. 12-6C). The mesophyll of angiosperms is commonly of two types. Beneath the uppersurface of the leaf, there are columnar cells with their longitudinal axes situated at right angles to the leaf surface (Fig. 12-6D). This is the **palisade mesophyll**. Below the palisade mesophyll is a tissue made up of cells of various shapes with many intercellular spaces (Fig. 12-6D), called the **spongy mesophyll**. Although intercellular spaces are conspicuous in spongy mesophyll, they are also present in palisade mesophyll. Because of the greater number of cells that constitute the palisade mesophyll, it has more total cell surface in contact with intercellular space than does spongy mesophyll.

Collenchyma Tissue

Collenchyma is a ground tissue consisting of collenchyma cells. Collenchyma cells, like parenchyma cells, are living when they perform the function for which they are adapted, and have only primary walls. However, these primary walls are unevenly thickened and unlignified, and the cells are elongated (Fig. 12-7A). The combination of thick walls and elongated cells indicates that collenchyma tissue functions in support. The special function of support is correlated with a unique feature of collenchyma cells: their walls are plastic—that is, after stretching they retain their new length without rebounding to their original dimensions. Hence, they can stretch to accommodate growth yet still offer support. Collenchyma tissue functions as support for primary tissues—tissues that are undergoing elongation.

There are three types of collenchyma cells as viewed in transverse section: angular, lamellar, and lacunar. Angular collenchyma cells have walls that are thickened in the corners (Fig. 12-7B, C). In a tissue, angular collenchyma has the appearance of beads on a string. Each "bead" is the thickened, angular, primary wall material of adjacent cells (Fig. 12-7B). Lacunar collenchyma results from the

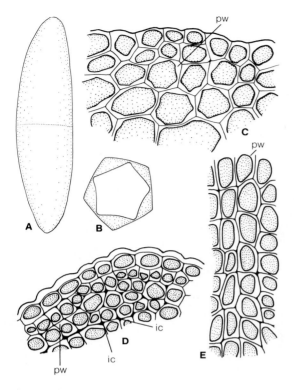

Figure 12-7 Types of collenchyma cell, × 500. **A, B,** angular collenchyma cells: **A,** diagram of single cell; **B,** transverse sectional diagram of single cell. **C,** tissue made up of angular cells. **D,** lacunar collenchyma from *Lactuca.* **E,** lamellar collenchyma from *Sambucus: ic,* intercellular space; *pw,* primary wall.

cells are the most common type of cell. The "typical" plant cell, as described in Chapter 10, is usually a parenchyma cell.

The primary functions of parenchyma tissue are photosynthesis, storage, and support (when in a turgid state), although parenchyma tissue can perform, with varied degrees of efficiency, practically all functions. Different kinds of parenchyma are distinguished by their cell morphology. For example, in *Juncus* (rush), a plant of wet habitats, the parenchyma in the center of the stem consists of lobed cells (Fig. 12-6B). As a result of this lobing, there is much **intercellular space** through which oxygen can diffuse to the roots. This could be viewed as an adaptation allowing the growth of roots in an thickening of the primary cell walls adjacent to an intercellular space (Fig. 12-7D). Lamellar collenchyma consists of cells that have their thickened walls on inner and outer tangential faces (Fig. 12-7E). Where collenchyma abuts on parenchyma or epidermis, intermediate cell types may form.

Sclerenchyma Tissue

Sclerenchyma is a more heterogenous tissue than either parenchyma or collenchyma; it is considered here as a type of ground tissue. Despite their heterogeneity, sclerenchyma cells have some features in common. They are dead when functioning and have a thick secondary wall that is usually impregnated with lignin, and an empty **lumen**. The lumen remains after the protoplast dies. The functions of sclerenchyma are support and protection.

Sclerenchyma can be classified into two major types, **sclereids** and **fibers**. Their differences are presented in tabular form below.

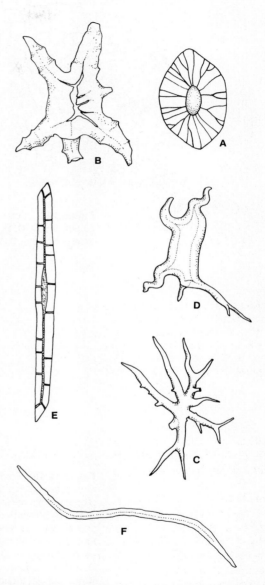

Feature	Sclereids	Fibers
Length	short	long
Shape	many different shapes	elongated, fusiform
Pits	large, often branching	smaller

Sclereids

Different types of sclereids are recognized on the basis of different cell forms:

1. Stone cells, or brachysclereids, which are little more than parenchyma cells with lignified secondary walls (Fig. 12-8A).
2. Branched sclereids with varied degrees of branching (Fig. 12-8B).
3. Astrosclereids, which are more or less star-shaped (Fig. 12-8C).

Figure 12-8 Sclerenchyma—types of sclereids, × 500. **A,** brachysclereid. **B,** branched sclereid. **C,** astrosclereid. **D,** columnar sclereid with ramified end. **E,** sclerified epidermal cell. **F,** fiber sclereid.

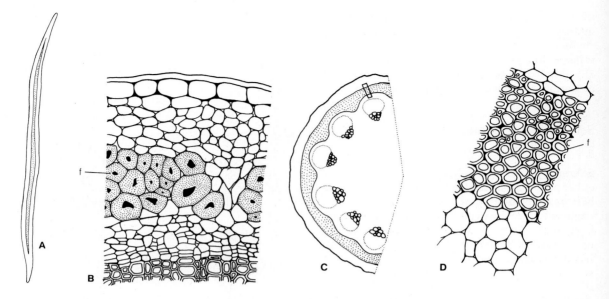

Figure 12-9 Sclerenchyma—fibers, ×100. **A**, fiber, ×100. **B**, cortical fibers in *Linum*, ×500. **C, D**, perivascular fiber in *Aristolochia:* **D**, enlarged view of rectangular area outlined in **C**, ×100; *f*, fiber.

Figure 12-10 Epidermis and epidermal cells. **A**, epidermis of *Aristolochia*. **B, C**, velamen of orchid root: **B**, transverse section of root; **C**, cellular detail of velamen. **D**, multiple epidermis of *Ficus*. **E**, surface view of epidermis of *Quercus* leaf. **F**, stoma of a grass. **G**, bulliform cells in grass. **H**, stoma of *Zebrina*. **I**, section of *Coleus* stoma. **J**, section of *Pyrus* stoma. **K**, section of *Pinus* stoma. **L**, subsidiary cells around stoma of *Pterophyllum*. **M**, mesogene development (*a–c*) of a stoma. **N**, perigene development (*a–d*) of a stoma. **O**, haplocheilic stoma. **P**, syndetocheilic stoma. **Q**, stellate trichome. **R**, peltate trichome. **S**, glandular trichome. **T**, branched trichome. **U**, simple trichome. **V**, multicellular trichome. **W**, root epidermis, showing root hairs; *bc*, bulliform cells; *c*, cortex; *cu*, cuticle; *e*, epidermis over vein; *gc*, guard cells; *gcmc*, guard cell mother cell; *i*, epidermis over area between veins; *lc*, long cell; *me*, multiple epidermis; *rh*, root hairs; *sc*, subsidiary cell; *scmc*, subsidiary cell mother cell; *shc*, short cell; *st*, stele; *v*, velamen. (**B**, × 100; **A, C-W**, × 500. **M**, redrawn after Payne; **N**, redrawn after Bonnet.)

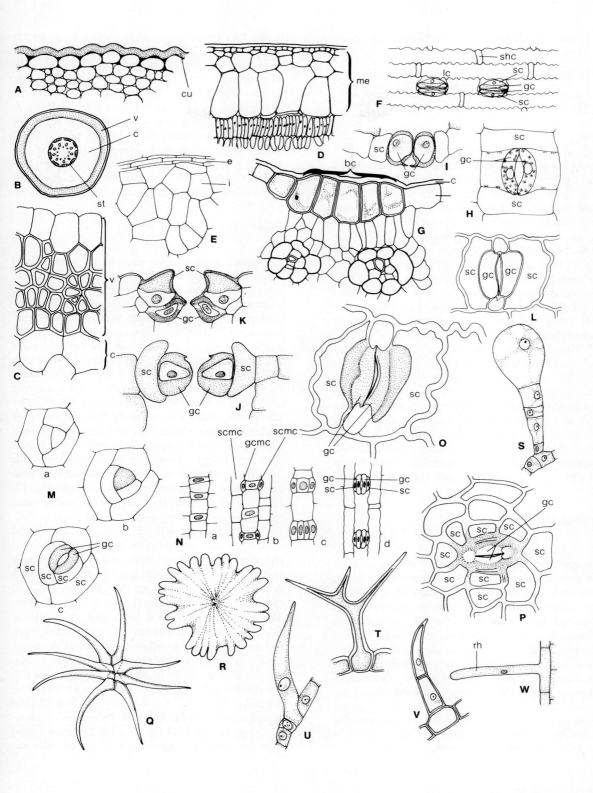

4. Columnar sclereids with ramified ends (Fig. 12-8D).
5. Sclerified epidermal cells (Fig. 12-8E).
6. Fiber-sclereids, considered to be too short for fibers but too long to be sclereids (Fig. 12-8F).

Fibers

Fibers are elongated cells that function as support and have thickened, usually lignified, secondary walls with small pits (Fig. 12-9A). Fibers occur in many tissues. In this section, we are concerned only with fibers that occur in ground tissues. There are two types of fibers, cortical and perivascular fibers.[1] Cortical fibers occur in the cortex (Fig. 12-9B). Perivascular fibers also occur in the cortex of the stem but only when the cortex is divided into inner and outer regions. This division is usually based on a ring of cells, as seen in transverse section, that have a high accumulation of starch, or show other cytological differences (Fig. 12-9C).

Fibers are unicellular, and in the primary plant body they can attain lengths up to 0.5 m. Two growth phenomena are associated with these extremely long cells—symplastic and apical intrusive growth. Symplastic growth is the growth of the developing fiber at a rate comparable to that of surrounding tissues. Apical intrusive growth occurs when a fiber grows faster than the surrounding tissues, so that the apex of the developing fiber intrudes between cells as it comes in contact with them.

Epidermis

The epidermis is a primary tissue that matures from the protoderm. In the shoot, it covers the plant; protects it against water loss, pathogen attack, and mechanical damage; and also provides for gas exchange. In the root, it has the primary functions of protection against pathogens and uptake of water and ions, with most uptake occurring through root hairs. As a primary tissue, the epidermis is usually not long-lived. There are, of course, exceptions. The epidermis of *Acer striatum* (striped maple) can persist for up to 20 years. In general, the epidermis is one cell-layer thick (Fig. 12-10A). However, in some instances the protoderm divides **periclinally** (see footnote, p. 373), to produce a multiple epidermis. For example, the roots of orchids have a multiple epidermis, the **velamen** (Fig. 12-10B, C), which seems specifically adapted to take up water from an aerial environment. Leaves of *Ficus* (fig) have a multiple epidermis (Fig. 12-10D) that has been implicated in water storage. The epidermis of the root system is different in structure, function, and ontogenetic origin from that of the shoot system.

Epidermis of the Shoot System

The shoot epidermis is made up of two general kinds of cells: epidermal cells and guard cells. Epidermal cells form the protective covering over the primary plant body. Guard cells are specialized epidermal cells involved in gas exchange. Most epidermal cells lack chloroplasts. They are barrel-shaped as seen in transverse section and have thickened outer walls (Fig. 12-10A) impregnated with cutin. They also have cutin on the surface, which forms a layer (the **cuticle**) over the entire tissue and is significant in reducing water loss. It functions by preventing water from escaping through the outer walls of the epidermal cells and also by reflecting light so that the leaf does not absorb as much heat in direct sunlight. In surface view, epidermal cells vary from elongate to equidimensional in outline and the end walls from sinuous to straight (Fig. 12-10E, F). Elongate cells are generally found over veins of leaves and on elongated organs.

Guard cells occur in pairs (Fig. 12-10F, H–J) and under certain physiological conditions undergo shape changes that result in an opening between them. The combination of guard cell and opening is referred to as the **stoma** (pl., **stomata**). Guard cells differ from epidermal cells in several ways. They usually have well-developed chloroplasts, whereas epidermal cells have no (or, at most, very reduced) plastids. Another difference is in shape: most guard cells are kidney-shaped as seen in surface view (Fig. 12-10H). Wall structure is also different: the only thickened wall seen in most epidermal cells is the outer wall, but in guard cells the distribution of thickened wall shows much variation (Fig. 12-10I, J). It has also been demonstrated that the cellulose microfibrils of guard cells are

---

[1]Since they occur in specific tissues other than cortical or perivascular, phloem and xylem fibers are discussed in the context of those specific tissues.

arranged along radii from the facing to the outer walls.

The nature of the guard cell wall has been implicated in its function. When light strikes a leaf, it initiates a series of physiological changes that are ultimately manifested as an increase in the potassium ion concentration in the guard cells. As a result of this change, the water potential of the guard cells lowers, and water enters them from surrounding cells. The resulting increase in internal pressure would normally result in swelling. However, the radially arranged microfibrils prevent the guard cells from increasing in girth. The increased pressure therefore tends to cause them to lengthen, but their thicker walls prevent any change in total length, so the pressure is manifested by bending.

A second mechanism causing guard cells to move and open stomata is correlated with another feature of guard cells: they share a common wall at the two ends as seen in surface view (Fig. 12-10H). The increase in internal pressure results in the wall directly opposite these common walls moving outward. As the walls opposite the facing walls move outward, radially arranged microfibrils transmit this movement to the facing walls and contribute to the opening of the stoma. The functional significance of this mechanism is obvious: the presence of light initiates a set of reactions that allow CO_2 to enter the plant, so that the essential process of photosynthesis can begin.

Other cells, called **subsidiary cells,** are associated with guard cells and also occur in some types of epidermis. These closely resemble typical epidermal cells but present a slightly different pattern in surface view (Fig. 12-10F, L). They are believed to be the source of the water that enters the guard cells and causes changes in turgor. The shape and arrangement of subsidiary cells vary greatly in angiosperms (class Magnoliopsida). We will mention only an ontogenetic aspect of this subsidiary cell/guard cell variation. Ontogenetically, there are two types of guard cell/subsidiary cell combinations. In one, guard cells and their subsidiary cells result from division of the same mother cell (Fig. 12-10M). In the other, the guard cell mother cell does not give rise to the subsidiary cells (Fig. 12-10N). The former is referred to as **mesogene development,** and the latter as **perigene development**. The terms mesogene and perigene apply only to angiosperms. A somewhat similar situation is encountered in gymnosperms, although the terminology is different. Based on a study of extant gymnosperms, the Swedish paleobotanist Florin described two basic types of stomata, **syndetocheilic** and **haplocheilic**. Developmentally, syndetocheilic stomata follow a perigene pattern and haplocheilic a mesogene pattern. Mature gymnosperm stomata of the two types form very distinctive patterns (Fig. 12-10O, P), and their designation as syndetocheilic or haplocheilic is often based on these patterns.

Various kinds of outgrowths, called **trichomes,** also occur in the epidermis. Trichomes are simple or branched, unicellular or multicellular, and glandular or nonglandular (Fig. 12-10Q, W). They are often important in taxonomic studies and are commercially important as well: cotton is from trichomes.

Epidermis of the Root System

The root system lacks stomata, and although some cutin is present, there is debate as to whether or not a cuticle is present. Two kinds of cells make up the root epidermis: epidermal cells and **root hair cells** (Fig. 12-10W). Root hairs are simply outgrowths of single epidermal cells (Fig. 12-10W). They increase the absorptive surface of the root and aid in water uptake by penetrating to new supplies of water. Root hairs are generally restricted to that part of the root where elongation has ceased and secondary growth, if it is to occur, has not yet started.

Vascular Tissues

The conducting tissues in the plant body, the xylem and phloem, make up the vascular tissue. Xylem transports water and ions; phloem conducts carbohydrates, water, and some ions.

Primary and secondary vascular tissues appear in developmental sequence. Primary tissues mature from cells derived from the apical meristem, whereas the secondary tissues are derived from vascular cambium. Primary vascular tissues are further classified into two types, designated by the prefixes *proto-* and *meta-*. Thus, a vascular plant will have **protoxylem** and **metaxylem** as well as **protophloem** and **metaphloem**. These terms refer to time of maturation: *proto-* indicates the first to mature, and *meta-* indicates the last. Being the first tissues to mature, protophloem and protoxylem do so in an area that is still undergoing elongation. Thus, after maturation, the cells of protophloem and protoxylem are subject to stretching and can be stretched to the point of obliteration.

The pattern of maturation in primary vascular tissue depends on the position of protoxylem and protophloem in the procambial bundle. Sometimes, protoxylem appears in the portion of the procambium that is near the center of the organ, and metaxylem develops externally to the protoxylem (Fig. 12-11A). This pattern of maturation is called **centrifugal,** and the primary xylem is referred to as the **endarch.** Conversely, protoxylem can appear in the portion of the procambium that is near the surface of the organ, and metaxylem develops internally to the protoxylem (Fig. 12-11B). Such a pattern is **centripetal,** and the primary xylem is **exarch.** A third type of primary xylem is **mesarch,** in which protoxylem appears in the center of the procambium, and metaxylem develops all around the protoxylem (Fig. 12-11C). All primary phloem undergoes centripetal maturation (Fig. 12-11A).

Stele

A broader term, the **stele,** is sometimes used to refer to vascular tissues plus some associated parenchyma. By definition, the stele occupies the central portion of the plant axis. Its outermost layer is called the **pericycle** and consists of parenchyma tissue. There is also parenchyma in the middle of some steles; it is referred to as **pith.** The stele is surrounded by cortex, the innermost layer of which is the **endodermis** (Fig. 12-12A–E).

The characteristic features of the endodermis vary somewhat. In roots of angiosperms and gymnosperms, and stems of lower vascular plants (class Filicopsida and more generalized subdivisions), the cells of the endodermis have a strip of **suberin,** the **casparian strip,** on all walls but their inner and outer tangential walls (Fig. 12-12C). In the stems of some angiosperms, the endodermis is characterized by the presence of a higher quantity of starch in certain cells (Fig. 12-12E). Because of this, the endodermis in stems is sometimes called a starch sheath.

Protosteles and Siphonosteles There are two basic types of steles, **protosteles** and **siphonosteles.** Pro-

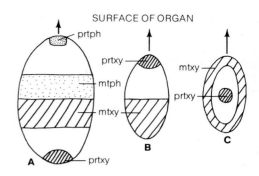

Figure 12-11 Types of primary vascular bundles. **A,** endarch primary xylem. **B,** exarch primary xylem, phloem not included. **C,** mesarch primary xylem, phloem not included: *mtph,* metaphloem; *mtxy,* metaxylem; *prtph,* protophloem; *prtxy,* protoxylem. *Clear* areas indicate primary vascular tissue that matures after protoxylem and phloem and before metaxylem and phloem.

tosteles have a central core of xylem (Fig. 12-13A–C), whereas siphonosteles have pith (Fig. 12-13E, G, I, K). There is an intermediate type of stele, called a *vitalized protostele;* it has a central core of xylem but contains a high percentage of parenchyma (Fig. 12-13D).

Protosteles. In general, protosteles are found in more generalized vascular plants (rhyniophytes and lycopods) and in roots of specialized vascular plants. Protosteles in lower vascular plants vary from the **haplostele** in *Psilotum* rhizomes (Fig. 12-13A) to **actinostele** in *Psilotum* stems (Fig. 12-13B), to **plectostele** in *Lycopodium* (club moss) stems (Fig. 12-13C). Roots of some higher vascular plants have a protostele like that seen in Figure 12-12A, B. Vitalized protosteles are found in the fern *Gleichenia* (Fig. 12-13D).

Siphonosteles. There are three different types of siphonosteles: **ectophloic,** with phloem external to xylem; **amphiphloic,** with phloem on both sides of the xylem; and **eusteles,** with phloem in various positions.

The ectophloic siphonostele may occur in various forms. It is a cylinder in the fern *Osmunda* (royal fern) (Fig. 12-13E). *Equisetum* (horsetail) has an ectophloic siphonostele consisting of separate

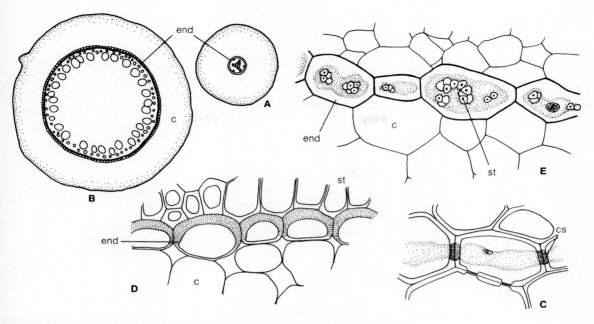

Figure 12-12 A stele and endodermis. **A,** stele of *Ranunculus* root. **B,** stele of *Zea* root. **C,** young endodermal cell. **D,** old endodermal cells. **E,** endodermis in stem of *Aristolochia*: *c*, cortex; *cs*, casparian strip; *end*, endodermis; *st*, starch grains. (**A, B,** × 100; **C-E,** × 500.)

bundles (Fig. 12-13F), with both internal and external endodermis. Most ectophloic siphonosteles have only an external endodermis.

Amphiphloic siphonosteles can also occur in many forms. One type, a **solenostele,** found in the fern *Dennstaedtia* (hay-scented fern) (Fig. 12-13G) has vascular tissue occurring in an unbroken ring. Some plants, such as the fern *Matonia* (Fig. 12-13H), have a stele called a **polycyclic solenostele,** which consists of a series of concentric rings, each having xylem surrounded on both sides by phloem, which is in turn surrounded on both sides by endodermis. A **dictyostele** consists of independent vascular bundles that can occur as one ring (*Polypodium,* Fig. 12-13I, J) or as more than one ring (*Pteridium,* Fig. 12-13K). The individual bundles of a dictyostele are called **meristeles.**

Eusteles are found in seed plants (classes Pteridospermopsida to Magnoliopsida). In transverse section, they are seen to have separate bundles, but these bundles have specific relationships to the longitudinal axis of the plant—relationships that are difficult to discern. One eustele has several series of vascular bundles that present a complex pattern in transverse section (Fig. 12-13L). These occur in monocotyledons, and Figure 12-14 shows their basic pattern. Figure 12-15 presents a summary of the various types of steles. The different steles are the result of evolution. The earliest vascular plants had protosteles and, with passing of time, vitalized protosteles and siphonosteles appeared in the fossil record.

Xylem

Xylem serves a dual function—the conduction of water and mineral ions, and support. Most of its cells have lignified secondary walls, and many are dead when functioning. Primary xylem develops from the procambium, and secondary xylem develops from the vascular cambium. Primary xylem is divided into protoxylem, the first to mature, and metaxylem, the last to mature.

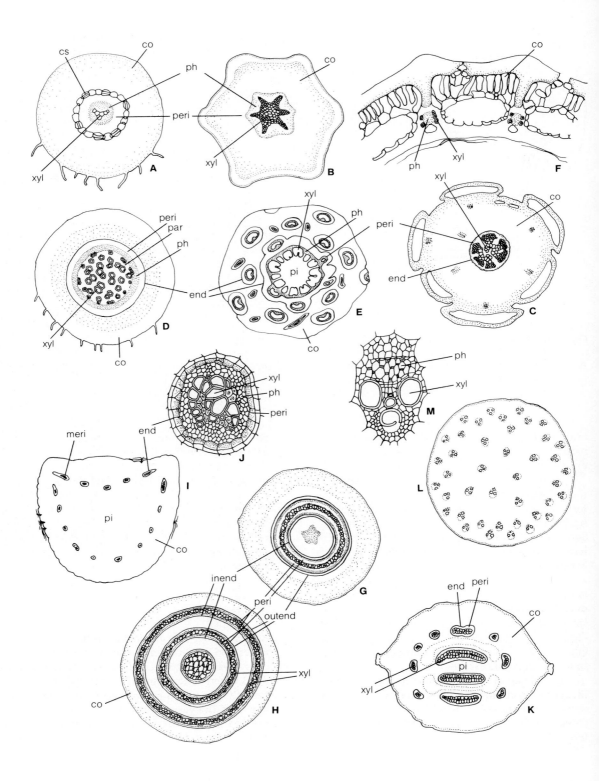

358   Chapter 12

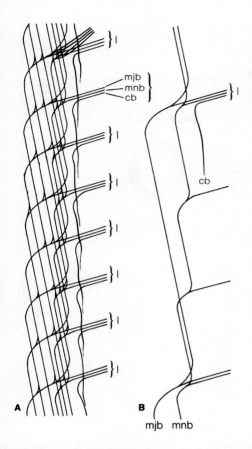

Xylem is composed of four different types of cells: **tracheids, vessel elements**, fibers, and parenchyma. In secondary xylem, these cells occur in two systems, **axial** and **radial**. The former is composed of cells with their longitudinal axes oriented parallel to the longitudinal axis of the organ in which they occur, and the latter is composed of cells with their longitudinal axis oriented along radii of the organ (Fig. 12-16). Cells in the radial system are said to occur in **vascular rays.** As seen in different sections of wood, cells of these two systems form distinctive patterns (Fig. 12-16).

Xylem parenchyma cells are living, but unlike other parenchyma cells, they usually have thick, secondary, lignified walls (Fig. 12-17A, B). When such walls are present, simple pits occur in them. Xylem parenchyma cells are found in both the axial and radial systems and function mainly in storage.

Tracheids and vessel elements are often classified together as **tracheary elements,** the conducting cells of the xylem. Tracheids are imperforate, elongated cells with numerous pits in the secondary wall (Fig. 12-17F). Water passes from tracheid to tracheid through the pits, but to do so it must traverse the pit membrane (Fig. 12-17H). The word *imperforate* means that a pit membrane exists between tracheids. Pits occur on all walls of tracheids but are most common on walls where tracheids overlap and on radial walls. Vessel elements differ from tracheids in the lack of pit membranes in the pits on certain walls, resulting in perforations (Fig. 12-17J). These perforations are found on the end walls of the vessel elements, which are called **perforation plates** (Fig. 12-17I–K, M, N). Vessel elements are also usually shorter than tracheids. Water is passed from one vessel element to the next through the perforation plates (Fig. 12-17J). Many vessel ele-

Figure 12-14 Diagram showing the longitudinal course of vascular bundles in a monocotyledon stem. **A,** several different sets of interconnecting bundles; each leaf is shown being supplied by one major, one minor, and one cortical bundle. **B,** a simplified diagram showing the three elements of **A**—major, minor, and cortical bundles: *cb,* cortical bundle; *l,* leaf base; *mjb,* major bundle; *mnb,* minor bundle. (From Zimmerman and Tomlinson.)

Figure 12-13 Types of steles. **A,** haplostele of *Psilotum* rhizome. **B,** actinostele of *Psilotum* stem. **C,** plectostele of *Lycopodium.* **D,** vitalized protostele of *Gleichenia.* **E,** ectophloic siphonostele of *Osmunda.* **F,** dissected siphonostele of *Equisetum.* **G,** solenostele of *Dennstaedia.* **H,** polycyclic solenostele of *Matonia.* **I, J,** dictyostele of *Polypodium:* **I,** diagram showing stelar arrangement; **J,** detail of a single vascular bundle. **K,** dictyostele of *Pteridium.* **L, M,** atactostele of *Zea* stem: **L,** diagram showing stelar arrangement; **M,** detail of a single vascular bundle: *co,* cortex; *cs,* casparian strip; *end,* endodermis; *inend,* inner endodermis; *meri,* meristele; *outend,* outer endodermis; *par,* parenchyma; *peri,* pericycle; *ph,* phloem; *pi,* pith; *xyl,* xylem. (**A, B,** × 250; **C-E, G-I, K, L,** × 200; **F, J, M,** × 500.)

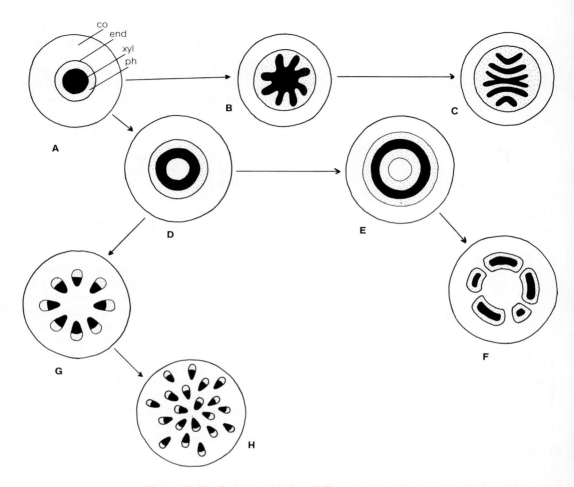

Figure 12-15 Evolution of steles. **A, B,** protosteles. **C,** plectostele. **D,** ectophloic siphonostele. **E,** amphiphloic siphonostele. **F,** dictyostele. **G,** eustele. **H,** monocotyledon stele: *co*, cortex (*clear*); *end*, endodermis; *ph*, phloem (*stippled*); *xyl*, xylem (*black*).

ments are joined end-to-end to constitute a vessel. Even though they are multicellular, vessels are of given lengths, and water must pass through pit membranes to reach another vessel.

Tracheary elements vary greatly. For example, the pattern of secondary wall deposition varies in the conducting cells (both tracheids and vessel elements) of the primary xylem. The different types of secondary wall deposition are **annular, helical, scalariform,** and **pitted** (Fig. 12-18A–D). In the fossil record, annular secondary wall depositions appear very early, followed later by helical scalariform and pitted depositions. Ontogenetically, conducting cells with annular or helical patterns of secondary wall deposition are in the protoxylem; scalariform and **reticulate** elements occur in the xylem, which matures just after protoxylem; pitted elements are seen in metaxylem and secondary

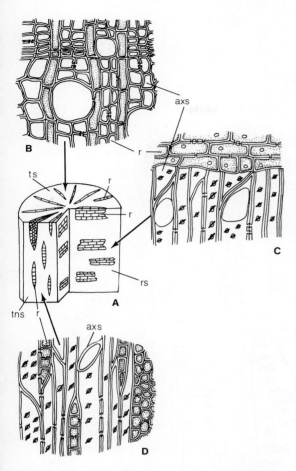

Figure 12-16 A–D, section of a stem showing different planes of sectioning: **A,** diagram showing orientation of planes; **B,** transverse section; **C,** radial section; **D,** tangential section: *axs,* cells of the axial system; *rs,* radial section; *r,* rays (cells of the radial system); *tns,* tangential section; *ts,* transverse sections. (**A,** × 100; **B-D,** × 250.)

xylem of specialized tracheophytes. Generalized tracheophytes usually have scalariform metaxylem elements.

Tracheids vary widely in types of pitting. In gymnosperms, the pits are **circular, bordered,** and often have a **torus** (see Fig. 10-8M, N). Ferns (class Filicopsida) usually have bordered pits, but individual pits are elongated and perpendicular to the longitudinal axis of the cell (see Fig. 10-8F): this is scalariform deposition (Fig. 12-17F). Other tracheids, including those of some angiosperms, have reduced, circular bordered pits (see Fig. 10-8J, K, L): this is pitted deposition (Fig. 12-17N).

Variation in vessel elements is extensive. They vary in

1. Angle of inclination of end walls (Fig. 12-17I, K, N)
2. Type of perforation plate, whether simple (with one opening) (Fig. 12-17N), scalariform (ladderlike) (Fig. 12-17I, J, K), or reticulate (Fig. 12-17M)
3. Length (Fig. 12-17I, K, N)
4. Width (Fig. 12-17L, O)
5. Shape of transverse section, from angular (Fig. 12-17G) to circular (Fig. 12-17O)
6. Pattern of secondary wall deposition, from scalariform (Fig. 12-17I) to pitted (Fig. 12-17K)
7. Pattern of pits, from opposite (Fig. 12-17K) to alternate (Fig. 12-17N)

There are correlations between these different characteristics. In general, long vessel elements have highly inclined end walls with scalariform perforation plates having many bars. They are also narrow and angled in transverse section and have a scalariform pattern of secondary wall deposition. Short vessel elements have essentially transverse end walls with simple perforation plates. These vessel elements are usually wide, circular in transverse section, and have a pitted pattern of secondary wall deposition with the pits being alternate. Between these extremes are series of intergrading types.

Such variation has initiated speculation concerning the evolutionary origins of vessel elements. Long vessel elements, with their associated features, are very similar to tracheids. Furthermore, vessels are found most commonly in angiosperm xylem, whereas tracheids occur mainly in xylem of the more generalized vascular plants. Tracheids also appear earlier in the fossil record than do vessel elements. These facts, coupled with the similarity between long vessel elements and tracheids, led I. W. Bailey and his associates to speculate that vessel elements evolved from tracheids through development of a perforation plate. Those

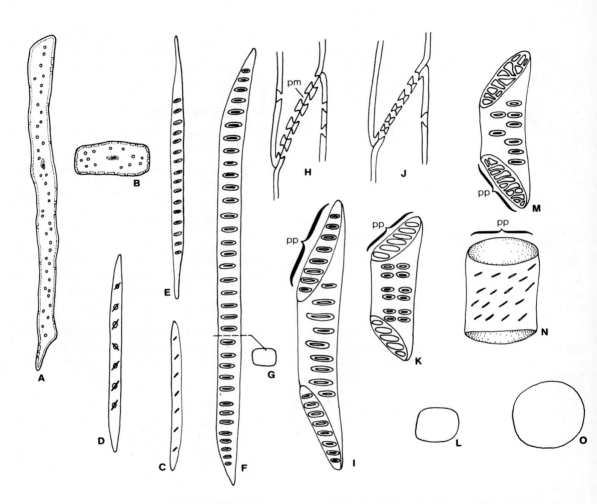

Figure 12-17 Cells of the xylem, × 500. **A,** xylem parenchyma cell. **B,** transverse section of xylem parenchyma cell. **C,** libriform fiber. **D,** fiber. **E,** fiber tracheid. **F,** tracheid. **G,** transverse section of a tracheid. **H,** overlapping walls between tacheids. **I,** vessel element with scalariform perforation plate. **J,** connection at perforation plate of two vessel elements. **K,** vessel element with scalariform perforation plate with fewer bars than **H. L,** transverse section of **K. M,** vessel element with reticulate perforation plate. **N,** vessel element with simple perforation plate. **O,** transverse section of **N**: *pm*, pit membrane; *pp*, perforation plate. (Redrawn after Bailey and Tupper.)

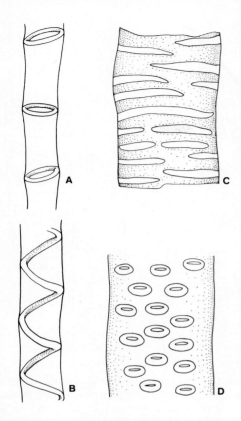

Figure 12-18 Kinds of secondary wall deposition in primary xylem, × 500. **A,** annular. **B,** helical. **C,** scalariform. **D,** pitted. (Redrawn after Bierhorst.)

vessel elements that are most similar to tracheids have undergone the fewest evolutionary changes from a tracheid and hence are most primitive. Vessel elements that are most unlike tracheids are more advanced. The most advanced vessel elements would be like those found in *Quercus* (oak) and grasses (Fig. 12-17N).

Fibers are also found in the xylem, often going by the name of **xylary fibers.** Like tracheary elements, fibers exhibit variation that allows them to be arranged in a morphological series (Fig. 12-17C–E), often interpreted as an evolutionary series. Xylary fibers also are considered to have evolved from tracheids through pit reduction and increase in wall thickness. Hence, fiber-tracheids, which resemble tracheids, are considered primitive. **Libriform fibers** (phloemlike), named for their similarity to phloem fibers, have much thicker walls and much smaller pits than fiber tracheids and are considered advanced (Fig. 12-17C). Intermediates are called *fibers* (Fig. 12-17D).

Vessels and vessel elements occur in diverse groups of plants and hence are interpreted as having evolved more than once. Vessels are found in *Pteridium* (bracken fern), *Marsilea*, *Selaginella*, *Equisetum* (horsetail), some gymnosperms [*Gnetum* and *Ephedra* (Mormon tea)],[2] and monocotyledons and dicotyledons of the angiosperms. Because patterns of intraplant vessel element variation are different in monocotyledons and dicotyledons (the most advanced vessel elements in monocotyledons are in the metaxylem of the root; the most advanced vessel elements in dicotyledons are in the stem), it is assumed that vessels evolved independently even in monocotyledons and dicotyledons.

Phloem

Phloem, the other conducting tissue of vascular plants, conducts products of photosynthesis, carbohydrates, water, and some ions. Because it contains phloem fibers, phloem also serves a support function. As in xylem, both primary and secondary phloem occur, developing from procambium and vascular cambium respectively. Primary phloem includes both proto- and metaphloem. There are radially oriented cells (rays) and longitudinally oriented cells in secondary phloem as there are in xylem. Phloem consists of three different types of cells: parenchyma, fibers, and **sieve elements.** Phloem fibers have thick secondary walls with very small pits (Fig. 12-17C).

Sieve Elements   **Sieve tube elements** and **sieve cells,** collectively referred to as **sieve elements,** are the characteristic cells of phloem and are involved

---

[2] A further indication of the independent evolution of vessel elements in *Gnetum* and *Ephedra* is the similarity between the vessel elements of these taxa and tracheids with circular bordered pits. Hence, it is assumed that vessel elements in these gymnosperms evolved from such tracheids. Vessel elements in other plants are assumed to have evolved from tracheids with a scalariform pattern of secondary wall deposition.

in conduction. Distinctive features of these cells are seen in both wall and protoplast. In the lateral and end walls, called **sieve areas** (Fig. 12-19), there are concentrations of pores. The two different kinds of sieve elements can be determined by the nature of the sieve areas. If all sieve areas have pores of equal size, the sieve element is called a sieve cell. This type is the longest of the sieve elements and occurs in gymnosperms and generalized tracheophytes. On the other hand, if sieve areas on the end walls have larger pores, the cell is called a sieve tube element, and the end wall is called a **sieve plate** (Fig. 12-19A). A sieve plate can be simple, with one sieve area (Fig. 12-19B), or compound, with more than one sieve area (Fig. 12-19A). Sieve tube elements are joined end-to-end to form a continuous tube called a **sieve tube.** The protoplast usually has no nucleus or vacuoles, and a proteinaceous substance, slime or P-protein, is present.

Sieve Tube Elements. The sieve area in a sieve tube element is more than just a cluster of pores. The protoplasts of adjacent sieve tube elements are connected through the pores of a common sieve area (Fig. 12-20A, B), which actually consists of the primary walls of adjacent sieve tube elements. Lining each pore, and outside the plasmalemma, is a carbohydrate, called **callose** (Fig. 12-20A, B). The amount of callose varies over the growing season. When phloem is young and functioning, there is relatively little callose present (Fig. 12-20B). As phloem ages, more and more callose develops, so that pores are gradually reduced in size (Fig. 12-20C, b-d). Finally, in the autumn (or at the end of the growing season), when phloem ceases to function, the pores become completely occluded with callose (Fig. 12-20C, e). After callose deposition, sieve tube elements can die, in which case all callose disappears, or they can reabsorb the callose and function again the following spring (or beginning of the growing season). The deposition of callose is manifested over the entire sieve area. The sieve plates of young phloem have a thin layer of callose. As phloem ages, the callose layer over the sieve plate becomes thicker and thicker (Fig. 12-20E).

In most plants, sieve tube elements function for only one or two years. In some ferns, palms, and some other monocotyledons—sieve elements have been said to function for many years. In these

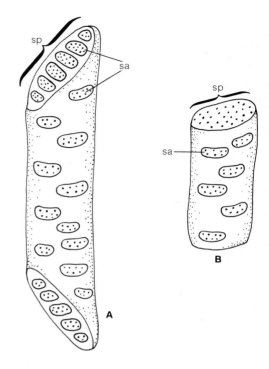

Figure 12-19 Sieve areas and sieve plates, × 500. **A,** sieve tube element with compound sieve plate. **B,** sieve tube element with simple sieve plate: *sa,* sieve area; *sp,* sieve plate. (Redrawn after Esau.)

plants, it would be suspected that definitive callose would be reabsorbed many times. However, such is not the case in palms, where definitive callose only appears with cessation of function.

Sieve areas become evident as the sieve tube element develops. An early step in development is the deposition of callose on opposite sides of common sieve tube element walls where a pore will develop (Fig. 12-21A). As the pore develops, the cell wall between opposing plates of callose becomes thinner, whereas the callose itself increases in thickness (Fig. 12-21B). Still later in development, the wall separating opposing callose layers is completely removed, so that the callose from different sieve tube mother cell elements is joined (Fig. 12-21C). The pore continues to enlarge until it reaches its mature size (Fig. 12-21D).

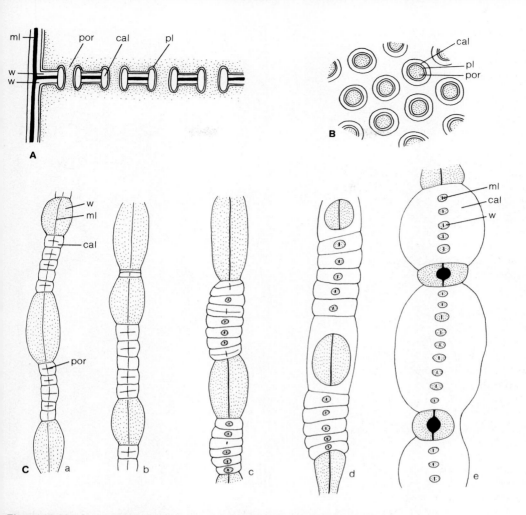

Figure 12-20 Sieve areas. **A,** sectional view of sieve plates of two adjacent sieve tube elements, showing callose and protoplasmic connections, × 750. **B,** surface view of part of sieve plate, × 750. **C,** buildup of callose over a growing season, × 1500: *a,* early in growing season; *b–d,* intermediate stages; *e,* end of growing season (*cal,* callose; *ml,* middle lamella; *pl,* plasmalemma; *por,* pore; *w,* wall). (After K. Esau.)

Protoplast. The protoplast of a mature sieve tube element lacks a nucleus and vacuoles (Fig. 12-23G) but has some organelles. Mitochondria and plastids are present, but both are somewhat modified (Fig. 12-22A). Endoplasmic reticulum is also present but in various forms, ranging from vesicles, to a **parietal** network, to stacked endoplasmic reticulum. Ribosomes are absent or rare, as are dictyosomes.

The slime of the sieve tube element protoplast is of great interest. Its exact configuration in undamaged elements is still open to debate. The sieve tube element has a very delicate protoplast and, following standard histological techniques, one usually sees slime as a mass around the sieve plate called a *slime plug* (Fig. 12-22B). Slime plugs are now considered to be artifacts. As viewed by electron microscopy, slime in at least one plant consists

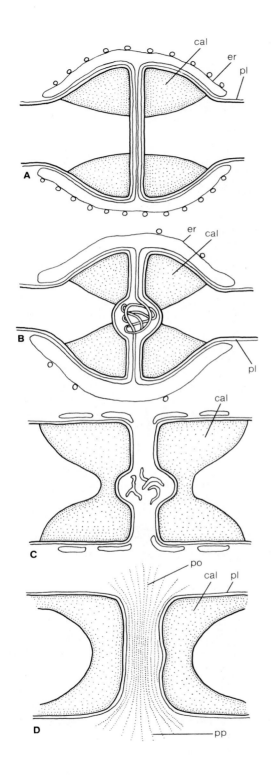
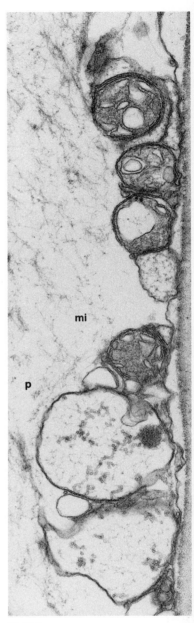

Figure 12-21 A–D, development of pores in sieve plates, × 3000: *cal,* callose; *er,* endoplasmic reticulum; *pl,* plasmalemma; *po,* pore; *pp,* strands of P-protein. (Redrawn after K. Esau.)

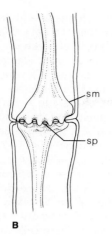

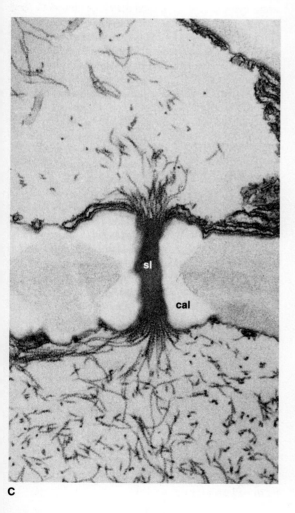

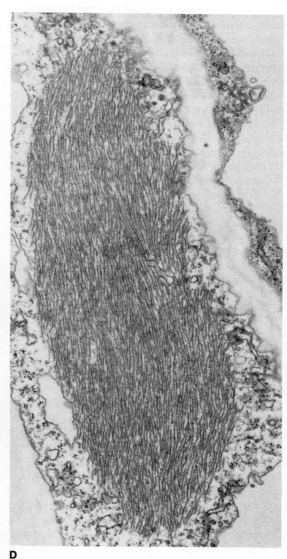

Figure 12-22 Sieve tube element protoplast. **A,** sieve tube element mitochondria and plastids. **B,** slime plug. **C,** slime in pore of sieve areas. **D,** slime in immature sieve tube elements: *cal,* callose; *mi,* mitochondria; *p,* plastid; *sl,* slime; *sm,* slime plug; *sp,* sieve plate. (**A, D,** from K. Esau and J. Cronshaw. Reproduction by permission of the National Research Council of Canada from the *Canadian Journal of Botany,* volume 46, 1968. **C,** from K. Esau, *The Phloem, Handbuch der Pflanzenanatomie,* vol. 5, part 2. Gebrüder Borntraeger, 1969.) (**A,** × 1000; **B,** × 200; **C, D,** × 2000)

of tubules that traverse the pores of the sieve area (Fig. 12-22C). Young sieve element protoplasts have slime that consists of parallel-oriented tubules (Fig. 12-22D).

The characteristic features of the sieve tube element protoplast are the result of developmental changes (Fig. 12-23). One of the first steps in the formation of sieve tube elements is the appearance of slime bodies in the cytoplasm (Fig. 12-23D). These slime bodies then begin to disperse, and the nucleus breaks down (Fig. 12-23E, F). Finally, the **tonoplast** disappears (Fig. 12-23F), and pores in the sieve areas reach their maximum size. The generalizations presented here apply also to sieve cells.

Parenchyma Phloem parenchyma cells are living, have thin primary walls, and primarily perform a storage function. Phloem parenchyma cells are elongated, the orientation of their longitudinal axis depending on their origin. Primary phloem parenchyma cells are elongated parallel to the longitudinal axis of the organ in which they occur. Some secondary phloem parenchyma cells are oriented in the same way; others, found mostly in vascular rays, have their longitudinal axes perpendicular to the longitudinal axis of the organ in which they occur.

There are also specialized parenchyma cells called **companion cells** and **albuminous cells**. In phloem, the usually enucleate sieve elements are closely associated with companion cells. In angiosperms, any one sieve tube element can have several such cells associated with it. Companion cells are ontogenetically related to the sieve tube elements with which they occur (Fig. 12-23A, B).

In gymnosperms and more generalized tracheophytes, the cell closely associated with sieve elements is not ontogenetically related to the sieve cell; in gymnosperms, it is referred to as an *albuminous cell*. Albuminous cells take their name from the deep staining they show with cytoplasmic stains, an indication of high protein level.

Companion cells in angiosperms have several distinctive features. As seen with the light microscope, the cytoplasm is denser than in a mature parenchyma cell (Fig. 12-24A). Electron microscopy reveals that this increased density results from the presence of many ribosomes (Fig. 12-24A). The cytoplasm of companion cells also contains well-developed mitochondria and few poorly developed plastids (Fig. 12-24A). Many protoplasmic connections appear between companion cells and their associated sieve tube elements. On the sieve tube element side of such a connection, there is a single large pore that is sometimes lined with callose. This pore opens into a chamber between the sieve tube element and its companion cell. From the companion cell, several **plasmodesmata** open into the same chamber (Fig. 12-24B).

A close physiological relationship between sieve elements and their associated companion or albuminous cells has been proposed. The lack of nuclei in sieve elements can be taken as evidence that companion and albuminous cells may perform nuclear functions. These cells may also carry out some cytoplasmic functions for sieve elements, and they have been implicated in the lateral movement of material into and out of sieve elements. Further evidence for a close physiological relationship between albuminous and companion cells and sieve elements is that when sieve elements cease functioning, their associated companion or albuminous cells die.

Growth in Plants

Two different aspects of growth exist in plants: growth in length (primary growth) and growth in girth (secondary growth). Growth in length is due to the activity of apical meristems, which are found at the tips of all stems and roots, including branch roots. Growth in girth results from the activity of lateral meristems that occur several cell layers under the surface in all portions of the plant except the very tips and parts of the leaves. There are two lateral meristems: vascular cambium and cork cambium. It should be noted that many nonwoody plants have neither a vascular cambium nor a cork cambium and therefore do not exhibit secondary growth. Others have a vascular cambium but no cork cambium.

Growth involves two phenomena: cell division and cell enlargement. Cell division is most rapid early in development. As noted previously, the actual increase in size during growth results from cell enlargement.

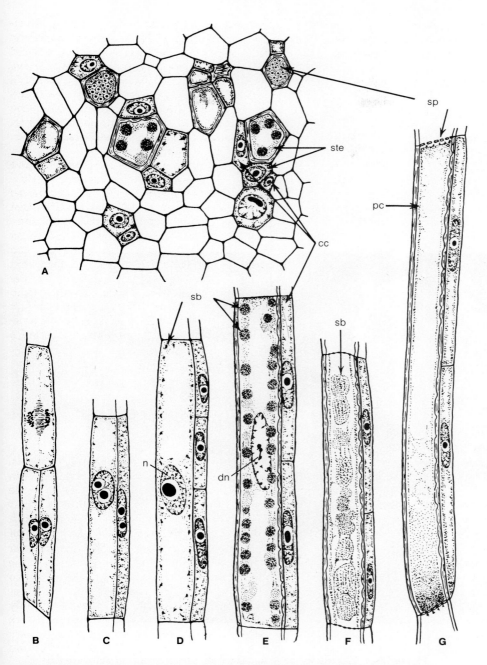

Figure 12-23 Development of a sieve tube element and some of its companion cells, × 1500. **A,** transverse section. **B–G,** longitudinal section: *cc,* companion cell; *dn,* degenerating nucleus; *n,* nucleus; *pc,* parietal cytoplasm; *sb,* slime body; *sp,* sieve plate; *ste,* sieve tube element. (Modified from K. Esau, copyright © 1965 by John Wiley & Sons, Inc. Reprinted by permission of John Wiley & Sons, Inc.)

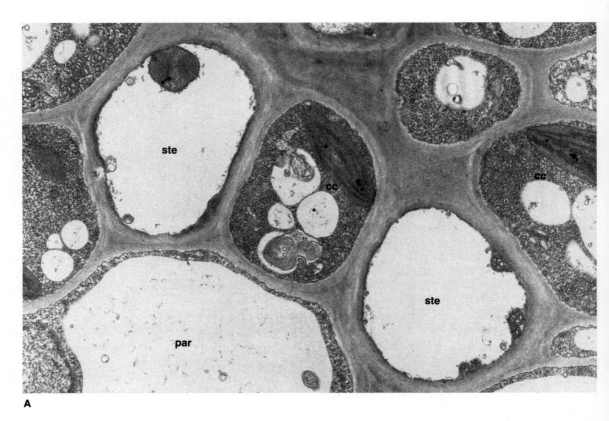

Figure 12-24 Companion cells. **A,** companion cell as compared with sieve tube element and phloem parenchyma, × 1000. **B,** connection between sieve tube element and companion cell, × 1500: *cal,* callose; *cc,* companion cell; *ml,* middle lamella; *par,* parenchyma; *pd,* plasmodesmata; *po,* pore; *ste,* sieve tube element. (**A,** from K. Esau, 1968; **B,** from K. Esau, *The Phloem, Handbuch der Pflanzenanatomie,* vol. 5, part 2, Gebrüder Borntraeger, 1969.)

Initials and Derivatives

All meristems consist of **initials** and **derivatives.** When an initial in a meristem divides, one daughter cell, the derivative, ultimately is incorporated into the mature plant body. The other remains in the meristem and continues to function as an initial. For a period of time, a derivative is still considered as part of the meristem and still undergoes cell division. Indeed, the rate of cell division in derivatives is commonly higher than that in initials. All daughter cells of a derivative also become part of the mature plant body.

Primary Growth

Many significant events are associated with primary growth. The apical meristems responsible for it are involved in the initiation and control of appendages in shoots. In turn, the type and position of appendages determine the form of the shoot system. For these reasons, a detailed understanding of the structure involved in primary growth is essential to an understanding of plant morphology.

Cytology of Apical Meristem Cells

Cells in apical meristems are cytologically diverse. This diversity can be ordered into a series of intergrading forms. At one end of the series are large cells, such as those found in the apical meristems of some generalized tracheophytes (Fig. 12-25A, C). At the other end are smaller cells (Fig. 12-25D, F) with nuclei that are large in relation to cell size. Many small vacuoles appear and make up only an insignificant part of the protoplast. In these cells, ribosomes are very common as are mitochondria, endoplasmic reticulum, and other cytoplasmic structures. These latter cells are found in part of the apical meristem in some angiosperms and gymnosperms. Between these two extremes are cells with intermediate characteristics. Such cells are found in parts of the apical meristems of lower tracheophytes, gymnosperms, and angiosperms. Generally, the larger, more vacuolate cells occur in more generalized tracheophytes.

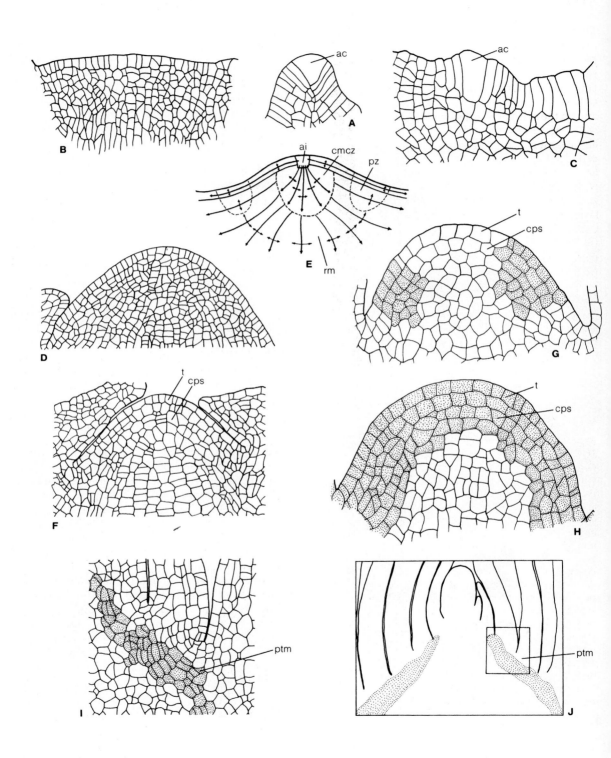

Shoot Apical Meristem

Three general types of shoot apical meristems are recognized: those found in generalized vascular plants, in gymnosperms, and in angiosperms. The shoot apical meristems of some generalized tracheophytes consist of a layer of rather large, vacuolated, rectangular cells that are bounded below and on the sides by smaller, cytoplasmically denser cells (Fig. 12-25B, C). In many such tracheophytes, an integral part of the apical meristem is an **apical cell** (see Fig. 12-25A, C). It is distinguished by its large size and pattern of cell division. Through **anticlinal divisions**[3], the apical cell gives rise to the surface layer of the meristem. Within this surface layer, both anticlinal and **periclinal divisions** occur that add to the surface and give rise to the rest of the plant (Fig. 12-25C). In some generalized tracheophytes, no apical cell seems to be present (Fig. 12-25B); in others the distinctiveness of the apical cell varies with the condition of the apical meristem. When relatively inactive, the apical cell is obvious; when very active, it is difficult to distinguish from others of the surface layer.

Most gymnosperms have a layer of initials at the apex, the *apical initials*, which give rise to both a surface layer of cells and a group of cells below the surface called the *central mother cells* (Fig. 12-25D, E). These are larger, more vacuolate, and have a slower rate of division than any other cells in the apical meristem. The area they occupy is called the *central mother cell zone*. Central mother cells give rise to the rib meristem in the center of the shoot, and the peripheral zone—the area between the rib meristem and the surface layer. The area between the central mother cell zone and these other two areas is referred to as the *transition zone*. Cells of the peripheral zone show the greatest amount of activity, have the densest cytoplasm, and are the smallest of the cells in the apical meristem. The distinctness of the various zones of the gymnosperm apex also varies somewhat with the activity of the apical meristem, showing slightly different patterns when it is dormant than when it is active.

Not all gymnosperms have apices like this. Some, such as *Araucaria* (monkey puzzle), a type of conifer, *Ephedra* (Mormon tea), and *Gnetum*, have a layer of discrete cells at the surface of the apical meristem overlying a body of cells (Fig. 12-25F). Such an arrangement is seen also in angiosperms and is referred to as a tunica-corpus organization. This is discussed with the angiosperms (see below).

It is possible to make some rather broad statements concerning the ultimate fate of cells in the various regions of the gymnosperm apex. Cells of the surface layers give rise to the epidermis and some tissues just internal to the epidermis. The rib meristem matures into pith, and the peripheral zone gives rise to most of the cortex and the procambium. Leaves also arise in the peripheral region.

The apical meristems of angiosperms show a different configuration from those of most gymnosperms (Fig. 12-25G). They consist of from one to as many as five surface layers, which form a core of tissue. Cells in the surface layers divide only anticlinally and hence show only surficial growth. Cells of the core divide in various planes, although anticlinal divisions occasionally predominate in an outer layer of the core. The surface layers are known as the **tunica layer,** and the core is called the **corpus.** As this terminology implies, each tunica layer

---

[3]An anticlinal division is one which results in a cell wall perpendicular to the nearest plant surface. The alternative, a periclinal cell division, results in a cell wall parallel to the nearest plant surface.

◀

Figure 12-25 Types of shoot apical meristems. **A,** *Equisetum* with large apical cell. **B,** *Lycopodium.* **C,** *Osmunda.* **D, E,** gymnosperm, apex of conifer: **D,** drawing showing cellular detail; **E,** diagram showing growth patterns in apex. **F,** *Araucaria,* a gymnosperm with a tunica-corpus organization. **G,** angiosperm vegetative apex (*stippling* indicates more densely stained cells). **H,** angiosperm reproductive apex (*stippling* indicates more densely stained cells). **I, J,** monocotyledon apical meristem with primary thickening meristem: **I,** diagram indicating region of primary thickening meristem (*stippled*); **J,** cellular detail of area outlined in **I** (*stippling* indicates primary thickening meristem): *ac,* apical cell; *ai,* apical initials; *cmcz,* central mother cell zone; *cps,* corpus; *ptm,* primary thickening meristem; *pz,* peripheral zone; *rm,* rib meristem; *t,* tunica. (**A-D, F-H, J,** × 500; **E, I** × 200) (**A,** redrawn after Golub and Wetmore, 1948; **B,** redrawn after Wardlaw, 1965; **C,** redrawn after Steeves, 1963; **D,** redrawn after Sacher, 1954; **E,** redrawn from Foster, 1938; **F,** redrawn after Griffith, 1952; **G, H,** redrawn after Gifford and Tepper, 1962; **I,** redrawn after K. Esau, 1977.)

has its own set of initials, and the corpus has one set of initials.

Superimposed on this cellular configuration in the shoot apical meristems of angiosperms is a type of cytohistological zonation. Apices in plants that undergo vegetative growth commonly have the most active, smallest, and cytoplasmically most dense cells in a ring below the tip (Fig. 12-25G). Cells at the tip of the apical meristem are larger, cytoplasmically less dense, and less active. Leaves are initiated in the region of higher activity, which shows a greater amount of cell division and synthesis of DNA and RNA. When the plant is induced to flower, the cells at the tip of the apical meristem are modified and come to appear cytologically like cells in the ring below them (Fig. 12-25H). This cytological change is accompanied by an increase in activity. The basic tunica-corpus organization of the meristem remains when flowering is initiated.

Another growth center is associated with the apical meristem of monocotyledons. This is an area below the apical meristem in which there is growth in girth (Fig. 12-25I, J); it is usually referred to as the *primary thickening meristem* and is responsible for much of the increase in girth in monocotyledons.

Root Apical Meristems

As in the shoot, variation occurs in the types of apical meristem in roots, both among different taxonomic groups and within them. In generalized vascular plants and ferns, there is an apical cell in the root apical meristem (Fig. 12-26A), although the patterns of growth that arise from it differ from those seen in the shoot. **Distal** to the apical cell is the **rootcap,** a protective structure common to all roots. **Proximal** to it arise tissues of the root proper. Patterns of growth are summarized in Figure 12-26A. The apical cell gives rise to tissues that in turn give rise to ground meristem, procambium, and rootcap. The protoderm differentiates from the ground meristem.

A different configuration is seen in the root apical meristem of gymnosperms (Fig. 12-26B, C). Here, a group of initials gives rise to a series of mother cell zones, which give rise to mature tissues of the root (Fig. 12-26B, C). One of these mother cell zones, proximal to the initials, gives rise to the central cylinder, which develops into vascular tissues. Two different mother cell zones arise distal to the initials. One of these zones is in the form of a short column; this column gives rise to the **columella** of the rootcap, which then gives rise to the rootcap itself. The other distally situated zone occurs as a ring around the rootcap mother cells; it gives rise to the ground meristem, the tissue that develops into the root cortex. As in generalized tracheophytes, the protoderm develops from the ground meristem.

The root apical meristem of angiosperms has a group of initials, but unlike that of gymnosperms, this group gives rise directly to the primary tissues and rootcaps. The initials in angiosperm root apical meristems show various configurations, three of which are presented in Figure 12-26D–F. In one, a single group of initials gives rise to the rootcap and all other tissues in the root (Fig. 12-26D). In this instance, the protoderm develops from the rootcap. Another type has three separate groups of initials: one gives rise to the rootcap, one to the ground meristem, and one to the procambium. In one such root apical meristem, the protoderm develops from the ground meristem cells (Fig. 12-26E). In another type, the protoderm develops from the rootcap (Fig. 12-26F). In some angiosperms, there is fusion of initials in the root apical meristem (Fig. 12-26G, H).

In many angiosperm roots, the center of the apex is occupied by a group of cells that divide at a much slower rate than the cells surrounding them. These slower-dividing cells also have a lower rate of DNA synthesis (Fig. 12-26H) and less cyto-

Figure 12-26 Types of root apical meristems. **A,** *Equisetum* with an apical cell. **B, C,** gymnosperm: **B,** drawing of cells; **C,** interpretative diagram of **B. D,** angiosperm, *Allium,* ground meristem stippled. **E,** angiosperm, *Zea,* ground meristem stippled. **F,** angiosperm, *Nicotiana,* ground meristem stippled. **G,** merger of outer ground meristem initials with root cap initials (*heavy stippling*); stellar initials lightly stippled. **H,** transversal meristem (*stippled*). **I,** root labeled with radioactive compound to show the lower rate of DNA synthesis in the quiescent center cells in division (*stippled*): *ac,* apical cell; *ccmc,* central cylinder mother cells; *clmc,* column mother cells; *cmc,* cortical mother cells; *iz,* initial zone; *prt,* protoderm; *rc,* root cap; *rci,* root cap initial. (**A, B, D-I,** × 500; **C,** × 200) (**A,** redrawn after Johnson, 1933; **B, C,** redrawn after Wilcox and Allen; **G-H,** redrawn after Seago and Heimsch; **I,** after Clowes.)

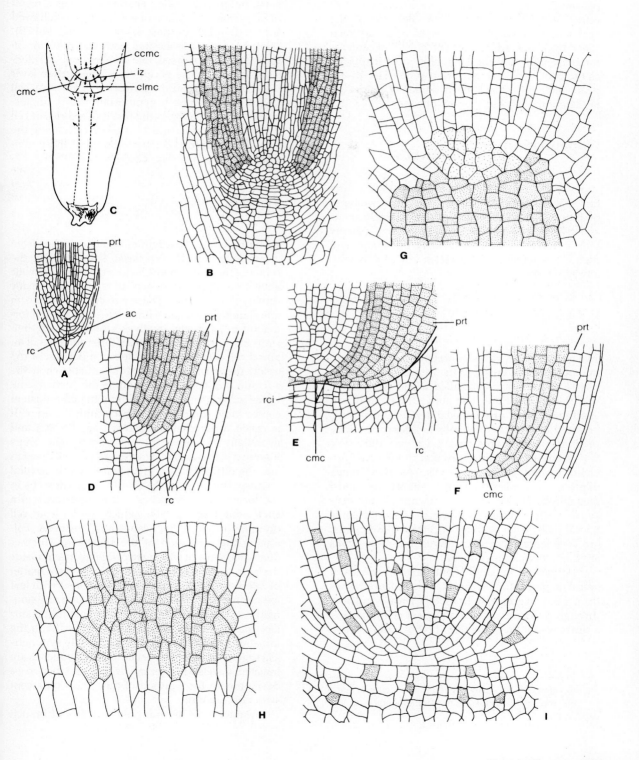

plasmic RNA and cytoplasmic protein than the more active cells surrounding them. Because of the apparent low activity of these cells, they have been called the **quiescent center**. The quiescent center is not present in young and initiating roots; it appears as the root ages. The quiescent center also disappears after roots have been damaged, either by radiation or cold, but then reappears.

Appendages

Types of appendages and their production are integral aspects of primary growth. There are two types of appendages in the shoot system, namely stems and leaves, and one in the root system, namely branch roots. Leaves are usually bilaterally symmetrical, whereas roots and stems are usually radially symmetrical.

Leaves

Two general types of leaves occur in vascular plants: **microphylls** and **megaphylls**. In plants with mega phylls, there is a break called a **leaf gap**[4] in the vascular tissue of the stem above the point of leaf attachment. Also, procambial differentiation is often oriented toward the megaphylls. Stems of microphyllous plants do not have a leaf gap, their stele is a protostele, and their vascular tissue is not oriented toward the microphylls. As their name implies, most microphylls are smaller than most megaphylls. However, some megaphylls are very small, such as those of some cedars, and some microphylls, such as those of certain fossil plants, are very large. Both microphylls and megaphylls are dorsiventral appendages.

Leaves are initiated at the sides of the shoot apical meristem in vascular plants. Initiation varies from group to group. In *Equisetum*, a leaf is initiated through differentiation of an apical cell following a series of cell divisions (Fig. 12-27A). In *Psilotum*, small outgrowths called **enations** arise as a result of tangential elongation of a ridge of cells followed by cell divisions in these same cells (Fig. 12-27B). The phenomena associated with leaf initiation in ferns are an increased rate of periclinal cell divisions in cells at the periphery of the surface layer (Fig. 12-27C, D) and the formation of an apical cell (Fig. 12-27D). In both gymnosperms and angiosperms, leaf initiation is the result of a series of cell divisions in surface and subsurface cells along the sides of the apical meristem. Some of these divisions are periclinal (Fig. 12-27E).

Branches

As with leaves, the origin of branches varies from group to group. In *Selaginella,* some cells in the center of the apical meristem lose their meriste matic ability, resulting in two centers of meristematic activity (Fig. 12-28A). These two centers then form apical meristems, and a branch results. Branch formation in *Equisetum* involves a series of cell divisions in a surface cell and then differentiation of an apical cell (Fig. 12-28B). In ferns, a portion of the apical meristem does not differentiate when the surrounding tissue does, and as the surrounding tissue grows, this bit of undifferentiated meristem comes to lie some distance back from the apex. It is called a **detached meristem** (Fig. 12-28C) and eventually develops into a branch. The steps involved in this are an increased rate of cell division and the differentiation of an apical cell. In several angiosperms, one of the first histological steps in the formation of branches is the appearance of a **shell zone** (Fig. 12-28D), which results from cell divisions in particular planes and positions. Following the formation of the shell zone, further divisions, some of which are periclinal, occur inside the area delimited by the shell zone. As a matter of interest, periclinal divisions associated with leaf initiation (and initiation of other bilaterally symmetrical organs) occur in tissues closer to the surface than do similar divisions associated with the initiation of branches (and other radially symmetrical organs). In many vascular plants, branches are initiated, develop to a certain degree, and then cease development for a period of time. This quiescent structure is a **bud**.

Variation occurs not only in types of branch

---

[4]Nambooderi and Beck have argued that leaf gaps in some conifers and angiosperms are not homologous with leaf gaps in ferns.

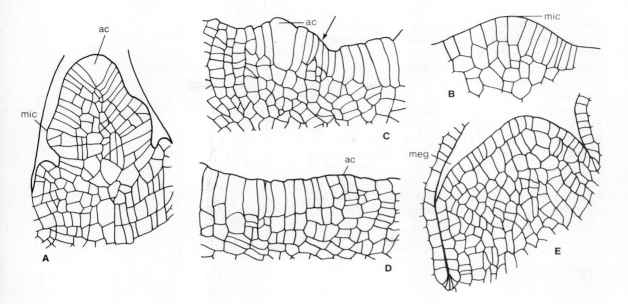

Figure 12-27 Initiation of microphylls (*Equisetum* and *Psilotum*) and megaphylls (*Osmunda* and an angiosperm), × 500. **A**, *Equisetum*. **B**, *Psilotum*. **C, D**, *Osmunda*: **C**, initiation of leaf (*arrow*); **D**, slightly older leaf with apical cell. **E**, angiosperm (*Ranunculus*): *ac*, apical cell; *meg*, initiating megaphyll; *mic*, initiating microphyll. (**A**, redrawn after Golub and Wetmore; **B**, redrawn after Bierhorst; **C, D**, redrawn after Steeves and Sussex; **E**, after Tepfer.)

initiation but also in branching patterns. One aspect of this variation is terminal versus lateral branching. The former describes branches that arise as a result of the apex dividing to form two branches, as in *Selaginella* (spike moss), *Psilotum* (whisk fern), and *Lycopodium* (clubmoss). Lateral branching refers to branches that arise from the side of the shoot and some distance behind the apical meristem, as in *Equisetum* (horsetail), ferns, gymnosperms, and angiosperms. Branching patterns also vary in the type of branch system they form. A plant with terminal branching can show **dichotomous branching**. This results from the formation of two branches that grow to be equal in size (Fig. 12-29A). Each of these branches can then divide again into two equal pairs (Fig. 12-29B). Dichotomous branching occurs in all planes.

Another type of branch system that can develop in a plant with terminal branching is a **sympodium**. In this branch system, two branches form, and then one outgrows the other and becomes predominant. At each point of branch formation alternate branches become the dominant ones (Fig. 12-29C).

In most instances of lateral branching, the branch arises in a specific position relative to a leaf. In *Equisetum*, branches alternate with scale leaves. In ferns, gymnosperms, and angiosperms, branches arise in the angle formed by a leaf and its stem, the axil. Buds in the axils, as in angiosperms, are called **axillary buds**. In most ferns and seed plants, branching is **monopodial**. Such a branch system consists of a main axis with lateral branches that developed from axillary buds (Fig. 12-29D). In some plants, the branching is **pseudomonopodial** or **sympodial**. In these systems, an axillary bud develops and its branch outgrows the main axis that produced it, taking on the aspect of a main branch (Fig. 12-29E). This event is repeated time and time again, producing a branch system that consists of a series of different lateral branches that combine to give a main axis (Fig. 12-29F). Alternate axillary buds predominate as development proceeds (Fig. 12-29F).

The initiation of branch roots in most vascular plants occurs some distance from the apical meristem. Unlike shoot appendages, root appendages initiate deep in the parent root (Fig. 12-30A) as a result of cell division in the pericycle, a tissue that is procambial in origin. In *Selaginella*, a plant with an apical cell in the root, branch roots originate when the root apical cell forms two such cells (Fig. 12-30B), which grow into roots.

Secondary Growth

Secondary growth results from the activity of lateral meristems and produces an increase in girth. Most secondary growth is from the activity of the vascular cambium, a meristematic region between xylem and phloem that gives rise to xylem to the inside and phloem to the outside (Fig. 12-31B).

Vascular Cambium

Vascular cambia in stems and roots have different origins. In those plants with vascular cambia, the shoot primary vascular tissue is a eustele. Vascular cambium initiates with cell division in the procambium between primary xylem and phloem. From there, cell division spreads to parenchyma cells in the pith ray (Fig. 12-31A, B). As a result of these developmental phenomena, a ring of vascular cambium is formed. Because of their different modes of origin, different names have been applied to regions of vascular cambium. The portion that develops from procambium is termed **fascicular cambium** (in reference to its development in the vascular bundle), and the portion that develops in the pith ray (between the bundles) is termed **interfascicular cambium**. Once the cambium has functioned for a period of time, it is impossible (in most plants) to distinguish between fascicular and interfascicular cambia. In some monocotyledons, a ring

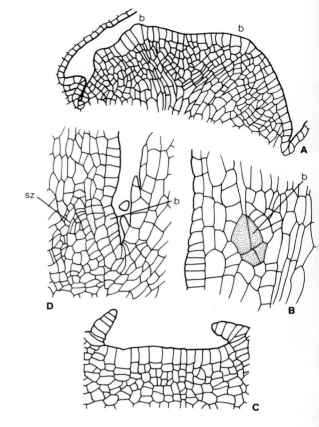

Figure 12-28 Initiation of branches, × 500. **A,** *Selaginella*. **B,** *Equisetum*. **C,** *Onoclea*. **D,** angiosperm: *b*, initiating branches; *sz*, shell zone. (**A,** redrawn after Webster and Steeves; **B,** redrawn after Johnson; **C,** after Wardlaw; **D,** redrawn after Sussex.)

of dividing cells external to mature vascular tissues produces parenchyma and new vascular tissue.

Unlike the vascular cambium in stems, root vascular cambium has no ground tissue involved in its formation. The first sign of cambial activity in roots is cell division in the procambium between primary xylem and primary phloem (Fig. 12-31D). After this portion of the cambium has functioned for a time, pericycle cells near the protoxylem begin to divide to form a ring of vascular cambium which gives rise to secondary tissue (Fig. 12-31E).

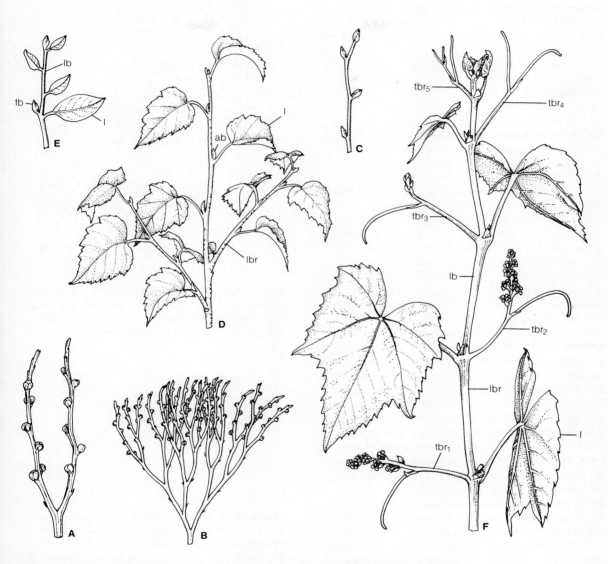

Figure 12-29 Types of branching systems. **A,** dichotomous branch. **B,** dichotomously branched system. **C,** sympodium of terminal branches. **D,** monopodial branching system. **E,** bases for a pseudomonopodial branch system. **F,** pseudomonopodially branched system (note the leaves opposite what appear to be lateral branches; the leaves are subtending lateral branches that develop in axillary buds): *ab,* axillary bud; *lbr,* lateral branch; *tb,* terminal bud; tbr_1-tbr_5, terminal branches that have been bypassed in sequence; *l,* leaf.

Vascular Plants 379

Another type of cambium is seen in the lycopod *Isoetes* (quillwort). In the **corm** of this plant, there is a group of cells that produce parenchyma to the outside and either xylem and phloem or parenchyma to the inside. The internal products of the *Isoetes* cambium depend on the age and species of the plant.

Phellogen

Another lateral meristem involved in secondary growth is cork cambium, or **phellogen.** This meristem gives rise to cork, or **phellem,** a protective tissue impregnated with suberin. Cork functions to protect the plant from mechanical damage and water loss. Phellogen also gives rise to a tissue internal to itself, the **phelloderm.** Unlike vascular cambium, the phellogen is not a permanent meristem. As the plant grows in circumference, the phellogen does not grow to keep pace; instead, new phellogen is initiated inside the old. This initiation of new phellogen results from dedifferentiation of parenchyma. In stems, the first phellogens develop from parenchyma in the cortex (Fig. 12-32A–D). As more phellogen appears, parenchyma in the secondary phloem will dedifferentiate and become phellogen. When new phellogen appears, the phellem produced cuts off the supply of water to the older, outer phellogen. Thus, older phellogen, and any other tissue external to the new phellogen, dies from lack of water and a source of energy. As a result of this phenomenon, an organ that has undergone much secondary growth will be clothed with a dead but complex tissue composed of layers of dead phellem, phellogen, phelloderm, and phloem (Fig. 12-32E). The phellogen and its derivatives are referred to as **periderm.** In organs with much secondary growth, all dead tissue outside the functioning phellogen is called the **rhytidome.**

The initiation of phellogen in roots is somewhat different from that in stems. It usually begins in the pericycle. New phellogens develop from pericycle cells, which are the result of old pericycle undergoing cell division, or from phloem parenchyma. Otherwise, periderm is formed as it is in stems.

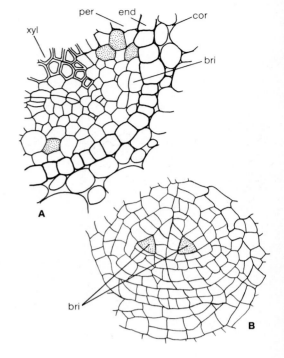

Figure 12-30 Branch root formation, × 500, **A,** angiosperm branch root initiation. **B,** *Selaginella* root branching: *bri*, branch initiation; *cor*, cortex; *end*, endodermis; *per*, pericycle; *xyl*, xylem. (**B,** after Webster.)

Figure 12-31 Vascular cambia. **A,** primary vascular tissue in dicotyledon. **B,** initiation of secondary growth in dicotyledon. **C,** vascular cambium of *Cordyline*. **D,** initiation of vascular cambium in root. **E,** secondary growth in root: *co*, cortex; *e*, endodermis; *fc*, fascicular cambium; *ic*, interfascicular cambium; *ifr*, interfascicular region; *nvb*, newly formed vascular bundle; *per*, pericycle; *ph*, phloem; *pph*, primary phloem; *pxy*, primary xylem; *sph*, secondary phloem; *sxy*, secondary xylem; *vc*, vascular cambium; *xy*, xylem. (**A-D,** × 500; **E,** × 200.)

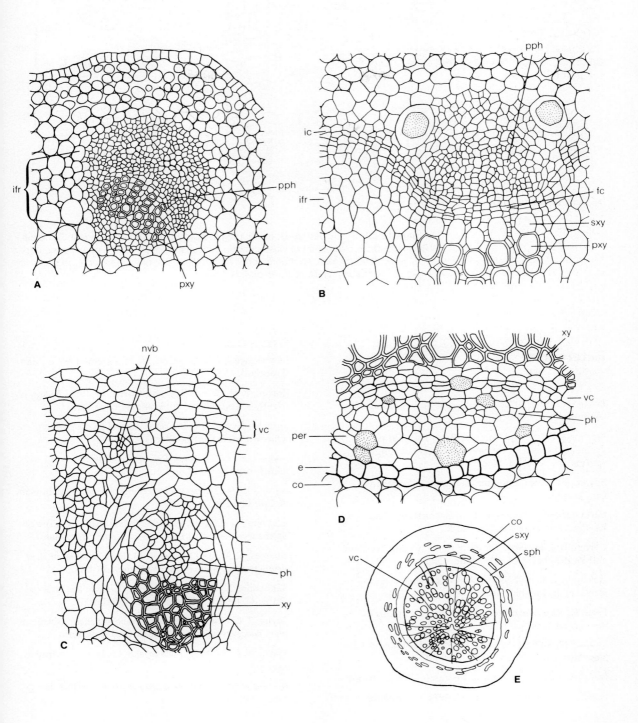

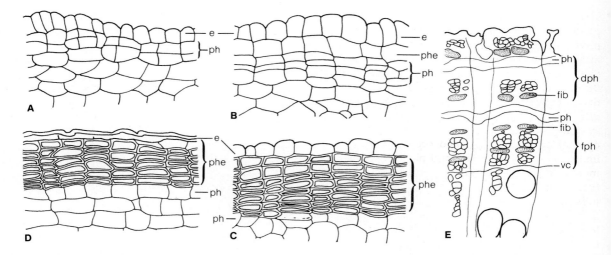

Figure 12-32 Phellogen. **A–D,** developing phellogen, × 500: **A,** youngest; **D,** oldest. **E,** phellogen in older stem of *Vitis*, showing successive layers of phellogen: *e*, epidermis; *dph*, dead phloem; *fib*, fibers; *fph*, living phloem; *ph*, phellogen; *phe*, phellem; *vc*, vascular cambium. (**A–D,** redrawn after K. Esau.)

References

Bierhorst, D. W. 1971. *Morphology of vascular plants*. Macmillan Co., New York.

Bierhorst, D. W., and Zamora, P. M. 1965. Primary xylem elements and element associations of angiosperms. *Amer. J. Bot.* 52:657–710.

Bower, F. O. 1935. *Primitive land plants*. Macmillan & Co., London.

Carlquist, S. 1961. *Comparative plant anatomy*. Holt, Rinehart and Winston, New York.

Corner, E. J. H. 1964. *The life of plants*. Collins, William, and World Publ. Co., Cleveland, Ohio.

Delevoryas, T. 1962. *Morphology and evolution of fossil plants*. Holt, Rinehart and Winston, New York.

Esau, K. 1960. *Anatomy of seed plants*. John Wiley & Sons, New York.

———. 1965. *Plant anatomy*, 2nd ed. John Wiley & Sons, New York.

Fahn, A. 1967. *Plant anatomy*. Pergamon Press, Oxford.

Foster, A. S., and Gifford, E. M. 1974. *Comparative morphology of vascular plants*. W. H. Freeman & Co., San Francisco.

Raven, P. H.; Evert, R.F.; and Curtis, H. 1981. *Biology of plants*. 3rd ed. North Publications, New York.

Sporne, K. R. 1970. *The morphology of pteridophytes*. Hutchinson & Co., London.

———. 1971. *The morphology of gymnosperms*. Hutchinson & Co., London.

———. 1974. *The morphology of angiosperms*. Hutchinson & Co., London.

Stebbins, G. L. 1974. *Flowering plants, evolution above the species level*. Harvard Univ. Press, Cambridge, Mass.

Steeves, T. A., and Sussex, I. M. 1972. *Patterns in plant development*. Prentice-Hall, Englewood Cliffs, N.J.

Takhtajan, A. 1969. *Flowering plants, origin and dispersal*. Trans. C. Jeffrey. Oliver & Boyd, Edinburgh.

Taylor, T. N. 1981. *Paleobotany: an introduction to fossil plant biology*. McGraw-Hill Book Co., New York.

Wardlaw, C. W. 1955. *Embryogenesis in plants*. Methuen, London.

———. 1965. *Organization and evolution in plants*. Longmans, London.

The three plant groups considered here, rhyniophytes (subdivision Rhyniophytina), trimerophytes (subdivision Trimerophytina), and zosterophyllophytes (subdivision Zosterophyllophytina), are of particular significance in understanding the origins of vascular plants. They contribute data that suggest the evolutionary pathways that vascular plants followed in originating **organs** and **organ systems.** The assumption is made that the development of these organs and organ systems has improved the potential of land plants to function more efficiently in diverse environmental conditions. Although these three plant groups are now extinct, it should not be concluded that they were unsuccessful, since they persisted for many millennia. Indeed, they existed for a longer period than the flowering plants have been on the earth, and the flowering plants are considered to be the most "successful" group of plants now existing.

Subdivision Rhyniophytina (Rhyniophytes)

13

RHYNIOPHYTES, TRIMEROPHYTES, AND ZOSTEROPHYLLOPHYTES

The Rhyniophytina comprise at least ten genera. They first appeared in late Silurian time, approximately 400 million years ago, and were extinct before the end of the Devonian, around 345 million years ago. Only sporophytic material has been identified with any confidence, although some gametophytic material may belong to this subdivision.

In all instances, the sporophyte consists of a leafless, dichotomizing axis with terminal sporangia. The vascular strand forms a central cylinder of xylem enclosed by a circle of cells, interpreted as phloem, and these elements are bounded by a thick cortex. This vascular organization is designated a **protostele.** The epidermis sometimes has stomata.

The sporangium has a thick, multistratose jacket, suggesting that it is **eusporangiate**—that is, derived from a multicellular primordium, as in bryophytes.

Each sporophyte produced numerous **homosporous** sporangia, which terminated the shoot apex. The erect shoot often branched equally (dichotomously), and each branch tip bore an apical sporangium. In some instances, the sporophyte had a **rhizomatous** system (i.e., the stem reclined and produced erect branches) and was attached to the substrate by **rhizoids.** This rhizomatous portion has occasionally been interpreted as a gametophyte, but since indisputable sex organs have not been demonstrated, this interpretation is questionable.

The nature of the sediments from which fossil material is always recovered indicates that these plants probably grew in swampy areas, sometimes partially submerged or in areas subject to inundation.

None of the plants is very large; their stems reach lengths of up to 20 cm but are generally much shorter. Axes are sometimes less than 1 mm in diameter and rarely exceed 4 mm, even in the largest representatives.

The earliest rhyniophyte is *Cooksonia* (Fig. 13-1A), of late Silurian age. This genus persisted into the early Devonian. The naked, equally dichotomizing axes are extremely slender (0.25–1.5 mm in diameter) and bear single, rounded to elongate sporangia at the apices of each of the dichotomies. Sporangia vary from 2–5 mm long, some of which contain spores, each showing a triradiate scar. Fossil specimens are very few, with material originating from various localities in Europe, Asia, and North America. No stomata have been demonstrated on the axes or sporangia, although the latter sometimes have elongate cells, a situation common in modern mosses. The central vascular strand consists of annular tracheids.

Rhynia (Figs. 13-2 and 13-3C) is the first rhyniophyte genus described in the literature. The material is remarkably well preserved and relatively abundant. Two species are described from the early Devonian deposits in a single locality in Scotland. Plants were apparently gregarious in *R. major* and had a rhizomatous portion (presumed to have been subterranean) that produced numerous unicellular rhizoids on its undersurface. From the **rhizome** arose erect cylindrical axes, each dichotomizing once or twice, with the apex sometimes terminated by a cylindrical sporangium. Axes of the largest plants probably reached 20 cm in height and 3 mm in diameter. The lower portion of the erect axis has a few scattered stomata in the epidermis, but the rhizome and upper portion of the axis lack stomata. The rhizomatous portion resembles the erect axis in anatomy, except that there are **mycorrhizae** among the cells of the rhizome. The

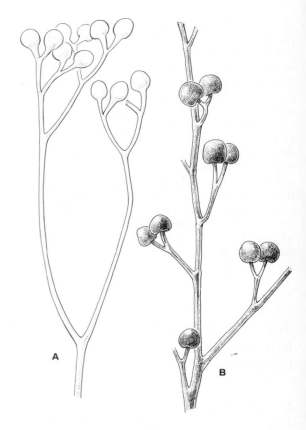

Figure 13-1 Rhyniophytina. **A,** *Cooksonia caledonica*, branch system with terminal sporangia, × 1. **B,** *Renalia hueberi*, reconstruction of a portion of an axis with lateral dichotomizing branches bearing terminal sporangia, × 3. (**A,** redrawn after D. Edwards; **B,** redrawn after Gensel.)

stem has a weak central stele, consisting of annular tracheids and surrounded by a cylinder of cells interpreted as phloem. The cortex is very broad, and the outer portion is presumed to have been **chlorophyllose** (Fig. 13-2C). Sporangia are up to 3 mm long and 1.5 mm wide, with a multistratose jacket that encloses spores arranged in tetrads. No dehiscence line has been observed. Features of cutinized spores and stomata in the epidermis of both stems and sporangia are cited as evidence that shoots were aerial rather than aquatic.

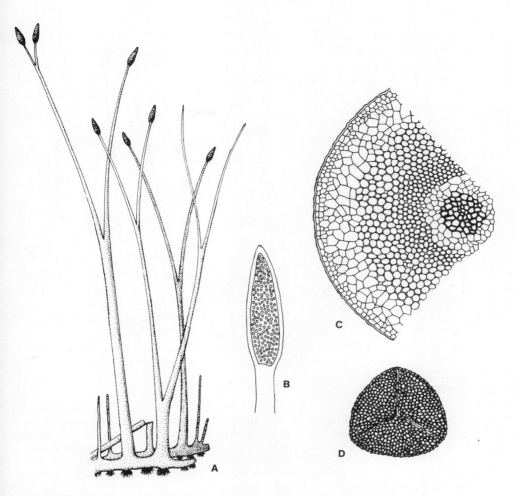

Figure 13-2 Rhyniophytina. **A–D,** *Rhynia major.* **A,** reconstruction of habit, showing rhizome, dichotomous branching, and terminal sporangia, × 0.2. **B,** longitudinal section through sporangium, × 5.5. **C,** transverse section of stem, showing protostele, cortex, and epidermis, × 16. **D,** single spore with reticulate ornamentation and trilete scar, × 400. (**A,** redrawn after Remy and Remy; **C,** redrawn after Andrews, *Studies in Paleobotany.* New York: John Wiley & Sons. Copyright 1961 by John Wiley & Sons.)

In *Rhynia gwynne-vaughnii* (Fig. 13-3C) the erect axes were up to 18 cm tall and produced hemispherical projections and presumably **adventitious** branches. The material has received a diversity of interpretations. The most recent, by D. S. Edwards (1980), describes the stems as 2–3 cm in diameter and the axis as multibranched and basically monopodial. It is assumed that this axis emerged from a horizontal system, which has not been described before. The aerial branches produced both dichotomous and adventitious branches. Some of the adventitious branches were produced following the production of terminal sporangia on the same axis. Some of the branches have small, rhizoid-covered

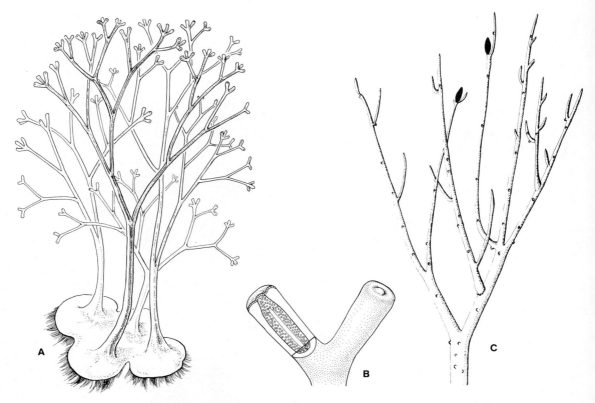

Figure 13-3 Rhyniophytina. **A, B,** *Horneophyton lignieri.* **A,** reconstruction of whole plant, × 0.5. **B,** reconstruction of a two-lobed sporangium with cutaway, showing columella, spores, and apical pore in one arm, × 5. **C,** *Rhynia gwynne-vaughnii,* reconstruction of aerial axis showing adventitious branches and two sporangia; small hemispherical projections are shown scattered on the axis, × 0.5. (**A, B,** redrawn after Eggert; **C,** redrawn after D. W. Edwards.)

pads on one surface. Sporangia were cylindric, had a multistratose jacket without stomata, and produced many spores. There appears to be a dehiscence line at the base of the sporangium, which was possibly shed as a unit.

Horneophyton (Fig. 13-3A, B) resembles *Rhynia* in many respects. It is also from the Lower Devonian and is associated with *Rhynia* specimens. Two main features that distinguish it from *Rhynia* are a columella in the forked sporangium and, in the rhizome, tuberlike protuberances from which erect axes emerge. Eggert (1974) interprets the sporangia as twice- or sometimes thrice-forked cylinders at the apices of branches. He suggests also that the sporangia may have released their spores through a pore at the apex of each fork. The sporangium-bearing apices appear to have been meristematic for an extended time and, as in hornworts (Bryophyta), the sporangium may have produced spores over an extended period.

Taeniocrada represents yet another growth form. This is also known from both the Lower and Upper Devonian. Of the few species known, *T. decheniana* (Fig. 13-4A, B) provides the most complete and

Figure 13-4 Rhyniophytina. **A, B,** *Taeniocrada decheniana*. **A,** reconstruction of the plant, × 0.3. **B,** reconstruction of sporangia, showing remarkable attachment, × 2. (**A,** from Schweitzer; **B,** redrawn from D. Edwards.)

of this structure, the plant is sometimes suggested to have been a submerged aquatic. No information is available concerning its attachment to its terrestrial substrate. Axes were often terminated by oval obtuse swellings, sometimes borne in clusters and each bearing a short **peduncle** that extended longitudinally up one wall (Fig. 13-4B). These structures are interpreted as sporangia, although no spores are known. The genus is known from both eastern North America and Europe.

Renalia hueberi (Fig. 13-1B), of Lower Devonian age, is a rhyniophyte that shows a number of remarkable structural advancements over other rhyniophytes. The plant consisted of erect axes at least 11 cm tall and 1–1.5 mm wide; these branched more or less equally. These branches also produced smaller, lateral, dichotomous branches terminated by round to **reniform** sporangia that dehisced along the distal margin. The basically **pseudomonopodial** axes and the dehiscence lines of sporangia resemble Zosterophyllophytina in some respects, and provide suggestive evidence that this subdivision was derived from Rhyniophytina.

The rhyniophytes appear to have been the progenitors of the zosterophyllophytes and the trimerophytes. Indeed, within the rhyniophytes, *Taeniocrada langi* has sporophytic axes much like those in the zosterophyllophytes, and *Renalia* has a branching pattern that suggests it is a precursor to the trimerophytes.

Subdivision Trimerophytina (Trimerophytes)

The trimerophytes (subdivision Trimerophytina) are restricted to Lower Devonian deposits, and have been the source of much dispute. Part of this arose because of the misinterpretation of the earliest described fossils of the genus *Psilophyton*. In his discussion of this genus in the mid-19th century, J. W. Dawson erroneously assumed an organic connection between two entirely different plants: *Psilophyton princeps* and the zosterophyllophyte, *Sawdonia ornata*. More recent work, especially that of H. P. Banks and F. Hueber and particularly their painstaking excavation of fossil material from its sandy matrix, has led to a great increase in our

abundant material. The axes are very long, with some fragments reaching lengths of 1.8 cm. The axes branch pseudodichotomously with many ultimate dichotomies; they are **terete,** up to 2.6 cm broad, and show a distinct central stele. The circinate tips of some vegetative axes give the species a distinctive appearance. The epidermis contains no stomata, and parenchyma tissue between the epidermis and stele is weakly developed. Because

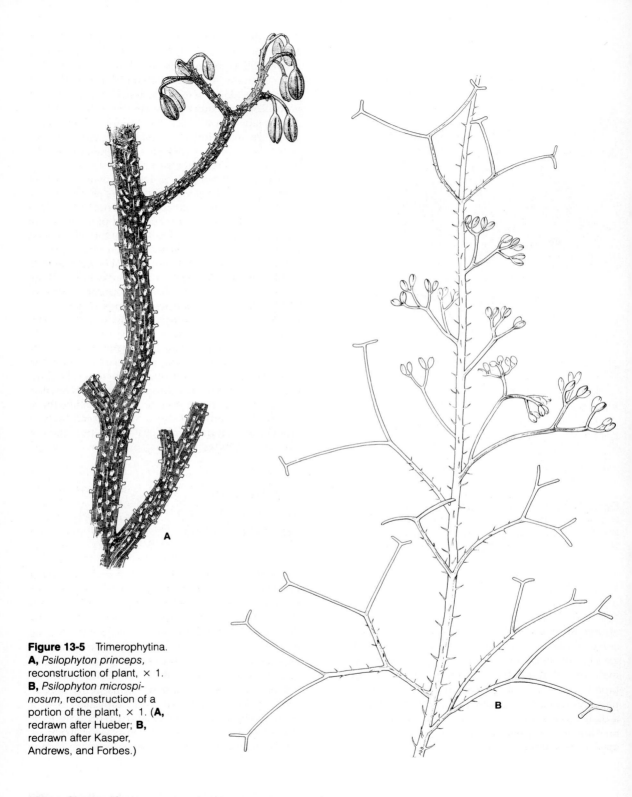

Figure 13-5 Trimerophytina. **A,** *Psilophyton princeps,* reconstruction of plant, × 1. **B,** *Psilophyton microspinosum,* reconstruction of a portion of the plant, × 1. (**A,** redrawn after Hueber; **B,** redrawn after Kasper, Andrews, and Forbes.)

Figure 13-6 Trimerophytina. *Psilophyton dawsonii*, reconstruction of plant fragment, showing determinate leaflike branches, mature sporangium-bearing branches, and an immature sporangium-bearing branch (near apex). (From Banks, Leclercq, and Hueber, 1975.)

knowledge of these plants. However, the taxonomic position of this group remains anomalous. One feature that differentiates it from the rhyniophytes is the pseudomonopodial axis that bears clusters of terminal sporangia at the tips of lateral systems of dichotomies or trichotomies. These sporangia are borne on a recurved branch tip in most specimens of *Psilophyton*, but in *Trimerophyton* and *Pertica*, they are erect on a shortened lateral branch from the main axis. Another significant feature is the roughly radial arrangement of determinate lateral branches, some of which dichotomize once or several times. Only erect axes are known for any of these genera; thus reconstructions of the entire plant are impossible. It is suggested that erect axes arose from a rhizome.

In most trimerophytes the axes are naked, but *Psilophyton microspinosum* (Fig. 13-5B) has radially arranged, widely spaced, spinelike emergences. In *P. princeps* (Fig. 13-5A), these emergences are peglike and rather irregularly distributed around the stem.

Psilophyton princeps possibly reached a height of a meter, and the pseudomonopodially branched main axis achieved a diameter of up to 10 mm. Lateral branching is dichotomous and three-dimensional (as contrasted to bilateral branching in a single plane). Sporangia are paired and elliptic, and are less than 1 mm long. Other species are smaller, *P. dapsile* being probably no more than 30 cm tall and bearing tiny sporangia.

In *Psilophyton dawsonii* (Fig. 13-6), the axes are naked. Some of its lateral branch systems are obviously determinate and terminated ultimate dichotomies (or trichotomies). These suggest leaflike systems. Thus the plant appears to have had a branch system specialized for sporangium production and other systems specialized for photosynthesis. The protostele occupied nearly one quarter of the diameter of the stem. It consists of **centrarch** xylem with several enlarged protoxylem areas and radially aligned, scalariform, pitted tracheids. The phloem and inner cortex are parenchymatous whereas the outer cortex is **collenchymatous.** Substomatal chambers were present in this outer cortex.

As in other species of *Psilophyton*, elliptical sporangia terminate ultimate dichotomies of axes that are restricted to sporangium production. Sporangia are $3–5 \times 1–1.5$ mm; they curve downward when young but are erect when mature. Dehis-

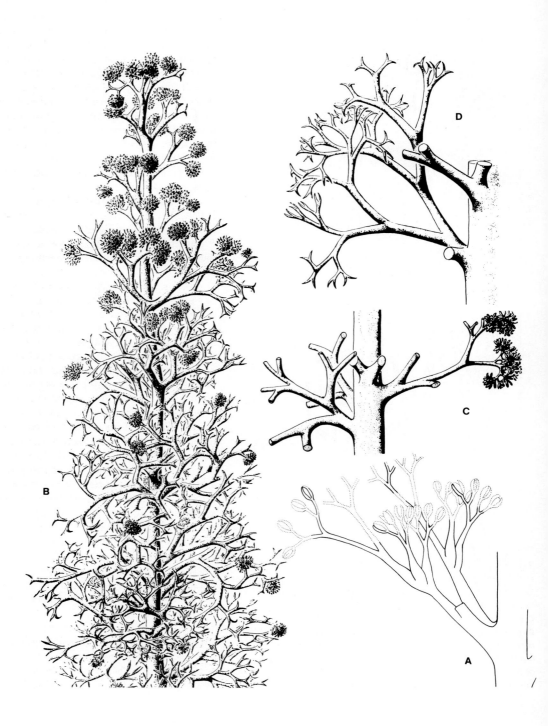

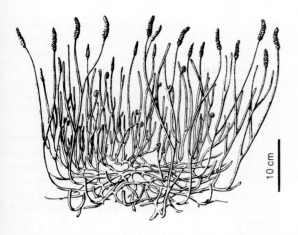

Figure 13-8 Zosterophyllophytina. *Zosterophyllum rhenanum*, reconstruction of plant with terminal sporangium-bearing spikes and circinate tips of vegetative shoots, × 12. (After Schweitzer.)

Figure 13-7 Trimerophytina. **A,** *Trimerophyton robustius*, showing branching axes with sporangia, × 0.2. **B–D,** *Pertica quadrifaria*. **B,** reconstruction of plant apex, showing admixture of leaflike branch systems and those bearing terminal clusters of sporangia, × 0.2. **C,** reconstruction of one unit of four side branches, showing their arrangement on the stem and dichotomous forking, × 1.2. **D,** reconstruction of sterile branch, × 0.8. (**A,** redrawn after Hopping; **B, D,** from Kasper and Andrews, with permission of *American Journal of Botany*.)

cence is along longitudinal lines. Spores are numerous and enclosed within a membranous sheath within the sporangium.

Pertica (Fig. 13-7B–D), of Lower Devonian age, also shows a certain degree of separation of photosynthetic and sporangium-bearing functions in the branches, although it is less marked than in *Psilophyton*. In *Pertica*, the axis was distinctly **monopodial**. It bears alternately arranged, short, much-dichotomized lateral branches. Some terminate by roundish clusters of small elliptical sporangia, and others are apparently determinate and vegetative. This plant is about 1 m tall, with a main axis of up to 1.5 cm in diameter.

Trimerophyton (Fig. 13-7A), of Lower Devonian age, is known only by fragments of axes with monopodial branching. The stem probably bears tiny bristles, because minute scars are present. Sporangium-bearing lateral branches fork several times; the final forkings are dichotomies, each terminated by a cluster of no more than three elliptical sporangia. The fragments available are approximately 20 cm long. Sporangium-bearing lateral branches are also short, and sporangia are 4–5 mm long. Tetrads of spores have been noted in sporangia.

Trimerophytes show a number of features indicating that they were derived from *Rhynia*like ancestors, including the nature of axes and position of sporangia. Short, lateral branches of *Pertica* suggest precursors of leaflike organs, and the clustering of sporangia is suggestive of the beginning of a sorus. Trimerophytes appear to be probable progenitors of cladoxylids and coenopterids (see Chapter 18), from which both ferns and progymnosperms possibly originated. This group of plants is one of great interest and will continue to attract the curiosity of paleobotanists attempting to probe deeper into the mystery of the progenitors of vascular plants.

Subdivision Zosterophyllophytina (Zosterophyllophytes)

The subdivision Zosterophyllophytina also consists entirely of fossil plants that are restricted to the Devonian and comprise two families with at least seven genera. The axes are protostelic and sometimes have **enations** (spinelike outgrowths on the stem); sporangia are borne laterally, rather than terminally, often on a short stalk, and dehisce by a transverse line along the distal margin.

Zosterophyllum, of Lower Devonian age, is represented by a number of described species, but the most completely known are *Z. rhenanum* (Fig. 13-8A) and *Z. myretonianum*. *Z. rhenanum* is presumed to have been an aquatic. Erect axes emerge from a complicated, entangled series of reclining, multibranched rhizomes. The axes are smooth and forked, and have a vascular strand. They are 1–2 mm in diameter and presumably exceeded 10–15 cm in length when living. Sporangia are borne spirally and laterally on an apical spike 1.5 cm long, and each sporangium dehisces by a transverse line. Sporangia are 2.5–4 mm broad, but no spores are known. Vegetative tips were circinate.

Zosterophyllum myretonianum is much smaller, and the curious, H-shaped branching pattern of the rhizome appears among several other zosterophyllophytes. Erect axes of *Z. myretonianum* are generally unbranched and bear apical spikes of sporangia. The sporangia have spores, and the epidermis of the axes bears occasional stomata.

The genus *Hicklingia* (Fig. 13-9) is based on two specimens of Middle Devonian age. They are of a bushlike plant about 17 cm tall, with many slender branches that arise from a thickened base. Naked axes branch either dichotomously or sympodially, and the upper parts of the shoots often bear spherical structures, presumed to be sporangia, arranged laterally and terminally on the upper portion of the axes. No cellular detail is available, but this fossil is considered a zosterophyllophyte because of the position of the presumptive sporangia.

Sawdonia (Fig. 13-10B, C), of late Devonian age, is known only by fragments of axes. These axes bear spirally arranged short spines, perhaps glandular in function. Interrupting the spines are nearly sessile sporangia that dehisce by a transverse line. The tips of the axes appear to be **circinately vernate**. This indicates that the shoot apex was coiled, and as the stem elongated it uncoiled behind the apex—a feature common in fern leaves. The axes are up to 5 mm in diameter, and multicellular spines are 0.5–3.9 mm long. The axes must have exceeded 30 cm in length; they branched monopodially, with a central axis producing smaller lateral branches. Stomata appear in the epidermis.

Drepanophycus spinaeformis has been treated as belonging to the lycopods (see Chapter 15), but it shows many features of the zosterophyllophytes, including the lateral position of sporangia and the H-shaped branching of the rhizomes. It has true radially arranged leaves, however, which characterize the lycopods. This genus, as well as *Kaulangiophyton*, are both intermediate between the lycopods and zosterophyllophytes.

Although somewhat fragmentary, *Kaulangiophyton* (Fig. 13-10D), is known from Devonian material from Maine in the United States. The plants achieved a height of several decimeters, with axes of 5–9 mm in diameter arising from a multibranched, reclining system. Sparsely distributed spines are scattered on the axes. In some upright axes, short-stalked, lateral sporangia are scattered among the spines. The sporangia opened by longitudinal dehiscence slits.

Gosslingia breconensis (Fig. 13-10A), of Lower Devonian age, also consists of slender pseudomonopodial and dichotomizing smooth naked axes up to 15 cm long and 4 mm in diameter. The living plant was probably up to 50 cm tall. The tips of the axes are conspicuously coiled. Globose to reniform sporangia are borne laterally on short stalks and on a single side, mainly on lateral branches. The stem has a cutinized epidermis and scattered stomata. The cortex is composed of two to four layers of thick-walled cells. The elliptical protostele is exarch and relatively broad compared to other genera.

Zosterophyllophytes, showing monopodial and dichotomous branching and lateral sporangia that are often restricted to an apical spike, generally suggest the possible appearance of the progenitors of more complex tracheophytes. The fact that sporangia were borne laterally and the presence of leaflike enations suggest a close relationship with the subdivision Lycophytina (see Chapter 15).

Ecologically, zosterophyllophytes show a diversity of growth forms that suggests an equal variety in habitats occupied. *Zosterophyllum* appears to have been aquatic. The somewhat smaller, tufted *Hicklingia* was apparently a plant of more open, drier sites. The morphology of *Sawdonia* suggests a humid environment. The rather elongate, slender axes of *Gosslingia* imply a sprawling habit.

A number of features originated among early land plants that persist in modern vascular plants: (1) a lignified vascular system with pitted tracheids; (2) an epidermis with cuticle and stomata; (3) an apparent suggestion of leaflike determinate

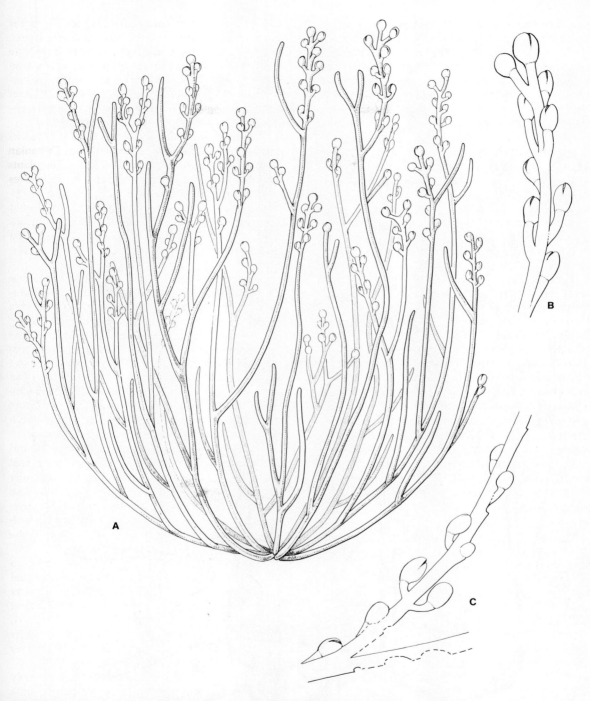

Figure 13-9 Zosterophyllophytina. **A-C,** *Hicklingia edwardi*. **A,** reconstruction of habit, × 1. **B, C,** lengths of sporangium-bearing branches, showing lateral arrangement of sporangia. (**B, C,** redrawn after D. Edwards.)

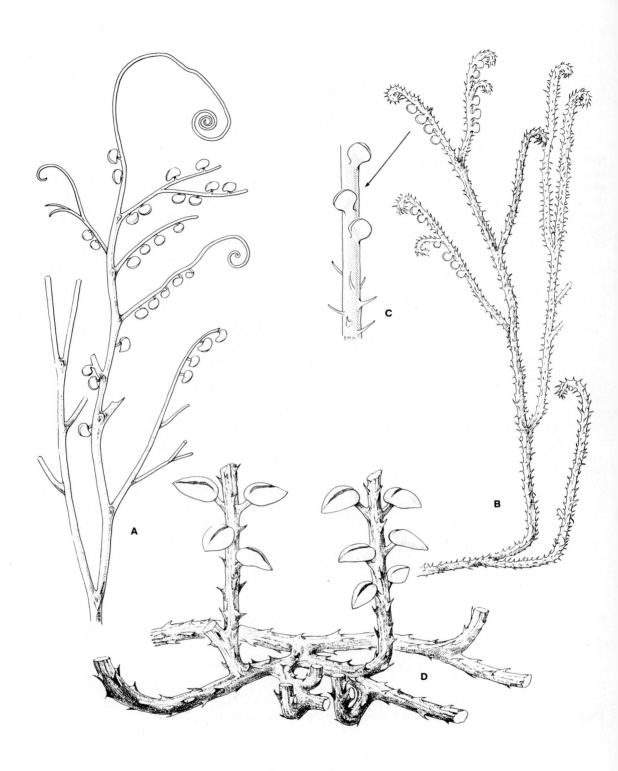

branches, but without the appearance of **planation;** (4) specialized determinate branches with terminal sporangia; (5) sporangia with single dehiscence lines, sometimes confined to a specialized organ; and (6) the beginning of a radial arrangement of appendages. It is from these innovations, as well as others, that the present structure of the rest of the vascular plants was probably derived.

Considering these subdivisions in sequence of specialization, Rhyniophytina are the simplest, and suggest progenitors of the others. The Zosterophyllophytina are more specialized, both in their diversity of form and in the specialization and position of their sporangia. The Trimerophytina are even more specialized, both in their branch systems and in the nature of their sporangia.

It has been postulated that during Devonian time, Europe and North America were part of one supercontinent (the Old Red Continent), whereas present Southern Hemisphere continents and the Indian subcontinent formed another. The Old Red Continent probably lay within an equatorial belt of tropical climates. This makes the presence there of such fragile plants as rhyniophytes, zosterophyllophytes, and trimerophytes rather more plausible than the present position of these land masses would indicate. It also makes credible the fact that very similar Devonian floras are found in areas that are now geographically distant from each other.

As further details are accumulated, fossilized fragments of these early land plants may help to support reconstruction of the habitats in which these plants evolved.

Figure 13-10 Zosterophyllophytina. **A,** *Gosslingia breconensis*, reconstruction of sporangium-bearing fragment, × 0.75. **B, C,** *Sawdonia ornata*. **B,** reconstruction of plant, × 0.5. **C,** reconstruction of sporangium-bearing stem (note dehiscence of sporangia), × 2. **D,** *Kaulangiophyton akantha*, reconstruction of a representative part of a plant, × 0.5. (**A,** redrawn after Edwards; **B,** redrawn after Ananjev and Stepanov; **C,** redrawn after Hueber; **D,** redrawn after Gensel, Kasper, and Andrews.)

References

Andrews, H. N. 1974. Paleobotany, 1947–72. *Ann. Missouri Bot. Garden* 61:179-202.

Andrews, H. N.; Kasper, A. E.; Forbes, W. H.; Gensel, P. G.; and Chaloner, W. G. 1977. Early Devonian flora of the Trout Valley Formation of northern Maine. *Rev. Palaeobot. Palynol.* 23:255–85.

Banks, H. P. 1968. The early history of the land plants. In Drake, E. T., ed. *Evolution and environment.* Yale University Press, New Haven, Conn.

———. 1973. Occurrence of *Cooksonia*, the oldest vascular land plant macrofossil, in the Upper Silurian of New York state. *Indian J. Bot.* 50A:227–35.

———. 1975. Palaeogeographic implications of some Silurian Early-Devonian floras. In Campbell, K. S. W., ed. *Gondwana Geology.* Australian National University Press. pp 75–97.

———. 1980. The role of *Psilophyton* in the evolution of vascular plants. *Rev. Palaeobot. Palynol.* 29:165–76.

Banks, H. P.; and Davis, M. R. 1969. *Crenaticaulis*, a new genus of Devonian plants allied to *Zosterophyllum*, and its bearing on the classification of the early land plants. *Amer. J. Bot.* 56:436–49.

Banks, H. P.; Leclereq, S.; and Hueber, F. M. 1975. Anatomy and morphology of *Psilophyton dawsonii* sp. n. from the late Lower Devonian of Quebec (Gaspé), and Ontario, Canada. *Palaeontogr. Amer.* 8:7–127.

Boureau, E. 1967. *Traité de paléobotanique.* II. *Bryophyta, psilophyta, lycophyta.* Masson et Cie., Paris.

Doran, J. B. 1980. A new species of *Psilophyton* from the Lower Devonian of northern New Brunswick, Canada. *Can. J. Bot.* 58:2241–62.

Edwards, D. 1970. Fertile Rhyniophytina from the Lower Devonian of Britain. *Palaeontology* 13:451–61.

———. 1970. Further observations on the Lower Devonian plants, *Gosslingia breconensis* Heard. *Phil. Trans. Roy. Soc.* London, Ser. 3, 258B:225–43.

———. 1973. Devonian floras. In Hallam, A., ed. *Atlas of Palaeobiogeography.* Elsevier Scientific Publishing Co., Amsterdam. pp 105–115.

———. 1976. The systematic position of *Hicklingia edwardi* Kidston and Lang. *New Phytologist* 76:173–81.

———. 1979. A Late Silurian flora from the Lower Old Red Sandstone of South-West Dyfed. *Palaeontology* 22:23–52.

———. 1980. Evidence for the sporophytic status of the Lower Devonian plants *Rhynia gwynne-vaughnii* Kidston and Lang. *Rev. Palaeobot. Palynol.* 29:177–88.

Eggert, D. A. 1974. The sporangium of *Horneophyton lignieri* (Rhyniophytina). *Amer. J. Bot.* 61:405–13.

Gensel, P. G. 1976. *Renalia hueberi*, a new plant from the Lower Devonian of Gaspé. *Rev. Palaeobot. Palynol.* 22:19–37.

Gensel, P. G.; Andrews, H. N.; and Forbes, W. H. 1975. A new species of *Sawdonia* with notes on the origin of microsporophylls and lateral sporangia. *Bot. Gaz.* 136:50–62.

Gensel, P. G.; Kasper, A. E.; and Andrews, H. N. 1969. *Kaulangiophyton*, a new genus of plants from the Devonian of Maine. *Bull. Torrey. Bot. Club* 96:265–76.

Hueber, F. M. 1964. The psilophytes and their relationship to the origin of ferns. *Mem. Torrey Bot. Club* 21:5–9.

———. 1972. *Rebuchia ovata*, its vegetative morphology and classification with the Zosterophyllophytina. *Rev. Palaeobot. Palynol.* 14:113–27.

Hueber, F. M., and Banks, H. P. 1967. *Psilophyton princeps*: the search for organic connection. *Taxon* 16:81–85.

Kasper, A. E., and Andrews, H. N. 1972. *Pertica*, a new genus of Devonian plants from Northern Maine. *Amer. J. Bot.* 59:897–911.

Lele, K. M., and Walton, J. 1961. Contributions to the knowledge of "*Zosterophyllum myretonianum*" Penhallow, from the Lower Old Red Sandstone of Angus. *Trans. Roy. Soc. Edinburgh* 64:469–75.

Lemoigne, Y. 1970. Nouvelles diagnosis du genre *Rhynia* et de l'espèce *Rhynia gwynne-vaughnii*. *Bull. Soc. Bot. France* 117:307–20.

Niklas, K. J., and Chaloner, W. G. 1976. Chemotaxonomy of some problematic Palaeozoic plants. *Rev Palaeobot. Palynol.* 22:81–104.

Schweitzer, H.-J. 1979. Die Zosterophyllaceae des rheinischen Unterdevons. *Bonner Paläobot. Mitt.* 3:1–32.

———. 1980. Die Gattungen *Taeniocrada* White and *Sciadophyton* Steinman im Unterdevon des Rheinlandes. *Bonner Paläobot. Mitt.* 5:1–38.

———. 1980. Die Gattungen *Renalia* Gensel und *Psilophyton* Dawson im Unterdevon des Rheinlandes. *Bonner Paläobot. Mitt.* 6:1–34.

———. 1980. Über *Drepanophycus spinaeformis* Göppert. *Bonner Paläobot. Mitt.* 7:1–29.

Taylor, T. N. 1981. *Paleobotany*. McGraw-Hill Book Co., New York.

———. 1982. The origin of land plants: a paleobotanical perspective. *Taxon* 31:155–77.

Walton, J. 1964. On the morphology of *Zosterophyllum* and some other early Devonian plants. *Phytomorphology* 14:155–60.

14
PSILOTOPHYTES (Whisk-Ferns and Their Relatives)

Tmesipteris and the whisk-ferns (*Psilotum*) make up the subdivision Psilotophytina. The relationships of these plants have baffled botanists for many years. Because they have many features common to fossils of the most ancient vascular plants, they were considered to be direct descendants of these fossil plants. The Rhyniophytina, for example, were considered to be in the same subclass as *Psilotum* and *Tmesipteris*. Now, however, these two modern genera are considered to belong to an entirely isolated group, showing some features in common with Rhyniophytina and sharing others with Pterophytina. The seemingly primitive features appear to be the result of reduction from an unknown but structurally more complex ancestor.

Psilotum, with two or possibly three species, is widespread throughout the tropics and subtropics but extends into temperate regions, for example New Zealand, and Arizona in the United States. *Tmesipteris*, with an estimated ten species, has a rather more restricted distribution. It is best represented in the Australasian area and extends southward to cool temperate areas in southernmost New Zealand. *Tmesipteris* is usually epiphytic in humid forests, but *Psilotum* grows also on soil and on cliff shelves in both forested areas and in rather open, shrubby vegetation. The epiphytic habit is a **specialized** rather than a **generalized** feature.

Psilotum

Psilotum nudum, the commonest species, has a rhizomatous system of many short branches (Fig. 14-1A), which bears no roots but is heavily encased by short, brown, multicellular rhizoids. From this tangled system, the chlorophyllose erect shoots arise. These shoots are unbranched at first, then fork at the apex, usually with two branches diverging in the same plane. Further forking, usually in pairs, but sometimes in threes, occurs in a different plane than that of the preceding branches, and makes the upper part of each shoot somewhat bushy with interwoven branches. The shoots are usually longitudinally grooved, with stomata abundant in the shallow grooves and infrequent on the ridges. The stomata are readily visible as whitish spots in the epidermis. The axes bear no leaves, but small, awl-shaped **enations** emerge in three rows. These

Figure 14-1 **A,** *Psilotum nudum*, single axis, showing corallike rhizomatous system with rhizoids and aerial system bearing sporangia and enations. **B,** *Tmesipteris*, single axis, showing sporangia on bifid leaves.

enations lack stomata and either lack or, on occasion, have a weakly developed vascular strand. The enations are usually yellowish and serve no known function; they are widely spaced in the lower part of the axis, becoming more frequent on smaller branches.

Bierhorst (1956) interprets the rhizomatous system as the stem and the chlorophyllose branch system as a leaf. Following the traditional interpretation, which is easier to defend and avoids confusion, we treat both the rhizomatous and erect systems as stems. Since the enations lack any vascular strands, this plant is considered leafless.

Rhizoids are evenly distributed over the rhizome; they are two or three cells long, and their walls, as well as the epidermal cells of the rhizome, have a conspicuous **cuticle**. The main absorptive surface of the rhizome is the mass of fungal hyphae that penetrates the cortex and interweaves among the rhizoids. The internal anatomy of the rhizome varies depending on its diameter. In larger rhizomes, most of the diameter is of parenchymatous cortical cells; the outer cells are invaded by fungi, and the inner cells contain starch grains. At the inner part of the cortex there is a conspicuous circle of tannin-impregnated cells, which often obscure the endodermis. The stele is **protostelic,** with a central, irregular mass of xylem surrounded by phloem. The phloem, in turn, is enclosed by several layers of parenchymatous cells that make up the **pericycle**. The smallest branches of the rhizome lack a vascular system, and are made up entirely of undifferentiated parenchyma cells. Somewhat larger rhizome branches have a central strand of elongate cells reminiscent of those in some bryophyte gametophores. As the diameter of the rhizome increases, the structure becomes more complex.

In different parts of the same rhizome system, therefore, there are portions that are anatomically identical to various stages of evolution of a vascular system, as exhibited in the fossil group Rhyniophytina. These same parts of the organ system also illustrate various stages in **ontogeny** potentially capable of producing a complete sporophyte.

The rhizomatous system can produce tiny, multicellular, ovoid gemmae. These arise at the tips of rhizoids, are starch-filled, and readily break free.

The chlorophyllose axes are usually 20–30 cm tall. They are always grooved, and have a conspicuous cuticle. In transverse section (Fig. 14-2A), the thicker axes have a structure similar to the rhizomatous portion, except that the epidermal cells are without chloroplasts and lack rhizoids, although they do have stomata. A narrow band of outer cells of the cortex is chlorophyllose and has intercellular spaces; this forms the main photosynthetic tissue of the plant. Internal to the chlorophyll-bearing tissue is a band of sclerenchyma cells that makes up the main supportive tissue. This band merges with thinner-walled parenchyma cells that lack intercellular spaces and usually contain starch grains. In the lower part of the main axis, there is a band of tannin-impregnated cells similar to those in the rhizome; this is absent in the upper branches. An endodermis forms the innermost layer of the cortex, enclosing the stele. The xylem is a several-lobed cylinder, with or without a **pith,** depending on whether it is nearer to the perpendicular stem base or to the apex. The xylem maturation is **exarch**; the xylem is made up of **scalariform** and **pitted** tracheids in the metaxylem and annular **spiral** and **helical tracheids** in the protoxylem. Between the xylem and endodermis there is a mass of cells interpreted as the phloem. The stem anatomy is similar in the finer branches, but the xylem strand is greatly reduced, and a pith is absent. Except for the diversity of tracheid elements and the presence of a pith, the stem is very like that of Rhyniophytina.

Sporangia are always borne on the uppermost branches (Figs. 14-1A and 14-2B) and are always associated with a forked appendage. In the axil of each appendage is a (usually) three-chambered structure, the **synangium** (Fig. 14-2C). It is so named because it is believed to have originated from the fusion of three sporangia. This interpretation is supported by the fact that from the short axis of the synangium, the vascular strand divides to send a strand to the base of each chamber. Further support is given by the rare appearance, in some aberrant plants, of three unfused sporangia borne terminally on their own short axes. The synangium is conspicuously lobed, with each lobe grooved by a longitudinal dehiscence line. The synangium opens from the center outward. After the spores are shed, they do not germinate for several months. Synangia are **eusporangiate** and **homosporous**.

The gametophyte of *Psilotum nudum* is small and subterranean, essentially cylindrical, and either irregularly or dichotomously branched (Fig. 14-3). It varies from 0.5–2 mm in diameter and is up to 18 mm long. It is colorless to yellowish to dark

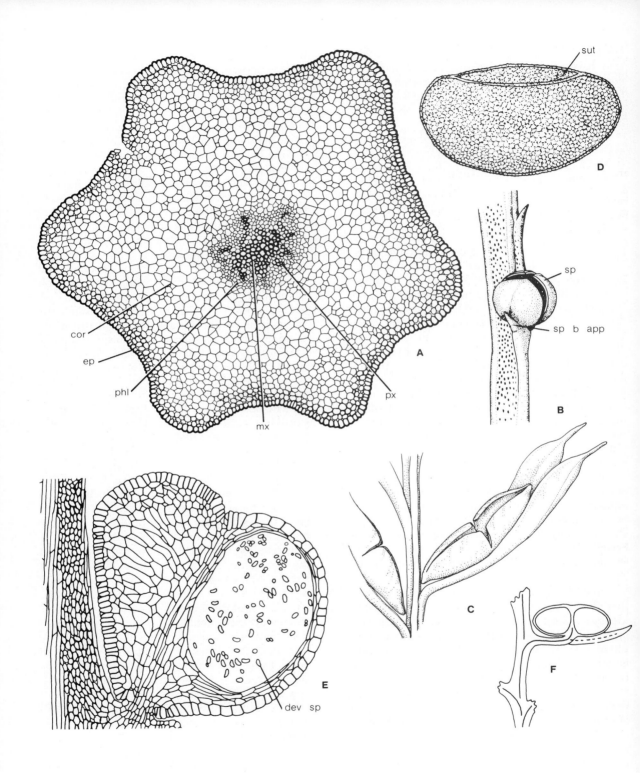

brown, depending on the density of the rhizoids and the abundance of the endophytic fungal mycelium on which it depends for its nutrition. The gametophyte usually lacks vascular tissue; rarely, however, slender vascular strands have been noted. In such instances, the gametophyte is indistinguishable from the rhizome of the sporophyte. Such a vascular system in the gametophyte is very rare in lignified plants, although equivalent structures are present in bryophytes. Antheridia and archegonia are intermixed over the entire surface of the gametophyte. The antheridia are hemispherical bulges on the surface of the gametophyte and have a **unistratose** jacket made up of almost a dozen cells. One of these cells disintegrates to release nearly 250 sperms from the antheridium. Each sperm is coiled and bears numerous flagella. The archegonium is embedded in the gametophyte and has a short, stout neck, which is usually considered a specialized feature. The gametophyte can produce gemmae indistinguishable from those of the rhizome.

Psilotum complanatum is an epiphytic species in which the axes are rather flattened and the enations have a vascular trace.

Figure 14-2 A–D, *Psilotum nudum*. **A,** transverse section of stem showing stele, cortex, and epidermis, ×55: *cor*, cortex; *ep*, epidermis; *mx*, metaxylem; *phl*, phloem; *px*, protoxylem. **B,** single synangium subtended by synangium-bearing appendage, ×2: *sp b app*, synangium-bearing appendage. **C,** vertical section through synangium, showing developing spores in one locule, ×20: *dev sp*, developing spores; *spmc*, spore mother cell. **D,** single spore, showing a single longitudinal suture and the faint reticulate ornamentation, ×500: *sut*, suture. **E, F,** *Tmesipteris tannensis*. **E,** synania, on sporangia-bearing appendage, ×3. **F,** vertical section through synangium and appendage, showing vascular bundle entering synangium, ×2. (**F,** redrawn after Eames.)

Tmesipteris

Tmesipteris (Fig. 14-1B) also has a rhizomatous system similar in structure to that of *Psilotum*, but it often has collenchymatous thickenings in the cortex. In contrast to *Psilotum*, however, the axes of *Tmesipteris* are usually unbranched. These axes bear relatively large, flattened, **lanceolate** leaves Fig. 14-1B). In many species, these are arranged bilaterally on the stem, giving the leafy shoot a flattened appearance. Each leaf has a **decurrent** base, and is oriented with the plane surface essentially parallel to the main axis. There is a single vascular strand that extends the length of the leaf and ends below the **mucronate** leaf apex. Stomata occur on both surfaces, but in flattened axes the stomata are more abundant on the "under" surface of leaves. Nearer the base of the axis there are enations rather like those of *Psilotum*, but in the upper part of the axis the leaves are replaced by **bifid** structures with an elongate, two-lobed synangium on the adaxial surface. As in *Psilotum*, each lobe has its own vascular strand. Dehiscence is by a simple longitudinal line. The whole sporophyte can reach lengths up to 40 cm but is usually smaller. The leaves are radially arranged, but not in three distinct rows. In epiphytic species, the axes are conspicuously flattened, and leaves are in two or three rows of **subopposite** pairs. In terrestrial species, subopposite rows persist, but the axes are not flattened, and the rows can increase to six. Thus the axis can roughly resemble that of some lycopods (e.g., *Huperzia*).

The Psilotophytina have an admixture of generalized features, including the nature of the stele, eusporangiate sporangia with numerous spores, rootlessness, little differentiation between the gametophyte and the rhizomatous system, and chlorophyllose, often leafless axes. Other features appear to be derived from a more complex ancestor, including synangia, the rather complex branching system in some, the adoption of an epiphytic habit in many, flattened axes in *Tmesipteris*, and simplified archegonia. This admixture of features has led to various interpretations concerning both the ancestry and possible relationships of these genera.

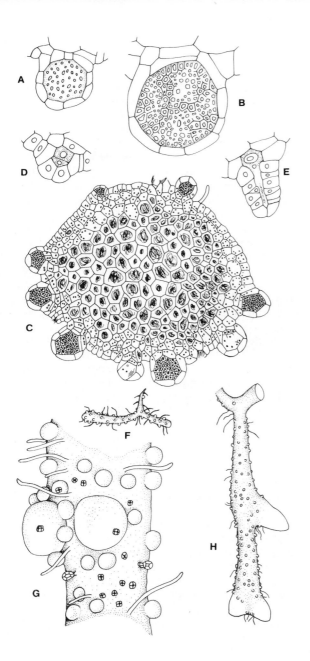

References

Bierhorst, D. W. 1956. Observations on the aerial appendages in the Psilotaceae. *Phytomorphology* 6:176–84.

———. 1968. On the Stromatopteridaceae (Fam. Nov.) and on the Psilotaceae. *Phytomorphology* 18:232–68.

———. 1971. *Morphology of vascular plants.* Macmillan Co., New York.

Holloway, J. E. 1939. The gametophyte, embryo, and young rhizome of *Psilotum triquetrum* Swartz. *Ann. Bot.* 3:313–36.

Rouffa, A. S. 1967. Induced *Psilotum* fertile-appendage aberration: morphology and evolutionary implications. *Can. J. Bot.* 45:855–61.

———. 1971. An appendageless *Psilotum*. Introduction to aerial shoot morphology. *Amer. Fern J.* 61:75–86.

Sykes, M. G. 1908. The anatomy and morphology of *Tmesipteris*. *Ann. Bot.* 22:63–89.

Figure 14-3 *Psilotum nudum* gametophytes. **A,** portion of well-grown gametophyte with antheridia on surface, ×4. **B,** nearly entire small gametophyte, ×4. **C,** portion of very large gametophyte, showing sex organs and three embryo protuberances (the antheridia are the smaller protuberances, whereas the apices of the archegonia are shown by the four cells in groups; two gemmae are also on the upper portion of the segment), ×16. **D,** transverse section of a gametophyte, showing antheridia and endogenous fungi, ×50. **E, F,** archegonia—two stages in development, ×150. **G, H,** two mature antheridia, one from a large gametophyte (**G**) and one from a small gametophyte (**H**). (Redrawn after Holloway.)

Plants in the subdivision Lycophytina are usually referred to as lycopods or lycophytes. The first indisputable evidence of their presence appears in the early Devonian, but the diversity of genera at that time implies that they were present earlier, possibly during the Silurian. The fossil record reveals several evolutionary lines within the group. The remarkable arborescent lycopods appeared in the Carboniferous and formed extensive forests. After the Carboniferous, they disappeared, and the lycopods as a whole declined in relative abundance among vascular plants. There are now at least six living **herbaceous** genera of lycopods with approximately 1000 species.

15

LYCOPHYTES (Lycopods)

General Morphology

A number of distinctive characteristics appear in lycopods that show selection for features that adapt the plant body for a terrestrial rather than an aquatic environment. These include the appearance of roots in some genera; true leaves, usually with a single vascular strand; and the first evidence of **heterospory** in some genera, in which **megaspores** produced megagametophytes with archegonia, and **microspores** produced microgametophytes with antheridia. Other features that appear first in lycopods are the frequent restriction of sporangia to **sporophylls,** which are sometimes confined to a special terminal organ, the **strobilus**; secondary growth of the stem in some instances; arborescent as well as herbaceous habit; deciduous leaves; and the evolution of a seedlike organ.

Besides these features, lycopods are frequently characterized by dichotomizing axes that bear spirally arranged leaves, which in heterosporous genera have **ligules** near their axils. A ligule is a small, membranous, tonguelike structure, the base of which is in a depression on the abaxial surface of the leaf. Its function is uncertain.

Early Lycopods

If *Aldanophyton* is indeed a vascular plant, then it is probably a lycopod. This material is of Middle Cambrian age, and consists of a dichotomizing axis with irregularly to spirally arranged spines. Neither cellular detail nor any structures that can be interpreted as sporangia are available, making the affinities of the material uncertain. It has been interpreted as an alga, but it may not be plant material at all.

The earliest lycopods were **homosporous** and presumably herbaceous. *Baragwanathia* (Fig. 15-1), of the early Devonian, produced dichotomizing axes that were probably up to 30 cm tall. The axes available are generally 1–2 cm in diameter, but they sometimes reach 6 cm, making it a relatively large plant. Slender leaves up to 4 cm long and 0.5–1 mm wide are spirally arranged on the axes. The sporangia are **reniform** and are associated with ordinary leaves, but their point of attachment is unclear. Possibly they were attached directly to the stem rather than to the leaves. A single **leaf trace** forms a central strand of vascular tissue in the leaves. The stele is a fluted protostele. It shows only primary growth and occupies about one-fifth of the stem; the remainder of the stem is cortex. Cellular detail of the leaves and of the stem epidermis is unknown.

Drepanophycus (Fig. 15-2A), which occurred from the Middle to the Upper Devonian, was a similar plant. Sometimes it grew with *Baragwanathia*. It bore shorter leaves that had conspicuously expanded bases. Sometimes, the leaves appear to be arranged in what resemble **whorls**. Stomata have been discovered in the stem epidermis. The sporangia are on short stalks and are scattered among the leaves; they usually occur along the main axis, not near the stem apices. The rhizomatous system was like that of some zosterophyllophytes, including *Kaulangiophyton* and some species of *Zosterophyllum*. There were rootlike organs on the rhizomatous system.

Protolepidodendron (Fig. 15-2B–D), of Lower to Middle Devonian age, was apparently herbaceous and its axes may have reached heights of 15 cm or more. In reconstructions, it is pictured with rhiomatous, leafy axes from which arise erect, dichotomizing axes. No fossil specimen is sufficiently complete to support such an assumption. The leaves are neatly arranged in spirals, and the leaf bases expand to completely cover the stem. The leaves have forked tips and are supplied by a single **vascular strand**. On some branches, the leaves are more widely spaced and have sporangia affixed to their

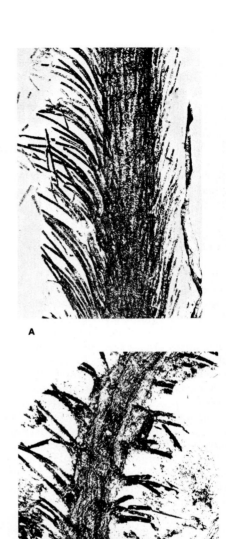

Figure 15-1 *Baragwanathia*. **A,** photograph of fragment of stem, showing crowded, strap-shaped leaves, ×0.4. **B,** closer view of stem and leaves with several sporangia in axils, ×0.7. (After Lang and Cookson, with permission of the Royal Society of London.)

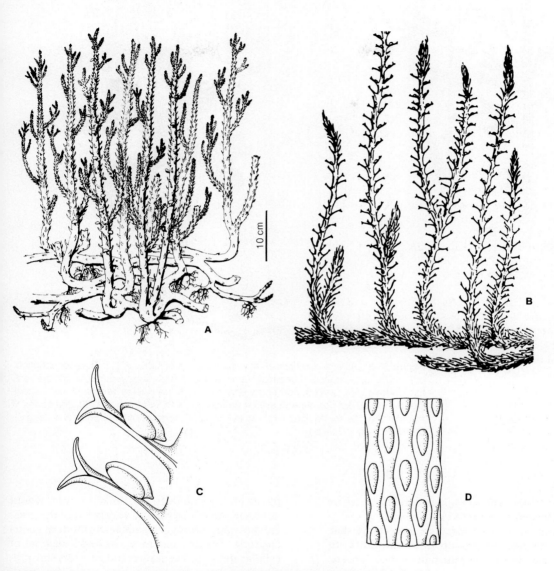

Figure 15-2 Early lycopods of the Devonian. **A,** *Drepanophycus spinaeformis,* showing rhizomatous system, sporangia among the sporophylls, and monopodial axes, ×0.13. **B,** *Protolepidodendron,* showing prostrate and upright dichotomous stems, forked leaves, and sporangia on adaxial surface of sporophylls, ×0.5. **C,** sporophylls with sporangia, ×10. **D,** detail of leaf-scar cushions. (**A,** after Schweitzer; **B,** from Krausel and Weyland, with permission of *Senckenbergische Naturforschende Gesellschaft;* **C,** redrawn after Krausel and Weyland; **D,** redrawn after Pontonié and Bernard.)

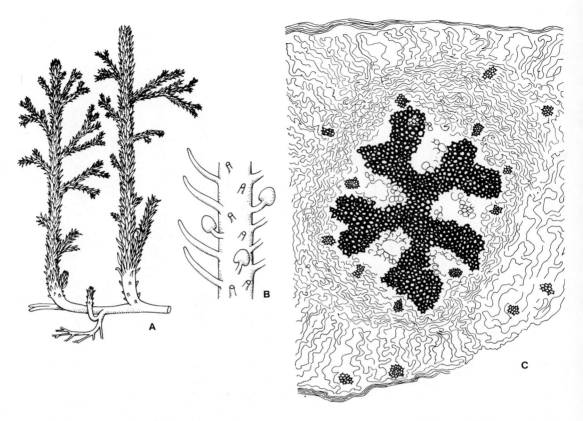

Figure 15-3 *Asteroxylon mackiei.* **A,** reconstruction of habit, × 0.5. **B,** reconstruction of possible appearance of one leafy fertile branch, much enlarged; sporangia are reniform in face view, stalked, and borne on the stem. **C,** transverse section of stem, showing five-armed actinostele with exarch protoxylem, and leaf traces, × 20: *cor,* cortex; *lf tr,* leaf trace; *mx,* metaxylem; *px,* protoxylem. (**A, C,** redrawn after Kidston and Lang; **B,** redrawn after Banks.)

adaxial surfaces. The main axis contains a triangular protostele surrounded by a cortex.

Of similar age, *Colpodexylon* has stems that probably reached lengths up to 50 cm. Since its stems are slender (12–15 mm in diameter) they are presumed not to have been rigid. The leaves are three-forked. As in *Protolepidodendron,* the sporangia are affixed to the adaxial surfaces of the leaves.

Leclercquia, also of the Middle Devonian, was somewhat smaller and bore unusual leaves with a central, recurved point and two lateral, once-forked branches. Ligules were also present.

All genera discussed so far are relatively small plants and, if one overlooks forked leaves, would not seem out of place in the vegetation of the present day; some strongly resemble the modern genus *Huperzia.* The absence in much fossil material of cellular detail in the leaves and of spores in presumptive sporangia makes one yearn for specimens with such details. When such information becomes available, more reliable reconstructions can be made.

Asteroxylon (Fig. 15-3A) and *Thursophyton* are also Devonian genera. Both have a rhizomatous leafless axis from which emerge erect, **monopodial** axes that give rise to monopodial or dichotomizing

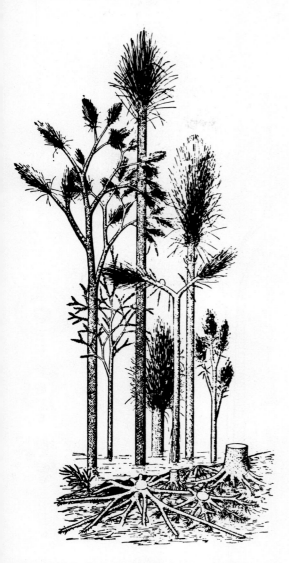

Figure 15-4 Reconstruction of *Lepidodendron*, two figures to *left*; and *Sigillaria*, five figures to the *right*. (After Grand'Eury.)

in *Asteroxylon* they do not. The rhizomatous system produced rootlike, dichotomizing branches in both genera. It is possible that *Thursophyton* had axillary sporangia, but the evidence is far from convincing. In *Asteroxylon*, the sporangia are also intermixed among scales (Fig. 15-3B); each terminated a small axis, was laterally flattened, and was homosporous. This type of sporangium position is more reminiscent of *Sawdonia* (Zosterophyllophytina) than of lycopods and suggests a close affinity between the groups. The stele in both genera is a much-fluted **actinostele** and is enclosed by a thick cortex (Fig. 15-3C).

Carboniferous Lycopod Forests

The remarkable arborescent lycopods (Fig. 15-4) appeared during the late Carboniferous. These represent the zenith of diversification in lycopods. During the Carboniferous, the lycopods influenced the evolution of many other groups of organisms. The multilayered forest community dominated by lycopods, cordaites, and calamites produced a somewhat shaded environment, which doubtless influenced the selection of plants that were tolerant of such conditions. This environment would also supply shelter for many plants and animals intolerant of the more exposed conditions that characterized most of the terrestrial environment at that time.

Like more ancient lycopods, the arborescent lycopods presumably occupied swampy sites, although there are **xeromorphic** modifications in morphology, especially of the leaves (e.g., embedded stomata). However, the limited root system suggests that these trees would be restricted to an environment where water was readily available.

Arborescent lycopods are considered collectively under the generic name *Lepidodendron* (Fig. 15-5), although several other form genera are commonly found among the inevitable fragments of such large organisms. Most arborescent lycopods are represented by **casts** of the trunks, and hence only the external morphology of the trunk is known for many species.

These trees often attained heights of more than 30 m, and their trunks achieved diameters of more than 60 cm near the base. Instead of a rhizomatous

branches. Erect axes probably reached heights near 1 m and bore numerous **imbricate,** roughly spirally arranged, leaflike scales. These lack any vascular tissue, although traces extend from the stele of the stem to the leaf scale base. The scales are **dorsiventrally** flattened and have stomata. In *Thursophyton*, the stem tips have **circinate vernation,** but

Lycophytes (Lycopods) **407**

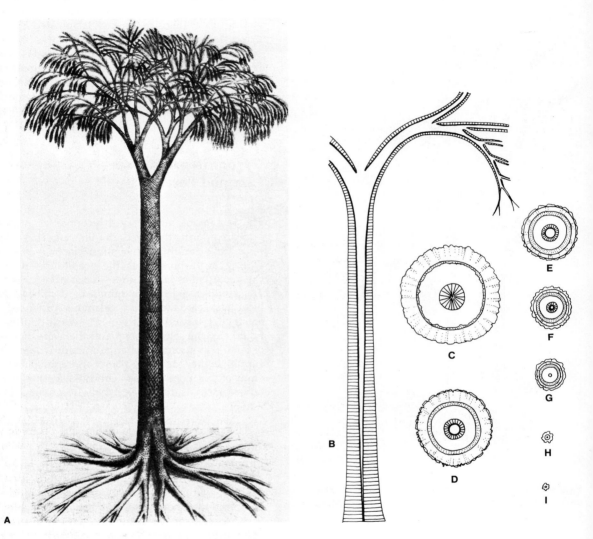

Figure 15-5 A, reconstruction of tree of *Lepidodendron,* showing roots, columnar stem, unequally dichotomous branching, and leafy twigs. **B,** schematic reconstruction of the xylem of the aerial portion of an arborescent lycopod of the *Lepidodendron* type; the primary xylem is black, whereas the secondary xylem is represented by horizontal lines. **C–I,** diagrams of transverse sections at several successive levels of an adult plant, primary xylem, and primary phloem shown as in **A**; the periderm is represented by discontinuous radial lines. **C,** base of trunk (level of maximum development in the upper part of the trunk). **D, E,** siphonostelic branches from the crown at successively higher levels. **H, I,** protostelic branches from the most distal part of the crown. (**A,** from Hirmer, with permission of R. Oldenbourg; **B–I,** redrawn after Eggert.)

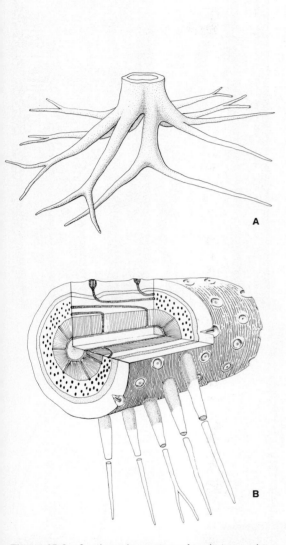

Figure 15-6 **A,** stigmarian system of scale-trees, showing characteristic dichotomy of the axes and spiral arrangement of the appendages. **B,** three-dimensional diagram of a stigmarian axis segment, showing the relationship of the appendages to the axis, and its internal anatomy: *V B*, vascular bundle; *P S*, protostele; *p x*, primary xylem; *phel*, phellogen; *2nd x*, secondary xylem; *camb*, cambium; *r scar*, root scar; *mid c*, middle cortex; *pheld*, phelloderm. (Redrawn after Stewart.)

system, the trunk arose from an expanded basal organ, the **stigmarian system** (Fig. 15-6A), which usually consisted of four main branches radiating outward from the tree base. The main branches dichotomized and sometimes reached lengths of up to 14 m. The axes of this system bore spirally arranged, circular root scars, from which short, simple dichotomizing roots emerged (Fig. 15-6B). Usually simple or with a single dichotomy, the roots were deciduous and left **root scars** on older portions of the stigmarian system.

The trunk was often unbranched up to 20 m or more, then produced a dichotomy that again forked several times, tapering to final dichotomies of less than 1 cm in diameter. Ultimate branches bore spirally arranged linear leaves that were 1–80 cm long. Like those in *Protolepidodendron,* the leaves tapered to the base, but in arborescent lycopods the leaf base formed a roughly diamond-shaped cushion. Leaves were **deciduous,** but their expanded bases persisted to produce a scalelike pattern on the surface of the trunk (Fig. 15-7A). Besides the leaf scar, cushions show **parichnos** scars (Fig. 15-7B–D). These scars mark the terminus of a series of channels of parenchymatous tissue continuous with the leaf and with cortical tissue of the stem. These channels possibly served in gas exchange.

Each leaf has a single midrib along the center of the blade. This vascular strand is continuous with the leaf trace, which departs from the primary xylem without a leaf gap. Numerous stomata are confined to two longitudinal bands on the adaxial surface of the leaf, one on each side of the thickened midrib region and sometimes in a deep groove. The leaves also show a degree of tissue organization that is not apparent in most older fossil lycopod genera (Fig. 15-7E). The epidermis has a cuticle, and beneath it are several layers of thick-walled cells. **Mesophyll** cells, with many intercellular spaces, make up much of the leaf. The transfusion tissue, part of the parichnos system, surrounds the vascular strand. Each leaf bears a diminutive, usually embedded **ligule** on the upper surface near its axil.

The vascular system is always preserved in isolation from the pattern of the trunk surface (Fig. 15-8), but from available material it is reconstructed as follows. There was a remarkably thick cortex, which presumably had its own cambium and continued growing throughout the life of the plant. The cortex was rather complex and consisted of several well-differentiated layers. The stele is an

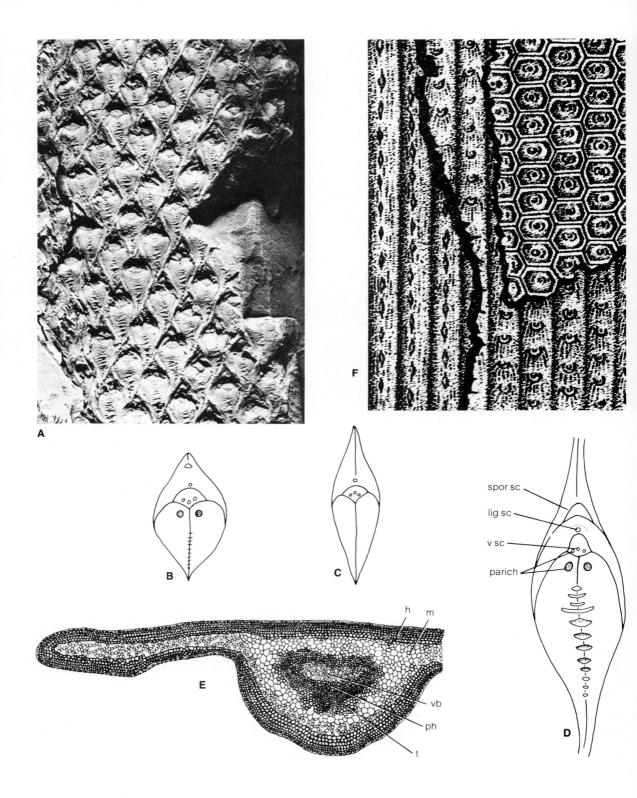

unbroken cylinder; near the base of the trunk it is protostelic, but further up there is a pith; hence the stele is **siphonostelic** (Fig. 15-5B–I). At the extremities of branches, the stele is again protostelic. The stele forms a very small fraction of the diameter of the trunk (none exceeded 10 cm). Although secondary growth occurred, there is no evidence of growth rings.

In most genera of scale-trees, the cambium of the **periderm** is located at the extreme outside of the stem, among the leaf bases. As a result, most periderm development was toward the inside, or **centripetal,** preserving the leaf cushions where the leaves were attached to the stem. At the trunk base, however, the cortex was probably irregularly ruptured, and the leaf scars broken.

Sporangia were borne at the tips of branches in a distinctive cylindrical cone, or strobilus (Fig. 15-9A). These varied from 1–7 cm in diameter, sometimes reaching lengths of up to 50 cm, but most were much smaller. Sporophylls were arranged spirally on the central axis of the strobilus, and sporangia were borne adaxially on each sporophyll. The apex of each sporophyll was often much expanded and overlapped the bases of sporophylls above it. Each sporophyll diverged at right angles to the axis of the strobilus. Most species of *Lepidodendron* were heterosporous, and each strobilus often contained both **megasporangia** and **microsporangia.** Microsporangia were usually concentrated on sporophylls toward the top of the cone, with megasporangia toward the base. Each microsporangium contained innumerable microspores; thus when strobili matured, vast dust clouds of microspores must have resulted. Microspores ranged in size from 20–40 μm, whereas megaspores ranged from 300 μm–2 mm in diameter. Megasporangia contained 16, 4, or sometimes 1 megaspore. **Gametophyte** development apparently was **endosporic.** In the few specimens that have been found, megagametophytes bore several archegonia.

Whole strobili apparently sometimes fell as a unit from trees, and sometimes the megasporophylls were deciduous, especially those with only a single megaspore in the megasporangium. Some of these exhibited a remarkable thickening and specialization of the sporangium wall. In these, the megaspore was almost completely enclosed; the gametophyte often germinated within the sporangium and was fertilized while attached to the sporophyll (Fig. 15-9D). This and more complex developments of megasporophylls in other lycopod genera represent the independent evolution of a seedlike habit in this group of plants.

The low absorptive area provided by the many small roots is suggested to have guided the selection for xerophytic adaptations to prevent excessive water loss. These include long, narrow leaves that were deciduous, the presence of stomata embedded in grooves, and other features.

The genus *Sigillaria* (Figs. 15-4 and 15-7F) is similar to *Lepidodendron*, but its leaf cushions are arranged in longitudinal rows. Evidently the trunk also dichotomized less frequently—sometimes apparently not at all. Strobili were borne on short, **adventitious** branches that were produced at different intervals and emerged high on the trunk. Strobili bore either microsporophylls or megasporophylls; this would assure **outcrossing** more than may have been possible in *Lepidodendron*. Furthermore, whereas in *Lepidodendron* the terminal strobili made the growth of the stems determinate, the adventitious strobili in *Sigillaria* meant that strobili could be produced in succeeding years, while the trunk continued to grow.

The Carboniferous record also contains herbaceous lycopods very like modern genera, such as *Selaginella* and *Lycopodium*.

Figure 15-7 A, part of branch cortex of *Lepidodendron*, showing rhomboidal leaf cushions in oblique rows, ×2. **B–D,** leaf cushions of three different species of *Lepidodendron*, showing details of structure: *lig sc,* ligule scar; *parich,* parichnos; *spor sc,* sporangial scar; *v sc,* vascular bundle scar. **E,** transverse section of leaf of *Lepidodendron*, showing vein, transfusion tissue, and mesophyll; *h,* hypodermal tissue; *m,* mesophyll; *ph,* phloem; *t,* transfusion cells; *vb,* tracheids of vein. **F,** bark of trunk of *Sigillaria;* note variation in appearance as layers are removed. (**B–D,** redrawn after Zeiller; **E,** after Andrews, *Studies in Paleobotany.* New York: John Wiley & Sons. Copyright 1961 by John Wiley & Sons; **F,** after Schimper.)

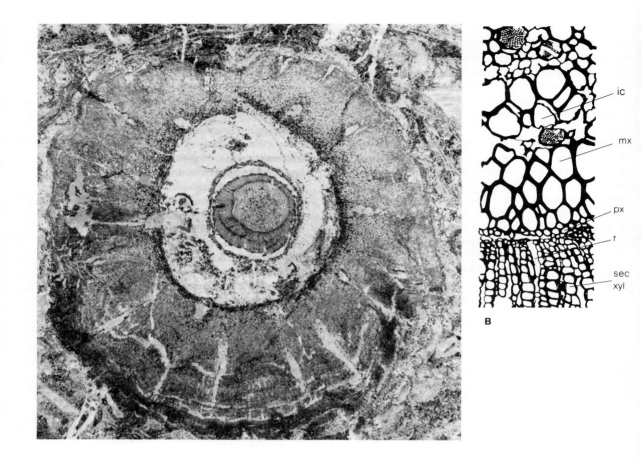

Figure 15-8 Transverse sections of stem of *Lepidodendron*. **A,** showing stele, cortex, and periderm, ×3.9. **B,** part of stele, showing pith, primary xylem, and secondary xylem, ×20: *ic,* inner cortex; *mx,* metaxylem; *oc,* outer cortex; *p,* pith; *per,* periderm; *px,* protoxylem; *r,* ray; *sec xyl,* secondary xylem.

Modern Lycopods

Modern lycopods are all herbaceous. The genera *Huperzia, Lycopodium,* and *Phylloglossum* are homosporous, whereas *Selaginella, Isoetes,* and *Stylites* are heterosporous. *Huperzia* and *Lycopodium* are widely distributed, occurring from frigid and temperate regions to the tropics. *Selaginella* and *Isoetes* occur in temperate to tropic regions, whereas *Stylites* and *Phylloglossum* are confined to temperate areas of the Southern Hemisphere. *Stylites* has two species, known only from the southern Peruvian Andes; *Phylloglossum* has a single Australian species.

Huperzia has perhaps 60 species, *Lycopodium* has approximately 50, *Selaginella* more than 200, and *Isoetes* about 60. Most genera are terrestrial or epiphytic, but *Isoetes* is generally a **littoral** aquatic, and *Stylites* is amphibious.

The genus *Huperzia* (Figs. 15-10 and 15-12) is the most generalized extant genus. The sporophyte consists of a simple, erect, dichotomizing axis of either indeterminate or determinate growth. This bears spirally arranged, evergreen, spiny leaves.

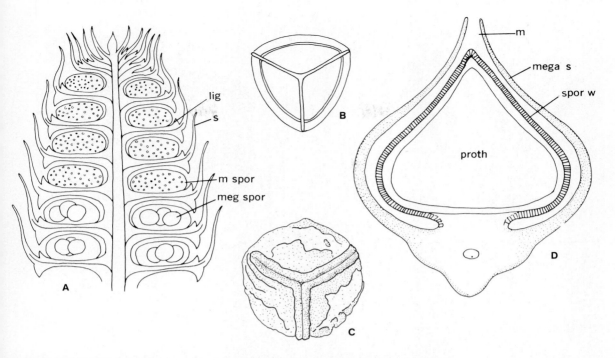

Figure 15-9 **A,** vertical section through cone of *Lepidostrobus*, showing segregaton of micro- and megasporangia on different sporophylls: *lig*, ligule; *m spor*, microsporangium; *meg spor*, megasporngium; *s*, sporophyll. **B, C,** spores of *Lepidodendron*. **B,** microspore, showing trilete scar and narrow flange *(fl)* encircling spore body, × 500. **C,** megaspore, showing trilete scar and roughened spore wall, × 40. **D,** vertical section through "seed" of *Lepidocarpon*, showing outer lamellae of megasporophyll *(l)*, spore wall *(spor)*, megaspore *(ms)*, prothallus *(proth)*, and micropyle *(m)*, × 4. (**A,** redrawn after R. Zeiller; **B, C,** redrawn after Potonié; **D,** redrawn after Scott.)

At intervals, annually in some species, are produced a series of sessile sporophylls that are succeeded by vegetative leaves on the same shoot. The sporophylls are green and closely resemble vegetative leaves. Sporangia are affixed to the axis of the adaxial surface of the sporophyll. Ligules are absent in *Huperzia*, *Lycopodium*, and *Phylloglossum*. Some terrestrial species of *Huperzia* are no taller than 1 dm, but some epiphytic, often pendulous species reach lengths of 3 dm or more. The leaves have entire or toothed margins and sometimes reach lengths of up to 3 cm. Some species produce reduced, leaflike **propaguliferous** branches that emerge near the shoot apex, and each bears a single, roughly heart-shaped propagule (Fig. 15-10D, E). The propagules can be caused to be ejected forcibly by a simple touch of the propaguliferous branch. Propagules root soon after they fall; an apical cell is initiated, and a new sporophyte grows.

Although roots are initiated near the shoot apex in *Huperzia*, they grow downward within the stem cortex and emerge near the base of the shoot. This sometimes gives the transverse section of the shoot a complex pattern of steles. Roots are dichotomously branched and have root hairs. The stele of the stem is a much-fluted actinostele, from which emerge leaf traces that extend unbranched to the apices of the leaves. The cortex is broad and the outer layers of cells contain chlorophyll. Stomata are in the stem epidermis and on one or both surfaces of the leaf epidermis. The leaves consist largely of chlorophyllose tissue and have a central vascular strand.

Sporangia are attached to the leaf axil or upper

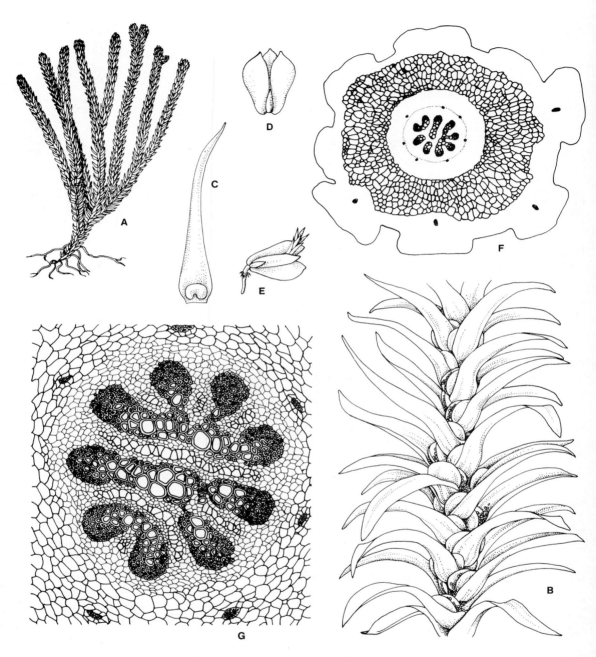

Figure 15-10 *Huperzia selago*. **A,** habit, showing dichotomizing axes, and apical clusters of gemma-bearing reduced branches, ×0.5. **B,** detail of a leafy stem segment, showing sporophylls with axillary sporangia, ×5. **C,** a single sporophyll, showing axillary sporangium, ×6. **D,** gemma, as seen from above, ×4. **E,** germinating gemma ×4. **F,** transverse section of stem, showing central actinostele, and also transverse sections of roots, ×8. (**D, E,** redrawn after Gibbs.)

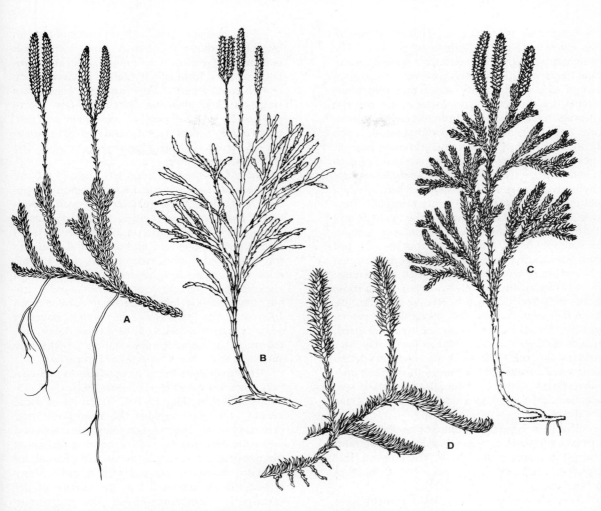

Figure 15-11 Diversity in the morphology in the genus *Lycopodium*. **A,** *L. clavatum*, ×1. **B,** *L. complanatum*, showing flattened leafy stems that dichotomize freely, ×1. **C,** *L. obscurum*, ×1. **D,** *L. inundatum*, ×1.

surface of the leaf by a stout stalk (Fig. 15-10B, C). The homosporous sporangium opens by a transverse line, and gapes open to expose the spores.

Spores are somewhat resistant to wetting, but they germinate within a few weeks under favorable moisture conditions. The spore produces an **exosporic** gametophyte. In nature, a **mycorrhizal** association is formed early, and it is **obligate**. Chlorophyll is formed when the gametophyte is on the soil surface rather than subterranean. The gametophyte, usually 2–4 mm long, is generally cylindrical, with either radial or bilateral symmetry and with a rim curving up over the margins of the upper surface. Rhizoids extend from the undersurface. Male and female sex organs are intermixed near the margin of the upper surface and are embedded in the thallus. Antheridia are produced first, followed by archegonia. Sperms are ovoid, with two apical whiplash flagella oriented back over the body of the sperm.

Lycopodium (ground pine, Fig. 15-11) resembles *Huperzia* (Fig. 15-12) in many respects. The evergreen sporophytes bear spiny leaves. *Lycopodium* frequently produces a rhizomatous or stoloniferous (see **stolon**) leafy shoot that sometimes reaches lengths of several meters. At irregular intervals, the roots emerge from this creeping stem, and upright monopodial, sympodial, or even dichotomizing branch systems also arise from it. These branch systems show determinate growth in some species, but others are perennial. Sporophylls are frequently colorless to brownish when mature and are confined to terminal strobili. They overlap somewhat and usually differ in shape and size from the vegetative leaves. As sporangia mature, the sporophylls dry out, diverge from the axis, and expose the **axillary** sporangia. The sporangia open by a transverse slit and scatter spores to air currents.

Leaf arrangement is always spiral, but in some more specialized species, vegetative branches are much flattened. Adaxial and **abaxial** leaves can be greatly reduced, and the bases of all leaves are fused; in such instances, leaves occur in four ranks. Stomata are on both surfaces of the leaves in species considered generalized, whereas those more specialized have stomata confined to the adaxial surface. Each leaf has a single vascular strand that forms a midvein.

The stele is generally much **dissected** and is termed a **plectostele,** from which a vascular strand emerges to each of the leaves. There are no leaf gaps (Fig. 15-13), and the cortex is broad, much as in *Huperzia*.

The xylem is all primary, and the maturation is exarch. The protoxylem tracheids have annular and helical thickenings; those of the metaxylem are scalariform, or with circular and bordered pits. The phloem consists of sieve cells with oblique end walls and scattered **sieve areas** interspersed with **parenchyma** cells.

The gametophyte (Fig. 15-14) is variable; sometimes it is a flattened **multistratose** thallus with a ruffled, infolding rim and rhizoids on the undersurface. If subterranean, it tends to lack chlorophyll and is mycorrhizal; when on the soil surface, it has chlorophyll, and mycorrhizal fungi are less abundant. Sometimes the gametophyte is shaped like a diminutive carrot, with the apex near the soil surface and the tapering portion subterranean. The upper surface contains the embedded sex organs. The sperms are like those in *Huperzia*.

Phylloglossum (Fig. 15-15) resembles an extremely reduced *Lycopodium* (about 2 cm tall at maximum), except that the minute, erect stem produces a **tuber,** from which a tuft of leaves and one or two roots emerge. This leafy tuft produces a single, stalked strobilus, bearing a few sporophylls. The leafy tuft persists one season, and produces another tuber near its base that gives rise to the next year's vegetative leaves and strobilus. The stele is a siphonostele.

In extant homosporous lycopods, there are conspicuous trends toward specialization in sporophylls and leaf arrangement, and even considerable reduction of the entire sporophyte, as in *Phylloglossum*. In stelar specialization, there is a trend from an actinostele to a plectostele and finally to the specialized siphonostele of *Phylloglossum*. This does not imply derivation of *Lycopodium* from *Huperzia*, although close relationship is indisputable. *Phylloglossum* may have been derived from *Lycopodium*. Its resemblance to juvenile stages of some species is too close to be a coincidence; furthermore, its spore ornamentation is extremely similar to that of some species of *Lycopodium*.

Heterosporous living lycopods include *Selaginella* (Figs. 15-16 and 15-17), *Isoetes* (Figs. 15-18–15-20), and *Stylites* (Fig. 15-21A, B). In all of these genera, the gametophyte is **endosporic** and extremely small. The megagametophyte is exposed when the triradiate scar of the spore is broken as the gametophyte enlarges beyond the bounds of the spore wall that encompasses it (Fig. 15-19). Many archegonia are embedded in the surface of the endosporic megagametophyte. The microspore contents essentially become a single antheridium, and sperms are shed only after the microspore wall is ruptured. A *Selaginella* microgametophyte produces numerous biflagellate sperms, sometimes as many as 256; *Isoetes* produces only 4 per microgametophyte, and each of these sperms has 15 flagella.

The leaves of heterosporous lycopods always bear an inconspicuous ligule near the axil (Fig. 15-17F). The sporophyte of *Selaginella* is generally evergreen and its leaves are spirally arranged. As in *Lycopodium*, species of *Selaginella* that are considered more generalized have all of the vegetative leaves alike (Fig. 15-17A). In more specialized species, they are in four ranks (Fig. 15-17B), and shoots are dorsiventrally flattened, with adaxial and abaxial leaves reduced in size. The sporophyte varies considerably in size, from mosslike species a few

Figure 15-12 *Huperzia* stem segments, showing diversity in leaf shape and their orientation on the shoots. **A,** *H. phlegmarioides,* ×1. **B,** *H. gnidioides,* ×1. **C,** *H. nummularifolia,* ×1.

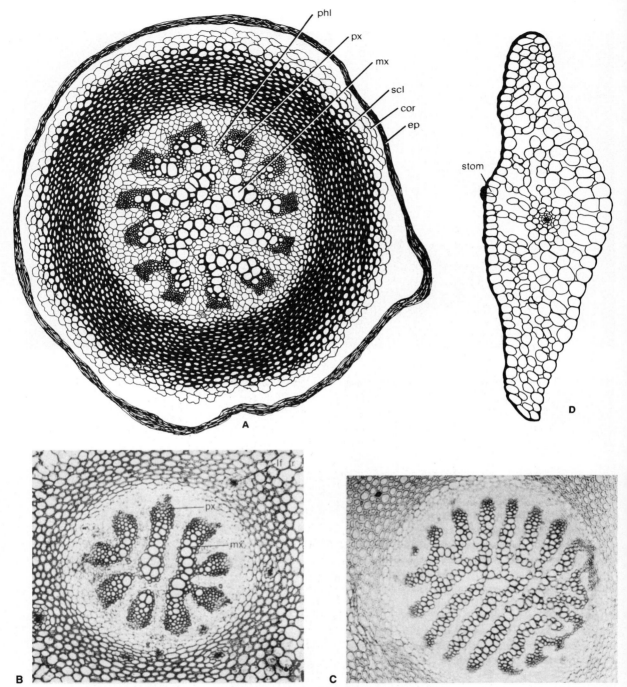

Figure 15-13 **A**, transverse section of root of *Lycopodium*, showing plectostelic arrangement of vascular tissue: *cor*, cortex; *ep*, epidermis; *mx*, metaxylem; *phl*, phloem; *px*, protoxylem; *scl*, sclerenchyma. **B, C,** transverse sections of stems of two species, × 50 (note actinostelic arrangement of xylem and phloem in **B** and plectostele in **C**): *lf tr*, leaf trace. **D**, transverse section of leaf, showing stoma *(stom)*, mesophyll and central vein.

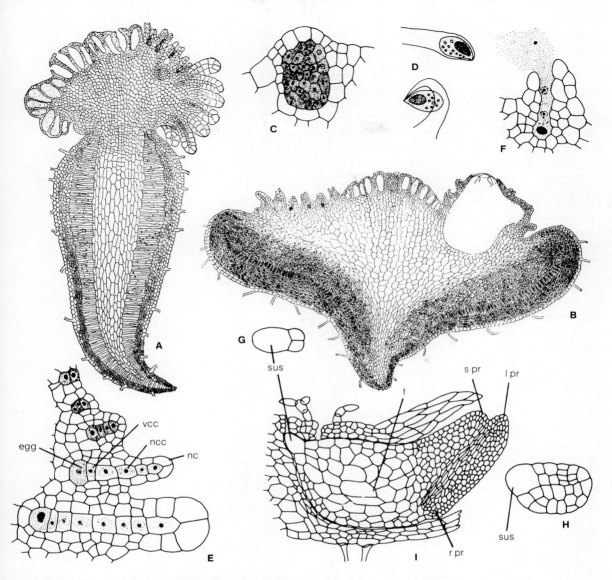

Figure 15-14 *Lycopodium* gametophyte and young embryo. **A,** longitudinal section through prothallus of *Lycopodium complanatum* (note the numerous antheridia on the *upper left,* many of which have shed the sperms; on the *right* are the archegonia), ×18. **B,** longitudinal section through prothallus of *Lycopodium clavatum* (note archegonia on *upper left* and antheridia in *central upper portion;* a young embryo is marked by the large white area; the mycorrhiza abound in the dense layers near the lower surface), ×37. **C,** immature antheridium showing central spermatogenous cells surrounded by jacket cells, ×50. **D,** mature sperms showing flagella, ×1000. **E,** stages of archegonial maturation with neck cells *(nc),* neck canal cells *(ncc),* ventral canal cells *(vcc),* and egg, ×50. **F,** mature archegonium which has disrupted prior to fertilization, ×50. **G–I,** stages in development of embryo, ×30; *f,* foot; *l pr,* leaf primordium; *r pr,* root primordium; *sus,* suspensor. (**A, B,** after W. W. Haupt from *Plant Morphology,* copyright 1953 by McGraw-Hill Book Co., reproduced with permission; **C, D,** redrawn after Bruchmann.)

Lycophytes (Lycopods)

centimeters high to some vinelike species that reach lengths of over 20 m. The branching pattern is highly variable, but chiefly in one plane from a main axis. The leaves are always small, usually less than 2 mm long. They usually have a single, unbranched vascular strand, but rarely the vascular strand is branched, making these leaves indistinguishable from megaphylls. Roots frequently arise perpendicular to the main axis, become somewhat chlorophyllose, and when their tips touch the soil, produce dichotomous branches. The root system is never extensive. In some tufted species of desert sites, the roots are brittle, and when the plant dries out, the branches curl inward, producing a hemispherical, wind-carried "tumbleweed." When moisture returns, the plant uncoils and begins to photosynthesize and grow.

The stele of *Selaginella* ranges from protostelic to siphonostelic and is often strap-shaped; in some species it is **polystelic** (Fig. 15-16E, F), with the vascular strand broken into several separate strands that fuse at the branch bases. Each stele has **exarch** protoxylem, with metaxylem formed centripetally. The xylem is surrounded by several layers of parenchyma, which in turn is surrounded by the sieve cells of the phloem. In most species the stele is sheathed by a single layer of parenchyma, separated from the cortex by a series of intercellular spaces, and suspended by uniseriate filaments, the **trabeculae** (Fig. 15-16F). The cortex is wide, and the epidermis is cutinized. In several species of *Selaginella*, there are true vessels—an unusual occurrence in archegoniate plants.

As in *Lycopodium*, the sporophylls are in strobili (Fig. 15-17A) that are arranged in four ranks. The lower ones tend to be megasporophylls, whereas the upper ones are microsporophylls; megasporophylls and microsporophylls may also be completely mixed in the same strobilus.

Microsporangia contain numerous microspres, but megasporangia normally contain four megaspores. In some species of *Selaginella*, the bivalved microsporangium has thicker-walled cells in the center of each of the valves. As the mature sporangium dries out, the valves gape open; cells of the central patch of the floor of the sporangium then cave inward against the spores, and the valves are abruptly clamped together, throwing spores into the air. In the megasporangium, a similar procedure throws the megaspores into the air.

Sometimes megagametophytes are retained in

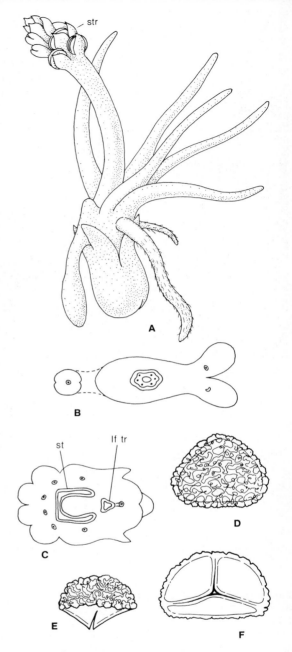

Figure 15-15 *Phylloglossum.* **A,** habit sketch, showing strobilus *(str)*, ×2. **B,** transverse section of base of strobilar axis with trace of petiole at left, ×4. **C,** transverse section toward base of stem, showing stele *(st)* and single leaf trace *(lf tr)*, ×4. **D–F,** three views of spores, ×500. **D,** distal surface, showing rugulate ornamentation. **E,** lateral view; **F,** proximal view, both showing the unornamented contact faces. (Redrawn after Bertrand.)

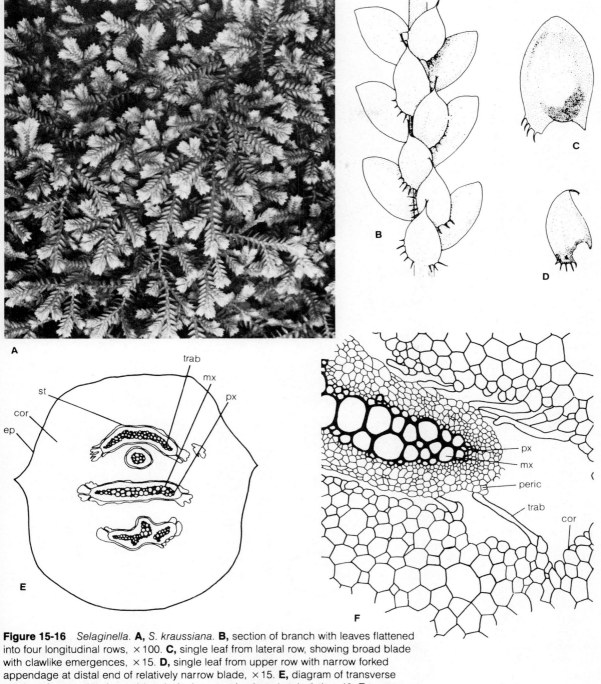

Figure 15-16 *Selaginella.* **A,** *S. kraussiana.* **B,** section of branch with leaves flattened into four longitudinal rows, ×100. **C,** single leaf from lateral row, showing broad blade with clawlike emergences, ×15. **D,** single leaf from upper row with narrow forked appendage at distal end of relatively narrow blade, ×15. **E,** diagram of transverse section of stem, showing epidermis *(ep),* cortex *(cor),* and stele *(st),* ×40. **F,** transverse section of part of stele and cortex of stem of *Selaginella,* showing exarch protoxylem *(px),* metaxylem *(mx),* pericycle *(peric),* trabeculae *(trab),* and cortex *(cor),* ×150.

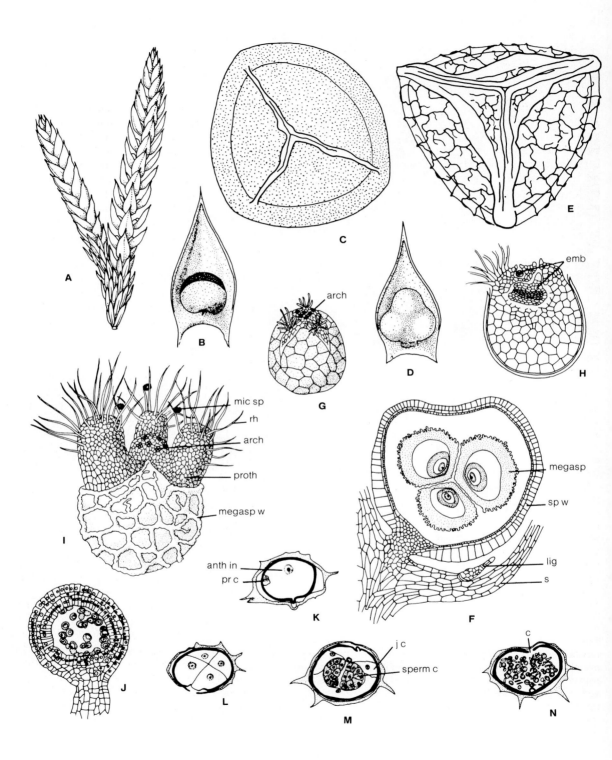

422 Chapter 15

the sporangium until fertilization occurs. This keeps the megagametophyte in a position where microspores are likely to fall, and resembles the seed habit. The young embryo can germinate while on the strobilus; the seedling ultimately falls to the ground to begin an independent existence.

In *Isoetes* (Figs. 15-18–15-20) and *Stylites* (Fig. 15-21A, B), the sporophyte has an erect stem like a **corm**. In *Isoetes* the stem is completely ensheathed by leaf bases and by roots, but in *Stylites* the stem is somewhat more elongate and sometimes dichotomously branched. Secondary growth occurs in both genera, but such growth is extremely slight each year.

The stem of *Isoetes* is composed of two to four lobes (Fig. 15-19B, D). The stem is very short (usually 1–3 cm tall), bears dichotomously branching roots between the lobes, and grows from both an apical and basal meristem.

Figure 15-17 *Selaginella.* **A,** four-sided strobili at tips of two branches, ×4. **B,** adaxial surface of single megasporophyll, showing megasporangium with three lobes reflecting three of enclosed megaspores, ×20. **C,** single microsporophyll with microsporangium split open to show mass of microspores, ×20. **D,** microspore showing well-defined trilete mark and outer flange on wall, ×100. **E,** megaspore, showing thick wall and thickenings along trilete tetrad scar, ×200. **F–H,** development of female gametophyte. **F,** vertical section through megasporangium and three megaspores, ×250. **G,** single megaspore showing archegonia on slightly protruding gametophyte, ×500. **H,** prothallus with archegonia protruding from megaspore, showing several microspores trapped on rhizoidal appendages, ×500. **I,** two embryos developing in archegonia (one on *left* is in two-celled stage; one on *right* has developed primordia and foot), ×500: *arch*, archegonia; *emb*, embryo; *lig*, ligule; *megasp*, megaspore; *megasp w*, megaspore wall; *mic sp*, microspores; *proth*, prothallus; *rh*, rhizoids; *s*, sporophyll. **J–N,** male gametophyte development in *Selaginella*. **J,** vertical section through mature microsporangium, showing wall and developing microspores in tapetal fluid, ×50. **K–N,** stages in development of microspores, ×500: *anth in*, antheridial initial; *c*, commissure; *j c*, jacket cell; *pr c*, prothallial cell; *sperm c*, spermatogenous cell. (**F,** redrawn after Haupt; **G–I,** redrawn after Bruchmann; **J–N,** from Slagg, with permission of *American Journal of Botany*.)

The arrangement of tissues within the stem is unique (Fig. 15-19B–D). A core of primary xylem in the central region consists of tracheids interspersed with parenchyma. This core is surrounded by a very thin layer of primary phloem. As in other lycopods, leaf traces depart from the vascular cylinder and leave no leaf gaps. In *Isoetes,* a cambium forms outside the primary phloem (Fig. 15-19C) and produces a small amount of secondary tissue annually. The cells toward the inside are sieve cells, parenchyma, and apparently occasional tracheids. Cells derived from the cambium outside are all parenchymatous and act as a food storage tissue. Each year, a new parenchyma is formed, and outer tissues are shed together with the leaf bases.

The leaves are spirally arranged and are long and narrow, but the expanded, flattened bases overlap each other. They may be up to 80 cm long in some species but are generally no more than 10 cm long. They form a rosettelike crown on the upper part of the corm or **rhizophore** (Fig. 15-19D). The earliest leaves formed each year are sterile, followed by megasporophylls, then microsporophylls. Sporophylls are morphologically like vegetative leaves. The leaf is made up of four longitudinal chambers arranged around the vascular strand. The chambers are partitioned along the length of the leaf. When submerged plants produce leaves, they lack stomata, but those produced when the plant is emergent have stomata.

Each megasporophyll produces 150–300 spores in its basal sporangium, whereas each microsporangium contains 10 times that number of microspores. Megasporangia are sometimes 5–10 mm long. They are enclosed by both the sporangial wall and a protective flap of leaf tissue called the **velum.** Sporangia are partitioned by transverse bands of sterile tissue.

During the unfavorable season, generally winter, sporophylls decompose and release the spores. The next season produces a new series of sporophylls. Water birds consume spores and even corms, and thus contribute to the dispersal of spores.

Stylites is similar in structure to *Isoetes*, except that the stem is cylindrical and more extensive, frequently producing a single dichotomy and sometimes dichotomizing twice. In *Stylites,* one species also produces propagules, which arise in the position of a leaf that is replaced by a small bud meristem. These persist on the parent plant for a time, but ultimately fall off, and begin a life of their own, forming new sporophytes.

Modern lycopods have a number of modest uses. Species of *Lycopodium* have long been used as Christmas decorations in temperate regions. Their spores were used to coat pills and prevent their adhering to each other. Spores were also used as baby powder by pioneer women. The Chinese used spores in the manufacture of firecrackers, since they ignite explosively. The same feature made them useful for spectacular flash effects in theatrical performances, and for "flash" photography before flashbulbs were available.

More Fossil Lycopods

In the fossil record, there are several species obviously closely related to *Isoetes* and *Stylites*, and they may have been ancestral to these genera. *Pleuromeia* (Fig. 15-21D, E), of Triassic age, had an upright, unbranched stem up to 2 m tall. The base was composed of an expanded, four-lobed rhizophore on which were borne, roughly spirally, simple or dichotomizing roots. Its leaves were linear and deciduous, up to 10 cm long, and spirally arranged. The apex of the shoot was terminated by an elongate strobilus in which sporophylls were either megasporophylls or microsporophylls. Although preservation of sporophylls is not satisfactory, sporangia appear to have been borne on the adaxial surface of the truncate sporophyll. The sporangia were partitioned, as in *Isoetes*.

Of Lower Cretaceous age, *Nathorstiana* (Fig. 15-21C) was a smaller plant consisting of a stem approximately 10 cm tall. It bore a tuft of spirally arranged leaves near its apex and a bluntly four-lobed rhizophore at the base, from which simple or once-dichotomizing roots emerged. Its leaves were deciduous, leaving scars on the stem and were approximately 5 cm long. Unfortunately, all known specimens are vegetative.

Therefore, in heterosporous lycopods there appears to be a very suggestive reduction series from the complex, branched system of *Lepidodendron*, with its stigmarian rhizophore, through the less-branched situation in *Sigillaria*, through the simple axis of *Pleuromeia*, with its reduced rhizophore, then to the very reduced *Nathorstiana*, the

Figure 15-18 *Isoetes*. **A,** *I. nuttallii*, ×1.

Figure 15-19 *Isoetes*. **A,** vertical section through single megasporophyll, showing ligule *(lig)*, velum *(v)*, and spores *(s)*, ×15. **B,** transverse section of stem, showing inner core of primary xylem *(pr xyl)* surrounded by primary phloem *(pr phl)*; cambium *(camb)*—the clear area—with secondary tissue *(sec t)* on inside, secondary parenchyma *(pa)* and cortex *(cor)* on outside, ×5. **C,** enlarged section of **B,** showing tracheids of primary xylem *(pr xyl)*; primary phloem *(pr phl)*; secondary tissue *(sec t)* inside cambium *(camb)*, and secondary parenchyma on outside *(par)*, ×15. **D,** three-dimensional drawing of the "corm," showing the relationship of roots of the rhizophore to the main axis, and essential details of the internal structure: *camb*, cambium; *pris 2nd*, "prismatic" layer made up of secondary xylem and secondary parenchyma; *mer*, meristem of rhizophore; *Prim st*, primary stem; *Prim x*, primary xylem; *V B*, vascular bundle. (**B,** redrawn after Haupt; **D,** redrawn after Stewart.)

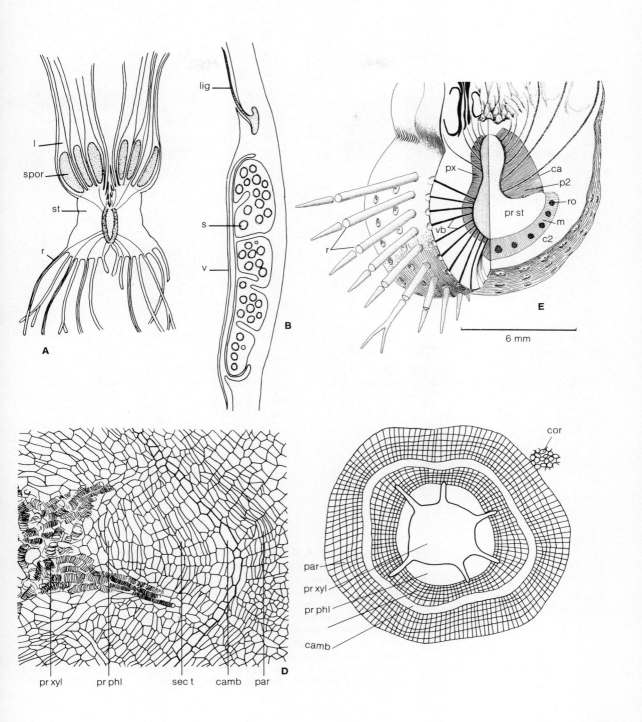

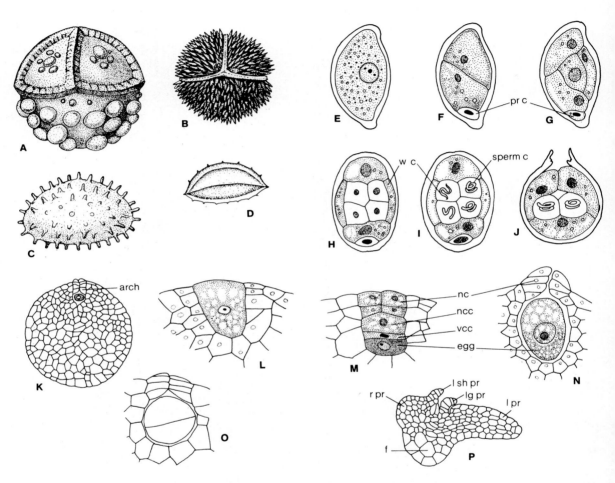

Figure 15-20 *Isoetes.* **A, B,** megaspores of two different species, ×5. **C, D,** microspores of two different species, **E–J,** development of male gametophyte, ×800. **E,** uninucleate cell. **F,** single prothallial cell *(pr c)* has formed at one end. **G,** antheridial initial in center surrounded by three wall cells. **H–J,** stages in formation of sperms from spermatogenous cells *(sperm c).* **K–N,** development of female gametophyte and archegonia. **K,** single archegonium on periphery of mature gametophyte, ×100. **L–N,** development of archegonium, ×400. **O,** two-cell stage of embryo development, ×400. **P,** a late embryo, ×300: *arch,* archegonium; *f,* foot; *l pr,* leaf primordium; *l sh pr,* leaf sheath primordium; *n c,* neck cell; *n c c,* neck-canal cell; *r pr,* root primordium; *v c c,* ventral canal cell. (**A–D,** redrawn after Motelay and Vendryes; **E–J, P,** redrawn after Liebig; **K–O,** redrawn after Campbell.)

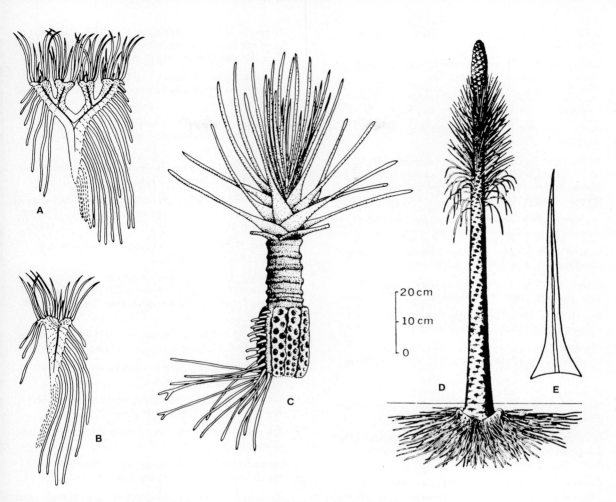

Figure 15-21 *Stylites.* **A,** side view, showing disposition of roots and leaves on stem, ×1.3. **B,** front view, showing double dichotomy of stem, ×1.3. **C,** reconstruction of *Nathorstiana,* showing rootstalk, very short stem, and spiral collar of leaves, ×0.5. **D,** reconstruction of *Pleuromeia,* showing rootstalk and roots, columnar stem, and leaves and cone toward top. **E,** single leaf. (**A, B,** after Rauh and Falk; **C,** redrawn after Mägdefrau; **D,** from Hirmer, with permission of *Paleontographica;* **E,** after Mägdefrau.)

more reduced situation in *Stylites,* and finally *Isoetes.* Unfortunately, it is not a neat series and should not be interpreted as a clear line of genera derived from each other. For example, it should be noted that the appearance of strobili in some ancient genera is a more derived condition than their absence in such modern genera as *Isoetes.*

References

Andrews, H. N., and Murdy, W. H. 1958. *Lepidophloios*—and ontogeny in arborescent lycopods. *Amer. J. Bot.* 45:552–560.

Arnold, C. A. 1960. A lepidodendrid stem from Kansas and its bearing to the problem of cambium and phloem in Paleozoic lycopods. *Contr. Mus. Paleo. Univ. Mich.* 15:247–67.

Balbach, M. K. 1962. Observations on the ontogeny of *Lepidocarpon*. *Amer. J. Bot.* 49:984–89.

———. 1965. Paleozoic lycopsid fructifications I. *Lepidocarpon* petrifications. *Amer. J. Bot.* 52:317–30.

Banks, H. P. 1960. Notes on Devonian lycopods. *Senckenbergiana Lethaea,* 41:59–88.

Banks, H. P.; Bonamo, P. M.; and Grierson, J. D. 1972. *Leclercquia complexa* gen. et sp. nov., a new lycopod from the late Middle Devonian of eastern New York. *Rev. Palaeobot. Palynol.* 14:19–40.

Bruce, J. G. 1976. Gametophytes and subgeneric concepts in *Lycopodium*. *Amer. J. Bot.* 63:919–24.

Danze-Corsin, D. 1962. Les Lepidophytes: historique et classifications. *Ann. Sci. Nat. Bot.* 12:293–319.

Eames, A. J. 1942. Illustrations of some *Lycopodium* gametophytes. *Amer. Fern J.* 32:1–12.

Eggert, D. A. 1961. The ontogeny of Carboniferous arborescent Lycopsida. *Palaeontographica* 108:43–92.

Freeberg, J. A., and Wetmore, R. H. 1967. The Lycopsida—a study in development. *Phytomorphology* 17:78–91.

Grierson, J. D. 1976. *Leclercqia complexa* (Lycopsida, Middle Devonian): its anatomy and the interpretation of pyrite petrifications. *Amer. J. Bot.* 63:1184–202.

Grierson, J. D., and Banks, H. P. 1963. Lycopods of the Devonian of New York State. *Palaeontogr. Americana* 4:217–95.

Grierson, J. D., and Hueber, F. M. 1968. Devonian lycopods from northern New Brunswick. International Symposium on the Devonian System, Calgary.

Holloway, J. E. 1919. Studies in the New Zealand species of *Lycopodium*. III. The plasticity of the species. *Trans. N. Z. Inst.* 51:161–261.

Jones, C. E. 1905. The morphology and anatomy of the stem of the genus *Lycopodium*. *Trans. Linn. Soc. London* (Ser. 2) 7:15–35.

Kidston, R. 1901. Carboniferous lycopods and sphenophylls. *Trans. Nat. Hist. Soc. Glasgow* n.s., 6:25–140.

Krassilov, V. A., and Zakharov, P. D. 1975. *Pleuromeia* from the Lower Triassic of the Far East of the U.S.S.R. *Rev. Palaeobot. Palynol.* 19:221–32.

Kräusel, R., and Weyland, H. 1932. Pflanzenreste aus dem Devon. IV. *Protolepidodendron* Krejči. *Senkenbergiana* 14:391–403.

Lemoigne, Y. 1961. Etudes analytiques et comparées des structures internes des Sigillaires. *Ann. Sci. Nat. Bot.* 12:473–578.

Lyon, A. G. 1964. Probable fertile region of *Asteroxylon mackiei* K. & L. *Nature* 203:1082–83.

Mägdefrau, K. 1932. Über *Nathorstiana*, eine Isoetaceae aus dem Neokom von Quedlinburg A. Harz. *Beih. Bot. Centralbl.* 49:706–18.

Nessel, H. 1939. *Die Barlappgewachse (Lycopodiaceae).* Gustav Fischer Verlag, Jena.

Neuberg, M. F. 1961. New data on the morphology of *Pleuromeia* Corda from the Lower Triassic Period of the Russian platform. *Doklady Akad. Nauk S.S.R.* no. 2:445–48.

Paolillo, D. J. 1963. *The Developmental Anatomy of Isoetes.* Illinois Biol. Monogr. 31. Univ. of Illinois Press, Urbana.

Retallack, G. J. 1975. The life and times of a Triassic lycopod. *Alcheringa* 1:3–29.

Rothmaler, W. 1944. Pteridophyten-Studien. I. *Feddes Report. Spec. Nov. Reg. Veg.* 54:55–82.

Schweitzer, H.-J. 1980. Über *Drepanophycus spinaeformis* Goeppert. *Bonn. Palaeobot. Mitt.* 7:1–29.

Sporne, K. R. 1975. *The Morphology of Pteridophytes.* Hutchinson Univ. Library, London.

Stewart, W. N. 1947. A comparative study of stigmarian appendages and *Isoetes* roots. *Amer. J. Bot.* 34:315–24.

Sykes, M. G. 1908. Notes on the morphology of the sporangium-bearing organs of the Lycopodiaceae. *New Phytol.* 7:41–70.

Wagner, W. H.; Beitel, J. M.; and Wagner, F. S. 1982. Complex venation patterns in the leaves of *Selaginella*: megaphyll-like leaves in lycophytes. *Science* 218:793–94.

Wilce, J. H. 1965. Section Complanata of the genus *Lycopodium*. *Nova Hedwigia* 19 (Beiheft):1–233.

Perhaps the most ancient living genus of vascular plants is *Equisetum*, the only survivor of the sphenophytes, subdivision Sphenophytina. There are 15 species and 7 hybrids of *Equisetum*, found in most parts of the world except for Australia and New Zealand, from subtropical to high arctic regions. They are popularly called horsetails or scouring rushes, and this name is also given to the whole subdivision. Although modern horsetails are predominantly modest, herbaceous perennials, some of those of the past were immense woody trees that formed extensive forests, presumably in swampy areas, and reached the pinnacle of their diversity and abundance some 300 million years ago. At the same time, *Equisetum*like plants were present in the understory of these forests.

16

SPHENOPHYTES
(Horsetails and Their Allies)

General Morphology

The sporophyte of most sphenophytes consists of a jointed stem, usually with longitudinal grooves. The leaves and branches are in whorls at the nodes. The roots also arise at the nodes. The sporangia are eusporangiate on a specialized, reduced branch system, termed the **sporangiophore.** Sporangiophores are in whorls aggregated into a terminal cone. Both homospory and heterospory are known, but homospory is most common. In *Equisetum* there is no secondary growth, but extinct arboreal genera had secondary growth. The gametophytes—at least those of *Equisetum*—are exosporic, usually small, thallose, and chlorophyllose. The sperms are multiflagellate.

Early Sphenophytes

The sphenophytes appeared first in the late Devonian. *Pseudobornia ursina* (Fig. 16-1) is known from sediments of this age from Spitzbergen and Alaska.
    Although subterranean portions are unknown and no detailed anatomy is preserved, the morphology of *Pseudobornia* places it in apparent relationship with the extant genus *Equisetum*. The plants had a monopodial axis with stems up to 20 m tall and 60 cm thick at the base. These articulated stems, with internodes (probably hollow within) up to 60

cm long, bore one or two upward-arching branches up to 3 m long just above each node. These branches, in turn, bore pairs of branches from each internode, and the somewhat fluted, smaller branches produced whorls of leaves. At each node there was probably a membrane within.

A whorl of four leaves emerged at each node. Each leaf had an elongate petiole that dichotomized twice and terminated in featherlike lobes up to 6 cm long and 0.8 cm wide. Their venation is too obscure to interpret.

Fertile shoots probably occurred at the distal ends of branches near the apex of the main stem; they appear to have been unbranched, with the fertile portion of condensed nodes and internodes. The whorls of sterile bracts appear to have differed from the vegetative leaves, and had several parallel veins that extended to the truncate, toothed tip of each bract. The sporangiophores were also in whorls, and each was forked and terminated by a cluster of approximately 30 sporangia. *Pseudobornia* trees apparently formed dense stands in the shallows of water bodies.

Sphenophyllum (Figs. 16-2 and 16-3) is abundant in the fossil record and persisted from the Devonian until the Triassic, reaching its greatest abundance and diversity during the Carboniferous. The probably herbaceous sphenophylls were slender plants with ribbed stems of usually less than 1 mm in diameter. Nodes and internodes were regularly spaced, and adventitious roots arose at the nodes. Some specimens have stems up to 85 cm long and bear branches up to 60 cm long; they probably reclined or grew intertangled among other plants. They grew in coal swamp forests, where they may have formed a significant part of the undergrowth.

The usually wedge-shaped leaves have a single vascular strand (Fig. 16-3) that dichotomizes several times in the flattened leaf blade. The leaves are in whorls of 6–18 leaves at the nodes and are from 9–20 mm long. Sometimes the leaves of one whorl are deeply dissected, whereas those in the whorl above and below are not.

The stem has a triangular stele composed of tracheids with protoxylem at the points. Some stems show secondary growth. The cortex of some specimens was comparatively thick (1.5–6 cm).

Slender cones terminate the lateral branches (Fig. 16-2A). *Sphenophyllostachys* is used as the form genus for cones of many species of *Sphenophyllum*.

The cones consist of shortened internodes. Attached at each node is a whorl of bracts, above which are the associated sporangiophores (Fig. 16-2B–D, G). These bracts often ensheathe the sporangiophores and give them some protection. The sporangiophores bear small terminal sporangia (0.5–3 mm in diameter). Specimens of spores are poorly preserved but appear to be smooth and homosporous. *Tristachya* (Fig. 16-4) is similar to *Sphenophyllostachys*, but the sporangia are globose.

Archaeocalamites (Fig. 16-5), of Devonian and Carboniferous time, appears to have been arborescent. The woody stems are sometimes up to 16 cm in diameter, and some stem fragments are several meters long. Grooves of the stem are continuous from node to node in this genus; in most other sphenophytes the grooves of one internode alternate with the grooves of the adjacent internodes. The leaves (Fig. 16-5B) emerge at the nodes in whorls (Fig. 16-5A, C), are up to 10 cm long, and dichotomize several times; the limb is usually little more than 1 mm broad. Branches and roots arise at the nodes. Fertile cones (Fig. 16-5D) are assumed to have been borne on lateral branches. The cones replaced several whorls of vegetative leaves of the stem at various intervals; thus the growth of the cone-bearing branch was indeterminate. These cones are made up of eight to ten sporangiophores per whorl, and several whorls make up each cone. Each sporangiophore has four sporangia. No sterile bracts are associated with the sporangiophores.

The most impressive of the sphenophyte genera is *Calamites* (Figs. 16-6 and 16-7), a codominant tree with *Lepidodendron* and *Cordaites* in the coal forests of the Carboniferous. The first representatives appeared in the Middle Carboniferous, and the last became extinct during the Lower Permian. They varied from immense trees 20–30 m tall and with trunks up to 30 cm in diameter down to small shrubs. As in most sphenophytes, the main shoot shows much the same diameter throughout its length. Its leaves and branches are in dense whorls at the nodes. When living, the shoots were probably densely clustered with the branches intermeshed so that the plants formed thickets. Since erect shoots arise from an extensive system of rhizomes, the clustered habit was further enhanced.

Much of the fossil material is preserved as casts of the interior of the trunk, which formed an extensive hollow cylinder. Other fossil material, however, permits a reasonable reconstruction of this

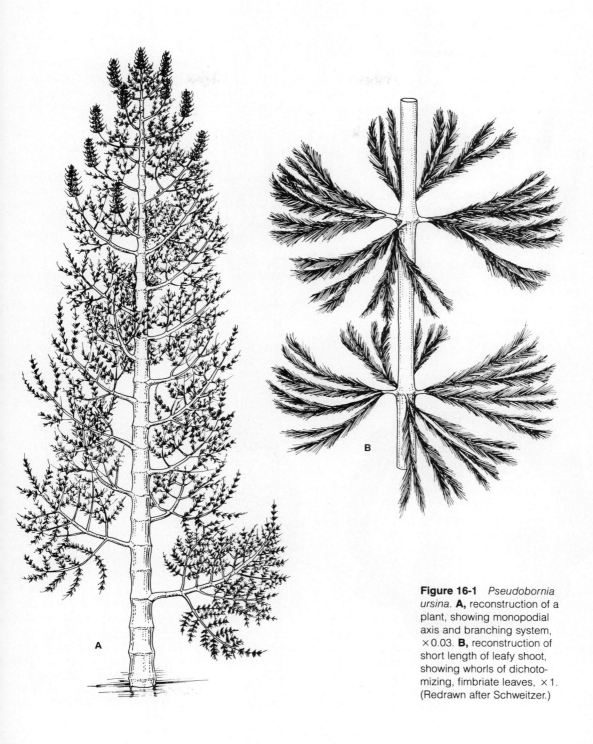

Figure 16-1 *Pseudobornia ursina*. **A,** reconstruction of a plant, showing monopodial axis and branching system, ×0.03. **B,** reconstruction of short length of leafy shoot, showing whorls of dichotomizing, fimbriate leaves, ×1. (Redrawn after Schweitzer.)

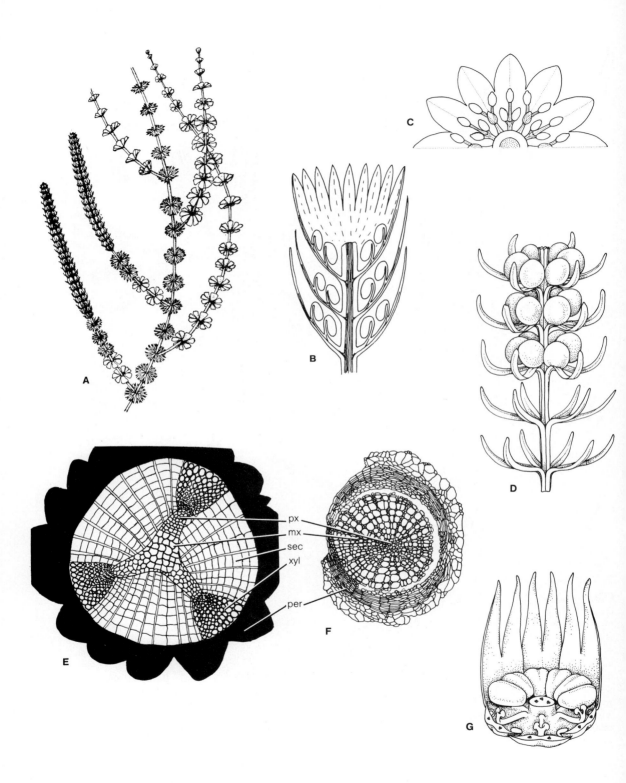

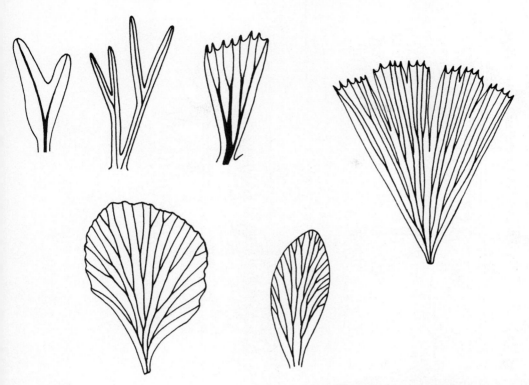

Figure 16-3 *Sphenophyllum* leaves, showing diversity in form.

Figure 16-2 **A-C,** *Sphenophyllum,* sporophyte and associated cones. **A,** reconstruction of shoot with several branches, showing whorls of leaves at nodes, and two terminal strobili on branches to left, × 0.2. **B,** median longitudinal section through cone, showing fused bracts and sporangiophores, × 10. **C,** plan view of one half of bract whorl, showing position of sporangia on sporangiophores, × 10. **D-F,** *Sphenophyllostachys.* **D,** reconstruction of a portion of a cone, showing whorls of sporangia and leaves. **E,** transverse section of stem, showing three-armed exarch protostele, radial rows of secondary xylem and rays, and periderm, × 50. **F,** transverse section of root with primary xylem and periderm, × 40. **G,** *Bowmannites,* reconstruction of a single whorl of bracts and associated sporangiophores. (**A,** after Smith, with permission of McGraw-Hill Book Co.; **B,** redrawn after D. H. Scott; **C,** redrawn after Hirmer; **D,** redrawn after Remy; **G,** redrawn after Andrews and Mamay.)

Figure 16-4 *Tristachya*, a sphenophyll, showing whorls of leaves and axillary strobili, ×0.5. (Redrawn after Remy.)

plant. It is suggested that the outer surface of the trunk was smooth.

The branching pattern varies from whorled branches that emerge from every node to produce a dense crown, to a trunk on which only two branches emerge from each node. The trunk sometimes branches only near the top. Little primary xylem is produced, but secondary wood is strongly developed.

The vascular tissue of the stem consists of a cylinder of xylem surrounding a large central pith (Fig. 16-7). The primary xylem occurs as regular

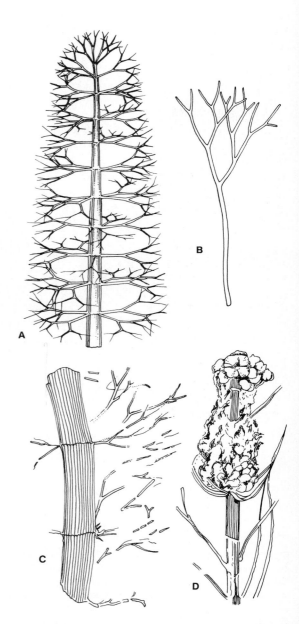

Figure 16-5 **A-D,** *Archaeocalamites*. **A,** reconstruction of a fragment of a stem, showing whorls of dichotomizing leaves, ×0.15. **B,** reconstruction of a dichotomizing leaf, ×0.5. **C,** a stem fragment with leaves, showing grooved internodes, ×0.25. **D,** a cone with sporangiophores, ×0.5. (Redrawn after Stur.)

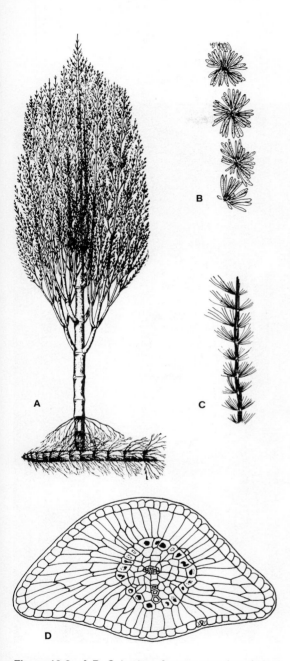

Figure 16-6 **A-D,** *Calamites*. **A,** reconstruction of plant, showing rhizome, roots, jointed aerial stem, and side branches bearing leaves. **B,** leaf whorls of *Annularia*, foliage of *Calamites*, ×0.5. **C,** leaf whorls of *Asterophyllites* (also foliage of some species of *Calamites*), ×0.5. **D,** transverse section of leaf. (**A,** redrawn after Hirmer; **B–C,** redrawn after Andrews; **D,** redrawn after Seward.)

bundles around the outside of the pith. Each bundle contains a hollow canal left behind by the disintegration of the endarch protoxylem. Secondary xylem, consisting of well-defined wood rays and tracheids with bordered pits, forms a cylinder outside the primary bundles. Adventitious roots emerged in whorls at the nodes, and some possibly formed prop roots to support the trunk. These have no nodes and contain exarch protoxylem.

Leaves are also whorled at the nodes of the ultimate branches (Fig. 16-6B, C), each whorl with 4–60 simple, single-nerved leaves. The leaves are either linear, lanceolate, or spatulate and vary from 0.5 cm to more than 5 cm long, although most are 0.5 cm long. Stomata are confined to the undersurface of the leaf and are not conspicuous. Sometimes the leaves are heavily cutinized.

The cones are borne singly at the nodes or, more usually, near the apex of the main shoot, or on upper branches (Fig. 16-8). Each of the cones is borne on a short lateral branch, and there are one or two at each node. They vary in length from 2 cm to more than 4 cm and are made up of a succession of whorls of sterile bracts alternating with whorls of sporangiophores. Bracts are fused into a disc at their bases, and sheathe the axis; the apex of the bract bends upward, offering some protection to the adjacent sporangiophores. Although sporangiophores are borne in vertical rows, bracts alternate in each successive whorl. Four sporangia are on each sporangiophore, attached to the inner surface of its peltate tip. The cone is usually homosporous, but in rare instances is heterosporous.

The calamites show parallel evolution of structures strongly resembling those of the arborescent lycopods: great stature, secondary growth, the nature of the fructifications and their position, and the appearance of heterospory. Selective pressures in a similar environment produced parallels in morphology that fitted entirely unrelated plants to this environment. Parallels in modern plants are numerous and impressive, but convincing parallels are less apparent in the fossil record.

The Modern Horsetails

Equisetum, the common horsetail or scouring rush, is the modest living representative of this once

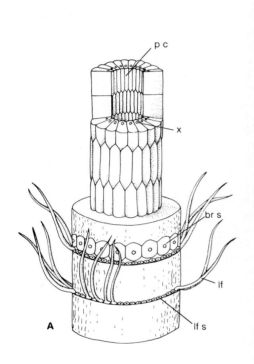

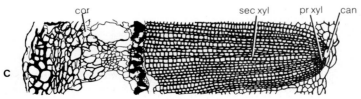

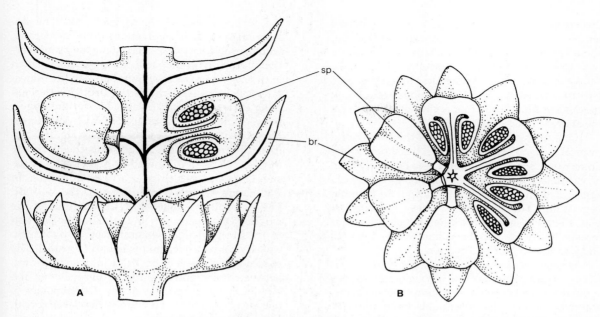

Figure 16-8 Reconstruction of a portion of a cone of *Calamostachys*. **A,** longitudinal view, showing whorls of bracts (*br*) and sporangiophores (*sp*), ×4. **B,** transverse sectional view of a single whorl of bracts and associated sporangiophores. (Redrawn after Zimmermann.)

Figure 16-7 A-C, *Calamites*. **A,** reconstruction of a stem segment, showing pith cavity (*p c*), xylem (*x*), branch scars (*br s*), whorls of leaves (*lf*), and leaf scars (*lf s*). **B,** photograph of transverse section of stem, showing pith cavity, primary xylem (*pr xyl*), and secondary xylem (*sec xyl*), × 1.6. **C,** part of stele, showing pith (*p*), canal (*can*) in the protoxylem, primary xylem (*pr xyl*), radiating rows of secondary xylem (*sec xyl*), and cortex (*cor*), × 10. (**A,** redrawn after Boureau.)

Sphenophytes (Horsetails and Their Allies) **437**

diverse group of plants (Figs. 16-9 to 16-12). In cool temperate climates of the Northern Hemisphere, some species of these rhizomatous perennials are widespread on sandy or gravelly soils of lake shores and roadsides. As invaders of gardens and pastureland, some species are noxious pests. Some species are found only in marginal shallows or on the shores of bodies of water.

One of the largest species is the New World tropical *E. myriochaetum*, in which the shoots sometimes reach 8 m in height and may be nearly 2 cm in diameter. Such slender shoots grow closely aggregated and are physically supported by other vegetation. The smallest species, *E. scirpoides* (Fig. 16-9H–J), is frequent in temperate coniferous forests; its shoots may be as small as 4–5 cm tall and less than 5 mm in diameter.

Several species of *Equisetum* are evergreen, and all species are perennial. The nonevergreen species sometimes have vegetative, chlorophyllose shoots, whereas the cone-bearing shoot lacks chlorophyll. For example, in *E. arvense* the unbranched, nonchlorophyllose, cone-bearing shoots emerge and pierce the soil surface in early spring. The vegetative, multibranched chlorophyllose shoots appear later; they mature after the spores have been shed, and the cone-bearing shoot dies back to the rhizome. Sometimes, later in the season, this species produces chlorophyllose shoots with terminal, miniature cones bearing well-formed sporangia.

All species of *Equisetum* have a markedly ridged stem. A thick septum is formed at each node, and the internodes are hollow. As in most sphenophytes, the ridges of each internode alternate with those of the internodes above and below. The ridges are usually heavily sclerified and impregnated with silica. At each node, there is an intercalary meristem, from which branches or roots can arise. The presence of this meristematic area somewhat weakens the shoots, and they break fairly easily. This fragmentation of the shoot facilitates vegetative reproduction, which occurs naturally in some aquatic species and is induced by zealous gardeners who unwittingly expand unwanted populations in their gardens.

Initially the stem is a simple protostele, but soon it becomes a siphonostele. A central pith is essentially absent, and the vascular bundles are in a single row around the circumference of the stele (Fig. 16-10). In transverse section, the internode of the stem reveals a cutinized epidermis. The stomata often occur in depressions in the epidermis, with the subsidiary cells overarching the guard cells. Longitudinally in the stem, the stomata are often in two vertical rows in regions between the ridges. Internally, the next layer is the strongly sclerified outer cortex, especially thickened over the ridges. Between the adjacent ridges are large air spaces, the **vallecular canals** (or cortical canals, Fig. 16-10B), that run vertically the length of each internode; these canals are surrounded by photosynthetic tissue. Opposite each ridge is a vascular bundle. In the mature stem is a **carinal canal** (Fig. 16-10B). This is another longitudinal **lacuna** running the length of the internodes and marking the position of the protoxylem. Frequently one or two thick-walled protoxylem elements bound this canal. The metaxylem consists of tracheids in two radial files; between these and toward the perimeter of the bundle is the primary phloem. Development is **mesarch**. Sometimes an endodermal layer is both external and internal to the vascular bundles; these layers are continuous around the circumference of the stem. Occasionally the endodermal layer surrounds each vascular bundle, or a single endodermal layer surrounds the stele.

The vascular system can be visualized as a ring of columns of vascular tissue that extend the length of the internodes, each column lying opposite an external ridge of the stem. These vascular bundles fuse at the node, and in the internode above, new columns of vascular bundles extend, alternating with those of the internode below. Tracheary elements

Figure 16-9 *Equisetum* sporophytes, showing diversity in habit. **A–E,** *E. sylvaticum*. **A,** strobilus-bearing shoot with whorls of chlorophyllose branches, ×1/6. **B,** vegetative shoot, ×1/6. **C,** strobilus with whorls of sporangiophores, ×2. **D,** detail of a whorl of branches of vegetative shoot, ×1. **E–G,** *E. hyemale*, an evergreen chlorophyllose species. **E,** strobilus, ×1. **F,** a single whorl of fused leaves, ×2. **G,** habit sketch, ×1/8. **H–J,** *E. scirpoides*, a small evergreen chlorophyllose species. **H,** habit sketch, ×1/2. **I,** strobilus, ×3. **J,** a single whorl of fused leaves, ×10.

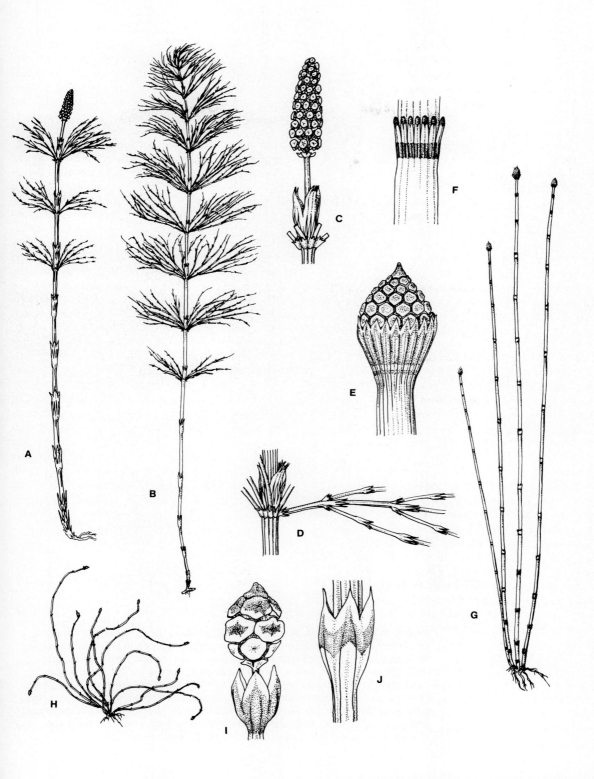

Sphenophytes (Horsetails and Their Allies)

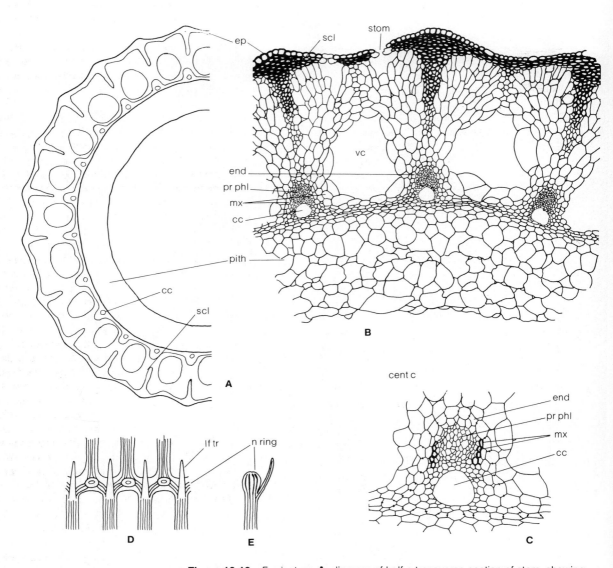

Figure 16-10 *Equisetum*. **A,** diagram of half a transverse section of stem, showing large central canal, carinal canals, vallecular canals, and bands of sclerenchyma, ×10. **B,** part of transverse section of *Equisetum* stem, showing details of tissues, ×80: *cc,* carinal canal; *cent c,* central canal; *end,* endodermis; *ep,* epidermis; *mx,* metaxylem; *pr phl,* primary phloem; *scl,* sclerenchyma; *stom,* stoma; *v c,* vallecular canal. **C,** part of transverse section of stem in area immediately surrounding vascular bundle, showing endodermis (*end*), carinal canal (*cc*), metaxylem (*mx*), and primary phloem (*pr phl*), ×250. **D, E,** drawing of part of decorticated node, showing nodal ring of xylem (*n ring*) and leaf traces (*lf tr*), ×30. (**D, E,** redrawn after Jeffrey.)

are variable. Those formed earlier are annular or helical; those formed later are often annular but have vertically to horizontally oriented secondary wall strands. Still others have circular bordered pits. Short vessels are found in the metaxylem elements of some species.

The leaves are always fused at the node to form a sheath that encloses the base of the internode just above the node (Fig. 16-9F, J). Leaf sheaths generally have little or no chlorophyll, and each leaf has a single, unbranched leaf trace. The leaf sheath protects the intercalary meristem. A sheath is composed of from 5 to 20 or more fused leaves. The number coincides with the number of ridges in the stem, and can vary somewhat from one node to the next. Along the midvein on the adaxial face of each leaf are **hydathodes,** through which water is secreted under high moisture conditions.

Equisetum usually has a subterranean rhizome, from which roots and erect shoots arise at the nodes. Occasionally, starch-rich tubers occur at the nodes of the subterranean stems. The roots are slender and fibrous and form a limited root system. Root anatomy differs conspicuously from that of the stem. The epidermis, one or two cells thick, surrounds a broad cortex. An endodermis, two cells thick, surrounds the central stele, which is a continuous cylinder made up of three to six protoxylem points that surround a single, central metaxylem element. Phloem occupies the area between the protoxylem points.

In most species of *Equisetum,* a single cone terminates the main shoot (Figs. 16-9C, E, I and 16-11A). Sometimes, if the main shoot apex is destroyed, short lateral branches arise in whorls from the nodes of the main shoot, and some of these lateral branches produce miniature apical

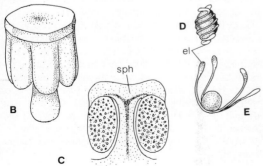

Figure 16-11 Reproductive organs of *Equisetum.* **A,** habit of cone-bearing plants of *E. telmateia.* **B,** single sporangiophore, showing axis, peltate disc, and recurved sporangia, ×12. **C,** vertical section through single sporangiophore *(sph),* showing masses of spores inside two sporangia, ×15. **D, E,** views of spores with elaters *(el)* tightly curled about endospore in **D,** and elaters expanded in **E,** × 120. (**B–E,** redrawn after Haupt.)

cones. In *E. myriochaetum*, all cones are normally borne on these lateral branches.

The cone is made up of several whorls of sporangiophores (Fig. 16-11A). Each sporangiophore has a short stalk, and the outer portion is expanded into a flattened disc lying at right angles to the stalk (Fig. 16-11B, C). From this disc, five to ten elongate sporangia hang downward, parallel to the stalk of the sporangiophore. When mature, each sporangium opens by a longitudinal line. Those that mature first appear to be in the widest part of the cone. The sporangium wall is two cells thick at maturity, and the cell walls of the outer layer bear spiral thickenings. The spores are rich in chlorophyll. Their outer coat is deposited in four strips, tightly coiled around the spore and attached at a common point (Fig. 16-11D). When the spore mass dries out, these strips of spore wall, termed **elaters,** uncoil and agitate the spore mass, loosening it for wind and gravitational dissemination. The intertangling of elaters also means that small groups of spores are released together. The spores have a brief span of viability, surviving only a few days.

The germinating spore ultimately produces a ruffled, chlorophyllose thallus, resembling some members of the hepatic order Metzgeriales (Fig. 16-12A). The thallus is usually less than 1 cm in diameter, but can be up to 3 cm in diameter. It consists of a multistratose horizontal body, from which chlorophyllose flaps emerge. Rhizoids attach the

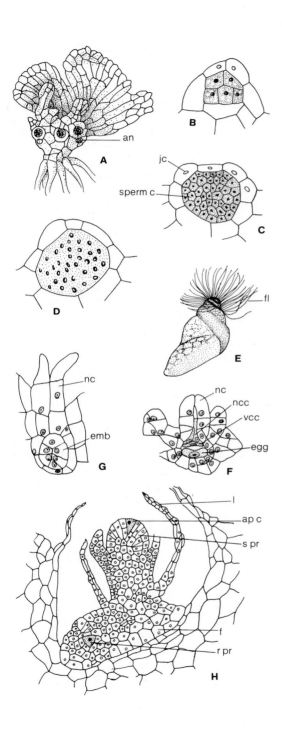

Figure 16-12 A–H, *Equisetum*. **A,** lateral view of prothallus, showing antheridia *(an)* toward base of lobes, ×25. **B–D,** stages of maturation of antheridia, showing jacket cells (*j c*) and spermatogenous cells (*sperm c*), ×200. **E,** single sperm with flagella (*fl*), ×1250. **F,** vertical section through archegonium, showing neck cell (*nc*), neck-canal cells (*n c c*), ventral canal cell (*v c c*), and egg, ×219. **G,** vertical section through archegonium, showing early stage of embryo *(emb)* in venter, ×210. **H,** late embryo, showing stem primordium (*s pr*) with juvenile leaves (*l*) and large apical cell (*ap c*), foot (*f*), and root primordium (*r pr*), ×120. (**A,** redrawn after Walker; **B–D,** redrawn after Smith; **E,** redrawn after Sharp; **F,** redrawn after Jeffrey; **G,** redrawn after Sadebeck; **H,** redrawn after Smith.)

thallus to the substrate. Usually the thallus initially produces either antheridia or archegonia. Generally, an antheridium-producing gametophyte produces only antheridia. However, a gametophyte that produces archegonia first usually produces antheridia later in the season. In nature, the gametophytes appear to be short-lived, but in culture some gametophytes persist for more than two years. Sex organs are embedded in the upper surface of the thallus (Fig. 16-12A). Each antheridium contains numerous sperms that are spirally coiled and multiflagellate (Fig. 16-12E).

The genus has few practical uses. Many gardeners abhor it as an aggressive weed, but Japanese gardeners use *Equisetum hyemale* as an ornamental that will survive even in small plots on the city sidewalks. Pioneer settlers used the silica-impregnated shoots for scouring dirty tinware and as a substitute for sandpaper for polishing wood. Some species are planted to stabilize dike banks. Aboriginal medicine men apparently threw shoots of *E. hyemale* into the fire near an ailing patient; the popping explosion of the shoots was thought to stimulate the patient.

The chromosome number in *Equisetum*, $n = 108$, is remarkably high. The uniformity in chromosome number enhances hybridization in the genus; thus natural hybrids are relatively frequent. Despite their sterility, such hybrids are often very widely distributed, both by vegetative propagation and by frequently recurring wherever the parents are together.

Further Fossil Sphenophytes

Schizoneura (Fig. 16-13), which probably reached heights of over a meter, appears to have been an herbaceous plant of swamps or shallow ponds. It occurred in Europe, Asia, and southern Africa from the late Permian to the Jurassic. The stems are relatively slender (1–2 cm in diameter) and grooved, with the grooves of one internode continuous with those of the next, as in *Archaeocalamites*. The branches are in whorls at the nodes, and there are from two to ten, depending on the species.

Near the ends of young shoots, leaves form a cylindrical sheath made up of 10–18 individual fused

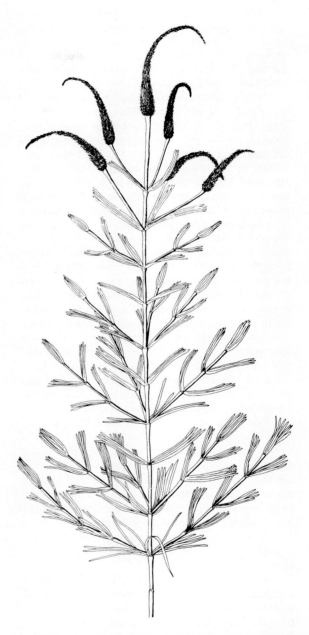

Figure 16-13 Reconstruction of the Triassic sphenophyte, *Schizoneura paradoxa*, ×0.2. (Redrawn after Mägdefrau.)

leaves, each with its own midrib. Further down the shoot, these leaves split into two flattened lobes that flare outward on opposite sides of the stem. On older portions of the stem, the leaves split into individual linear blades that form a widely flaring whorl at the node. The linear leaves are up to 10 cm long, and each blade is 2–4 mm wide; thus in the two-lobed sheath, each lobe is sometimes more than 2 cm wide.

In *S. paradoxa*, the fructifications were terminal on branches. Each is catkinlike and up to 15 cm long. They appear to have been made up of whorls of linear sterile bracts, with sporangiophores in the internodes above them, but none has been preserved in sufficient detail to be certain of their structure. On the other hand, in *S. manchuriensis*, the cones are remarkably large and relatively well preserved. These cones were several centimeters long and were thus rather massive. It is therefore suggested that, as in some extant species of *Equisetum*, the cones were borne on a shoot separate from the vegetative shoot and arose directly from the rhizome. These cones were made up of whorls of cuplike sterile sheaths composed of fused leaves; on the internodes between the sheaths, longitudinal rows of crowded sporangiophores were arranged in six or more whorls per internode. Each sporangiophore was made up of a polygonal disc that bore up to six homosporous sporangia. In *S. gondwanensis*, the cones show no evidence of any sterile bracts among the whorls of sporangiophores; they resemble *Equisetum* in this respect.

The great diversity in cone structure in *Schizoneura* strongly suggests that this form genus is very complex. In this group of species, all of which share the general growth form, there are similarities to *Equisetum* (e.g., fused leaf bases) and to more ancient sphenophytes (e.g., nature of the ribbing; large, presumably chlorophyllose leaves; sterile bracts among the sporangiophores).

As further details become available, the evolutionary significance of the genus will become more apparent.

The Sphenophytina, then, form an isolated evolutionary line, showing many morphological features not repeated in either ferns or fern allies: the jointed stems, intercalary meristems, nature of the siphonostele, and nature of the sporangiophores, as well as other features.

Within the sphenophytes, there are several notable evolutionary trends. In the leaves, for example, the complex dichotomizing leaves of *Archaeocalamites* are replaced in *Sphenophyllum* by a flattened leaf, still with a dichotomizing vascular system but forming a broad blade. In *Calamites*, the leaf is greatly simplified, with an unbranched vascular system; in *Schizoneura*, the simple leaves are initially fused and presumably chlorophyllose; in *Equisetum*, the simple leaves still have a single trace, but the leaves have fused and have virtually lost their photosynthetic function. There are similar evolutionary trends exhibited in the nature of the cones and their compound sporangiophores, in the details of stem structure, and in the habit of the plants.

References

Abbot, M. L. 1958. The American species of *Asterophyllites, Annularia* and *Sphenophyllum. Bull. Amer. Paleont.* 38:289–90.

Anderson, B. R. 1954. A study of American petrified *Calamites. Ann. Missouri Bot. Gard.* 41:395–418.

Baxter, R. W. 1948. A study of the vegetative anatomy of the genus *Sphenophyllum* from American coal balls. *Ann. Missouri Bot. Gard.* 35:209–31.

Bonamo, P. M., and Banks, H. P. 1966. *Calamophyton* in the Middle Devonian of New York State. *Amer. J. Bot.* 53:778–91.

Boureau, E., ed. 1964. *Traité de paléobotanique.* Vol. 3. *Sphenophyta, Noeggerathiophyte.* Masson & Cie., Paris.

Dawson, J. W. 1871. On the structure and affinities of *Sigillaria, Calamites* and *Calamodendron. Quart. J. Geol. Soc.* 27:147–61.

Eggert, D. A. 1962. The ontogeny of Carboniferous arborescent Sphenopsida. *Palaeontographica* 110B:99–127.

Hauke, R. L. 1963. A taxonomic monograph of the genus *Equisetum*, subgenus *Hippochaete. Nova Hedwigia* (Beihefte) 8:1–123.

———. 1966. A systematic study of *Equisetum arvense. Nova Hedwigia* 13:81–109.

———. 1969. The natural history of *Equisetum* in Costa Rica. *Rev. Biol. Trop.* 15:269–81.

Leclercq, S. 1961. Strobilar complexity in Devonian sphenopsids. *Adv. Bot. IX Int. Bot. Congress.* 2:968–71.

Page, C. N. 1972. An interpretation of the morphology and evolution of the cone and shoot of *Equisetum*. *Bot. J. Linn. Soc.* 65:359–97.

Schaffner, J. H. 1925. Main lines of evolution in *Equisetum*. *Amer. Fern. J.* 15:8–12, 35–39.

———. 1930. Geographic distribution of the species of *Equisetum* in relationship to their phylogeny. *Amer. Fern J.* 20:89–106.

Schweitzer, H. J. 1967. Die Oberdevon-flora der Båreninsel, part 1: *Pseudobornia ursina* Nathorst. *Palaeontographica* 120B: 116–37.

The pterophytes, or Pterophytina, comprise the largest subdivision in the plant kingdom. It contains all ferns and their allies as well as all seed plants. The Pterophytina first appeared in the Middle Devonian period. Although innumerable species, genera, and even families of Pterophytina have become extinct, at the present time it shows its greatest diversity and abundance and occupies nearly every environment where living things can survive.

Characteristics of Pterophytina

Sporophyte

The subdivision Pterophytina is characterized by the following features of the sporophyte:

1. They bear, in most instances, a complex flattened leaf (or leaves) consisting of a blade and **petiole** in most, and supported by a network of vascular strands. The leaves are generally the main photosynthetic organ. They usually show much tissue differentiation and are equipped for gas exchange, controlled mainly by the presence of characteristic pores, termed **stomata**. Leaves vary greatly in size and complexity. They reach lengths of many meters in some ferns and seed plants, and the blade is simple or much-divided.

2. The vascular strand, as it enters the leaf, leaves a gap of nonvascular tissue in the **stele**, termed a **leaf-gap**.

3. Many pterophytes show secondary growth, the sporophyte is perennial, and the stem is often **woody**. There are also many annual flowering plants in which there is little secondary growth and the stems are **herbaceous**.

4. The stele shows much diversity in structure, especially among ferns, in which virtually every known stele type is represented. Among seed plants, the stele shows a

17

PTEROPHYTES

remarkable constancy in general design, although detailed cellular structure is diverse. Among flowering plants, steles of **monocotyledons** and **dicotyledons** are conspicuously different.

5. Sporangia are often borne on **sporophylls** or modified sporophylls. Although there is little dispute concerning ferns, sporangia of seed plants are borne on structures that are generally not interpreted as sporophylls.

6. Most pterophytes have a root system; its primary functions are anchorage and absorption of water and dissolved nutrient minerals.

7. Many pterophytes, particularly seed plants, have an elaborate **embryo** with much structural differentiation; in ferns and allies, the embryo is less elaborate.

8. In most ferns, **homospory** predominates; in seed plants, sporophytes are exclusively **heterosporous**.

9. Vegetative reproduction of the sporophyte is common; diverse structural modifications enhance vegetative reproduction.

10. Sporophyte size varies from nearly microscopic to trees more than 100 m tall.

Gametophyte

In pterophytes, gametophytic features are less easy to generalize:

1. The gametophyte is generally annual and of brief duration.

2. Occasionally it is **exosporic**, but in seed plants it is **endosporic**.

3. It shows little structural differentiation; among ferns and allies it is generally a photosynthetic thallus, but in seed plants it is reduced to a few cells and is nonphotosynthetic.

4. Among ferns, the gametophyte generally bears both male and female sex organs; in seed plants, gametophytes are exclusively unisexual.

5. Sperms are flagellate in ferns and more primitive seed plants, but in most seed plants sperms lack any motile organ whatsoever, and thus their transfer involves direct contact between male and female gametophytes.

Diversity

The pterophytes form the conspicuous vegetational cover of the earth. Their structural and physiological diversity permits them to occupy a wide range of environments. Pterophytes are the vital link in the food chains of most terrestrial organisms since they are the major primary producers on land. They also supply shelter for most organisms against both environmental extremes and predation by other organisms. The vast diversity in sporophytic design has profoundly influenced the cultural development of mankind.

Leaves Compared to other land plants, pterophytes have undergone a number of remarkable specializations. The concentration of photosynthetic activity largely in the leaves has brought vast diversification to the nature of the leaf, both in the complexity of its gross morphology and in its internal anatomy. This has shown several lines of selection:

1. **Dichotomous** venation to **netted** venation

2. Segregation of two main trends in netted venation: (a) those in which the main veins are parallel, and (b) those in which the main veins are strongly **divergent** from a main central axis of the leaf

3. Compound leaves or simple leaves

4. Leaves restricted to photosynthesis versus those in which water and food reserves can be stored

5. Long-persistent **evergreen** leaves versus early **deciduous** leaves

6. Leaves that can serve as vegetative propagants

7. Leaves that make up the bulk of the sporophyte, and some which bear the sporangia

8. Leaves that are scalelike and protect the apical and lateral meristems of the stem

Numerous specializations appear in leaves, particularly in those of the flowering plants, in which they are sometimes insectivorous organs or soil-and-water traps. Others are brightly colored and serve as bracts that attract animals to the sporangia.

Sporangia There has also been considerable elaboration of the sporangia, which has generally favored the dissemination of spores. These elaborations are not conspicuous in homosporous pterophytes, since gametophytes are exosporic, usually **monoicous** (bisexual), autotrophic, and have flagellated sperms. In these plants, the spores are usually airborne, and structural modifications enhance exposure of the dehiscing sporangium to moving air.

Gametophytes In heterosporous pterophytes, where the gametophytes are essentially endosporic and **dioicous** (unisexual), the gametophytes must be close together so that male gametes can reach female gametes.

Selection has tended to favor a completely dependent and protected female gametophyte that remains within enveloping structures while attached to the sporophyte. On the other hand, male gametophytes are released from the sporangia and must be borne to a receptive surface near the female gametophyte before male gametes are released. In heterosporous ferns, such elaborations also involve getting the entire male gametophyte to the female gametophyte and trapping it there until sperms are released. In seed plants, this process of carrying male gametophytes and depositing them on a receptive surface near female gametophytes is termed **pollination.** Wind, water, and animals are variously involved in pollination. Structural modifications of sporangia and their enveloping organs influence the behavior of the pollinator as well as the reception of the pollen and the way male gametes are ultimately released.

Since seed plants have female gametophytes attached to the sporophyte when fertilization occurs, the new sporophyte that arises from each of these gametophytes must be released from the parental sporophyte. Diverse modifications favor release of these young sporophytes in their enclosing **seed**; many involve water, wind, fire, animals or simply gravity.

Other Modifications

Many modifications permit the sporophyte to survive an unfavorable period of time. These modifications allow pterophytes to occupy a great diversity of environments. They usually involve storage of food reserves in stems, roots, or leaves. However, in pterophytes the life history is sometimes annual and is best represented in flowering plants. The **herbaceous habit,** although not confined to pterophytes, is frequent.

Compared to that of other vascular plants, the vascular system of pterophytes has undergone much diversification. Although there are woody vascular plants in other subdivisions, it is unlikely that the sporophytes of any fossil vascular plants had a life-span as extended as do extant pterophytes; some can live 4000 years or more, and many live for several centuries.

Reduction of the gametophyte in size and structure has reached its extreme in the pterophytes. All heterosporous pterophytes have endosporic gametophytes, and are often of few cells. In angiosperms, the entire male gametophyte can be reduced to as few as two cells, one of which divides to produce two gametes. The female gametophyte can be reduced to as few as four cells, one of which serves as an egg.

The Pterophytina include the most successful plants that occupy the land, occurring in nearly every habitat available, including marine and freshwater aquatic habitats. It is the pterophytes that are the major agricultural crops and on which most research has been concentrated in the plant kingdom. Despite this, our knowledge of these fascinating organisms remains imperfect. Enrichment of this knowledge will challenge botanists for many centuries.

References

Andrews, H. N. 1961. *Studies in paleobotany.* John Wiley & Sons, New York.

———. 1963. Early seed plants. *Science* 142:925–31.

Banks, H. P. 1970. *Evolution and plants of the past.* Wadsworth Publ. Co., Belmont, Calif.

Bierhorst, D. W. 1971. *Morphology of vascular plants.* Macmillan, New York.

Bold, H. C.; Alexopoulos, C. J.; and Delevoryas, T. 1980. *Morphology of plants and fungi.* 4th ed. Harper & Row, New York.

Bower, F. O. 1908. *Origin of a land flora.* Macmillan & Co., London.

Campbell, D. H. 1940. *The evolution of the land plants (Embryophyta).* Stanford Univ. Press, Stanford, Calif.

Delevoryas, T. 1962. *Morphology and evolution of fossil plants.* Holt, Rinehart and Winston, New York.

Eames, A. J. 1936. *Morphology of vascular plants: lower groups.* McGraw-Hill Book Co., New York.

Foster, A. S., and Gifford, E. M. 1974. *Comparative morphology of vascular plants.* W. H. Freeman and Co., San Francisco.

Goebel, K. 1905. *Organography of plants, especially of the archegoniatae and spermatophyta* (English ed., transl. L. B. Balfour). Oxford Univ. Press, London.

Ingold, C. T. 1939. *Spore discharge in land plants.* Clarendon Press, Oxford.

Lawrence, G. H. M. 1951. *Taxonomy of vascular plants.* Macmillan, New York.

Meeuse, A. D. J. 1965. Angiosperms—past and present. *Adv. Frontiers Plant Sci.* (New Delhi) 11:1–228.

Spore, K. R. 1969. *The morphology of gymnosperms.* Hutchinson Univ. Library, London.

———. 1970. *The morphology of pteridophytes.* Hutchinson Univ. Library, London.

Taylor, T. N. 1981. *Paleobotany.* McGraw Hill Book Co., New York.

18

CLADOXYLIDS AND COENOPTERIDS (Preferns)

The cladoxylids (class Cladoxylopsida) and coenopterids (class Coenopteridopsida) are known only as fossils. Preserved material consists of fragments of axes and leaflike organ systems. The two classes are highly artificial and are probably not closely related. As further details of the structure of these organisms become known, it is likely that many will be placed among the Filicopsida, and others transferred to other classes. They appear to be derived from the Trimerophytina, showing particular resemblances to the morphology of *Pertica*. The internal anatomy of their axes is highly distinctive: Cladoxylopsida has a much-dissected stele, and Coenopteridopsida has a single, three- or five-lobed stele of striking morphology.

Class Cladoxylopsida

The class Cladoxylopsida includes at least ten genera and appeared first in Middle Devonian, represented by the unusual plant *Cladoxylon scoparium* (Fig. 18-1). Regrettably, the material is fragmentary; thus the growth form of the entire plant remains unknown. The fragments that exist suggest a plant that was at least 22 cm tall, and probably taller. The main axis was 1.5 cm broad, and the narrowest axes were less than 3 mm broad. The branching of the most robust axes shows that several branches arose simultaneously. These axes bear radially arranged, flattened appendages that are interpreted as either **pinnules** or leaves (Fig. 18-1B, C). Near the ends of some narrower branches are fan-shaped sporophylls or sporangium-bearing pinnules (Fig. 18-1D). These sporophylls are widely spaced and bear sporangia embedded in the tips of each division. The spores are assumed to have been homosporous.

Stem anatomy consists of a rather complex, dissected stele (Fig. 18-1E). In transverse section, it shows 10–33 curved or branched bundles with free ends radiating outward. These are made up of metaxylem and include tracheids that are scalariform, reticulate, or (rarely) pitted. There is a well-developed cortex.

The stele and form of the plant are similar in *Calamophyton* (Fig. 18-2), except that the ultimate branches of this plant appear to dichotomize. As in *Cladoxylon*, leaflike appendages appear to be spirally arranged; they are terete, dichotomize several times, reach lengths up to 1 cm (Fig. 18-2B), and have a single vascular strand. Fertile appendages are aggregated on a specialized branch and are also near the apices of main shoots. They, too, are radially arranged, but each fertile structure forks once or twice and bears a recurved pair of sporangia laterally on the main axis of the appendage (Fig. 18-2C). These are homosporous and dehisce by a longitudinal line. *Calamophyton* shows some similarity to the structurally simplest Sphenophytina, but despite the sporangiophorelike structure, it is included in the Cladoxylopsida.

The next most completely understood cladoxylid is the genus *Pseudosporochnus* (Fig. 18-3). The elegant reconstruction by Leclercq and Banks (1962) shows *P. nodosus* as a tree 2–3 m tall with a trunk 5–8 cm in diameter. The trunk surface was covered by small **nodules** of sclerenchyma, compacted over each other on the bark. The top of the trunk was crowned by several branches, each of which had an unbranched stem for approximately 30 cm. These branched several times simultaneously, and final branches were up to 32 cm long and sometimes forked. Branches also bore nodules on the bark as well as leaflike appendages radially arranged. These leaves were compound, with narrow segments, and were about 4 cm long. They bore two to four pairs of subopposite **pinnae** and were terminated by a series of dichotomous segments. The fertile segments terminated in ovoid sporangia (Fig. 18-3C). The main stem bore a basal series of roots.

The branches of *P. nodosus* resemble fragments of other cladoxylids, and the polystelic anatomy of the stele is also similar. The main branches suggest precursors of complex compound leaves, as in coenopterids; these resemble the large **fronds** of ferns.

Figure 18-1 *Cladoxylon scoparium*. **A,** reconstruction of plant, showing radially arranged appendages, and sporangium-bearing "pinnules", × 2.5. **B, C,** sterile "appendages" or "pinnules", × 0.3. **D,** sporangium-bearing appendage with "pinnules", × 15. **E,** xylem outline of stem in transverse section, × 8. (Redrawn after Krausel and Weyland.)

452 Chapter 18

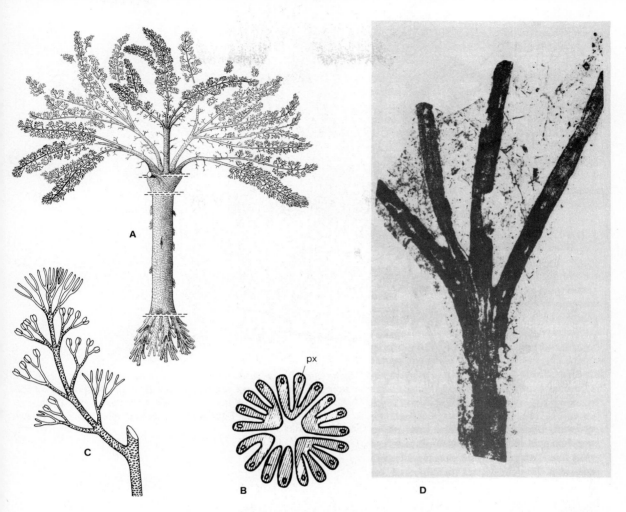

Figure 18-3 *Pseudosporochnus.* **A,** reconstruction of whole plant, showing roots, columnar stem, digitate first-order branches, and leafy appendages, × 0.003. **B,** pattern of vascular bundles as found in a third-order branch, showing the location of protoxylem *(px),* × 4; **C,** third-order branch, showing leafy appendages and sporangia, × 0.3; **D,** stem and digitate branching, × 0.75. (After Leclercq and Banks, with permission of *Palaeontographica.*)

Figure 18-2 *Calamophyton.* **A,** showing digitate branching of first order, dichotomous or monopodial branching of higher orders, and sterile and fertile appendage, ×1. **B,** sterile appendage, showing first dichotomy and three secondary branches that develop on each primary, ×10. **C,** fertile appendage, showing two orders of branching and six sporangia on each secondary branch, ×10. (After Leclercq and Andrews, with permission of Missouri Botanical Garden.)

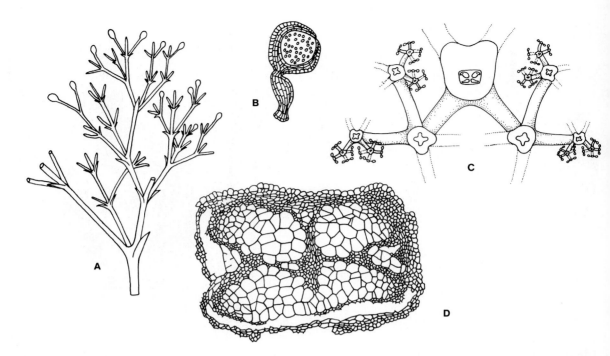

Figure 18-4 *Stauropteris oldhamia.* **A,** reconstruction, showing branches arising in pairs, scalelike aphlebiae, and terminal sporangia, × 1. **B,** longitudinal section of sporangium, × 10. **C,** branching pattern of frond, as viewed from above, × 7. **D,** stele of the major axis, × 30. (**A,** redrawn after Delevoryas; **B,** redrawn after Scott; **C,** redrawn after Hirmer; **D,** redrawn after Tansley.)

Figure 18-5 **A–E,** details of the coenopterid *Rhacophyton ceratangium.* **A,** restoration of a representative portion of an erect axis with determinate appendages; the pairs of circles shown near the bases of fronds represent the position of sporangium-bearing portions, shown in detail in **C,** ×0.25. **B,** reconstruction of a single unit of a frond, showing vegetative and sporangium-bearing pinnae on the *right* and two aphlebiae on the *left,* × 2/3; **C,** detail of sporangium-bearing portion of "pinnule," × 2. **D,** detail of paired sporangia, × 8. **E,** a single sporangium, × 17. (From H. N. Andrews and T. L. Phillips, by permission of the Council of The Linnaean Society of London.)

Cladoxylids and Coenopterids (Preferns) 455

Class Coenopteridopsida

Coenopteridopsida are represented by more than 15 genera that occurred from Upper Devonian until Permian. Most are fernlike in general morphology, with megaphyllous leaves, sporangia on some of these leaves, **circinate vernation,** and **adventitious** roots. The stems are typically solid protostelic, but there are some siphonostelic forms. Secondary xylem is known in a single genus.

In coenopterids, as in cladoxylids, it is difficult to distinguish stem from leaf. Unfortunately, the fossil material in coenopterids in usually too fragmentary to reconstruct the gross morphology of the whole plant. It is customary to treat determinate branches that bear pinnalike organs as **fronds,** comparing them to the leaves of ferns. They are suggested to be precursors of fern fronds, but the single, vascular leaf trace emerges from the stem without leaving a leaf gap. Sometimes the vascular strand of the main stem is different in morphology from that of the axis (or **rachis**) of the frond. The transverse sectional anatomy of the stele of the rachis is used to separate various families. This results in a classification that cannot be compared with fern classification, since in ferns the sporangial structure is of paramount significance in classification. When sporangial information is discovered that relates to the stelar anatomy types, the fossil group can be transferred to the appropriate class. This has occurred recently with the families Botryopteridaceae, Tedeleaceae, and Anachoropteridaceae, which were previously considered to be coenopterids but are now treated as fossil families of true ferns.

These fossils are particularly abundant in **coal balls** found in deposits of coal. These are concretions of calcium carbonate that often contain abundant plant remains. Most of the fossil remains are vegetative parts, but some fertile parts are known.

Coenopterids are a large and diverse group; some may have been precursors of true ferns, and others apparently led to evolutionary dead ends. In any event, the group occupies a critical position among vascular plants, and as further knowledge accumulates, their relationships become more apparent. The most comprehensive recent studies have been by Eggert (1964), Phillips (1974), and Dennis (1974), whose papers should be consulted for details.

Stauropteris (Fig. 18-4), of Carboniferous age, in many respects resembled Rhyniophytina in general appearance. The plants were probably bushy and small, with main axes attaining a diameter of no more than 3 mm. Delevoryas has reconstructed *S. oldhamia* (Fig. 18-4A, C) as an axis trichotomizing at various intervals and bearing at each **trichotomy** a pair of **aphlebiae** (scalelike structures with a single vascular strand). Terminating the ultimate branches are elongate sporangia. This plant was apparently homosporous; the sporangia lacked an annulus, but a **stomium** occurred at the sporangial apex (Fig. 18-4B). *S. burntislandica* was similar, except that it was heterosporous. The microsporangia contained many spores, but the megasporangia contained two large megaspores plus remnants of two aborted spores.

In *S. oldhamia*, the stele of the main axis was divided into four parts. Further up the smaller branches, the stele was four-lobed or three-lobed, and in the finest divisions, it was reduced to a few tracheids (Fig. 18-4D).

In *Rhacophyton ceratangium*, the axis is shown as a slender, elongate stem with spirally arranged leaves of two types: sterile and fertile (Fig. 18-5). Sterile fronds consist of a petiole bearing two ranks of primary pinnae, and an elongate, vegetative frond bearing pinnae, each of which, in turn, bears small, dichotomously branching systems helically arranged around the pinna (Fig. 18-5B). The fertile frond is larger and more complex, and it bears four ranks of primary pinnae arising in pairs from opposite sides of the petiole. These primary pinnae are of two types: two are elongate with bilaterally arranged secondary pinnae, each of which is a reduced dichotomizing system; the other two are shorter dichotomizing systems, in which the ultimate dichotomies bear numerous small, **fusiform** sporangia (Fig. 18-5C–E).

The stem was approximately 1 cm thick and probably attained a meter in height, with the longest fronds up to 30 cm long. The plant was probably **procumbent** or scrambling, since the stele is too slender to support such an assortment of large fronds on an erect stem.

Zygopterid ferns represent the largest array of coenopterids and are found from Devonian to Permian. Some species of *Zygopteris* are visualized as rhizomatous plants producing numerous dichotomous branches, with numerous roots emerging randomly. The rhizome was clothed by spirally

Figure 18-6 Reconstruction of a zygopterid fern, *Zygopteris berryvillenis*, × 0.15. (From Dennis, with permission of *Palaeontographica*.)

borne, **digitate** aphlebia pairs, and the petioles emerged in two lateral ranks. These also bore paired aphlebiae and probably produced primary pinnae in four ranks. The sporangia were elongate and apparently eusporangiate and were produced in clusters on the abaxial surface of the pinnules or at the frond tips. The plants were rather small, with the rhizome usually less than 1 cm in diameter and the frond rachis often less than 1 cm in diameter. The reconstruction of *Z. berryvillensis* (Fig. 18-6) presented by Dennis (1974) illustrates this morphology. *Z. primaria* was apparently arborescent, with a trunk up to 20 cm in diameter, but this consisted mainly of petioles and roots. *Zygopteris* produced secondary wood, as in the extant grape fern *Botrychium*.

Fragmentary rachis material is described under the form genus *Etapteris*, and the vascular strands are H-shaped in transverse section. Since the **stelar** morphology varies in different ontogenetic stages and in different parts of the stem and rachis, this leads to problems in interpretation, and has caused considerable inflation in the names applied to specimens.

A number of remarkable sporangia are associated with coenopterid fossils. Some of these are in **synangia,** and others are borne in terminal clusters on axes (Fig. 18-7).

Coenopterids were clearly a diverse group of plants; the available evidence merely hints at this diversity. It is little wonder that they have fascinated many paleobotanists. Each new fragment discovered further improves the reconstruction proposed. Among others, this group of fossil plants has tested the knowledge and ingenuity of botanists seeking insight from the known morphology and anatomy of extant ferns to assist in interpretation of fossil groups. They are especially important in understanding the origin of megaphyllous leaves.

References

Andrews, H. N., and Phillips, T. K. 1968. *Rhacophyton* from the Upper Devonian of West Virginia. *J. Linn. Soc. Bot.* 61:37–64.

Banks, H. P. 1964. Putative Devonian ferns. *Mem. Torrey Bot. Club* 21:10–25.

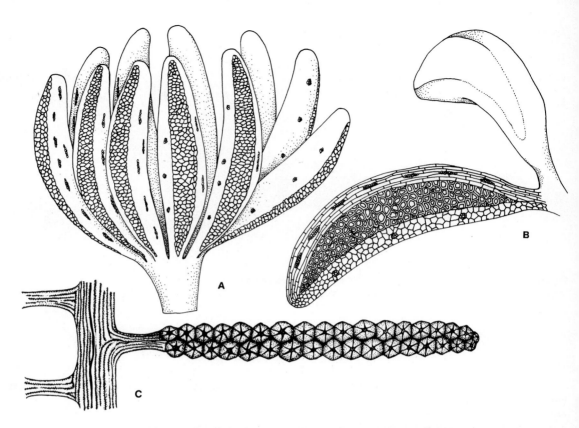

Figure 18-7 Sporangium-bearing organs of coenopterids. **A,** *Monoscalitheca*, reconstruction of a group of terminal sporangia, showing longitudinal annulus on each sporangium, × 30. **B,** *Biscalitheca* sporangia, reconstruction, showing longitudinal annulus, × 28. **C,** *Corynepteris*, pinna with sporangia, × 5. (**A,** redrawn after Abbott; **B,** redrawn after Mamay; **C,** redrawn after Bailey.)

Bonamo, P. M., and Banks, H. P. 1966. *Calamophyton* in the Devonian of New York state. *Amer. J. Bot.* 53:778–91.

Boureau, E., ed. 1970. *Traité de paléobotanique.* IV (I). *Filicophyta.* Masson & Cie., Paris.

Delevoryas, T. 1962. *Morphology and evolution of vascular plants.* Holt, Rinehart and Winston, New York.

Dennis, R. L. 1974. Studies of Paleozoic ferns: *Zygopteris* from the Middle and Upper Pennsylvanian of the United States. *Palaeontographica* 148B:95–136.

Eggert, D. A. 1964. The question of the phylogenetic position of the Coenopteridales. *Mem. Torrey Bot. Club* 21:38–57.

Leclercq, S. 1951. Étude morphologique et anatomique d'une fougère du Devonien Supérieur. Les *Rhacophyton zygopteroides* nov. sp. *Ann. Soc. Geol. Belgique Mém.* 9:1–92.

Leclercq, S., and Andrews, H. N. 1960. *Calamophyton bicephalum* a new species from the Middle Devonian of Belgium. *Ann. Missouri Bot. Gard.* 47:1–23.

Leclercq, S., and Banks, H. P. 1962. *Pseudosporochnus nodosus* sp. nov. a Middle Devonian plant with cladoxylalean affinities. *Palaeontographica* 110B:1–34.

Leclercq, S., and Lele, K. M. 1968. Further investigation on the vascular system of *Pseudosporochnus nodosus* Leclercq and Banks. *Palaeontographica* 123B:97–112.

Leclercq, S., and Schweitzer, H. J. 1965. *Calamophyton* is not a sphenopsid. *Bull. Acad. Roy. Belgique, Sci.* 11A:1395–1403.

Long, A. G. 1967. Some specimens of *Cladoxylon* from the calciferous sandstone series of Berwickshire. *Trans. Roy. Soc. Edinburgh* 68:45–61.

Matten, L. C. 1974. The Givetian flora from Cairo, New York: *Rhacophyton, Triloboxylon* and *Cladoxylon. Bot. J. Linn. Soc.* 68:303–18.

Phillips, T. C. 1974. Evolution of vegetative morphology in coenopterid ferns. *Ann. Missouri Bot. Gard.* 61:427–61.

19
FILICOPSIDS (TRUE FERNS)

General Features

True ferns (filicopsids) belong to the class Filicopsida. Fern sporophytes have leaves, stems and, generally, roots. The leaves are usually evergreen.

Fern leaves are generally spirally arranged and usually much divided; a pinnate leaf is common and is usually termed a **frond.** The midrib of the frond is an extension of the petiole, and is termed the **rachis.** If the blade is subdivided, the leaflets are termed **pinnae.** These pinnae can be further dissected into **pinnules,** and the pinnules themselves may be subdivided. The fronds are usually bilaterally symmetrical and are often parchment-like in texture, but can be filmy or leathery. **Vernation** varies from open dichotomous to netted. The leaf is usually tightly coiled when immature and uncoils as it grows. Since uncoiling usually coincides with spring in temperate regions, this type of frond is said to have **circinate vernation** (Fig. 19-1). Sometimes pinnae and even pinnules are also coiled. It sometimes takes up to four years from the time of leaf initiation until it is fully mature.

In primitive ferns, the sporangia are **eusporangiate,** but in most ferns they are **leptosporangiate** (Fig. 19-2E–I). A leptosporangiate sporangium originates from a single, superficial initial cell rather than a mass of cells, as is the case in the **eusporangium.** Also, in the **leptosporangium,** the sporangial jacket is unistratose, rather than multistratose, as in most eusporangia (Fig. 19-2A–D). The distinctions between eusporangiate and leptosporangiate are sometimes blurred in some ferns. Most ferns are homosporous, but a few aquatic ferns are markedly heterosporous.

Sporangia are usually clumped in masses called **sori** on the **abaxial** surface of the frond. Often the **sorus** is protected by a flap of tissue termed the **indusium.** A ring (**annulus**) of thick-walled cells is usually differentiated in the wall of the sporangium. The annulus (Figs. 19-12F, 19-14E, F, and 19-15) is sometimes important in initiating dehiscence of the sporangium and throwing spores into the air.

Fern spores of the more generalized groups are usually roughly spherical with a trilete face. The surface is often elaborately ornamented with warts or reticulations. The spores of more specialized ferns are more or less kidney-shaped and have a **monolete** scar. Spores are unicellular and usually lack chlorophyll.

In most generalized homosporous ferns, the spore germinates to produce a gametophyte of a plate of cells, but in more specialized ferns a short filament precedes the formation of the thallus. In heterosporous ferns, the spore germinates endosporously.

Many ferns can propagate vegetatively, particularly in the local expansion of a **clone;** in rhizomatous taxa a single clone can occupy an area of several square meters. Some ferns produce deciduous vegetative **bulbils** on the leaves; in others adventitious roots form at or near the apex of the frond, and a new active shoot apex differentiates to produce a new sporophyte.

Most ferns are terrestrial. Sometimes they grow in rather arid environments, but generally the gametophyte can survive only under humid conditions, and water is necessary to transport the sperms. In consequence, the sporophyte does not readily establish itself unless liquid water is available at least part of the year. Most ferns grow best under somewhat humid, shaded conditions.

Ferns show a wide distribution, both latitudinally and altitudinally. Although few kinds of ferns occur in arctic and alpine climates, some occur on the northernmost land masses and at alpine elevations. Ferns are abundant in both coniferous and deciduous forests in temperate climates; they reach their greatest abundance and diversity in tropical and subtropical forests in South America and Southeast Asia, where they can carpet the forest floor, grow on the trunks and branches of trees, and flourish on cliffs and boulders. A few ferns are strictly aquatic.

Unquestionable fern fossils are known from the Carboniferous. Many fernlike fossils are also present in the earlier geological record, but their true relationships are frequently uncertain because only their foliage characteristics are known.

The gametophyte in most ferns is a chlorophyllose thallus, and both male and female sex organs are borne on the same plant (Fig. 19-3). Occasionally, however, the gametophyte is filamentous, and sometimes it is endosporic, unisexual, and has little or no chlorophyll. The gametophyte is generally an annual.

Figure 19-1 Crozier of *Cibotium* sp., showing circinate vernation of the frond and pinnae.

The gametophyte in most ferns is exosporic and chlorophyllose. Generally it is attached to the substrate by rhizoids and produces both antheridia and archegonia. Sperms are multiflagellate and must have free water in order to swim to the egg in the archegonium and effect fertilization.

The fern embryo is relatively simple. As it grows, it differentiates root and shoot apical meristems early and becomes photosynthetic and independent of the gametophyte.

The mature stem varies greatly in structure among ferns, but it is generally relatively short and rhizomatous. However, sometimes it is erect and forms tall trees. Roots are usually present.

The Filicopsida can be separated into seven distinctive evolutionary lines that coincide with six subclasses of extant ferns: Ophioglossidae, Marattiidae, Osmundidae, Filicidae, Marsileidae, and Salviniidae.

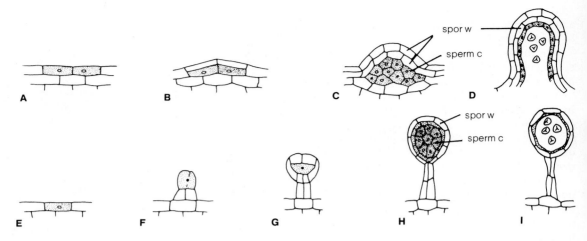

Figure 19-2 Two types of sporangial development. **A–D,** eusporangiate development, with several wall layers (*sp w*) surrounding sporogenous cells (*sp c*)., ×20. **E–I,** leptosporangiate development, in which outer initials form both wall and sporogenous tissues on stalk, ×120.

Subclass Ophioglossidae

Ophioglossidae, the grape ferns, contain three genera in a single family, Ophioglossaceae. *Helminthostachys*, with a single species, is endemic to Southeast Asia in the Indo-Malayan area and Australia, whereas *Botrychium* (Fig. 19-4) (about 20 species) and *Ophioglossum* (Fig. 19-5) (about 40 species) are found throughout the world from tropical to arctic climates. *Cheiroglossa* (tropical) and *Rhizoglossum* (South Africa) are two genera sometimes extracted from *Ophioglossum*. Each has a single species. Most species are terrestrial, but a number of species of *Ophioglossum* are **epiphytes.**

The subclass is not closely related to any other fern or fern ally, and its origins are obscure. The fossil record is not very helpful; only spores of *Botrychium* and *Ophioglossum* have been identified from as early as the Lower Jurassic.

The leaves of Ophioglossidae lack circinate vernation, and each consists of a sterile blade plus a sporangium-bearing branch, the sporangiophore. The blade is simple and undivided in most species of *Ophioglossum*, but in *Botrychium* (Fig. 19-4D, F) and *Helminthostachys* it is fundamentally pinnately divided. The leafy portion of the sporophyte is usually less than 30 cm tall. *Helminthostachys* and some species of *Botrychium* reach heights of 50 cm, but *Ophioglossum pendulum*, a pendulous epiphyte, can reach lengths up to 2.7 m. The lower portion of the leaf petiole expands into sheathing **stipules,** and the whole leaf is somewhat fleshy. All Ophioglossidae are herbaceous, and each plant tends to mature a single leaf annually in temperate climates, although the former year's leaf blade sometimes persists while the current year's leaf is present. A leaf can take up to four years to develop after it has been initiated at the rhizome apex.

In *Ophioglossum*, the leaf venation is reticulate (Fig. 19-5A, B), but in the other two genera it is typical of most ferns: open and dichotomous. The sporangiophore in all genera usually emerges adaxially near the junction of the petiole and the leaf blade, or sometimes it diverges further down the petiole. In *Ophioglossum palmatum*, several sporangiophores emerge bilaterally on both sides of the petiole (Fig. 19-5B).

The sporangia are eusporangiate and comparatively large (0.5–3 mm in diameter) and thick-

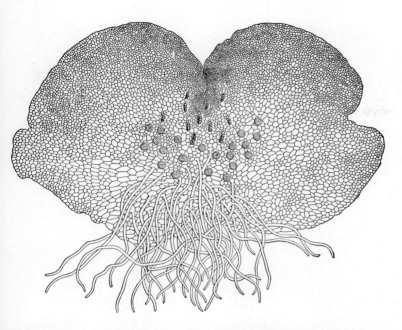

Figure 19-3 Lower surface of leptosporangiate fern gametophyte (prothallus), showing numerous rhizoids, rounded antheridia, and archegonia with necks pointing toward base. (From *The Plant Kingdom* by W. H. Brown, copyright © 1935, by William H. Brown. Used by permission of the publisher. Ginn & Company [Xerox Corporation].)

walled (4–8 cells thick), with stomata in the outer wall; dehiscence is by a transverse slit. In *Botrychium* (Fig. 19-4D), the sporangia are terminal on ultimate branches of the sporangiophore. In *Ophioglossum* the sporangiophore is usually unbranched, and sporangia form lateral rows on either side of it (Fig. 19-5). They are fused laterally, making the whole **spike** resemble a miniature rattle of a rattlesnake—hence the popular name, rattlesnake fern. All genera produce many spores in each sporangium.

In *Ophioglossum* species that grow on soil, the stem is subterranean, but in epiphytic species it is exposed. The stem is soft and fleshy, and lacks sclerenchyma. The young stem is protostelic, but as it matures, it becomes a dictyostele. *Botrychium* produces secondary vascular tissues and a periderm, features that are absent among other living ferns. The stem apex consists of a single, pyramidal apical cell with three cutting faces, a feature also typical of mosses and leafy hepatic gametophytes. In Ophioglossidae, as in early bryophytes, each segment derived from these cutting faces gives rise to a leaf. The stem is usually unbranched and short (normally less than a centimeter long). In *Helminthostachys*, however, it is often a rhizome several centimeters in length. Roots arise singly near the base of the leaf petiole. They are somewhat fleshy, lack hairs, are little-branched, and are heavily infected by fungal hyphae.

Terrestrial species have a subterranean gametophyte that lacks chlorophyll and is strongly mycorrhizal, whereas epiphytes have a surficial gametophyte that sometimes has chlorophyll and mycorrhizae. The gametophyte is annual or, occasionally, perennial. It is usually only a few millimeters long, and is either elongate and branched or thick and tuberlike. Each parenchymatous gametophyte produces antheridia and archegonia embedded in its surface layers (Fig. 19-6). The

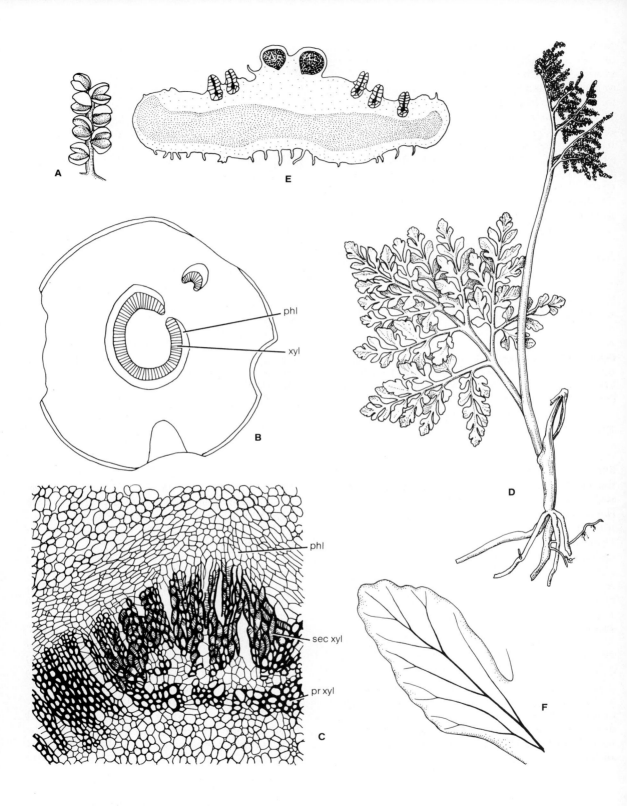

464 Chapter 19

antheridia are relatively large (up to 200 μm, but usually less than 100 μm). Each produces many sperms with numerous anterior flagella. The neck of the embedded archegonium forms a conical projection. Antheridia appear first in the gametophyte, and archegonia appear later. The sexes are not intermixed but are confined to specific areas, usually on the upper surface of the gametophyte.

Although relationships of the Ophioglossidae are unclear, they appear to be closest to the coenopterids or perhaps even to the trimerophytes. Details of the sporangia and stem anatomy are cited as features that relate these groups of plants.

Subclass Marattiidae

The Marattiidae contain a single order with a fossil record extending back to the middle Carboniferous. Although numerous fossil genera have been recognized, the extant flora probably includes only five genera, although seven are sometimes identified. These ferns occur in tropical and subtropical humid forests, with most of the nearly 200 species confined to the western Pacific area.

The subclass is characterized by stems of remarkable complexity, often made up of a series of concentric **meristeles** embedded in a ground tissue richly supplied by mucilage ducts and tannin-filled cell clusters (Fig. 19-7E, F). In extant genera, the stem is greatly thickened by the persistence of succulent sheathing stipules of the petiole bases. Further thickening results from the thick fleshy adventitious roots that emerge from near the leaf bases and from an entangling mass that sheathes the stem and the associated structures that envelop it. Generally, the leaves are pinnately divided and circinately vernate. Sporangia are large, borne abaxially on the pinnae and clustered in sori that are frequently fused into synangia; dehiscence is either by a longitudinal slit or a terminal pore (Fig. 19-7B, D, G). Numerous spores are produced in each sporangium.

In the fossil genus *Psaronius*, the unbranched trunk probably reached heights of up to 10 m and was supported largely by **persistent** leaf bases and masses of fleshy adventitious roots (Fig. 19-8). In young stems, the stele is protostelic, but with continued production of leaves and roots, the stele is broken into a complex **siphonostele.** The internal outline of the stem is conic, with the narrow point downward. The external form of the stem has the narrower portion uppermost, built up of an accumulation of thickened leaf bases and particularly by the continual production of persistent adventitious roots growing downward to envelop the stem and the associated leaf bases.

The stems in modern genera are usually relatively short and partially embedded in the soil, although some erect species have fleshy stems (with their accompanying persistent leaf stipules and roots) that reach a height (and sometimes breadth) of a meter (Fig. 19-8A).

One or two (or more) fleshy roots emerge from each leaf base and have multicellular root hairs. They are mycorrhizal and arise endogenously, piercing the stipules as they emerge. The stele is **polyarch,** with long radiating xylem arms. It is embedded in a thick cortex, which, like the stem, contains **mucilage ducts** and tannin cells.

The leaves are usually pinnately compound. In the fossil genus *Psaronius*, the leaves are twice-pinnate, a situation that persists in some modern species of *Angiopteris*. Leaf size varies; *Psaronius* bore leaves up to 3 m long, giving the tree a crown of fronds, and in some extant genera the leaves are up to 6 m long with a petiole up to 6 cm in diameter. Venation is usually open dichotomous, but in *Christensenia* it is netted (Fig. 19-7G). The leaves are spirally arranged on erect stems, but on rhizomatous stems they are roughly two-ranked. They are leathery and evergreen and persist several years. When they break off, the thickened stipular bases persist on the stem. The stipular bases have dor-

Figure 19-4 *Botrychium.* **A,** dehiscing sporangia on branch of sporangiophore, ×5. **B,** transverse section of stem, showing xylem (*xyl*) surrounded by phloem (*phl*) in stele and one leaf trace toward upper right, ×7. **C,** section of stele, showing primary xylem (*pr xyl*), secondary xylem (*sec xyl*), and phloem (*phl*), ×60. **D,** habit sketch of *B. multifidum*, ×1/2. **E,** section through subterranean gametophyte, showing mycorrhizal lower portion, archegonia, and antheridia, ×20. **F,** pinna of *B. multifidum*, showing open dichotomous venation, ×15. (**E,** redrawn after Campbell.)

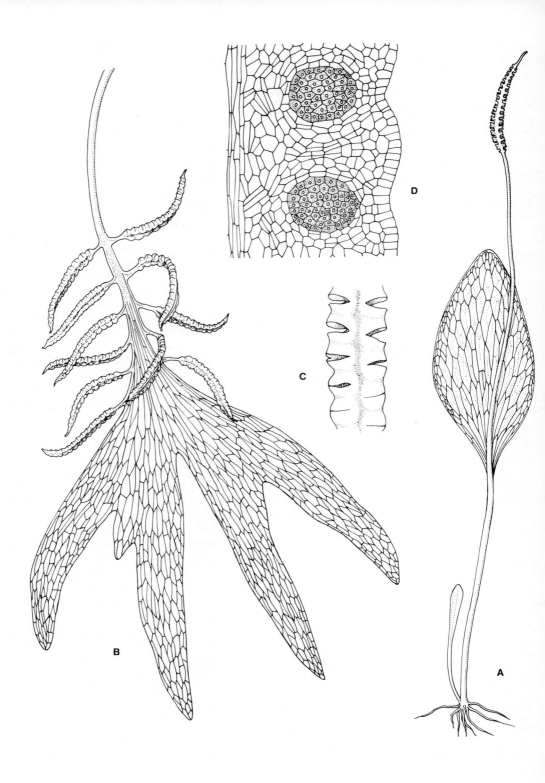

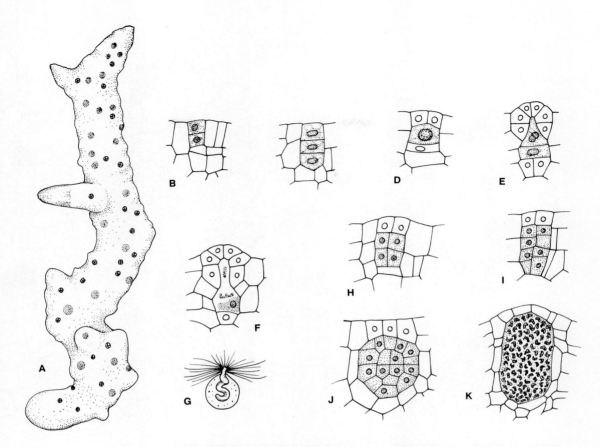

Figure 19-6 Gametophyte of *Ophioglossum*. **A,** habit sketch of whole gametophyte, × 30. **B–F,** development of archegonium, × 215. **G,** sperm, × 500. **H–K,** development of antheridium: **H–J,** × 215; **K,** × 140. (**A, G,** redrawn after Bruchmann; **B–F, H–K,** redrawn after Smith.)

Figure 19-5 *Ophioglossum*. **A,** habit sketch of *O. palmatum*, ×1.5. **B–D,** *O. vulgatum*. **B,** habit sketch showing netted venation, ×1/2; **C,** portion of sporangiophore, × 3; **D,** longitudinal section through sporangia, × 160. (**D,** redrawn after Smith.)

Filicopsids (True Ferns) 467

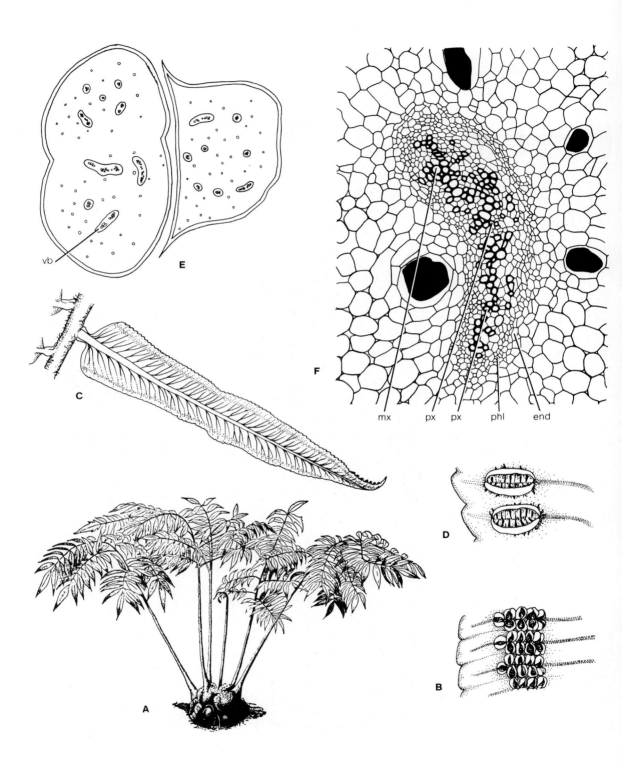

468   Chapter 19

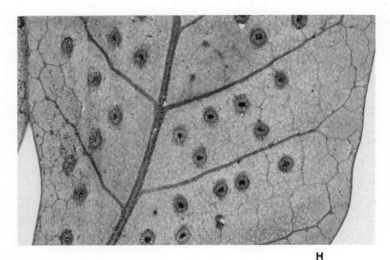

Figure 19-7 Subclass Marattidae. **A–C,** *Angiopteris evecta*. **A,** habit sketch, ×1/50; **B,** sori on veins of pinnule, ×0.5; **C,** pinna, showing venation. **D–F,** *Marattia*. **D,** synangia on pinnule, ×1.0; **E,** transverse section of stem and one frond, showing dictyostelic arrangement of vascular bundles (*v b*), ×5; **F,** anatomy of one vascular bundle from stem in **E,** showing protoxylem (*px*) with small tracheids, metaxylem (*mx*), phloem (*phl*), and endodermis (*end*), ×40. **G,** *Marattia* pinnules, showing sporangia, ×1.0. **H,** *Christensenia*, portion of pinna, showing netted venation and synangia, ×1.0. (**A,** from Wettstein.)

Figure 19-8 Fossil Marattidae. **A,** *Psaronius,* habit sketch (reconstruction) (height about 8 m). **B,** *Psaronius,* transverse section of stem, showing arrangement of vascular bundles and leaf scars, ×0.5. **C,** *Eoangiopteris andrewsii,* portion of pinnule, showing sori, ×10. **D,** *Scolecopteris incisifolia,* portion of pinnule, showing sori, ×10. (**A,** redrawn after Morgan; **B,** redrawn after Stidd; **C, D,** redrawn after Mamay.)

mant buds at their axils; when these bases fall off, they serve as vegetative propagants.

The sporangia are borne abaxially on the pinnae and are exposed. They are large, sometimes exceeding 2 mm in diameter, and are clustered into sori that are frequently fused together into synangia (Fig. 19-7D, G). Some synangia are up to 3 cm long and consist of more than 100 sporangia. All sporangia in a sorus mature at the same time and contain thousands of spores. As the sporangial walls dry out after maturation, the sporangium gapes open to release spores.

Gametophytes are thallose, chlorophyllose, perennial, and attached to the substrate by rhizoids. Both sexes are on a single thallus that can reach lengths up to 3 cm. Sex organs are sunken in the thallus, with archegonia on the upper surface and confined mainly to the thickened centre of the fleshy thallus. Antheridia are scattered on both surfaces of the thallus.

Although their specific relationships are obscure, the Marattiidae show a number of features that relate them to other ferns, for example, the nature of the leaves, including the circinate vernation, and the position of the sporangia. They resemble the Ophioglossidae in their possession of stipules and in their massive eusporangiate sporangia, but bear no close relationships to that subclass.

Subclass Osmundidae

The Osmundidae contain a single family, Osmundaceae, consisting entirely of herbaceous ferns. They first appeared during the Permian and persist as three genera widely distributed in temperate to tropical regions. *Osmunda* is the most widespread genus and contains about 12 species. *Todea*, with a single species, is found in Australasia and South Africa, whereas *Leptopteris* (6 species) is found in Australasia and many Pacific islands. The Osmundidae are intermediate in morphology between eusporangiate and leptosporangiate ferns, but are not closely related to any known representatives of either group and appear to be an evolutionary dead end. Features that characterize them are (1) the eusporangiate origin of some sporangia and essentially leptosporangiate origin of others; (2) the sporangia are never grouped in a sorus, and the wall is unistratose; (3) sporangia lack a functional annulus and open by a longitudinal line; (4) the spores are chlorophyllose; (5) the stem is erect, and the bases of the petioles are expanded and persist after the leaf is shed; and (6) leaves are always pinnately compound and circinately vernate.

The stems of most Osmundidae are relatively short, although a New Caledonian species of *Leptopteris* has a trunk up to 3 m tall. The stem is formed by a small dictyostele surrounded by a thick, dark-colored cortex of sclerenchyma, outside of which are the numerous persistent leaf bases, each with its C-shaped vascular bundle (Fig. 19-9C, G). The roots are coarse and hard; two arise from the base of each leaf, and they may be multibranched.

The leaves, up to 3 m long, are borne spirally and are essentially annual in most species. Leaf venation is dichotomous (Fig. 19-9F). In *Todea* and *Osmunda*, the leaves are leathery, but in *Leptopteris* the fronds are very thin and filmy. Sporangial position varies among the genera and even among species within a genus. In several species of *Osmunda*, some leaves have the pinnae converted largely to sporangia, whereas in other species of *Osmunda* (Fig. 19-9A), as well as in *Leptopteris* and *Todea*, only a portion of the frond is converted to sporangium-bearing pinnae. In both *Leptopteris* and *Todea*, the sporangia are confined mainly to the abaxial surface of the frond, but in many species of *Osmunda* the main axis of the pinna and the axes of the pinnules become stalks on which sporangia are borne (Fig. 19-9D). Sporangia are comparatively large, sometimes up to 1 mm in diameter, and contain numerous trilete spores (Fig. 19-9E).

The gametophytes are chlorophyllose, simple thalli, and are sometimes perennial, reaching lengths of up to 5 cm (Fig. 19-9B). Archegonia are on the undersurface of the thallus near the thickened longitudinal midrib. Antheridia project on the surface of the thallus margin and on the undersurface between the margin and the midrib. The rhizoids are on the undersurface of the midrib.

Fossil material of osmundaceous fern stems of Late Permian age make this one of the most ancient of fern families with a continuous record up to the present time. Identification of the leaves relies on the presence of sporangia. Such information indicates that the extant subgenera of *Osmunda* were present in early Tertiary times and probably earlier.

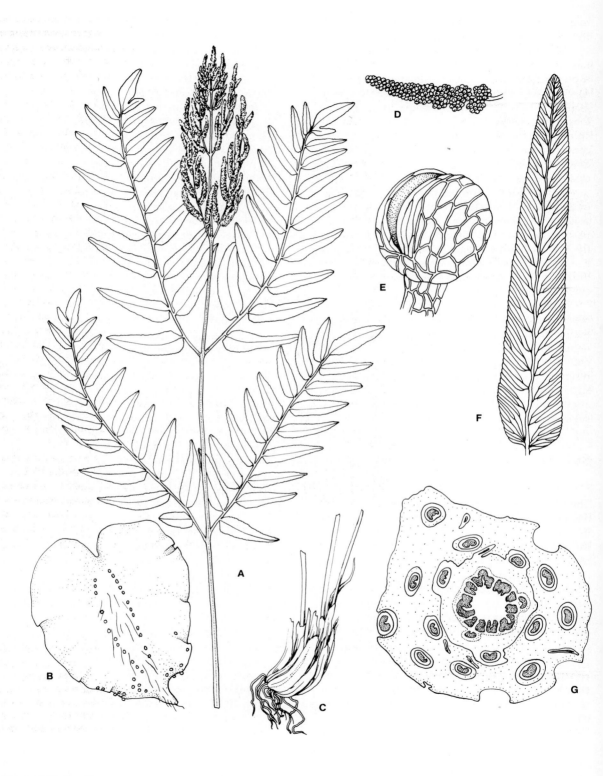

Todea shows great similarity to fossils of a Jurassic fern of widespread occurrence.

The Osmundaceae have been used extensively to probe physiological and morphogenetic problems. It appears that leaf primordia, very early after their initiation, determine internally all succeeding morphogenesis of the frond. Stem and leaf primordia are often extremely similar in *Osmunda,* and the precise mechanism involved in determining whether a stem or a leaf arises from a specific primordium is not understood.

Subclass Filicidae

Most living ferns are in the subclass Filicidae, which is distributed from arctic to subantarctic regions and from sea level to alpine elevations. It is represented in the fossil record of the Carboniferous, when ferns related to the modern family Gleicheniaceae were present. At present there are approximately 300 genera with nearly 9000 species included in 32 families. Concepts at all levels of classification vary greatly among different researchers, partly because careful research has been confined to relatively few taxa.

The Filicidae occupy most habitats, from dry sites on boulders and tree trunks to humid forest floors and tree trunks; a few taxa are submerged aquatics. No Filicidae are marine, although a few are restricted to maritime sites.

The vast array of genera exhibits much diversity in morphology and anatomy. Features that characterize the group include (1) the homosporous leptosporangiate condition, (2) the usual restriction of the sporangia to a sorus (usually on the abaxial surface of the leaf), (3) the usual presence of a well-defined annulus on the sporangium, (4) the usual possession of hairs or scales on the stems, and (5) leaves that are usually circinately vernate and frequently pinnately compound. The gametophyte is usually a chlorophyll-rich, flattened, thin thallus, with sex organs protruding on the surface. Both antheridia and archegonia are on the same thallus in most taxa and, as a rule, gametophytes are annual.

Evolutionary trends in living ferns are highly suggestive, and are generally supported by the time sequence of the appearance of different fern structures in the fossil record. The trends are usually interpreted as tending from a generalized to a more specialized condition. In the following list, generalized features are stated first, and comparative specialized features are stated second:

1. Stem slender, creeping, protostelic, covered with hairs, fronds in two ranks versus stems erect, thick, with complex dictyostele, covered with scales, fronds spirally arranged

2. Fronds large, branched dichotomously, with single leaf trace in stipe, and with open venation versus fronds reduced, simple, pinnately branched or unbranched, with many leaf traces in stipe, and with netted venation

3. Sporangia few, terminating a vein, and marginal versus sporangia numerous, not terminating vein, and abaxial

4. Sporangia relatively large, with stout stalk and poorly defined annulus, maturing simultaneously within a sorus, and producing a large number of spores (128, 256) versus sporangia relatively small, with slender stalk and well-defined annulus, maturing at different times within a sorus, and producing a small number of spores (32, 64)

5. Gametophyte thallus with thickened midrib and often perennial versus uniformly thin thallus that is generally annual

6. Antheridia large, with jacket consisting of more than four cells, and containing several hundred sperms versus antheridia small, reduced to jacket consisting of three cells, and containing few sperms (14, 16, or 32)

7. Archegonia with long, straight neck versus archegonia with short, curved neck

Figure 19-9 *Osmunda regalis.* **A,** habit sketch of sporangium-bearing frond, ×0.5. **B,** gametophyte, ×10. **C,** rhizome with persistent bases of leaves and roots, ×0.5. **D,** sporangium-bearing pinna, ×2. **E,** dehiscing sporangium, ×30. **F,** pinna, showing open dichotomous venation, ×1.5. **G,** transverse section of rhizome, showing stele in center and leaf traces in cortex, ×0.5.

In the following discussion, several distinctive orders are discussed, illustrating some of the structural diversity and evolutionary trends among Filicidae. These cannot be arranged in a linear series to illustrate the trend from generalized to specialized morphology, since, within some orders, there is a considerable range of degrees of specialization. Furthermore, although some taxa show several generalized features, they also have some specialized features. These taxa lead to controversy in classification, but they also stimulate further research that attempts to establish their relationships more confidently.

Order Schizaeales

This order appeared first during the Jurassic. There are five modern genera containing approximately 160 species. Three families are generally recognized: Schizaeaceae with *Schizaea* (about 28 species) and *Actinostachys* (about 13 species); Anemiaceae with *Anemia* (about 90 species) and *Mohria* (3 species); and Lygodiaceae with *Lygodium* (about 40 species). The order is largely tropical and subtropical throughout the world, but a single species of *Schizaea* is found in cool, temperate bogs on the Atlantic coast of North America, and a number of species of *Schizaea* are found in temperate New Zealand and South America. *Actinostachys* is predominantly tropical and subtropical. *Anemia* is most richly represented in tropical South America, *Mohria* is African, and *Lygodium* is widely distributed in the tropics and subtropics, with a single species extending northward to New England in the United States. An obvious distinctive feature of the order is the sporangium, which has a thick stalk and an annulus that forms an apical cap. Sporangia arise simultaneously, are borne singly on the leaf margin rather than in sori, and open longitudinally. In *Lygodium*, each sporangium is sheathed by an indusiumlike flap, but in the other genera sporangia are exposed.

The leaves of Schizaeales are diverse in morphology. In *Schizaea*, vegetative leaves are usually dichotomously divided or sometimes simple and grasslike (Fig. 19-10F). The leaves of *Schizaea* and *Actinostachys* are usually relatively small, often less than 3 cm long. However, in the genus *Lygodium* the leaf is sometimes almost indeterminate in length; in some species it continues to add new pinnae each year, and reaches a length up to 30 m (Fig. 19-10A). In *Lygodium*, the leaves are pinnate, and the pinnules are sometimes much-divided. Some pinnules are strictly vegetative, whereas others bear sporangia (Fig. 19-10B). Venation in *Lygodium* is usually dichotomous.

The leaf of *Anemia* is pinnately compound, with the pinnae further divided (Fig. 19-10G). The two lowermost pinnae on sporangium-bearing fronds are much elongated, sometimes diverging almost perpendicularly from the rest of the frond leaf, which is nearly horizontal. The ultimate divisions of the pinnules bear two rows of sporangia on the abaxial surface. Leaves sometimes reach lengths of 30 cm or more.

Schizaeales have relatively small stems; most are rhizomatous, and leaves arise from the upper surface (Fig. 19-10C). The stem has hairs or scales. However, in *Schizaea* and *Actinostachys* stems are short and erect, and leaves are spirally arranged (Fig. 19-10F). Steles vary among genera, from protosteles to dictyosteles (Fig. 19-10D, E).

Gametophytes of some species of *Schizaea* are filamentous, with sex organs forming swollen, short, lateral branches. In *Actinostachys* and some species of *Schizaea*, the gametophyte is cylindric, much resembling that of the Psilotophytina; in other genera, the gametophytes are flattened, chlorophyllose thalli.

As treated here, Schizaeales contains several independent evolutionary lines, distinctive in both sporophytic and gametophytic structure. The uniseriate, filamentous gametophyte is among the most

Figure 19-10 Schizaeales. **A–F,** *Lygodium palmatum*, a fern with a perennial climbing frond. **A,** habit sketch of portion of vegetative frond, ×1.0. **B,** portion of sporangium-bearing frond, showing converted pinnules, ×25. **C,** rhizome, ×1.0; **D,** transverse section of part of stele, showing sclerenchyma fibers (*scl*), endodermis (*end*), protoxylem (*px*), metaxylem (*mx*), and phloem (*phl*), ×170. **E,** three-armed protostele, showing protoxylem (*px*) at poles and metaxylem (*mx*) forming core, ×30; **F,** habit sketch of *Schizaea pusilla*, ×1. **G,** photograph of *Anemia* sp., ×0.5.

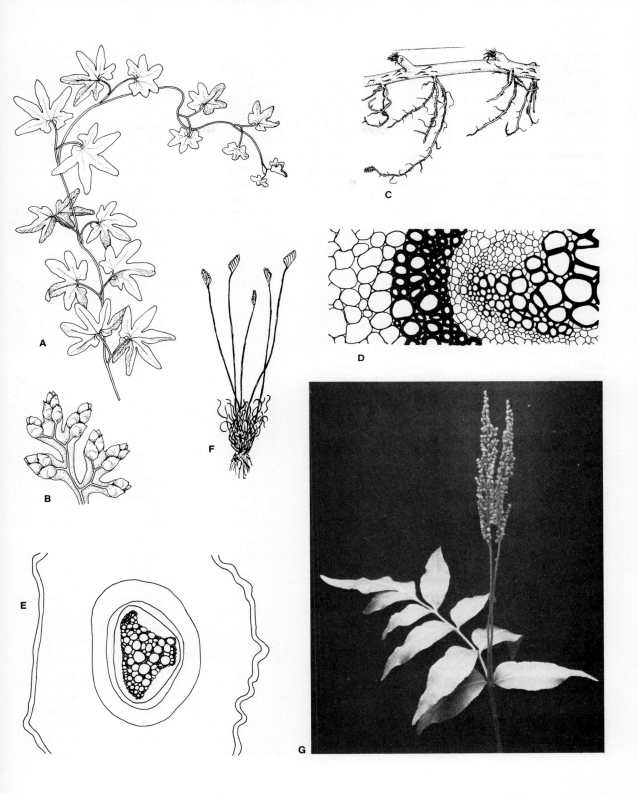

Filicopsids (True Ferns) 475

reduced and specialized known in ferns, but separate marginal sporangia, each with an apical annulus, suggest a rather generalized condition. All genera produce a large number of spores within each sporangium, which is also a generalized feature.

The Schizaeales exhibit a mixture of generalized and specialized features in both the gametophyte and sporophyte. These characteristics place them in a critical position for interpreting the interrelationships of ferns and their nearest allies. They also contribute to our understanding of the presumptive pathways of the origin of organs and organ systems in ferns.

Order Pteridales

Possibly derived from schizaealean ancestors, the Pteridales comprise a large assortment of ferns, probably of relatively unrelated families. The sporangia are in marginal sori and are not protected by an indusium, and **paraphyses** are rare. The sporangium has a distinct, longitudinal annulus and a slender stalk that characterizes most leptosporangiate ferns. Dehiscence is by a transverse slit through a definite stomium. The sporangia of a single sorus mature at different times, a condition referred to as mixed.

The order includes the widely distributed, large tropical genera *Pteris* (about 280 species) and *Adiantum* (about 200 species). *Adiantum* (Fig. 19-11A–D) is also widespread in temperate and subarctic regions. Pteridales are essentially terrestrial ferns, although some are epiphytes, and they show great ecologic amplitude, from relatively **xeric** sites (*Pellaea*, Fig. 19-11E), to humid sites (*Adiantum* and *Pteris*), to aquatic sites (*Ceratopteris*). Some taxa extend into high elevations.

The order contains approximately eight families with about 60 genera and more than 1000 species. They have a wide range of leaf types, from pinnately compound and usually less than 2 cm tall, to very small, wedge-shaped leaves (*Hecistopteris*). Usually, the leaves are pinnately compound, and in rare instances, as in *Jamesonia*, they have indefinite growth. They are usually annual in temperate climates, but in some genera they are evergreen for several years. Usually the vegetative fronds are morphologically like the sporophylls, but the fronds of some genera are dimorphic, with the pinnules of the sporophylls conspicuously slender and the whole margin reflexed to protect the sori of the abaxial surface. In *Adiantum*, sori are restricted to interrupted portions of the margin of a pinnule, and each of these portions is recurved to form an indusiumlike covering (Fig. 19-11B–D). The venation of leaves is usually open and dichotomous, but it is sometimes netted. Leaves are borne alternately or spirally on the rhizomes, and sometimes have a wiry rachis.

Stems vary from short to long rhizomes, and steles also vary greatly in structure. Hairs or sometimes broad scales clothe the rhizomes. Gametophytes are **cordate, reniform,** or asymmetrical chlorophyllose thalli.

In this order, there begins a greater consistency in general morphology that characterizes most of the true ferns: elaborate fronds with protected sori, sporangia with effective mechanisms for scattering spores, and annual, chlorophyllose gametophytes.

Order Hymenophyllales

This order of filmy ferns is richly represented in the Southern Hemisphere and in the tropics and subtropics. Copeland (1947) recognized 33 genera and at least 650 species. Most botanists are more conservative, and some treat the species very broadly by including them in two genera, *Hymenophyllum* and *Trichomanes*. A few species occur in temperate, humid climates in the Northern Hemisphere. In New Zealand, nearly one-fifth of the total, rich fern flora belongs to this order.

Filmy ferns are most common in humid forests, abounding as epiphytes, on rocks, and on the forest floor. Most species are small; some are less than a centimeter tall and closely resemble gametophytes of the liverwort family Metzgeriales (Fig. 19-12).

The leaves are usually thin, often unistratose except at the veins, and lack stomata. The vascular system is extremely reduced in most species; the protostele is often only two or three tracheids in circumference. Because of this reduction, leaf gaps are sometimes absent. Leaves are palmately or finely

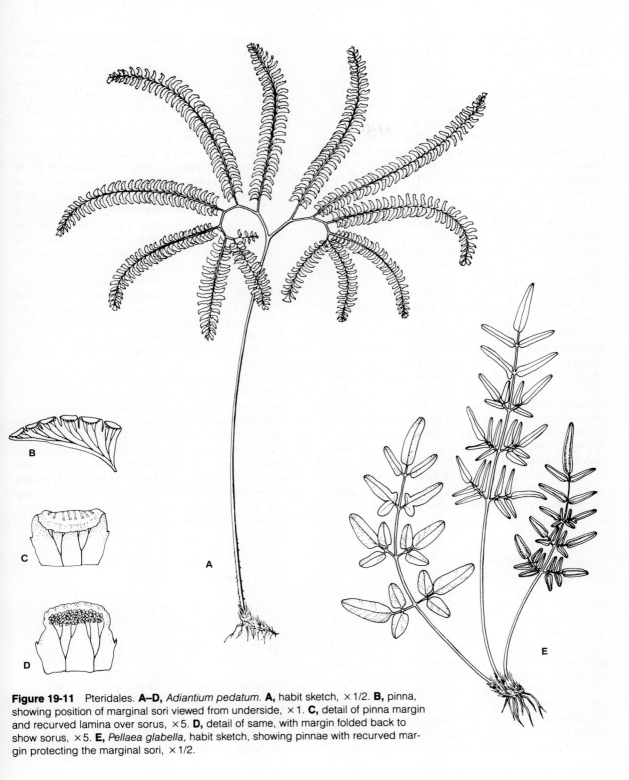

Figure 19-11 Pteridales. **A–D,** *Adiantium pedatum.* **A,** habit sketch, ×1/2. **B,** pinna, showing position of marginal sori viewed from underside, ×1. **C,** detail of pinna margin and recurved lamina over sorus, ×5. **D,** detail of same, with margin folded back to show sorus, ×5. **E,** *Pellaea glabella,* habit sketch, showing pinnae with recurved margin protecting the marginal sori, ×1/2.

pinnately divided, or are simple. Venation is usually open and dichotomous. Sori are generally marginal at the tips of veins, and are surrounded by a cup-shaped or bivalved indusium (Fig. 19-12A–E). Often sporangia are on a cylindric receptacle that protrudes beyond the indusium mouth (Fig. 19-12A, D, E). The maturation of sporangia is gradate in many taxa; they mature first at the apex and then toward the base of the cylindric receptacle. In other taxa, all sporangia mature simultaneously. The sporangia are short-stalked, and the annulus is oblique (Fig. 19-12F).

Often the plant has no roots but is attached to the substrate by rhizoids. In such instances, water and nutrient uptake is like that in most bryophyte gametophytes—over the entire surface. This feature tends to restrict these plants to areas of high precipitation or moisture. Most taxa are rhizomatous, and their leaves emerge in two rows; however, some have a short, erect stem and leaves that are spirally arranged.

The gametophyte is often a multibranched, uniseriate protonema, much resembling that of some Schizaeales. Sometimes gametophytes produce marginal gemmae that are effective for vegetative reproduction.

The Hymenophyllales are an unusually specialized order of ferns in many respects, showing extreme reduction in both sporophyte and gametophyte. However, some retain generalized features, for example, the simultaneous maturation of sporangia in a sorus. The protostele and lack of leaf gaps appear to result from extreme reduction rather than retention of generalized features.

Order Gleicheniales

The order Gleicheniales is an ancient one, possibly dating back to the Carboniferous. There are six genera in the extant flora with approximately 130 species. The order flourishes in tropical to subtropical areas, both in open sites and in forests. It extends to temperate regions in the Southern Hemisphere. During the Cretaceous, the order was abundant in the Northern Hemisphere.

The perennial leaves are characteristically dichotomous and occasionally reach lengths of 6 m, adding new growth annually after a period of dormancy. Often the leaf has a long rachis in which the development of the apical meristem is arrested; two lateral meristems that differentiate on either side of it tend to grow to produce a false dichotomy. Pinnae frequently emerge below this false dichotomy, and they are sometimes further subdivided (Fig. 19-13).

There is no separation of sporophylls and vegetative leaves. Sori occur on the abaxial surface along a vein of the pinna and lack an indusium (Fig. 19-13E, F). Sporangia develop simultaneously, are pyriform, and have a short, stout stalk and a well-defined, oblique annulus. The sporangium opens by a longitudinal slit (Fig. 19-13F), aided by contraction of the annulus, which is interrupted at the sporangium apex. Many trilete spores are produced in each sporangium.

The stem is usually rhizomatous and slender; it may be shallowly buried or on the soil surface (Fig. 19-13C). The stele is usually protostelic but is sometimes solenostelic. The leaves arise alternately from the upper surface of the rhizome.

The gametophytes are chlorophyll-rich, flattened thalli with a thickened midrib.

Gleicheniales show numerous generalized features that they retain from their ancient origin: (1) protostelic stem, (2) stout-stalked sporangium, (3) simultaneous development of sporangia in a sorus, (4) production of numerous spores, (5) thickened thallose gametophyte, (6) long-necked archegonia, and (7) enormous antheridia that produce vast numbers of sperms. Among the more specialized features are the false dichotomies of leaves and their indeterminate growth, the annulus-bearing sporangia, and the presence of sori.

Figure 19-12 Hymenophyllales. **A,** *Hymenophyllum affine*, habit sketch, ×8. **B,** *Trichomanes saxifragoides*, habit sketch, × 5. **C,** *Trichomanes* (*Microgonium*) *tahitiense*, habit sketch of sterile fronds. **D,** *Trichomanes* (*Cardiomanes*) *reniforme*, habit sketch, × 1/2. **E,** vertical section through sorus of *Trichomanes*, showing indusium (*ind*), and sorus (*sorus*), × 100. **F,** sporangium with oblique annulus, × 120. (**A–C,** redrawn after Brownlie; **E–F,** redrawn after Bauer.)

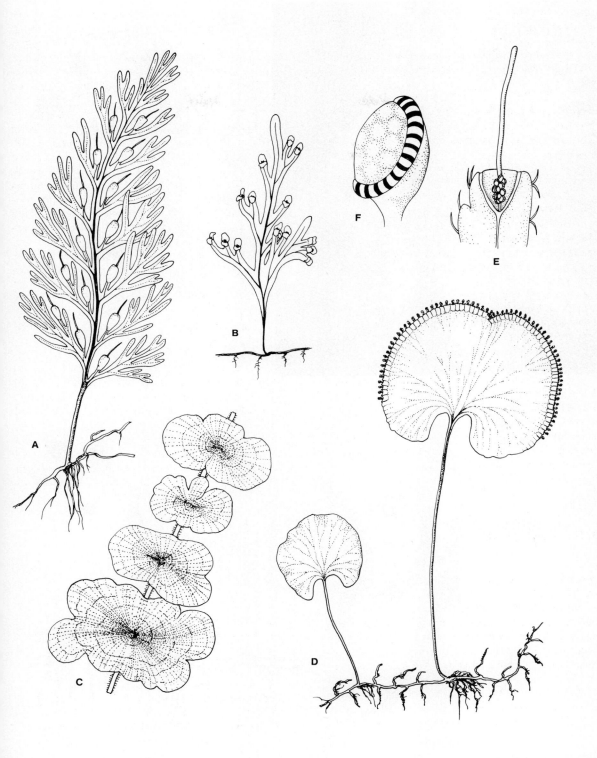

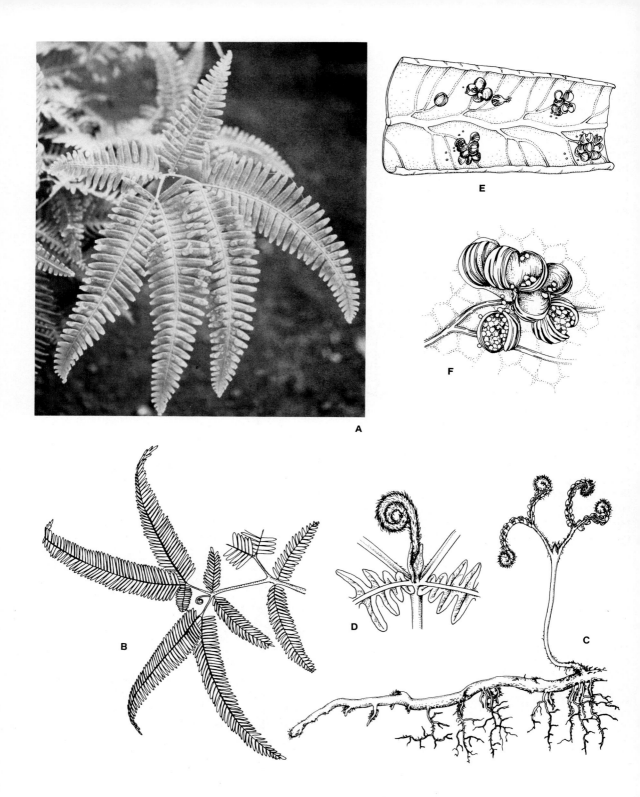

Order Cyatheales

The order Cyatheales consists entirely of tree ferns. In the fossil record as early as Jurassic time, it is represented by fragments of leaves and stems. In the extant flora, the order includes approximately eight genera and around 750 species, most of which belong to the genus *Cyathea*. The order is essentially tropical and subtropical, but it extends to temperate regions in New Zealand and South America.

These tree ferns abound near streams in forest openings and as understory species in forests. The trunks are usually unbranched and vary from less than a meter in height to over 20 m (Fig. 19-14A). The trunk is often ornamented by leaf scars and scales or hairs (Fig. 19-14B) that are often much overgrown by a tangle of roots. The tall trunks can reach a diameter of about 50 cm near the base and taper somewhat toward the apex. The stele varies from a single siphonostele to a complex **dictyostele** (Fig. 19-14G). No cambium is present. In some species of *Cyathea*, the trunk grows up to 30 cm in a year, and probably lives less than 70 years.

The pinnate leaves sometimes reach a length of nearly 7 m, and widths of nearly 2 m. Venation is open, sori are rounded and on the veins and are borne on a receptacle. The sori are often completely sheathed by a globose indusium when young, which opens when mature (Fig. 19-14C, D). Sometimes a flap of tissue overlaps the sorus; sometimes an indusium is absent, and the sporangia are intermixed with hairs. Each sporangium has a short, slender stalk and complete, oblique annulus (Fig. 19-14E, F). Dehiscence is by a transverse slit, and few spores are produced in each sporangium. Maturation within a sorus varies from gradate to mixed. Gametophytes are chlorophyllose and thallose, sometimes with a thickened midrib.

Tree fern trunks are often used as fence posts and as house-building timber, since they are remarkably resistant to decomposition. The delicate and graceful beauty of the plants makes them attractive ornamentals.

Order Aspidiales

Aspidiales are represented by approximately 75 genera and more than 3000 species. They are the predominant order of ferns in arctic and temperate floras, although they reach their greatest diversity in the tropics and subtropics. The largest genera are *Asplenium* (about 700 species), *Athyrium* (about 600 species), *Lastraea* (more than 500 species), and *Elaphoglossum* (more than 400 species). The order is so large that it is difficult to characterize satisfactorily.

The order contains most of the indusiate ferns. These are predominantly terrestrial and rhizomatous and bear pinnate leaves. The sporangia have a longitudinal annulus that is interrupted by a stomium and stalk (Fig. 19-15); thus dehiscence is transverse. Despite these *usually* common features, some representatives are epiphytic, and a few are climbers. Spore discharge in these ferns is by a catapult mechanism. As the sporangium dries out, the annulus shrinks in length, gaping open at the stomium. This tension reaches a level that the annulus cannot maintain, and it snaps back to its original position, in the process catapulting the spores into the air (Fig. 19-15).

The order shows great variety of habitat diversity; some taxa tolerate xeric sites, whereas others have the rhizomes in very wet substrata. Most taxa grow in humid, somewhat shaded sites.

The creeping, rhizomatous stem is commonly dictyostelic, relatively short, and clothed with scales. The leaves emerge either alternately or spirally. They are relatively small and rarely exceed 1 m in height; some epiphytic forms have leaves a few centimeters long. In most temperate genera, the leaves are deciduous, but in tropical regions they are often

---

Figure 19-13 Gleicheniales. **A,** photograph of pinnae of *Dicranopteris*, ×0.5. **B–E,** *Dicranopteris*. **B,** part of frond, showing accessory branches, ×1/5. **C,** end of rhizome, with one young frond and numerous roots, ×3/5. **D,** main fork of frond, showing bases of accessory branches, stipular pinnae, and circinate young rachis, ×1. **E,** detail of a pinnule, showing venation and sori, ×4. **F,** detail of sorus, showing dehiscing sporangia, ×15. (**B–D,** redrawn after Holttum; **E, F,** redrawn after Campbell.)

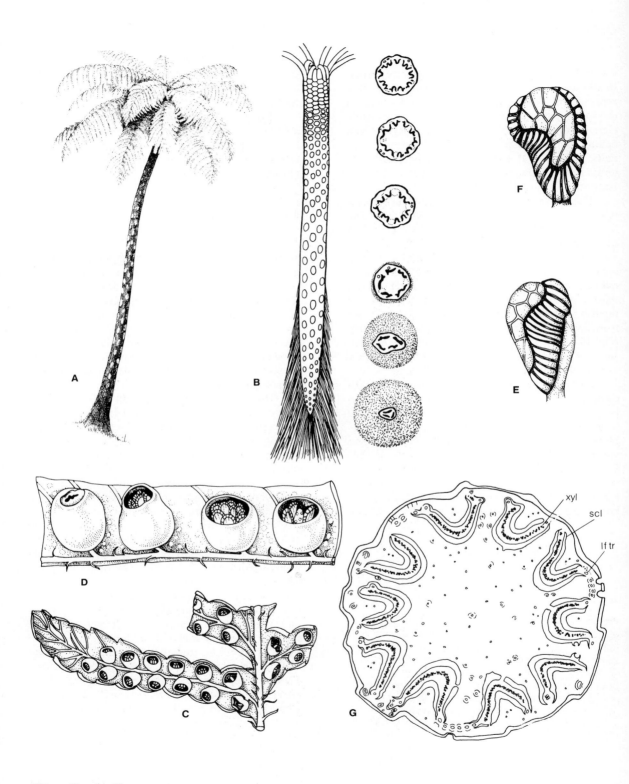

482 Chapter 19

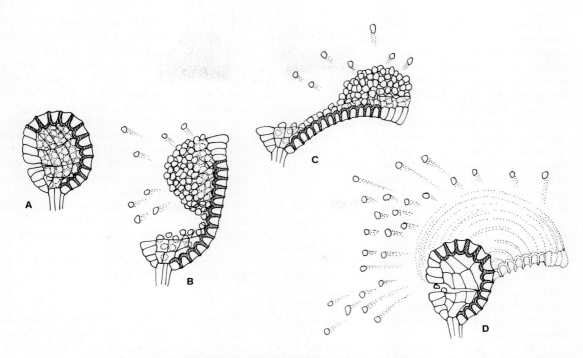

Figure 19-15 Sporangium dehiscence in leptosporangiate Filicidae, showing "catapult" mechanism of throwing out spores. (Redrawn after Atkinson.)

Figure 19-14 Cyatheales. **A,** *Cyathea,* habit, ×100. **B,** *Cyathea,* detail of stem with leaves removed, diagrammatic representation of transverse section of stem, showing nature of stele, ×1/10. **C–D,** *Cyathea elegans.* **C,** portion of sorus-bearing frond, ×1.0. **D,** detail of sori, showing various stages in opening of indusium, ×4. **E–F,** views of sporangium of *Cyathea,* showing oblique annulus, ×30. **G,** transverse section of *Cyathea* stem, showing U-shaped meristeles and numerous leaf traces (*lf tr*), ×0.5: *scl,* sclerenchyma, *xyl,* xylem. (**B, G,** redrawn after Ogura; **C, D,** redrawn after Bauer; **E, F,** redrawn after Goebel.)

Filicopsids (True Ferns) 483

evergreen. The petiole usually has two or more vascular strands that fuse in the rachis. The leaf is usually pinnate and is often much subdivided (Fig. 19-16A, D). Venation is usually open but can be netted. Occasionally, the leaves are dimorphic, with the sporophylls having constricted pinnae. In some instances (e.g., *Matteuccia, Onoclea*), pinnae are remarkably reduced and infolded to form a berrylike structure that encloses the sori. The sori are round to elongate and are usually covered by an indusium; they are along veins or expand over the abaxial surface of the pinnae. The indusium is usually fixed below the sorus and overlaps it, but it is sometimes fixed within the middle of the sorus and expands over it as a peltate flap (Fig. 19-16E). Occasionally an indusium is lacking; in such instances paraphyses are usually intermixed among the sporangia. Maturation of sporangia is mixed.

The gametophyte is generally a thin, annual, chlorophyllose, cordate or reniform thallus. Sex organs vary from somewhat complex in generalized taxa to reduced in cell number and size in more specialized taxa.

The order appears first in the Jurassic record, but fossil material is frequently so incomplete that it is difficult to interpret affinities. Commercially this order contains a number of garden ornamentals, and the leaves of some members are frequently used for decoration by florists. A number of taxa are used as food when the leaves are in the **crozier stage.**

Order Polypodiales

The Polypodiales form a large order with approximately 60 genera and 1500 species. Most taxa are epiphytic, but some are terrestrial. The order is widely distributed in temperate to tropical latitudes, but it reaches its greatest diversification in the tropics, particularly in the Eastern Hemisphere. The order appears first in the Upper Cretaceous.

All taxa are rhizomatous, and the rhizome is covered with hairs or scales (Fig. 19-17A). They are dictyostelic or solenostelic (Fig. 19-17C, D). The leaves tend to be two-ranked and vary from simple to simple-pinnate; sometimes they are further divided. Occasionally, the leaf is dichotomously divided. Leaves are usually relatively small, rarely reaching more than 50 cm, but some species of *Dipteris* are more than 2 m tall. Leaves are sometimes strikingly **dimorphic,** but occasionally sporophylls and vegetative leaves are alike. Venation is either open dichotomous or netted; sometimes both types occur on the same sporophyte. The sori always lack an indusium and are frequently round (Fig. 19-17A, B). Occasionally, they are elongate along the vein, and sometimes they form an extensive **coenosorus** over the abaxial surface of the pinna. Elaborate paraphyses are occasionally intermixed among the sporangia. Maturation is mixed. The sporangia are like those of Aspidiales.

Gametophytes are diverse, from generalized, thickened, midrib-bearing thalli with large sex organs to thin, annual thalli with small sex organs.

Discussion of Filicidae

The origins of Filicidae are difficult to determine. It is possible that they arose from ancestors that were similar to Schizaeales or Gleicheniales but not actually from either of these orders. It has been suggested that the ancestor had marginal sporangia, open dichotomous venation, and a creeping rhizome. It is possible that Filicidae are polyphyletic in origin—that is, derived from several separate progenitors. The many features that they have in common, in both sporophytic and gametophytic features, argue against this.

Even in small orders, there are clear evolutionary trends within the order, so that some mem-

Figure 19-16 Aspidiales. **A–C,** *Polystichum lonchitis*. **A,** habit sketch, ×1/4. **B,** detail of pinna with sori, as viewed from underside, ×2. **C,** detail of single sorus, ×20. **D–F,** *Woodsia ilvensis*. **D,** habit sketch, ×1. **E,** detail of pinna, viewed from underside, showing sori, ×10. **F,** detail of sorus, showing sporangia and paraphyses, ×20. **G,** vertical section through a fern sorus, showing receptacle (*rec*) containing scalariform tracheids, indusium (*ind*) emanating from receptacle, and stalked sporangia with annuli (*an*); mesophyll (*mes*) and epidermis (*ep*) of leaf are shown above, ×25.

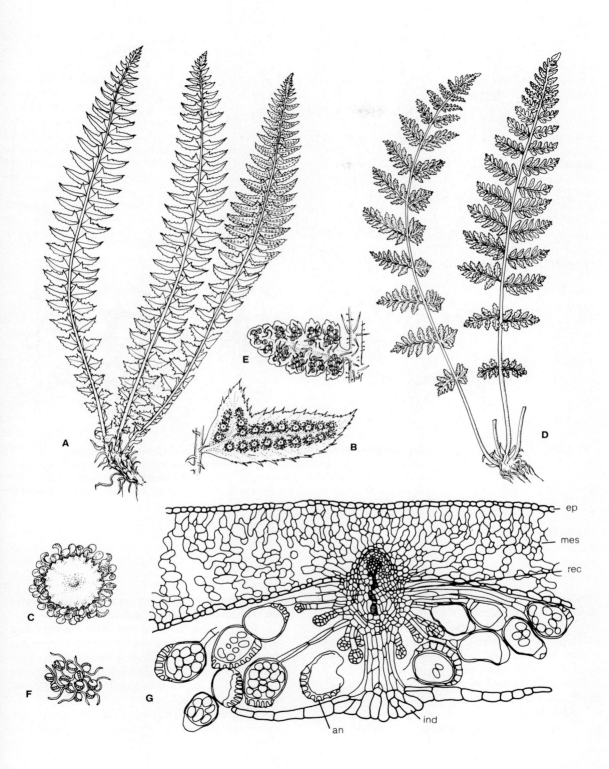

Filicopsids (True Ferns) 485

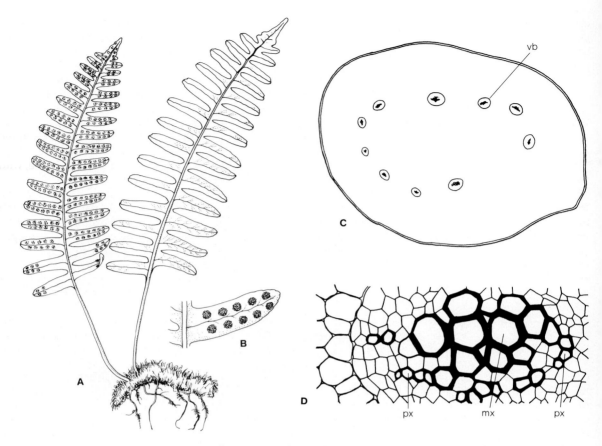

Figure 19-17 Polypodiales—*Polypodium vulgare*. **A,** habit sketch, showing fronds viewed from above and below, ×1. **B,** detail of pinna, showing exposed sori, ×2. **C,** transverse section of stem, showing dictyostelic arrangement of vascular bundles (*v b*), ×2. **D,** single vascular bundle of **C,** showing protoxylem (*px*), and metaxylem (*mx*), ×25.

bers are more specialized in most features than the most generalized member of another order in which large numbers of highly specialized taxa occur. This overlapping of characteristics used to segregate orders and families has presented difficulties in classification and confusion in interpretation. Greater stability will not result until much more detail is available concerning the controversial taxa.

Apogamous production of sporophytes on gametophytes and **aposporous** production of gametophytes on sporophytes further complicates the reproductive biology of Filicidae.

Tryon (1964) has suggested that, among living ferns, the pinnate, determinate leaf form has given rise to all other principal forms, the most derivative condition being the simple blade. Another type of specialization, rather infrequent, is the indeterminate leaf with vinelike habit.

Wagner (1964b) has suggested that the nature and ornamentation of spores may be helpful in clarifying the interrelationships among ferns, particularly Filicidae. However, this procedure must be used with caution and always in correlation with other features.

Subclass Marsileidae

The Marsileidae have a single order, Marsileales, with three closely related genera: *Marsilea* (about 70 species), *Pilularia* (6 species), and *Regnellidium* (one species). *Marsilea* and *Pilularia* are widespread in both tropical and temperate areas, whereas *Regnellidium* is endemic to Brazil. Fossil material is scarce, but *Rodeites*, of late Cretaceous sediments in India, appears to be closely related to *Marsilea*. All genera are aquatic, growing in freshwater marginal shallows with the vegetative pinnae floating on the water surface. The leaves are circinately vernate. All genera are heterosporous, and the sporangia are completely enclosed within a specialized, nutlike structure termed a **sporocarp**. All genera have a creeping rhizome from which leaves emerge in two rows. Roots emerge opposite the leaf bases. The rhizome is irregularly branched, and the stele is a solenostele. The stem usually has a concentric circle of longitudinal air chambers in the cortex.

In *Marsilea*, the leaves have an elongate petiole terminated by four pinnae, rather like a four-leaf clover (Fig. 19-18A). These pinnae apparently arise as three successive dichotomies of the leaf-tip primordium. Venation is essentially dichotomous, although veins frequently fuse along their lengths. Near the base of the petiole, and on a branch from the petiole, are the sporocarps. These vary in number from 1 to 20, depending on the species. Sometimes they occur as pairs on a dichotomizing branch of the petiole; they may arise singly in a row along one side of the petiole. Sporocarps are often covered with hairs. In *Regnellidium* and *Pilularia*, the sporocarps are solitary and on a stalk at or near the petiole base.

The sporocarps are bilaterally symmetrical and very hard in all genera and can remain viable up to 30 years or more. Since most taxa grow in shallow water that appears and disappears seasonally, the existence of sporocarps is clearly important in survival. The sporocarp is generally interpreted as a modified pinna, with sporangia borne in sori along open dichotomous veins. Megasporangia are borne along a ridge, whereas microsporangia are borne along the flanks (Fig. 19-18B, D). Bierhorst (1964) disputes this interpretation of the sporocarp, considering it to be an entirely newly synthesized organ, since early ontogenetic stages in the sporocarp are entirely unlike those that initiate pinnae. In any event, the elongate sori are arranged in rows lining sporocarp walls and are sheathed in indusia (Fig. 19-18B–F, H). Development is generally gradate, with megasporangia initiated first, followed by microsporangia. Each megasporangium contains only a single megaspore at maturity, and it is large enough to be clearly visible without magnification (slightly less than 1 mm long in some species). Microsporangia are more numerous than megasporangia, and each contains 32–64 microspores. Overarching the sori is a mucilaginous structure that runs the length of the sporocarp and extends down either end of it. When the sporocarp "germinates," this mucilaginous structure, with paired sori attached, swells up and extrudes as a tongue, pushing the sori out into the water (Fig. 19-18E, F). The sporangial walls have no special lines of dehiscence; they simply disintegrate when exposed to water and release their spores. Spores develop gametophytes endosporically soon after they are released. The mature megagametophyte ruptures the upper end of the megaspore and exposes the single archegonial neck and the cells surrounding it. Most of the megagametophyte is made up of a single, starch-rich cell (Fig. 19-18I). The megaspore wall imbibes water and sheathes the megagametophyte in a mucilaginous mass.

When mature, the microgametophytes contain 16 spermatocytes surrounded by jacket cells. They are released as two masses, which break up on dissolution of their enclosing membranes, allowing the sperms to swim freely. Sperms are essentially spherical and have a long, corkscrew tail with numerous flagella. When trapped by the gelatinous covering of the megagametophyte, the flagellate sperms burrow through it, and one unites with the egg, forming a zygote that undergoes cell divisions soon afterward. The young embryo is nourished by the starch-rich cell of the megagametophyte and is initially enclosed by a **calyptra** of gametophytic origin, much as in the bryophytes. In time, a leaf and root are initiated, and the young sporophyte is independent.

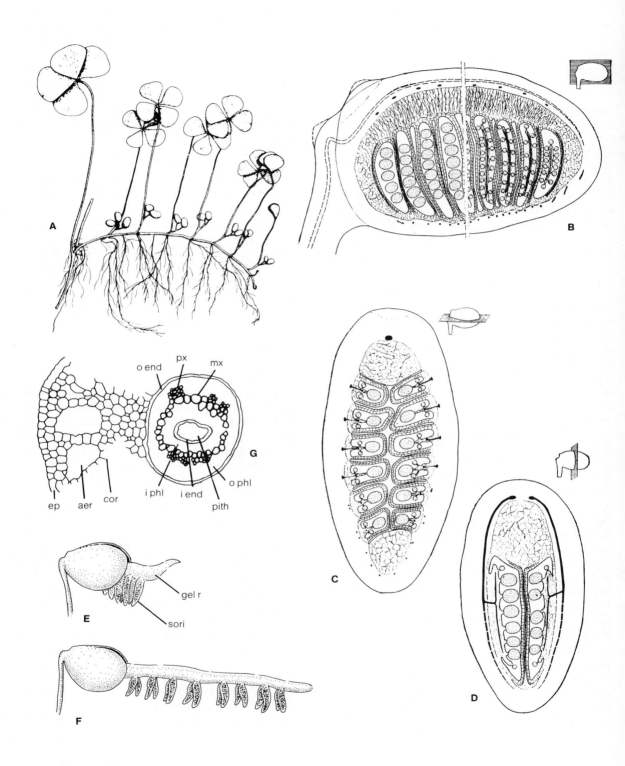

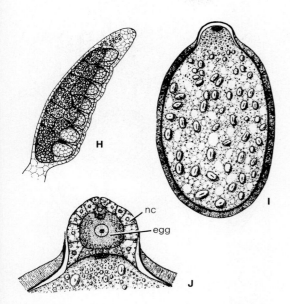

Figure 19-18 *Marsilea quadrifolia.* **A,** habit sketch, showing rhizome with leaves and sporocarps, ×1/2. **B–D,** diagrams illustrating structure of the sporocarp. **B,** section in dorsiventral plane, somewhat to one side of the median line, one portion slightly farther away than the other. **C,** longitudinal section at right angles to **B**. **D,** transverse section at right angles to **C**. The vascular strands are indicated by heavy black. The mucilaginous ring is shown with irregular, cell-like structure. **E–F,** various stages in germination of the sporocarp, showing extrusion of the mucilaginous ring with attached sori, ×1. **G,** transverse section of stem, showing stele and cortex with aerenchyma, ×100; **H,** detail of a sorus, ×4. **I,** female gametophyte in megaspore, showing starch grains and archegonium, ×25. **J,** detail of archegonium, ×75: *aer,* aerenchyma with large cavities; *cor,* cortex; *ep,* epidermis; *i end,* inner endodermis; *i phl,* inner phloem; *mx,* metaxylem; *o end,* outer endodermis; *o phl,* outer phloem; *px,* protoxylem. (**A–D,** from A. J. Eames, *Morphology of Vascular Plants,* © 1936; used with permission of McGraw-Hill Book Co. **E, F, H,** redrawn after Eames; **G,** redrawn after Smith; **I, J,** from A. W. Haupt, *Plant Morphology,* © 1953; used with permission of McGraw-Hill Book Co.)

In *Pilularia,* the leaf is a simple cylinder and lacks any leaflets. The single sporocarp at the leaf base contains only four sori. The sporangial walls vaguely resemble those of Schizaeales, but the annulus is **vestigial.** When ruptured, the sporocarp breaks into four parts and exposes the sporangia. In *Regnellidium,* the leaf has two fleshy leaflets, and the single sporocarp is similar in construction to that of *Marsilea. Regnellidium* produces latex; it is the only nonflowering plant that does so.

In Marsileidae, then, there are numerous specializations: leptosporangiate heterosporous condition, remarkably specialized sporocarp, extreme reduction of sporophyte and gametophytes, and extreme simplification of sporangial dehiscence. However, associated with these features are generalized characteristics, including the persistence of a calyptra reminiscent of that in bryophytes, dichotomous venation, and the basically marginal position of the sori. The subclass is suggested to be closest to Schizaeales, but this relationship is very distant.

Subclass Salviniidae

The Salviniidae form another subclass of aquatic ferns, all of which float on the water surface. The single order, Salviniales, has two genera: *Salvinia* (10 species) and *Azolla* (18 species). *Salvinia* is found mainly in tropical America and Africa, but it does extend northward in Europe to the Netherlands and Czechoslovakia. *Azolla* is widely distributed in temperate and tropical areas. *Azolla* is known in the fossil record from as early as late Cretaceous time; during the Tertiary, the genus was widespread in the Northern Hemisphere, far north of its present natural range. *Salvinia* is also known as fossil material of Tertiary age, both in the northern United States and northern Europe. Salviniidae lack circinate vernation, have a horizontal, irregularly branched rhizome, are heterosporous, and have sporocarps that are modified sori.

Azolla is small and mosslike, with minute, bilobed leaves (usually less than 0.5 mm long) that are arranged in alternating helical rows (Fig. 19-19A, B). The upper lobe is photosynthetic and floats on the water surface, whereas the lower, submerged lobe is colorless and more delicate. Roots generally arise at the junctions of branches that

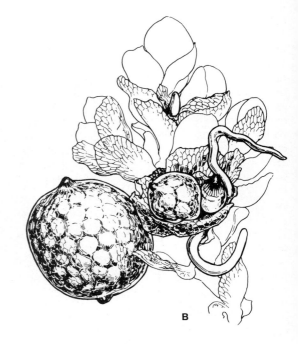

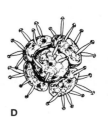

appear irregularly, but bilaterally, on the stem. A cavity at the base of each leaf contains the nitrogen-fixing bluegreen alga *Anabaena*. The stems are fragile, and the stele is essentially a protostele, presumably a result of extreme reduction.

The sporocarps are in pairs (or fours). They are modified indusia and arise as outgrowths of submerged leaf lobes. At first, each sporocarp contains both microsporangia and megasporangia. If the megasporangium develops, the microsporangia abort; if the megasporangium does not develop, microsporangia develop. Consequently, one of each pair of sporocarps contains only megasporangia or microsporangia. Each megasporangium produces 32 spores, one of which matures, resulting in a single megasporangium within the sporocarp. The megaspore enlarges to fill the sporangium and the sporocarp.

The formation of the megaspores is unusual. The eight **sporocytes** are scattered in a multinucleate mass formed by the breakdown of the cell walls of the sporangium's inner walls. These sporocytes undergo meiosis, and the 32 megaspore nuclei are free within the sporangium. This multinucleate mass ultimately becomes vacuolate, and all but one of the megaspores abort. The remaining one enlarges and fills the lower portion of the sporocarp. The vacuoles above the spore become pear-shaped and compress against each other to crown the megaspore with a cap.

The microsporangia contain 16 sporocytes that ultimately produce 64 microspores embedded in a vacuolar mass similar to that of megaspores in their early development. As vacuoles enlarge, the microspores move into peripheral vacuoles, several occupying each vacuole. This vacuolar mass with the included spores is termed a **massula**. Sometimes the outer surface of the massula produces tubular outgrowths with anchor-shaped tips, called **glochidia** (Fig. 19-19D).

Megaspores are shed by the breakdown of the sporocarp and sporangial wall. The gametophyte is endosporic, with one or more archegonia embedded in the end facing the pear-shaped vacuoles. The remainder of the spore contains nutrient material. The irregular surface of the lower face of the spore catches the glochidia of the massulae. Male gametophytes form after rupturing the microspore wall; thus they are partially exosporic. They produce eight sperms that are released from the massula when the vacuolar walls surrounding them decay.

Salvinia has leaves in whorls of three. Two are grooved, boat-shaped, and floating, and one is submerged and finely dissected. There are no roots (Fig. 19-19E). The stem is siphonostelic. The sporocarps (4–20) are borne on the inner segments of the submerged leaves. The first-formed sporocarps contain megasporangia and, as in *Azolla*, ultimately contain a single one. As in *Azolla*, the sporangial contents ultimately produce a single spore enclosed by the accumulated material of aborted spores. A thick coat is deposited, and its upper end contains a chamber similar to the pollen chamber in seed plants. Microspores ultimately produce a massula much as in *Azolla*, but no glochidia.

Both genera have brittle stems; thus vegetative reproduction is common, and colonies often expand in mats over stagnant water. Flotation is enhanced by the slow wetting properties of the floating leaves. Salviniales are suggested to be closest to Hymenophyllales, with which they share terminal indusia, gradate development of sporangia, and a similar residual annulus. Both orders consist of extremely reduced representatives, and it is possible that these shared features have resulted from parallel evolution. Salviniales show numerous specializations in the vegetative sporophyte, including reduction of the leaves and reduction of the stele.

In the fossil record, besides *Azolla* and *Salvinia*, three strictly fossil genera are recognized, based largely on distinctive megaspores and massulae: *Azollopsis*, *Parazolla*, and *Glomerisporites*. The abundance of fossil material suggests that the order was highly successful in the past.

Figure 19-19 Salviniidae. **A–D**, *Azolla filiculoides*. **A,** photograph of a population of fronds shown floating on water surface, ×2. **B,** detail of plant, shown dissected out and viewed from below to show roots, a microsporocarp (*left*, shown in detail to its left), and a megasporocarp to its right, ×20. **C,** young stalked microsporangium, ×40. **D,** separating massulae, showing glochidia, ×40. **E,** photograph of *Salvinia*, shown floating on water surface. (**B–D**, after H. L. Mason, in *A Flora of the Marshes of California*, with permission of University of California Press.)

References

Arnold, C. A. 1964. Mesozoic and Tertiary fern evolution and distribution. *Mem. Torrey Bot. Club* 21(5):58–66.

Bierhorst, D. W. 1964. Suggestions and comments on teaching materials of the non-seed bearing plants. *Amer. Biol. Teacher* 26:105–107.

Bower, F. O. 1923–28. *The ferns (Filicales) treated comparatively with a view to their classification.* 3 vols., Cambridge University Press.

Campbell, D. H. 1918. *The structure and development of mosses and ferns.* Macmillan & Co., London.

Copeland, E. B. 1947. *Genera Filicum, the genera of ferns.* Chronica Botanica Co., Waltham, Mass.

Dyer, A. F. 1979. *The experimental biology of ferns.* Academic Press, New York.

Goebel, K. 1930. *Organographie der pflanzen.* Teil 2. *Bryophyten-Pteridophyten.* G. Fischer, Jena.

Hall, J. W. 1974. Cretaceous Salviniaceae. *Ann. Missouri Bot. Gard.* 61:354–67.

Hewitson, W. 1962. Comparative morphology of the Osmundaceae. *Ann. Missouri Bot. Gard.* 49:57–93.

Holttum, R. E. 1947. A revised classification of leptosporangiate ferns. *Bot. J. Linn. Soc.* 53:123–58.

———. 1949. The classification of ferns. *Biol. Rev.* 24:267–96.

———. 1958. Morphology, growth, habit, and classification in the family Gleicheniaceae. *Phytomorphology* 7:168–84.

———. 1965. The evolution of the vascular system in ferns with special reference to dorsiventral rhizomes. *Phytomorphology* 14:477–80.

Jermy, A. C.; Crabbe, J. A.; and Thomas, B. A., eds. 1972. The phylogeny and classification of the ferns. *Bot. Journ. Linn. Soc.* 67 (Suppl. I).

Klekowski, E. J. 1971. Ferns and genetics. *Bioscience* 21:317–22.

Lloyd, R. M. 1974. Reproductive biology and evolution in the Pteridophyta. *Ann. Missouri Bot. Gard.* 61:318–31.

Manton, I. 1955. Pteridology. In *A Century of Progress in the Natural Sciences, 1853–1953.* Calif. Acad. Sci., San Francisco.

Mickel, J. T. 1974. Phyletic lines in the modern ferns. *Ann. Missouri Bot. Gard.* 61:474–82.

Nayar, B. K., and Kaur, S. 1971. Gametophytes of homosporous ferns. *Bot. Rev.* 37:295–396.

Ogura, Y. 1972. Comparative anatomy of vegetative organs of the pteridophytes. In Zimmerman, W., ed., *Handbuch der Pflanzenanatomie.* ed. 2. Band VII, Teil 3. Gebrüder Borntraeger, Berlin.

Phillips, T. L. 1974. Evolution of vegetative morphology in coenopterid ferns. *Ann. Missouri Bot. Gard.* 61:427–61.

Pichi-Sermolli, R. E. G. 1958. *The higher taxa of the pteridophyta and their classification.* In O. Hedberg, ed., *Systematics of today.* Uppsala Univ. Årsskrift, 1958.

———. 1959. Pteridophyta. In Turrill, W. B., *Vistas in Botany*, Vol. 1.

Sporne, K. R. 1966. *The morphology of pteridophytes.* Hutchinson and Co. (Publ.), London.

Steeves, T. A. 1966. *On the determination of leaf primordia in ferns.* In Cutter, E. G., ed., *Trends in plant morphogenesis.* Longmans, London.

Stidd, B. M. 1974. Evolutionary trends in the Marattiales. *Ann. Missouri Bot. Gard.* 61:388–407.

Tryon, R. M. 1964. Evolution in the leaf of living ferns. *Mem. Torrey Bot. Club* 21(5):73–85.

Verdoorn, F., ed. 1938. *Manual of pteridology.* Nijhoff, The Hague.

Wagner, W. H. 1964a. Paraphyses: Filicineae. *Taxon* 13:56–64.

———. 1964b. The evolutionary patterns of living ferns. *Mem. Torrey Bot. Club* 21(5):86–95.

Wardlaw, C. W. 1963. Apical organization and differential growth in ferns. *Bot. J. Linn. Soc.* 58:385–400.

White, R. A. 1974. Comparative anatomical studies of the ferns. *Ann. Missouri Bot. Gard.* 61:379–87.

Gymnosperms (from the Greek for *naked seed*) are plants of diverse habit and structure, but they have similar reproductive patterns. Although they belong to several distinct classes, we now believe that the different lineages of gymnosperms have a common ancestry within a group of Devonian plants, the progymnosperms. Extant classes of gymnosperms are: Cycadopsida, Ginkgopsida, Coniferopsida, and Gnetopsida. Two other classes, Pteridospermopsida (pteridosperms, or seed-ferns) and Progymnospermopsida (progymnosperms), are extinct.

Class Progymnospermopsida

The concept of a group of plants ancestral to the several gymnosperm groups was established in 1960 when Beck discovered leaf-bearing branches called *Archaeopteris* connected with woody stems of the genus *Callixylon* in Upper Devonian rocks of New York state. Up to the time of this discovery, botanists had usually considered *Archaeopteris* an early fern because it had pinnate arrangement of parts and reproduced by spores. On the other hand, *Callixylon*, with bordered-pitted tracheids in the secondary xylem, was classified as a gymnosperm. When Beck found *Archaeopteris* leaves actually attached to *Callixylon*, he realized the significance of the combination of pteridophytic reproduction by spores and secondary xylem of gymnosperms, and proposed a new group, progymnosperms. Since then, Beck and others have reinvestigated several Devonian plants, concluding that they are more closely related to *Archaeopteris* than to strictly pteridophyte groups. As a result, two main orders of progymnosperms are now recognized, Aneurophytales and Archaeopteridales. Progymnosperms are important not only because they represent early evolutionary stages of gymnosperms, but also because they reveal some of the earliest stages of leaf development, bordered pitting of gymnosperm tracheids, and the formation of periderm.

Order Aneurophytales

Of the two groups, Aneurophytales are more primitive, occurring first in early Middle Devonian, some

20

PROGYMNOSPERMS

370 million years ago. They appear to represent one evolutionary offshoot from the Rhyniophytina-Trimerophytina lineage.

Aneurophyton appears to have been a small tree with a central trunk and branches up to 1 m long (Fig. 20-1A). The branches are arranged spirally to give a bushy, two-dimensional appearance. Ultimate branching is dichotomous, producing two fine appendages that are believed to have functioned as leaves. The sporangia are borne in clusters on tiny recurved branches, and (as far as is known) are **homosporous.** The spores have a **trilete mark,** and are classified in the **dispersed-spore genus** *Aneurospora.*

The stem has a three-armed **protostele** that is **mesarch** (Fig. 20-2A). This is surrounded by secondary xylem consisting of tracheids with numerous rows of **circular bordered pits** and tall **uniseriate rays.** This is a gymnospermous characteristic shared with other progymnosperms. The presence of **vascular** and **cork cambia** is significant, as they permitted continuing lateral growth, resulting in additional strength and support.

Another interesting genus of Aneurophytales is *Tetraxylopteris,* a good-sized, woody and shrubby plant with **decussate branching** (Fig. 20-1B). The ultimate tips also branch decussately, with three different patterns of ultimate branching on the same axis (Fig. 20-1B, a–c). In *Tetraxylopteris,* ultimate appendages are interpreted as leaves, because vascular tissue in the tips is a simple circular strand compared to the four-armed protostele of other parts of the stem (Fig. 20-2B). The sporangia are borne on special fertile appendages (Fig. 20-3B) that are more leaflike than sterile branches (Fig. 20-3A). Each fertile stalk is four-lobed and divides twice dichotomously to produce four major branches. Each of the four then further divides three times to give an irregular, pinnate arrangement of small branches (Fig. 20-3B). Each ultimate branch tip bears one sporangium. Sporangium-bearing branches arise from the upper surface of the pinnate branches and are flattened *in one plane* like a leaf. The spores are trilete, spherical, and **pseudosaccate,** with the outer part of the **exine** expanded into a bladder or **pseudosaccus** (Fig. 20-2F). They are identical with spores of the genus *Rhabdosporites,* which is widely dispersed in Devonian rocks.

The secondary xylem is similar to that in *Aneurophyton* and shows a gymnospermous pattern of circular bordered pits crowded on both the radial and tangential walls of tracheids.

These examples of Aneurophytales show the essential characteristics of early progymnosperms: (1) three-dimensional branching; (2) forking, leaflike branch tips; (3) sporangia terminal on ultimate divisions, with sporangia borne on pinnately arranged branches in *Tetraxylopteris;* and (4) circular bordered pits on the radial and tangential walls of tracheids of the secondary xylem.

Order Archaeopteridales

The Archaeopteridales include several genera ranging from late Middle Devonian to early Carboniferous. Although it was previously thought that Archaeopteridales were derived from Aneurophytales, Beck (1976) has proposed that the two orders of progymnosperms represent two different evolutionary pathways from earlier ancestral stock (Fig. 20-6).

By far the best-known genus of Archaeopteridales is *Archaeopteris* (Fig. 20-4A), a large tree forming one of the dominants of Upper Devonian plant communities. Beck's discovery of *Archaeopteris* foliage attached to woody stems of gymnosperms, uniting fernlike and gymnospermous characteristics, was enormously significant for our understanding of the early evolution of gymnosperms and the phylogeny of the various classes. It also opened up vistas of the relationships of Aneurophytales to gymnosperms rather than to early fernlike groups, such as coenopterids and cladoxylids.

Beck reconstructed *Archaeopteris* with a habit similar to that of modern conifers: a straight main trunk and numerous branches. Specimens with the largest trunk are over 1.5 m in diameter and 9 m long, but the trees undoubtedly reached greater heights. The main branches are horizontal or angled slightly upward; they bear frondlike branches in the axils of leaves. The frondlike secondary branches bear opposite, leaf-bearing branches that are **planated**—that is, flattened into one plane (Fig. 20-4B). Significantly, studies of the vascular system of the leaf-bearing branches indicate that vascular strands originate spirally and that these branches apparently became planated by unequal growth and twisting during maturation. Also, it is noteworthy that each leaf-bearing branch occurs between two **rachial** leaves of the secondary branch, again suggesting that they represent branches. The clinching evidence that all branches of the axis are really **cauline** in origin was provided by critical studies indicating that the stele was radially structured in all parts except the terminal laminated blades. As a result, we now know that the *Archaeopteris* "frond" is really a planated series of branches bearing leaves on their outermost regions. Each leaf receives strands that leave the radial stele spirally. The leaf is currently interpreted as representing an evolutionary modification of small, dichotomous, ultimate branch systems through **reduction, planation,** and **lamination** (webbing) (see Chapter 27).

The woody stem contains a narrow pith surrounded by mesarch primary xylem bundles (Fig. 20-2D). Branch traces were produced tangentially and in spiral sequence. Secondary wood occurs outside of the primary, and consists of radial rows of tracheids and thin, high rays. It is remarkably gymnospermous in character: tracheids have groups of circular bordered pits on the radial walls (Fig. 20-2C); the tracheid wall is thickened between the groups of pits; and the narrow vascular rays have characteristic **ray tracheids,** which are otherwise known only in the order Coniferales. Thus the secondary xylem of *Archaeopteris* indicates that plants with very definite gymnospermous anatomical features had evolved by Upper Devonian time.

The reproductive structures of *Archaeopteris* consist of sporangia borne on small stalks on the leading edges of fertile branchlets that occur in place of leaves (Fig. 20-5). It is interesting that the ends of sporangial branchlets have one dichotomy and several uneven dichotomous projections, suggesting an origin by reduction from **telomes.** At least

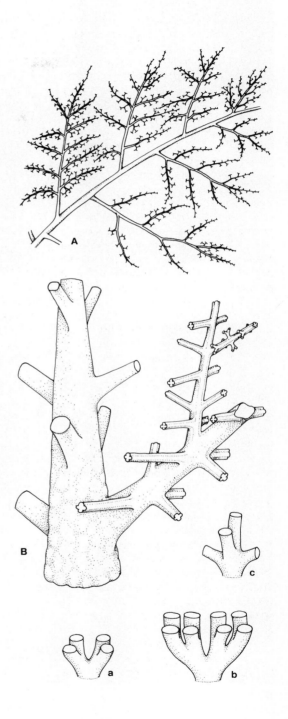

Figure 20-1 A, lateral branch system of *Aneurophyton,* showing spirally arranged branches and small, dichotomously forked branch tips, × 1/2. **B,** arrangement of several orders of branching in *Tetraxylopteris schmidtii,* showing three different arrangements of branch endings in **a, b,** and **c,** × 10. (**A,** after Krausel and Weyland, 1925; **B,** after S. E. Scheckler and H. P. Banks, *Amer. J. Bot.* 58:737-51, 1971, with permission of the authors and *American Journal of Botany.*)

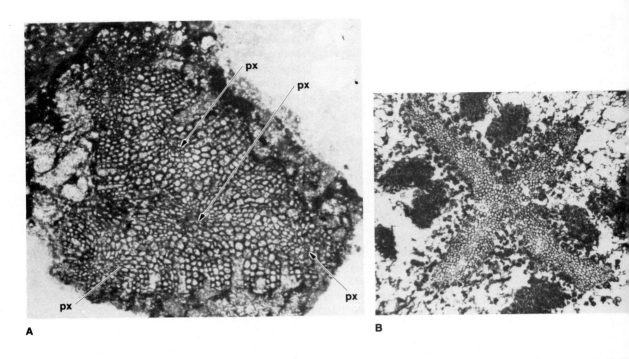
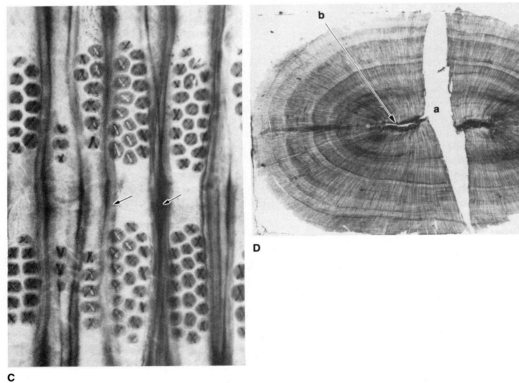

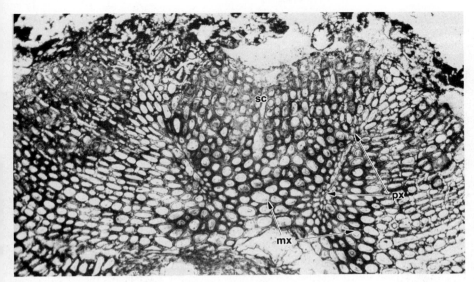

E

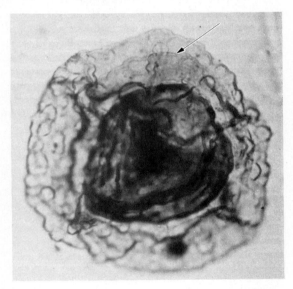

F

Figure 20-2 **A,** transverse section of petrified branch of *Aneurophyton,* showing the three-armed xylem with four mesarch protoxylem strands (*px*) and some secondary xylem cells on the outside, ×42. **B,** transverse section of branch of *Tetraxylopteris,* showing four-armed protostele, × ca.40. **C,** pits of *Archaeopteris (Callixylon)* tracheids, showing variation from elliptic to stiltlike apertures, and crassulae *(arrows),* ×280. **D,** transverse section of *Archaeopteris* axis, showing well-developed growth rings and radial files of tracheids and rays in secondary xylem (where *a* shows wedge of cells destroyed by secondary mineralization; arrow *b* shows pith tissue crushed during preservation), × ca.1/4. **E,** transverse section of vascular bundle of *Archaeopteris,* showing small protoxylem cells (*px*), polygonal metaxylem (*mx*), and files of rectangular secondary tracheids (*sc*) just beyond the metaxylem, ×88. **F,** spore of *Tetraxylopteris,* ×500: note dark central spore body and the lighter surrounding bladder-like layer—the pseudosaccus; arrow points to contact of two layers that may be a split in wall of pseudosaccus. (**A,** photograph courtesy of Harlan P. Banks, with permission of B. S. Serlin and H. P. Banks and *Palaeontographica Americana.* **B,** after H. P. Banks, *Evolution and Plants of the Past,* Wadsworth Publishing Co., 1970. **C,** after C. B. Beck, with permission of the author and Cambridge University Press, from "On the appearance of gymnospermous structure," *Biol. Rev.* 45:379–400, 1970. **D,** after H. P. Banks, with permission of the author and Wadsworth Publishing Co. **E,** from C. B. Beck, with permission of the author and *American Journal of Botany,* from *Amer. J. Bot.* 58:758-84, 1971. **F,** after P. M. Bonamo and H. P. Banks, with permission of the authors and *American Journal of Botany,* from *Amer. J. Bot.* 54:755-68, 1967.)

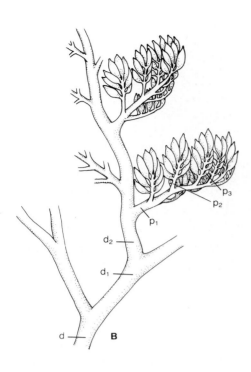
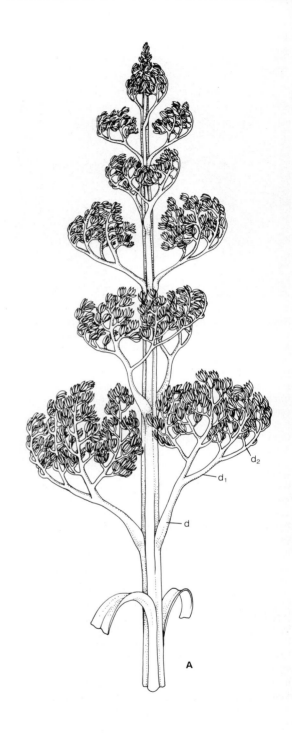

Figure 20-3 Reconstructions of fertile branches of *Tetraxylopteris*. **A,** whole fertile branch system, showing three orders of dichotomous branches (d, d_1 and d_2). **B,** three orders of smaller branches (p_1, p_2, p_3) on larger branches; p_3 branches are obscured by sporangia and p_1 branches have been much reduced in these reconstructions. **A,** × ca. 1; **B,** × 2. (After P.M. Bonamo and H. P. Banks, with permission of the authors and *American Journal of Botany*, from *Amer. J. Bot.* 54:755-68, 1967.)

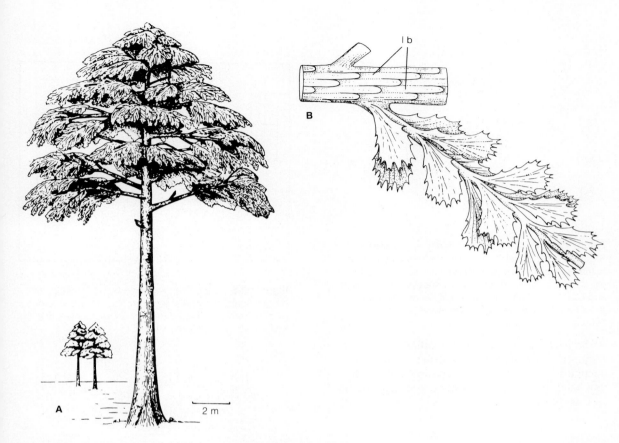

Figure 20-4 A, reconstruction of *Archaeopteris*. **B,** restoration of part of lateral branch system of *Archaeopteris macilenta*, showing part of main axis ensheathed in leaf bases (*l b*) as if leaves had abscissed, and part of lateral branch bearing leaves, × ca.2. (**A,** after C. B. Beck, with permission of the author and *American Journal of Botany*, from *Amer. J. Bot.* 49:374, 1962; **B,** after C. B. Beck, with permission of the author and *American Journal of Botany*, from *Amer. J. Bot.* 58:758-84, 1971.)

one species is known to be **heterosporous,** and several others are considered likely to be heterosporous, with the possibility that **microsporangia** and **megasporangia** were on different fertile branches or even on different plants. Furthermore, it has been postulated that several species of *Archaeopteris* may have produced seeds with **cupules** of the *Archaeosperma* type, which is the earliest known gymnospermous seed that occurs with *Archaeopteris* in some Upper Devonian sediments. Although this cannot be corroborated until seeds are actually found attached to *Archaeopteris*, frequent association has led Beck and others to be very watchful for such a union.

The results of all this critical research on progymnosperms has resulted in new concepts regarding evolution and the relationships of gymnospermous groups. The interpretation becoming increasingly accepted is that evolution progressed from rhyniophyte-psilotophyte-trimerophyte ancestors of the Lower Devonian to Aneurophytales of Middle and Upper Devonian in one branch, and

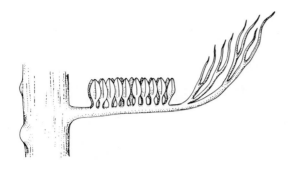

Figure 20-5 Reconstruction of fertile appendage of *Archaeopteris*, showing two rows of sporangia attached to the adaxial surface and unequal divisions of the terminal region, ×2.5. (After L. M. Carluccio, F. M. Hueber, and H. P. Banks, with permission of the authors and *American Journal of Botany*, from *Amer. J. Bot.* 53:721, 1966.)

to the Archaeopteridales in the other branch (Fig. 20-6). *Archaeopteris* represents structural and reproductive advancement, with development of a true tree habit, true laminated leaves, and heterospory (at least in some species).

Evolution of the Seed

One of the most significant events in the history of vascular plants was the evolution of the **seed habit**. The earliest seed found to date is from late Devonian, some 35 million years after the first vascular plants appeared. Following the evolution of the seed, the several groups of gymnospermous seed plants became dominant in floras of successive geologic periods, culminating with conifers, cycads, and ginkgos of the Mesozoic Era. Because of the seed's significance, it is important to note some key developments in this evolutionary progression.

In extant gymnosperms, the seed develops from an **ovule** following fertilization and subsequent development of the embryo. The gymnosperm ovule consists of an **outer integument** that surrounds a megasporangium containing (usually) one functional megaspore. This enclosure is complete except

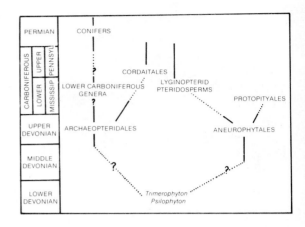

Figure 20-6 Suggested evolutionary relationships of progymnosperms and several groups of Carboniferous gymnosperms. (Adapted from Andrews, Gensel, and Kasper, 1975; and Beck, 1976.)

Figure 20-7 A, probable main stages in evolution of seed (**d**) from heterosporous ancestors (**b, c**), which in turn probably evolved from a homosporous ancestor (**a**); **c** shows a single functional megaspore surrounded by three aborted megaspores: *arch*, archegonia; *f gam*, female gametophyte; *int*, integument; *mega*, megaspores; *meio*, meiospores; *micro*, microspores; *nuc*, nucellus (= megasporangial wall); *spor*, sporangium. **B, a–e,** a series of early Carboniferous seeds attributed to pteridosperms, showing degrees of fusion of integumentary lobes leading to true micropyle in **e**; **b** shows development of pollen chamber in nucellus of seed in **a**: *int*, integument; *int l*, integumentary lobe; *m*, micropyle; *nuc*, nucellus; *p ch*, pollen chamber; *spor*, sporangial wall. (**A**, after H. N. Andrews, with permission of the author and *Science* 142:926, 15 November 1963. Copyright 1963 by the American Association for the Advancement of Science. **B, a, c–e,** after H. N. Andrews, with permission of the author and *Science* 142:927–28, 15 November 1963; copyright 1963 by the American Association for the Advancement of Science; **b**, after A. G. Long, with permission of the author and the Royal Society of Edinburgh.)

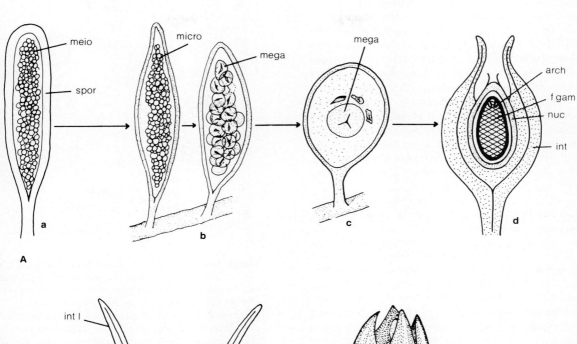

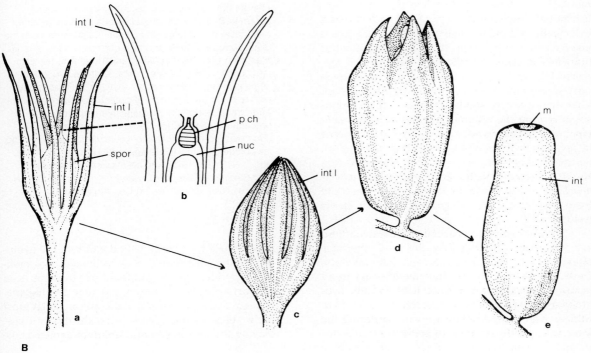

Progymnosperms

for one point on the distal end, an opening called the **micropyle.** The **female gametophyte** develops completely *inside* the integument. Following fertilization *inside* the ovule, the embryo develops within the nutritive tissue of the female gametophyte. The integument usually becomes harder after fertilization and develops into a protective **seed coat.**

The evolution of the seed appears to have followed evolutionary progression from homosporous to heterosporous ancestors, and thence to seed plants (Fig. 20-7A, a–d). **Palynological** studies of Devonian spores have shown that early vascular plants were almost certainly homosporous. Increases in the size range of spores occurred from Lower to Middle Devonian, suggesting that this marked the inception and development of heterospory. The evolution of two spore sizes was probably accompanied by separation of the two sexes onto the respective gametophytes.

The next stage in the evolution of the seed appears to have been the development of **integumentary lobes** from tissues surrounding the megasporangium. Such lobes are expressed in the earliest seed, *Archaeosperma,* from the Upper Devonian. They suggest that the seed did evolve by enclosure of the megasporangium by integumentary lobes, accompanied by reduction of the functional megaspores to one. Such enclosure can be visualized if the fertile branch of *Archaeopteris* is taken as an example (Fig. 20-5). Five dichotomous branches at the end of the sporangial branch represent telomelike segments that could have enveloped a megasporangium by reduction and fusion. That this or a very similar process occurred gains support from the different degrees of integumentary enclosure observed in early Carboniferous seeds believed to be those of seed ferns (Fig. 20-7B, a–e). Although this series has not been shown to represent a **lineal evolutionary progression,** it suggests one way in which the megasporangium could have become progressively enveloped by sterile tissue.

It is interesting to speculate on the evolutionary advantages provided by the seed. We have already noted that during this same interval of the Devonian, there was a gradual evolutionary trend toward the woody, **arborescent** habit in some evolutionary lines. It is almost certainly more than coincidence that the seed was evolving during the same interval. In addition to better protection for

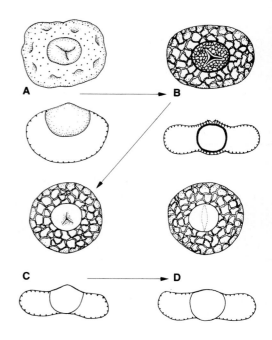

Figure 20-8 **A–D,** proposed evolutionary sequence for cordaite pollen. In each figure, *upper view* is polar and proximal, *lower* is an equatorial side-view. Evolutionary change from **A,** a pseudosaccate progymnosperm spore, to **B,** a saccate prepollen of a cordaite required pseudosaccus stabilization by saccus attachment to the distal surface as in **B,** and also change in saccus ornamentation from the monoform-uniform of **A** to the reticulate pattern of **B.** In a later stage, **C,** there is a loss of some external ornamentation, thinning of the corpus wall, and the development of potential distal germinal aperture, although the proximal trilete aperture is still present. In pollen of late Carboniferous cordaites, **D** (*Florinites*-type), there is a lenticular distal aperture, and loss of the proximal trilete aperture. (After M. A. Millay and T. N. Taylor, 1976, with permission of the authors and *Review of Palaeobotany and Palynology* 21:65–91, Elsevier Scientific Publishing Co., Amsterdam.)

both female gametophyte and **embryo,** the evolution of the seed would have facilitated both pollination and fertilization in the air, free from the need for surface water shared by **vascular cryptogams.** Indeed, fossil evidence to date indicates that most early seed-bearing plants were shrubs or trees. Having the production and union of gametes take

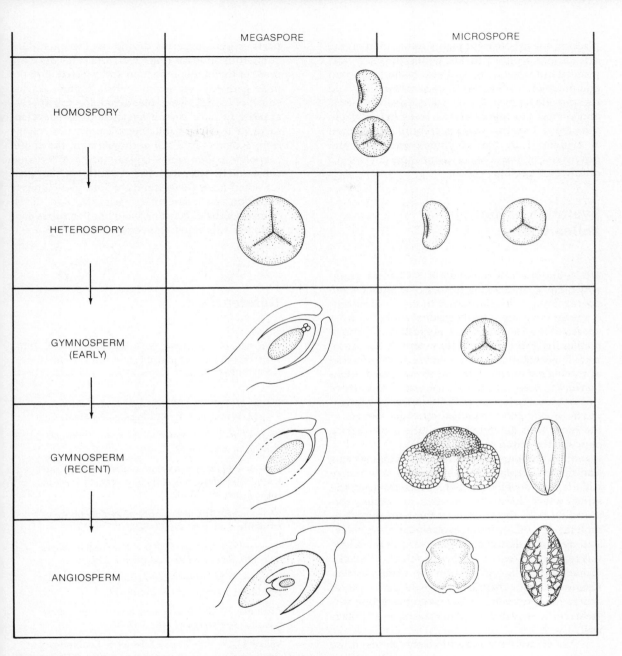

Figure 20-9 Diagram of spore evolution from early homosporous stage to angiosperm condition. Proximal hemispheres are shown by *open stippling;* distal hemispheres are *clear.* Megaspores retained in the seed are indicated by *close stippling.* Note that in the "Gymnosperm (early)," prepollen still germinated through the proximal trilete aperture, even though the megaspore had already become enclosed in the seed. (After W. G. Chaloner, 1970, with permission of the author and *Geoscience and Man,* copyright the School of Geoscience, Louisiana State University, Baton Rouge.)

place on the arborescent plants rather than on the soil surface would provide better efficiency and protection for gametes, and also better protection and nutrition for the developing embryo. It would also provide for the relatively wide dispersal of seeds and embryos by wind currents or by free-fall from the plants. That this was a successful combination is evidenced by the continued evolution and expansion of the several gymnosperm lineages during later geologic epochs.

Evolution of the Pollen Grain

The counterpart of the ovule is the **pollen grain,** which has an interesting history of evolution from spores. From examples found in the fossil record, it appears that there was a gradual evolution from spores of the Devonian into **prepollen** of the Carboniferous and hence to true pollen in the Mesozoic. This evolutionary series involved a reduction of the male gametophyte and its increased retention *inside* the protective spore wall. Later, there was a gradual change in the location of gametophytic germination from the **proximal surface** of the spore with the **tetrad scar** to the distal surface opposite the tetrad mark.

In evolutionary branches of gymnosperms leading to Cordaitales and Coniferales, this change included the development of a protective coat, the exine, which formed a bladder, or **saccus,** surrounding both the spore and the reduced gametophyte (Fig. 20-8). In the evolutionary branch(es) leading to cycads and some seed ferns, no bladder was formed; instead, a furrow, or **sulcus** (*pl.* **sulci**), formed as a thin region(s) on the distal surface. This was a major change in the male gametophyte generation, providing for increased protection and greater efficiency of energy supply *inside* the pollen wall.

As yet, we have not put all the pieces of the evolutionary story together. However, it is interesting, intriguing, and likely very significant that similar stages of reduction and enclosure were occurring in both the male and female reproductive parts of gymnosperms during the Devonian-Carboniferous interval (Fig. 20-9). As with development of the ovule, reduction and protection of the male gametophyte inside pollen grains—at least some of which were bladdered for transport—appears to have been a response to the development of the tree habit. Significantly, the change from proximal to distal germination in the spore-prepollen-pollen series appears to have occurred substantially later than the evolution of the seed. As noted later, these trends toward gametophyte reduction, enclosure of megaspores, and development of distal germination by **pollen tubes** continued in subsequent geologic epochs.

References

Andrews, H. N.; Gensel, P. G.; and Kasper, A. E. 1975. A new fossil plant of probable intermediate affinities (trimerophyte-progymnosperm). *Can. J. Bot.* 53:1719–1728.

Banks, H. P. 1970. *Evolution and plants of the past.* pp. 77–112. Wadsworth Publishing Co., Belmont, Calif.

Beck, C. B. 1960. Connection between *Archaeopteris* and *Callixylon. Brittonia* 12:351–68.

———. 1962. Reconstruction of *Archaeopteris,* and further considerations of its phylogenetic position. *Amer. J. Bot.* 49:373–82.

———. 1970. The appearance of gymnospermous structure. *Biol. Rev.* 45:379–400.

———. 1976. Current status of the Progymnospermopsida. *Rev. Palaeobot. and Palynol.* 21:5–23.

Bonamo, P. M. 1975. The Progymnospermopsida: building a concept. *Taxon* 24:529–79.

Chaloner, W. G. 1970. The evolution of meiospore polarity. *Geoscience and Man* 1:47–56.

Millay, M. A., and Taylor, T. N. 1976. Evolutionary trends in fossil gymnosperm pollen. *Rev. Palaeobot. and Palynol.* 21:65–91.

Pettitt, J. 1970. Heterospory and the origin of the seed habit. *Biol. Rev.* 45:401–15.

21

PTERIDOSPERMS (SEED FERNS)

The pteridosperms, or seed ferns, are an extinct group with frondlike compound leaves bearing both seeds and pollen-producing organs. As with progymnosperms, the abundant leaves of the Carboniferous were once believed to be those of true ferns. The concept of the pteridosperms as a separate phylogenetic lineage arose near the beginning of this century, when it was shown that a genus of fossil seed belonged to the same plant as a genus of fernlike fronds.

The pteridosperms, class Pteridospermopsida, are usually classified into six orders. Three of these are limited to the Paleozoic, whereas the other three occur in the Mesozoic. Although all the orders are generally considered to be true seed ferns, they show great diversity in structure, and their exact phylogenetic relationships are not known. The current belief among those working closely with the seed ferns is that they appear to have been derived as one evolutionary offshoot of the progymnosperms, possibly from the Aneurophytales. Evidence also suggests that the compound leaf of the seed ferns, as with the progymnosperms, was formed by the **planation** of extensive branch systems.

As a group, the seed ferns are characterized by frondlike compound leaves, some of which bear seeds and pollen-producing organs in place of leaflets (pinnules). The seeds show structural similarities to those of modern cycads, whereas many of the pollen organs consist of **exannulate** sporangia coalesced into a compound structure called a **synangium**. Other features include secondary xylem with bordered pits on the radial walls of tracheids, a prominent cortex with fiber strands, and large **leaf traces** consisting of one to several vascular strands. Some of the pteridosperms reached a height of 5 m, whereas others were apparently smaller, somewhat sprawling plants.

Paleozoic Pteridosperms

Order Lyginopteridales

Lyginopteridales is the order of Carboniferous pteridosperms that contains the classical *Lyginopteris oldhamia*, the first plant in which seeds were shown to have been borne by the compound leaves that occur in many Carboniferous plant-bearing rocks (Fig. 21-1). For some years, there had been a strong suspicion among botanists that the fernlike leaves were not actually true ferns; among other features, they had never been found bearing sporangia, as had many true ferns. The actual proof was obtained in 1903 by Oliver and Scott, who demonstrated that the distinctive secretory glands on the stems, leaves, and seed-bearing **cupule** of closely associated specimens were identical. Since that important breakthrough, the pteridosperms have been shown to be one of the most important groups of Carboniferous and Mesozoic seed plants.

The stem of *Lyginopteris*, with a diameter up to 3 cm, has a central pith, surrounded by bundles of primary xylem and phloem (Fig. 21-2). The protoxylem is usually mesarch, and in some bundles is located towards the outside. The leaf traces depart from the primary xylem and pass through the secondary xylem into the cortex.

The secondary xylem and phloem form a narrow cylinder outside the primary bundles. The secondary xylem has large tracheids with irregularly arranged bordered pits and numerous rays. Outside the secondary phloem is a parenchymatous inner cortex consisting of **anastomosing** fibre cells with interspersed parenchyma. Stalked glands with globular heads cover the outer surface of smaller stems. There is a periderm in larger stems that appears to have differentiated from cells of the inner cortex.

Compound leaves of *Lyginopteris* are up to half a meter long, with pinnate branching (Fig. 21-1). The leaflets have several lobes, are constricted at the base, and have dichotomous venation. The seeds of *Lyginopteris* were borne on the tips of the leaves. They are barrel-shaped and approximately 5 mm long, and the innermost gametophyte is enclosed by the megasporangial wall, or **nucellus** (Fig. 21-3A). This in turn is enclosed by a single integument, except for the narrow opening of the **micropyle** at the distal tip. Here the nucellus projects as a beak, forming a flask-shaped **pollen chamber** between the beak and the overarching integument. It has been suggested that the pollen grains found in the pollen chamber produced sperms that fertilized the egg in the archegonium directly, as in extant cycads. The whole seed is enclosed in a cupule, consisting of lobes covered with globular glands (Fig. 21-3B). The glands are identical to those

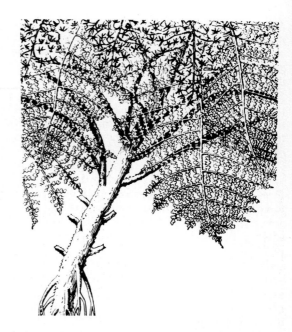

Figure 21-1 Reconstruction of part of the Paleozoic pteridosperm *Lyginopteris oldhamia*, showing roots, trunks, and frondlike compound leaves, ×0.5. (From L. Emberger, 1968, *Les Plantes Fossiles*, with permission of Masson et Cie, Paris.)

on the stems and petioles and hence provide important evidence for linking isolated fragments of seeds, leaves, and stems.

The pollen organs have never been found attached to *Lyginopteris*, but may be those of *Crossotheca* (Fig. 21-3C), a genus that is often associated with the leaves. With the same general structure as the compound leaves, they consist of a central axis with pinnate lateral branches terminating in slightly flattened appendages. The closely appressed sporangia hang from the margin of the lower surface of each appendage. In at least one species, the spores are trilete and smooth-walled, and they measure from 43–58 micrometers.

Other stems, leaves, and seeds found in Carboniferous rocks are very similar to *Lyginopteris* and hence are classified in the same order.

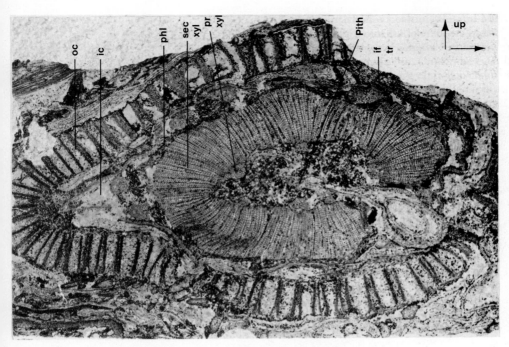

Figure 21-2 Transverse section of stem of *Lyginopteris*, showing stele and cortex, ×4: *ic*, inner cortex; *lf tr*, leaf trace; *pr xyl*, protoxylem; *oc*, outer cortex; *phl*, phloem; *sec xyl*, secondary xylem.

Order Medullosales

Among Carboniferous pteriodosperms, the order Medullosales is distinguished by **polystelic** stems. Most of the stems are classified in a single genus, *Medullosa*, of which *M. noei* is a good example.

As reconstructed, *Medullosa noei* consists of a trunk about 5 m in height, with prop roots around the base, and spirally arranged fronds toward the top (Fig. 21-4A). Numerous leaf bases cover the central regions of the stem. The fronds are dichotomously branched and have a pinnate arrangement of both secondary pinnae and pinnules.

One of the most abundant leaves found in Carboniferous beds of Europe and North America is *Alethopteris*, and there is evidence that at least one species of the leaf-genus *Alethopteris* represents the leaves of *Medullosa*. Recently, well-preserved specimens were examined for internal anatomical features, illustrated in Figure 21-4B. It is interesting that they show several adaptations for living in dry conditions, including abundant multicellular epidermal hairs, a hypodermal layer of large cells, and a thick mesophyll. In some, the stomata appear sunken in furrows, which may have served to cut down air flow and hence the diffusion rate from stomata. In general anatomy, *Alethopteris* leaves are comparable with those of the extant cycad *Cycas revoluta*, which also has many **xeromorphic** characters (see Chapter 22, p. 520).

The anatomy of the stem is very characteristic (Fig. 21-5A). The vascular tissue is in the form of a polystele; the number of steles varies from a few

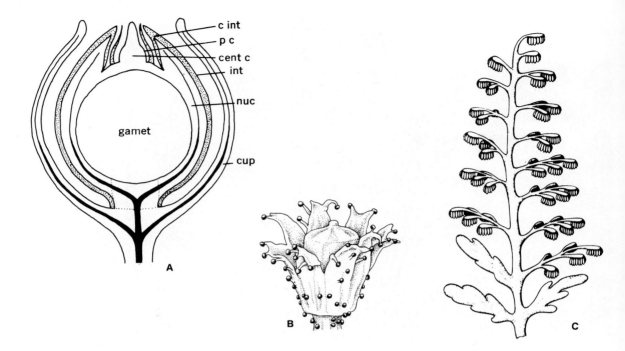

Figure 21-3 **A,** reconstructed vertical section of the seed and cupule of the pteridosperm *Lyginopteris* ×1: *c int*, canopy of integument; *pc*, pollen chamber; *cent c*, central column; *int*, integument; *nuc*, nucleus; *cup*, cupule. **B,** reconstruction of the seed and cupule of the pteridosperm *Lyginopteris*, showing the stalked glands, ×5. **C,** *Crossotheca*, a probable pollen-bearing organ of *Lyginopteris*, showing the sporangia suspended from the flattened appendages, ×2. (**A, B,** after Oliver and Scott; **C,** from H. N. Andrews, Jr., *Studies in Paleobotany*, New York and London: John Wiley & Sons, 1961, with permission of the author and John Wiley & Sons.)

in young stems to many in older stems. Each stele contains a central core of primary xylem with mesarch protoxylem and is surrounded by a cylinder of secondary xylem and phloem. The steles are surrounded by parenchyma tissue containing secretory cells and are bounded on the outside by a parenchymatous cortex. Periderm occurs on the inner face of the cortex. In older stems, the cortex was sloughed off during development, with the periderm forming the outer limits of the stem.

Seeds named *Pachytesta* have been found closely associated with *Medullosa noei* (Fig. 21-5B). *Pachytesta* has a nucellus that contains a vascular strand and has a somewhat flattened pollen chamber at the **distal** end. The integument has several layers and, except at the base, is completely free from the nucellus; it arches over the pollen chamber, forming the micropyle. Seeds of the general form of *Pachytesta* have been found in place of a terminal or lateral pinnule on some fernlike fronds.

The pollen-bearing organ of *Medullosa* is called *Dolerotheca* (Fig. 21-5C); it is shallow and cup-shaped, with many elongate sporangia fused together into a synangium. The pollen grains are bilaterally symmetrical and up to 500 μm in length. They have a single **monolete suture** on the **proximal** surface, and two longitudinal furrows on the distal (Fig. 21-6). There has been some uncertainty and controversy in recent years as to the surface on which germination and expulsion of the gametes would have occurred. One opinion is that germination in the grains from *Dolerotheca* may have been distal,

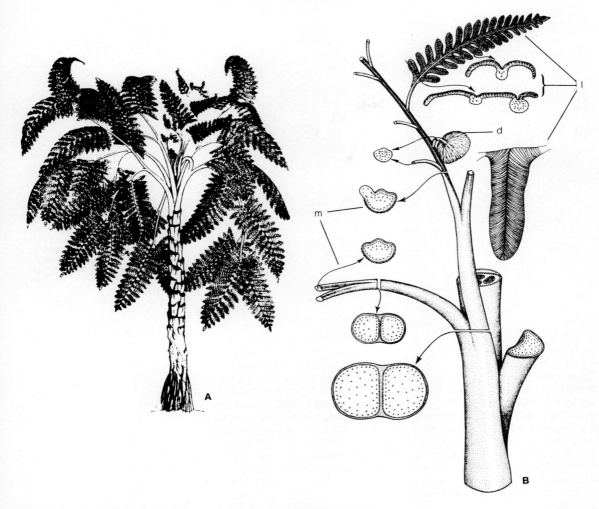

Figure 21-4 **A,** reconstruction of *Medullosa,* showing roots, leaf bases, and compound leaves, ×0.03. **B,** reconstruction of a medullosan frond with *Alethopteris* leaves, rachises called *Myeloxylon,* and a *Dolerotheca*-type pollen organ. The morphology and venation patterns at several levels of the frond are shown in the transverse sections. (**A,** from W. N. Stewart and T. Delevoryas, 1956, with permission of the authors and the New York Botanical Garden, publisher of *Botanical Review;* **B,** from C. G. K. Ramanujam, G. W. Rothwell, and W. N. Stewart, 1974, with permission of the authors and *American Journal of Botany.*)

whereas in pollen of other genera of the Medullosaceae, it was most likely proximal. It has also been suggested that this type of large pollen (or **prepollen**) may have been ancestral to the type found in the modern cycads, with a marked reduction in size during evolution.

Order Callistophytales

The order Callistophytales was erected in 1970 by Stidd and Hall of the University of Minnesota when it was discovered that several genera of detached

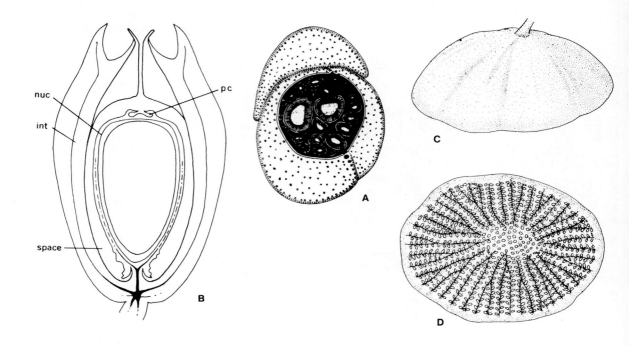

Figure 21-5 A, a transverse section of *Medullosa* stem, showing the polystelic arrangement of vascular tissues, and secretory cells in the outer parenchyma, ×1. **B,** vertical section through *Pachytesta*, showing nucellus (*nuc*), integument (*int*), and pollen chamber (*pc*), ×1. **C,** *Dolerotheca,* the pollen-bearing organ of some medullosan pteridosperms. **A,** top view of organ, ×2; **B,** view of underside, showing numerous pollen sacs opening to the outside, ×2. (**A,** from W. N. Stewart and T. Delevoryas, 1956, with permission of the authors and the New York Botanical Garden, publisher of *Botanical Review*; **B,** after W. N. Stewart, 1954, with permission of the author and the *American Journal of Botany*; **C,** after J. M. Schopf, 1949, with permission of the author, the *Journal of Paleontology* and the Society of Economic Paleontologists and Mineralogists.)

organs contained similar secretory cavities; also, pollen grains similar to those produced by the pollen organ were found in the pollen chamber of the seed. The various genera are listed in Table 21-1. On the evidence to date, the Callistophytaceae are more closely related to the lyginopterid than the medullosan pteridosperms.

The stem and leaf genus *Callistophyton* was the first to be described by Delevoryas and Morgan from the University of Illinois in 1954 from a **coal ball** in Carboniferous deposits in Illinois (Fig. 21-7). The largest of the petrified stems averages 1.8 cm grading down to less than 3 mm. Internally, (Fig. 21-8), it has a solid parenchymatous pith surrounded by isolated **mesarch** primary xylem strands that represent the leaf traces. Secondary xylem and phloem are well-developed as a cylinder, interrupted only where leaf traces pass outward into petioles. The secondary xylem has tracheids with circular bordered pits and wide rays. Other noteworthy anatomical features of the stem are the rays interspersed with sieve cells and parenchyma in the secondary phloem, an inner cortex, a periderm, and an outer cortex with noteworthy secretory cavities. These secretory cavities are also found in the pith, in laminae of leaves, and in the developing leaves of the shoot apex. Significantly, identical cavities have also been found in both the pol-

Figure 21-6 View of the distal surface of a pollen grain from *Dolerotheca*, showing two longitudinal grooves, ×180. (From M. A. Millay and T. N. Taylor, 1976, with permission of the authors and Elsevier Publishing Co., Amsterdam, publishers of *Review of Palaeobotany and Palynology*.)

Table 21-1 List of Genera of Detached Organs of the Pteridosperm Family Callistophytaceae. (Names of the describers of the genera and the dates are given in brackets.)

Structural part	Genus
Stem and leaf	*Callistophyton* (Delevoryas & Morgan, 1954)
Seed	*Callospermarion* (Eggert & Delevoryas, 1960)
Pollen organ	*Callandrium* (Stidd & Hall, 1970)
Pollen	*Vesicaspora* (Schemel, 1951, as a dispersed pollen in rock)

len organ and seed, providing part of the evidence for considering all three as representing the same genus.

Callospermarion seeds are of the pteridosperm type (Fig. 21-9A). They are small, with a three-layered integument, and the nucellus (megasporangial wall) is free from the integument except at the base. Secretory cavities identical to those of *Callistophyton* occur in the outer integument. Pollen grains with a single bladder surrounding the inner body (which contained the male gametophyte) and a distal germinal **sulcus** have been found inside the pollen chamber.

The pollen organ is *Callandrium* (Fig. 21-9B). It consists of six or seven individual sporangia arranged radially in a synangium. *Callandrium* differs from the synangia of other pteridosperms in occurring superficially on the abaxial surface of a laminated fertile frond. Pollen grains found within the sporangia are identical to those found in the pollen chamber of the seed. The bladdered pollen grains are identical with dispersed pollen grains in rocks of the same age called *Vesicaspora*. In having a bladder they are similar to pollen grains of other groups of gymnosperms, such as some conifers, but different from the **trilete** prepollen of *Lyginopteris* and the **monosulcate** grains of *Medullosa*. It has recently been proposed that *Vesicaspora* pollen evolved from trilete-monolete pteridosperm prepollen by the formation of a **saccus**, and the development of a distal germination area (Fig. 21-10).

The chief characteristics of the three main Paleozoic pteridosperm families are summarized in Table 21-2.

Phylogeny of Paleozoic Pteridosperms

The three families described above are the most completely known of the Paleozoic pteridosperms. However, other fossils are known from the late Devonian and early Carboniferous that represent the earliest records of the seed fern lineage. All of these (e.g. the Calamopityaceae) are known from petrified stems or leaf branches only; we lack information of other parts that could fill in our knowledge of the evolution of the group.

From all the available evidence, the evolution of the Paleozoic pteridosperms from progymnospermous ancestors such as the Aneurophytales appears very likely (Fig. 21-11). As outlined in the last chapter, it also seems probable that the

Figure 21-7 Reconstruction of *Callistophyton* as a scrambling shrub growing on a coal swamp near a dead tree stump, ×0.1. (After G. W. Rothwell, 1975, with permission of the author and E. Schweitzerbart'sche, Stuttgart, publisher of *Palaeontographica*.)

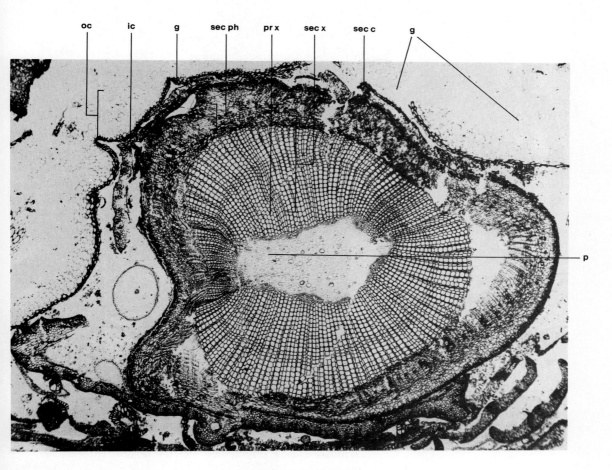

Figure 21-8 Transverse section of the stem of *Callistophyton*, showing the arrangement of vascular and cortical tissues: *p*, pith; *pr x*, primary xylem; *sec x*, secondary xylem; *sec ph*, secondary phloem; *i c*, inner cortex; *o c*, outer cortex; *s c*, secretory canal; *g*, gland, × 13. (From T. Delevoryas and J. Morgan, 1954, with permission of the authors and E. Schweitzerbart'sche, Stuttgart, publisher of *Palaeontographica*.)

Archaeopteris lineage was separated from the pteridosperm line. Figure 21-11 also indicates the close relationship of the Calamopityaceae to the Lyginopteridaceae and the affiliation of the Lyginopteridaceae with the Callistophytaceae. The rather uncertain relationships of the Medullosaceae are reflected in the broken line.

Thus the pteridosperms, with their planated compound leaves, seeds, cupules, and synangiate pollen organs, represent one evolutionary pathway from the progymnosperms of the Devonian. They flourished during the Upper Carboniferous, possibly from advantages derived from the seed-unit, prepollen, and leafy segments adapted for **xeric** conditions. Several lines appear to have survived the drastic evolutionary upheaval of the

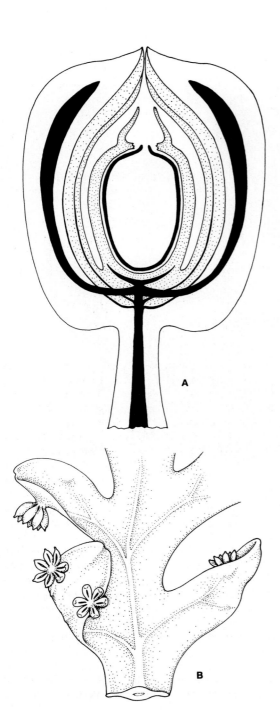

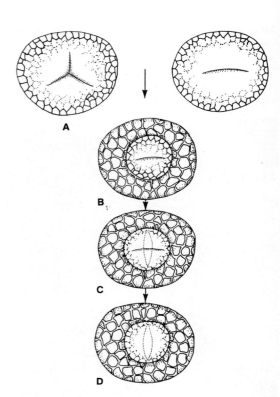

Figure 21-10 Proposed evolution of pollen of the Callistophytaceae. The grains of **A** represent prepollen of pteridosperms with external ornamentation; the second stage (**B**) is hypothetical, showing exine separation into a saccus with internal ornamentation. **C** shows the formation of a distal sulcus and complete internal ornamentation of the saccus, such as found in mid-Pennsylvanian species of *Vesicaspora*. **D** shows the loss of the proximal tetrad scar, as in later Pennsylvanian species of *Vesicaspora*. (From M. A. Millay and T. N. Taylor, 1976, with permission of the authors, and Elsevier Scientific Publishing Co., Amsterdam.)

Figure 21-9 **A,** diagram of longitudinal section through *Callospermarion*, the ovule of *Callistophyton*, ×48: *mg m*, megaspore membrane; *nuc*, nucellus; *in int*, inner integument; *m int*, middle integument; *o int*, outer integument; *p c*, pollen chamber; *v t*, vascular tissue. **B,** reconstruction of the fertile pinnules of *Callandrium*, the probable pollen-bearing organ of *Callisotophyton*, approx. ×805. (**A, B,** from B. M. Stidd and J. W. Hall, 1970, with permission of the authors and *American Journal of Botany*.)

Table 21-2 Main Characteristics of Three Families of Paleozoic Pteridosperms.

Structural Part	Medullosaceae	Lyginopteridaceae	Callistophytaceae
Stems	Polystelic	Monostelic	Monostelic
Fronds	Large; petioles with many traces	Small; petioles with one V-shaped trace; bifurcate rachis	Small; petiole with one flat vascular trace
Seeds	Large; noncupulate; simple pollen chamber; free nucellus; radial symmetry	Small; cupulate; elaborate pollen chamber; fused nucellus; radial symmetry	Very small; noncupulate; simple pollen chamber; free nucellus; bilateral symmetry
Pollen organs	Large; complex synangia; sporangia tubular. *Dolerotheca, Aulacotheca*	Small; nonlaminar. *Telangium, Crossotheca*	Small; fernlike synangia borne on laminae. *Callandrium*
Pollen	Prepollen; monolete	Prepollen; trilete	Pollen; saccate; sulcate

(From B. M. Stidd and J. W. Hall, 1970, with permission of the authors and *American Journal of Botany*.)

Permian; they occur as several families of Mesozoic pteridosperms.

Mesozoic Pteridosperms

Plants affiliated with the seed ferns occur in Permian, Triassic, and Jurassic rocks of both the Northern and Southern Hemispheres. They are usually classified into three distinct orders with no apparent relationship to one another. They have seeds, cupulate structures, pollen, and pollen organs, however; hence they are more closely related to the Paleozoic pteridosperms than to any other lineages.

Their possession of seeds, cupulate structures, and pollen, and their position in time make the Mesozoic pteridosperms potentially significant in providing clues to the origin of **angiospermy**. This is particularly true of the plants of the Caytoniales, which have characteristics sufficiently close to angiospermy that they are considered by some prominent botanists to represent possible stages in angiosperm evolution.

Order Corystospermales

Several genera from early Mesozoic rocks in Africa and South America have been classified in the order Corystospermales. The plants are small, with fernlike leaves and unisexual reproductive organs arranged on branched axes. Seed-bearing branches arise from the stem in the axil of a bract and have recurved cupules on the branch tips (Fig. 21-12A). The helmet-shaped cupules contain seeds; micropyles project beyond the lips of the capsule. The pollen-bearing organs, *Pteruchus*, have sporangia clustered on the expanded tips of the branch endings (Fig. 21-12B). Bladdered pollen grains of the same type as in the Paleozoic Callistophytaceae have been found in some of the sporangia (Fig. 21-12C). However, in gross morphology, the pollen organ *Pteruchus* appears to be closer to the lyginopterid than to callistophyte pteridosperms.

Order Peltaspermales

The Peltaspermales are represented by plants from the Permian and Triassic of South Africa and

Greenland. The best-known genus is *Lepidopteris*, which has twice-pinnate compound leaves with open venation in the **pinnules** (Fig. 21-12D). Seeds about 15 mm in diameter have been found attached to the lower surface of a **peltate,** stalked disc (Fig. 21-12E). Each seed is oval, about 7 mm long, with a micropyle formed by a curved integument (Fig. 21-12F). The oval pollen grains are contained in pollen sacs arranged in two rows on the lower surface of branches resembling reduced pinnae.

Order Caytoniales

The Caytoniales are represented by leaves, seed-bearing organs, and pollen-producing organs in rocks ranging in age from Triassic to Lower Cretaceous. They were originally believed to be angiospermous, but later investigations indicated that the plants are not true angiosperms, although they show the angiospermous character of ovules enclosed in cupules. Based mainly on this character, Dr. V. Krassilov (1977) of Vladivostok has proposed that the Caytoniales and two other Mesozoic groups be categorized as "Proangiosperms," allied with the present Gnetopsida (see Chapter 25).

The leaves of the caytonias, called *Sagenopteris*, usually have two pairs of elongated leaflets emanating from the tip of a petiole (Fig. 21-12G). Each leaflet has a central midvein and branching reticulate venation. Although the venation resembles that of many dicotyledon leaves, the veins lack free endings except at the margin.

Most significant is the seed-bearing organ called *Caytonia*, consisting of a central axis with two lateral rows of circular sacs (Fig. 21-12H). Each of these is a cupule that opens by a lip near the base (Fig. 21-12I), with seeds attached to the wall and projecting into the cavity. The resemblance of this cupule to the **carpel** of an angiosperm originally prompted several botanists to assign *Caytonia* to the angiosperms. Later work, however, disclosed pollen grains not only inside the cupule, but also inside the micropyle of the seed, indicating that pollination took place in each ovule rather than on the lip of the cupule—not a true angiospermous condition, although several angiosperms have a similar type of pollination. The pollen grains contained in *Caytonanthus* (Fig. 21-12K) are small and **bisaccate,**

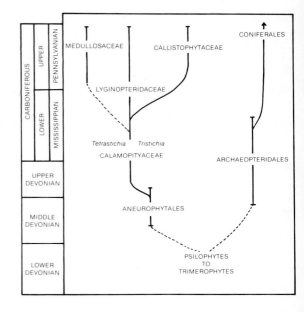

Figure 21-11 Suggested evolutionary relationships of the Paleozoic pteridosperms to the progymnosperm orders. (Adapted from Stidd and Hall, 1970, and Beck, 1970, 1976.)

of the same general type as those of *Vesicaspora* of the Paleozoic pteridosperm family Callistophytaceae (Fig. 21-10D). These same grains found dispersed in rocks of the Mesozoic are classified in the form-genus *Vitreisporites*.

Other Pteridosperms

A large assemblage of plants called the *Glossopteris* flora occurs in Permian and early Triassic rocks of Australia, South Africa, and South America. Its name is derived from *Glossopteris*, one of the main genera of leaves. Poorly preserved reproductive structures have been discovered attached to some of the leaves. Although these are interpreted as cupulate structures bearing seeds, there is insufficient evidence to date. Hence *Glossopteris* and

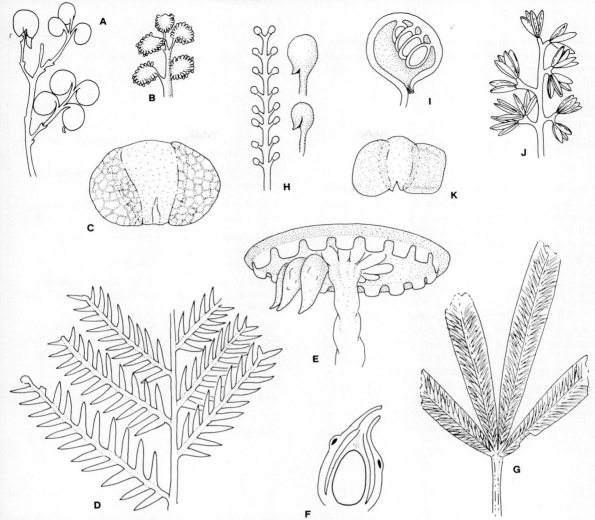

Figure 21-12 Mesozoic pteridosperms. **A–C,** Corystospermales. **A,** seed-bearing branches of *Umkomasia,* showing typical shield shape of seed in upper left, ×1. **B,** *Pteruchus* pollen-bearing organ, ×0.75. **C,** bisaccate pollen grain isolated from *Pteruchus,* ×500. **D–F,** Peltaspermales. *Lepidopteris.* **D,** part of frond, ×0.3. **E,** peltate disc with seeds attached to underside, ×4. **F,** vertical section through seed, showing nucellus (*nuc*) and integument (*int*), ×15. **G–K,** Caytoniales. **G,** *Sagenopteris,* leaf of plant bearing *Caytonia* seeds, ×0.3. **H,** *Caytonia* seed-bearing axis, showing two helmet-shaped cupules, ×0.6. **I,** vertical section through cupule of *Caytonia,* showing four seeds attached to inner wall, ×5. **J,** *Caytonanthus,* pollen-bearing organ, ×0.6. **K,** bisaccate pollen grain isolated from pollen sacs of *Caytonanthus,* ×1000. (**A, B** after Thomas, from *Philosophical Transactions,* 1933, with permission of the Royal Society, London; **C,** from J. A. Townrow, 1962, with permission of the author and Almqvist & Wiksell, Stockholm, publishers of *Grana Palynologica;* **D–F,** after T. M. Harris, 1932, with permission of the author and the Commission for Scientific Research in Greenland, publishers of *Meddelelser om Grønland;* **G,** after Seward; **H–J,** after Thomas, with permission of the Royal Society, London; **K,** after J. A. Townrow with permission of the author and Almqvist & Wiksell, Stockholm, publishers of *Grana Palynologica.*)

related plants are usually classified provisionally as pteridosperms, although it has been suggested recently that they show some relationships with the Gnetales (Chapter 25).

Phylogeny of Mesozoic Pteridosperms

The evidence of evolutionary relationships is based largely on similarities of pollen and pollen organs between Paleozoic and Mesozoic genera; the vegetative parts and cupules are very different in the older and younger groups. To date, relationships of the Caytoniales with the Paleozoic Callistophytales, and also of *Pteruchus* (Corystospermales) with lyginopterid lines have been suggested, based on pollen organ and pollen comparisons.

All the evidence suggests that the seed ferns were diminished in numbers during the relatively rapid evolution of the Permian, but survived in greatly modified forms to provide reasonable populations of the Mesozoic families. As a group, the pteridosperms appear to have declined progressively through the Mesozoic, becoming extinct in later intervals of the Cretaceous. They appear to be the evolutionary lineage that was most likely ancestral to the angiosperms as well as the next group to be described, the cycadophytes.

References

Banks, H. P. 1970. *Evolution and plants of the past*. Wadsworth Publishing Co., Belmont, Calif.

Beck, C. B. 1976. Current status of the Progymnospermopsida. *Rev. Palaeob. and Palynol.* 21:5–23.

Krassilov, V. A. 1977. Contributions to the knowledge of the Caytoniales. *Rev. Palaeob. and Palynol.* 24:155–78.

Long, A. G. 1977. Some Lower Carboniferous pteridosperm cupules bearing ovules and microsporangia. *Trans. Roy. Soc. Edinburgh* 70:1–11.

Millay, M. A., and Taylor, T. N. 1976. Evolutionary trends in fossil gymnosperm pollen. *Rev. Palaeob. and Palynol.* 21:65–91.

Stidd, B. M. and Hall, J. W. 1970a. *Callandrium callistophytoides*, gen et. sp. nov., the probable pollen-bearing organ of the seed fern, *Callistophyton*. *Amer. J. Botany* 57(4):394–403.

———. 1970b. The natural affinity of the Carboniferous seed, *Callospermarion*. *Amer. J. Botany* 57(7):827–36.

Taylor, T. N. 1981. *Paleobotany*. McGraw-Hill Book Co., New York.

Until very recently, the interesting group of gymnosperms collectively called cycadophytes were classified as two separate orders of the single class Cycadopsida (cycads), based largely on morphological similarities of stem, leaves, and pollen. Now, however, the strictly fossil class Cycadeoidopsida (cycadeoids) is generally recognized as a distinct evolutionary lineage. The record of the cycads extends from the late Carboniferous to the present with nine extant genera. The cycadeoids, on the other hand, have a history from the early to late Mesozoic; they became extinct before the end of the Cretaceous period some 75 million years ago. Although not closely related, the cycads appear linked with Paleozoic pteridosperms. It is probable that the cycadeoids are also descended from pteridosperms, although the connection is less certain.

22

CYCADOPHYTES

Class Cycadopsida

The Cycadopsida form a small but most interesting group of living vascular plants. From Mesozoic times, when they were widespread and relatively abundant, they declined in number of species and individuals during the Cretaceous and Tertiary periods. The nine extant genera are found in relatively limited regions of the tropics and subtropics, usually on arid and well-drained sites (Table 22-1). The relatively restricted ranges suggest that many if not all of the cycads are heading toward extinction. This is particularly true of the genus *Microcycas* in Cuba, some species of *Macrozamia* in Australia, and *Encephalartos* in central and southern Africa.

Resembling palms, many cycads are pleasant to the eye; they are cultivated in greenhouses, gardens, parks and in some of the famous conservatories and botanical gardens. At least one cycad, *Cycas revoluta*, is marketed through retail outlets in North America under the popular name of Easter palm. Cycads are usually propagated from buds on the stem; wounding causes prolific bud production, and this method is often used by gardeners. The starch from the pith and cortex of the stem of some species is used locally for food (sago), and cycas gum is produced commercially from species of *Cycas* in Asia. The leaves of many cycads are poisonous to foraging animals, and at least one

Australian species was almost exterminated by arsenic to prevent the paralysis and death of cattle. A report that some cycads produce aphrodisiacs is interesting but unconfirmed.

Morphology

In general habit, the extant cycads resemble tree ferns as well as some of the pinnately-leaved palms (Fig. 22-1). The stem ranges from very short and bulbous to tall and columnar. Some cycads are very small, with mostly subterranean stems (*Bowenia* and *Stangeria*), whereas others have a trunk reaching 18 m (*Macrozamia*). Most range between these extremes, seldom reaching over 2 m.

The leaves of the extant cycads are pinnately compound, with *Bowenia* having a bipinnate configuration. The leaves are arranged helically at the stem tip, and most plants form a new set each year. The leaves are tough, and some species have well-developed spines that make them dangerous to touch. With age, the leaves develop an **abscission layer** that develops in the petiole a short distance out from the stem (Fig. 22-2). After abscission, the leaf bases form a formidable cover on the stem. The number of leaf bases has been used to calculate the age in several species; some plants of *Dioon* in Mexico have been estimated to be as old as 1000 years using this method.

The pinnate leaflets are usually entire and leathery; they are **xeromorphic,** with thick **cuticle, sclerified hypodermal** layers, and **sunken stomata** (Fig. 22-3A). The leaflets of most species have well-marked palisade and spongy mesophyll zones. They also have prominent stomatal rows on the underside, with well-developed cuticle on the outer edges of both the guard and subsidiary cells (Fig. 22-3B). Although the venation pattern appears at first glance to be parallel, it is instead steeply dichotomous, with the veins branching obliquely and then assuming a parallel orientation. In some species, the veins join together near the margins; in others they end blindly. *Cycas* leaflets have only one vein, which forms a prominent midrib. Most cycad leaflets have cells surrounding the vein and in the mesophyll that are specialized into tracheidlike cells with **scalariform** or **reticulate** pitting. These **transfusion cells** apparently function in conducting water

Table 22-1 Distribution of Extant Cycads

Genus	Distribution by Countries
Cycas	Australia; East Indian Islands; India; China; southern Japan
Macrozamia, Bowenia	Australia
Encephalartos	central and southern Africa
Stangeria	southern Africa
Zamia	Florida; West Indies; Mexico; Central America; northern South America; Andes Mountains south to Chile
Microcycas	Western Cuba
Ceratozamia	Mexico
Dioon	Mexico

between the veins and mesophyll, and are particularly well developed in *Cycas revoluta* leaflets.

Stem Anatomy

The apical meristem consists of a cap of five zones of meristematic cells rather than a large apical cell, as in ferns, lycopods, and *Equisetum*. Derivatives differentiate into a ring of vascular bundles. The cycads differ from most other arborescent gymnosperms in having very limited secondary tissue developed from the vascular cambium (Fig. 22-4). The small amount of secondary xylem develops **centrifugally** in radial rows, but without forming well-defined growth rings. The pitting of the tracheids is bordered in all except *Stangeria* and *Zamia*, which have scalariform pits. Rays vary from one to several cells in width, and are up to 20 cells in depth; they often contain starch grains or crystals of calcium oxalate. Both the pith and cortex contain prominent mucilaginous canals and large amounts of starch. Transfusion cells similar to those in the leaf mesophyll are found in the pith and cortex of many species. The cortex is bounded on the outside by a well-marked but thin periderm.

One of the most interesting features of cycad anatomy is the arrangement of the **leaf traces.** The

Figure 22-1 *Cycas revoluta* in a park in Japan, showing the columnar stem studded with leaf bases, and a rosette of pinnate leaves at the top, ×0.08.

number of traces varies with the species but always shows a pattern called **girdling** (Fig. 22-5). Individual traces arise from **leaf gaps** that are located some distance around the stele from the leaf insertion. Thus, in passing to the leaf base, the traces girdle the stele. Each trace passes upward slightly through the cortex, approaching closer to other traces before entering the leaf base. This process is also found in many angiosperms, particularly in monocotyledons with parallel leaf venation. However, it is not present in the cycadeoids, which have traces passing directly from the stele to the leaf base.

The primary root of cycads is large, especially in the seedling. Adult roots attain great lengths; in one example, a *Dioon* root has a diameter of 3 cm some 12 m distant from the stem. The primary vascular tissue is an exarch protostele, ranging from **diarch** to **tetrarch**. Secondary growth is common. A noteworthy feature is the negative **geotropism** exhibited widely in the roots. Many of these **apogeotropic** roots have a region containing endosymbiotic bluegreen algae of the genera *Nostoc* and *Anabaena*. In most cycads, these bluegreen cells occur in intercellular spaces in both the inner and outer cortex, but in some species they are concentrated in an algal zone in the cortex.

Life History

The life cycle of the cycads, like that of other gymnosperms, has the sporophyte generation dominant, with the gametophytes much reduced and completely dependent on the sporophyte (Fig. 22-6). The male gametophyte develops inside the pollen grain and at maturity is reduced to four cells. The female gametophyte is multicellular and devel-

Figure 22-2 Trunk of *Dioon spinulosum*, showing the thick cover of leaf bases, ×0.3. (Photo courtesy of Chicago Natural History Museum.)

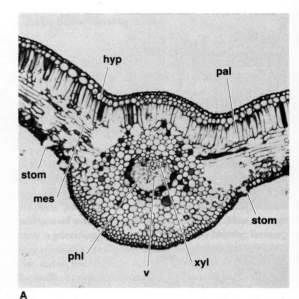

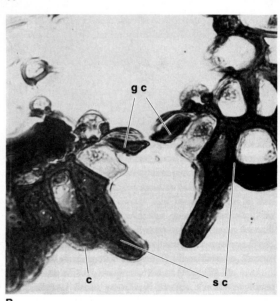

Figure 22-3 A, transverse section of leaflet of *Cycas revoluta*, showing the large central vein, mesophyll, and stomata, × 5: *v*, vein; *xyl*, xylem; *phl*, phloem; *mes*, mesophyll; *pal*, palisade parenchyma; *hyp*, hypodermis; *stom*, stomata. **B,** a single sunken stoma of *Cycas revoluta*, showing guard cells (*g c*), overarching subsidiary cells (*s c*), and cuticle (*c*), × 645.

ops within the protection of the megasporangium (= nucellus) and the integument. These sporophytes grow for several to many years before producing pollen and seeds and normally continue to produce reproductive units throughout the rest of their lives.

Reproduction

In all the cycads, the sexes are segregated on separate plants; that is, they are **dioecious** (from the Greek for two houses). The reproductive units consist of compact cones in all except the female reproductive units of *Cycas*, in which the seeds are borne in two rows along the **rachis** of what is considered a reduced fertile leaf (see Fig. 22-11).

Most of the cycads bear the cones at the apex of the stem, but in at least two genera they are borne laterally and in the axils of the leaves. Many of the cones are extremely large and heavy; female

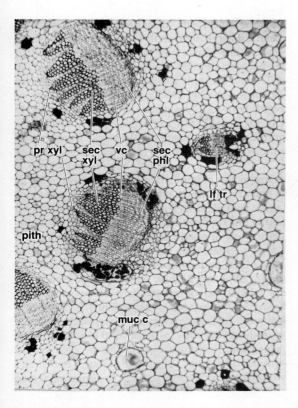

Figure 22-4 Transverse section of a *Zamia* stem, showing pith, primary xylem (*pr xyl*), secondary xylem (*sec xyl*), vascular cambium (*vc*), secondary phloem (*sec phl*), mucilage canals (*muc c*), and oblique leaf traces (*lf tr*) in cortex, ×40.

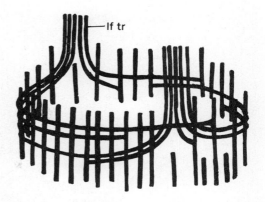

Figure 22-5 Vascular pattern of cycad stem, showing "girdling" of leaf traces (*lf tr*), ×0.3. (After H. A. Dorety, 1919, *Botanical Gazette* 67:251-57, with permission of the University of Chicago Press.

cones (seed-cones) exceeding 70 cm in length and weighing over 30 kg are typical for the Australian *Macrozamia denisonii*. In most cycads, the female cones are conspicuously larger than the male (pollen) cones.

The male cones are borne singly or in small groups at the stem tip (Fig. 22-7). Each cone consists of many sporophylls spirally arranged on a central axis (Fig. 22-8A). The sporophylls are usually oriented in vertical rows; they are narrow at the point of insertion and flare out distally (Fig. 22-8B). Each sporophyll has clusters of pollen sacs congregated on both sides of a median keel. These clusters have been called sori and contain from one to five sporangia, depending on the genus. The total number of sporangia per sporophyll varies from 28 in *Zamia* to over 1000 in *Cycas*.

Each pollen sac is essentially **sessile,** and develops in the **eusporangiate** manner (see p. **000**). After repeated divisions of initials, several layers of cells are produced. The inner cells become the microsporocytes, and the outer layers develop into the **tapetum** and the sporangial wall. During development of the microsporocytes, the tapetum and inner wall cells break down, forming a multinucleate protoplasm that serves as a source of food for the maturing pollen grain.

Following meiosis in the microsporocytes, the microspores differentiate **endosporally** into the male gametophyte (Fig. 22-9A–F). In the first stage, the microspore nucleus divides into a small **prothallial cell** and a large central **antheridial initial.** The antheridial initial then divides to form a small **generative cell** that lies against the prothallial cell and a larger tube nucleus that is centrally located (Fig. 22-9B).

At this stage of maturation, the immature gametophytes are shed from the sporangia as pollen grains, which are the immature male gametophytes within the pollen walls. The cone axis grows slightly to separate the sporophylls sufficiently for the pollen to escape. The pollen grains of most cycads are circular or elliptical, with a single ger-

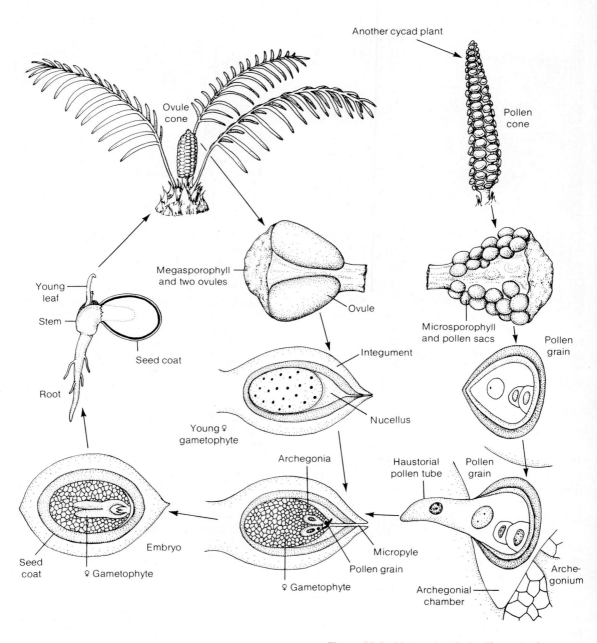

Figure 22-6 Main stages in the life cycle of a cycad.

Figure 22-7 Single large pollen cone of *Cycas* attached to the top of the stem. Note the cluster of erect new leaves in front of the cone that are surrounded by sterile appendages, ×1/6.

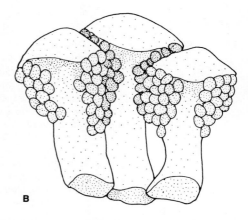

Figure 22-8 A, pollen cone of *Dioon*, showing vertical rows of microsporophylls and sporangia on the abaxial surface of the sporophylls, ×0.4. **B,** three microsporophylls of *Zamia*, showing the clustered pollen sacs, ×5.

minal furrow on one surface (Fig. 22-9E). With this form, they reflect the prepollen of pteridosperm ancestors. They are produced in large numbers (up to 30,000 per pollen sac) and are mainly dispersed by wind. Significantly, the sporophylls of the female cone separate slightly for a few days during pollen production so that ample pollen grains are available for fertilization.

When a pollen grain lands on the **pollen drop** at the micropyle of an ovule, further elaboration of the gametophyte takes place. A pollen tube begins to evaginate from the germinal furrow of the grain; it gradually grows into the nucellus of the ovule,

acting as an **haustorium** (Fig. 22-9C). At the same time, the tube nucleus migrates out into the basal part of the pollen tube. The swollen base extends into the pollen chamber formed from the dissolution of nucellar cells. While the tube is growing

Cycadophytes **525**

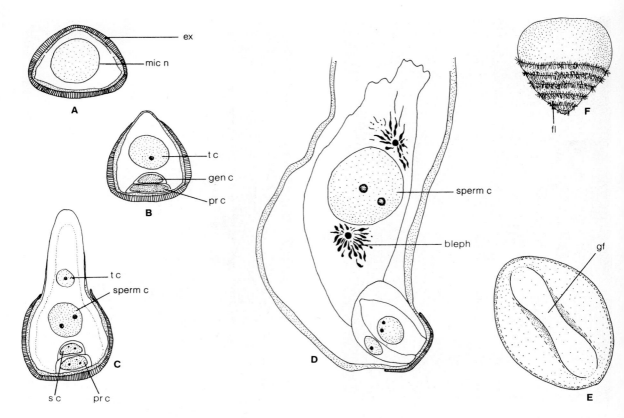

Figure 22-9 Male gametophyte of cycads. **A,** single-cell stage in a young pollen grain, showing the exine (*ex*) and the microspore nucleus (*mic n*), ×900. **B,** pollen grain containing prothallial cell (*pr c*), generative cell (*gen c*), and tube cell (*t c*), and the pollen tube just starting to form (*top*), ×900. **C,** late stage of male gametophyte, with prothallial cell (*pr c*), sterile cell (*s c*), spermatogenous cell (*sperm c*), and tube cell (*t c*), ×900. **D,** later stage of male gametophyte, showing two gamete nuclei forming inside the spermatogenous cell (*sperm c*), and two blepharoplasts (*bleph*), ×350. **E,** external view of *Zamia* pollen grain, showing a single germinal furrow (*g f*) on the distal surface, ×1000. **F,** single cycad sperm showing the helix of flagella (*fl*), ×1000. (**A–D,** after C. J. Chamberlain, 1909, *Botanical Gazette* 47:215-236, with permission of the University of Chicago Press; **F,** after Webber.)

through the nucellus, the generative cell divides to form a **sterile cell** that develops next to the prothallial cell, and a **spermatogenous cell** near the base of the pollen tube (Fig. 22-9C). Finally the spermatogenous cell undergoes a single division to form two sperms (Fig. 22-9F) which approach $\frac{1}{3}$ mm in diameter, the largest plant sperms known. The period between pollination and fertilization ranges from several days to at least eight months, during which the stages outlined above develop.

All but two genera of cycads have compact female cones with spirally arranged sporophylls (Fig. 22-10A, B). In contrast, *Cycas* has a rosette of sporophylls that resembles reduced pinnate leaves (Fig. 22-11A, B). In *Dioon*, the female cone contains loosely compacted entire sporophylls. This reduction in complexity of sporophylls is accompanied by a reduction in ovule number from six to eight

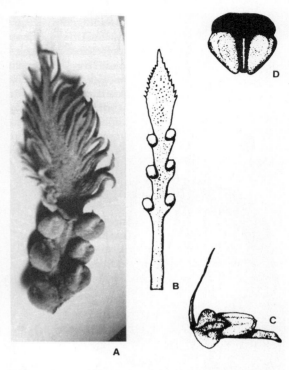

Figure 22-11 Ovulate sporophylls of cycads. **A,** two lateral rows of seeds below leaflike blade of *Cycas revoluta* sporophyll, ×0.3. **B,** lateral seeds on sporophyll of *Cycas circinalis*, ×0.2. **C,** single sporophyll of *Macrozamia* with terminal upturned spike, ×0.2. **D,** sporophyll of *Zamia*, showing two seeds attached to peltate sporophyll, ×0.6. (**B–D,** after A. W. Haupt, 1953, with permission of McGraw-Hill Book Co.)

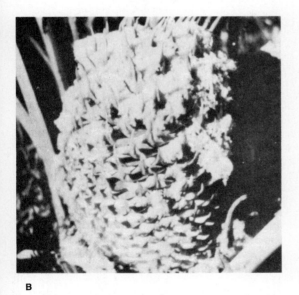

Figure 22-10 Female cones of cycads. **A,** single cone of *Encephalartos*, attached to the top of the stem amongst leaf bases, ×0.14. **B,** cone of *Macrozamia*, showing the sharp pointed tips of the sporophylls, ×0.5.

in *Cycas* to two in most other genera. This is generally considered to represent a reduction series that reflects an evolutionary trend.

The ovules are usually sessile or borne on short stalks on the adaxial surface of the sporophylls (Fig. 22-11C, D). In most instances, the ovules are oriented with the micropyle toward the central axis. Ovules vary in size from 6 cm in some species of *Cycas* and *Macrozamia* to as small as 5 mm in *Zamia pygmaea*.

The essential structures of an ovule can be observed in median longitudinal section (Fig. 22-12A, B). In very young stages, the white central mass is the megasporangium (nucellus). This is surrounded by an integument that comprises three

Cycadophytes

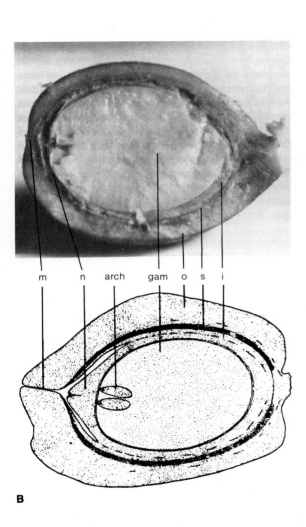
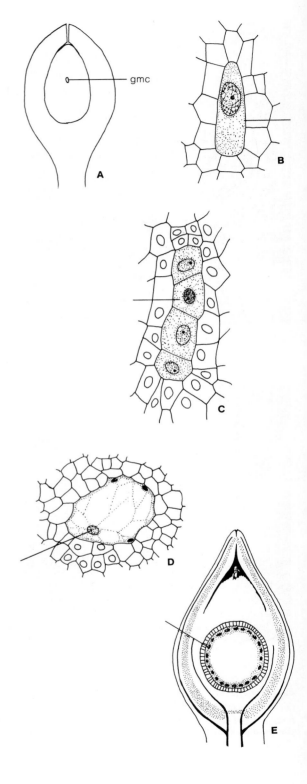

Figure 22-12 Longitudinal median sections of a cycad ovule. **A,** section of ovule of *Cycas revoluta*, ×2.5. **B,** diagram of longitudinal section of ovule of *Zamia*, ×4: *m*, micropyle; *n*, nucellus; *arch*, archegonia; *gam*, female gametophyte; *o*, outer fleshy integumentary layer; *s*, middle stony layer of integument; *i*, inner fleshy layer of integument appressed to the nucellus. (**B,** after A. W. Haupt, 1953, with permission of McGraw-Hill Book Co.)

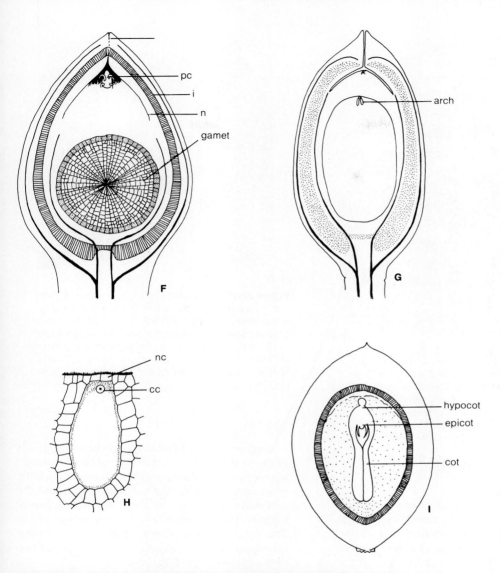

Figure 22-13 Development of the female gametophyte in cycads. **A,** megaspore mother cell in the nucellus, ×150. **B,** enlarged megaspore mother cell, ×250. **C,** linear tetrad of megaspores, ×900. **D,** four nuclei of free nuclear stage, ×600. **E,** free nuclei around periphery of gametophyte, ×4. **F,** female gametophyte after walls have formed around free nuclei, ×6. **G,** two archegonia at micropylar end of gametophyte, ×6. **H,** single archegonium, showing large central cell, two small neck cells, and surrounding archegonial jacket, ×50. **I,** single embryo, showing two cotyledons, two leaf primorida, epicotyl, and hypocotyl, ×4: *arch,* archegonium; *cc,* central cell; *cot,* cotyledon; *epicot,* epicotyl; *f n,* free nuclei; *m c,* megaspore mother cell; *gamet,* female gametophyte; *ms,* megaspore; *hypocot,* hypocotyl; *i,* integument; *m,* micropyle; *n,* nucellus; *nc,* neck cell; *pc,* pollen chamber; *arch j,* archegonial jacket. (**B–D,** after F. G. Smith, 1910, *Botanical Gazette* 50:128–41, with permission of the University of Chicago Press; **E–I** from *The Living Cycads* by Charles Joseph Chamberlain, New York: Hafner Press/Macmillan, 1965; previously published by the University of Chicago Press, 1919.)

Cycadophytes   **529**

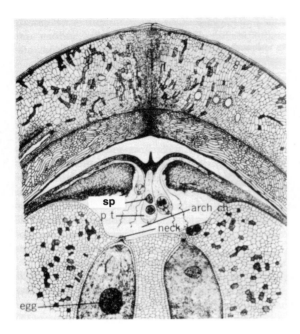

Figure 22-14 Vertical section of ovule of cycad just prior to fertilization, showing several pollen tubes and two archegonia ready to receive sperms: *sp*, sperm; *arch ch*, archegonial chamber; *neck*, neck; *pt*, pollen tube. (From *The Living Cycads* by Charles Joseph Chamberlain, New York: Hafner Press/Macmillan, 1965; previously published by the University of Chicago Press, 1919.)

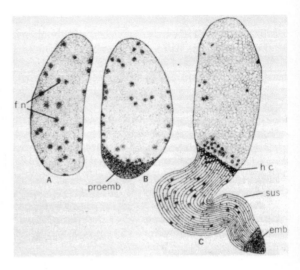

Figure 22-15 Early embryogeny of a cycad. **A,** free-nuclear stage, ×16. **B,** cells at chalazal end forming proembryo, ×16. **C,** embryonic cells at tip, suspensor cells in middle, and haustorial cells next to free nuclei, ×16: *emb*, embryonic cells; *f n*, free nuclei; *h c*, haustorial cells; *proem*, proembryo; *sus*, suspensor. (From A. W. Haupt, 1953, with permission of McGraw-Hill Book Co.)

distinct layers: a middle stony layer and two fleshy layers, one on either side. The inner fleshy tissue is largely resorbed during development of the gametophyte, but part remains as a thin, papery layer. Two vascular strands enter the base of the integument; each branches to give strands into both the outer and inner fleshy layers.

Within the central region of the megasporangium, a megasporocyte undergoes meiosis to form four megaspores, thereby initiating the gametophyte generation (Fig. 22-13A). The four megaspores are normally arranged in a linear tetrad within the megasporangium. Usually only the basal megaspore remains functional, and the other three degenerate. In their stages of germination, the megaspores are very similar to those in a **free-sporing** plant, such as *Selaginella* (see p. 416). The functional megaspore enlarges markedly while undergoing many free-nuclear divisions (Fig. 22-13D). At the same time, the cells of the surrounding megasporangium are digested, providing both space and nutrient for the enlarging megaspore. At maximum development, the free-nucleate gametophyte fills most of the original megasporangium. The many free nuclei are suspended in watery cytoplasm, and are congregated at the periphery of the gametophyte (Fig. 22-13E).

Following the completion of free-nuclear divisions, cell walls begin to form around the nuclei. Wall formation starts at the periphery and continues centripetally until all or most of the nuclei are walled (Fig. 22-13F). The resulting tissue is the female gametophyte, which produces archegonia toward the micropylar end and later acts as a nutritive tissue for the developing embryo (Fig. 22-13G).

In most genera, archegonia develop from superficial archegonial initials at the micropylar end of the gametophyte; from one to four large archegonia mature. At maturity, each archegonium has two neck cells (a characteristic of all cycads), a ven-

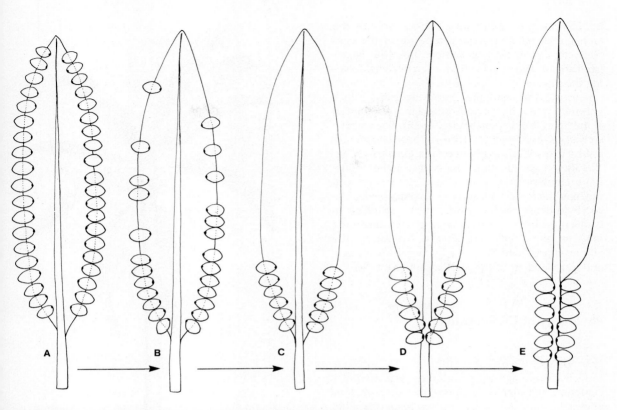

Figure 22-16 Hypothetical evolutionary development of primitive *Cycas*like megasporophyll (**E**) from *Spermopteris*like pteridosperm (**A**). **A,** a dorsal view of a *Spermopteris*like megasporophyll, with a row of abaxial ovules attached (at the black dots) on either side of the midrib, and the micropyles protruding beyond the edges of the sporophyll lamina. **B,** a derivative stage with a reduced number of seeds. **C,** further reduction in seed number, and seeds restricted to basal part of lamina. **D,** seed number essentially as in **C,** but with reduction of basal lamina, and basal seeds attached to stalk of megasporophyll (*see* the Lower Permian *Archaeocycas*). **E,** the ultimate *Cycas*like derivative, with basal lamina completely reduced, and all seeds attached to megasporophyll stalk. (After S. H. Mamay, 1976, "Paleozoic Origin of the Cycads," U.S. Geological Survey Prof. Paper 934, with permission of the author and the U. S. Geological Survey.)

tral canal nucleus, and an egg nucleus (Fig. 22-13H). In most instances, no wall forms between the ventral canal and egg nuclei. The layer of gametophytic cells enclosing the egg forms an archegonial jacket, and probably functions in transferring nutrients from the gametophyte to the egg and young embryo. This is facilitated by haustorial threads extending from the egg cytoplasm through plasmodesmata into the cells of the jacket. When fully expanded, the egg measures up to 3 mm long and is extremely turgid.

The ventral canal nucleus is short-lived and begins to disintegrate soon after formation. The egg nucleus then moves to a central position and expands to 0.5 mm in some instances (Fig. 22-14). At fertilization, the swollen proximal wall of the pollen tube bursts, discharging the two large, flagellated sperms and the liquid contents into the pollen chamber. A sperm then swims through the archegonial neck into the cytoplasm of the egg. Here the sperm nucleus escapes from the flagellated sheath and fuses with the egg nucleus, forming the zygote.

In most genera, the zygote develops into a proembryo that, in turn, develops into an embryo at the tip of the suspensor (Fig. 22-15). The embryo consists of two large cotyledons; these encompass a short axis comprising **hypocotyl** and **epicotyl** (Fig. 22-13I). At germination, the bases of the cotyledons elongate, forming a "new" hypocotyl and pushing the **radicle** out through the micropyle. The cotyledons remain for several weeks or longer in contact with the female gametophyte, absorbing food for the developing shoots. Eventually, the young sporophyte becomes anchored in the soil and independent of the food supply in the seed.

Phylogeny of Cycadopsida

Prior to 1969, the oldest cycads in the fossil record were Triassic in age. Although there was no direct evidence, investigators suspected an ancestral link of cycads with Carboniferous pteridosperms. This was based mainly on similarities in the pinnately compound leaves, stomatal pattern, and seeds. In the period 1969 to 1976, the definitive link of the cycads with the Paleozoic pteridosperms was provided by Mamay (1976) of the U.S. Geological Survey. Using recently discovered megasporophylls from the Lower Permian, Mamay proposed a hypothetical evolutionary series from the Upper Carboniferous *Spermopteris* type (Fig. 22-16A) through intermediate stages shown in parts *B* and *C* of Figure 22-16 to a stage represented by a Lower Permian genus *Archaeocycas* (Fig. 22-16D). The last stage in part *E* of Figure 22-16 is represented by another Lower Permian genus, *Phasmatocycas,* which is essentially equivalent to the megasporophyll of the extant *Cycas*. This series shows reduction in the number of seeds, reduction of the extent of indusiumlike covering of the seeds by the sporophyll lamina, and a concentration of seeds to the sporophyll stalk below the lamina. From the stages of development in the early Permian, the cycads evolved into one of the main groups of vascular plants of the Mesozoic Era. One of the oldest and most completely reconstructed is *Leptocycas* from the Upper Triassic of North Carolina (Fig. 22-17).

Figure 22-17 Reconstruction of the Upper Triassic cycad *Leptocycas*. (After T. Delevoryas and R. C. Hope, with permission of the authors and *Postilla*, Peabody Museum of Natural History, Yale University.)

Class Cycadeoidopsida (Cycadeoids)

Order Cycadeoidales (Bennettitales)

The plants of the Cycadeoidales form one of the striking and dominant elements of mid-Mesozoic

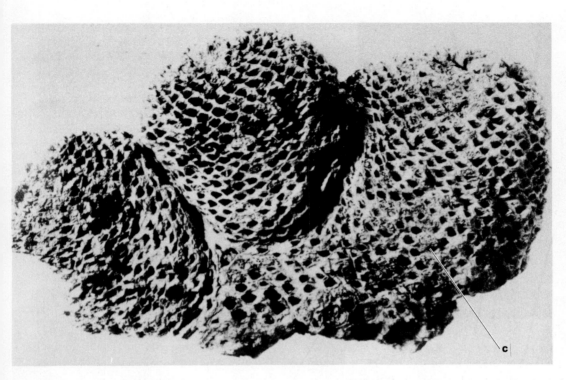

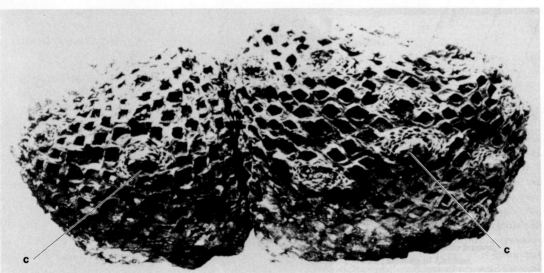

Figure 22-18 A, B, petrified trunks of *Cycadeoidea*, showing rhomboidal leaf bases and embedded cones (*c*), ×0.33. (After Wieland.)

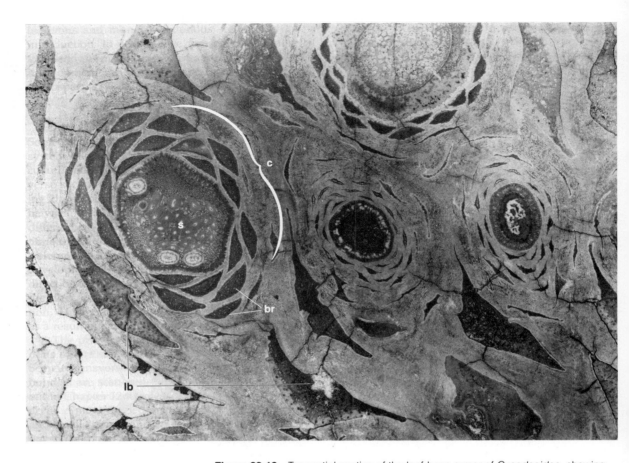

Figure 22-19 Tangential section of the leaf-base armor of *Cycadeoidea*, showing transverse sections of several cones. The larger cone at the left is mature and seed-bearing: *lb*, leaf base; *br*, bract; *c*, cone; *s*, seed-bearing receptacle, ×2.7. (From T. Delevoryas, 1968, with permission of the author and E. Schweizerbart'sche, Stuttgart, publisher of *Palaeontographica*.)

floras; together with the cycads, they were so numerous that some investigators have called the Mesozoic Era the "Age of Cycads." The cycadeoids apparently evolved in the late Carboniferous or Permian, reached a zenith of development in the Jurassic, and declined dramatically to extinction in the late Cretaceous. Thus, in general evolutionary pattern, they appear to parallel the rise and fall of the dinosaurs. This has led to the suggestion by some that the cycadeoids and dinosaurs were to some degree interdependent, although no direct evidence for this has been found.

The Cycadeoidales comprise two families, the Cycadeoidaceae and Williamsoniaceae. Although obviously related, plants of the two families show distinct differences in habit, anatomy, and reproductive structures.

Cycadeoidea is the best-known genus of the Cycadeoidaceae (Fig. 22-18). It has been found in many parts of the world but especially in the Black Hills of South Dakota. The stems consist of short, often barrel-shaped trunks superficially resem-

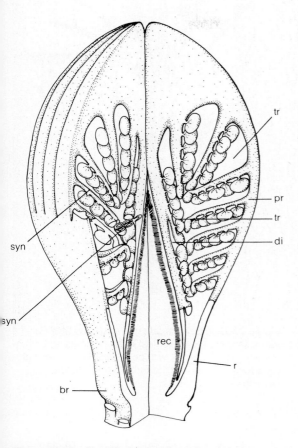

Figure 22-20 Simplified three-dimensional reconstruction of the cone of *Cycadeoidea*, showing the outer sporophylls surrounding the central conical ovulate receptacle. All of the surrounding bracts have been removed. The stippling shows those surfaces covered with epidermis. The right-hand side of the cone is a radial view of a sporophyll; the left side shows a transverse sectional view of another sporophyll. Note the trabeculae supporting the synangia, and that the trabeculae on the proximal rachis are adnate to those on the distal rachis: *br*, bract; *rec*, ovulate receptacle; *r*, rachis of sporophyll; *di*, distal portion of rachis; *pr*, proximal portion of rachis; *tr*, trabeculae; *syn*, synangium. (From W. L. Crepet, 1974, with permission of the author and E. Schweitzerbart'sche, Stuttgart, publisher of *Palaeontographica*.)

bling pineapple fruits. Some species have branched trunks, but many are unbranched. A close examination reveals a dense covering of leaf bases arranged spirally around the stem. These bases are surrounded by a thick mat of long, flat, multicellular hairs or scales. No mature leaves have been found attached to the stems, but young leaves indicate a pinnate arrangement of leaflets. This is generally interpreted to mean that when mature, the pinnate leaves were arranged in a crown at the top of the trunk.

The stems have a large pith and cortex, and a narrow cylinder of xylem. This arrangement resembles that in the stems of the extant species of the Cycadales. The primary xylem is endarch, with tracheids and rays of secondary xylem arranged in radial rows. The tracheids have bordered pitting of a circular or **scalariform** pattern. Leaf traces are numerous, and they are usually conspicuously C-shaped, with the concave side pointing inward.

The reproductive cones of *Cycadeoidea* are very interesting and have been the object of a great deal of critical research, particularly by Prof. T. Delevoryas (University of Texas) and his students (1968, 1973). The cones are deeply embedded in some trunks among the armor of leaf bases (Fig. 22-19). In most cases, the cones are **bisporangiate,** containing both ovules and pollen grains. They consist of a long stalk (**peduncle**) with a terminal, ovule-bearing receptacle (Fig. 22-20). Bracts with a dense **ramentum** are attached helically to the peduncle below the receptacle. The ovules are surrounded by tightly packed, polygonal, interseminal scales with swollen tips; only the micropyles project beyond the scales.

The microsporophylls in a **bisporangiate cone** are attached to the peduncle just below the receptacle. Each sporophyll is similar in form to the wedge-shaped section of an orange. The pollen-bearing synangia are suspended from extensions of the sporophyll rachis called **trabeculae**. Trabeculae from the proximal parts of the rachis develop towards the distal rachis and eventually meet and join trabeculae from the distal parts of the rachis (Fig. 22-20). The sporophylls are fused laterally and partly lengthwise, with only the outer regions free. The **monosulcate** pollen grains are similar to those of the Cycadales and the medullosan pteridosperms, although considerably larger. Ovules are attached to the receptacle by relatively long stalks (Fig. 22-21).

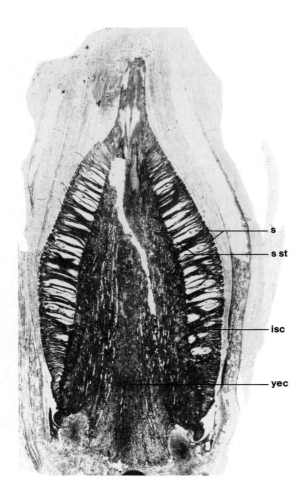

Figure 22-21 Longitudinal section of an ovulate receptacle (*rec*) of a cone of *Cycadeoidea*, showing ovules (*ov*) on stalks (*st*) surrounded by interseminal scales (*isc*), ×3. (From W. L. Crepet, 1974, with permission of the author and E. Schzweitzerbart'sche, Stuttgart, publisher of *Palaeontographica*.)

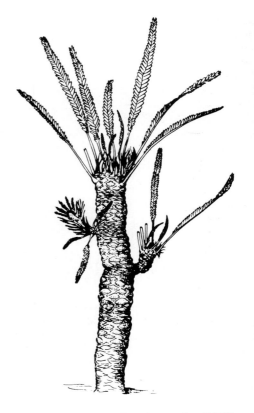

Figure 22-22 Reconstruction of a species of *Williamsonia*, showing branches, leaf bases, and leaves, ×0.03. (From B. Sahni, with permission of the Geological Survey of India and the Birbal Sahni Institute of Palaeobotany.)

Some cones of *Cycadeoidea* appear to be ovulate cones with no trace of pollen-bearing sporophylls. In some cones, it would have been difficult for the pollen from other cones to reach the ovules on the deeply embedded receptacle. There is evidence that several possible mechanisms may have been operating for a limited amount of cross pollination. Some evidence indicates that insects and possibly other animals may have chewed both pollen-sporophylls and receptacles; the pollen-sporophylls may have abscissed from the peduncle, dropped from the cone and trunk, and been blown around on the ground, releasing pollen; or the pollen-sporophylls may have split open at the tips on drying, thereby allowing the pollen grains to discharge and blow over the ovulate cones. However, the generally closed structure of the cone suggests a high level of self-pollination. The possibility of getting additional information on these basic processes makes it exciting to collect and do research on the cycadeoids.

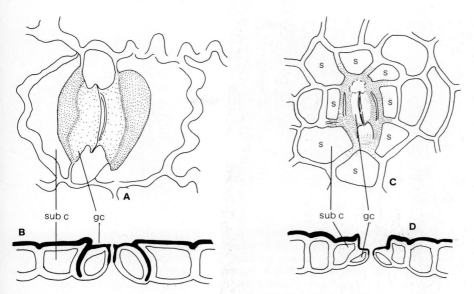

Figure 22-23 Stomatal patterns in cycadophytes. **A, B,** syndetocheilic stomatal pattern of the cycadeoid leaf *Pterophyllum,* showing two subsidiary cells (*sub c*), guard cells (*g c*), and cuticle in black, ×600. **C, D,** haplocheilic stomatal pattern of the cycad leaf *Pseudoctenis,* showing a ring of subsidiary cells around the guard cells, and cuticle in black, ×500. (After T. M. Harris, 1932, with permission of the author and of the Commission for Scientific Research in Greenland, publishers of *Meddeleser om Grønland.*)

Order Williamsoniales

Genera of the order Williamsoniales are well represented in mid-Mesozoic rocks; most of the williamsonias appear to have had slender trunks and side branches, and pinnate leaves. One of the better known of these is *Williamsonia* (Fig. 22-22). Although the arborescent habit contrasts with that of the short, bulbous *Cycadeoidea,* the anatomy and reproductive organs are similar. A few bisporangiate cones have been found, but most specimens appear to have been unisexual. The reproductive unit of a species recently described from the Jurassic of Mexico consists of an ovulate cone, subtended by poorly preserved bracts. The circular cone consists of seeds and surrounding interseminal scales attached to a receptacle, and the seeds of some species contain embryos.

Detached Cycadophyte Leaves

Detached leaves of both the cycadeoids and the cycads are widespread and locally abundant in Mesozoic rocks. On morphology alone, the leaves of the two groups cannot usually be distinguished. However, critical research has shown that the **stomatal apparatus** and epidermal pattern of cycadeoid leaves are markedly different from those of cycads, a difference used widely to identify and classify many detached leaves.

The most important diagnostic feature in the stomatal pattern is the arrangement of the **guard** and **subsidiary cells.** In cycadeoid leaves, the development of the stomatal apparatus is called **syndetocheilic.** In this type, both the guard and

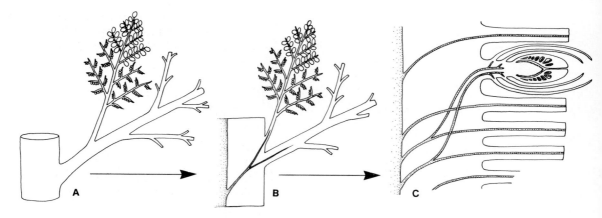

Figure 22-24 Possible pathway for the evolutionary development of the cycadeoid cone from Paleozoic pteridosperm ancestors. **A,** part of an idealized Paleozoic pteridosperm, showing a base of a leaf bearing fertile structures. **B,** hypothetical intermediate stage between the Paleozoic pteridosperm and the cycadeoids, showing a basal pinna of a leaf partially embedded within the fleshy stem. **C,** longitudinal section of part of a cycadeoid stem, showing the reduced fertile appendage (= cone) embedded within the stem. [Reproduced with permission from T. Delevoryas, 1968, "Some Aspects of Cycadeoid Evolution," *Botanical J. Linnean Soc.* 61:137–46. Copyright by Academic Press Inc. (London) Ltd.]

the subsidiary cells originate from the same mother cell (Fig. 22-23A, B). Another characteristic feature of this pattern is the presence of thick cuticle layers on the outer walls of the guard cells.

The other type of development occurring in both extant and fossil Cycadales is called **haplocheilic**. In this type, the two guard cells develop from a single initial, whereas the three to nine subsidiary cells originate from ordinary epidermal cells that later become modified (Fig. 22-23C, D). The subsidiaries usually form a ring around the two guard cells, which have no appreciable cuticular thickening.

Phylogeny of Cycadeoidales

The cycadeoids, with their earliest undoubted record in the early Mesozoic, appear to be most closely related to Paleozoic pteridosperms. With their pollen organs in the form of a synangium, and their sporophylls probably representing portions of a reduced pinnately compound leaf, they are similar to seed ferns of the Medullosaceae; they could have been derived from that or a closely related family. The chief feature in which they differ from both the seed ferns and the Cycadales is the structure of the ovulate cone. To date, nothing similar has been found in any group that can be interpreted as ancestral to the receptacle of the cycadeoid cone. However, assuming a pteridosperm ancestry, Delevoryas (1968) has presented several hypothetical pathways for the evolution of the cycadeoid cones. One of these, shown in Figure 22-24, derives the cone from one part of a Paleozoic pteridosperm (Fig. 22-24A); this is followed by telescoping of the whole leaf, so that the fertile branch is first encompassed within the fleshy stem (Fig. 22-24B), and leads to the cycadeoid cone embedded in leaf bases (Fig. 22-24C). The other hypothetical pathway views the cone as originating from the reduction of the fertile branch in the axil of a leaf (Fig. 22-25).

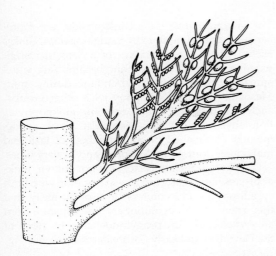

Figure 22-25 Diagram of an axillary fertile branch that could represent the ancestral structure of the cone of cycadeoids. [Reproduced with permission from T. Delevoryas, 1968, "Some Aspects of Cycadeoid Evolution," *Botanical J. Linnean Soc.* 61:137–46. Copyright by Academic Press Inc. (London) Ltd.]

reached a peak in the mid-Mesozoic and became extinct in the late stages of that era. In this, they match the general evolutionary pattern of the dinosaurs.

References

Banks, H. P. 1970. *Evolution and plants of the past.* Wadsworth Publishing Co., Belmont, Calif.

Crepet, W. L. 1974. Investigations of North American cycadeoids: the reproductive biology of *Cycadeoidea*. *Palaeontographica* B, 148:144–69.

Cridland, A. A., and Morris, J. E. 1960. *Spermopteris*, a new genus of pteridosperms from the Upper Pennsylvanian of Kansas. *Amer. J. Bot.* 47:855–59.

Delevoryas, T. 1968. Some aspects of cycadeoid evolution. *Bot. J. Linn. Soc.* 61:137–46.

Delevoryas, T., and Gould, R. E. 1973. Investigations of North American cycadeoids: williamsonian cones from the Jurassic of Oaxaca, Mexico. *Rev. Palaeobot. Palynol.* 15:27–42.

Harris, T. M. 1969. The Yorkshire Jurassic Flora. III. Bennettitales. *Br. Mus. (Nat. Hist.), London, Publ.* 675:1–186.

Mamay, S. H. 1976. Paleozoic origin of the cycads. *U.S. Geol. Survey, Prof. Paper* 934:1–48.

Norstog, K. 1974. Fine structure of the spermatozoid of *Zamia*: the vierergruppe. *Amer. J. Bot.* 61:449–56.

Taylor, T. N. 1973. A consideration of the morphology, ultrastructure and multicellular microgametophyte of *Cycadeoidea dacotensis* pollen. *Rev. Palaeobot. Palynol.* 16:157–64.

Thus the Cycadeoidales represent another evolutionary line of gymnosperms for which precise phylogenetic connections are unknown. With likely ancestry in the Paleozoic pteridosperms, they parallel the Cycadales in geologic time span as well as in many structural similarities. In having an ovulate cone with a receptacle and the stomatal-cuticular pattern, they stand as a distinct group that

23
GINKGOPHYTES

The ginkgophytes, class Ginkgopsida, are represented in the extant flora by a single species, *Ginkgo biloba*. It is the sole survivor of a class that originated in the later Paleozoic and became very widespread and moderately abundant during mid-Mesozoic times. In company with other groups, such as cycadophytes and conifers, ginkgophytes flourished during the Mesozoic. They subsequently diminished in both numbers of taxa and individuals during later Mesozoic and Tertiary times.

Although their ancestors are not known, ginkgophytes have characteristics in common with both cycadophytes and conifers. It appears likely that they represent an evolutionary line separate from an *Archaeopteris*like, progymnospermous stock of the Devonian and early Carboniferous (see also Fig. 23-14).

Although several fossils suggesting ginkgophytean relationships have been reported in the Carboniferous, the earliest definite ginkgophyte in the geologic record is *Trichopitys* from the early Permian of France (Fig. 23-1A). It consists of a small axis that bears leaves in which the blade is dichotomously divided into five filamentous laminae. Fertile, branched shoots bearing ovules occur in the axils of some leaves. Usually there are 4–6 ovules per fertile shoot, but as many as 20 have been recorded.

Almost all remains of ginkgophytes in Mesozoic rocks are leaf compressions. In some sites, leaves are so abundant they overlap each other on the rock surface. Some 16 genera have been described, but the two most common are *Baiera* and *Ginkgoites*. Leaves of *Baiera* have narrow, dichotomously segmented blades (Fig. 23-1B). Depending on species, the *Ginkgoites* leaf is more or less deeply cleft (Fig. 23-1C), or it may be entire with lobes, as in some leaves of the extant *Ginkgo biloba* (Fig. 23-6). In general, the deeply incised leaves are more common in the early and mid-Mesozoic, and the entire forms are more common in the later Mesozoic and Tertiary. However, these forms occur together in many early and mid-Mesozoic deposits.

Ginkgo biloba

Ginkgo biloba, popularly called the maidenhair tree because its leaves are similar to those of *Adiantum*,

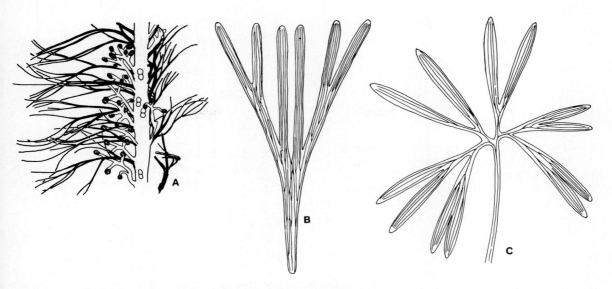

Figure 23-1 Leaves of fossil ginkgophytes. **A,** *Trichopitys*, from the Permian, with sporangia-bearing axes in axils of finely dissected leaves, ×0.3. **B,** *Baiera*, with long dichotomous leaf segments, ×0.5. **C,** *Ginkgoites*, with primary, secondary, and tertiary dichotomies of blade from petiole, ×0.5. (**A,** after Florin with permission of Bergianska Stiftelsen, Stockholm, Sweden, publisher of *Acti Horti Bergiani;* **B, C,** after Heer.)

the maidenhair fern, is often referred to as a "living fossil" because of its position as the sole surviving member of an ancient lineage. Originally known only from gardens in eastern Asia, particularly China and Japan, it has been found in recent times in what appear to be wild stands in a small mountainous area of southeastern China. The trees are deciduous and very attractive, with young leaves a light yellow-green in spring, turning to a vivid yellow in fall. The starchy seeds, called sal-nuts, are eaten in Japan and China. *Ginkgo* trees are widely cultivated for shade and as ornamentals; they are now becoming more popular because of their resistance to pollution as well as disease and fire. The species is dioecious, and male trees are preferred because the outer fleshy layer of the female seeds decays and produces butyric acid, which pollutes the air with the smell of rancid butter! It was formerly impossible to tell whether a seedling was male or female, and horticulturists had to wait for a plant to mature before selecting male trees. Now sex can be determined by identifying the X and Y chromosomes in somatic cells (Fig. 23-2).

Morphology and Anatomy

In younger trees of *Ginkgo*, the strong central stem and ascending branches produce a marked conical outline (Fig. 23-3A). However, as trees mature, the main trunk loses its prominence, and side branches grow relatively larger and longer, eventually producing a rounded crown (Fig. 23-3B). Some trees reach a height approaching 30 m and a trunk diameter of over 1 m.

As do some conifers, *Ginkgo* has two kinds of branches, the **long shoot** and the **spur** (or short) **shoot.** Long shoots have indeterminate growth with scattered leaves and form the main branches of the tree. On the other hand, spur shoots have limited growth, increasing only a few millimeters in length each year; they form on the long branches during the second year of growth and bear a cluster of leaves at the tip (Fig. 23-4). Abscission scars of former petioles form a prominent spiral around the spur shoots. Although spurs usually remain short

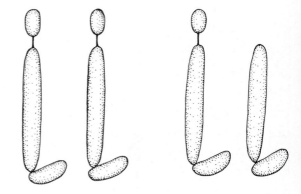

Figure 23-2 Diagram of the pairs of sex chromosomes in somatic cells of the female and male *Ginkgo* trees ($2n = 24$). Note the absence of the satellite on the male chromosome at the right. (Based on Lee and Pollock.)

Figure 23-3 *Ginkgo biloba*. **A,** young tree, showing characteristic pyramidal branching. **B,** mature tree with rounded crown. (Photographs courtesy of Chicago Natural History Museum.)

Figure 23-4 Part of a branch of *Ginkgo biloba*, showing the long shoot (*ls*), the spur shoots (*ss*) with petioles spirally arranged at the tips, and abscission scars (*as*) marking positions of previous petioles, ×0.5.

and leaf-bearing, in some instances a spur suddenly develops into a long shoot with scattered leaves, often as a result of injury to a neighboring long shoot. Similarly, long shoots sometimes become retarded and change into spur shoots.

There is a marked difference in the relative amounts of tissues produced by long and spur shoots of the same age. Long shoots have a narrow pith, wide vascular cylinder, and thin cortex. In contrast, spur shoots have a wide pith, very narrow xylem cylinder, and wide cortex. The secondary xylem forms a cylinder of irregular radial rows of tracheids, rays, and weakly defined growth rings. The tracheids of primary xylem have helical thickenings, whereas those of secondary xylem have one to two rows of bordered pits with **crassulae** between them. Except for rays, there is no xylem parenchyma in the wood. The pith and cortex are studded with mucilage cavities and cells containing calcium oxalate crystals and tannins. A periderm originates in the outer cortical layers and later, with secondary phloem, forms the bark (Fig. 23-5).

The distinctive leaves of *Ginkgo biloba* have a long petiole that fans out into a broad blade (Fig. 23-6). On long shoots, the leaf blades are usually notched in the middle (Fig. 23-6B); hence, the specific epithet *biloba*. In contrast, the leaves of spur shoots are mainly entire (Fig. 23-6A). The petiole contains two leaf traces that divide dichotomously in the base of the blade. The two traces come from two completely separate primary vascular strands in the stem; this is the only gymnosperm with separate traces. Additional dichotomies occur in the blade as far as the margin, giving a fanlike appearance to the leaf. Internally, the mesophyll consists of a relatively weak palisade layer and a spongy mesophyll below (Fig. 23-7A). The xylem and phloem strands of veins are surrounded by rounded and somewhat thickened parenchyma cells. Cells containing tannins and oxalate cells are also common. Both upper and lower epidermis are coated with a cuticle, which is thickened on top of some cells (Fig. 23-7A). Stomata are found mostly on the lower surface; they have four to seven subsidiary cells flanking the slightly sunken guard cells (Fig. 23-7B).

Reproduction

Like cycads, *Ginkgo biloba* is consistently **dioecious**. Male trees produce lax **microstrobili** on the tips of spur shoots interspersed among leaves (Fig. 23-8A). Each consists of a branching axis, and each branch ends in a knob and bears two hanging microsporangia. Each **microsporangium** has four to seven wall layers, a **tapetum**, and a number of central **microsporocytes** that undergo meiosis to form tetrads of **microspores** in early spring. Soon after their formation, microspores continue dividing to form the male gametophytes, the pollen grains.

The pollen grains of *Ginkgo biloba* are similar to those of cycads (Fig. 23-8B). They are circular to elliptical, about 25–37 μm, with the **exine** deposited on the **intine** except along the area of the germinal furrow on the concave distal surface. Dis-

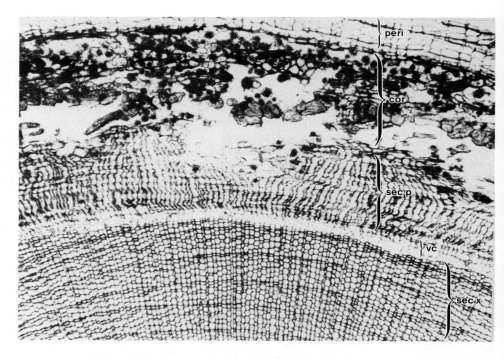

Figure 23-5 Transverse section of stem of *Ginkgo biloba*, showing periderm (*peri*), cortex (*cor*), and secondary phloem (*sec p*) in relation to the vascular cambium (*vc*) and secondary xylem (*sec x*), ×67.

persal and pollination occur in April–May in middle northern latitudes, depending on location and weather.

The development of the immature male gametophyte in the pollen grain is very similar to that of cycads (Fig. 23-9). In the first stage, the **androsporal cell** divides twice to produce two small **prothallial cells** and a larger **androgenous initial** (Fig. 23-9A–C). The androgenous initial then divides to produce a generative cell next to the second prothallial cell, and a tube cell (Fig. 23-9D). Pollen grains are shed at this stage; their large numbers yield a yellow cloud when the limbs are shaken by wind or by hand.

Further development of the male gametophyte occurs after the pollen grain has lodged in the pollen chamber of the ovule. At this time, the pollen tube begins to penetrate the integumentary wall of the outer part of the pollen chamber. Concurrently, the generative cell in the pollen divides to form a stalk cell next to the prothallial cells and a body cell (Fig. 23-9E) located near the base of the pollen tube. Shortly before fertilization, the body cell divides to form two sperms. Particles that densely surrounded the body cell now align themselves around each sperm and appear to form spiral bands of flagella at one end. After the pollen wall ruptures, the relatively large (*ca.* 80 μm) sperms swim through fluid in the pollen chamber to fuse with the large egg.

The interval between pollination and fertilization ranges from four to seven months, depending on locale, weather conditions, and probably on time of pollination. See Figure 23-13 for the timing of this and other events in the **ontogenetic** cycle of *Ginkgo*.

Two ovules are borne on each peduncle, one at the tip of each dichotomous branch (Fig. 23-10A, B). Beneath each ovule is a collar, which some botanists believe to be homologous to a leaf because occasional freaks occur in which the ovules are associated with leaflike structures. As in cycads,

Figure 23-6 **A, B,** *Ginkgo biloba.* **A,** photograph of a leaf of *Ginkgo biloba* attached to a spur shoot (*sp s*), showing petiole (*pet*) and blade (*bl*), ×1.2. **B,** drawing of a cleared leaf from a long shoot, showing the prominent notch (*n*) separating the two lobes of the blade and the dichotomous veins fanning out in the lobes, ×0.8.

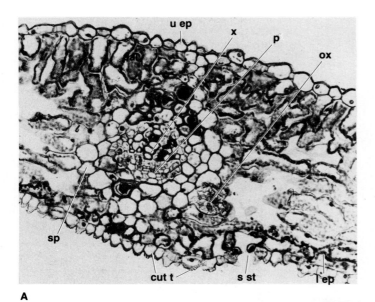

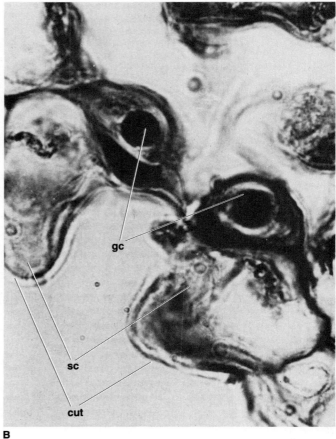

Figure 23-7 Leaf of *Ginkgo biloba*. **A,** vertical section through leaf blade, showing upper epidermis (*u ep*), palisade (*pal*) and spongy mesophyll (*sp*), a vein with xylem (*x*) and phloem (*p*), oxalate crystals (*ox*), and lower epidermis (*l ep*) with cuticular thickenings (*cut t*) and sunken stomata (*s st*), ×175. **B,** vertical section through a stoma, showing the sunken guard cells (*gc*), subsidiary cells (*sc*), and cuticle (*cut*), ×1600.

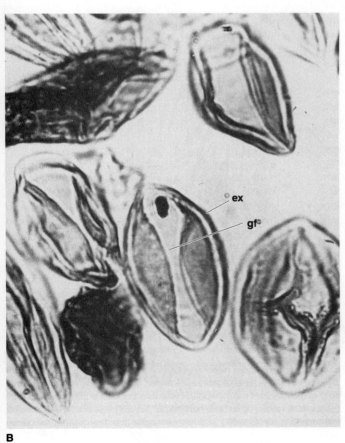

Figure 23-8 *Ginkgo biloba*. **A,** lax microstrobili, with spirally arranged microsporophylls (*ml*) and microsporangia (*ms*) that have all split open and discharged the pollen grains, ×1.5. **B,** several pollen grains, showing the exine (*ex*) and single elongated germinal furrow (*gf*) on the concave distal surface, ×1250.

each ovule has a single integument with three distinct tissue regions (Fig. 23-10C). Following pollination, the three layers mature into an inner fleshy layer that dries out to a papery skin, a central stony layer, and an outer fleshy layer. The outer layer contains the butyric acid that causes the foul odor of seeds in autumn.

The female gametophyte in the ovule is similar to that of cycads and begins developing in May. The megasporocyte in the megasporangium undergoes meiosis to form a linear tetrad of megaspores (Fig. 23-11A); usually only the one at the **chalazal** end develops. The functional megaspore wall expands, and the nucleus undergoes many free-nuclear divisions (Fig. 23-11B). At maximum expansion, numerous nuclei are packed within the periphery of the megaspore wall. Cell wall formation proceeds centripetally, so that eventually all nuclei become walled to form the female gametophyte (Fig. 23-11C). A noteworthy feature is the pale green color resulting from the development of chlorophyll in cells of the gametophyte—a condition that also occurs in gametophytes of some **free-sporing** plants.

In most instances, two or three archegonia differentiate from outer cells at the micropylar end of the female gametophyte. Each archegonium has two neck cells, a ventral canal cell, and a large egg (Fig. 23-12A). As in cycads, a jacket layer one cell thick forms on the outside.

The zygote nucleus undergoes a series of free-nuclear divisions, forming some 256 nuclei around which cell walls develop almost simultaneously to form a proembryonic tissue. The cells at the micropylar end elongate slightly, and those in the middle

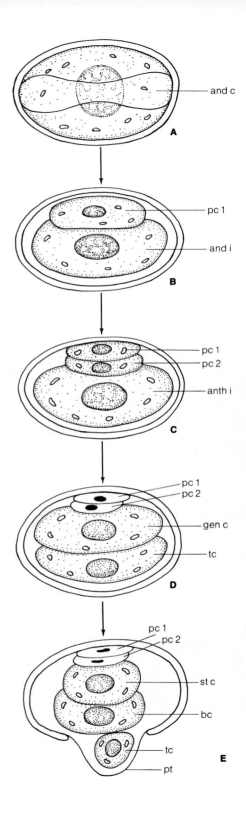

Figure 23-9 A–E. Stages in the development of the male gametophyte within the pollen grain of *Ginkgo biloba*. **A,** androsporal cell (*and c*), which divides to produce one prothallial cell (*pc-1*) and androgenous initial (*and i*) in **B. C,** after the next division with a second prothallial cell (*pc-2*) and antheridial initial (*anth i*). **D,** after the next division, in which the antheridial initial produces a generative cell (*gen c*) and tube cell (*tc*). **E,** after the division of the generative cell to give a stalk cell (*st c*) and body cell (*bc*), and the persistence of the tube cell (*tc*); note the developing pollen tube (*pt*), ×1000.

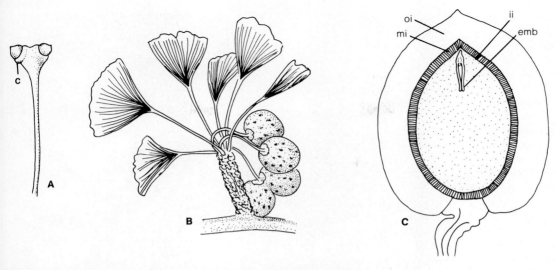

Figure 23-10 Female reproductive organs of *Ginkgo biloba*. **A,** two young ovules developing on dichotomous branches, and subtending collar (c), ×1. **B,** several mature seeds borne at tip of spur shoot, ×1. **C,** longitudinal median section of seed with young embryo, ×2: *oi,* outer integument; *mi,* middle integument; *ii,* inner integument; *emb,* embryo. (**C,** from J. M. Coulter and C. J. Chamberlain, *Morphology of Gymnosperms,* Chicago: University of Chicago Press, copyright 1910 by The University of Chicago Press.)

regions enlarge moderately; those at the bottom or chalazal end divide rapidly but do not enlarge (Fig. 23-12B). The elongating cells at the micropylar end do not form a definite suspensor as in cycads. Cells at the chalazal end continue to divide, eventually forming **primordia** of the cotyledons and young shoot apex (Fig. 23-12C). The primordia enlarge and absorb the cell contents of the female gametophyte. Eventually, the young root grows out through the micropyle to establish itself in the soil. The cotyledons remain attached to the seed and continue to absorb nutrients from the gametophyte until the young leaves of the seedling can provide sufficient energy from photosynthesis.

The ontogenetic events for *Ginkgo biloba* are summarized in Figure 23-13.

Phylogeny of Ginkgophytes

The ancestors of the ginkgos are unknown. *Ginkgo* is similar to cycads in many respects, especially in pollen, ovule, and female gametophyte development. However, the leaves are distinctive and generally unlike the pinnate leaves of cycads and pteridosperms. In addition, the stem, with extensive wood, small pith and cortex, and bordered pitting, is similar to that of coniferophytes (Chapter 24). The most likely hypothesis is that ginkgophytes are a separate line that evolved from the same ancestral stock as Cordaitales, most likely from progymnosperms such as *Archaeopteris*. Such a phylogeny is shown in the flow diagram in Figure 23-14, in which Meeuse (1966) has indicated possible evolutionary pathways (**semophylesis**) for both the male and female organs of ginkgophytes. In presenting this hypothesis, Meeuse has deliberately used only reproductive organs of *Cordaites* to illustrate organ phylogeny, deliberately omitting the leaves and anatomical details of *Cordaites* plants. This allows for the possibility that it was indeed a plant other than *Cordaites*—but one with *Cordaites*like reproductive organs—that was a developmental stage between *Archaeopteris* and *Trichopitys-Baiera-Ginkgo* (Fig. 23-14).

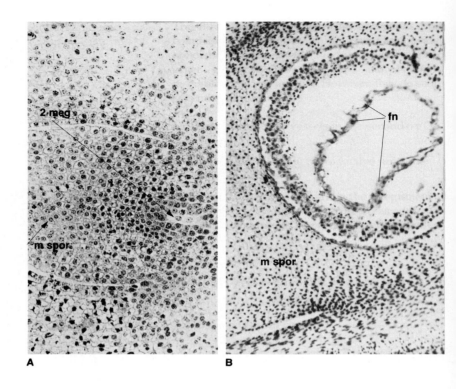
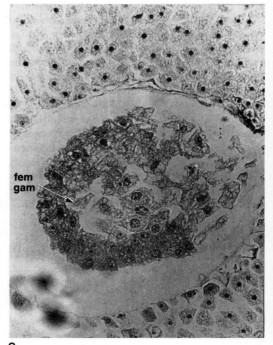

Figure 23-11 Female gametophyte development in *Ginkgo biloba*. **A,** the first division of meiosis in the megasporocyte (*2 meg*) within the megasporangium (*m spor*), ×150. **B,** free-nuclear stage (*fn*) of female gametophyte development, ×180. **C,** formation of female gametophyte cells (*fem gam*) by wall formation around free nuclei, ×200.

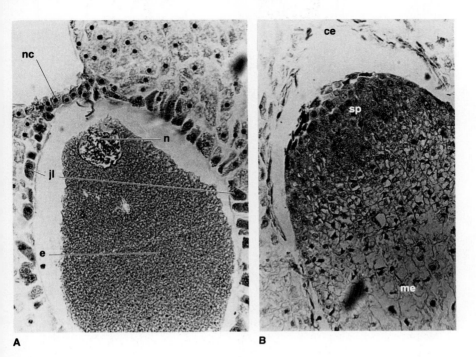

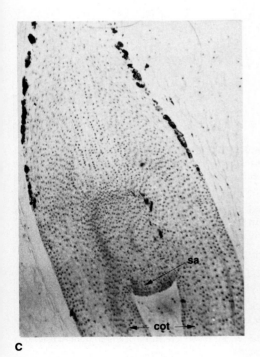

Figure 23-12 *Ginkgo biloba.* **A,** archegonium with large egg cell (*e*), prominent nucleus (*n*), jacket layer (*jl*), and neck cells (*nc*), ×150. **B,** proembryo, showing elongating cells at the micropylar end (*me*) and the early shoot primordium (*sp*) at the chalazal end (*ce*), ×180. **C,** embryo with stem apex (*sa*) and cotyledons (*cot*), ×160.

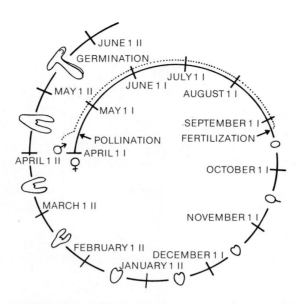

Figure 23-13 Diagram of the ontogenetic cycle of *Ginkgo biloba*. (From M. Favre-Duchartre, 1958, with permission of the author and the International Society of Plant Morphologists.)

Although this hypothesis appears reasonable, there is also the possibility that ginkgophytes descended from Carboniferous pteridosperms.

References

Chamberlain, C. J. 1963. *Gymnosperms: structure and evolution*. Hafner Publishing Company: University of Chicago Press (1935).

Dorf, E. 1958. The geological distribution of the *Ginkgo* family. *Bull. Wagner Free Inst. Sci.* 33(1):1–10, Philadelphia.

Florin, R. 1949. The morphology of *Trichopitys heteromorpha* Saporta, a seed plant of Paleozoic age, and the evolution of the female flowers in the Ginkgoinae. *Acta Horti Bergiani* 15(5):79–109.

Gifford, E. M. Jr., and Corson, G. E. Jr. 1971. The shoot apex in seed plants. *Bot. Rev.* 37:143–229.

Li, H. L. 1956. A horticultural and botanical history of *Ginkgo*. *Morris Arbor. Bull.* 7:3–12.

Meeuse, A. D. J. 1966. *Fundamentals of phytomorphology*. The Ronald Press, New York.

Pollock, E. G. 1957. The sex chromosomes of the maidenhair tree. *J. Heredity* 48:290–94.

Seward, A. C. 1938. The story of the maidenhair tree. *Sci. Progress* 32:420–40.

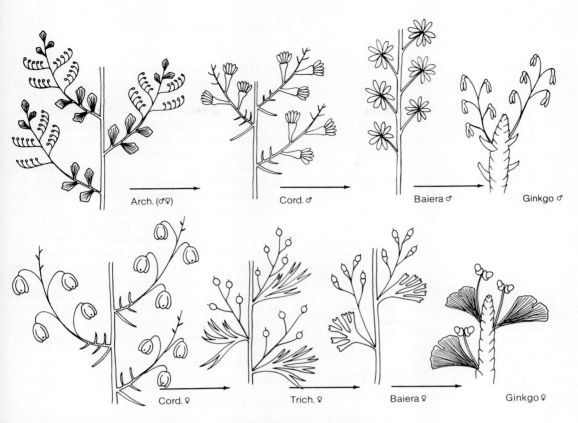

Figure 23-14 Tentative semophylesis (organ derivation) of the reproductive regions of the Ginkgoales. *Arch, Archaeopteris* (Devonian); *Cord, Cordaianthus* (Carboniferous); *Trich, Trichopitys* (early Permian); *Baiera* (Triassic-early Cretaceous); *Ginkgo* (Mesozoic-Recent). Note the major trend in both male and female semophylesis in reduction in both sterile and fertile appendages. (From A. D. J. Meeuse, *Fundamentals of Phytomorphology*. Copyright © 1966, by the Ronald Press Company. Reprinted by permission of John Wiley & Sons, Inc., and the author.)

24

CONIFEROPHYTES

Class Coniferopsida

From their earliest appearance in the late Paleozoic, the conifers, class Coniferopsida, evolved into one of the major elements in the vascular floras of the Mesozoic era. Although they diminished in number after the angiosperms assumed dominance in the later Cretaceous, the conifers today form dense forests over large expanses of earth, particularly in cool and wet regions. In extant forests, conifers are among the tallest (*Sequoia*), largest (*Sequoiadendron*), and oldest (*Pinus aristata*) plants. Historically, the conifers have been one of our most important sources of wood and wood products. Also, many have a pleasing appearance and are hardy; hence, they are cultivated extensively in gardens and parks. The main distinctive feature in most is that both the pollen and seeds are borne in cones.

The coniferopsid class encompasses four orders: the Cordaitales of the Carboniferous and the Voltziales of the Permian are both extinct, whereas the Coniferales and Taxales contain both fossil and extant taxa.

Order Cordaitales

The Cordaitales contain three families: the Eristophytaceae, the Poroxylaceae, and the Cordaitaceae. The first two families comprise genera of petrified wood, and the seed is known as well for the single genus of the Poroxylaceae. The best known is the Cordaitaceae, particularly the genus *Cordaites*, in which stems, leaves, and cones have been found attached. The cordaites, as they are commonly called, flourished during the Upper Carboniferous, forming forests of tall, majestic trees (Fig. 24-1).

Judging from petrified stems, *Cordaites* trees reached a height of over 20 meters. Their upper reaches supported branches bearing long, strap-shaped leaves. Lax cones were borne in the axils of some of the leaves.

The stem anatomy of *Cordaites* is known in detail (Fig. 24-2). The pith has a diameter of up to 1.5 cm and contains lens-shaped cavities within the parenchyma tissue. Primary xylem lies immediately outside the pith and is endarch (Fig. 24-2). The tracheids show a progression outward from helical to scalariform to bordered. The primary xylem grades into secondary xylem, which has tracheids with one to three rows of alternating and hexagonal bordered pits. Rays of the secondary wood are variable in height and usually one cell wide. The secondary xylem forms a thick cylinder of wood. In most Carboniferous specimens, the wood shows no growth rings, suggesting the absence of any well-defined seasonal or climatic changes during the interval of formation of extensive coal swamps of the Carboniferous.

The secondary phloem contains much parenchyma and forms a thin layer on the inner edge of the cortex of smaller branches (Fig. 24-2). Leaf traces, either single or double, angle through the cortex to the leaf bases. The traces divide in the cortex, with several strands entering each leaf.

The leaves are thin, strap-shaped, and sometimes over a meter long and 15 cm wide. Many lengthwise veins that appear to be parallel are actually acutely dichotomous. Stomata are arranged in bands on the under surface of the leaf. Stomatal development is **haplocheilic,** with four to six subsidiary cells surrounding the sunken guard cells. The leaves apparently derived much of their support from bands of sclerenchyma above and below the veins, as well as hypodermal intermediate bands (Fig. 24-3).

The reproductive organs of *Cordaites* are loose, compound cones borne in the axils of some leaves on outer branches. Both male (pollen-bearing) and female (seed-bearing) cones are placed in the same genus, *Cordaianthus* (Figs. 24-4A and 24-5). The compound cones consist of two rows of awl-shaped bracts, one on either side of a central axis. In the axil of each bract is a short shoot bearing spirally arranged, flattened scales. In the male cones, some of the outer scales bear pollen sacs at the tips (Fig. 24-4B). Several have yielded pollen grains (Fig. 24-4C) that show a central body, or **corpus,** with the wall folded into chambers. The corpus is surrounded by a single bladder, or **saccus,** that formed an expansion of the outer **exine** from an inner **intine.** The saccus undoubtedly acted as a floating apparatus, assisting in wind dispersal. In having a saccus, *Cordaites* pollen is similar to the disaccate pollen of many conifers, such as pine, spruce and fir. This appears to represent an evolutionary development: the progressive reduction of a single large saccus into two, with connecting stages shown in the pollen of the early conifers (see p. 561).

Figure 24-1 Reconstruction of *Cordaites*. Note compound strobili interspersed among leaves of side branches, ×0.003. [From *Studies in Fossil Botany*, Pt II, 3rd edition by D. H. Scott by permission of C. Black (Publishers) Ltd., London.]

The female cone of one species is shown in part in Figure 24-5. The bracts occur in two lateral rows, with a **dwarf shoot** in the axil of each. The dwarf shoot has spirally arranged scales, both sterile and fertile. The fertile scales are usually longer, with one or more terminal ovules. In a later species, each fertile scale is very short, bears but one ovule, and is embedded deeply among the surrounding sterile scales.

The work of Florin (1938–1945) and others has suggested that the seed cone of the Coniferales was probably derived by the reduction and elaboration

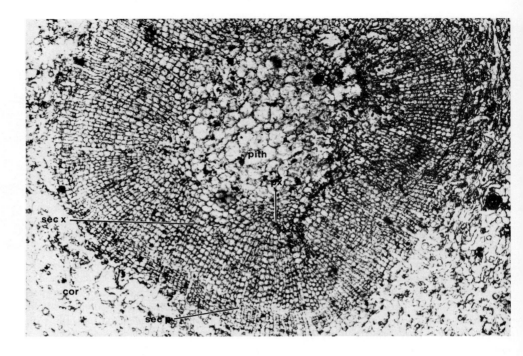

Figure 24-2 Part of a transverse section of a stem of *Cordaites*, showing pith, primary xylem (*px*) extending into pith, radial rows of tracheids and rays of the secondary xylem (*sec x*), secondary phloem (*sec p*), and cortex (*cor*), ×35.

of a compound cone such as *Cordaianthus*. This will be considered in some detail under the extant conifers; probable intermediate stages can be noted in the ovule-bearing shoots of the family Lebachiaceae described below. Other features showing relationships to the conifers are the bordered pitting of the tracheids, similar to those of the araucarian conifers, the saccate structure of the pollen grains, and the similarities of leaves to those of some of the araucarias (e.g. *Agathis*).

Order Voltziales

The Voltziales contain the two fossil families, Lebachiaceae and Voltziaceae, ranging from the late Paleozoic to the early Mesozoic. It includes genera and species that appear to be the immediate ancestors of the six extant families of conifers.

These two families consist of several genera found on several continents of the Northern Hemisphere from the late Paleozoic and early Mesozoic, thereby overlapping the time span of the Cordaitales. The earliest and best-known genus is *Lebachia* of the Lebachiaceae (Fig. 24-6A). It was a tree of uncertain size, with whorls of five or six main branches on the trunk. The secondary branches were arranged in two rows on the primaries—a **distichous** arrangement. The pattern is very similar to that of many modern Araucariaceae, particularly *Araucaria heterophylla*, the Norfolk Island pine.

Lebachia has two types of leaves. Those on the trunks and larger branches are longer, broader, and more flattened than those of smaller shoots.

Internally, the well-developed secondary xylem has araucarian-type tracheids with one to three rows

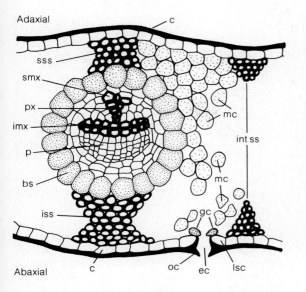

Figure 24-3 Transverse section of part of a leaf of *Cordaites*, showing the main tissues and cells in the vicinity of a vein: *c*, cuticle; *oc*, overarching cuticular extension; *lsc*, lateral subsidiary cell; *ec*, epistomal chamber; *gc*, guard cells; *iss*, inferior hypodermal sclerotic strands; *sss*, superior hypodermal sclerotic strands; *int ss*, intermediate sclerotic strands; *bs*, bundle sheath; *p*, phloem; *imx*, inferior metaxylem arc; *smx*, superior metaxylem; *px*, protoxylem, ×350 (approx.). (Redrawn, with modifications, from Harms and Leisman.)

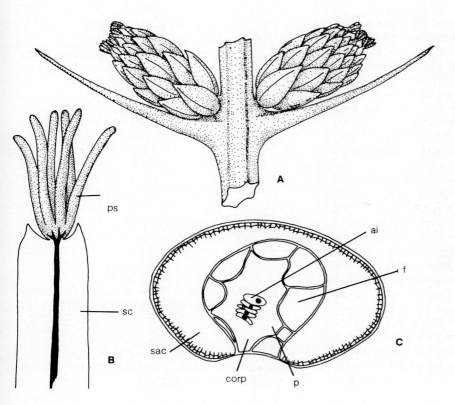

Figure 24-4 Male reproductive organs of *Cordaites*. **A**, two male (micro-) strobili in the axils of bracts on primary strobilar axis, ×6. **B**, single scale (*sc*) of a male strobilus, showing radial attachment of pollen sacs (*ps*) at the tip, ×25. **C**, pollen grain (*Florinites*) of *Cordaites*, with a single saccus (*sac*) surrounding a central corpus (*corp*); the cell-like chambers on the inside (*f*) are actually folds in the wall of the corpus; the four lenticular bodies on the inside (*p*) are considered remnants of prothallial cells; the large oval body is the antheridial initial (*ai*), otherwise called the embryonal cell, ×800. (**A**, after Delevoryas, 1953, with permission of the author and *American Journal of Botany*; **B, C,** after R. Florin, with permission of *Palaeontographica* and *Svensk Botanisk Tidskrift*.)

Coniferophytes

of alternate bordered pits. There are no resin canals and only a trace of xylem parenchyma. The xylem rays are **uniseriate,** another araucarian characteristic. Growth rings are weakly delimited.

Both seed-bearing and pollen-bearing cones are attached terminally on some of the side branches of *Lebachia* (Fig. 24-6B, C). The pollen-bearing cones have an axis with spirally arranged and scalelike **microsporophylls.** Two pollen sacs are borne on the abaxial surface of each microsporophyll and are partly covered by an outgrowth of the sporophyll itself.

It must be emphasized that the male cone is a simple cone, like that of modern conifers, and very unlike the compound cones of *Cordaites* and *Lebachia*. Hence there is a significant gap in our knowledge of the evolution of the conifer male cone, for which only future discoveries will provide evidence.

Pollen grains of *Lebachia* have a single saccus surrounding a central corpus (Fig. 24-7A). The bladder is interrupted on the distal surface by the germinal furrow. In several genera, two sacci are attached to a corpus, a disaccate condition occurring in some genera of the modern conifer families, namely, the Pinaceae and Podocarpaceae.

It has been suggested that the disaccate condition evolved from the **monosaccate** of *Lebachia* by a suppression of the single saccus along the **proximal** surface of the grain (Fig. 24-8). The evolutionary trend toward the disaccate condition, which started in the progymnosperms and continued with the cordaitalean pollen (see Fig. 20-8), has been linked with the inversion of the ovules that occurred in several genera of the Voltziaceae (e.g., *Ullmannia*) found somewhat later in the geologic record than *Lebachia*.

The female cone of *Lebachia* is elliptical to circular in outline, with many compact bracts spirally arranged on the central axis (Fig. 24-7B). Each bract is two-forked, with a single vein that divides to provide one vein for each of the two lobes. In the axil of each bract there is a dwarf shoot with spirally arranged, **decurrent,** and upright scales. Usually, only one scale is fertile and bears a single ovule at the tip (Fig. 24-7C). The seed-bearing scale occurs next to the main axis of the cone, in the same spiral sequence as the sterile scales. The micropyle of the ovule is directed outward. The integument appears to be a continuation of the tissue of the seed scale that overarches the nucellus, forming a relatively deep micropyle.

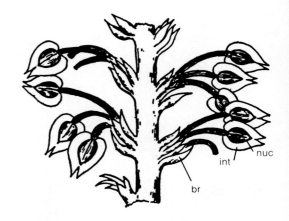

Figure 24-5 Part of a female cone of *Cordaites*, showing dwarf shoots in axils of bracts, with both sterile and fertile scales emanating from the dwarf shoots. Seeds are flat and heart-shaped, with a winglike integument surrounding the central nucellus, ×20: *br*, bract; *int*, integument; *nuc*, nucellus. (After Carruthers.)

Figure 24-6 *Lebachia*. **A,** sterile branch with five twigs, showing pinnate arrangement of ultimate branches, ×0.33. **B,** single branch supporting both foliage twigs and pollen-bearing cones (*pbc*); note forked leaves on main branch and unforked falcate leaves on ultimate branches, ×0.33. **C,** branch with seed-bearing cones and leaf-bearing twigs, ×0.33. (After R. Florin, with permission of *Palaeontographica*.)

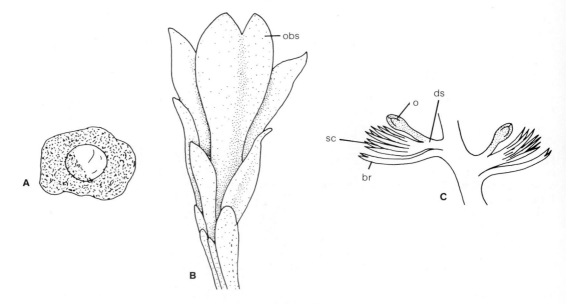

Figure 24-7 *Lebachia.* **A,** single pollen grain, showing central corpus and surrounding saccus, ×185. **B,** dwarf shoot of seed-bearing cone, showing spirally arranged scales and single ovule-bearing scales (*obs*) at top, ×10. **C,** longitudinal section of seed-bearing cone, showing bracts (*br*), sterile scales (*sc*), and single ovule (*o*) attached to dwarf shoot (*ds*), ×4. (After R. Florin, with permission of *Palaeontographica*.)

The overall structure of the compound female cones in the Lebachiaceae and Voltziaceae is similar to that of both the Cordaitales and the more modern conifers. In the Lebachiaceae, there is a general reduction in the length of the dwarf shoot and in the numbers of sterile scales and ovules. This may represent an early stage in the evolution of Mesozoic and Cenozoic conifers, which resulted in further reduction of parts in several directions in the conifer families.

Order Coniferales

The extant conifers are grouped into six families, with about 50 genera and over 500 species. The families, distribution, and main morphological characteristics of each are shown in Table 24-1. Several families, such as the Pinaceae, Taxodiaceae, and Cephalotaxaceae, occur more abundantly in the Northern Hemisphere; others, such as Araucariaceae, are more widespread in the Southern Hemisphere.

In terms of the absolute number of extant species, there are only about one-twentieth as many conifers as ferns, and only about one-sixtieth as many conifers as monocotyledons. Thus, although the number of genera and species is relatively low, some conifers, such as *Pinus* and *Picea*, are conspicuous because of the large numbers of individuals that occur in certain regions. As a group, the conifers are among the dominant forest trees of the world, particularly in the boreal forests of the Northern Hemisphere. Depending on the species, they are adapted for various conditions, and occupy most ecological niches except areas such as extreme deserts and the tundra. They provide a main source of lumber and wood products, including pulp and paper.

Modern conifers are the result of a long period

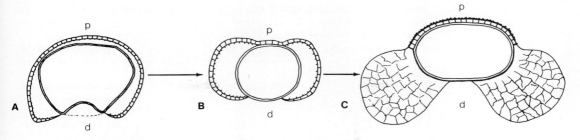

Figure 24-8 Apparent evolution of coniferalean saccate pollen. **A,** *Lebachia* (Lower Permian). **B,** *Ullmannia* (Upper Permian). **C,** extant grain of *Pinus*: *d*, distal pole and surface; *p*, proximal pole and surface. (**A, B,** after R. Florin, with permission of *Palaeontographica*.)

of evolution during the Mesozoic and Tertiary. During this long interval, conifers became a predominant part of the floras.

During the late Mesozoic and early Tertiary, almost all of the extant genera had become differentiated, but apparently at different times. During this interval, many of the genera also began to appear in the regions of their present distribution, which in some instances entailed a retreat from a more widespread distribution. Such restrictions are almost certainly an indication of senescence or even of impending extinction. Some of the most dramatic examples are found in the family Taxodiaceae, in which such genera as *Metasequoia* (dawn redwood), *Sequoia* (California redwood), and *Taxodium* (bald cypress) were widespread across North America, Siberia, and Greenland during the earlier Tertiary. Today, the same genera are restricted to very narrow geographical and ecological ranges and are almost certainly facing extinction. The main factors leading to extinction appear to be competition, particularly from angiosperms, and genetic senescence that prevents speciation with changing climatic and other ecological conditions.

Life History

The conifers have a life history essentially similar to that of the other gymnosperms; the sporophyte is dominant, and the gametophyte is reduced, being enclosed by and dependent on the sporophyte (Fig. 24-9). The male gametophyte at maturity ranges from zero to many prothallial cells and develops within the spore wall into a pollen grain. As in the cycads and *Ginkgo*, the female gametophyte consists of a mass of cells embedded in the nucellus. In most conifers, the sporophyte grows for several to many years before maturing and producing gametes. However, under normal conditions, the sporophyte continues to produce pollen and ovules for the rest of its life.

Morphology

In habit, the conifers are nearly all trees, some reaching gigantic proportions. The largest is *Sequoiadendron giganteum*, the "big tree" of California, with some specimens over 10 meters in diameter, 90 meters in height, and close to 4000 years old (Fig. 24-10). A few genera, such as *Juniperus*, are characteristically shrubby; none is herbaceous.

Most of the conifers are monopodial, with excurrent and whorled or spiral branching. Some conifers, such as the Pinaceae, lose the leaves on older branch regions, whereas others, such as the Araucariaceae, retain the leaves on all but the oldest branches. Several genera, such as *Pinus*, *Larix*, and *Cedrus*, have both long and **spur shoots** as in *Ginkgo*. In *Pinus*, one to eight leaves are arranged at the tip of the spur, spirally when more than one, which occurs in the axil of a nonphotosynthetic scale leaf of the long shoot (Fig. 24-11A). Each dwarf

Table 24-1 Main Morphological Characteristics and Distribution of Coniferales and Taxales

Family (number of genera in parentheses)	Range	General Characteristics
Coniferales		
Pinaceae (10)	Almost entirely Northern Hemisphere	Leaves needlelike; leaves and cone scales spirally arranged; bract and ovuliferous scale distinct; pollen grains mostly saccate
Taxodiaceae (10)	China, Japan, Formosa, Tasmania, California, southern United States, and Mexico	Leaves and cone scales spirally arranged; bract and ovuliferous scale almost completely fused; pollen small, with a papilla
Cupressaceae (16)	Widespread in both Hemispheres	Leaves and cone scales cyclic; bract and ovuliferous scales strongly fused; pollen small, with a small pore
Araucariaceae (2)	Almost completely Southern Hemisphere	Leaves and cone scales spirally arranged; bract and ovuliferous scale completely fused; ovules solitary; pollen large and nonsaccate
Podocarpaceae (7)	Mostly Southern Hemisphere, Central America, West Indies, and the Orient	Leaves flat and broad, with either a single vein or many veins; ovules terminal and single; pollen saccate
Cephalotaxaceae (1)	China, Japan, tropical Himalya	Leaves flat and narrow; spiral; cone scales decussate; 2 ovules on ovuliferous scale; pollen circular and nonsaccate
Taxales		
Taxaceae (5)	Mostly Northern Hemisphere	Leaves flat and pointed; spiral; ovule terminal and solitary, with fleshy aril; pollen round and nonsaccate

Figure 24-9 Life history of *Pinus: arch*, archegonium; *br*, bract; *emb*, embryo; *f c*, female cone; *f g*, female gametophyte; *int*, integument; *m*, micropyle; *m c*, male cone; *o s*, ovuliferous scale; *p g*, pollen grain; *s c*, seed coat; *sl*, seedling; *w*, wing.

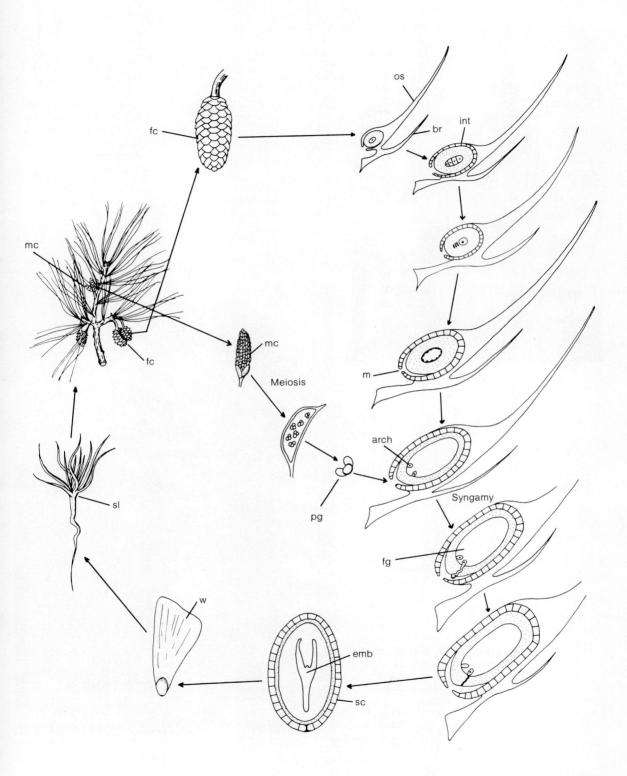

Figure 24-10 Habit views of conifers. **A,** *Sequoiadendron giganteum,* the Big Tree, native of the western Sierra Nevada in California, ×0.04. **B,** *Picea abies,* the Norway spruce, native of Europe, ×0.05.

Figure 24-11 Arrangement of dwarf shoots in conifers. **A,** twig of *Pinus,* showing scale leaves (*sl*) spirally arranged on the twig and dwarf shoots (*ds*) emanating from the axils of the scale leaves. In this species, each dwarf shoot supports three needle leaves, ×1.2. **B,** twig of *Cedrus,* showing the scale leaves (*sl*) and dwarf shoots (*ds*) with clustered needle leaves, ×0.5.

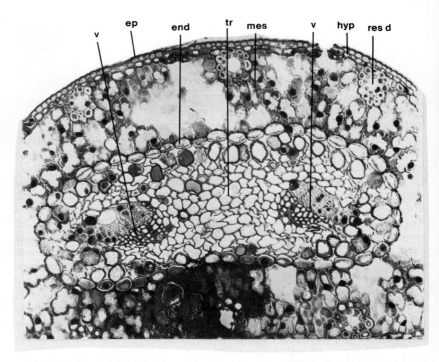

Figure 24-12 Transverse section of part of a two-veined pine leaf, showing the heavily cutinized epidermis (*ep*), compact mesophyll (*mes*), resin ducts (*res d*), transfusion tissue (*tr*), endodermal layer (*end*), and the two veins (*v*), ×155.

shoot in a pine is surrounded by a membranous sheath of bud scales. The spur falls off after several years, leaving a scaly long shoot. The leaves of *Larix* and *Cedrus* vary in number and are spirally arranged on the spur shoot (Fig. 24-11B).

The leaves of conifers vary in shape, texture, and size from long needle leaves in *Pinus* to lanceolate leaves in *Sequoia* and *Metasequoia*, scale leaves in Cupressaceae, and flat, ovate leaves in *Agathis* and some species of *Araucaria* and *Podocarpus*.

Most of the conifers are evergreen, but several genera are deciduous (*Larix, Taxodium,* and *Metasequoia*). In *Larix*, the needles fall from spur shoots; in *Taxodium* and *Metasequoia*, complete leafy twigs fall each autumn. Almost all conifer leaves are adapted for **xeric** conditions, with compact tissues and reduced surface area (Fig. 24-12). Typically, there is a heavy cuticle on a thick epidermis with sunken stomata. The stomata have prominent subsidiary cells with cuticular thickenings that sometimes overarch the stomatal opening, thereby providing a constricted aperture. The stomata occur in longitudinal bands on the leaves of many conifers.

Immediately inside the epidermis, there is a hypodermis of one to three layers of thick-walled cells (Fig. 24-12). This surrounds the mesophyll tissue; it consists of convoluted cells in the needle and scale leaves but differentiates into palisade and spongy layers in many of the broad, flat leaves. The vascular bundles vary from one to many; there are usually one or two in the needle and scale leaves and many in the broad-leaf forms. In most leaves, the bundle (or bundles) is surrounded by a well-defined **endodermis** (Fig. 24-12). Often, cells with secondary walls and bordered pits extend from the vascular bundles to the endodermis, presumably

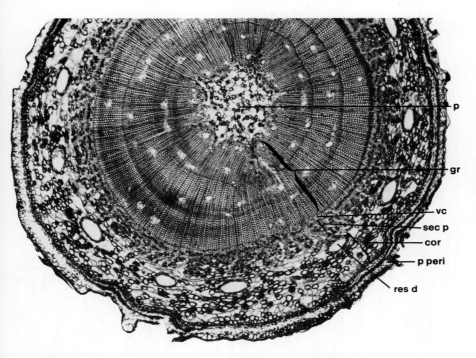

Figure 24-13 Transverse section of young stem of *Pinus*, showing pith (*p*), three growth rings (*gr*), vascular cambium (*vc*), secondary phloem (*sec p*), cortex (*cor*), resin duct (*res d*), and primary periderm (*p peri*) in the outer cortex, ×30.

functioning in conduction between the mesophyll and vascular bundles. These are called transfusion tracheids, and the tissue the transfusion tissue.

Stem Anatomy

The apical meristem in the conifers conforms with the general gymnosperm pattern. It resembles the five-zoned apex described for the cycads and *Ginkgo biloba*, except that only four zones are identifiable in *Pinus* and most genera of the Pinaceae. Another difference, noted particularly in *Araucaria* and some other genera, is the development of a well-defined **tunica layer** (one or two cell layers) as well as a subapical initial flanking the rib meristem zones.

The stems of all conifers have a narrow pith, primary vascular cylinder, and cortex (Fig. 24-13).

In contrast, many have a very wide cylinder of secondary xylem or wood that reaches diameters up to 11 meters. The bark is often as thick as 30 cm, and in *Sequoiadendron* it has been reported to reach 100 cm! The bark includes both secondary phloem and **periderm,** with large resin ducts in many genera. The first periderm develops usually in the outer region of the cortex.

As in cycads and *Ginkgo*, the annual rings of secondary xylem (wood) and similar increments of secondary phloem result from the annual activity of the thin vascular cambium. The cells comprising the secondary xylem include tracheids, ray parenchyma, and, in some genera, xylem parenchyma. The arrangement of cells and tissues is best observed in sections of wood that have been cut along the transverse (Fig. 24-13), radial (Fig. 24-14A), and tangential planes (Fig. 24-14B).

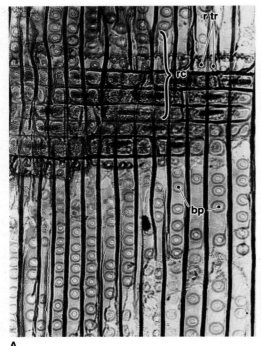

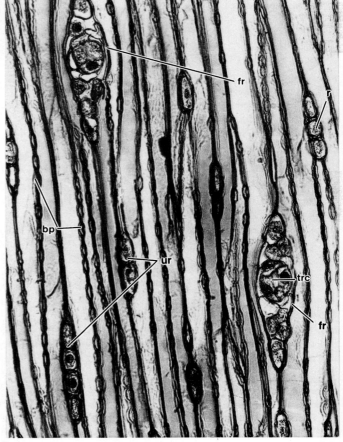

Figure 24-14 Longitudinal sections of secondary xylem of *Pinus*. **A,** radial section, showing tracheids with single rows of bordered pits (*bp*), ray cells (*rc*) cutting across toward the top, and ray tracheids (*r tr*) at the top of the ray, ×488. **B,** tangential section, showing bordered pits in vertical section, lenticular uniseriate rays, and large fusiform rays containing transverse resin canals, ×375: *bp*, bordered pit; *r*, ray; *fr*, fusiform ray; *ur*, uniseriate ray; *trc*, transverse resin canal.

cells are elongated, and have sieve areas on the radial walls. Sieve areas are particularly numerous on the overlapping end walls. Certain of the phloem parenchyma cells are often filled with dense cytoplasmic contents that stain deeply. These are called **albuminous cells.**

The periderm in conifers, as in cycads and *Ginkgo*, develops from a cork cambium (**phellogen**) that arises normally in the cortex within the first year. The cork cambium divides to give **phelloderm** cells to the inside and **phellem** (cork) cells to the outside. In many conifers, the periderm forms as **lenticular** or **conchoidal** plates in successively deeper layers, and the older and outer layers split off and fall away. Internal to the phelloderm is nonfunctional (conducting) phloem containing distorted and crushed sieve cells (Fig. 24-15).

Reproduction

The conifers have both pollen sacs and ovules aggregated into cones. Most genera are **monoecious;** that is, both male and female cones are borne on the same tree. A few, such as *Araucaria*, are **dioecious** like the cycads and *Ginkgo*. Some families (e.g., Cupressaceae) contain both monoecious and dioecious genera. Although direct evidence is lacking, it is generally believed that the dioecious condition was derived from the monoecious, mainly because in most plant groups from algae to angiosperms, evolution has apparently tended toward a separation of sexes.

The male cones are borne on smaller branches, often in lower reaches of the tree. They vary in length from about 2 mm in the common juniper to over 12 cm in some species of *Araucaria*. In many conifers, the cones occur singly; in others, such as the pines (*Pinus*) and firs (*Abies*), the cones are tightly clustered about the shoot (Fig. 24-16). In all except one family, the microsporophylls are arranged spirally on the axis. The pollen sacs, varying from 2 to over 15, are borne on the abaxial surface of the microsporophylls (Fig. 24-17).

The young male cones develop during the winter or dormant months, and mature during the spring months. In midnorthern latitudes, maturation of most conifers occurs during mid-May to early June, depending on elevation and climatic conditions. At maturity, the microsporophylls dry out

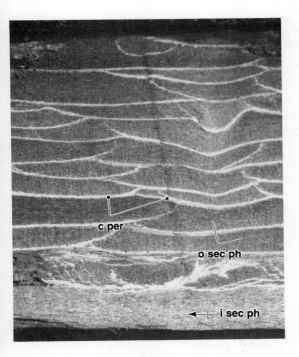

Figure 24-15 A radial section of bark of *Pseudotsuga* (Douglas fir), showing the innermost (youngest) secondary phloem (*i sec ph*) and the conchoidal plates of periderm (*c per*) in older secondary phloem (*o sec ph*) that consists mainly of fibers, × 2.4.

The xylem rays vary in length, width, and depth; they are only a single cell wide in many genera (Fig. 24-14B). In addition to having ray parenchyma cells with simple pits, the xylem rays of most genera of the Pinaceae have one to three rows of cells called **ray tracheids,** with bordered pits and without protoplasts, at the bottom and top of the ray (Fig. 24-14A). **Fusiform** rays are several cells wide and contain resin ducts (Fig. 24-14B). These develop from the separation of resin-producing cells, forming a central duct. The resin-producing cells are called **epithelial cells;** they often enlarge to close the duct, and are then called tylosoids.

Each year the secondary phloem is incorporated as an inner layer to the bark. It contains sieve cells, parenchyma, and phloem fibers. The sieve

Figure 24-16 Male cones of *Pinus* clustered around stem with the shoot extending beyond the cone cluster, ×0.8.

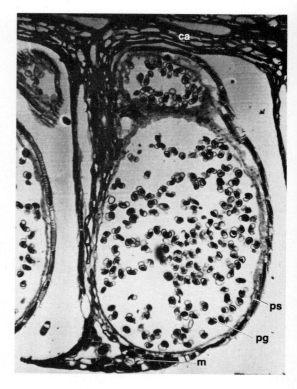

Figure 24-17 Single pollen sac (*ps*) of *Pinus* with pollen grains (*pg*) attached to abaxial surface of a microsporophyll (*m*), which in turn is attached to the cone axis (*ca*), × 60.

and separate, and the pollen sacs split to disgorge the pollen grains.

All of the conifers are mainly **anemophilous,** or wind-pollinated. The grains are shed at different times of the year, depending on latitude, altitude, and climate. Some travel over several hundred miles from the producing trees, but most appear to settle within shorter distances. Production is often prodigious, with an estimated several million grains from a single cone of *Araucaria*. In heavy coniferous forests, yellow clouds of pollen occur during the peak of pollen dispersal, and pollen scums form on ponds and lakes. In general, only a small number of the millions of grains shed ever reach a micropyle to effect pollination and subsequent fertilization.

The pollen grains of conifers are quite variable, ranging from 20 to about 150 micrometers, with various sculpturing patterns on the **exine** (Fig. 24-18). Most have a thin germinal aperture, the **leptoma,** through which the intine bulges out as the pollen tube. In the bladdered grains of the Pinaceae and Podocarpaceae, the leptoma occurs between the two sacci (Fig. 24-18).

The development of the male gametophyte inside the pollen grain is essentially the same as that in the cycads and *Ginkgo* (Fig. 24-19). Division of the microspore nucleus results in the production of a variable number of prothallial cells and an antheridial initial. As noted in Figure 24-19, the Pinaceae have 2 prothallial cells, whereas up to 8 have been recorded in the Podocarpaceae and up

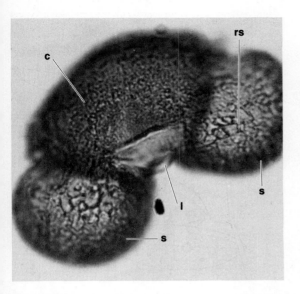

Figure 24-18 *Pinus* pollen, showing the two sacci (*s*) attached to the corpus (*c*), the reticulate sculpturing (*rs*) of the exine, and the leptoma (*l*) at the distal pole between the sacci, ×1540.

to 40 in the Araucariaceae. No prothallials are produced in Taxodiaceae, Cepholotaxaceae, or in most genera of Cupressaceae. Following this, the antheridial initial divides into a generative cell and a tube nucleus. In many conifers, the pollen is shed at this stage.

If a grain lands on a micropyle and conditions are favorable, development of the male gametophyte continues. The tube nucleus migrates towards the tip of the extending pollen tube, and the generative cell divides to form a sterile cell and a spermatogenous cell (Fig. 24-19E). In most instances, the spermatogenous cell lies next to the sterile cell. As the tube lengthens, the sterile cell loses its wall and, together with the spermatogenous cell, begins to move into the pollen tube. Just before fertilization, the spermatogenous cell divides to form two nonmotile sperms (Fig. 24-19F).

Conifer ovules are borne in compound cones, as in the Cordaitales and Lebachiaceae. They consist of bracts with an ovuliferous scale attached to the basal part of the adaxial surface of each bract (Fig. 24-20).

One or two ovules are attached to the adaxial surface of the ovuliferous scale. There is only one integument in the ovule, either free or partly fused to the nucellus. Unlike the cycads and *Ginkgo*, conifer ovules have no pollen chamber or nucellar beak. The integument is divided into an outer fleshy layer, a middle stony layer, and an inner fleshy layer. The outer layer is usually sloughed off. The ovules of most conifers lack vascular tissue, but some have bundles at the base. Only *Podocarpus* has vascular bundles extending to the tip of the integument.

In some species of conifers, the seed has an attached wing. In *Pseudotsuga*, it is long and flat and assists in wind dispersal of the seed (Fig. 24-21A). The wing is not actually a part of the seed but is the upper surface of the ovuliferous scale that abscises at the time of seed maturation.

The development of the female gametophyte of conifers is very similar to that in the cycads and *Ginkgo*. The megasporocyte develops deeply or shallowly within the ground meristem of the nucellus. Four megaspores are formed at meiosis, nearly always in a linear tetrad, but usually only one develops. The functional megaspore expands greatly, undergoing free-nuclear division (Fig. 24-22A). Walls then form centripetally, producing a central mass of gametophytic cells, with the cells toward the **chalazal** end charged with nutrient materials.

In all conifers, the archegonia arise from the differentiation of superficial cells of the female gametophyte at its micropylar end along the flanks of the nucellus (Fig. 24-22B). The number of archegonia that develop is variable, ranging from 1 to 200.

The archegonia usually develop two tiers of neck cells, each tier with four cells. Occasionally each tier has eight cells, and sometimes as few as two. In the Pinaceae, the ventral canal nucleus is separated from the egg by a wall. In the other conifers, no intervening wall is formed. The conifers with a cell wall surrounding the ventral canal nucleus also have thick membranes around the sperms; the converse also is true.

Following fertilization, the zygote undergoes a period of free-nuclear division in all conifers except *Sequoia* (Fig. 24-23A), thereby initiating the first stage of the proembryo. In most families, 4 to 8 nuclei are formed, but as many as 64 are reported in the Araucariaceae. Again, except in the Araucariaceae, the nuclei migrate to the basal end of the egg cell, where they form into a single plane. The nuclei

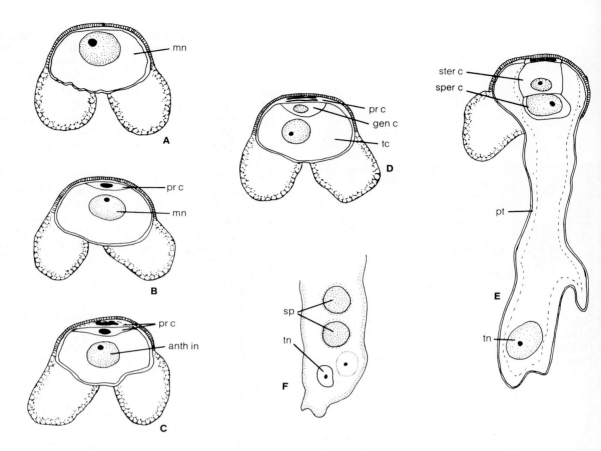

Figure 24-19 Development of male gametophyte of *Pinus*, ×500. **A,** pollen grain with microspore nucleus (*mn*). **B,** after first division, showing one prothallial cell (*pr c*) and the microspore nucleus (*mn*). **C,** after second division, with two prothallial cells and antheridial initial (*anth in*). **D,** antheridial initial has divided to produce the generative cell (*gen c*) and tube cell (*tc*). **E,** tube nucleus (*tn*) has migrated to tip of pollen tube (*pt*), and generative cell has divided to form a sterile cell (*ster c*) and spermatogenous cell (*sper c*). **F,** spermatogenous cell has divided to produce two nonmotile sperms (*sp*), while tube nucleus remains at tip of pollen tube. (Redrawn from *Morphology of Gymnosperms* by J. M. Coulter and C. J. Chamberlain, 1910, by permission of the University of Chicago Press. Copyright 1910 by The University of Chicago Press.)

then undergo several additional divisions accompanied by wall formation, resulting in three tiers of four cells each in the proembryos of the Taxodiaceae, Cupressaceae, and Araucariaceae (Fig. 24-23B), and four tiers of four cells in the Pinaceae. The four tiers are called progressively backward: the embryonal (t_1); the suspensor (t_2); the rosette (t_3); and the upper (t_4). The rosette tier serves as a suspensor in the proembryos of conifers with only three tiers of cells.

In the Pinaceae, the cells of the suspensor tier elongate first rather quickly to form primary suspensors (Fig. 24-23C, D). Soon after, the cells of the embryonal tier divide to produce secondary suspensor cells. The secondary suspensors elongate to variable extents and serve in pushing the embry-

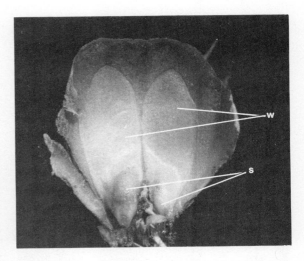

Figure 24-21 Two seeds (*s*) with wings (*w*) attached to top surface of ovuliferous scale of *Pseudotsuga* (Douglas fir), ×2.5.

onic cells even deeper into the gametophyte (Fig. 24-23E).

The young embryos produce from 2 to 18 cotyledons at the chalazal end of the seed, and a **radicle** at the micropylar end (Fig. 24-24). As development continues, the radicle emerges and anchors into the soil, thereby assuming independent nourishment. In many seedlings, the cotyledons remain attached to the seed coat by their tips for a short time before becoming detached (Fig. 24-24).

Order Taxales

The Taxales contain one family and five extant genera, of which *Taxus*, the yew, is the most widespread and best known. Although considered here and by some other botanists to be a separate order, or even a distinct subdivision, the Taxopsida, the only basis for the separation is the isolated terminal ovule surrounded by an **aril** in the taxads, versus the female cone with reduced ovule-bearing dwarf shoots and bracts in the other conifers.

Taxus consists of nine species of shrubs or trees, some reaching up to 25 m (85 ft). In the Himalayan

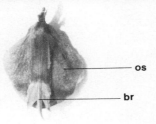

Figure 24-20 Female cone of *Pseudotsuga* (Douglas fir). **A,** two female cones, showing spiral arrangement of ovuliferous scales and bracts, ×0.8. **B,** single ovuliferous scale and bract extending beyond, ×1: *os*, ovuliferous scale; *br*, bract.

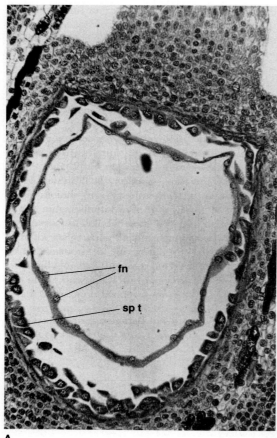
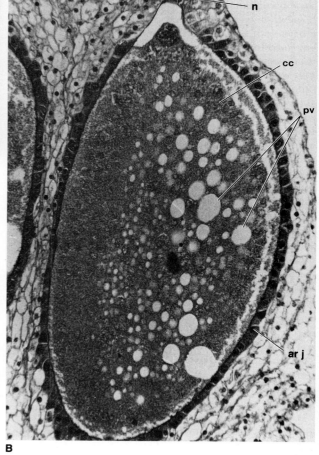

Figure 24-22 Female gametophyte of *Pinus*. **A,** free-nuclear division, showing free nuclei (*fn*) appressed against a membranous ring of cytoplasm that has separated from the spongy tissue (*sp t*) forming the outer layers of the gametophyte, ×125. **B,** single archegonium, showing the archegonial neck (*n*), the large central cell (*cc*), conspicuous protein vacuoles (*pv*), and archegonial jacket (*ar j*), ×230.

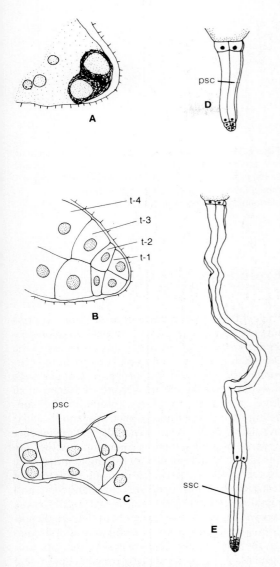

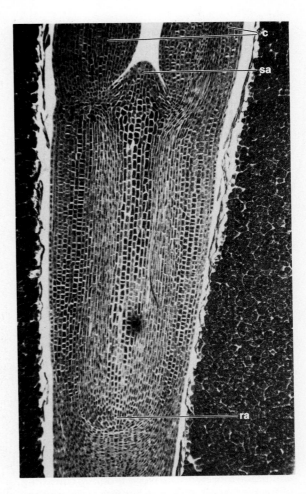

Figure 24-24 Embryo of *Pinus*, showing two cotyledons (*c*), stem apex (*sa*), and radicle apex (*ra*), ×150.

Figure 24-23 Embryonic development in conifers. **A,** free-nuclear stage with two proembryonic nuclei at chalazal end, ×1000. **B,** longitudinal section of proembryo, showing four tiers of cells (*t-1, t-2, t-3, t-4*), each with four cells, ×100. **C,** cells of central tier elongated to form primary suspensor cells (*psc*), ×100. **D,** primary suspensor cells further elongated, and cells of first tier divided to form several tiers of embryonic initials, ×40. **E,** primary suspensors greatly elongated, and the secondary suspensor cells (*ssc*) have formed to push embryonic cap even deeper into gametophytic tissue, ×50. (**A–C** redrawn after Johansen; **D, E,** after J. T. Buchholz, with permission of Illinois State Academy of Science.)

mountains and Guatemala, they extend from sea level to about 3000 m (9800 ft) either singly or in small clumps among other trees in moist and shaded sites. The wide geographical range of *Taxus* is attributed in part to birds that eat the seeds, which have a bright red and somewhat sweet aril.

Taxus and the other taxads are evergreen, with the leaves spirally arranged, mostly two-ranked and flattened; their blades vary from linear to lanceolate (Fig. 24-25). The stem is usually multibranched and reaches a girth of 6 m (20 ft). The bark is thin, scaly,

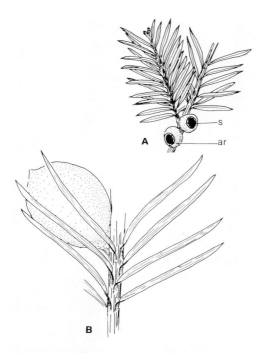

Figure 24-25 Foliage and seeds of the Taxaceae, ×0.5. **A,** *Taxus*, showing two ranked leaves and seed (*s*) surrounded by aril (*ar*). **B,** *Torreya* with several needle leaves and single seed.

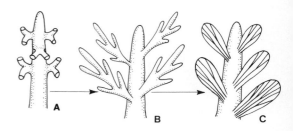

Figure 24-26 Hypothetical evolutionary progression in leaf development in gymnosperms, from a three-dimensional branch system in **A,** to a system of branching in one plane following planation in **B,** to a flat, veined blade following webbing, in **C.**

and reddish-brown. The wood is dense, consisting of tracheids with uniseriate bordered pits, and has many tertiary spirals on the secondary walls. These thickenings have been credited with giving yew wood the unusual elasticity required for archery bows.

Most trees of *Taxus* are dioecious, although a few bear both ovules and male cones. The male cone is unusual and striking, with three to nine pollen sacs reflexed from a **peltate** microsporophyll. The pollen grains are small and spherical, with a moderately thick granular exine and a weak leptoma.

The ovules are solitary and terminal on short shoots that emanate from the main branches (Fig. 24-25B). The aril remains as an inconspicuous bud until after fertilization. Then it grows up around and eventually beyond the seed, turning the bright red that attracts birds.

Evolution of Conifers

The evidence from studies of fossil gymnosperms suggests to some investigators that evolution proceeded from progymnospermous ancestors of the Devonian to the cordaites and conifers of the Carboniferous. The main supporting evidence for such an evolutionary link is the secondary xylem. In all three groups, it is typically coniferophytic, consisting of bordered-pitted tracheids and vascular rays. Other features are admittedly less convincing evidence of phylogenetic relationship, but they are considered by at least several leading gymnosperm researchers (e.g., Beck, 1970) to be indicative of descendancy. One concerns the derivation and homology of the leaves. In *Archaeopteris*, the leaves vary with species. They range from slender, dichotomous segments to those with laminae and deep sinuses and to others that are entire. This morphologic series suggests that the progymnosperm leaf may have evolved by modifications of dichotomous branch tips through telomic processes (see **telome concept,** Chapter 27). Such a hypothetical progression is outlined in Figure 24-26.

The leaves of the cordaites and most conifers are simple, arising in spiral succession on the twigs; they appear to be homologous to the leaf of *Archaeopteris*. However, the leaf-bearing frond of *Archaeopteris*, believed to be equivalent to the pteridosperm and cycad frond, also has an equivalent in the conifer *Metasequoia*. Thus both simple and compound leaves were apparently derived during

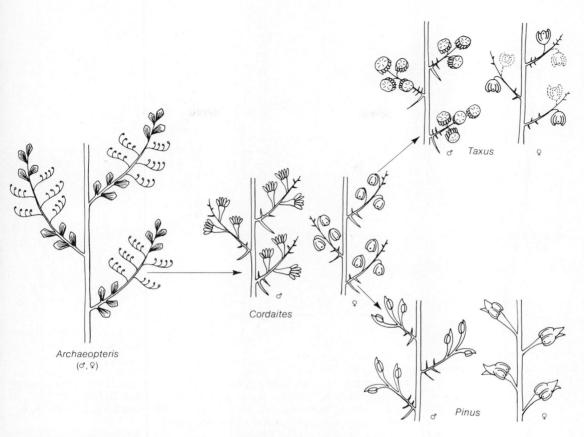

Figure 24-27 Proposed evolution of male and female reproductive organs of cordaites and conifers from ancestral *Archaeopteris*. (From A. D. J. Meeuse, Fundamentals of Phytomorphology. Copyright © 1966, by the Ronald Press Company. Reprinted by permission of John Wiley & Sons, Inc., and the author.)

the evolution of the gymnosperm groups from the progymnosperms.

It has also been postulated that both the female and male reproductive units of the coniferophytes and cordaites were derived from progymnospermous ancestors (Fig. 24-27). Such a scheme would have entailed the reduction of the fronds of *Archaeopteris*like ancestors, accompanied by modifications of the leaves into bracts and sterile scales, and of sporangia into ovules and pollen sacs. According to this scheme, the dwarf fertile shoot of the cordaites and early conifer *Lebachia* are the direct homologues of a secondary axis and subtending leaf of *Archaeopteris*.

One main problem is whether the conifers descended from the cordaites, or whether both lineages evolved along separate lines from the progymnosperms (see Fig. 20-6). Both the early conifers and *Cordaites* occur essentially in the same time interval of the later Carboniferous, some 40 million years later than *Archaeopteris* of the Upper Devonian. However, there are some genera that occur in this 40-million-year interval that have been related to the cordaites, suggesting that at least the cordaites may have a significantly longer history and possibly a closer link with the progymnosperms than do the conifers.

The cordaites and early conifers such as *Leba-*

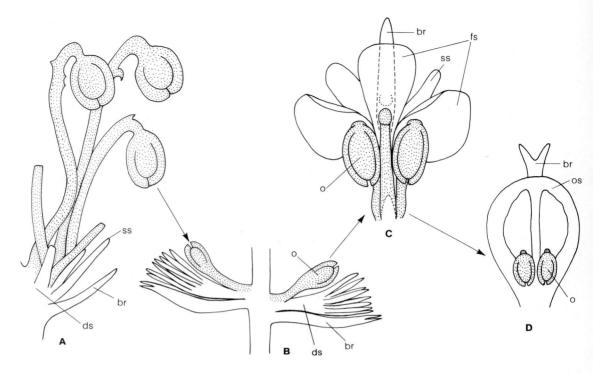

Figure 24-28 Probable stages in the evolution of the female reproductive organ of coniferophytes. **A,** *Cordaites.* **B,** *Lebachia.* **C,** *Pseudovoltzia.* **D,** *Pinus: br,* bracts; *ds,* dwarf shoot; *ss,* sterile scale; *fs,* fertile scale; *os,* ovuliferous scale; *o,* ovule. (**A, B,** after R. Florin, with permission of *Palaeontographica.*)

chia (see p. 556) persisted throughout the later Carboniferous and early Permian. The cordaites appear to have become extinct in the middle part of the Permian, possibly in response to the tectonic activity and increasing aridity in the Northern Hemisphere plates, and the glaciation in the Southern (Gondwana) plates. On the other hand, the conifers survived these disrupting conditions, probably because they were better adapted for the changes. Such features as needle leaves, sunken stomata, and compact reproductive units, as well as physiological adaptations, would provide protection against water loss under increasing cold or aridity. Thus the conifers survived the upheaval of the Permian and evolved into one of the dominant groups of vascular plants in the Mesozoic.

The most definitive outline of the evolution of the cordaites and conifers is found in a long series of classic studies by the late Rudolf Florin from Sweden. Among other landmarks, Florin (1938–1944; 1951) catalogued and synthesized the main events that took place in the evolution of the female reproductive unit. Starting with the cordaites and earliest conifers of the Carboniferous, Florin documented how the female strobilus of the cordaites is directly homologous with both the lax female cone of the early conifers and the compact cone of the Mesozoic and modern conifers. Some of the more notable events of this reduction series are outlined in Figure 24-28.

Another noteworthy contribution is a probable evolutionary development of pollen structure and pollination mechanisms (Fig. 24-29). The basic type is shown in genera of the Voltziaceae with inverted ovules and disaccate pollen, which floated the grains upward in the pollination fluid and

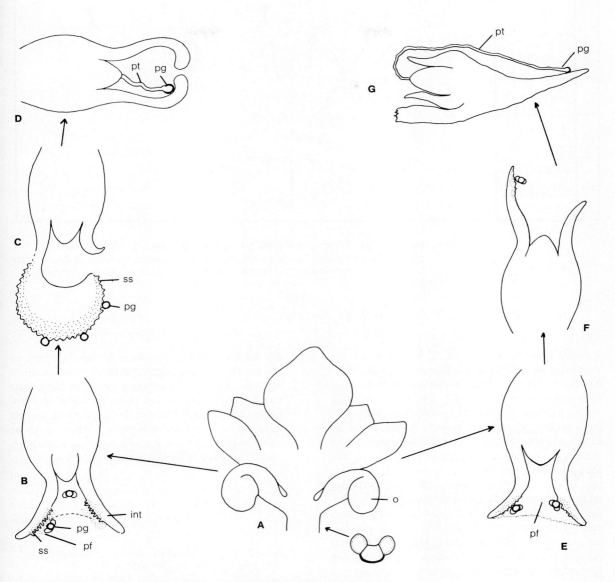

Figure 24-29 Probable evolution of pollination mechanisms in pinaceous conifers. One pathway is shown in the series *Pseudovoltzia*-type (**A**) to *Pinus* (**B**) to *Larix* (**C**) to *Pseudotsuga* (**D**); a second pathway is shown from *Pseudovoltzia*-type (**A**) to *Abies* (**E**) to *Cedrus* (**F**) to *Tsuga* (**G**): *p f*, pollination fluid; *p g*, pollen grain; *p t*, pollen tube; *o*, ovule; *int*, integument; *s s*, stigmatic surface. (After Doyle, with permission of the Royal Dublin Society.)

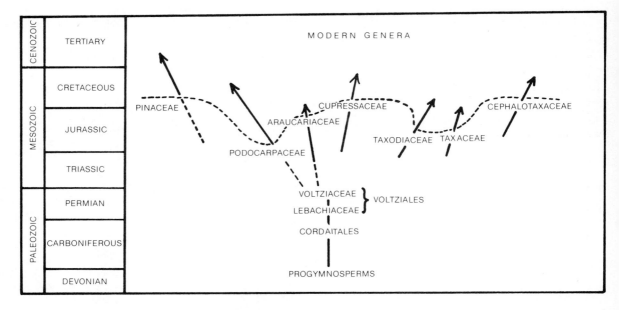

Figure 24-30 Flowchart depicting the probable evolutionary pattern of coniferophytes. The bars below the dotted line indicate the earliest occurrence of modern families; lines with arrows above the dotted line indicate the earliest record of modern genera.

thereby oriented the germinal area toward the nucellus (see p. 578). In the Pinaceae, Doyle (1945) has derived two evolutionary series, each showing four major modifications (Fig. 24-29): loss of pollination fluid exudate; development of stigmatic surfaces on the integument; reduction and loss of air sacs; and pollen germination elsewhere than on the nucellus. The most specialized mechanisms are found in *Larix* and *Pseudotsuga* (in the first series, Fig. 24-29B–D), which have well-developed stigmatic surfaces and bladderless pollen; and in species of *Tsuga* (second series) with bladderless pollen and vigorous pollen-tube growth (Fig. 24-29E–G). According to Doyle, the most specialized mechanism in conifers occurs in *Araucaria*, in which the apex of the nucellus protrudes from the micropyle, and the completely bladderless pollen produces a very active and long pollen tube.

The main picture of conifer and taxad evolution is shown in Figure 24-30. From the Voltziaceae, the modern families appear at different times in the Mesozoic; the Podocarpaceae occur first, and the Pinaceae last. In the later Mesozoic, many of the modern genera became distinct. Most of those have persisted to the present, but some have their numbers and areal extent reduced. Some of these genera, particularly those of the Taxodiaceae, appear to be heading toward extinction. Others, such as some of the Cupressaceae, Pinaceae, and Podocarpaceae, appear to be still actively evolving.

References

Beck, C. B. 1970. The appearance of gymnospermous structure. *Biol. Rev.* 45:379–400.

Chamberlain, C. J. 1963. *Gymnosperms: structure and evolution.* Hafner Publishing Co.; University of Chicago Press (1935).

Dallimore, W., and Jackson, A. B, 1966. *A handbook of Coniferae and Ginkgoaceae.* 4th ed. rev. by S. G. Harrison. Edward Arnold, London.

Doyle, J. 1945. Developmental lines in pollination mechanisms in the Coniferales. *Scient. Proc., Roy. Dublin Soc. N.S.* 23:43–62.

Florin, R. 1938–1945. Die Koniferen des Oberkarbons und des Unteren Perm. *Palaeontographica.* 85B, Pts. 1–8.

———. 1951. Evolution in Cordaites and Conifers. *Acta. Hort. Berg.* 15:285–388.

———. 1963. The distribution of conifer and taxad genera in time and space. *Acta. Hort. Berg.* 20(4):121–321.

Foster, A. S., and Gifford, E. M., 1974. *Comparative morphology of vascular plants.* 2nd ed. W. H. Freeman and Co., San Francisco.

Harris, T. M. 1976. The Mesozoic gymnosperms. *Rev. Paleobot. and Palynol.* 21:119–34.

Miller, C. N. 1977. Mesozoic conifers. *Bot. Review* 43:217–80.

Mirov, N. T. 1977. The genus *Pinus*. Ronald Press, New York.

25

GNETOPHYTES

The three orders, each with but one genus, that comprise the gnetophyte group, class Gnetopsida, are distinctive plants with no close relationships to other gymnosperms and little morphological resemblance to one another. However, they do share several characteristics:

1. The occurrence of vessels in the secondary xylem
2. The male and female reproductive units in **compound cones**
3. Extra structures surrounding the integument of the ovule, variously interpreted as additional integuments, sporophylls, or **bracteoles**
4. Extra bractlike appendages surrounding the pollen sacs in the male cones
5. A micropyle that projects as a long tube
6. Opposite or whorled leaves
7. Absence of resin canals.

The presence of vessels in the three genera of gnetophytes deserves emphasis because of the parallel with angiosperms. A great deal of research has shown that the origin of the **perforation plate** of the vessel elements is fundamentally different in the two groups. The vessels of angiosperms, as well as those of *Selaginella, Equisetum,* and the ferns *Pteridium* and *Marsilea*, evolved by a series of changes in the sloping end walls of tracheids with elongated bordered pits to produce scalariform pitting. Gnetophyte vessels, by contrast, form by the dissolution of circular bordered pits in the end walls of elongated tracheids. Thus, it appears that the occurrence of vessels in angiosperms and gnetophytes is an interesting example of convergent evolution and not the result of phylogenetic connections between the two groups.

Order Ephedrales

The genus *Ephedra*, with about 40 species, is generally shrubby; some reach a height of 2 m (Fig. 25-1A). It is multibranched and spreads by rhizomes from underground buds. *Ephedra* shrubs are used as a ground cover on dry sandy banks and some-

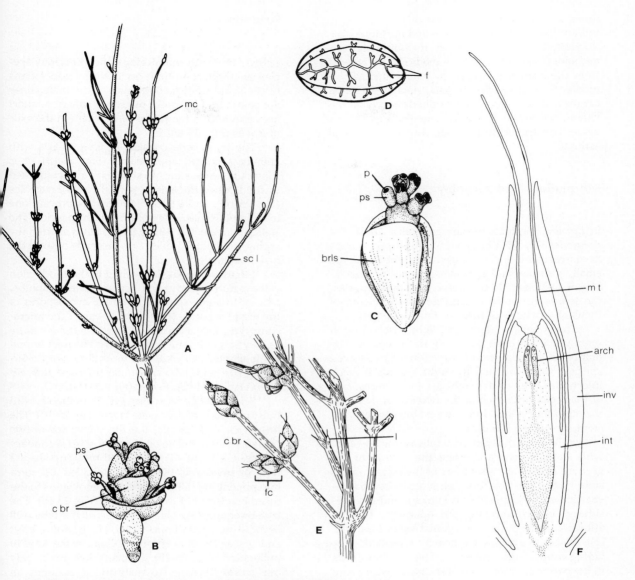

Figure 25-1 *Ephedra*. **A,** several branches bearing very tiny scale leaves (*sc l*), and whorls of male cones (*mc*) at nodes, ×1. **B,** single male cone, showing whorls of cone bracts (*c br*) and protruding clusters of pollen sacs (*ps*), ×5. **C,** single secondary strobilar axis with two bracteoles (*brls*) subtending pollen sacs (*ps*) with pores (*p*), ×5. **D,** pollen grain in lateral view, showing sinuous branched sulci or furrows (*f*), ×500. **E,** several female cones (*fc*) at nodes of twigs with leaves (*l*) and cone bracts (*c br*), ×1. **F,** median longitudinal section of ovule, showing integument (*int*) projecting as micropylar tube (*m t*), involucre (*inv*) comprising fused bracteoles, and two archegonia (*arch*) at tips of nucellus, ×42. (**A–C,** redrawn after Metro and Sauvage; **E,** from C. J. Chamberlain, *Gymnosperms: Structure and Evolution,* Chicago: University of Chicago Press, 1935. Copyright 1935 by The University of Chicago Press. **F,** after W. J. G. Land, 1904, by permission of The University of Chicago Press.)

Gnetophytes 583

times provide graze on open rangeland. Species in the southwestern United States and northern Mexico are called by several names, including Mexican tea, joint pine, and poptillo. A medicinal tea brewed from the branches is used to relieve urinary ailments. Several species in Asia, notably *E. sinica*, are the source of the alkaloid ephedrine, called Ma Huang in Chinese. It is used in medicine as a blood vessel constrictor and in relieving symptoms of asthma.

Morphology and Anatomy

In *Ephedra*, the branches are generally whorled at the **nodes** (Fig. 25-1A, E), although several species have irregular branch arrangements. The leaves are small, scalelike, and **ephemeral;** hence the resemblance to horsetails (*Equisetum*). In most species, the leaves are opposite and **decussate;** in others they are in whorls of three or four per node.

The stems are slight and lightly ribbed, with most photosynthesis occurring in the cortex. The number of vascular strands varies; in the two-leaved species, there are usually eight strands in each internode. Four of the eight are foliar traces; two of them pass into each of the two leaves, whereas the other four persist into the next internode as the stem, or **cauline bundles.**

In the stem, the epidermal cells are thick-walled, with stomata arranged along the bottom of the grooves and on the shoulders of the ridges on the stem. As in the conifers, the stomata are **haplocheilic,** with the guard cells sunken, and the subsidiary cells overarching the openings, thereby affording extra protection against water loss. The cells below the ridges are heavily **sclerified,** providing support and protection. The remaining cells of the cortex are mainly chlorenchyma, with many intercellular spaces, thus forming a photosynthetic tissue resembling leaf mesophyll. The primary vascular bundles are scattered inside the cortical chlorenchyma and outside the pith.

A vascular cambium develops between the primary xylem and the phloem, and also between the bundles. Secondary xylem develops on the inside, with radial rows of bordered-pitted tracheids and vessels. The secondary phloem consists of **sieve cells, albuminous cells** and phloem parenchyma.

Reproduction

Ephedra has both **monoecious** and **dioecious** species, and both conditions occur within populations of the same species. Both male and female cones are compound. In both, the cones consist of short axes attached to the nodes of the stem in the axils of leaves (Fig. 25-1A, E).

The male cone consists of a short axis with spirally arranged appendages (Fig. 25-1B). It is compound, consisting of a short axis, a bract attached to the base of the axis, a pair of fused bracteoles (Fig. 25-1C), and a short central stalk bearing one to eight pollen sacs, depending on the species. The stalks elongate just before the pollen is shed in the spring, so that the pollen sacs protrude beyond the fused bracteoles (Fig. 25-1C).

Microsporogenesis, leading to the production of the mature pollen grains, begins in winter months, at least in some species. The four microspores of the tetrad separate from the cell wall of the **microsporocyte,** increase in size, and develop an **exine** as they form into the pollen grain. The mature pollen is elliptical to **fusiform** in outline, with alternating ribs and **sulci** (germinal furrows) that are often elaborately branched (Fig. 25-1D).

Female cones also consist of short shoots, with two to six pairs of decussate bracts (Fig. 25-1E). The lower bracts are sterile, but the upper one of two contains ovules borne on the very short secondary shoots. Each ovule is subtended by a cup-shaped disc, or **involucre,** that has been called a second integument, but which appears homologous to the fused bracteoles of the male cone (Fig. 25-1F). The true **integument** projects as a micropylar tube well beyond the upper limits of the surrounding bract and bracteoles; it is chlorophyllous at the time of pollination (Fig. 25-1F). In many instances, only one ovule develops to maturity; it assumes an apparent terminal position, although it is attached laterally.

Gametophytes

The development of the male gametophyte in the young pollen grain begins in the pollen sac, and is essentially the same as in cycads, *Ginkgo,* and the Pinaceae. Division of the microspore nucleus yields a prothallial cell and an antheridial initial. A second division produces a second prothallial nucleus without a cell wall. The antheridial initial then divides to form a generative cell and tube nucleus. In some species, the generative cell apparently divides once again to produce stalk and body nuclei, with the body nucleus then producing two sperms. However, in at least one species, the generative cell divides directly into two sperm nuclei. At this time, the pollen grains are shed from the pollen sacs and are wind-blown to the ovule with its **pollen drop** and **pollen chamber.**

The female gametophyte develops very much as in the conifers, going through the main stages of the linear megaspore tetrad, the functional megaspore that enlarges while undergoing free-nuclear division, wall formation to produce a female gametophyte, and the formation of one to three archegonia (Fig. 25-1F). The archegonial neck keeps dividing to keep pace with the extension of the gametophyte tissue into the pollen chamber.

Several hours after the lodging of the pollen grain in the micropyle, the pollen tube emerges, grows into the archegonial neck, and releases the two flagellated sperms close to the egg. One sperm fertilizes the egg; occasionally, the other sperm fuses with the ventral canal nucleus, but no embryo or tissue develops from this second union, as occurs in the formation of angiosperm endosperm.

The zygote divides three times to produce eight proembryonic derivatives. Each of the eight is capable of forming an embryo, but usually only a few do so. Those proembryonic cells that do develop divide to produce an embryonic initial and a **suspensor.** The suspensor then elongates and pushes the proembryo into the female gametophyte tissue, where the nutrients are absorbed by the embryo. As the embryo reaches maturity, the integument hardens into a seed coat, and the bracts and bracteoles turn a bright red. The seed can germinate whenever conditions become favorable.

Relationships

The general consensus is that *Ephedra* is virtually unrelated to either of the other two gnetophyte genera. *Ephedra* differs in having haplocheilic stomata, different primary stem and nodal anatomy, wood structure, and especially in having the ovule terminal on a *lateral* branch of the fertile axis, instead of terminal on the shoot axis itself. Although direct evidence of phylogeny is lacking, it has been suggested that *Ephedra* is related to the cordaites or cordaitean ancestors. According to this hypothesis, the cones of *Ephedra* are homologous with the fertile short shoots of the cordaites' strobilus (see Fig. 24-9), which have been much reduced; the bracteoles and involucre parts represent the sterile scales of the cordaite strobilar short shoot.

Order Welwitschiales

The Welwitschiales are represented by the single species *Welwitschia bainesii*. Named in honor of Dr. Friedrich Welwitsch, who discovered it in Angola in 1860, it is indeed a remarkable plant. Presently restricted to a small area along the southwestern coast of Africa, it is well adapted for growth in the desert conditions prevailing in that region. Because there are so few individuals, the plant is protected against unauthorized collecting.

The stem of *Welwitschia* is bowl-shaped and up to 1 m in diameter, with the rim extending a few centimeters above the surface of the soil (Fig. 25-2A, B). The stem tapers sharply below, merging with a large tap root that extends several meters toward the water table.

The vascular structure of the young stem consists of a ring of individual vascular bundles. Older stems have a disc-shaped collar of vascular tissue, from which traces lead to the leaves, cones, and root. In still older stems, there are concentric layers of xylem and phloem, as in some species of the cycad genus *Cycas*.

The two leaves emanating from the rim of the stem are broad, flat, and **coriaceous** (Fig. 25-2A, B). With a width of over 20 cm and length exceeding 2 m, the leaves split at the tips and become tattered by the action of wind and sand. New growth occurs from a basal meristem, thus compensating for terminal abrasion. Stomata are numerous on both surfaces of the leaf; they are both sunken and **syndetocheilic,** as in *Gnetum*, the cycadeoid cycadophytes, and some angiosperms. The main collateral vascular bundles are arranged essentially parallel in the leaf, with numerous oblique bundles splitting and uniting with each other between the main bundles. Each vein is surrounded by well-developed transfusion tissue containing shortened tracheids with small pits. Also, each bundle derives additional support from strands of fibers above and below. There are several layers of palisade parenchyma in the outer regions of mesophyll, a normal development in leaves exposed to intense and direct sunlight. The parenchyma tissue of both the stem and mature leaves contains numerous **sclereids** (Fig. 25-3A).

Reproduction

The reproductive organs of *Welwitschia* are male and female compound cones produced on different plants. In both instances, cones are borne on branched stalks emanating from the rim of the stem (Fig. 25-2B), with parts arranged in an opposite and decussate pattern. When mature, the decussate **cone-bracts** turn a bright red.

The male cone consists of beautifully overlapping cone-bracts tightly packed at the end of a short branch and subtended by a **nodal bract** (Fig. 25-3B). The usual arrangement is three (or two) male cones borne at the node. Each cone-bract subtends an aggregate of scales and pollen-bearing structures that can be called, for want of a better term, a **secondary strobilus** (Fig. 25-3C, D). Each secondary strobilus consists of two pairs of opposite bracteoles surrounding a central column that supports six lobed pollen sacs and a central sterile ovule. Each pollen sac has three lobes and is stalked, forming a unit sometimes called a **synandrium.** Because the central ovule is always sterile, the secondary strobilus is functionally unisexual (male).

The male gametophyte within the pollen grain is reduced from those in other gymnosperms. The first division of the antheridial initial produces a generative and tube cell. At this stage, the boat-shaped, ridged pollen grain is discharged (Fig. 25-3E). When conditions for germination are favorable, the prothallial cell disintegrates, and the generative cell divides to form two sperm nuclei. The female cones are also borne on branched stalks and subtended by nodal bracts (Fig. 25-4A). As with the male, each female cone has four rows of overlapping cone-bracts, with protruding integuments. Each cone-bract subtends and encloses two pairs of bracteoles, which surround the single central ovule with its extended integument and micropyle.

A

B

Figure 25-2 *Welwitschia.* **A,** habit view of single plant, showing two large, spreading leaves and bowllike stem supporting strobili, ×0.05. **B,** closeup photo showing strobili (*str*) attached to stem (*st*), ×0.1.

The inner bracteoles are fused to form an envelope around the integument (Fig. 25-4B, C). At maturity, this envelope extends into a wing that assists in seed dispersal.

As in other gymnosperms, a single megaspore mother cell develops directly into four megaspores; however, in *Welwitschia* no linear tetrad is formed. The megaspores develop into a free-nuclear stage with well over 1000 free nuclei, which then form

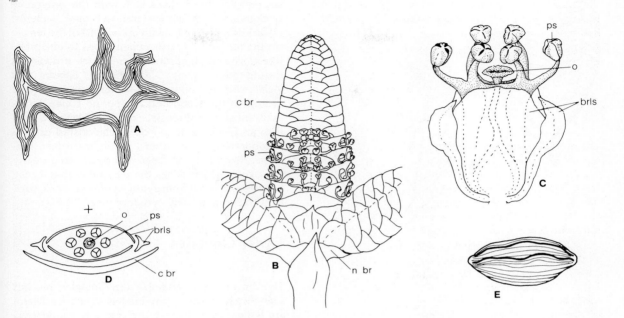

Figure 25-3 *Welwitschia.* **A,** single sclereid from male cone, showing concentric layers of secondary wall, ×225. **B,** male cone, showing four rows of overlapping cone-bracts (*c br*), nodal bract (*n br*), and protruding pollen sacs (*ps*), ×5. **C,** single secondary strobilus, showing three of four bracteoles (*brls*), central sterile ovule (*o*), and six pollen sacs (*ps*), ×12. **D,** plan view of single secondary strobilus from a male cone, showing the cone-bract (*c br*), bracteoles (*brls*), pollen sacs (*ps*), and central sterile ovule (*o*), ×12. **E,** single pollen grain, showing five longitudinal ribs, and a single sulcus down the middle, ×500. (**A, D,** from *Morphology of Gymnosperms*, by J. M. Coulter and C. J. Chamberlain, by permission of The University of Chicago Press. Copyright 1910 by The University of Chicago Press. **B, C,** after A. H. Church, *Phil. Trans. Roy. Soc. Lond.* B,V. 205:115–51, with permission of the Royal Society of London.)

walls to produce a massive female gametophyte. Unlike other gymnosperms, *Welwitschia* forms no archegonia on the gametophyte, and many of the prothallial cells are polyploid as a result of one to several free nuclei being enclosed by cell walls.

The process of fertilization in *Welwitschia* is different from that in all other plants, with both the male and female units taking an active part. The pollen tubes carrying the male gamete nuclei grow down through the nucellus in the usual way. However, coincidentally, some of the cells at the apex of the female prothallus form archegonial tubes that extend upwards to meet the pollen tubes (Fig. 25-5A). When these tubes meet, one male gamete nucleus unites with an archegonial nucleus to form the zygote. The zygote then divides to form a two-celled proembryo. The upper cell functions as a suspensor (Fig. 25-5B); the lower cell divides sev-

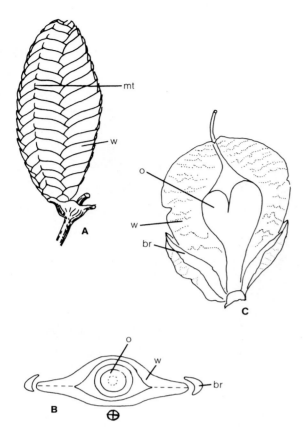

Figure 25-4 *Welwitschia.* **A,** single female cone, showing rows of overlapping wings (*w*), nodal bract (*n br*), and micropylar tubes (*mt*), ×5. **B,** single secondary female strobilar unit from cone in **A,** showing ovule (*o*), wing (*w*), and bract (*br*), ×12. **C,** diagrammatic view of secondary female strobilar unit, showing wing (*w*), bracteoles (*brls*), and ovule (*o*), ×12. (Modified from Martens.)

eral times to form secondary suspensor cells as well as a multicellular embryo that is pushed into the prothallial tissue (Fig. 25-5C, D). Many embryos are produced from many independent fertilizations, but only one embryo matures in each seed.

Relationships

It has been noted that *Welwitschia* differs rather significantly from *Ephedra* and appears more closely allied with the third genus, *Gnetum.* Differences from *Ephedra* include (1) syndetocheilic stomata (those of *Ephedra* are haplocheilic); (2) a lack of prothallial cells in the male gametophyte and suppression of the stalk cell, with the generative nucleus dividing directly into two male gametes; (3) lack of archegonia, with apical cells of the female prothallus uniting with pollen tubes to effect fertilization; and (4) absence of free-nuclear divisions during embryogeny. Nevertheless, *Ephedra* and *Welwitschia* do show some similarities: both have vessels in the wood and compound cones. However, most morphologists regard *Welwitschia* as a distinct genus, with no clear relationship or evolutionary link with either *Ephedra* or *Gnetum.* *Welwitschia*like pollen has been reported from the early Mesozoic, which raises the possibility that it may have been derived from gymnospermous stock, such as seed ferns or early cycadophytes.

Order Gnetales

The third order of gnetophytes, Gnetales, has but a single genus and approximately 40 species. Many are **lianas,** some are shrubs, and a few are trees. Most species are native to Indonesia, South China, Malaysia, and India, but others inhabit wet tropical forests of parts of West Africa and the Amazon River basin. One of the best known is *Gnetum gnemon.* It is cultivated in Malaysia for its seeds and in Java for the young leaves and cone axes, which are eaten, and for the strong fibers of its bark, which are used to make rope.

One of the most noteworthy features of *Gnetum* is the structure of its leaves. They resemble the leaves of dicotyledonous angiosperms, with broad blades and reticulate venation (Fig. 25-6A). The stomata are syndetocheilic as in *Welwitschia. Gnetum* has both long and short shoots; in liana species, leaves are borne only on short shoots.

The apex of the stem is very similar to that of the angiosperms, with **tunica** and **corpus** regions. The primary vascular tissues consist of a ring of vascular bundles with endarch xylem. In the tree species, a single cambium produces secondary xylem and phloem, forming wood and secondary phloem in the regular manner. In liana species of *Gnetum,* there are several concentric rings of bundles and associated cambia, giving a pattern similar to that of many angiosperm lianas. The vessels are well developed in the secondary xylem. The secondary phloem in some species differs from that of other gymnosperms in having sieve tubes and companion cells, as in angiosperms.

Reproduction

The reproductive structures are compound cones, similar in basic construction to those of *Ephedra* and *Welwitschia*. The male cone consists of collars evenly spaced along an axis (Fig. 25-6A). Each collar bears a ring of secondary male strobili in its axil (Fig. 25-6B). Each male strobilus consists of a disc of almost completely fused bracteoles surrounding a stalk that supports two pollen sacs (Fig. 25-6C). The pollen grains are small and circular; they are bedecked with many coarse papillae and have no apparent germinal aperture (Fig. 25-6D).

The male gametophyte within the pollen of *Gnetum* differs from that of *Ephedra* and *Welwitschia* in lacking prothallial cells. First the antheridial initial divides to form a generative cell and a tube nucleus. Next, the generative cell divides to produce a spermatogenous cell and a sterile cell; the spermatogenous cell divides into two nonmotile male gametes, one of which fertilizes the egg.

The ovules are encompassed in compound cones that are whorled at the nodes, in the axils of nodal bracts (Fig. 25-7A). Each cone consists of three distinct envelopes surrounding the ovule. The innermost extends distally some distance, forming an integumentary beak containing a micropylar tube and pollen chamber (Fig. 25-7B). The middle envelope becomes sclerified and has been variously interpreted as a second integument or fused inner bracteoles, as in *Welwitschia*. The outer envelope is fleshy and has been called a third integument, a **perianth**, or fused outer bracteoles.

The female gametophyte of *Gnetum* is similar to that of *Welwitschia* and different from that of the other gymnosperms. It develops from several megaspore mother cells that undergo meiosis without forming transverse walls between the four megaspores. Nuclei from two or three of the mother cells divide freely, forming a free-nucleate stage. The nuclei at the micropylar end of the gametophyte remain free, whereas those at the chalazal end become walled (Fig. 25-7C).

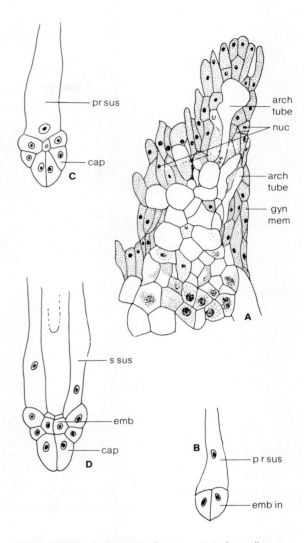

Figure 25-5 *Welwitschia*. **A,** upper part of nucellus (*nuc*) in which archegonial tubes (*arch tube*) are expanding upward to meet pollen tubes growing downward (*meg*, megaspore membrane), ×70. **B,** early stages of embryonic development, showing two embryonic initials (*emb in*) and single primary suspensor (*pr sus*), ×300. **C,** later stage of embryonic development, showing cells at tip forming cap and single primary suspensor cell, ×300. **D,** later stage, still with embryonic cells forming behind cap and two secondary suspensors flanking primary suspensor, ×300. (After H. H. W. Pearson, *Phil. Trans. Roy. Soc. Lond.* B,V. 200:331–402, with permission of the Royal Society of London.)

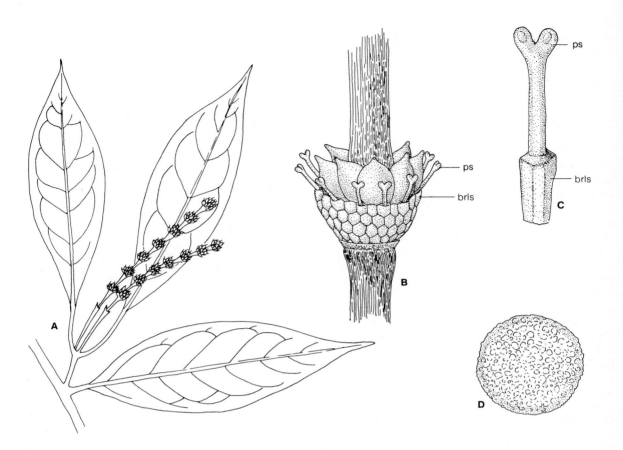

Figure 25-6 *Gnetum*. **A,** several leaves and compound male cones, ×0.5. **B,** a male cone at a node, showing ring of secondary male strobili with pollen sacs (*ps*) protruding from fused bracteoles (*brls*), ×15. **C,** single male strobilus, showing pollen sacs (*ps*) atop central stalks and bracteoles (*brls*) ensheathing the base, ×15. **D,** single pollen grain, showing coarse papillae on wall, ×500. [**A–C,** adapted from *The Plant Kingdom* by William H. Brown, copyright © 1935, by William H. Brown. Used by permission of the publisher, Ginn and Company (Xerox Corporation.)]

When pollen grains land on the micropyle, they are pulled into the pollen chamber by a pollen-drop mechanism. Several pollen tubes can germinate and penetrate the prothallus. As in *Welwitschia*, there are no archegonia; several of the free nuclei become female gametes, apparently in response to touching the pollen tube. Actually, many zygotes are often formed from multiple gamete fusions. This process is somewhat closer to that of angiosperms, in which no gametophytic tissue is formed, and the polar nuclei combine directly with male gametic nuclei to form the endosperm.

The early embryo of *Gnetum* consists of two primary suspensor cells. Each of these elongates and forms branches. These primary branches further extend into secondary branches, each of which is multicellular and develops a proembryo at the tip. The formation of several female gametophytes

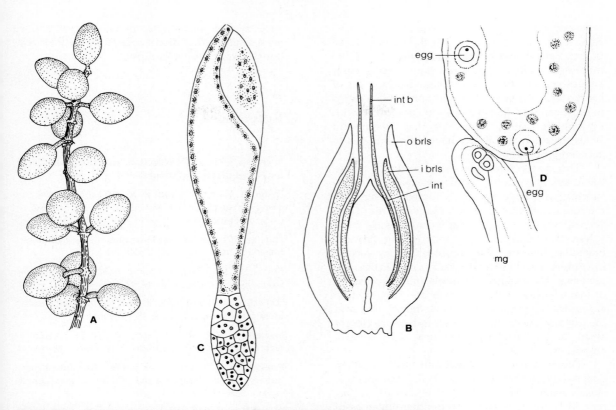

Figure 25-7 *Gnetum.* **A,** twig with seeds, ×0.4. **B,** median longitudinal section of single ovule, showing integument (*int*) extended into integumentary beak (*int b*), inner bracteoles (*i brls*), and outer bracteoles (*o brls*), ×40. **C,** female gametophyte with free nuclei in upper part and cells below, ×50. **D,** part of female gametophyte with free nuclei and two egg cells above and two male gametes (*mg*) below, just prior to fertilization, ×700. (**A,** adapted from *The Plant Kingdom* by William H. Brown, copyright © 1935, by William H. Brown. Used by permission of the publisher, Ginn and Company [Xerox Corporation]; **B, C,** after Lotsy; **D,** after Thompson.)

with several fertilizations in each, together with many embryonic branches with proembryos, makes *Gnetum* extensively polyembryonic. This undoubtedly is a mechanism to ensure that at least one embryo matures to propagate.

Gnetum has several features in common with *Welwitschia*, which also serve to set it apart from *Ephedra*. These include syndetocheilic stomata, a lack of prothallial cells in the male gametophyte, absence of archegonia, and absence of free-nuclear divisions during embryogeny. On the other hand, *Gnetum* differs from both *Welwitschia* and *Ephedra* in the reticulate broad leaves, the extra envelope surrounding the ovule, and the circular, spiny pollen. As with the other two genera, *Gnetum* appears to be a gymnosperm apart, with no close links to any other gymnosperm.

Phylogeny of Gnetophyta

The origin and relationships of these three very distinct genera of gymnosperms are obscure. Pollen grains similar to those of *Ephedra* and *Welwitschia* have been found as early as the Permian, whereas pollen assigned to *Gnetum* has only been reported from the Tertiary. On the basis of fossil pollen, *Ephedra* and *Welwitschia* appear to have had more species in former times than they do now, suggesting that they have declined in number of species since the early Mesozoic. Some pollen grains in Mesozoic and Tertiary rocks are associated with mesophytic plants, suggesting that *Ephedra* and *Welwitschia* may not always have been restricted to **xeric** environments.

The gnetophytes have been cited as a transitional group between gymnosperms and angiosperms, and have been considered by some as the immediate forerunners of angiosperms. Their angiospermous characteristics include vessels in the xylem, compound strobili that resemble flowers, bracteoles that are often compared with a second integument, and the time of fertilization in *Gnetum*. However, the absence of **carpels** and the presence of many nuclei in the free-nucleate gametophyte are typically gymnospermous characteristics, and the gnetophytes are almost certainly more closely allied to gymnosperm ancestors than to any angiosperm stock.

The evidence from fossil and modern pollen grains suggests that *Ephedra* and *Welwitschia* are more closely related to one another than either is to *Gnetum*. In addition, evidence from other pollen associated with the Permian pollen of *Ephedra* and *Welwitschia* suggests that these two genera were most likely derived from the early coniferophyte line. But the pollen of *Gnetum*, with its numerous small spines, suggests that it might be a much-reduced angiosperm. This view is supported by the general morphology of the *Gnetum* leaf, the lack of archegonia, and the typically angiospermous embryogeny. However, for all three genera the fossil ancestry and lines of descent are obscure, and we must await further discoveries to clarify their relationships with other groups.

References

Bornman, C. H. 1972. *Welwitschia mirabilis:* a paradox of the Nanul-desert. *Endeavour* 31:95–99.

Foster, A. S., and Gifford, E. M. 1974. *Comparative morphology of vascular plants.* 2nd ed. W. H. Freeman and Co., San Francisco.

Hus, S. 1969. *Ephedra* (Ma-Huang) in the new Chinese Materia Medica. *Econ. Bot.* 23:346–51.

Kers, L. E. 1967. The distribution of *Welwitschia mirabilis* Hook. F. *Svensk. Bot. Tidskr.* 61:97–125.

Maheshwari, P., and Vasil, V. 1961. *Gnetum.* Botanical Monograph, No. 1, C.S.I.R. New Delhi.

Martens, P. 1971. *Les gnétophytes.* Vol. 12 (2). Encyclopedia of Plant Anatomy. Gebrüder Borntraeger, Berlin.

Peterson, R. L., and Vermeer, J. 1980. Root apex structure in *Ephedra monosperma* and *Ephedra chilensis* (Ephedraceae). *Amer. J. Bot.* 67:815–23.

Rodin, R. J. 1963. Anatomy of the reproductive bracts in *Welwitschia. Amer. J. Bot.* 50:641–58.

Vasil, V. 1959. Morphology and embryology of *Gnetum ula* Brong. *Phytomorphology* 9:167–215.

Waterkeyn, L. 1960. Études sur les Gnétales. IV. Le tube micropylaire et la chambre pollinique de *Gnetum africanum* Welw. *La Cellule* 61:79–96.

Wilson, L. R. 1959. Geological history of the Gnetales. *Oklahoma Geol. Notes* 19(2):35–40.

The most structurally complex of all extant plants are the angiosperms (class Magnoliopsida), or flowering plants. They show the greatest diversity in leaves, stems, and roots, more elaboration in reproductive structures, and the most highly modified cell types in the vascular plants.

Distinctive Features

An obvious feature of the angiosperms is the **flower,** composed of an aggregation of **mega-** and **microsporangia, sepals** and **petals** (sterile appendages) associated in a predictable fashion (Fig. 26-1). Not all flowers have all of these parts; some are reduced to a single sporangium.

The word *angiosperm* means "enclosed seed." This emphasizes another distinctive feature of this group: the **ovules** are enclosed in one of the flower parts, the **carpel.** After fertilization, part or all of the carpel, sometimes with its surrounding tissues, undergoes changes to produce a **fruit.** In most angiosperms, the **seeds,** which develop from fertilized ovules, are enclosed in the fruit. Ovules and seeds are not always completely enclosed; sometimes the carpels are open, or partially open at maturity. *Drimys* (a tropical tree), *Pereskia* (Barbados gooseberry, a leafy cactus) and *Akebia* (akebia) all have carpels that are completely enclosed by intermingled hairs, and all carpels are open during development.

Angiosperms also have distinctive **gametophytes** and a unique process of fertilization. The female, or **megagametophyte** (the embryo sac), has a briefer **coenocytic** stage in its development than is shown by other tracheophytes. A characteristic seven-celled, eight-nucleate structure is usually produced (Fig. 26-2R and 26-4A). The male, or **microgametophytes,** of angiosperms commonly have fewer sterile cells than do those of other tracheophytes (Fig. 26-3 and 26-24). The details of fertilization are unique, because two **sperm cells** participate. One sperm fuses with the egg to form the zygote, and the other fuses with a pair of megagametophyte nuclei to form the primary **endosperm nucleus** (Fig. 26-4B). The primary endosperm nucleus ultimately gives rise to the **endosperm** (Fig. 26-4F), the nutritive tissue of the seed. This process, called **double fertilization,** occurs only in angiosperms.

26

ANGIOSPERMS
(Flowering Plants)

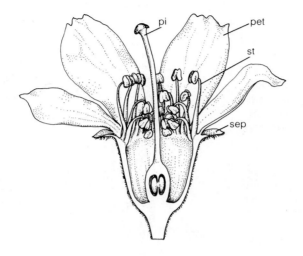

Figure 26-1 Longitudinal section through a flower of *Prunus*, showing sporangia surrounded by sterile appendages, ×3: *pet*, petal; *pi*, pistil; *sep*, sepal; *st*, stamen.

Figure 26-2 Ovule development. **A–E** and **K–N**, ovules. **F–J** and **O–R**, longitudinal sections of **A–E** and **K–N**, respectively, showing megasporogenesis (**G–I**) and megagametogenesis (**J–R**) resulting in female gametophyte (at micropylar end, two synergids and egg nucleus; at chalazal end, three antipodal cells; at center, two polar nuclei fusing to form primary endosperm nucleus). [Redrawn from *The Plant Kingdom* by W. H. Brown, © copyright, 1935, by William H. Brown, used by permission of the publisher, Ginn and Company (Xerox Corporation).]

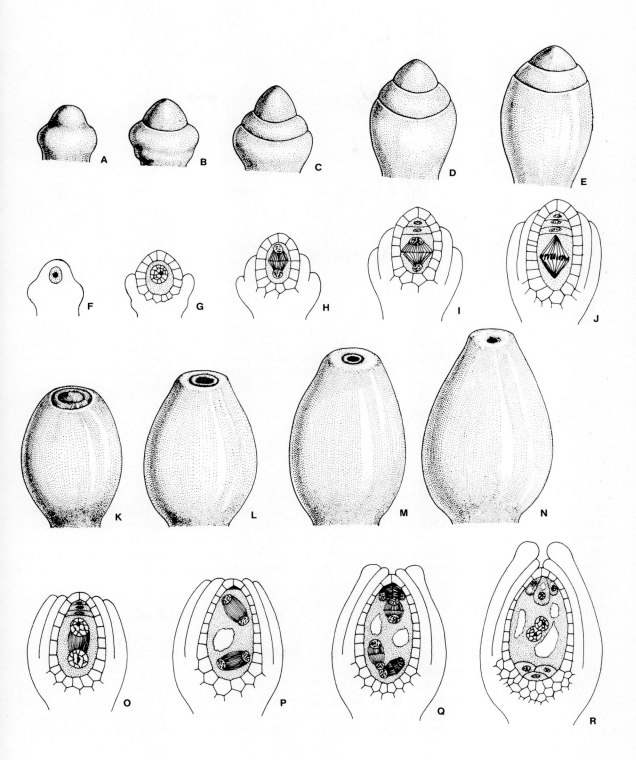

Angiosperms (Flowering Plants) 595

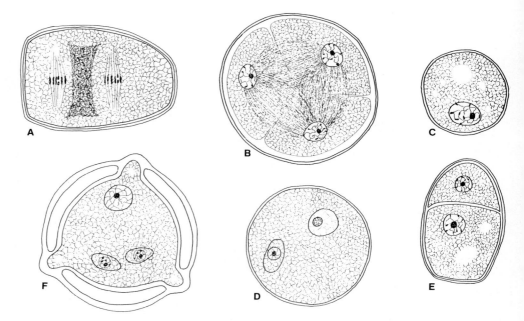

Figure 26-3 Microsporogenesis in *Lobelia cardinalis*. **A,** second meiotic division within the spore mother cell wall, ×900. **B,** microspores nearly formed, partition walls developing, ×900. **C,** mature microspore, ×900. **D,** spore enlarged, first cell division completed, ×900. **E,** vegetative and reproductive nuclei mature, ×900. **F,** mature pollen grain with sperm nuclei, ×1800. (Redrawn after Eames.)

Distinctive angiosperm features in the vegetative plant body include the **apical meristems** (Chapter 12), which differ from those of most other tracheophytes, **vessel elements** (Chapter 12), which appear more frequently in angiosperms than in other vascular plants, and sieve tube elements associated with companion cells in the phloem.

Major Angiosperm Groups

Two major groups are in the class Magnoliopsida: the subclasses Magnoliidae (dicotyledons) and Liliidae (monocotyledons). Although modern research indicates that perhaps the Magnoliidae are not a natural group, they are recognized here for historical reasons. The differences between these two subclasses are presented in Table 26-1. Familiar dicotyledons are carnations, daisies, oak and eucalyptus trees, cacti, sunflowers, and most shrubs. Representative monocotyledons are grasses (barley, wheat, rye, rice, and corn), palms, lilies, pineapples, and orchids.

Vegetative Plant Body

Shoot System

The general aspects of the shoots of vascular plants are considered in Chapter 12, to which the reader is referred for information on primary and secondary growth, an anatomical comparison of dicoty-

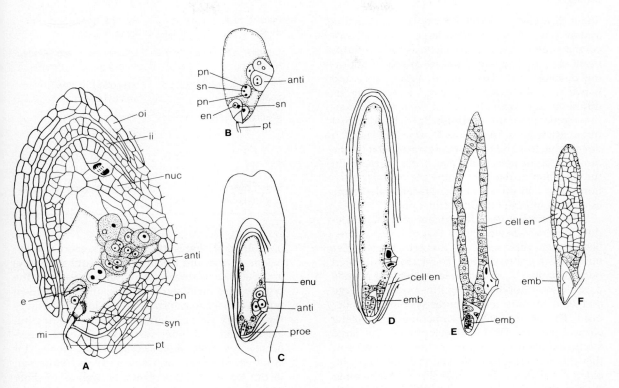

Figure 26-4 Stages in the development of the embryo and cellular endosperm in *Agrostis interrupta* (Poaceae). **A,** longitudinal section of ovule just prior to fertilization. **B,** embryo sac with pollen tube, zygote, fused polar nuclei beside male nucleus. **C,** zygote divided once, to give proembryo; four free endosperm nuclei. **D,** embryo several-celled; many free endosperm nuclei. **E,** embryo multicellular with endosperm becoming cellular. **F,** embryo with cotyledon; endosperm cells completely walled: *cell en,* cellular endosperm; *en,* endosperm; *mi,* micropyle; *pt,* pollen tube; *anti,* antipodals; *e,* egg; *emb,* embryo; *en,* egg nucleus; *enu,* endosperm nucleus; *ii,* inner integument; *nuc,* nucellus; *oi,* outer integument; *pn,* polar nuclei, *proe,* proembyro; *sn,* sperm nucleus; *syn,* synergids.

ledonous and monocotyledonous stems, and an introduction to leaves. We present here more detailed information on stem variation and on leaf development and morphology as shown in angiosperms.

Stems

Some dicotyledons exhibit a primary conducting system (Fig. 26-5) with both large and small **vascular bundles** arranged in a circle as seen in transverse section. The large bundles appear first during development. The smaller appear later and between the large ones. The large vascular strands differentiate acropetally toward the apex as a continuation of older **procambium.** Small bundles first differentiate as isolated strands near the base of a developing leaf and then undergo **acropetal** (in the direction of the apex of the plant) and **basipetal** (in the direction of the base of the plant) differentiation. They join up with other **procambial strands** as a result of basipetal differentiation.

The primary vascular system of monocotyledons differs somewhat from that of dicotyledons. Seen in transverse section of the stem, the vascular bundles are scattered (Fig. 26-5A, B). As pointed out in Chapter 12, two vascular systems of strands occur in monocotyledons: an inner system of large and small strands and an outer system of smaller ones. The outer system is sometimes termed the cortical system. During development, the large strands of the inner system are the first to appear; they differentiate acropetally and in connection with older procambium. The smaller strands of the inner system appear at a given point and then differentiate acropetally and basipetally. The stems of both monocotyledons and dicotyledons probably reflect modifications of the same basic pattern; that is, both consist of (1) large strands that differentiate acropetally early and (2) smaller, later-developing strands that differentiate both acropetally and basipetally. The basic difference between monocotyledons and dicotyledons is that the smaller strands in the monocotyledons are *outside* the larger strands and they have a cortical system; in dicotyledons, the smaller strands *alternate* with the larger ones and there is no cortical system.

Stem Variations In most dicotyledons, the primary vascular bundles of the stem are in a ring and

Table 26-1 Comparison of Characteristics Found in Subclasses of Class Magnoliopsida

Subclass Liliidae (monocotyledons)	Subclass Magnoliidae (dicotyledons)
One cotyledon in embryo	Generally two cotyledons in embryo
Plan of flower usually in threes	Plan of flower usually in fours or fives
Usually parallel venation in leaves	Usually reticulate venation in leaves
Vascular bundles in stem scattered	Vascular bundles in stem in a cylinder
Vascular cambium absent	Vascular cambium commonly present (produces secondary xylem and secondary phloem)
Pollen basically monocolpate	Pollen basically tricolpate

are **collateral;** that is, the phloem is external to the xylem. But in the Cucurbitaceae (squash family) and Solanaceae (tobacco family), for example, the bundles are **bicollateral;** that is, there is phloem both internal and external to the xylem. Sometimes the primary vascular bundles are not in a ring but are in inner and outer systems and hence superficially resemble monocotyledonous stems, for example, *Macropiper* and *Apium* (celery). However, in the stems of these plants, as well as of other dicotyledons of similar stem anatomy, the vascular systems differ from monocotyledons either in the time and pattern of development, or in the course of the vascular bundles in the inner and outer systems.

Most monocotyledons lack **secondary growth.** The secondary growth that does occur in monocotyledons results from the proliferation of cells external to the vascular tissue. As these cells divide, a parenchyma tissue is produced in which new vascular bundles differentiate. Vascular bundles that differentiate in the "secondary plant body" of monocotyledons are treated as part of the cortical system.

Variations also exist in the pattern of secondary growth in dicotyledons. In many **lianas** (woody vines), such as *Clematis* (virgin's bower), the **interfascicular cambium** produces parenchyma, whereas

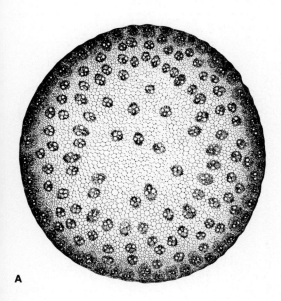

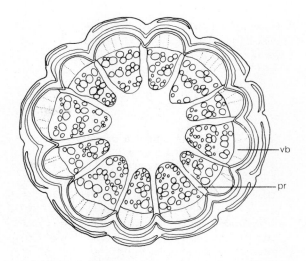

Figure 26-6 Transverse section of stem of the woody vine *Clematis*, showing vascular bundles separated by large pith rays, ×30: *pr*, pith ray; *vb*, vascular bundle.

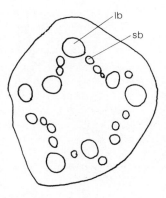

Figure 26-5 Transverse sections of monocotyledon and dicotyledon stems. **A,** the monocotyledon *Zea mays;* the dark oval areas are vascular bundles, × 20. **B,** the dicotyledon *Populus* (diagrammatic), showing small and large procambial strands, × 100; *lb,* large strands; *sb,* small strands. [**A,** from *The Plant Kingdom* by William H. Brown, © copyright, 1935, by William H. Brown, used by permission of the publisher, Ginn and Company (Xerox Corporation).]

the **fascicular cambium** produces the xylem and phloem. This results in large vascular bundles that are separated from each other by large **pith rays** (Fig. 26-6).

Although most angiosperm stems are typically elongated axes, there are some striking modifications. In some cacti, for example, the stem is flattened and thickened into a photosynthesizing and water-storing structure (Fig. 26-7A), features that permit the plant to tolerate extreme aridity. The spines on a cactus stem are interpreted as modified leaves. The needlelike "leaves" on *Asparagus* (asparagus) are also modified stems (Fig. 26-7B). Some stems are modified into **thorns.** For example, *Crataegus* (hawthorn) has branches that produce leaves at one stage of their development; these branches later develop hardened tips and become thorns (Fig. 26-7C).

A less striking stem modification is the **stolon,** which is a stem that grows horizontally along the surface of the ground (Fig. 26-7D). Another modified stem, the **rhizome,** is an underground stolon.

Two general types of modified upright stems are bulbous. One is a true **bulb,** which has an abbreviated stem enclosed by fleshy leaves (Fig. 26-7E). The other bulbous upright stem is a **corm,** which is a thickened stem sheathed in scalelike leaves (Fig. 16-7F, G).

Angiosperms (Flowering Plants) **599**

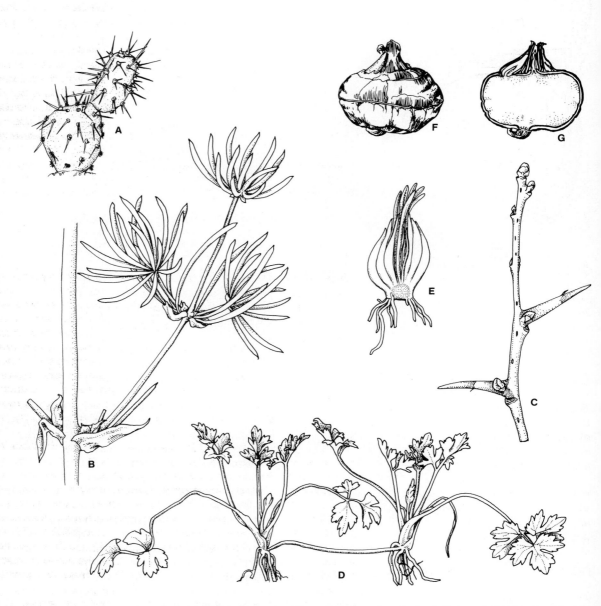

Figure 26-7 Stem modifications in flowering plants. **A,** flattened photosynthesizing and water-storing stems of the cactus *Opuntia*, ×1. **B,** a branch of *Asparagus*, showing stems forming needle-shaped, leaflike structures, ×2. **C,** stems modified as thorns in *Crataegus*, ×1. **D,** stoloniferous stems in *Ranunculus repens*, ×1. **E,** a bulbous stem (longitudinal section) enclosed by fleshy leaves in *Crocus*, ×1; **F, G,** corm of *Eleocharis tuberosa*, ×3: **F,** surface view; **G,** view of longitudinal section: *st*, stem.

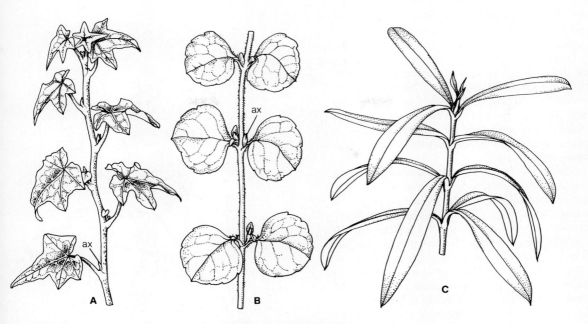

Figure 26-8 Leaf arrangements in flowering plants. **A,** alternate leaves in *Hedera helix;* note also axillary buds at leaf bases, × 0.5. **B,** opposite leaves in *Linnaea borealis;* note axillary buds, × 3. **C,** whorled leaves in *Nerium oleander,* × 0.5: *ax,* axillary bud.

Leaves

The leaves are the lateral appendages on a vegetative stem and are the major photosynthetic organs of most angiosperms. They are usually bilaterally symmetrical, determinate in growth, and associated with a bud in their **axils** (Fig. 26-8A, B).

The leaves of an angiosperm appear in a definite and predictable pattern of attachment in most plants. In the mature shoot, they can occur singly at **nodes** (leaves alternate; Fig. 26-8A), two on opposite sides of the stem at a node (leaves opposite; Fig. 26-8B), or three or more at a node (leaves whorled; Fig. 26-8C).

Leaves often consist of three parts: the **lamina** (or blade), the **petiole** (or leaf stalk), and an expanded leaf base (Fig. 26-8A). Sometimes two structures, the **stipules,** are at the base of the petiole (Fig. 26-9B). Leaves vary greatly in the shape of the lamina, petiole, or stipules. These various structural differences permit plants to occupy specific environments.

Stipules, when present, vary from small (Fig. 26-9B) to large foliate structures (Fig. 26-9E), which sometimes sheathe the stem. Stipules are occasionally modified into **spines** (Fig. 26-9A), or tendrils (Fig. 26-9C).

Petioles may also be present or absent. A leaf without a petiole is described as **sessile.** Petioles usually consist of only the stalk, but sometimes they have expansions of photosynthetic tissue.

Most morphological variation shown by leaves is in the lamina, which is simple or compound, and varies in size, shape, margin, and pattern of venation. In a simple leaf, the lamina consists of one unit, regardless of how deeply it is incised (Fig. 26-10A). In a compound leaf, the lamina is clearly divided into subunits (Fig. 26-10B-C), each of which is called a **leaflet.** Each leaflet is sometimes attached to the main axis of the leaf by a stalk called a **petiolule.** There are two different types of compound leaves: **pinnate,** in which the leaflets are attached along the extension of the petiole (the **rachis**) (Fig. 26-10C) and **palmate,** in which the leaflets are all attached to one point on the petiole (Fig. 26-10B). Terms used to describe the lamina of simple leaves are also applied to the leaflets of compound leaves.

Leaf blade (and individual leaflet) outlines vary from **linear** (Fig. 26-8C) to **orbicular** (Fig. 26-8B), with many intermediate types. The perimeter of the leaf, the **margin,** also shows many different forms. The margin may be toothless (**entire**) or toothed in various ways, for example, **serrate** or **dentate,** or it may be lobed, divided, **dissected,** or **spinose** (Fig. 26-11).

Conducting tissues in the leaves are called **veins,** and their outline is the **venation** pattern. There are two major venation patterns: **netted,** or **reticulate,** in which a great deal of interconnection exists between the veins (Fig. 26-12A), and **parallel,** in which most of the prominent veins are parallel to each other, and there are fewer interconnecting veins (Fig. 26-12B). Reticulate venation exhibits two main patterns: pinnate, in which the secondary veins form a featherlike pattern (Fig. 26-12C), and palmate, in which they form a pattern with all major veins attached to one point (Fig. 26-12D). Generally speaking, leaves with reticulate venation are found in the dicotyledons, and those with parallel venation are found in the monocotyledons.

Some monocotyledons have what are called **unifacial leaves.** In these, the base sheathes the stem, and an upper portion tends to be radially symmetrical or flattened in a plane perpendicular to that in which most leaves are flattened and has a ring of vascular bundles (Fig. 26-13). Somewhat similar leaves appear in some dicotyledons, including species of *Acacia* (wattle; Fig. 26-14).

The variants cited are exhibited by different kinds of plants. Variation occurs also in the form of leaves produced on a single plant. **Cotyledons** (Fig. 26-15B) occur on plant embryos in seeds and

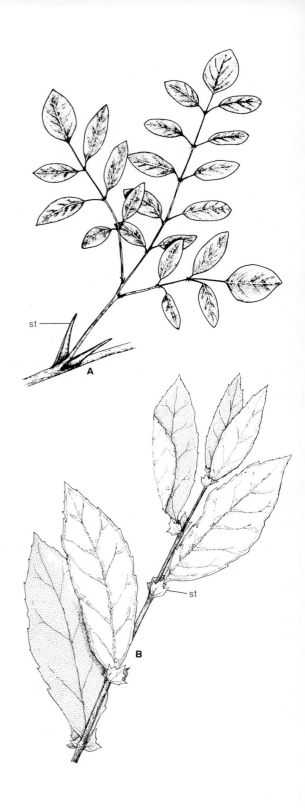

Figure 26-9 Leaves with stipules in flowering plants. **A,** *Robinia pseudoacacia,* with stipular thorns, × 0.5. **B,** *Salix scouleriana,* with persistent leaflike stipules, × 0.5. **C,** *Smilax laurifolia,* with stipular tendrils, × 0.5. **D,** *Liriodendron tulipifera,* with stipules functioning as bud scales, × 0.5. **E,** *Pisum sativum,* with leaflike stipules; note also the terminal leaflets modified as tendrils, × 1: *st,* stipule.

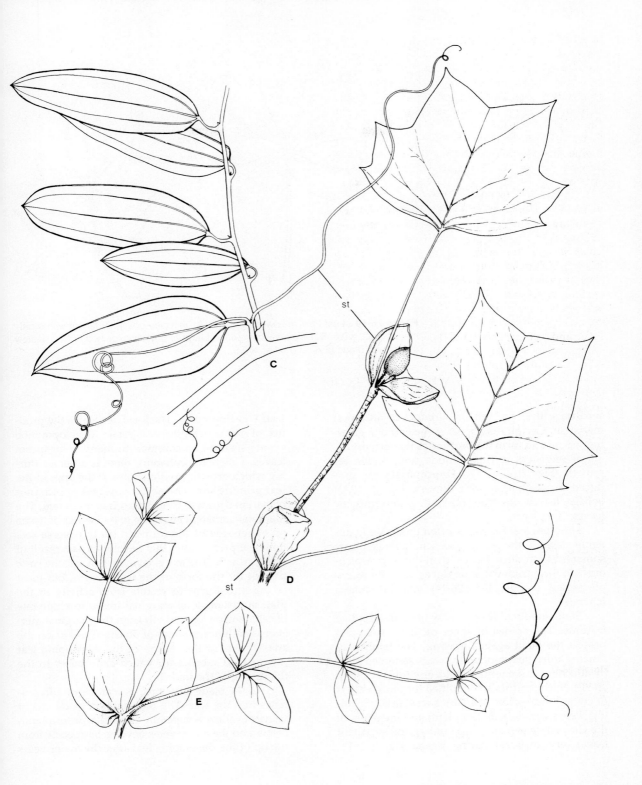

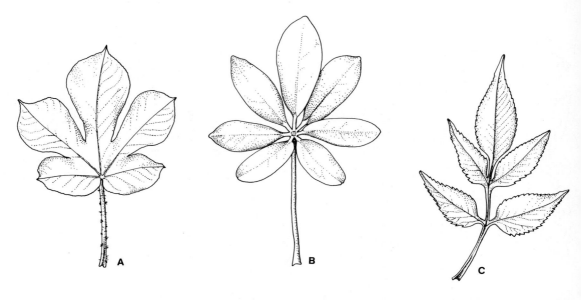

Figure 26-10 Leaf form in flowering plants. **A,** *Jatropha gossypifolia* leaf with a palmate blade, ×0.5. **B,** palmately compound blade of *Brassaia* sp., ×0.5. **C,** pinnately compound blade of *Bidens alba*, ×0.5.

seedlings; they are not the product of an apical meristem. **Cataphylls** appear at the end of a growing season and form the **bud scales** that protect the apical meristem during periods unfavorable for growth (Fig. 26-15C, D). **Hypsophylls** are produced in association with flowers (Fig. 26-15A). Foliage leaves are the "normal" photosynthetic leaves of a plant.

Different leaf forms are often produced at different stages in the growing season. Juvenile plants sometimes produce leaves that differ conspicuously in form from those produced later, for example *Acacia* (wattle; Fig. 26-14) and *Adenostema* (chemise).

Some modified leaves show obvious adaptive features. The spines of some cacti, for example, protect the plant against grazing. The **tendrils** of *Pisum* (garden pea) support the weak-stemmed plant by attaching to something in its environment (Fig. 26-9E). Such tendrils are modified terminal leaflets of compound leaves. In many **succulents,** including *Mesembryanthemum* (ice plant) and members of the Crassulaceae (stonecrop family), the succulent leaves serve as water-storing organs (Fig. 26-15E).

Leaf Development The form of a leaf is the product of its development. Certain developmental events seem to be common to many angiosperm leaves. Following initiation, there is a period during which growth predominates at the apex of the **leaf primordium.** The next stage is the appearance of marginal growth from a **marginal meristem.** The marginal meristem gives rise to two wings of tissue on either side of the rounded (in transverse section) leaf primordium that developed as a result of apical growth. The marginal meristem sets the basic pattern for the form of the leaf margin, but most of the actual growth results from activity in the **plate meristem,** a submarginal area with a high rate of cell division. As growth from the marginal meristem continues, a series of divisions occurs on the **adaxial** side of the midrib of the developing leaf (the **adaxial meristem**), which contributes to the thickening of the midrib.

As the blade develops, procambium differentiates into the leaf. The procambium of the major veins develops acropetally from the procambium below, and the minor veins develop basipetally from the tip of the developing leaf after the major veins

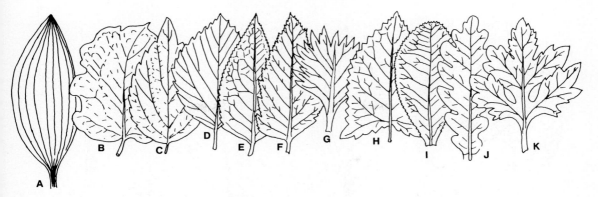

Figure 26-11 Leaf margins in flowering plants. **A,** entire; **B,** undulate; **C,** crenate; **D,** serrate; **E,** serrulate; **F,** double serrate; **G,** incised; **H,** dentate; **I,** denticulate; **J,** lobed; **K,** parted. (After Lawrence.)

have been delimited. The difference between the **palisade** and **spongy mesophyll** (Chapter 12) of the leaf results from the fact that the spongy layer ceases dividing sooner than the palisade layer. The leaf expands as a result of an increase in cell number and enlargement in the palisade layer, but only cell enlargement in the spongy layer. However, cell enlargement in the spongy layer is insufficient to fill the space below the palisade layer, and spaces are left between the cells of the spongy layer.

The petiole is the last structure to appear in a developing leaf. It is the result of intercalary growth between the sheathing leaf base (which results from the young leaf primordium growing around the apex) and the lamina.

Leaf Morphology The leaf can be considered to consist of three portions: the base (the lower leaf zone), the lamina (the upper leaf zone), and the petiole. The work of D. R. Kaplan (1973) has demonstrated that the various shapes of many angiosperm leaves result from varying degrees of development in these three areas. For example, in **compound leaves,** the marginal meristem of the lamina is active in restricted areas and thus gives rise to leaflets. The unifacial leaves found in certain monocotyledons result from activity in the adaxial meristem, with no activity in the marginal meristem (Fig. 26-13B). These unifacial leaves are equivalent to the lamina of most dicotyledons and do not represent expanded, bladeless petioles.

Another type of leaf has a well-marked blade and petiole and is found in certain monocotyledons, such as *Zantedeschia* (calla lily) and *Hosta* (plantain lily; Fig. 26-12B). The widened leaf blades of such plants develop from the basal portion of the leaf rather than from the upper portion, as in the dicotyledonous leaf. A small tip that terminates the leaf blade in these monocotyledons is the equivalent of the blade of dicotyledonous leaves.

The different types of leaves produced by a single plant can also be traced to such developmental patterns. Hypsophylls (Fig. 26-15C, D) and cataphylls (Fig. 26-15A) in some plants result from the development of the lower leaf zone and suppression of the upper leaf zone.

Root System

Collectively, all roots of a plant are called the **root system.** Angiosperms have two general types of root systems: the **taproot** system, which consists of a main root and some secondary **branch roots** (Fig. 26-16A); and the so-called **fibrous system,** composed of many roots approximately equal in size (Fig. 26-16B). The general features of angio-

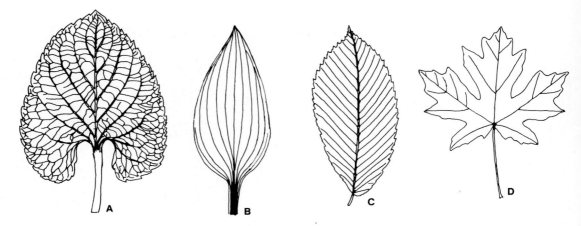

Figure 26-12 Venation in angiosperm leaves. **A,** netted venation in *Viola* sp., ×1. **B,** parallel venation in *Hosta* sp., ×0.5. **C,** main veins pinnately arranged in *Ulmus americana*, ×0.5. **D,** main veins palmately arranged in *Acer saccharum*, ×0.25.

sperm roots are presented in Chapter 12. Here we consider variations.

Root Variation

In most dicotyledonous roots, xylem occupies the center of the root, and there can be two, three, four, or more **protoxylem** poles, the variants being called **diarch, triarch, tetrarch,** and **polyarch,** respectively (Fig. 26-17A-C). In most monocotyledonous roots, parenchyma instead of xylem occupies the center of the root. The xylem is polyarch and patches of phloem alternate with patches of xylem (Fig. 26-17D).

Roots are typically underground and enlarged, and they serve the primary functions of anchorage, absorption, and storage. There are variations in this pattern. One root variant includes the thorns that develop from roots in some palms. These develop from an apical meristem wherein cells elongate and harden. Another modification is the **prop root,** a support root that develops from the lower **internodes** of some plants (Fig. 26-18B). Some tropical trees produce modified roots called **buttress roots,** a name that reflects their supposed function of additional support (Fig. 26-18A).

Aerial roots (Fig. 26-18D) that form on some tropical **epiphytes** hang down like long cords from the trees. On reaching the soil, these behave as normal roots. Some corm-bearing plants produce large, thick, fleshy roots that usually have few **root hairs.** Such roots have a large store of carbohydrates in the cortex; under certain conditions, this is rapidly absorbed by the plant, and the cortex collapses so that the root contracts downward and pulls the corm deeper into the soil. Such roots are called **contractile roots.** Certain storage roots have considerable economic significance. In these roots, a large amount of parenchyma is produced either by a normal vascular cambium, as in carrots (*Daucus*, Fig. 26-18C) or by **anomalous cambia.** One type of anomalous cambium is seen in beets (*Beta*), in which the normal cambium is surrounded by concentric rings of anomalous cambia. Another type is seen in sweet potato (*Ipomaea*), wherein vessel elements in the secondary xylem are surrounded by anomalous cambia that produce large amounts of parenchyma.

Reproductive Plant Body

Flowers

Flowers characterize the diverse evolutionary lines within the angiosperms and are involved in repro-

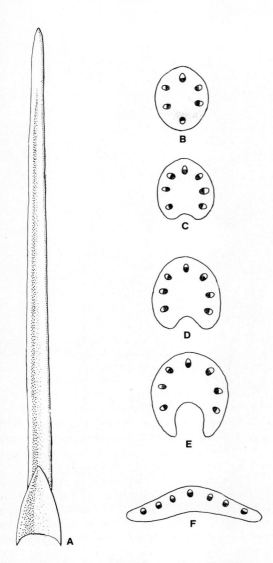

Figure 26-13 **A,** unifacial leaf of *Allium*, with diagrams (**B–F**) at right showing arrangement of vascular bundles. (Redrawn after Kaplan.)

Figure 26-14 Dimorphic leaves in *Acacia* sp. Most of the leaves are compound, but two leaves are simple and unifacial, × 0.5.

duction. They consist of appendages—sepals (outermost), petals, **stamens,** and carpels (innermost) (Fig. 26-1)—attached to a modified portion of the stem (the **receptacle**).

Sepals

Sepals are collectively termed the **calyx** and function primarily in protecting the immature flower. Sepals are commonly green, but they are sometimes colored and form the showy part of the flower (e.g., *Clematis*). In the Asteraceae (sunflower family), the calyx is highly modified and is referred to as the **pappus.** The "parachutes" of dandelion fruits are actually a modified calyx, and this enhances air dispersal of the fruits.

Sepals can be free (not attached to each other) or fused in varying degrees into one unit.

Petals

Petals are enclosed above the sepals and alternate with them (Fig. 26-1). Collectively termed the **corolla,** petals are usually brightly colored and form the showy part of the flower. The most common

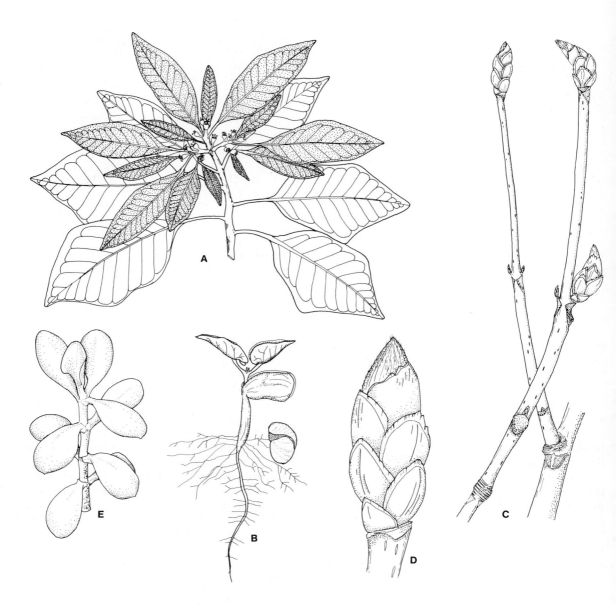

Figure 26-15 Leaf modifications in flowering plants. **A,** leaves modified as colored bracts around the flowers in *Euphorbia pulcherrima,* ×1. **B,** leaves modified as storage cotyledons in *Phaseolus vulgaris,* ×1.5. **C, D,** leaves modified as bud scales in *Aesculus hippocastanum.* **D,** enlarged view of bud scales: C, ×1; D, ×2. **E,** leaves modified for storage in *Crassula* sp., ×1.

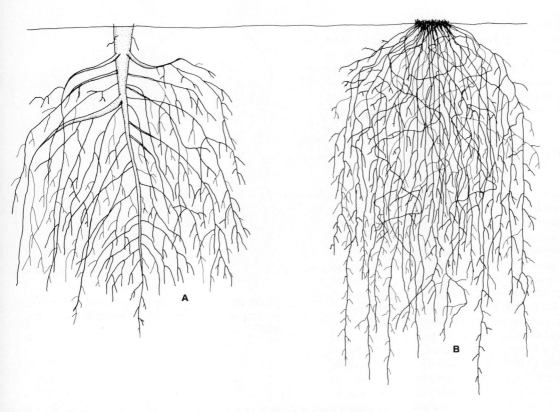

Figure 26-16 Root systems in flowering plants. **A,** taproot, showing large main root and many smaller lateral roots, ×1. **B,** fibrous roots, in which many small roots emerge from a single point, ×1.

function attributed to them is to attract animal **pollinators**—insects, birds, and bats.

The corolla varies in size, color, degree of fusion, symmetry, distinctness from the calyx, and may be present or absent. Petals range from minute, as in some Juncaceae (rush family) to large, as in many Magnoliaceae (magnolia family, Fig. 26-19). They vary in color from white to deep crimson. The color and surface pattern appearance of the petals results from the reflection of light waves. Because the pattern of reflection differs between visible and ultraviolet light, organisms that can perceive ultraviolet light will perceive flowers differently than those that cannot. This is of great significance, since insects, the common pollinators, can perceive ultraviolet light. The reflection and absorption of ultraviolet light are determined by the nature of the reflecting surface and the presence of certain types of chemicals called *phenolics*.

Viewed head on, the corolla of a flower can be **actinomorphic** (radially symmetrical), **zygomorphic** (bilaterally symmetrical), or **irregular** (essentially asymmetrical). Some botanists use the term *irregular* as a synonym for *zygomorphic*. Although a flower is referred to as actinomorphic, zygomorphic, or irregular, it is usually the structure of the corolla that is described. When the corolla is zygomorphic, the calyx and the other flower parts are usually also zygomorphic.

Collectively, sepals and petals are termed the **perianth,** a term used when it is difficult to distinguish between sepals and petals, as in magnolias

(Fig. 26-19) and lilies. The individual units in such an undifferentiated perianth are called **tepals**. The term **perianth** is used also when either petals or sepals are lacking. Even when the petals are lacking, the single whorl of appendages is often referred to as a perianth rather than as **sepals**. Some flowers lack a perianth (e.g., *Euphorbia*; Fig. 26-15A).

Stamens

Stamens are equivalent to microsporophylls; they are appendages enclosed by the petals and usually alternate with them (Fig. 26-1). The collective term for the stamens is the **androecium**. The stamens are composed of the **filament** that supports the **anther**, and the anther itself, which bears the microsporangia (Fig. 26-19 and 26-20). The stamen produces **pollen** and occasionally forms the showy part of the flower, as in *Mimosa* (sensitive plant).

The androecium shows much variation. The individual stamens can be free from each other and from other flower parts or fused to the corolla, to each other, or both. In the pea family, for example, stamens are fused to each other but not to the corolla (Fig. 26-21B). Sometimes the stamens and **gynoecium** are fused and form an elaborate pollinating mechanism, as in milkweed and orchid families (Fig. 26-22D, E).

Individual stamens also vary. Many consist of the typical filament and anther. Others are leaflike; this feature is especially apparent in dicotyledons with large flowers, such as *Nymphaea* (water lily) and *Magnolia* (Fig. 26-19).

The filament can be attached to the middle of the anther, in which case the attachment is called **versatile** (Fig. 26-20E). If the filament is attached to one end of the anther, it is called **basifixed** (attached to the base) (Fig. 26-20B).

Anthers open to release pollen by slits, pores, or flaps (Fig. 26-20A, H, L). If the opening is toward the inside of the flower, dehiscence is called **introrse**; if toward the outside, it is called **extrorse**.

In some flowers, there is a bowl-shaped structure that surrounds the gynoecium and to which the sepals, petals, and stamens are attached (Fig. 26-23A–C); this is the **hypanthium**, or *floral cup*. The nature of the hypanthium can often be determined by the direction the vascular tissue follows (Fig. 26-23E–I).

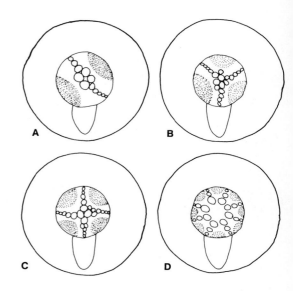

Figure 26-17 Variation in pattern of root growth in flowering plants (diagrammatic). **A,** diarch; **B,** triarch; **C,** tetrarch; **D,** polyarch.

Carpels

Carpels are the uppermost (innermost) floral appendages and form the female part of the reproductive axis. They are the equivalent of megasporophylls and, collectively, are referred to as the **gynoecium**.

One or more carpellary units can be distinguished in the gynoecium of the flower. Each unit consists of three regions: the **stigma**, the **style**, and the **ovary** (Fig. 26-24A). The stigma (pl. **stigmata**) is the enlarged portion at the free end of each gynoecial unit where pollen is deposited. The style is an elongated portion that elevates the stigma to a level that enhances pollination. The ovary, usually the largest portion of each unit of the gynoecium, is attached to the receptacle. Within the ovary, there are one or more chambers, called **locules**, that enclose ovules. Ovules are attached to areas of the carpels called **placentae** (*sing.* **placenta**) by stalks called **funiculi**. After fertilization, the ovules develop into seeds, and the ovary develops into the fruit.

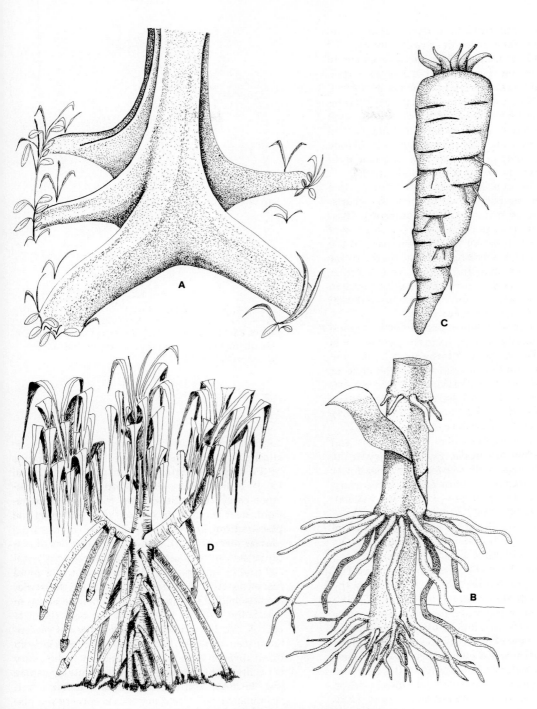

Figure 26-18 Functional and structural variation in roots of flowering plants. **A,** buttress roots of tropical rain forest tree, × 0.02. **B,** prop roots, *Zea mays,* × 0.5. **C,** storage root, *Daucus carota,* × 1. **D,** aerial roots, *Pandanus,* × 0.05.

Each of the compartments of the gynoecium can be made up of one carpel, or of two or more fused carpels. If there is more than one component in the gynoecium each is made up of a single carpel. The old term *pistil* was used to refer to the carpellary units of the gynoecium because of their similarity to the pestles of the druggists of old. Although outdated, this term is still encountered in botanical literature, where one may read of a *simple pistil* (one composed of only one carpel) or a *compound pistil* (one composed of two or more fused carpels).

When the gynoecium consists of many free carpels, they can be attached in spirals or whorls. A gynoecium of free carpels is **apocarpous.** Gynoecia with fused carpels show varying degrees of fusion. Some are barely fused at the base of the ovaries (some Ranunculaceae); in others the ovary is fused, but the style and stigma are free (other Ranunculaceae and some Hypericaceae); and in others the stigma, style, and ovary are fused. Fused gynoecia are said to be **syncarpous.**

To understand the form of any flower, knowledge of the number of carpels in the gynoecium is important. The number of carpels can be derived from (1) the number of branches in the style or stigma; (2) the number of placentae; (3) the number of locules; and (4) the vascular bundle pattern leading to the placentae. Each of these features generally reflects the carpel number.

Besides varying in number, arrangement, and degree of fusion of carpels, gynoecia vary in the position of the ovary with respect to the other flower parts. When other flower parts are below the ovary, the flower is termed **hypogynous** (Fig. 26-25A). When the sepals, petals, and stamens are around the ovary on a bowllike structure (the hypanthium or floral cup), the flower is **perigynous** (Fig. 26-23C). If the flower parts seem to arise from the top of the ovary, the flower is termed **epigynous** (Fig. 26-25B). When no tissue is **adnate** to the ovary, the ovary is said to be *superior* (Fig. 26-25A, C); if a tissue is adnate to the ovary, the ovary is said to be *inferior* (Fig. 26-25B). Thus, a hypogynous flower has a **superior ovary,** whereas an epigynous flower has an **inferior ovary.** A perigynous flower can have either a superior or an inferior ovary. It is also possible to have tissue adnate to only the lower portion of the ovary, in which case the ovary is termed *half-inferior* (Fig 26-23C).

Gynoecia vary in their type of **placentation,** or attachment of the ovules to the ovaries. When gyn-

Figure 26-19 Flower of *Magnolia* sp., showing gradual change from petals to stamens to carpels: *car,* carpel; *pe,* petal; *st,* stamen.

oecia consist of only one carpel, ovules can be attached in four different positions: at the margins of the carpel (**marginal placentation**) (Fig. 26-26, I.a, J); to the surface of the carpel and some distance away from the margin (**laminar placentation**); near the margin but not at it (**submarginal placentation**); and to the dorsal midrib of the carpel (**dorsal placentation**). Two additional types of placentation can be found in a one-carpellary gynoecial unit or in gynoecia consisting of two or more fused carpels. These are **apical,** in which the ovules are attached at the apex of the locule, or **basal,** in which the ovule is attached at the base (Fig. 26-26D, E, M, N). Three additional types of placentation can occur in the ovaries of gynoecia with fused carpels: **axile placentation,** when the ovary is partitioned into more than one locule by septae (Fig. 26-26Aa, Ab), **free central placentation,** which is similar to axile placentation, except that the placenta is not attached to the apex of the locule, and there is only one locule (Fig. 26-26Ca, Cb); and **parietal placentation,** in which the ovules are attached

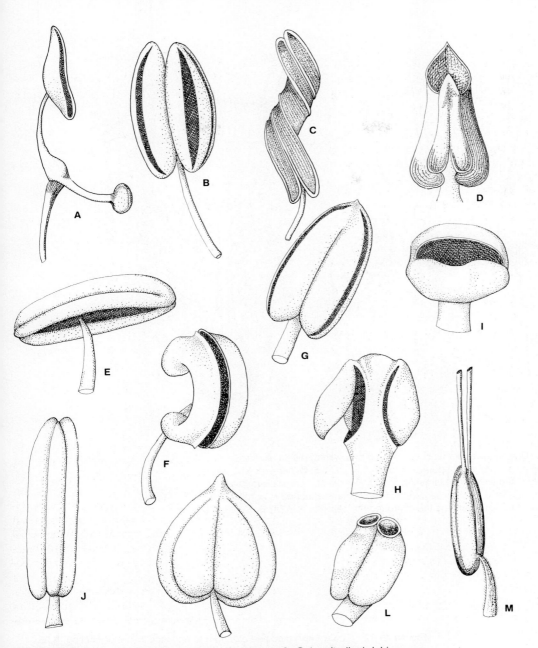

Figure 26-20 Diversity of stamens in flowering plants. **A–C,** longitudinal dehiscence: **A,** *Salvia officinalis,* ×25; **B,** *Calandrinia compressa,* ×35; **C,** *Erythraea centaurium,* ×45. **D,** *Cyclamen europaeum,* apical pore, ×20. **E, F,** dorsifixed, longitudinally dehiscing with one chamber at anthesis: **E,** *Lilium canadense,* ×45; **F,** *Globularia cordifolia,* ×35. **G,** *Juglans regia,* basifixed, longitudinal dehiscence, ×35. **H,** *Hernandia ovigera,* with valvular flaps, ×30. **I,** *Sibbaldia procumbens,* transverse dehiscence, ×30. **J,** *Beta vulgaris,* ×45; **K,** *Lilaea scilloides,* ×35; **L, L–M,** apical pores: *Rhododendron* sp., ×30; **M,** *Oxycoccus* sp., ×60.

Angiosperms (Flowering Plants)

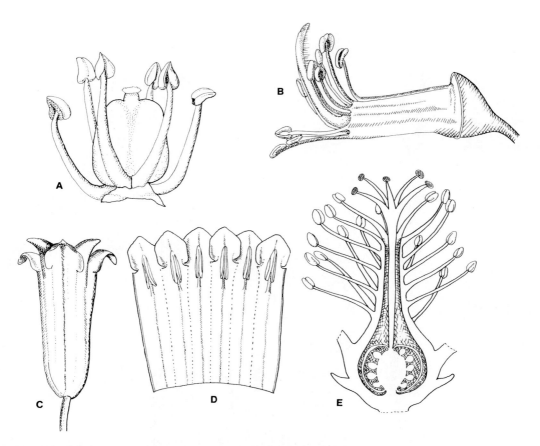

Figure 26-21 Diverse arrangements of stamens in flowering plants. **A,** *Thlaspi* sp., with stamens 6 in number and separate from each other, ×8. **B,** *Lathyrus* sp., with stamens mainly fused along the filaments, ×5. **C,** *Primula* sp., corolla with calyx removed, ×4; **D,** *Primula* sp., dissected to show stamens attached to corolla tube, ×4. **E,** *Hibiscus* sp., with stamens fused around the carpels, ×5 (**A, B, E,** after Lawrence.)

Figure 26-22 Flowers with modified structure to enhance insect pollination. **A, B,** *Aristolochia sipho:* **A,** flower erect with stiff, downward-pointing hairs on inner wall to prevent escape of trapped small flies, ×2; **B,** flower limp and hairs wilted so that flies can escape after pollination has been effected, ×2. **C,** *Stapelia grandiflora,* with flower showing small masses of larvae near bases of petals, ×0.5. **D–F,** *Asclepias speciosa:* **D,** flower with five scoop-shaped nectaries and translator as dark spot between two, ×2; **E,** pair of pollinia suspended from translator, ×15; **F,** dead insect with pollinia caught in its legs, ×5. (**A, B,** redrawn and adapted from B. J. D. Meeuse, with permission of John Wiley & Sons, Inc., New York.)

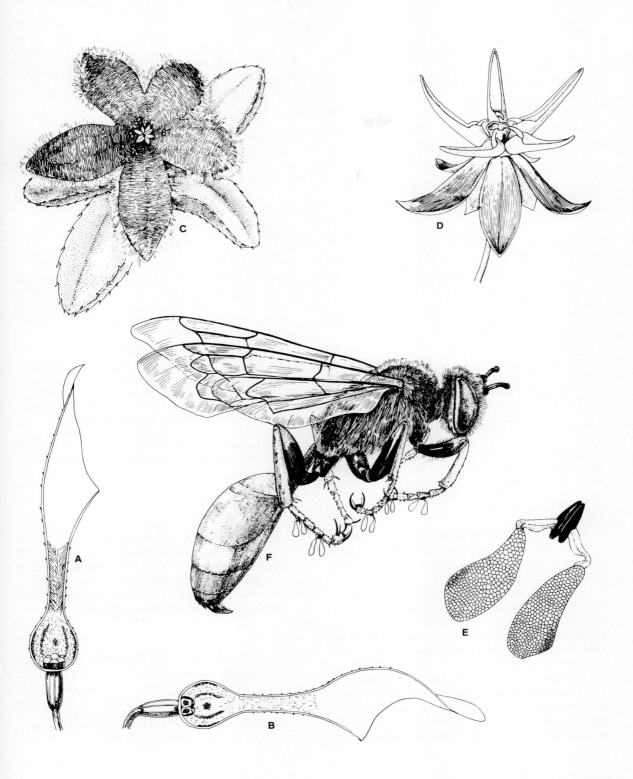

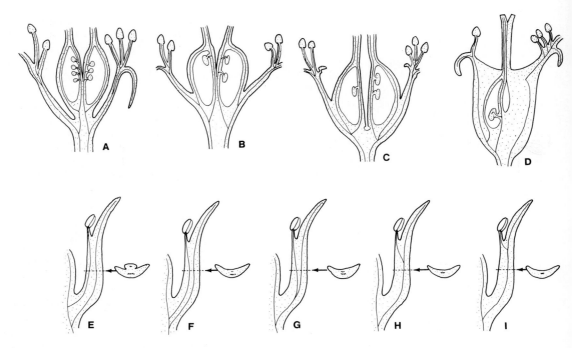

Figure 26-23 Diagrams of vascularization in a flower (longitudinal sections in upper row; bottom row with longitudinal sections to the left and transverse sections to the right of each figure). **A,** *Physocarpus opulifolius*, ovary superior, stamens adnate to hypanthium. **B,** *Sorbus sorbifolia*, with floral cup fused to the carpels. **C,** *Spiraea* sp. with floral cup adnate to carpels halfway to top of ovary. **D,** *Malus pumila*, with floral cup adnate to the ovary, making the ovary inferior. **E–I,** stages in fusion of vascular bundles under adnation in stamens and petals: **E, F,** organs only lightly fused; **G,** fusion more complete; **H–I,** fusion for various distances from base. (Redrawn after Eames and MacDaniels.)

to the walls of the ovary (Fig. 26-26H, La, Lb). Plants in which the ovules seem to be attached to tissue derived directly from the apical meristem are said to have **cauline placentation.**

An aid to understanding types of placentation is provided in Figure 26-26 that shows the interrelationships between the different types of placentation. Although this figure suggests lines of evolution, each type of placentation could have arisen through several possible origins.

Placentation is usually based on mature ovule attachment. The mature type of placentation is usually reflected in the development of the gynoecium. For example, in axile placentation, ovule primordia (developing ovules) generally appear on a central column; in marginal placentation, ovules develop on the margins of the carpel. However, in the Poaceae (grass family), the ovule is initiated in a basal position, but the basal part of the gynoecium becomes shifted to the side; the mature ovary then appears to show parietal placentation.

Perfect and Imperfect, Complete and Incomplete Flowers

Not all flowers have a complete complement of parts. If either petals or both sepals and petals are lack-

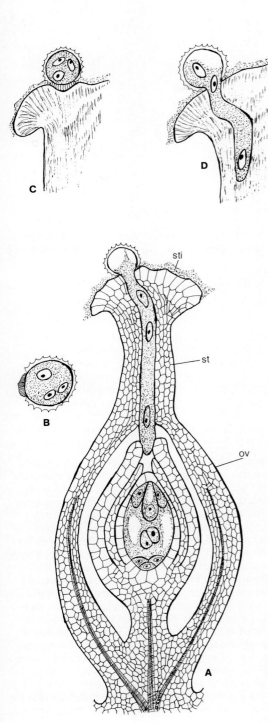

ing, flowers are referred to as **incomplete**. Flowers that lack either stamens or carpels are called **imperfect**. Flowers that have only stamens are termed **staminate**, whereas those with only gynoecia are termed **pistillate**. In a plant with imperfect flowers, both staminate and pistillate flowers can occur on one plant, in which case the plant is called **monoecious**. Conversely, staminate and pistillate flowers can occur on separate plants, in which case the plants are said to be **dioecious**. Sometimes, staminate flowers have reduced gynoecia, and pistillate flowers sometimes have rudimentary stamens.

Inflorescences

Flower clusters on a single plant are referred to as **inflorescences**. The stalk of an inflorescence is called a **peduncle**, whereas the individual flower stalks within the inflorescence are called **pedicels**. An inflorescence is a branch system. Each branch in an inflorescence develops from an **axillary bud**, the **subtending** leaf of which is usually highly reduced (or absent) and is called a **bract**.

The two major types of inflorescences are **determinate** and **indeterminate**. Determinate inflorescences have the oldest flower at the top (Fig. 26-27F, I), whereas an indeterminate inflorescence has the youngest flower at the top (Fig. 26-27E, J). These two types of inflorescences show several different forms that can be distinguished on the basis of branching pattern. These are illustrated in Figure 26-27.

Figure 26-24 Diagrammatic representation of an ovary with a straight, erect ovule, showing development of the pollen tube down the style. **A,** longitudinal section, showing embryo sac within ovule, at the eight-nucleate, seven-cell stage—the egg surrounded by its two synergids at the upper end, the fusion nuclei in the center, and three antipodals at the distal end; the pollen grain in the stigma has extended a pollen tube that is entering the micropylar end of the ovule. **B,** pollen grain at the three-nucleate stage. **C,** pollen grain on stigma. **D,** formation of pollen tube. (Redrawn from *The Plant Kingdom* by William H. Brown, © copyright, 1935, by William H. Brown, used by permission of the publisher, Ginn and Company (Xerox Corporation).)

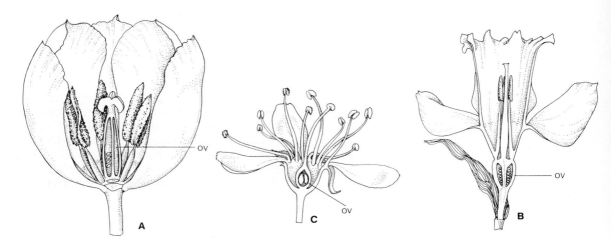

Figure 26-25 Varying position of the ovary in flowers (diagrammatic). **A,** *Tulipa* sp., hypogynous, ×1. **B,** *Narcissus* sp., epigynous, ×1. **C,** *Prunus* sp., perigynous, ×2: *ov,* ovary.

Figure 26-26 Relationships between ovary and placentation types (diagrammatic). **A,** trilocular ovary with axile placentation: **Aa,** schematic view of ovary; **Ab,** transverse section of ovary, showing carpellary vascularization (xylem elements blackened). **B,** intermediate stage between **Ab** and **C. C,** unilocular ovary with free-central placentation, derived from **Aa: Ca,** schematic view of ovary; **Cb,** transverse section of ovary, differing from **B** only in complete loss of septation. **D,** compound ovary with basal placentation, derived from **Ca** by reduction of central placentae. **E,** compound ovary with single basal ovule, derived from **D** by ovule reduction. **F,** transverse section of unilocular tricarpellate ovary with parietal placentation, derived from **Ab. G,** same as **F** but with septa reduced. **H,** advanced stage of **G** (no placental intrusion). **I,** hypothetical, primitive situation of axis with three open carpels: **Ia,** schematic view; **Ib,** transverse section of same. **J,** compound ovary (unilocular, tricarpellate) with parietal placentation, derived by fusion of adjoining carpel margins. **K,** compound ovary with parietal placentation, derived from **J: Ka,** schematic view of ovary; **Kb,** transverse section of same, showing union of adjoining ventral carpellary strands. **L,** compound ovary (parietal placentation), derived from **K; La,** schematic view of ovary; **Lb,** transverse section of ovary. **M,** compound ovary with single basal ovule, derived from **M** by ovule reduction. (Redrawn after Lawrence, reprinted with permission of Macmillan Publishing Company from *Taxonomy of Vascular Plants,* by G. H. M. Lawrence, copyright 1951 by Macmillan Publishing Company.)

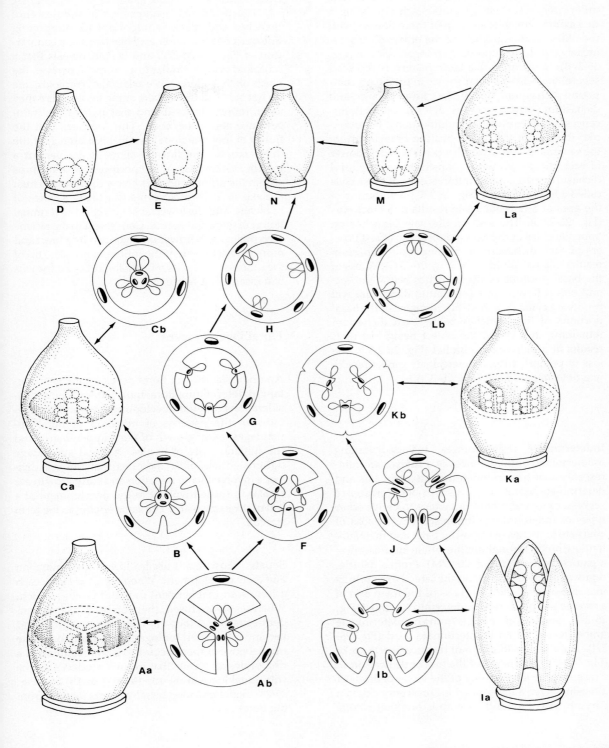

Angiosperms (Flowering Plants) 619

Determinate Inflorescences The determinate inflorescence consists of units, each one being a **dichasium,** which is a group of three flowers with the oldest in the middle and the others of approximately equal age oppositely arranged below it (Fig. 26-27B). Sometimes this basic pattern is repeated several times in an inflorescence to produce a **compound dichasium** (Fig. 26-27A). From a compound dichasium, other types of determinate inflorescences can be derived by modifications. For example, the loss of the flower-bearing branches from the same sides of the axes that bear the oldest flower of each dichasium which constitute a compound dichasium results in an inflorescence of loose coils called a **helicoid cyme** (Fig. 26-27C). Shortening of the axes of a helicoid cyme results in a **cincinnus** (Fig. 26-27D). The loss of alternate branches on opposite sides of the axes that bear the oldest flowers of each dichasium in a compound dichasium results in a **scorpioid cyme** (Fig. 26-27G). A loss of the basal portions of the branches below the second oldest flowers in a compound dichasium will produce a **cyme** (Fig. 26-27H). Removal of the basal portions of the branches below each dichasium remaining in a cyme, the oldest being lost first, results in a determinate **umbel** (Fig. 26-27F). The loss of pedicels from a determinate umbel results in a determinate head (Fig. 26-27I).

Indeterminate Inflorescences Included among indeterminate inflorescences is a **thyrse,** an inflorescence that has an indeterminate main axis and determinate lateral axes. This type of inflorescence serves as a logical point from which to derive other types of indeterminate inflorescences. By loss of alternate branches of opposing pairs and their subtending bracts throughout the whole inflorescence, a **panicle** is formed (Fig. 26-27M). From a panicle, removal of the branches on the lateral axes results in a **raceme** (Fig. 26-27K). Loss of the main axis of a raceme will produce an indeterminate umbel (Fig. 26-27E), and loss of pedicels from an indeterminate umbel results in an indeterminate head (Fig. 26-27J). Another modification of the raceme is a **spike** (Fig. 26-27L), the result of the loss of pedicels from a raceme. The lengthening of the lower pedicels of a raceme gives a flat-topped inflorescence called a **corymb.** A compound corymb is one that results from pedicel lengthening in a panicle (Fig. 26-27M).

The positional relationships of inflorescence branches and bracts is important in interpreting inflorescence morphology. For example, a comparison of Figures 26-27G and 26-27K reveals that a scorpioid cyme is similar to a raceme. However, the bracts on the inflorescence axis of the scorpioid cyme do not subtend the flowers at the node where they arise; rather, they subtend that part of the inflorescence above the node. This indicates that the flower at any one node terminates a branch of the inflorescence; the flowers above it are part of a branch system that, in the positional relationships within the inflorescence, is below the flower. Thus, the axis of a scorpioid cyme is actually a series of branches, and the lowermost flower is the terminal flower of the inflorescence. Conversely, in the raceme each pedicel and flower is subtended by a bract and represents a branch of the inflorescence. The tip of the inflorescence is its morphological apex; the axis of a raceme is a true axis.

Floral Development

An understanding of floral development is important in determining floral morphology. In our considerations of floral development, we will concentrate on the patterns of initiation of floral appendages, early growth of floral appendages, and the initiation site of ovules. A floral appendage designated as *leaflike* indicates that the cell divisions involved in its initiation and early growth are similar to those involved in the development of a leaf of the same plant. The same applies to the term *stemlike*.

Sepals Most sepals are leaflike in their initiation and early development. When sepals are free, each arises as an independent unit and grows as such. When sepals are fused, they can begin as independent sepals, but then (1) their margins grow and become interlocked (**ontogenetic fusion**), or (2) the tissue between them begins to grow, resulting in a ring of dividing cells that in turn produces a tube with sepal lobes at the top (**toral growth**) (see Fig. 26-28). Sepal initiation usually occurs directly from the **floral apex.**

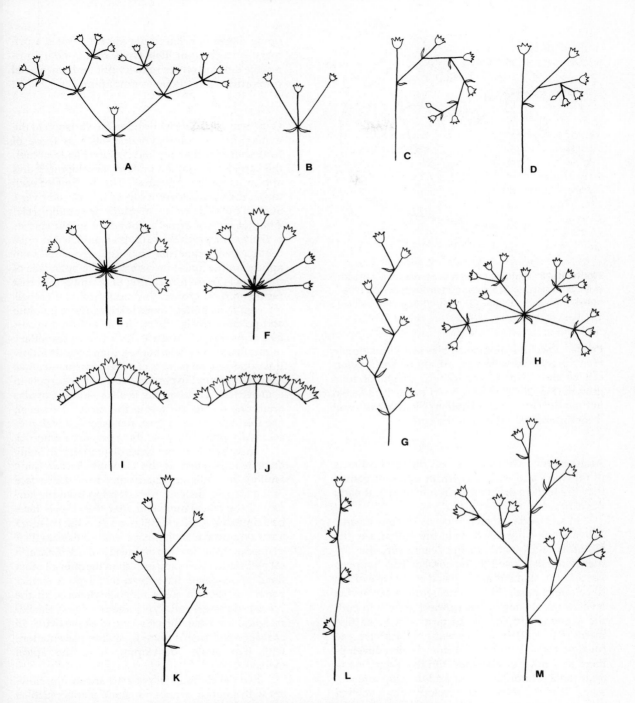

Figure 26-27 Inflorescence types (diagrammatic) in flowering plants (diagrammatic). **A,** compound dichasium; **B,** simple dichasium; **C,** helicoid cyme; **D,** cincinnus; **E,** indeterminate umbel; **F,** determinate umbel; **G,** scorpioid cyme; **H,** compound cyme; **I,** determinate head; **J,** indeterminate head; **K,** raceme; **L,** spike; **M,** panicle.

Angiosperms (Flowering Plants)

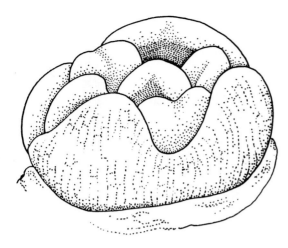

Figure 26-28 *Lantana camara*, side view of floral bud (with calyx removed), showing developing corolla tube between petal primordia, ×146. (Redrawn after Sattler.)

Petals Petal initiation resembles sepal initiation in many plants. Furthermore, fusion between petals can be the result of ontogenetic fusion or toral growth (Fig. 26-28). The form of petals is reflected in their development. Zygomorphic corollas result from differential growth in the corolla.

Androecium The stamens have different patterns of initiation and early growth in different plants; they are leaflike or stemlike in initiation and early development.

In androecia, stamens fuse by various means. When anthers are fused, as in the Asteraceae, the fusion is ontogenetic. In the Asteraceae, the filaments are also fused to the corolla. This happens because the stamens are initiated at the base of the developing petals; then zonal growth below the level of the stamens gives upward growth in both the stamens and corolla. In taxa with fused filaments (Fig. 26-21E), such as many of the Malvaceae (mallow family), the androecium begins development as a ridge on the apex. On this ridge, many androecial primordia appear and develop into stamens. It is the ridge that develops into a **stamen tube** that is characteristic of the Malvaceae. In some members of the Alismataceae (water-plantain family), an interesting developmental variant involves the petals and stamens. On the apex of a developing flower, a primordium appears, called a **primary primordium.** Two types of secondary primordia develop on the primary primordium: a petal primordium and stamen primordium.

Gynoecium There is considerable variation in the gynoecium. In some gynoecia, such as those of *Ranunculus* (buttercup) and *Aquilegia* (columbine), the carpels are leaflike in early development and appear as folded structures. The margins of each carpel then grow together to form a closed ovary. Such a carpel is called **involute** or **conduplicate.** Another type of carpel is found in certain species of *Drimys,* in which the carpels are initiated as crescent-shaped primordia, but cell divisions soon appear on the apical meristem on the open side of the carpel. These form a ring of dividing cells that results in toral growth to produce tubular carpels referred to as **peltate** or **ascidial.** In these involute and peltate carpels, the ovules arise from the surface of the carpels and are described as **carpellary** or **phyllosporous** (leaf-borne). Many plants follow this same general pattern of development—leaflike carpels and phyllosporous ovules. The carpels in some species of *Alisma* are leaflike, and the ovules are borne on the surface of the carpel. However, like the stamens in *Alisma,* the carpels develop as secondary primordia from three primary primordia.

There are also other types of gynoecia. In many dicotyledons, such as the Caryophyllaceae (pink family), the ovules do not develop from the surface of the carpels; instead, they develop from the surface of the apical meristem after the carpels have been initiated. Other families, such as the Polygonaceae (knotweed family), have single ovules in their gynoecia; these develop as a result of the floral apical meristem developing into an ovule after all other floral appendages have been initiated. A similar pattern is seen in some monocotyledons. In the Poaceae (grasses) and Cyperaceae (sedges), the floral apical meristem develops into a single ovule. In *Luzula* (wood rush), there is cauline placentation, with two ovules developing from the apical meristem.

Two developmental patterns are seen in flowers with inferior ovaries. In some plants, such as *Downingia* (downingia) and *Lactuca* (lettuce), the sepals, petals, and stamens are initiated and then carried upward by toral growth below the level of initiation. The base of the portion that develops as a result of such toral growth develops into the ovary wall. In these taxa, the stigma and style develop as units separate from the ovary wall. On the other hand, the inferior ovary of some Cactaceae results from growth phenomena in the receptacle by which the sepals, petals, stamens, and sometimes part of the gynoecium are elevated upward and ultimately lie directly above the placentae. When a part of the gynoecium lies above the placentae, the tissue to which the seeds are attached is derived from the central portion of the flower. Hence, what appears to be parietal placentation in some Cactaceae is, in fact, a highly modified form of axile placentation.

The type of inferior ovary seen in *Lactuca* and *Downingia* is referred to as **appendicular**; that of the Cactaceae and other similar types is called **receptacular**. Some plants, including certain Rosaceae, have inferior ovaries that are receptacular in their bases and appendicular at their tops.

Floral Interpretation

Floral interpretation is centrally concerned with the question: What is a flower? Most botanists would be satisfied with a broad definition of a flower as "an appendage-bearing axis that serves in reproduction." Such a broad definition allows some disagreement concerning the nature of the appendages. The classical theory considers the flower to be a modified stem that bears modified leaves (sepals and petals), microsporophylls (stamens), and megasporophylls (carpels), and interprets the diversity in floral form as having evolved from a flower much like that seen in modern day ranalean plants. This view considers the "primitive" flower to consist of many perianth segments that are undifferentiated into calyx and corolla, many laminar stamens, and many large, leaflike carpels which bear ovules on their surface. All of these floral appendages are on an elongated receptacle and are spirally arranged.

The evolution of diverse floral types from this generalized flower is interpreted as the result of an evolutionary loss of parts, modification of form, and shortening of the receptacle. An appendicular inferior ovary is considered to be the result of the evolutionary fusion of sepals, petals, and stamens. A similar process is suggested to account for some fusions seen within the calyx and corolla and between the androecium and corolla. The various

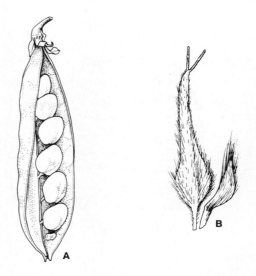

Figure 26-29 Carpels of flowering plants. **A**, *Pisum* sp. pod, showing seeds within carpel, ×1. **B**, *Salix* sp., carpel, ×5.

types of gynoecia are interpreted as modifications of the putatively generalized type. In cases of cauline placentation (the attachment of ovules to stem tissue), such as that found in the Caryophyllaceae and other families, it is assumed that the surface of the floral apical meristem is clothed in carpellary tissue. This interpretation receives support from the fact that some vascular bundles in the placentae of the Caryophyllaceae and other plants with cauline placentation show the kind of orientation one would expect if the floral apical meristem were clothed with carpellary tissue.

The classical theory of the flower offers a theoretical framework within which to compare all flowers; it emphasizes the unity of all flowers and hence the unity of the angiosperms. It is also an excellent learning tool, since it leads to an understanding of floral form by focusing first on one type (the ranalean) and then treating other types as modifications of it.

Some features that support the classical theory include the leaflike carpels found in *Drimys* (Fig. 26-30) and the leaflike stamens found in *Magnolia* and other ranalean plants (Fig. 26-19). Furthermore, the floral appendages of many plants initiate and grow like leaves, at least during early stages of development, and many plants have gynoecia in

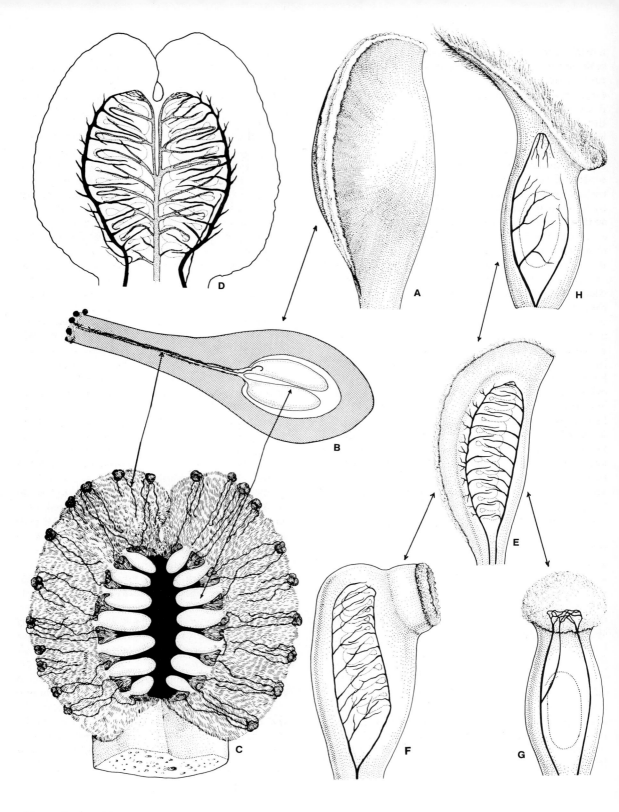

which ovules develop on the surface of carpels. However, certain floral structures cannot be interpreted readily by the classical flower theory. A prime example is cauline placentation, in which the carpels cannot be interpreted as megasporophylls, since they do not bear ovules. Proponents of the classical flower theory have evoked hypothetical evolutionary processes to explain how certain structures that refute the theory could have evolved from the generalized flower type. One of the most commonly utilized hypothetical evolutionary processes is congenital fusion: fusion that occurred in the past so that what were once free structures now appear as a single unit. Cauline placentation is supposed to be an example of congenital fusion between carpellary tissue and stem tissue.

Not all botanists adhere to the classical flower theory, and other theories of floral construction have been presented. A. D. J. Meeuse (1966), for example, envisions the flowers, as well as the reproductive structures of all gymnosperms, as having evolved from a modification of a frond of *Archaeopteris* (Fig. 20-4B), a fossil progymnosperm. In Meeuse's view, the several different types of flowers evolved from different portions of the *Archaeopteris* fronds that were modified into the various flower parts. It is possible, according to Meeuse, for a flower to be the result of evolutionary modification of one frond or of several fronds. In the latter case, each flower part may represent one highly reduced frond. However, this theory is based largely on hypothetical evolutionary modifications and is difficult to apply, since there are no precise criteria whereby one can determine the true nature (*sensu* Meeuse) of floral appendages. Furthermore, where some of Meeuse's interpretations have been subjected to rigorous testing, they have been found to present inconsistencies.

A theory proposed by Melville (1962) views flowers as evolving from units called **gonophylls** that consist of a bract and a stalked sporangium. Like Meeuse, Melville believes that there are several different types of flowers, depending on the hypothetical evolutionary pathways. Melville's theory has been applied, and sometimes floral structure has been revealed to differ from what would have been predicted by the theory.

H. J. Lam (1963) also departed from the classical theory. He viewed the flowers as either **stachysporous** (with stem-borne sporangia) or **phyllosporous** (with leaf-borne sporangia) and attempted to relate stachysporous and phyllosporous flowers to proposed ancestors. These attempts to classify angiosperms according to how their sporangia were borne resulted in division of the angiosperms into two very different groups, and the angiosperms were thus viewed to be **polyphyletic.** Most botanists reject this view, since angiosperms do not fit the criteria that characterize a polyphyletic taxon.

The various theories concerning floral morphology clearly show that the interpretation of floral structures is strongly influenced by speculation about the evolution of the flower. Some morphologists consider it undesirable to interpret floral structure according to evolutionary speculations, since further hypotheses, some of which stretch even the most vivid imagination, must sometimes be erected to explain away facts inconsistent with the theory.

Another approach to floral morphology attempts to interpret flower structure not by comparing flower parts with vegetative structures or hypothetical evolutionary units, but by comparing the units with each other; sepals with sepals, petals with petals, stamens with stamens and gynoecia with gynoecia. If flowers in different plants are observed to have dissimilar structures, then it is assumed that the flowers are indeed different and have not evolved from the same basic type. The grass flower, for example, has stamens that are different from the stamens of the Alismataceae, and the ovules in grasses are cauline, whereas those of the Alismataceae are leaf-borne. In the dicotyle-

Figure 26-30 Simple carpels of flowering plants. **A,** side view, showing paired stigmatic crests. **B,** transverse section, showing pollen grains and penetration of pollen tube. **C,** unfolded lamina, showing placentation, distribution of glandular hairs, and course of pollen tubes. **D,** cleared, unfolded lamina, showing vasculature. **E,** simple form of conduplicate carpel. **F,** lateral and terminal closure, with stigmatic crests restricted to projecting unsealed part. **G,** laterally sealed carpel with capitate stigma. **H,** laterally sealed carpel, with expanded, terminal, stigmatic crests. (After Bailey and Swamy, with permission of American Journal of Botany; and adapted from Irving W. Bailey and B. G. L. Swamy, *The Conduplicate Carpel of Dicotyledons and Its Initial Trends of Specialization*, by Irving W. Bailey, 1954, John Wiley & Sons, Inc., New York.)

dons, the ovules of *Datura* and the Caryophyllaceae are cauline, whereas in many other dicotyledons they are borne on the surfaces of carpels.

This approach to the interpretation of floral structure emphasizes function. For example, the angiosperm gynoecium functions in **pollination**, in the protection of developing ovules, and in the dispersal of seeds. The prerequisite for these functions is enclosure of the ovules; it is immaterial whether the ovules are leaf-borne or cauline. Similar arguments could be applied to stamens. One implication of the functional approach to floral interpretation is that flower parts have evolved from various structures. This general concept is not new. It is well-accepted that petals show two different origins: they have evolved from bracts in some plants and from stamens in others.

To assume that there are different types of flowers—or, more specifically, different types of androecia and gynoceia—is, for most botanists, to assume that angiosperm flowers are polyphyletic. This assumption implies that angiosperms themselves are polyphyletic and, as such, form an unnatural taxon. However, whether or not flowers are polyphyletic has no bearing on the naturalness of the subdivision Magnoliopsida (see the discussion of polyphyletic taxa in Chapter 1). The Magnoliopsida meet the criteria that characterize a monophyletic taxon.

Floral Function

Flowers are the reproductive units of angiosperms and, as such, are involved in pollination, fertilization, and all the subsequent events that lead to the formation of mature seeds and fruits.

Pollination

Pollination in angiosperms is defined as the transfer of pollen from an anther to a stigma. Many adaptations in flowers enhance pollination, and these are the primary subject of this section. A detailed review that is also a source for literature in the field is the fascinating book by B. J. D. Meeuse, *The Story of Pollination* (1961); the recent and more technical book by Faegri and van der Pijl (1979) is also worthwhile.

The relationship between pollen and the formation of the fruit was known to the ancient Mesopotamians in the ninth century B.C. Bas-reliefs of this period illustrate a ceremonial dusting of the fruit-bearing date palms with the inflorescences from pollen-producing plants. However, this particular folklore was without any scientific basis until the end of the seventeenth century A.D., when Nehemiah Grew wrote that the grains within the stamens represent male parts and the seed-producing parts were female. The first experiments on sex in plants were those of Camerarius in 1684. He discovered that unless pollen came into contact with a stigma, fruit would not develop.

Since these early considerations of the relationship between pollen and reproduction, the study of pollination has advanced greatly, resulting in a classification of types of pollination based on the pollinating agent involved. The most important agents of pollination are wind (the type called **anemophily**), insects (**entomophily**), and birds (**ornithophily**). Pollination by bats (**chiropterophily**) and by water (**hydrophily**) also occurs, but less frequently than the other types mentioned. Although we do not intend to discuss pollination in depth, we should point out that the features associated with various pollinating mechanisms have evolved numerous times in independent lines.

Entomophily

Although the first seed plants were probably wind-pollinated, irregular insect visits to flowers may have occurred even among the seed ferns. The insects of the Carboniferous had biting jaws, rather than mouth parts adapted for sucking nectar. Part of their food may have been fleshy sporophylls and possibly even spores. It was not until the later Cretaceous and Tertiary, when flowering plants predominated, and most modern plant families appeared, that most contemporary orders of insects also appeared. It is almost certain that the intimate association between particular flowers and their insect visitors began during several epochs of the later Cretaceous and Tertiary.

An insect visits a flower to obtain food, either in the form of **nectar** or pollen. Nectar is a watery fluid containing sugar and other dissolved materials, secreted by special glands known as **nectaries**. Nectaries on a plant may be in the flower (flo-

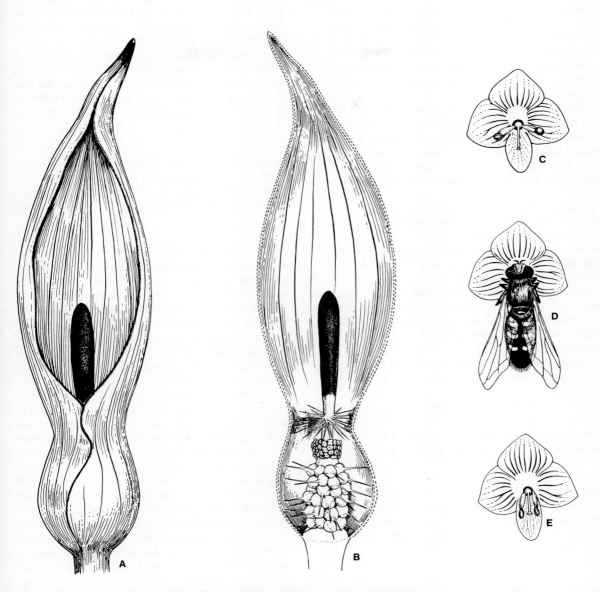

Figure 26-31 Dipterous flowers and inflorescences. **A, B,** *Arum maculatum:* **A,** inflorescence enclosed by spathe, upper part of spadix visible, ×2; **B,** portion of spathe removed to show sterile male flowers at constriction of spathe, functional male flowers immediately below them, and female flowers at base of spadix, ×2. **C–E,** *Veronica,* flowers before (**C**), during (**D**), and after (**E**) visit of crescent fly, ×2. (**C–E,** redrawn and adapted from *The Story of Pollination,* by B. J. D. Meeuse, copyright © 1961, John Wiley & Sons, Inc., New York.)

Angiosperms (Flowering Plants) 627

ral) or elsewhere (extrafloral). Floral nectaries can occur on different organs and in a number of positions. They sometimes occur at the base or apex of the ovary, and the nectar is produced by modified stems. Petals are commonly a source of nectar; even the receptacle sometimes produces nectar. *Prunus* (cherry), which has perigynous flowers, produces nectar in the tissue that lines the inside of a cup formed by the fusion of sepals, petals, and stamens (Fig. 26-1).

Insects are attracted to flowers chiefly by color and scent, the relative attractiveness of which varies with the type of insect visitor. Bees can discriminate among and are apparently attracted to yellows, blues, and purples, which explains why bee-pollinated flowers characteristically have these colors.

Various means of guiding the insect to the nectar (nectar guides) have evolved in different kinds of flowers. Sometimes the guide is a vivid patch of contrasting color on the background color of the corolla (Fig. 26-31C–E) or a set of darker stripes, streaks, or dots (Fig. 26-32).

Many insects perceive light at the ultraviolet end of the spectrum. Some flowers have patterns of reflection and absorbance of ultraviolet light, undetectable by the human eye, that often serve as nectar guides.

Many details of floral structure can be related to the form, size, and habits of particular insects, all of which influence the way they visit the flower. Entomophilous flowers have been segregated into a number of classes, partly according to their structure and partly according to their insect visitors. A few of these types of flowers are discussed here.

Pollen Flowers Lacking nectar, the pollen flowers have an abundance of pollen that is gathered for food, especially by certain bees and certain beetles. Pollen flowers include the poppy *(Papaver)* and the rose *(Rosa)*.

Hymenoptera Flowers Visited almost exclusively by bees, Hymenoptera flowers have a specialized form that permits the visit of only one particular kind of insect or of a few kinds of insects of about the same size. Some of these flowers have the nectar located so deeply in a tube that only an insect with a very long tongue, or **proboscis,** can reach it. In red clover *(Trifolium),* for example, the nectar is situated about 9 mm from the mouth of the flower; among the bees, only the bumblebees can reach it. In other flowers, the parts of the corolla are so rigid that only a heavy insect can open them (Fig. 26-33B). In markedly zygomorphic flowers, a landing stage for the insects is present along with nectar at the base of a long corolla tube, or at the base of **spurs.** In many such flowers, small flies and beetles may be prevented from emerging by hairs or scales at the throat of the corolla.

Lepidoptera Flowers Most of the Lepidoptera flowers are actinomorphic, and their insect visitors hover while feeding (Fig. 26-34). A common characteristic is a somewhat pungent, but **aromatic** scent.

Diptera Flowers Visited largely by flies, the Diptera flowers are generally less specialized than the previous two types. The starflowers *(Stapelia)* of South African deserts attract flies by an odor reminiscent of rotting meat; the flowers not only smell like but also resemble rotten meat (Fig. 26-22C). Female carrion flies are attracted to the blossoms, pollinate them, and are so completely fooled that they deposit their eggs (or larvae) on the perianth. Finding no food, the larvae perish.

Carrion flies are also attracted by the odor of the small European spotted arum lily *(Arum maculatum)*. What is often called the flower of this species is an inflorescence consisting of a **spadix,** bearing numerous flowers, enveloped by a large bract, the **spathe** (Fig. 26-31A, B). There are two sets of flowers: female at the bottom, and male above. The topmost flowers are sterile and modified into a set of bristles. The upper part of the inflorescence is the naked, reddish-brown extension of the inflorescence axis and is the source of a **fetid** odor. Small flies attracted to the spadix slide down the smooth surface of the spathe, past the male flowers to the female flowers below. Their attempts to escape up the spadix are thwarted by the bristly palisade of sterile male flowers, and they become trapped for the day. During this time, they crawl over the stigmata, dusting them with any pollen they may have already gathered. On the second day, the anthers of the male flowers dehisce, the bristles wilt, and the flies are free to leave and visit another inflorescence, continuing their role as pollinators. Dutchman's pipe *(Aristolochia)* has a somewhat similar trap, although only a single flower is involved rather than an inflorescence, and a gnat, rather than a fly, is caught.

Ornithophily

Various kinds of birds are important pollinating agents for a number of tropical and subtropical species of plants. Birds active in this respect are sunbirds, honeysuckers, and humming birds—some of which are no larger than moths. Bird flowers and insect flowers often resemble each other and are certainly more similar to one another than they are to wind-pollinated flowers. Bird flowers have bright, intense colors; reds predominate in more than 80% of 159 species studied. Brilliant color contrasts are common. Most bird flowers lack scent. B. J. D. Meeuse states that approximately 2000 species of birds belonging to about 50 families visit flowers more or less regularly; about two-thirds of these rely on flowers as their most important source of food. Since these bird visitors vary so much in size and feeding habits, there is no single type of flower that is particularly adapted to bird visits; however, some do have special features.

Anemophily

Most gymnosperms and other more generalized seed plants are characterized by anemophily, or wind pollination. Because of this predominance—almost to the exclusion of other types of pollination in the gymnosperms—wind pollination is commonly considered to characterize the first seed plants, although this is conjectural. The many cases of anemophily found among angiosperms are interpreted as having evolved from entomophilous flowers. One reason for this conclusion is the presence of anemophilous flowers in (or closely related to) many different families of angiosperms that normally have brightly colored entomophilous flowers. Another reason why anemophily is thought to be derived rather than a primitive characteristic is that some predominantly anemophilous plants are distantly related. Implicit in this reasoning is that if anemophily were primitive in the Mag-

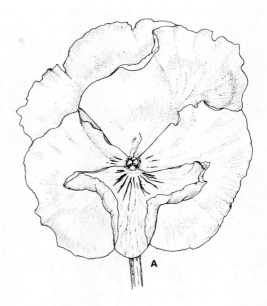

Figure 26-32 Honey guides. **A,** *Viola,* × 1; **B,** *Tropaeolum,* × 1.

Angiosperms (Flowering Plants) **629**

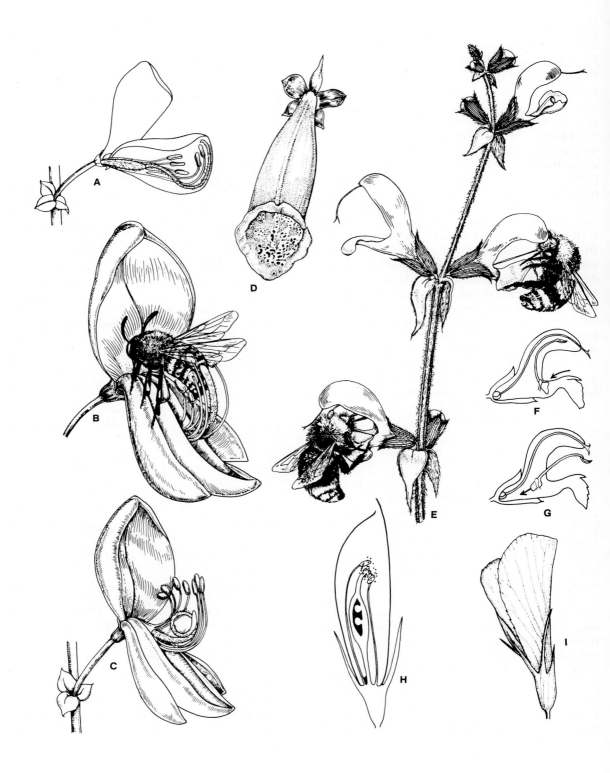

noliopsida, it would occur also in closely related groups.

All anemophilous flowers have certain features obviously associated with wind pollination. They are commonly aggregated in elongated inflorescences (Fig. 26-35A), which results in a concentrated release of pollen and a relatively large area where stigmata are concentrated. The anthers are often suspended from long filaments so that they hang free from the body of the flower (Fig. 26-35B, C). Such an arrangement allows the anther to swing in the wind and thereby release pollen to be caught up by air currents. The stigmata, like the stamens, are freely exposed. The many branches of the stigmata give them a feathery appearance and a large surface area for intercepting airborne pollen (Fig. 26-35B, C). The pollen grains are generally small and smooth, features that supposedly enhance their buoyancy in the air. They are also produced in enormous numbers, which compensates for the low probability that any single grain will contact a receptive stigmatic surface. Wind-pollinated flowers also commonly lack nectar, scent, or a brilliantly colored perianth.

Chiropterophily

Some tropical trees are pollinated by bats. Two distinct types of bats are involved. The Central and South American flower bats are related to **insectivores**, whereas those of Africa and Asia are related to fruit-eating bats. Bat flowers open at dusk, have copious nectar, are dull in color, and frequently have a mouselike odor that is apparently attractive to their visitors. As is to be expected, bat flowers are located so that bats can get at them easily (Fig. 26-36).

Hydrophily

Plants that grow immersed in water have particular problems in pollination, since they are denied access to normal pollinating agents. Such plants utilize water, in some fashion, as a pollinating agent. Some genera, such as *Ceratophyllum*, are monoecious; the ovulate flowers remain below the water level, and the staminate flowers shed their pollen into the water. The pollen is of the same specific gravity as the water and so is dispersed through it; some of the pollen eventually comes into contact with the submerged stigmata. A more complicated arrangement has evolved in the dioecious ribbon grass *(Vallisneria)*. The ovulate flower buds are initiated well below the surface of the water. As these mature, the pedicel becomes elongated and loosely coiled, eventually raising the flower to the surface of the water, where it floats (Fig. 26-37A, C). The flower then opens and rests, with its stigmas recurved, in a dimple on the water surface. The minute staminate flowers (Fig. 26-37B), only about 3 mm in diameter, are produced by the hundreds from a single inflorescence. Each consists of two stamens enclosed by a perianth of two large and one very small segment. The whole staminate flower is released under water and rises to the surface. Once it reaches the surface, it opens, and the perianth recurves to form a little **bipod** for the flower. The staminate flowers float until they come near the dimple caused by the pistillate flower (Fig. 26-37B). They are then drawn into the dimple by surface tension and effect pollination when they tip over and touch the stigmata of the ovulate flowers.

Ovule

The angiosperm ovule consists of the female gametophyte (megagametophyte), the **nucellus** in which the megagametophyte is embedded, and two (rarely

Figure 26-33 Hymenopterous flowers. **A–C,** *Sarcothamnus scoparius.* **A,** median section of flower showing stamens and style concealed within the keel, ×1. **B,** visit of heavy insect depresses keel and triggers explosive upward movement of stamens so they come in contact with belly and back of visitor, with stigma touching back, ×2. **C,** a "tripped" flower showing further coiling of style, ×1. **D,** *Digitalis purpurea,* ×1. **E–G,** *Salvia pratensis,* ×1: **E,** bumblebees visiting flowers; **F, G,** median sections of flowers, showing pivot mechanism at base of filament, which brings anther down on back of bee, ×1. **H, I,** *Trifolium*-type flower in sectional and side views, nectary shown at base of carpel, ×5. (**A–G,** redrawn and adapted from *The Story of Pollination* by B. J. D. Meeuse, copyright © 1961, John Wiley & Sons, Inc., New York; **H–I,** redrawn after Hagerup and Petersson.)

Figure 26-34 Lepidopterous flowers. **A,** *Anagraecum sesquipedale,* showing long, slender spur, ×1. **B,** *Lilium martagon,* showing hawk moth taking nectar from petal pouch, ×1. **C,** *Epiphyllum oxypetalum,* night-blooming cactus, ×1. (**B,** after Ross and Morin.)

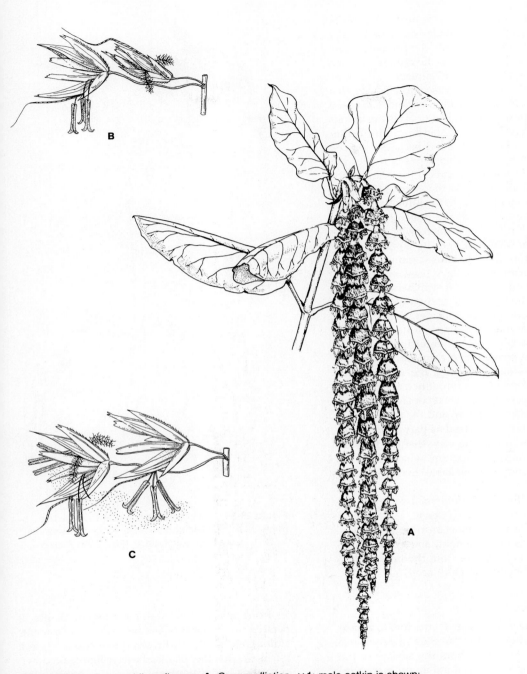

Figure 26-35 Anemophilous flowers. **A,** *Garrya elliptica,* ×1: male catkin is shown; female flowers are in similarly pendent catkins, × 1; **B, C,** *Avena* sp. flowers: **B,** spikelets on a calm day with glumes distended and anthers pendulous, × 1; **C,** spikelets in a wind, pollen escaping from the pendulous anthers on the right, while those on the left have shed their pollen, × 1: *gl,* glume. (**B, C,** after Kerner.)

one) investing structures, integuments (Fig. 26-2), around the nucellus. The ovule is attached to the ovary at the placenta by a stalk, the **funiculus** (Fig. 26-24A). Part of the funiculus (the **raphe**) is adnate to the ovule in some plants. The funiculus and raphe (when present) carry the vascular bundles that supply the ovule. Classification of ovules is based on the orientation of the funiculus. At the extremes of ovule type are those that stand upright (**orthotropous**) and those which bend back 180° on the funiculus (**anatropous**). Other types of ovules are intermediate, in varying degrees, between orthotropous and anatropous. At the apex of the ovule is a small, circular opening in the integuments, the **micropyle,** through which the pollen tube grows (Fig. 26-24A). Opposite the micropylar end of the ovule is the **chalazal** end, the site where the vascular supply from the funiculus commonly ends.

Many interpretations have been made of the angiosperm ovule. Some researchers view the nucellus and integuments as homologous to a megasporangium. Others consider the inner integument and the nucellus to constitute the megasporangium and consider the outer integument to be an investing structure. A third opinion is that the nucellus alone constitutes the megasporangium. In another view, an undefined layer of cells around the megagametophyte is considered to represent the megasporangium, and the remainder of the nucellus is not considered as part of the megasporangium. We prefer the view that the nucellus represents the megasporangium, which the integuments surround; the available fossil evidence supports this assumption.

The preceding comments refer to an angiosperm ovule in the broadest sense. However, in order to deal with the ovule in more detail, one must consider its development, since it is continually changing in structure from the time of initiation to its final maturation as a seed.

Figure 26-36 Bat pollinating flower of sausage tree *(Kigelia)*, ×0.5. (Adapted from *The Story of Pollination*, by B. J. D. Meeuse, copyright © 1961, John Wiley & Sons, Inc., New York.)

Nucellus In the ovule, the first structure to appear during development is the nucellus. It arises as a bump (Fig. 26-2A), either from the surface of the carpel or directly from the floral apical meristem. As it grows, any change in orientation results from a change in the growth patterns in the ovule.

The function of the nucellus is to furnish energy for the developing megagametophyte. Its fate varies from one plant to another. In some plants, it persists until relatively late in ovule development, although it eventually breaks down and is left as a thin tissue. In other plants, the nucellus breaks down very early in development.

Integuments The second structures to appear on the developing ovule are the integuments (Fig. 26-

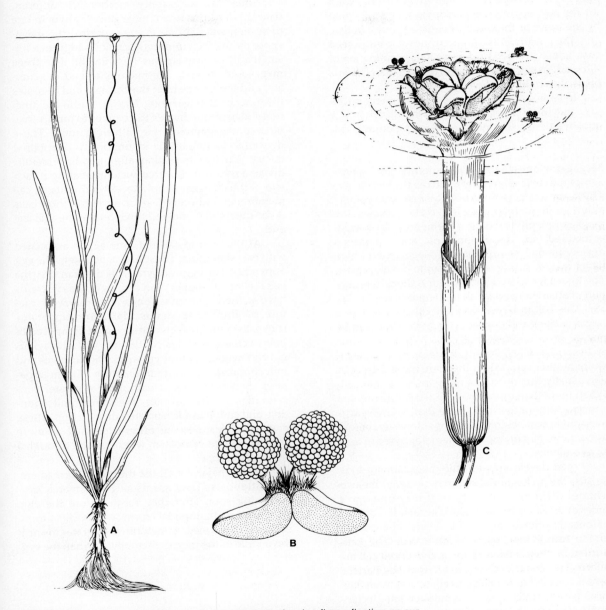

Figure 26-37 *Vallisneria americana*. **A,** female plant, showing flower floating on surface of water, ×0.2. **B,** male flower, ×5. **C,** boatlike male flowers coming in contact with stigmas, ×15. (**B,** after Gleason with permission of New York Botanical Garden.)

2B). They arise first on one side of the ovule and spread around to the other side. Two integuments, the inner and the outer, arise one after the other; the inner usually arises first. Once the integuments have been initiated, they grow faster than the nucellus and eventually completely invest it, except for the micropyle. The micropyle may be formed by the inner or the outer integument or by both.

The integuments commonly develop into a **seed coat,** which protects and/or inhibits germination of the seed. In some seeds, however, the integuments break down during development.

In some plants that have an ephemeral nucellus, the inner integuments are modified into a nutritive tissue, the **integumentary tapetum.**

Megagametophyte In angiosperms, the megagametophyte is often referred to as the *embryo sac*. This term was borrowed from animal embryology; although it is inaccurate (at certain stages the megagametophyte bears no embryo), it is well-embedded in discussions of angiospermous developmental morphology. Megagametophyte development begins early in ovule development. The first stage in its development is the differentiation of an **archesporial cell** in the nucellus (Fig. 26-38). This cell is larger than the others and has a larger nucleus and denser cytoplasm. There can be one or more archesporial cells. When only one archesporial cell is present, it functions as a **megaspore mother cell.** When there are more than one, one usually functions as a megaspore mother cell. When more than one archesporial cell functions as a megaspore mother cell, more than one megagametophyte will begin to develop. However, before maturation, all but one megagametophyte are usually eliminated.

When development of the megagametophyte begins, the archesporial cell may undergo meiosis without dividing mitotically, or it may undergo a series of divisions before meiosis occurs. If it divides mitotically prior to meiosis, it forms cells exterior to the future megaspore mother cell. The cells formed by the division of the archesporial cell are referred to as **parietal.** An ovule that has parietal cells is referred to as **crassinucellate.** An ovule that lacks parietal cells is called **tenuinucellate.** In tenuinucellate ovules, the megaspore mother cell is usually immediately beneath the nucellar **protoderm.** Occasionally, the nucellar protoderm undergoes a series of periclinal divisions to form a **nucellar cap** over the tip of the nucellus. When an ovule has such a cap, the megaspore mother cell is buried several cell layers below the surface of the ovule. Such an ovule is called **pseudocrassinucellate.**

Once the megaspore mother cell differentiates, it develops into a megagametophyte in one of several ways. In most angiosperms, the megaspore mother cell undergoes meiosis to form a linear tetrad of **megaspores** (Fig. 26-38). The three megaspores nearest the micropylar end degenerate, and the one nearest the chalazal end becomes the functional megaspore. The nucleus in the functional megaspore divides three times to form a free-nuclear megagametophyte with eight nuclei. These divisions do not result in uniformly distributed nuclei. Nuclear migration after the first meiotic division results in four free nuclei appearing at each end of the megagametophyte. After this stage, cell membranes and walls develop and form three cells at the micropylar end and three cells at the chalazal end.

At the micropylar end, one egg is associated with two **synergids.** These are, collectively, the **egg apparatus.** The egg and synergids differ in size (the egg is generally larger) and cytology. The synergids have at their micropylar ends a specialized structure, the **filiform apparatus.** The filiform apparatus is usually a highly convoluted region composed of polysaccharides and is continuous with the synergid and megagametophyte walls. Dictyosomes and mitochondria are common near the filiform apparatus. Synergids also usually have a greater concentration of dictyosomes, ribosomes, mitochondria, and endoplasmic reticulum than do eggs. These cytological features are often expressed as more heavily stained cytoplasm when viewed through a light microscope.

At the chalazal end, the three cells formed are the **antipodals.** They are usually ephemeral, disappearing soon after they form. Before the egg apparatus and antipodals form, one nucleus from each end of the megagametophyte migrates toward the middle of the megagametophyte. When the egg

---

Figure 26-38 Important types of female gametophytes in flowering plants. (After Maheshwari.)

Type	Gynosporogenesis			Gynogametogenesis			Mature female gametophyte
	Gynospore mother cell	Division I	Division II	Division III	Division IV	Division V	
Monosporic 8-nucleate Polygonum type							
Monosporic 4-nucleate Oenothera type							
Bisporic 8-nucleate Allium type							
Tetrasporic 16-nucleate Peperomia type							
Tetrasporic 16-nucleate Penaea type							
Tetrasporic 16-nucleate Drusa type							
Tetrasporic 8-nucleate Fritillaria type							
Tetrasporic 8-nucleate Plumbagella type							
Tetrasporic 8-nucleate Plumbago type							
Tetrasporic 8-nucleate Adoxa type							

apparatus and antipodals form, a cell membrane also appears around the cell containing these two nuclei to form a large, bi-nucleate central cell. The nuclei in the central cell are the **polar nuclei** and are ultimately involved in the formation of **endosperm,** the nutritive tissues of the angiosperm ovule. As a matter of interest, the polar nucleus that comes from the micropylar end is a sibling of the egg nucleus.

Megagametophyte development in angiosperms varies. Figure 26-38 presents a chart that shows the different types of megagametophyte development. The variation exists in (1) the number of meiotic products involved in the formation of the megagametophyte (one, two, or four); (2) the number of cell divisions noted in development up to the stage when the megagametophyte is ready for fertilization; (3) the number of antipodals; (4) the ploidy level of the polar nuclei; and (5) the position of the functional megaspore in the megaspore tetrad. In spite of this developmental variation, two cells are always present: the egg and the central cell. It is interesting to note that a plant that lacks synergids, *Plumbago* (leadwort), has a filiform apparatus in the egg.

Anther

Microgametophytes are produced in the anthers. Early in anther development, the primordia are more or less four-cornered structures as viewed in transverse section. In each corner, the cells in the layer just below the protoderm enlarge and develop larger nuclei. These enlarged cells form an **archesporium** consisting of one to several cells. After differentiation, each cell of the archesporium divides to form a **primary parietal cell** to the outside and a **primary sporogenous cell** to the inside. The primary parietal cells undergo perclinal divisions to produce a layer three to five cells thick, including the protoderm. This layer will develop into the anther wall. When the anther wall is mature, the epidermis is stretched and lost. The outer layer of the anther wall, which originally was the layer below the epidermis, consists of cells with fibrous bands. This layer is called the **endothecium** and is involved in anther dehiscence. In flowers with anthers that do not dehisce, or in which dehiscence is by a pore rather than a slit, no endothecium develops. Internal to the endothecium are middle layers that are crushed by the time meiosis occurs in the developing anther. The innermost wall layer is the tapetum, the source of nutrients and energy for developing microspore mother cells. There are two types of tapetum: glandular (or secretory), in which the cells remain in place, and amoeboid, in which cells break down and "wander" around in the developing microspores.

Microgametophyte The primary sporogenous cells undergo meiosis to produce the tetrads of **microspores.** Before undergoing meiosis, they may or may not divide several times. During meiosis, two types of cytokinesis can occur: simultaneous or successive. In the former, cytokinesis is delayed until after both meiotic divisions and then occurs as a result of the cytoplasm pinching in from the outside. In the latter, cytokinesis occurs after each meiotic division and does so by the formation of a cell plate in the center of the cell; the dividing membrane develops from the middle of the cytoplasm to the wall (Fig. 26-4D). After meiosis, most tetrads are either **tetrahedral** or **isobilateral.**

Meiosis occurs while the flower is still in the bud stage. The microspores break apart after meiosis or, sometimes, are shed as a tetrad. After meiosis, each microspore nucleus divides once to produce a large tube cell and a smaller generative cell (Fig. 26-3 and 26-4), and a characteristic wall is deposited around the microgametophyte. After the wall is deposited, the microgametophyte is commonly called a pollen grain, and this structure is shed from the anther.

When a pollen grain reaches a compatible stigma, it germinates and sends out a **pollen tube,** an extension of the inner wall layer of the pollen grain (Fig. 26-24). The pollen tube grows down through the stigma and style to the locule of the ovary. The pollen tube usually grows within a specialized tissue in the style, the **stigmatoid tissue.** In the locule, the pollen tube grows to the micropyle of the ovule, and fertilization (see following paragraph) is effected. At some time between pollen formation and fertilization, the **generative cell** divides to form two sperm cells.

Fertilization

The process of fertilization is an event of considerable biological significance. In angiosperms, fertilization involves the egg apparatus and the central cell. The pollen tube enters the embryo sac at the base of one synergid, the penetrated synergid (Fig. 26-4A, B). The reaction of this synergid varies. The synergid degenerates in some plants, and the pollen tube enters the base of the degenerated synergid. In other plants, the penetrated synergid does not degenerate until after the pollen tube enters. There have been reports that the penetrated synergid does not degenerate at all. The control of the direction of pollen tube growth seems to involve the filiform apparatus. In all cases, the pollen tube enters the synergid at the filiform apparatus; in one plant without synergids (*Plumbago capensis*), the egg has a filiform apparatus, and the pollen tube enters there. It is speculated that as a result of its convoluted nature, the filiform apparatus offers a greater surface area through which pollen-tube-directing growth substances can diffuse.

Once within the synergid, the pollen tube opens and discharges its contents, two sperms and some cytoplasm. The sperms are cells consisting of nuclei and some cytoplasm, the whole structure being surrounded by a cell membrane. The functional part of the sperm cell is the nucleus. One sperm nucleus enters the egg and fuses with the egg nucleus to produce the zygote. The other sperm nucleus fuses with the nuclei in the central cell to form the primary endosperm nucleus. This phenomenon is called **double fertilization.** The primary endosperm nucleus ultimately develops into the endosperm, the nutritive tissue of the angiosperm seed.

Endosperm The endosperm usually begins to develop before the embryo. There are three different types of endosperm in angiosperms: **free-nuclear, cellular,** and **helobial.** Free-nuclear endosperm undergoes a series of nuclear divisions without the development of cell wall (Fig. 26-4C, D); eventually cell walls form (Fig. 26-4E, F), resulting in cellular endosperm. In helobial endosperm, the division of the primary endosperm nucleus is accompanied by cytokinesis to form a chalazal and micropylar cell. In each of these cells, the nuclei undergo free-nuclear division. There are usually more divisions in the micropylar chamber than in the chalazal chamber. The cells at the chalazal end are usually crushed after a period of time, and the endosperm in the micropylar part behaves like nuclear endosperm. In some plants, the endosperm forms **haustoria** from either the micropylar or chalazal end, or both (Fig. 26-4E, F). Usually haustoria are found in plants with cellular endosperm.

Embryo The embryo develops from the zygote. Many types of embryos have been described in plants, based on the pattern of cell division noted in the earliest stages of embryo development. No specific type of embryo is restricted to monocotyledons or dicotyledons; similar patterns of early development are found in both dicotyledons and monocotyledons. For example, *Poa annua* (annual bluegrass) has an embryo that is classed as **asterad**; it undergoes the same type of development as that in members of the Asteraceae, a dicotyledon family. An outline follows of the embryo types

I. The zygote divides by a wall which is either vertical or so obliquely oriented that it is essentially longitudinal. Piperad type; named after *Peperonia* of the Piperaceae.

II. Zygote divides by a transverse wall to give a basal (toward micropyle) cell and a terminal (away from micropyle) cell.
   A. The first division of the terminal cell is longitudinal.
      1. The basal cell contributes little or nothing to the embryo. Onagrad type; named after the Onagraceae, in which it is common.
      2. The basal cell contributes to the embryo. Asterad type; named after the Asteraceae, in which it is common.
   B. The first division of the terminal cell is transverse.
      1. The basal cell contributes little or nothing to the embryo.
         a. Basal cell becomes large suspensor without division. Caryophyllad

type; named after the family Caryophyllaceae.
   b. The basal cell forms a suspensor of two or three cells. Solanad type; named after the family Solanaceae.
2. The basal cell contributes to the embryo. Chenopodiad type; named for the Chenopodiaceae.

Despite these differences, some features are common to most angiosperm embryos. Early in development, a short filament is produced; then at the chalazal end of this filament, a globular structure appears. This globular structure then develops the cotyledons (seed leaves), shoot apical meristem, and root apical meristem.

Differences between monocotyledon and dicotyledon embryos begin to appear at the globular stage. The apex of the globule in a dicot develops two cotyledons with the shoot apical meristem in between them. The apex of the monocot globule develops into the single cotyledon and then a shoot apical meristem appears laterally.

The differences between monocotyledon and dicotyledon embryos are **organographic**: differences in the pattern and appearance of the complex structures, such as cotyledons and shoot apices (Fig. 26-39).

Asexual Reproduction

Asexual reproduction is not especially common in angiosperms but assumes great significance where it is present. For example, many weeds reproduce asexually and hence are difficult to eradicate. Many horticultural plants are also asexually propagated in an attempt to maintain genetic uniformity.

Asexual reproduction is of two types, vegetative reproduction and **apomixis** (or **agamospermy**). Vegetative reproduction refers to asexual reproduction that involves plant parts other than the gynoecium: for example, the stolons of *Fragaria* (strawberry); the rhizomes of *Agropyron repens* (quack or couch grass); the "eyes" of *Solanum tuberosum* (potato), which are actually small buds; or the production of plantlets on the leaves of *Kalanchoe*. A type of vegetative reproduction that resembles apomixis occurs in the grass *Poa bulbosa*, in which the flower is replaced by a plantlet.

Apomixis is asexual reproduction that involves seemingly normal formation in the gynoecium. There are two general types of apomixis: nonrecurrent (occurring once), and recurrent (occurring more than once, generation after generation). In nonrecurrent apomixis, a cell in a normal haploid megagametophyte develops into an embryo that gives rise to a haploid plant. The embryo may develop directly from the egg or from other cells of the megagametophyte. The former situation is called **haploid parthenogenesis** and is found in *Solanum* (nightshades, potato). The latter is called **haploid apogamy** and has been reported in *Lilium* (lily).

Recurrent apomixis is of different types, depending on the origin of the embryo. In one type, the embryo forms from a megagametophyte that has two possible origins: an archesporial cell that has not undergone meiosis or a nucellar cell. The former is called generative apospory and is seen in *Hieracium* (hawkweed); the latter is called somatic apospory and is found in *Eupatorium* (eupatorium). In these two types of apomixis, the embryo forms from the egg (**diploid parthenogenesis**) or from a cell other than the egg (**diploid apogamy**). In the other general type of recurrent apomixis, the embryo forms from either nucellar or integument cells. The megagametophyte may or may not form and may be $1n$ or $2n$. This type of apomixis is called **adventive embryony** or sporophytic budding and is found in *Citrus* (orange, lemon, grapefruit).

Some plants are **facultative apomicts**: they reproduce sometimes by apomixis and sometimes by sexual means. These are of special interest to taxonomists and evolutionists, since they produce variation patterns that are difficult to assess. Apomictic reproduction results in large numbers of very similar plants, which gives the impression that these populations represent an independent species. Periodic sexual reproduction produces many new variants, each of which can reproduce apomictically and give rise to many clusters of plants. Each cluster is uniform and apparently distinct from others, and yet they tend to intergrade.

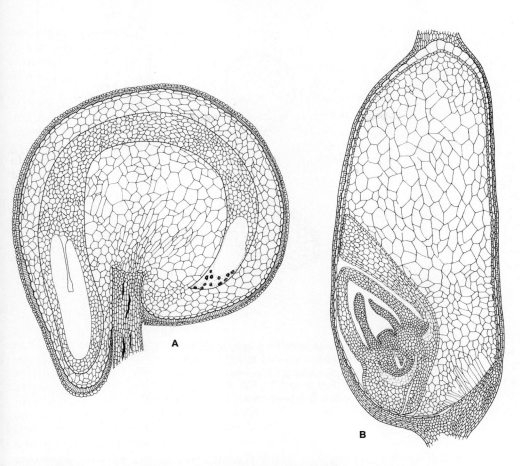

Figure 26-39 Longitudinal sections of seeds. **A,** a dicotyledon, *Beta vulgaris*, ×50. **B,** a monocotyledon, *Sorghum* sp., ×30. (After Eames.)

Other Considerations of the Reproductive Structures

Studies of megagametophyte, endosperm, and ovule development are sources that contribute to a significant understanding of reproduction and add greatly to our knowledge of relationships among angiosperms. Embryology provides very convincing evidence that the Empetraceae (crowberry family) are closely related to the Ericaceae (heath family); such evidence was also used in placing the Garryaceae in the same order as the Cornaceae (dogwood family). The structure of the mature embryo has proven to be very valuable in studies of relationships in the Poaceae (grass family). Undoubtedly further studies of all aspects of embryology will contribute greatly to an understanding of relationships. It is possible that the development of the embryo proper also offers valuable information. Unfortunately, studies of embryo development are currently out of vogue in botany, and much valuable information is simply not being gathered.

The study of pollen grains, the science of palynology, likewise assumes significance beyond its

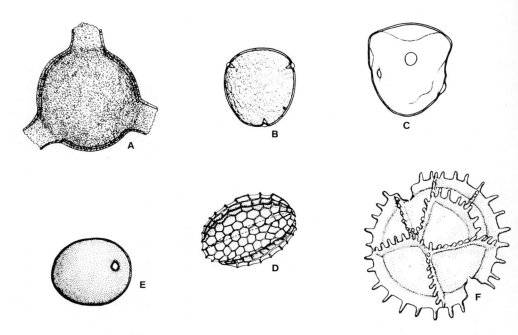

Figure 26-40 Angiosperm pollen grains. **A,** *Epilobium angustifolium* (Onagraceae), ×400. **B,** *Fagus sylvatica* (Fagaceae), ×600. **C,** *Carex* sp. (Cyperaceae), ×800. **D,** *Salix sitchensis* (Salicaceae), ×1600. **E,** *Bromus inermis* (Graminae), ×500. **F,** *Taraxacum* sp. (Compositae), ×1200.

role in reproduction. The wall of a pollen grain has two properties that make it valuable for many purposes: it can occur in many different yet consistent ornamentation patterns, and it is highly resistant to decay. Because wall ornamentation and structure are distinctive in some plants, it is possible to identify them from their pollen. Because it is resistant to decay, fossil pollen retains its distinctive features for millions of years. Hence, **palynologists** can reconstruct past vegetational patterns from the pollen record, and this information helps in recognizing past climatic patterns, since there is a marked correlation between vegetation and climate. Furthermore, oil companies make extensive use of palynological data to identify rock strata in which oil deposits may occur.

From the many constant and characteristic forms of pollen grains, palynology has also offered clues concerning taxonomic relationships. Palynologists who work closely with paleobotanists have been able to present a more comprehensive picture of the nature of ancient angiosperms.

Pollen grains are of numerous types (Fig. 26-40). Pollen grains may be globose, ellipsoid, or **fusiform.** Superimposed on this shape are further features: **porate** and **colpate.** Porate grains have from one to many pores in the outer covering, the exine. Colpate grains have one to several furrows, or **colpi** (sing. **colpus**), in the exine that run from one pole of the pollen grain to the other. Many angiosperms have grains with both colpi and pores and are called **colporate.**

Many botanists consider the primitive pollen type for angiosperms to be the **monocolpate** type with its single colpus on the distal side, that is, the side away from the point of contact with other pollen grains of the same tetrad. Such pollen grains characterize the monocotyledons, most of the Ranales, and are found also in cycads, bennettites, and some pteridosperms (see Fig. 27-31).

Embryo

The embryo represents the earliest stage in the new sporphyte and develops from the zygote. Within the seed, it is usually in an arrested stage of development, which persists until the seed germinates, and the embryo begins to grow. The embryo consists of **cotyledons, epicotyl, radicle** and **hypocotyl** (Fig. 26-41C, F, H). The cotyledons are the embryonic leaves and serve one or more functions. In some plants, cotyledons serve as storage organs (beans; Fig. 26-41A–C); in others they serve as photosynthetic structures (cabbage); and in still others, they are structures that absorb the food reserves in the seed (grasses). In some plants, such as *Ricinus* (castor bean), cotyledons serve as absorbing structures while still within the seed, and when they emerge from the seed, they function in photosynthesis (Fig. 26-41D, F).

The epicotyl or **plumule** is above the cotyledonary node and consists of rudimentary leaves and the apical meristem. It is this structure that grows into the shoot system of the plant.

The radicle (embryonic root) is at the end of the embryonic axis opposite the epicotyl and forms the root system in most plants. In others (e.g., grasses), the root that develops from the radicle is short-lived, and the root system of the plant consists of adventitious roots that arise above the radicle.

The part of the embryonic axis located between the cotyledons and the radicle is the hypocotyl. This is the region that first begins to grow at germination and forms the connecting link between the root and stem. The primary vascular tissues of roots and stems differ markedly (Chapter 12). At the base of the hypocotyl, the vascular tissue is similar to that found in the root. At higher levels in the hypocotyl, the rootlike pattern gradually changes into a stemlike pattern at the top of the hypocotyl. The region where this change in pattern of vascular tissue occurs is called the **transition region**.

Before the seed germinates, the cells of the embryo present a cytological picture that has been interpreted to indicate inactivity. The features seen are (1) the presence of reserve protein and lipids; (2) poorly defined organelles, with some critical ones possibly absent; and (3) lack of membrane

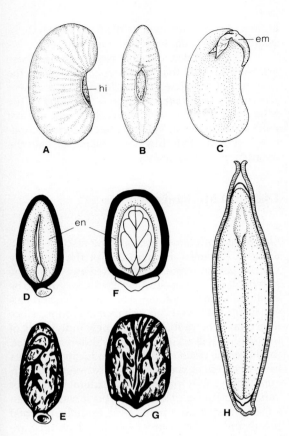

Figure 26-41 Seeds of flowering plants. **A–C**, lima bean (*Phaseolus lunatus*), ×2: **A**, lateral view, with hilum on right; **B**, ventral view, with hilum in center; **C**, with one cotyledon removed, showing embryo. **D–G**, *Ricinus communis*. **D, F,** longitudinal sections of **E** and **G**. **D, E,** lateral view, embryo embedded in endosperm; **F, G,** dorsal view; **H,** squash (*Cucurbita maxima*), longitudinal section, micropylar end uppermost, × 4; *em*, embryo; *en*, endosperm; *hi*, hilum. (Redrawn from *The Plant Kingdom* by William H. Brown, used by permission of the publisher, Ginn and Company (Xerox Corporation).)

Seeds

Seeds are the end product of fertilization and develop from the ovule in the ovary. In angiosperms, as in other seed-bearing plants, the seed consists of an embryo (the new sporophyte), some nutritive tissue, and a covering, the seed coat, or **testa** (Fig. 26-41). Each of these is considered in turn.

clarity. When germination occurs, several changes occur in the cells of the embryo: (1) reserve material is lost, (2) organelle and membrane clarity increase, (3) organelles appear that were lacking in the dormant stage, and (4) structures develop that were present in a highly modified state in the dormant embryo.

Nutritive Tissue

In most angiosperms, the nutritive tissue develops from the primary endosperm nucleus and is referred to as the **endosperm.** There are, of course, exceptions. Orchids have neither endosperm nor any other nutritive tissue. In the pea family (Fabaceae), there is no endosperm, and cotyledons have taken over the function of food storage. In one plant, *Bertholletia excelsa* (Brazil nut), the stored food reserves are found in the hypocotyl. Still another source of stored food is the nucellus, which undergoes considerable growth. This storage tissue is called **perisperm.** The nutritive tissue of angiosperm seeds is of particular significance to humans. Most of the food value we obtain from the cereal grains (wheat, oats, rice, rye, barley, and corn) is in the endosperm contained in the seeds of these plants. Seeds that have an endosperm are called **albuminous**; seeds that lack an endosperm are called **exalbuminous.**

Testa

Angiosperm seeds are covered with a testa, or seed coat. In most plants the testa is derived from the integuments. It may develop from both integuments, from parts of both integuments, from one integument, or from part of one integument. There are even cases in which the testa is not derived from the integuments; sometimes, it develops from the nucellus, and the integuments disappear. In rare cases, the testa develops from the outer layer of the endosperm. In the cactus *Opuntia*, the protective covering of the seed develops from an outgrowth of the funiculus. Sometimes the testa does not develop at all, or develops weakly. For example, in the Poaceae and Asteraceae, the usual functions of the testa are taken over by the fruit wall; in these plants, the testa does not disappear entirely.

Despite different developmental histories, there are two features that characterize all testas: the presence of a **cuticle** and a sclerified layer. These features are associated with two functions of the testa: protection against desiccation and against mechanical damage. The testa has also been implicated in two other functions: dispersal and the maintenance of dormancy.

Dispersal Mechanisms

One aspect of plants that has long fascinated botanists is dispersal mechanisms, or the means by which plant propagules are carried away from the parent plant. Dispersal allows the propagules to be excluded from the influence of the established parent plant. Dispersal can enhance evolution, since it increases the probability that new individuals of one species will occupy a slightly different habitat. Seeds are the result of sexual reproduction and thus involve a genetic recombination between two parents; hence, they present a considerable array of variants. Because of dispersal, these variants can fall into new environments away from the parental plants and may encounter one to which they are better adapted.

Dispersal has another advantage in that it removes propagules from each other's influence. When propagules are dispersed from the parent plant, they are more widely scattered than if they were not. When they germinate and begin to grow, there is less competition among more thinly dispersed seedlings than among tightly packed ones.

Wind Dispersal There are two major types of adaptations for wind dispersal of seeds. One is the presence of hairs on the seeds, in the Malvaceae (mallow family), *Epilobium* (fireweeds), and *Asclepias* (milkweeds) (Fig. 26-42F). The other major adaptation for wind dispersal is apparent in the Orchidaceae, the orchid family, in which the integument is thin and made up of inflated cells. This structure of the integument, along with the very reduced form of the embryo and lack of endosperm, result in a seed that has very high buoyancy in air.

Animal Dispersal Animal dispersal is usually associated with a structure on the seed that serves as an attractant (source of food) to an animal. This food source may be a fleshy integument, as in *Magnolia*, the fleshy outgrowth of an integument, funiculus, or both (such an outgrowth is called an **aril**), as in *Viola* (violet). Many of these structures attract birds, which eat the aril and attached seed, or ants which eat the aril, but leave the seed undisturbed. In seeds that are ingested, the testa of the seed protects it in the digestive tract, and the seed is ultimately excreted. The distance of dispersal from the parental plant is often great.

Water Dispersal In the genus *Nymphaea* (water lily), the aril is an air-filled sac. This sac keeps the seed of *Nymphaea* afloat and hence serves as a means of dispersal for this water plant.

Seed Dormancy

Dormancy is the physiological state during which the embryo in a seed will not grow and germinate, even though environmental conditions may be amenable to germination. Dormancy may result from extra-hard seed coats, the lack of oxygen, the necessity for a chemical change (referred to as afterripening), or the presence of growth inhibitors. Regardless of the mechanism of dormancy, it is significant for several reasons. First, it allows a seed to survive adverse conditions. Second, it prevents seed germination until favorable conditions have persisted long enough to assure at least some survival of the seedlings. Third, it prevents seed germination during the time of dispersal. Fourth, it helps to maintain a population by ensuring that not all the seeds of one species germinate in one year, all to be eliminated by a common catastrophe. The breaking of dormancy may require (1) a period of cold temperature, (2) a period of cold and a period of warm temperature, (3) rotting or other means of removal of the seed coat, or (4) removal of growth inhibitors.

The duration of seed viability is associated with dormancy; it enables the seed to withstand the condition of dormancy for long periods of time. The time that seeds remain viable varies from a brief period in such genera as *Quercus* (oak), *Acer* (maple), *Populus* (poplar), *Salix* (willow), *Citrus* and some grasses, to hundreds of years in other cases. It is reported that seeds of *Nelumbo* and *Albizzia julibrissen* that had been in the British Museum for 150 years germinated after becoming wet during the Battle of Britain. The record is apparently held by the lotus, *Nelumbo nucifera*. Seeds of this species were found by a Japanese botanist in a peat deposit in an old lake bed in Manchuria. Radiocarbon dating showed the seeds to be about 1000 years old, although other evidence suggests an even greater age. When the seed coats were filed to permit water to enter, every one of them germinated.

Correlations between Seeds and Habitat

Salisbury (1942) has carried out some very extensive investigations on the seed size of plants that grow in various kinds of habitats. Generally speaking, he discovered that plants that grow in closed habitats (forests) have larger seeds than those found in more open habitats. This caused him to speculate that the larger seeds in closed habitats would allow seedlings to survive under conditions of reduced light intensity until they grew large enough to support themselves with their own photosynthetic apparatus. In open habitats, light is less of a limiting factor. The seedlings produced from small seeds would have an immediate source of light for photosynthesis and would depend less on stored energy for growth. Small seeds are much more effectively dispersed than larger ones, which gives their plants a selective advantage in open habitats: they are more capable of reaching more potential sites of seed germination. This is important because open areas tend to become closed with the passage of time. Some plants of open habitats have large seeds; in these, the large seed serves as a source of energy for the establishment of a large root system.

Salisbury notes exceptions to these generalizations that also illustrate selective advantages. Parasites and saprophytes tend to have small seeds, regardless of the kind of environment in which they live. Such plants do not need large food reserves in seeds, since the seedling from the time of its germination extracts nutrition from the material on which it is growing. In the heath family (Ericaceae) and orchid family (Orchidaceae), for example, there

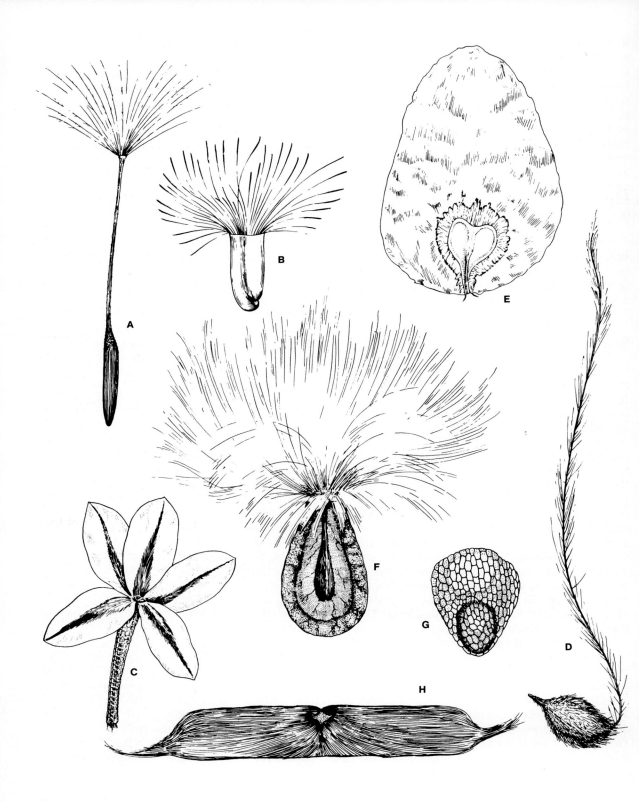

are plants that form mycorrhizal associations with a fungus and derive from it the energy necessary for the developing seedling. Evidence is being accumulated that the angiosperm in a few mycorrhizal associations is, in fact, parasitizing the fungus. Some parasites, including *Cuscuta* (dodder), have large seeds, but *Cuscuta* must survive as an independent plant for a period of time before the host-parasite relationship is established.

Fruits

Fruits, like seeds, are usually the end result of fertilization. Unlike seeds, however, fruits can be produced in which the seeds are inviable and usually not formed. Such production of fruits is called **parthenocarpy**. Several familiar parthenocarpic fruits are bananas, seedless oranges, and seedless varieties of grapes (Fig. 26-43).

The term **fruit** has been loosely applied; one often hears, incorrectly, of the "fruit" on a gymnosperm. Like many biological units, fruits defy a rigorous definition. In botany, the term *fruit* is defined as a plant structure that develops from the ovary (and sometimes associated structures) after fertilization, encloses the seeds, and is found only in angiosperms.

To facilitate communication and understanding, the many kinds of fruits are classified into various types. The features used in classifying fruits are

1. The number of ovaries and flowers involved in the production of the fruit
2. The nature of the ovary wall, the pericarp (fruit wall); when mature, whether dry or fleshy
3. When fleshy, the nature of the three layers of **pericarp**, the **exocarp** (outer layer), **mesocarp** (middle layer), and **endocarp** (inner layer); which of these are fleshy and which, if any, are stony
4. Where pericarp is dry, whether the fruit opens by a predefined opening (is **dehiscent**) or remains closed (is **indehiscent**) at maturity
5. The number of carpels involved in the fruit
6. Whether the ovary is superior or inferior
7. Whether or not extra ovary tissue is involved.

In classifying fruits, three main types can be recognized. The first group is comprised of **simple fruits** formed from a single, one-to-many carpellary ovary. The second type includes **aggregate fruits**, which are formed from a single flower that has several free carpels; each carpel forms a fruit, but the whole aggregates as a unit, as in raspberry (*Rubus*), strawberry (*Fragaria*), and *Magnolia*. In the third type, **multiple fruits**, the ovaries of several flowers form fruits, and these occur as a unit, as in flowering dogwood (*Cornus nuttallii*), plane tree or sycamore (*Platanus*), fig (*Ficus*), and pineapple (*Ananas*).

In multiple and aggregate fruits, there are two different components to consider: the entire unit, and the individual fruits that make it up. For example, a strawberry consists of a fleshy receptacle into which many small **achenes** are embedded. A raspberry, on the other hand, consists of many small **drupes**. The fruit of *Ficus* (fig) is made up of a fleshy, urn-shaped inflorescence that bears many small achenes on the inside. In the pineapple, the axis of the inflorescence becomes part of the fruit. Table 26-2 gives some concept of the diversity of fruit types.

When applying names to different kinds of fruits, botanists use various criteria. Some researchers emphasize the nature of the mature fruit and place little emphasis on its ontogenetic origin or the taxonomic group to which it belongs. Other botanists are more explicit, and use fruit names that carry ontogenetic and taxonomic implications. For example, the **nut** of *Quercus* (oak) is distinguished from the nut of *Tilia* (basswood), since the former is derived from an inferior ovary and the latter from a superior ovary; the **berry** found in some Liliaceae is distinguished from the berry found in some Rosaceae, since they occur in widely different taxonomic groups; and the **capsule** of the Orchidaceae is considered to be different from that

Figure 26-42 Airborne fruits and seeds. **A–D**, fruits: **A**, *Hypochaeris*, ×5; **B**, *Centaurea*, ×5; **C**, *Ursinia*, ×5; **D**, *Clematis*, ×5. **E–H**, seeds: **E**, *Spathodea*, ×5; **F**, *Asclepias*, ×5; **G**, *Castilleja*, ×20; **H**, *Catalpa*, ×2.

of the Primulaceae, since they differ in ontogenetic origin (the orchid fruit is derived from an inferior ovary), and these two families are very distantly related. Kaden and Kirpicnikow (1965) have proposed a system of fruit nomenclature that takes into account the ontogeny and taxonomic relationships within which various fruit types occur. This classification recognizes 84 fruit types. There is much merit in proposals such as those of Kaden and Kirpicnikow, but a simpler system of fruit classification, based primarily on the nature of the pericarp, is more useful to a greater range of people.

Biologically, fruits are of tremendous significance; they offer some measure of protection to the developing seeds, and surround the seeds with sterile tissue that is sometimes modified to aid in their dispersal. It may be argued that dispersal is a more important factor in the evolution of flowering plants than their many and ingenious adaptations for pollination. Apparently few species are deprived of pollination by the absence of an essential pollinator; on the other hand, the migration of plants to new areas is highly dependent on modifications for dispersal. In view of the undoubted biological importance of dispersal and of the variety of related adaptations that have evolved in the flowering plants, it is surprising that this study has not attracted the attention of more investigators. The encyclopedic work of Ridley (1930) should be consulted by anyone wishing to pursue this topic further.

Dissemination by Wind

Tumble-weeds The whole plant, or the fruiting portion, breaks off and is blown by wind across open country, scattering seeds as it goes. Such plants are herbaceous and are usually annuals.

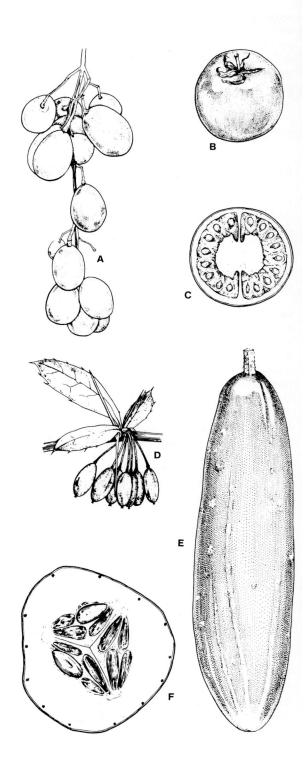

Figure 26-43 Berries. **A,** grape *(Vitis)*, × 1. **B, C,** tomato *(Lycopersicum esculentum)*, × 0.5. **D,** barberry *(Berberis)*, × 1. **E, F,** cucumber *(Cucumis sativus)*, a special type of berry known as a pepo; **E,** × 0.5, **F,** × 1.

Table 26-2 A Classification of Fruits

I. Fruits dry, pericarp dry when fruit is mature.
   A. Indehiscent fruits in which the pericarp remains closed at maturity.
      1. *Achene*—a small single-seeded fruit with a relatively thin pericarp; except for its attachment by the funiculus, the seed lies free in the cavity of the ovary: buttercup *(Ranunculus),* crowfoot *(Potentilla),* water plantain *(Alisma),* buckwheat *(Fagopyrum).*
      2. *Caryopsis*—achenelike, except that it is derived from a compound ovary, and the seed coat is firmly united to the pericarp: characteristic fruit of the Poaceae (grass family).
      3. *Cysela*—achenelike but derived from an inferior, compound ovary: characteristic fruit of the Asteraceae (sunflower family).
      4. *Nut*—like an achene but derived from two or more carpels, with a hard or stony pericarp: hazel nut *(Corylus),* basswood *(Tilia),* acorn *(Quercus).*
      5. *Samara*—a winged achene: elm *(Ulmus),* ash *(Fraxinus),* tree-of-heaven *(Ailanthus),* hop tree *(Ptelea).*
      6. *Schizocarp*—the product of a compound ovary that splits apart at maturity into a number of one-seeded portions termed *mericarps:* maple *(Acer),* many Apiaceae, Laminaceae, Malvaceae, Geraniaceae.
   B. Dehiscent fruits in which the fruit splits or opens in some manner to release the seeds.
      1. *Follicle*—derived from a single carpel that splits at maturity down one side, usually along the ventral suture: columbine *(Aquilegia),* peony *(Paeonia),* larkspur *(Delphinium),* milkweed *(Asclepias).*
      2. *Legume*—also from a single carpel, but dehiscing down both the dorsal and the ventral sutures to form two valves: characteristic fruit of Fabaceae (pea family; Papilionaceae).
      3. *Capsule*—from a compound ovary; various types of dehiscence are found, e.g., longitudinal, porous, or circumscissile; in general the dehiscence is from top downward, and the separated portions (valves or teeth) remain attached: characteristic of numerous families.

II. Fruits fleshy, pericarp partly or wholly fleshy or fibrous.
      1. *Drupe*—carpels one or more but usually single-seeded; mesocarp fleshy but endocarp hard and stony: cherry, peach, plum *(Prunus),* coconut *(Cocos),* olive *(Olea).*
      2. *Berry*—one to several carpels, usually many-seeded; both mesocarp and endocarp are fleshy. One-seeded: nutmeg *(Myristica),* date *(Phoenix).* Single carpel and several seeds: baneberry *(Actaea),* barberry *(Berberis).* Others with more than one carpel: grape *(Vitis),* tomato and potato *(Solanum), Asparagus.*
      3. *Pome*—derived from a compound inferior ovary; much of the fleshy portion is the enlarged base of the perianth tube, with only the central part composing the pericarp; both exocarp and mesocarp are fleshy, while endocarp (the core) is stony or cartilaginous: characteristic fruit of the Pomoideae, apple *(Malus),* pear *(Pyrus),* quince *(Cydonia),* mountain ash *(Sorbus).*

Winged Fruits and Seeds Wings on fruits may be formed by persistent bracts, perianth parts or may form the pericarp. Winged fruits are most common in trees and shrubs such as maple *(Acer)* and dipterocarps *(Dipterocarpus;* Fig. 26-42) and ash *(Fraxinus;* Fig. 26-44B). Winged fruits are found also on some herbaceous plants of open areas, e.g., *Halogeton.* Wings on seeds also offer a means of wind dispersal. Examples include *Spathodea* (Fig. 26-42E), *Castilleja* (paint brush; Fig. 26-42G), and *Catalpa* (Fig. 26-42H).

Plumed Fruits and Seeds Two different morphological types are found in plumed fruits: fruits in which the pericarp bears the plume, and fruits borne in an accessory structure that, in turn, bears the **plume.** Examples of the latter are found in the Asteraceae (sunflower family) (Fig. 26-42A, B). Seeds bearing plumes are found in *Asclepias* (milkweed; Fig. 26-42F).

Dispersal by Water

Rain Rainwash in the plains, both in temperate and tropical regions, is of great importance in dispersal. The action of rain is accentuated in the humid mountain forests of the tropics. Periodically, the rush of water from mountains is tremendous. At

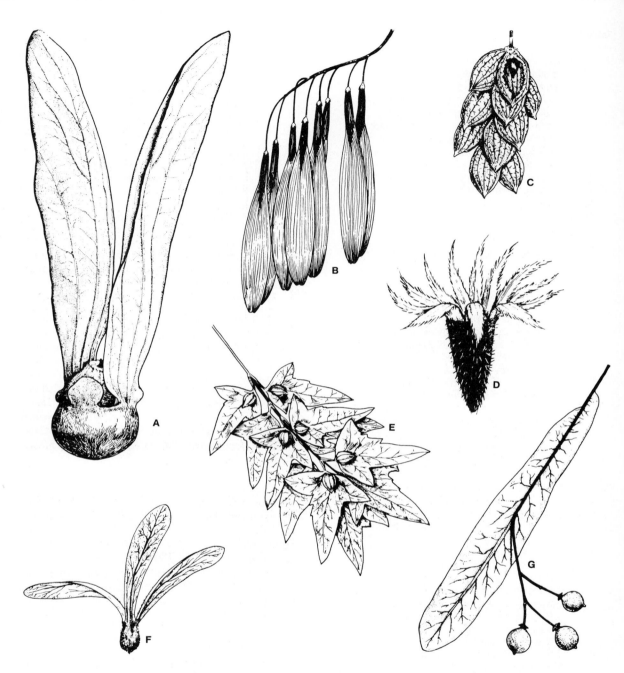

Figure 26-44 Winged fruits. **A,** *Dipterocarpus retusus* (the two wings are part of persistent calyx), ×0.5. **B,** *Fraxinus,* samara, ×1. **C,** *Ostrya virginiana,* with each fruit contained in inflated, involucral sac, ×1. **D,** *Galinsoga,* with cypsela fruit and wings of pappus scales, ×50. **E,** *Carpinus caroliniana,* with nutlike fruit subtended by single, involucral bract, ×1. **F,** *Triplaris,* with wings of elongated perianth lobes, ×1. **G,** *Tilia,* with petiole adnate to a large, strap-shaped bract, ×2.

gold *(Caltha)* and several aquatic Apiaceae that have corky ribs in the pericarp.

Sudd The dense masses of living vegetation that grow out from river banks and often block the channels are called **sudd**. The whole mass, or a portion thereof, may be torn loose by a sudden flood and carried downstream to new sites or into a lake. Sudd is usually found in slow-moving rivers of flat open country, such as the Nile or Ganges.

Oceans Ocean currents disperse plants over very long distances, sometimes over 1000 miles. There are two conditions that are imposed on plants that are ocean dispersed. First, the seeds must be able to float for very long periods of time without becoming waterlogged and without germinating. Second, the seeds must be adapted to establishing themselves on sand or mud banks, the kind of habitat in which they are most likely to be deposited. A good example of ocean dispersal is the coconut *(Cocos)*.

Dispersal by Animals

Dispersal by animals may be either internal or external. The former is referred to as **endozoochory**, the latter as **exozoochory**.

Dispersal by Ingestion Birds and mammals eat fleshy fruits of many kinds and pass the seeds through the alimentary tract unharmed. Sometimes germination is enhanced in seeds that have passed through the alimentary tract. It appears that the majority of fruits with a fleshy pericarp are adapted to this mode of dissemination (Fig. 26-43). Another type of dispersal by internal transport is illustrated in plants that produce small, hard fruits, such as nuts and achenes.

Adhesion to Fur and Feathers The adaptations in these cases are well known and consist of the development of prickles, hooks, spines, hairs, and sticky or viscid fruits or other structures (Fig. 26-46). The whole inflorescence in many fruits and other structures that are adapted for dispersal by adhering to animals also creates problems for live-

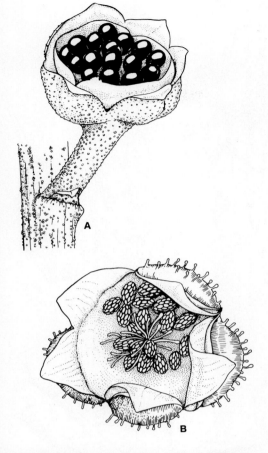

Figure 26-45 "Splash-cups." **A,** mitrewort *(Mitella),* ×5. **B,** pearlwort *(Sagina),* ×5. Under damp conditions, seeds are splashed out of capsules by rain. (**B,** after Gibbs.)

such times, vast numbers of **disseminules** are carried down the mountain slopes and strewn in the plains. In a few cases, as in pearlwort *(Sagina)* and mitrewort *(Mitella)* (Fig. 26-45), the open capsule with seeds lying in it is very similar to the "splash-cup" found in many fungi and may function in the same way.

Streams The fruits of many streamside and aquatic plants are adapted for floating for short periods of time. Examples of these include the marsh mari-

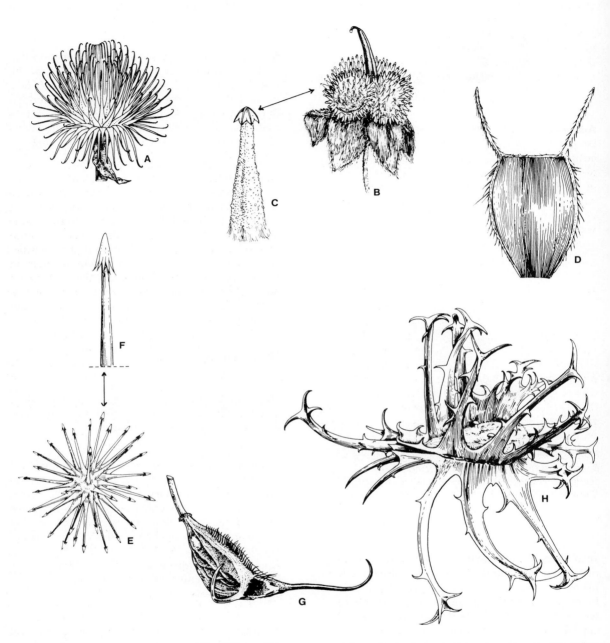

Figure 26-46 Animal-dispersed fruits. **A,** *Arctium minus,* involucral bracts with stiff, hooked tips, ×2. **B,** *Cynoglossum officinale,* nutlets covered with short, barbed spines, ×2. **C,** single spine of *Cynoglossum,* ×20. **D,** *Bidens,* pappus represented by two stiff, barbed awns, × 10. **E,** *Acaena,* with sepals sharp and spiny, × 7.5. **F,** *Acaena,* showing grapnellike tips of sepaline spines, × 25. **G, H,** capsular fruits; **F,** *Proboscidea,* × 0.5; **G,** *Harpagophytum,* × 0.5.

stock ranchers. Such structures can cause an animal to become sick and lose weight (weight means money to a rancher). They may become embedded in the wool of sheep and decrease its value, and they have been known to cause the death of animals by working their way into the body cavity.

Adhesion of Small Disseminules A large number of plants probably owe their distribution to the adherence of disseminules to the fur, feathers, or feet of passing animals. These disseminules are generally picked up in mud and show no particular adaptations aside from their small size. This form of dispersal is especially likely for marsh and other water plants.

Man as a Dispersal Agent

In modern times, people have assumed an important role as disseminators of seeds. Most weeds of the world have been transported by people, either intentionally or accidentally. Intentional introduction is usually associated with agriculture or horticulture. Accidental introduction usually results from weed disseminules being carried inadvertently by travelers. They may be carried as impurities in agricultural crops, on the feet, on cars, or as ballast in ships. As the efficiency of modern transportation increases, people become more efficient as dispersal agents over great distances.

Mechanical Dispersal

Some plants throw seeds away from themselves. In *Impatiens* (touch-me-not), for example, the fleshy capsules are often dilated somewhat at the end where the seeds are borne. Upon maturation, the fruit wall develops a high turgor pressure in specialized cells, the pericarp separates from the ovary, and the valves of the fruit become only slightly coherent. Then the valves separate violently, either spontaneously or as a result of motion or touch, and very rapidly curl upwards (Fig. 26-47D), throwing the seeds a great distance. *Oxalis* has a similar mechanism for dispersal.

In many legumes the dehiscence of fruits is related to a hygroscopic mechanism. The different layers of the pericarp contract to different extents as they dry out, so that considerable tensions are set up between them. When these tensions overcome the cohesion of the cell walls, the walls suddenly separate. Often, the valves of the legume flick into a spiral (Fig. 26-47A) and frequently eject the seeds a meter or more; the separation is usually violent and rapid.

Quite a different kind of dehiscing mechanism is found in the squirting cucumber (*Ecballium elaterium*), a trailing herb of Mediterranean regions. The fruit is an oblong berry about 5 cm in length, borne on a stout pedicel. The apex of the pedicel projects into the base of the fruit like a stopper. When the fruit is ripe, the tissues around the stopper break down, and its connection with the fruit becomes loosened. The seeds, by this time, are surrounded by a mass of semiliquid mucilage. A considerable turgor then builds up within the fruit, distending the outer layers and placing them in a state of tension. Finally the tissue at the base of the pedicel completely disintegrates and is blown out like a cork (Fig. 26-47C). At that time, the walls of the fruit contract and discharge the juicy contents, including the seeds, to a considerable distance.

Long-range Dispersal

Many puzzling distributional patterns show the same or closely related species to occur on widely separated land masses. One such pattern is exemplified by *Carex magellanica* that occurs in the north and south temperate zones but not in intervening areas. Distribution patterns of a somewhat similar nature are to be seen on oceanic islands and seem best explained as a result of long-distance dispersal, possibly by migratory birds.

This is not to imply that long-distance dispersal explains all unusual distributional patterns. Other possibilities include continental drift, the growth and disappearance of mountain ranges, and the breakup of a once continuous distribution.

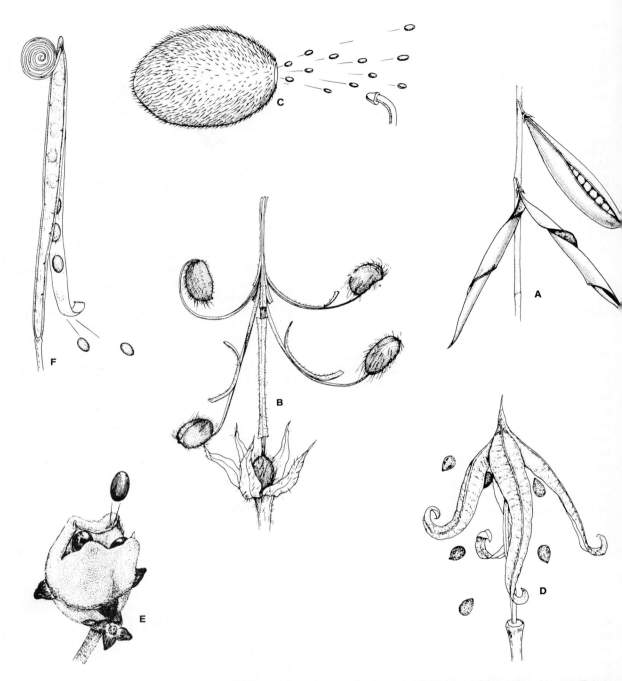

Figure 26-47 Examples of mechanically dispersed seeds. **A,** *Vicia gigantea* (Fabaceae), ×1. **B,** *Geranium bicknellii* (Geraniaceae), ×4. **C,** *Ecballium elaterium* (Cucurbitaceae), ×0.5. **D,** *Impatiens aurella* (Balsaminaceae), ×6. **E,** *Hamamelis virginiana* (Hamamelidaceae), ×3. **F,** *Cardamine hirsuta* (Cruciferae), ×4. (**C,** after Kerner von Marilaun and Oliver.)

Change in Dispersability

A feature of great evolutionary significance is an apparent adaptation that results in a change in the mechanism of dispersability. Plants that occupy restricted habitats with unique ecological conditions are usually well adapted to these conditions, and an efficient means of dispersal would transport **propagules** to sites for which the plants were not adapted. Plants growing on oceanic islands, for instance, might suffer (in the evolutionary sense) if their disseminules were dispersed into the ocean. Carlquist (1974) has presented several examples of apparent changes in dispersability among plants that occur on oceanic islands in the Pacific Ocean.

Classification of the Angiosperms

Since very early times, people have attempted to classify the variations observed in angiosperms. The classifications produced have varied with the level of knowledge about plants—a corollary of available scientific technology—and with the general objectives that influence the process of classification, as dictated by prevailing philosophies of biology and current interpretations of angiosperm relationships.

The earliest published angiosperm classifications were based on growth habit: herb, shrub, or tree. As levels of technology and knowledge increased, the structure of flowers and other reproductive organs became primary criteria for classification. At the present time, taxonomists consider as many features as possible when they erect classifications, ranging from biochemistry and chromosome morphology to features of vegetative anatomy, floral anatomy, floral and vegetative morphology, and the surface structure of various plant organs as viewed with a scanning electron microscope.

Objectives, general philosophies, and interpretations of angiosperm relationships have also varied among taxonomists over time. Before Darwin's *Origin of Species* was published, it was generally believed that biological diversity was the result of Divine Creation. Classification schemes developed during this time had two purposes: (1) to understand God's plan and (2) to design a filing system for the many different organisms recognized. Many of the earliest classifications were artificial, in that they used few features (e.g., number of stamens) to characterize major groups. In spite of that, some of the taxa recognized in these earliest classifications were natural. Before the publication of Darwin's theory of evolution, classification schemes were based on more characteristics than were earlier schemes, and many of the taxa recognized in these classifications are still recognized. With the publication and ultimate acceptance of Darwin's theory of evolution, taxonomists attempted to incorporate evolutionary principles into their classifications. This attempt did not drastically alter any of the preceding classifications but simply resulted in evolutionary explanations to support taxa and the construction of their presumed phylogenies.

The construction of phylogenies is based on theories of angiosperm evolution. Since the publication of Darwin's work, there have been two major schools of thought concerning the evolution of angiosperms. Engler (1904) and some other European botanists considered the angiosperms to have evolved from the conifers. Hence, the angiosperms considered to be primitive had structures similar to those of the conifers. Their flowers would occur in clusters, lack a perianth, and be imperfect. Among the angiosperms, the oaks (*Quercus*), alders (*Alnus*), willows (*Salix*), and cattails (*Typha*), among others, fit this description. The other major school of thought, the one accepted by most contemporary botanists, is that angiosperms evolved from pteridospermous ancestors, and the angiosperm flower is a modified stem that bears modified leaves. This interpretation of the flower is known as the *classical flower theory*, which considers the most primitive angiosperm flower to be one that most closely resembled a stem with leaves. It would have an elongate receptacle and many free flower parts arranged in a spiral. The perianth would not be differentiated into a calyx and corolla, and the stamens and carpels would be laminar. Since such flowers are found among the extant Ranales, many researchers consider this taxon to resemble closely the primitive stock from which all other angiosperms evolved.

Much attention has been directed toward an attempt to understand the interrelationships among angiospermous orders and families. A number of researchers have proposed a series of interrelated evolutionary trends, and on this basis have constructed a scheme that suggests particular interrelationships among the major taxa of angiosperms. To support these schemes further, the researchers have incorporated biochemical, cytological, and genetic data. The difficulties encountered have been considerable, and the resulting systems, although usually similar in general terms, often differ drastically in particulars. Furthermore, the detailed data available remain limited to a very restricted sample from the known angiosperms, and even the descriptive stage in this accumulation of information remains very incomplete. Until the body of data available becomes reasonably comprehensive, an unstable scheme of interpretations of interrelationships will persist.

The nature of the flower has been basic to all systems of classification that attempt to reflect interrelationships. A careful assessment of the evidence of the fossil record should supply a picture of the fundamental generalized floral structure from which the modern diversity of types was derived. David Dilcher (1979) has presented an analysis of the available fossil data. He has pointed out the serious shortcomings encountered because the fossil material is very fragmentary and the general morphology and biology of the whole plants that bore specific reproductive organs cannot be reconstructed with any degree of confidence. Furthermore, the data available are extremely limited, and it is premature to draw any broad generalizations.

It appears that the angiosperms were derived from Mesozoic pteridosperms (see Chapters 21 and 27; Fig. 27-30). Since this group is so broadly defined, and contains so many apparently unrelated elements, this provides a rather shadowy beginning for the angiosperms. Furthermore the time of this radiation and the mechanisms that controlled the evolution remain uncertain. The fossil record bears evidence of both **unisexual** and **bisexual** flowers. Among these fossils, there is no suggestion that one particular type of receptacle is the most primitive. Magnolialean flowers, long considered to be the basic type of primitive flower, appear to be only one of several primitive types, and certainly not the most primitive. Both anemophilous and entomophilous flowers appear to have developed in the early stages of angiospermy. Although recent work indicates anemophily developed somewhat later, from these lineages both anemophily and entomophily led to diverse radiant evolutionary lines, within each of which both entomophilous and anemophilous descendants were derived. It should be pointed out that Dilcher's (1979) results are not in agreement with those of Doyle and Hickey (1976), who offer support for the opinion that the Magnoliales are closest to primitive angiosperm stock based on a study of leaves and pollen (see Chapter 27, p. 691).

In summary, the angiosperms are a highly diverse group of plants occupying a wide range of habitats. They are important, indeed, for continued human existence and have been the subject of both intensive and extensive research. Despite this interest in the angiosperms, our knowledge is still very small compared to our ignorance.

References

Axelrod, D. I. 1952. A theory of angiosperm evolution. *Evolution*, 6:29–60.

———. 1961. How old are the angiosperms? *Am. J. Sci.* 259:447–59.

———. 1970. Mesozoic paleogeography and early angiosperm history. *Bot. Rev.* 36:277–319.

Bailey, I. V., and Swamy, B. G. L. 1951. The conduplicate carpel of dicotyledons and its initial trends of specialization. *Am. J. Bot.* 38:373–79.

———. 1954. *Contribution to plant anatomy.* Chronica Botanica Co., Waltham, Mass.

Beck, C. B., ed. 1976. *Origin and early evolution of angiosperms.* Columbia Univ. Press, New York.

Brown, W. H. 1935. *The plant kingdom.* Ginn & Co., Boston.

Carlquist, S. 1974. *Island biology.* Columbia Univ. Press, New York.

Constance, L. 1955. The systematics of angiosperms. In *A century of progress in the natural sciences, 1853–1954.* California Academy of Science, San Francisco.

Corner, E. J. 1964. *The life of plants.* Cleveland World Publishing Co.

Cronquist, A. 1961. *Introductory botany.* Harper & Row, New York.

———. 1968. *The evolution and classification of flowering plants.* Houghton-Mifflin Co., New York.

———. 1981. *An integrated system of classification of flowering plants*. Columbia University Press, New York.

Darwin, C. 1859. *The origin of species*. John Murray (Publishers) Ltd., London.

Davis, P. H., and Heywood, V. H. 1963. *Principles of angiosperm taxonomy*. Oliver & Boyd, Edinburgh.

Dilcher, D. L. 1979. Early angiosperm reproduction: an introductory report. *Rev. Palaeobot. and Palynol.* 27:291–328.

Douglas, G. E. 1957. The inferior ovary. II. *Bot. Rev.* 23:1–46.

Doyle, J. A., and Hickey, L. J. 1976. Pollen leaves from the mid-Cretaceous Potomac Group and their bearing on early angiosperm evolution. In Beck, C. B., ed., *Origin and Evolution of Angiosperms*. Columbia University Press, New York and London.

Eames, A. J. 1961. *Morphology of the angiosperms*. McGraw-Hill Book Co., New York.

Engler, A. 1904. *Syllabus der Pflanzenfamilien*. Gebrüder Borntraeger, Berlin.

Esau, K. 1953. *Plant anatomy*. John Wiley & Sons, New York.

Faegri, K., and van der Pijl, L. 1979. *The principles of pollination biology*. Pergamon Press, London.

Foster, A. S., and Gifford, E. M., Jr. 1974. *Comparative morphology of vascular plants*, 2nd ed. W. H. Freeman and Co., San Francisco.

Gibbs, R. D. 1950. *Botany, an evolutionary approach*. Blakiston Co., Philadelphia.

von Goethe, J. W. 1970. *Versuch die Metamorphose der Pflanzen zu erklären*. Gotha.

Grant, V. 1971. *Plant speciation*. Columbia Univ. Press, New York.

Gustafsson, A. 1946. Apomixis in the higher plants. I. The mechanism of apomixis. *Lunds Univ. Årsskr.* 42:1–66.

Heywood, V. H. ed. 1978. *Flowering plants of the world*. Mayflower Books, New York.

Hutchinson, J. 1973. *The families of flowering plants*, 3rd ed. Vols. 1 and 2. Oxford University Press, London.

Johansen, A. 1950. *Plant embryology*. Chronica Botanica Co., Waltham, Mass.

Jones, S. G. 1939. *Introduction to floral mechanisms*. Blackie and Son, London.

Kaden, N. N., and Kirpicnikow, M. E. 1965. A possible contemporary system of fruit terminology. *Taxon*. 14:218–23.

Kaplan, D. R. 1973. The problem of leaf morphology and evolution in the monocotyledons. *Quart. Rev. Biol.* 48:437–57.

Lam, H. J. 1963. Taxonomy: general principles—angiosperms. In Turrill, W. B. (ed.), *Vistas in botany*. Pergamon Press, London.

Lawrence, G. H. M. 1951. *Taxonomy of vascular plants*. Macmillan Co., New York.

McLean, R. C., and Ivimey-Cook, W. R. 1956. *Textbook of theoretical botany*. Vol. 2. Longmans, Green & Co., London.

Maheshwari, P. 1950. *An introduction to the embryology of the angiosperms*. McGraw-Hill Book Co., New York.

Meeuse, A. D. J. 1966. *Fundamentals of phytomorphology*. The Ronald Press Co., New York.

Meeuse, B. J. D. 1961. *The story of pollination*. Ronald Press Co., New York.

Melville, R. 1962. A new theory of the angiosperm flower. I. The gynoecium. *Kew Bull.* 16:1–50.

Pijl, L. van der. 1969. *Principles of dispersal in higher plants*. Springer Verlag, Berlin.

Puri, V. 1952. Floral anatomy and inferior ovary. *Phytomorphology*, 2:122–29.

Radford, A. E. et al. 1974. *Vascular plant systematics*. Harper & Row, New York.

Ridley, H. N. 1930. *The dispersal of plants throughout the world*. L. Reeve and Co., Ashford, Kent.

Salisbury, E. J. 1942. *The reproductive capacity in plants*. G. Bell & Sons, London.

Sporne, K. R. 1974. *The morphology of angiosperms*. Hutchinson Univ. Library, London.

Stebbins, G. L., Jr. 1974. *Flowering plants: evolution above the species level*. Harvard Univ. Press, Cambridge, Mass.

Takhtajan, A. L. 1969. *Flowering plants: origin and dispersal*. Oliver & Boyd, Edinburgh.

Taylor, T. N., and Millay, M. A. 1979. Pollination biology and reproduction in early seed plants. *Rev. Palaeobot. and Palynol.* 27:329–55.

Walker, J. W., ed. 1975. The bases of angiosperm taxonomy. *Ann. Missouri Bot. Garden.* 62:515–16.

Went, F. W. 1957. *Control of plant growth*. Chronica Botanica Co., Waltham, Mass.

The various groups described in preceding chapters illustrate the diversity within plants of such main features as size, morphology, anatomy, cytology, genetics, physiology, biochemistry, ecology, and reproductive processes. As stressed throughout, variations in structure and function of modern plants are the end result of evolutionary change occurring through at least 3 **aeons**, or **Ga** ($= 10^9$ years). During this time, the main events in botanical evolution occurred, such as the formation of prokaryotes, evolution of eukaryotes, syngamy and meiosis, alternation of generations, and adaptations to terrestrial habitats.

Early Evolution

The origin and evolution of prokaryotes and early eukaryotes is linked intimately with events that occurred during the earth's geologic history. Recent estimates of the earth's age range from 4–6 Ga. Relative to this, the earliest fossils, resembling bacillar bacteria (Fig. 27-1) and coccoid bluegreen algae, have been found in rocks in South Africa dated at approximately 3 Ga. By contrast, the first apparent eukaryotes have been found in rocks of the late Precambrian, about 1.3 Ga old. Thus the fossil record demonstrates clearly that prokaryotes were the earliest organisms to evolve on earth and supports the belief that eukaryotes evolved from prokaryotes prior to the late Precambrian interval.

An interesting and informative model showing main events in the evolution of primitive earth and early life forms is presented in Figure 27-2. Drafted by Preston Cloud (1976), a leading student of the Precambrian, it shows some of the main events between 4.6 and 0.7 Ga. One significant event proposed by Cloud was a period of gravitational heating after the origin of the earth. This could have resulted from frictional heat generated by tides produced when the moon was captured by the earth's gravitational field, from meteorite showers, from the natural production of heat by radioactive decay, or from a combination of these factors.

In the interval between approximately 4.5 and 3.5 Ga, the earth's atmosphere most likely contained gases such as H_2O, CO_2, CO, N_2, SO_2 and HCl, but little or no molecular oxygen. This was the period of **biogenesis,** leading to the evolution

27

EVOLUTION AND PHYLOGENY

of the first prokaryotes, including the first **autotrophs**. We do not know whether the first prokaryotes were more like bacteria or bluegreen algae, as both occur together. Presumably the early prokaryotic forms were anaerobic and heterotrophic, and derived their energy from fermentation. Once the photosynthetic process had become established in the first **photoautotrophs**, at around 3.5 Ga, oxygen would have been produced. As indicated in Figure 27-2, oxygen would have been dispersed first into the hydrosphere; it would have taken a long time to establish atmospheric oxygen and the prerequisite ozone layer in the outer atmosphere.

During this same time interval, beds of banded iron formation (*BIF* in Fig. 27-2) are found extensively in sediments. Banded iron consists of alternating layers of iron-rich and iron-poor silica. It has been suggested that this alternation resulted from periods when ferrous iron (unoxidized) was deposited in sites of bluegreen algal growth, where it was oxidized to ferric iron by oxygen that the algae released. This process lasted until about 2.0 Ga, at which time there was a marked change in deposition from banded ironstones to red beds. This appears to have marked the time when sufficient oxygen had been produced and accumulated in the hydrosphere that it began to escape to the atmosphere. Here it would have produced a high rate of oxidation in detrital terrestrial and marginal marine sediments, resulting in the extensive red beds of the later Precambrian. The buildup of an oxygenated atmosphere and ozone layer appears to coincide with the origin of eukaryotes; many believe these to be two of the main factors in eukaryotic evolution.

A prominent feature in the Precambrian is the presence of **stromatolites**. These are domed or columnar formations in carbonate rocks and cherts, particularly well developed in the Pongola and Bulawayan rock series of South Africa (Fig. 27-2). Bluegreen algal cells have been found in some stromatolites, and from the study of modern stromatolitic analogs, such as those from beaches in Australia (Fig. 27-3), it is generally believed that most Precambrian stromatolites were formed by the growth of carbonate-secreting bluegreen algae.

It appears that most morphological and (by inference) physiological features of bluegreen algae had evolved by 2.0 Ga. Any additional evolution in later times appears to have been at or below the family level, although the latest group to evolve, the Stigonematales, did not appear until the Devonian Period, about 370 Ma (= years × 10^6).

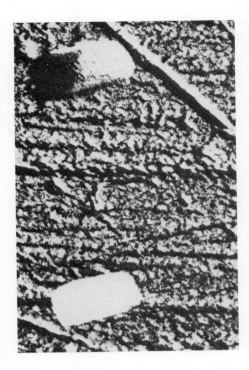

Figure 27-1 Electron micrograph of bacillar bacteria from the Precambrian of South Africa. (From E. S. Barghoorn and J. W. Schopf, "Microorganisms Three Billion Years Old from the Precambrian of South Africa," *Science* 152:758-63, 6 May 1966, with permission of the authors and the American Association for the Advancement of Science. Copyright 1966 by the American Association for the Advancement of Science.)

The relative conservatism in the evolution of bluegreen algae may be explained by their physiological flexibility, as evidenced by their wide ecological tolerance, and their genetic stability. Being photosynthetic and requiring few nutrients, they can exist in relatively impoverished ecological niches. Their genetic stability may be due to the presence of several sets of genetic material; these provide protection against mutational changes that would be expressed if they had only a single set of genes. Furthermore, being completely asexual, they lack the means of increasing variability that results from

Evolution and Phylogeny

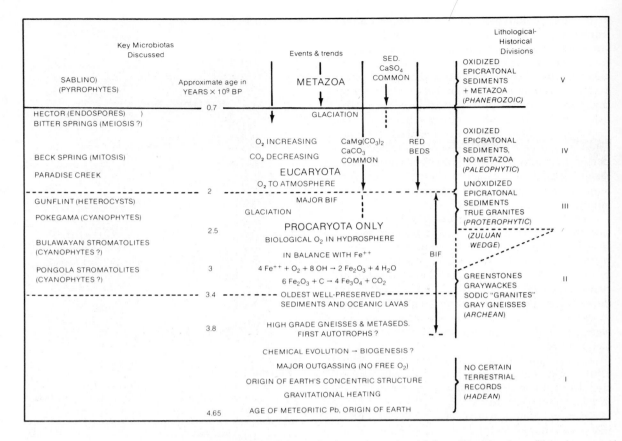

Figure 27-2 Proposed scheme for early evolution of the biosphere, lithosphere, and atmosphere: *BIF,* banded iron formation. (From P. Cloud, 1976, with permission of the author and the Paleontological Society, publishers of *Paleobiology.*)

meiosis and syngamy in eukaryotes. The net result of these factors has been a relatively slow rate of evolutionary change in bluegreen algae.

The record of the bacteria is so meager that no evolutionary trends can be identified. The presence of bacillar forms at 3 Ga, together with bacterialike filaments and rodlike structures in the mid-Precambrian, suggest a diversity of form in early periods. Bacteria have also been credited with the formation of some of the extensive iron ore deposits around the western Great Lakes in the late Precambrian. Recently, Woese (1981) has proposed a completely new lineage, the Archaeobacteria, which resemble the Eubacteria morphologically but differ radically in their biochemistry.

Evolution of Eukaryotes

As mentioned in Chapter 3, there is a sharp evolutionary break between the prokaryotes and the eukaryotes that first evolved some time in the middle to late Precambrian. The essential features that set eukaryotes apart so distinctly are the double-membrane-limited organelles, including nucleus,

Figure 27-3 Stromatolitic formation on a beach at Shark Bay, Western Australia. (Courtesy of Australian Information Service.)

mitochondria, and photosynthetic plastids, together with differentiated **endoplasmic reticulum, golgi bodies, lysosomes, microtubules, microbodies,** and **flagella.** These differences in structure are so sharp that many evolutionists and taxonomists believe the division of living organisms into these two groups to be more fundamental and essential than distinguishing green plants, animals, fungi, and protists into separate kingdoms!

Although there is no direct proof from either fossil or living members, the general belief is that eukaryotes evolved from prokaryotic ancestors. Without direct evidence as to how this came about, two main schools of thought have developed to account for the likely evolutionary pathways. Until the 1960s, most researchers adhered to the "autogenous" school. According to several **autogenous** hypotheses, the various membrane-bounded organelles of eukaryotes arose from step-by-step changes in the nonbounded photosynthetic, respiratory, and genetic units of the prokaryotic cell. In this progression, there was gradual compartmentalization, condensation, and specialization of the functional units until the discrete membrane-bound organelles of eukaryotes were evolved (Fig. 27-4A). This involved the engulfing of the func-

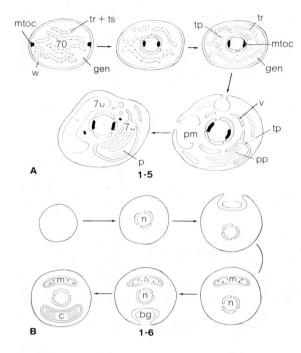

Figure 27-4A 1–5, hypothetical model of autogenous evolution of a eukaryotic cell from a prokaryotic ancestor by condensation, compartmentalization, and specialization. **1,** a prokaryotic cell showing a wall (*w*), thylakoid lamellae performing both photosynthesis and respiration (*tr + ts*), several areas of genome material (*gen*), and early stages of microtubule organizing centers (*mtoc*). **2,** a later stage, showing some of the genome material concentrated into a nucleoidlike mass surrounded loosely by membranes and containing the microtubule organizing centers. **3,** still later stage, showing a nuclear envelope around the pronucleus, and the specialization of membranes into photosynthetic thylakoid (*tp*) and respiratory thylakoids (*tr*): note the continued presence of cytoplasmic genome. **4,** a stage in which the cell wall has been lost; the membrane is undergoing a series of endocytoses, forming vacuoles (*v*), and surrounding some of the cytoplasmic genomes with membranes to form protoplastids (*pp*) and protomitochondria (*pm*). **5,** compartmentalization of organelles is essentially as in the eukaryotic cell, with nucleus (*n*), mitochondria (*m*) and plastids (*p*) delimited by membranes. (After F. J. R. Taylor, 1976, with permission of the author and the International Association of Plant Taxonomists.)

tional units by a process known as **cytosis.** In such a scheme, mitochondria would have evolved by invaginations of the **plasma membrane** encircling cytoplasmic DNA particles. Similarly, the chloroplasts of eukaryotic cells would have evolved by a gradual envelopment of both photosynthetic pigments and cytoplasmic DNA particles, possibly by budding and branching **thylakoid membranes.** Additional details can be found in the papers of Cavalier-Smith (1978, 1982) and F. J. R. Taylor (1976, 1979).

The other main view on the origin of the eukaryotes is that they obtained their organelles, particularly the DNA-containing mitochondria and chloroplasts, by a series of endosymbioses of prokaryotic cells (Fig. 27-4B). In contrast, this has been called a **xenogenous** hypothesis. Although not a new idea, the general xenogenous hypothesis, as well as its detailed steps, have been amplified in recent years, notably by Margulis. In general, Margulis (1970) and others postulate that a series of **endosymbioses** occurred in the middle part of the Precambrian. They believe that the first step was symbiosis of **protomitochondria,** most likely in the form of an early bacterium that had already differentiated protomitochondria. Likewise, the second main step was the endosymbiotic incorporation of bluegreen algal cells that then acted as **protoplastids;** these later evolved step by step into full-fledged chloroplasts of eukaryotic algae (Fig. 27-4B, step 5). Margulis also derived the 9 + 2 flagellum by endosymbiosis of flagellated cells. This last step has not gained the general acceptance that mitochondrial and plastid symbioses have, mainly

Figure 27-4B 1–6, hypothetical model of the xenogenous formation of the eukaryotic cell by a series of endosymbioses. **1,** ancestral, nonphotosynthetic prokaryote. **2,** the formation of a nucleus (*n*) by concentration of genome material surrounded by a nuclear envelope. **3,** endosymbiosis of a bacterium (*b*) to provide mitochondrial component (*m*) shown in **4. 5,** endosymbiosis of the bluegreen algal cell (*bg*) to provide the chloroplast component. **6,** eukaryotic plant cell with the three essential membrane-delimited organelles. (After F. J. R. Taylor, 1974, with permission of the author and the International Association of Plant Taxonomists.)

because DNA has not been found associated with either **centrioles** or basal bodies. However, there is some evidence favoring the derivation of eukaryotes by endosymbioses, including numerous examples of the endosymbioses of algae in many invertebrates. Although these and other supporting features cannot be detailed here, this exciting concept can be followed up in the excellent papers by Margulis (1970), Cavalier-Smith (1978, 1982), Pickett-Heaps (1974), and F. J. R. Taylor (1976, 1979).

The different pattern of chromosome structure in the dinoflagellates has been suggested as representing an evolutionary stage intermediate between the prokaryotic **nucleoid** and the eukaryotic nucleus. Lacking histones, and having complex whorls of DNA fibrils distinct from the nucleoplasm, the dinoflagellates have been called **mesokaryotes**. Although this opinion may have some merit, the general cell organization is typically eukaryotic, and the nucleus is regarded as a variant of the basic eukaryotic type.

Mitosis, Sex, and Alternation of Generations

The evolution of eukaryotes in an oxygenated atmosphere in the middle to late Precambrian was almost certainly accompanied by the development of mitotic cell division, meiosis and syngamy, and the alternation of generations. These processes are of utmost significance to plants. Their development provided for the duplication of cells (mitosis) to give relatively large and complex plant bodies, the introduction of more variability through the sexual process (meiosis and syngamy), and the elaboration of different kinds of life histories in plants (alternation of generations).

The development of mitosis apparently originated with the evolution of eukaryotes. The formation of DNA and histones into discrete linear chromosomes would permit replication of duplicate sets of genetic material for each daughter cell. Although fossil evidence is lacking, it has been speculated by Cavalier-Smith (1978) that the evolution of the mitotic process was accompanied by the development of **spindle** microtubules, **nuclear envelopes**, and a **cleavage furrow** for cell division

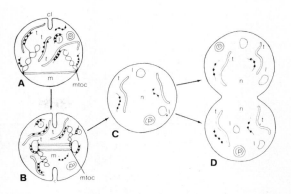

Figure 27-5 Probable evolutionary stages in the development of mitosis. **A,** an amoeboid "prealga," showing the evolution of spindle microtubules (*m*) as a device for separating the two daughter chromosomes into separate cells following cleavage (*cl*) along the furrow: *mtoc*, microtubule organizing center; *t*, thylakoid membrane; *l*, protolysosome; *p*, phagosome. **B,** chromosome attachment sites, initially at the cell membrane, have moved to more central regions by endocytosis; this would permit the two daughter nuclei to segregate into two daughter cells following cleavage at *cl*. **C,** origin of sexual process by fusion of wall-free cells to double the chromosome complement. (After T. Cavalier-Smith, 1975. Reprinted by permission from *Nature*, Vol. 256, No. 5517, pp. 463–68. Copyright © 1975, Macmillan Journals Limited.)

(Fig. 27-5). It has also been suggested that the sexual process started very early in the evolution of eukaryotic cells, along with the mitotic process, and before development of the nuclear envelope. Cavalier-Smith (1978) postulated that the early amoeboid **prealga** would have had one circular chromosome. This could have divided by a primitive mitosis to provide segregation. Recombination could have resulted from cell fusion by cytosis, a process that would be selectively advantageous by providing twice the internal food supply of unfused cells, particularly during times of starvation (Fig. 27-5).

Regardless of the details of their evolutionary development, the inception of meiosis and syngamy was a very significant step in plant evolution. The variability introduced by segregation and recombination followed by natural selection acting on new combinations resulted in the evolution of diverse eukaryotes. One general scheme showing presumed evolutionary developments following the establishment of eukaryotes is shown in Figure 27-6.

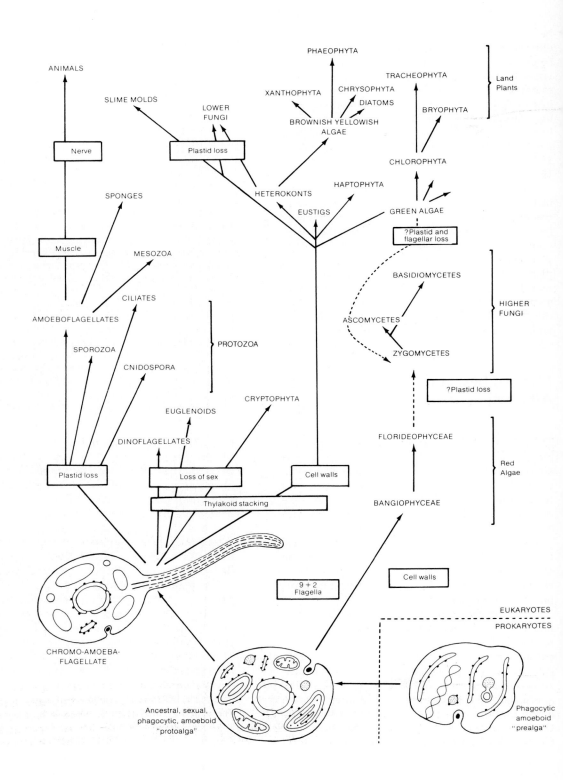

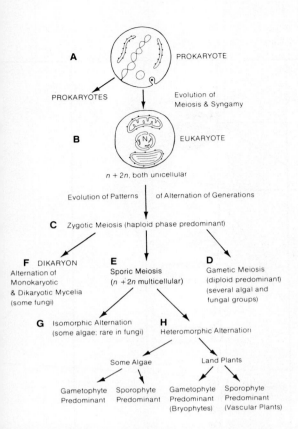

Figure 27-7 Probable evolutionary developments (**A–F**) of the main types of alternation of generations in plants.

Figure 27-6 A scheme showing possible diversification of eukaryotes following evolution of the ancestral phagocytic amoeboid "protoalga" from the amoeboid phagocytic "prealga." (After T. Cavalier-Smith, 1975. Reprinted by permission from *Nature*, Vol. 256, No. 5517, pp. 463–68. Copyright © 1975, Macmillan Journals Limited.)

The inception of the diploid phase or generation as part of the sexual cycle was an important step in the evolution of eukaryotes. In addition to increasing the reservoir of genetic variability, it also provided protection against detrimental mutations. This in turn could account for the fact that diploid-dominant organisms have evolved relatively complex tissues, organs, and physiological processes for adaptation to rigorous ecological conditions. Notable examples are the vascular plants and vertebrates on land, and the larger algae of exposed intertidal zones, such as the Laminariales and Fucales.

The alternation of the haploid and diploid generations in the meiotic-syngamy cycle has been expressed in four main patterns, based on the timing of meiosis in the life history (Fig. 27-7). We do not have a record of the actual evolution of the four patterns. However, from our knowledge of life histories in modern plants and our ability to link these to the fossil records, it seems probable that the pattern of **zygotic meiosis** was the earliest to evolve (Fig. 27-7C). It also seems likely that the **gametic** (Fig. 27-7D) and **sporic** (Fig. 27-7E) types are two separate patterns that evolved from the zygotic pattern by delay in meiosis until the multicellular diploid generation had developed, later producing either gametangia or sporangia. Although we do not know precisely when these two patterns developed, it was most likely in the interval of later Precambrian to early Cambrian, from which we have the first record of plants whose modern descendants have the same pattern of alternation. The fourth pattern of alternation, the **dikaryon** (Fig. 27-7F), is different from the others and is limited to some fungi. Of obscure origin, it consists basically of a fusion of gametic cytoplasm without nuclear fusion, resulting in binucleate cells.

Evolutionary Patterns

In attempting to recognize true evolutionary relationships among groups of plants, botanists are hampered by a lack of good fossil series to show various stages of the evolution of taxa. In a few instances, we can catch the main stages of evolution because the fossil record is reasonably good. Examples of **phylogenetic** series can be found in the Dasycladales (Chlorophyta) (Fig. 27-8). Even with

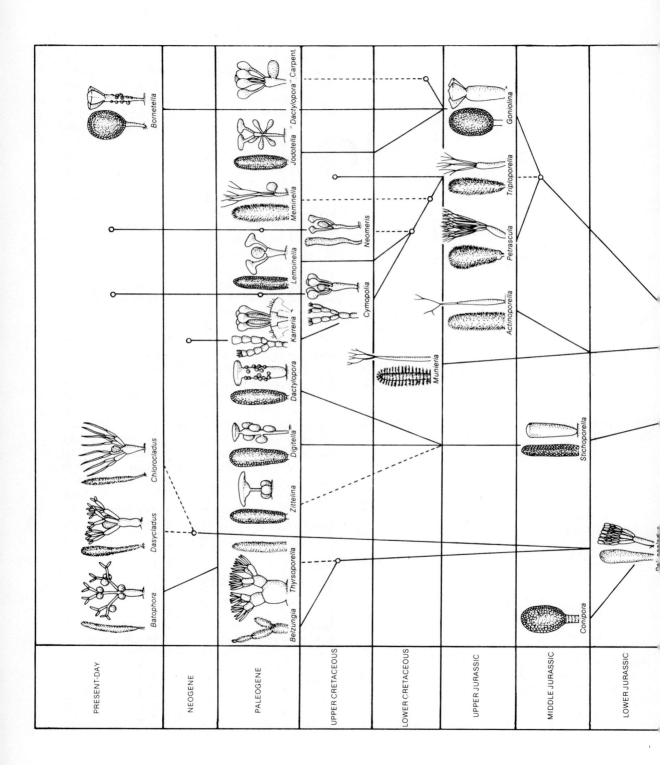

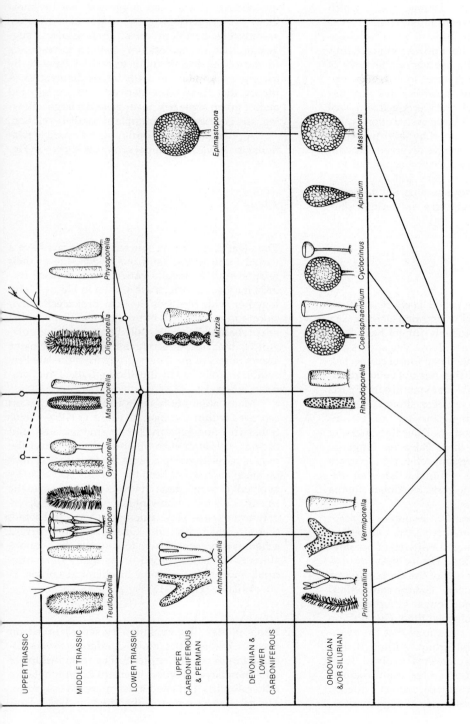

Figure 27-8 Phylogeny of genera of dasyclad algae. Circles within or at the end of solid lines indicate the occurrence of the genus within the geologic time zone. Broken lines indicate a measure of uncertainty in the evolutionary link between the genera. (Modified after Pia.)

such series, however, we do not know from which group or groups the Dasycladaceae arose, or which other groups (if any) arose from that family. Hence the phylogenetic relationships of the Dasycladaceae to other groups are not known with certainty. Because of this, botanists use various features of modern members in an attempt to establish relationships as closely as possible. For this they use biochemical, physiological, and reproductive, as well as structural criteria. The basic assumption is that the more characteristics shared by taxa, the closer they are related; hence they have diverged less with evolution. If the taxa have become separated to a greater extent by evolution, they will share fewer characteristics. Such series showing the degree of relationships among living taxa are termed **phenetic series.**

Fungi

Probable fungal hyphae called *Eomycetopsis* have been found from the late Precambrian (about 1 Ga). Also, spores that are probably fungal have been found in beds of similar age. However, spores and hyphal remains have never been found connected; hence the fungal relationship has not been proved beyond a doubt. Funguslike filaments in earlier parts of the Precambrian have not been shown unequivocally to be those of fungi, and the earliest undoubted fungi are found in Paleozoic rocks at about 400 Ma. Many of these Paleozoic fungi are very similar to modern zygomycetes. Examples are hyphal vesicles and cysts within the tissues of several vascular plants in the early Devonian Rhynie chert from eastern Scotland.

Although several possible examples of ascomycetes have been reported from the Carboniferous, there are no undoubted remains of ascomycetes until the Cretaceous. The earliest basidiomycete is from the latter half of the Carboniferous and has a well-developed **clamp connection,** suggesting that this major fungal group evolved before that period of geological time. Given these records, it is reasonable to assume that the zygomycetes are the ancestral group. Although it is often assumed that ascomycetes evolved from zygomycetes, and basidiomycetes from ascomycetes, the fossil record provides no direct evidence other than the earlier occurrence of zygomycetes.

However, on comparative structural and biochemical criteria, it has been suggested that the most likely evolutionary progression was zygomycetes-ascomycetes-basidiomycetes. There is also a suggestion that this evolutionary pattern corresponds to the colonizing of more terrestrial habitats by ascomycetes and particularly by basidiomycetes. A theory that fungi were derived from certain red and/or green algae has never gained a large following, and is probably an example of parallel evolution.

A phenetic scheme outlining the possible relationships of various fungal groups is shown in Fig. 27-9.

Algae

The algal divisions recognized in this text have a varied fossil record, and our knowledge of their relationships is equally varied. Known fossil records for algal divisions are shown in Figure 27-10, together with those of the two prokaryotic divisions and microfossils of unknown affiliations that are probably plant cells, called **acritarchs.** It is noteworthy that next to acritarchs, the oldest remains are members of the Chlorophyta from the late Precambrian, followed by the Rhodophyta of the Cambrian interval.

Many attempts have been made to draw up schemes within the algal divisions to show evolutionary relationships and trends among classes and orders. Although such schemes clearly reflect the degree of relationship among the taxa, they cannot be taken as proof of how the various taxa actually evolved through time. However, such phenetic schemes do tell us how much evolutionary divergence has taken place among the taxa, and sometimes in which direction. One of the more current schemes is presented here to illustrate the relationships of divisions, based on as many ultrastructural, morphological, and biochemical criteria as are available (Fig. 27-11).

Rhodophyta The evolutionary record based on fossil members of the family Corallinaceae was discussed in detail in Chapter 9. The relationships of other families and orders of extant Rhodophyta are rather obscure and need much more investigation. The general consensus is that the two classes Ban-

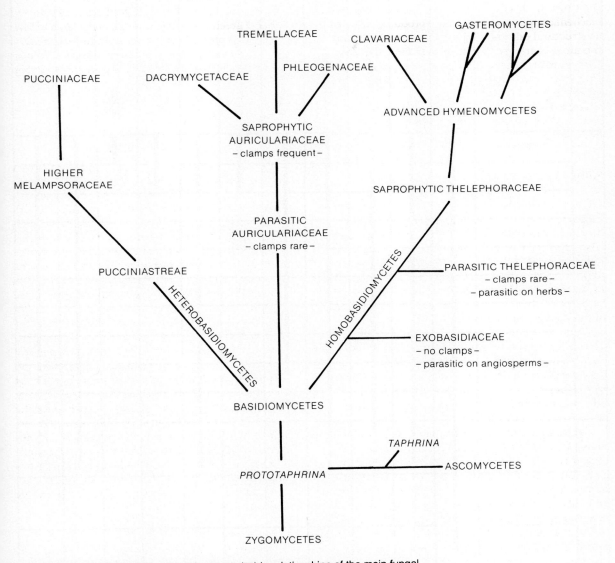

Figure 27-9 Phenetic series showing the probable relationships of the main fungal groups. (After D. B. O. Savile, with permission of the author. Reproduced by permission of the National Research Council of Canada from the *Canadian Journal of Botany* 33:60–104, 1955, and from Ainsworth and Sussman, ed., *The Fungi*, Vol. 3, *The Fungal Population*. Copyright © 1968 by Academic Press.)

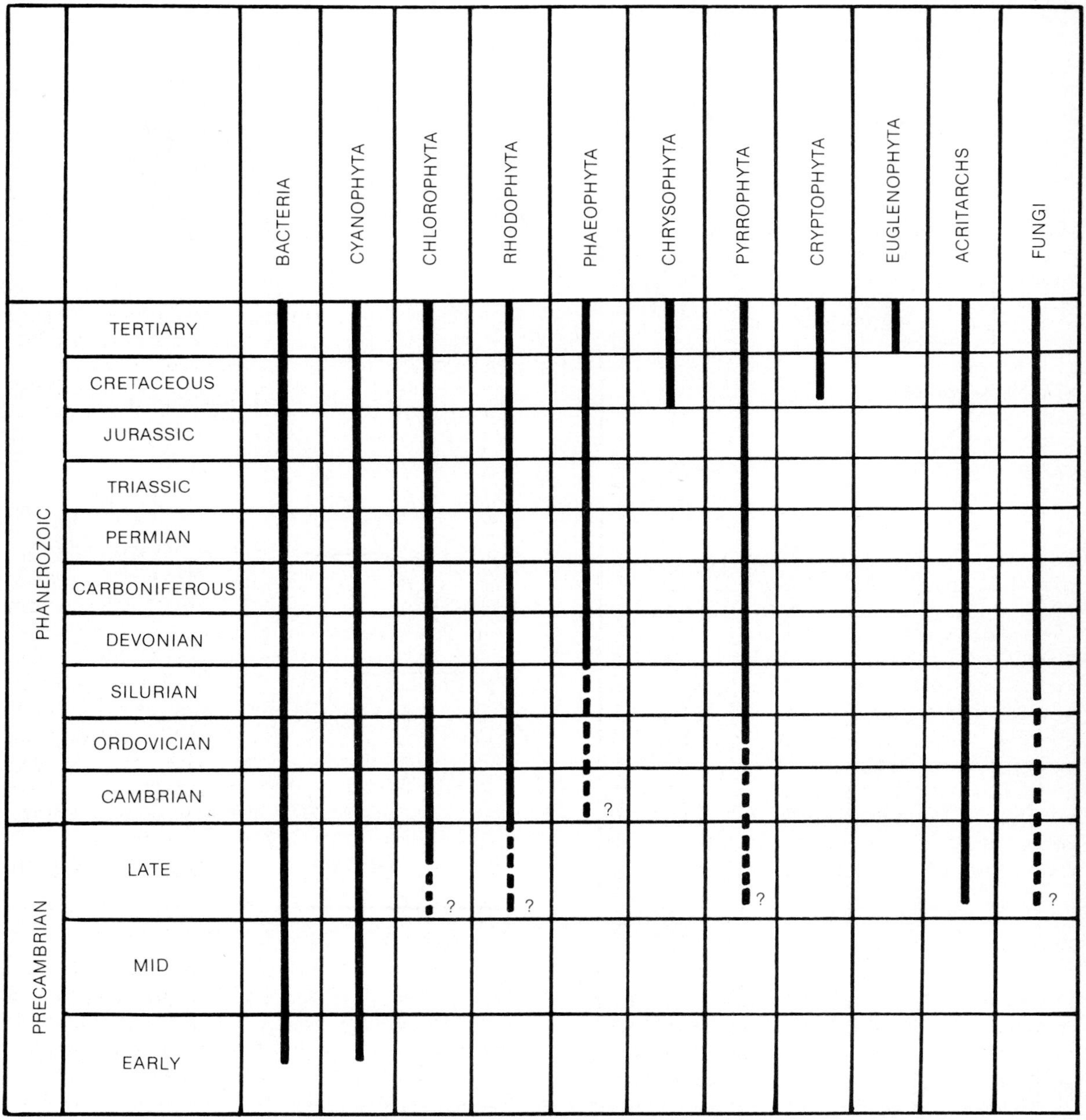

Figure 27-10 Geologic ranges of the main prokaryotic and eukaryotic divisions. Broken lines connect questionable or unverified occurrences with established ranges.

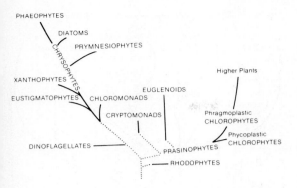

Figure 27-11 Contemporary scheme showing relationships of algal groups based on ultrastructural, morphological, and biochemical criteria. (After F. J. R. Taylor, reprinted with permission of the author and the Society of Protozoologists, from the *Journal of Protozoology* 23:28–40, 1976.)

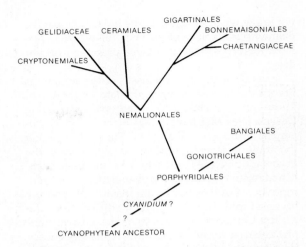

Figure 27-12 Phenetic series showing the relationships of the main groups of the Rhodophyta and their possible connection with the cyanophytes through *Cyanidium*. (After K. Fan, 1961, "Morphological Studies of the Gelidiales," Univ. of Calif. Publ. in Bot. 32:315–68, with permission of University of California Press.)

giophyceae and Florideophyceae are reasonably related by similarities in biochemical and ultrastructural characteristics. The main difference is the apparent lack of a sexual process in many of the Bangiophyceae. It has been postulated that the filamentous **conchocelis stage** of *Porphyra* with pit connections may represent a link with the Florideophyceae. Also, the Bangiophyceae, apparently lacking a sexual cycle in many instances, may be an evolutionary dead end. Although it is purely provisional, the scheme outlined in Figure 27-12 approximates the phenetic relationships of some of the main orders of the group.

Phaeophyta It was pointed out in Chapter 8 that the fossil record is spotty, although remains ascribed to brown algae extend from the Paleozoic to the present. Also, ancestors of the group are unknown. In assessing the relationships of orders, it is generally agreed that the Ectocarpales represent a group to which all others are closely related except the Fucales. Hence, a phenetic scheme such as that shown in Figure 27-13 represents a reasonable picture of what are believed to be the relationships in the division. The Fucales are considered to represent a group only distantly related to the others, mainly because of marked differences in the form and flagellation of the sperm.

There are two distinct morphological forms in the phaeophytes, excluding the Fucales. One consists of filamentous forms, ranging from simple filaments to filament clusters and complex **pseudoparenchymatous** forms. The other is **parenchymatous**, ranging up to the largest and most complex aquatic plants with meristems and differentiated tissues. Within both of these, the reproductive pattern ranges from isogamy to anisogamy to oogamy.

Chlorophyta The geologic record of the Chlorophyta is the oldest and one of the best-documented for the algal groups. Certain chlorophytan fossils extend from the latest Precambrian to the present. It is interesting, and perhaps significant, that the oldest known chlorophytes are thalloid and lime-encrusting, and that they antedate the earliest undoubted filamentous chlorophytes from the later Devonian Period by about 230 million years. We do

Evolution and Phylogeny

not know whether this is a true indication that the filamentous habit evolved later, or whether filamentous types have not been preserved and/or discovered.

Significantly, the first fossils of the siphonous Bryopsidophyceae occur in the late Precambrian. These are genera of the carbonate-encrusting Dasycladales that have a reasonably good phylogenetic history (see Fig. 27-8). This indicates that the multinucleate condition characteristic of the Bryopsidophyceae probably had become established relatively early in the overall history of Chlorophyta. Together with the record of another bryopsidophycean order, the Codiales, from the earliest Cambrian, it indicates that the Bryopsidophyceae have formed a discrete evolutionary lineage since late Precambrian times.

The record of the Charophyceae extends from the Devonian to the present. Our knowledge of evolution in this group is limited to carbonate-encrusted oogonia and their encircling sheath-cells, collectively called **gyrogonites**. These are often well preserved and occur in sufficient numbers to be very useful as geologic-time and stratigraphic indicators. The type of arrangement of sheath cells in the gyrogonites appears to reflect the evolution of several main lines of charophytes (Fig. 27-14).

Although the fossil record of some orders of the Chlorophyta is relatively good, it cannot shed much light on the degree of relationships among other orders, particularly those with a short fossil history. For this it is necessary to group the classes and orders in phenetic series that reflect our present knowledge of similarities. Such a scheme is presented in Figure 27-15.

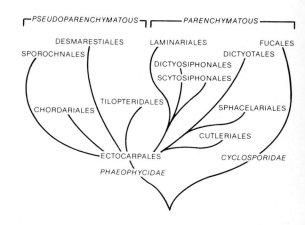

Figure 27-13 Phenetic relationships of the orders of the Phaeophyceae. (After Scagel, R. F., "The Phaeophyceae in Perspective," *Oceanogr. Mar. Biol. Annual Rev.* 4:123–94, 1966, with permission of George Allen & Unwin Ltd., London.)

Evolution of Land Plants

The origin of land plants, bryophytes and vascular plants, is very problematical and has stimulated much interest among botanists for many decades. As mentioned in Chapter 6, the general consensus is that land plants evolved from algae; the chlorophyta are usually chosen because of similarities in pigmentation, cell wall formation, ultrastructure of motile cells, food manufacture, and storage prod-

Figure 27-14 Distribution of charophyte gyrogonites from Devonian to recent times. Note the relatively short ranges of the Devonian families and the Clavatoraceae in the late Jurassic-Cretaceous interval, and the dextral (right hand) spiral in the Devonian Trocholiscaceae. (After L. J. Grambast, 1974, with permission of International Association of Plant Taxonomy.)

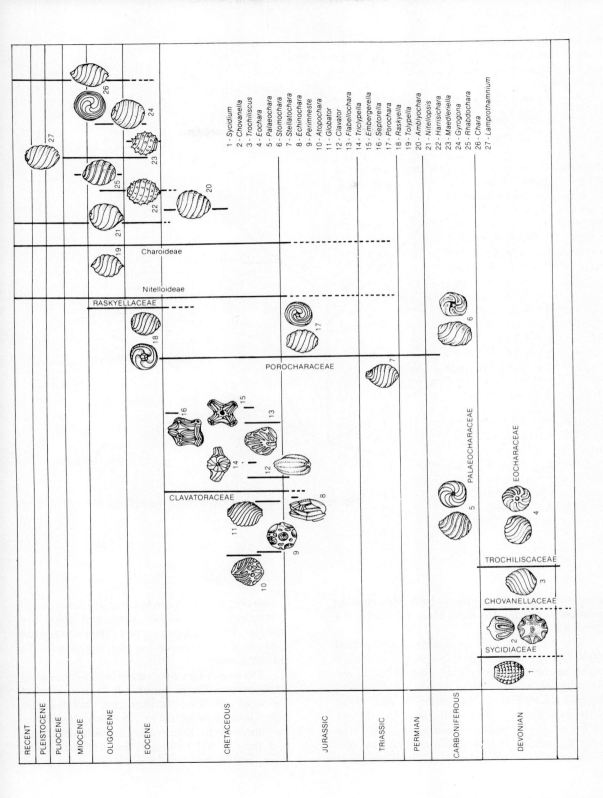

Evolution and Phylogeny 673

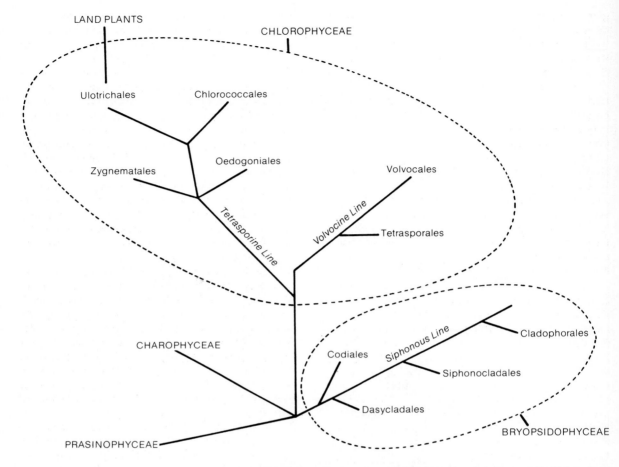

Figure 27-15 Phenetic relationships of main groups of the Chlorophyta.

ucts. Although several groups have been cited as probable ancestors, the most likely are among the Ulotrichales.

The transition from an aquatic to a terrestrial mode was one of the most important evolutionary events in the history of plants, comparable to the evolution of terrestrial from aquatic vertebrates in the animal kingdom. In both instances, derivatives of the evolutionary succession had to be adapted for the relatively hostile environment on land. In order to survive, it was absolutely essential that they evolve features for obtaining, distributing and conserving water. These features are found in the sporophytes of vascular plants and some bryophytes. They include a surface cuticle and stomata to regulate water loss and conducting cells for distributing water.

The search for evolutionary links between algal ancestors and land plants usually focuses on selecting algae with the most likely ancestral characteristics. In a hypothetical model, such algae would be chlorophytan, filamentous, and **heterotrichous,** and they would have two multicellular generations. Fertilization would occur *within* the female gametangium, and the zygote would germinate directly into a diploid sporophyte that would eventually produce meiospores in sporangia. Some of these features are found in the modern Ulotrichales, especially in the genus *Coleochaete*.

The steps in the hypothetical model of evolution on to land are illustrated in Figure 27-16A–G, based largely on Jeffrey (1962). The model postulates that the ancestral algae were growing in shallow waters of sites such as embayments, ponds, or lakes. They were essentially similar to freshwater chlorophytes such as *Coleochaete*, with a haploid, heterotrichous plant body (Fig. 27-16A). Sexual reproduction was likely oogamous, with fertilization occurring in the oogonium. Maturation occurred in the $2n$ zygote, which acted as a resting stage. On resumption of favorable conditions, the zygote underwent meiosis to form zoospores that developed into haploid plants.

It seems likely that one of the earliest evolutionary developments would have been the retention of the zygote on the haploid plant body, essentially as happens in *Coleochaete*, where sterile cells form around the oogonium after fertilization (Fig. 27-16B). This would have provided some protection for the zygote against periods of desiccation when water levels dropped. Once this was accomplished, the stage would be set for a multicellular diploid generation to develop, and to become attached to the gametophyte for nourishment. Although there are no known instances of this in *Coleochaete*, or other chlorophytes, there are red algae (e.g., *Griffithsia* and *Callithamnion*) that have analogous attachment of (carpo-) sporophyte to gametophyte. With further evolution, the sporophyte would have become larger and assumed its main function of spore dispersal (Fig. 27-16C). As habitats became more emergent, some sporangium-bearing branches would have become exposed for prolonged periods to air and the effects of drying. The evolution of a cuticle on such emergent sporophytes would protect against desiccation and provide a selective advantage for aerial spore dispersal.

In this evolutionary progression, there would have been different **selection pressures** exerted on the two generations. As a result, some plants would remain adapted for an aquatic existence to fulfill the main function of producing gametes for fertilization. At the same time, selection pressure on others would result in progressive adaptation of the sporophyte for its main function of dispersing spores. In order for spores to be disseminated on land, selection pressures would favor emergent sporophytic filaments (Fig. 27-16D, E).

Continuation of the two sets of selection pressures through long periods of time would result in increased emergence of the sporophyte, cutinization of emergent parts, development of conducting tissues, and formation of water control structures, the stomata (Fig. 27-16F, G). Thus a sporophyte adapted for existence in the relatively hostile environment on land and carrying out its prime function of spore dispersal for increasing the population could have become established. At the same time, the gametophyte remained tied to water, or at least to habitats with sufficient water to fulfill the prime function of gamete production and fertilization.

In this process of evolution on to land, there probably was a major split into two main evolutionary pathways (Fig. 27-16F). One of these, leading to the bryophytes, would see the gametophyte maintain its characteristics and independence (Fig. 27-16F*1*). This would be a natural response of plants living in water, in very wet or humid sites, or in shade provided by topographic features. In these sites, there would be sufficient water to supply the gametophyte and to ensure the prime functions of gamete production and union.

The other evolutionary pathway, leading to vascular plants, expressed the opposite trend in generation development (Fig. 27-16F*2*). There was increased selection for larger sporophytes that were able to obtain and conduct water and nutrients, to photosynthesize, and to disperse spores over drier sites. In this evolutionary pathway, the trend was toward increased independence of the sporophyte, accompanied by better conduction and protection of tissues against water loss. At the same time, selection worked against largely flimsy gametophytes requiring water; it favored smaller and fast-maturing gametophytes that could produce gametes quickly during periods of available water, such as during or after rains. The gametophytes of the earliest vascular plants would probably have been submerged. Those of later vascular plants on terrestrial sites may well have been wholly or partially subterranean, like those of some modern species of *Lycopodium*. Some may have been infected with **mycorrhizal** fungi, and **saprobic**. Whatever the case, the trend would have been toward decreasing size

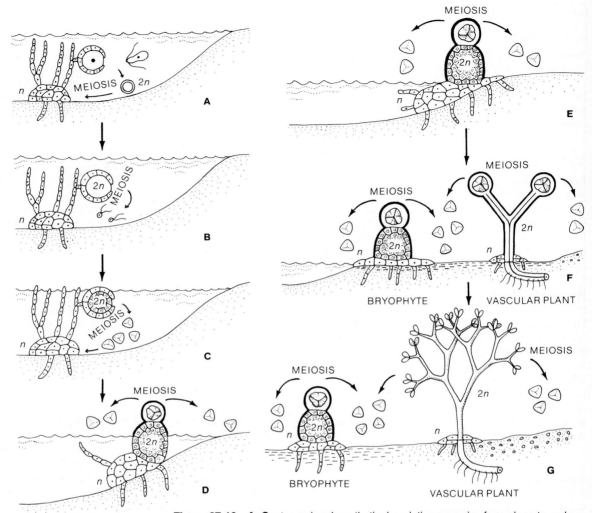

Figure 27-16 **A–G,** stages in a hypothetical evolutionary series from algae to early land plants (based in part on Jeffrey). The first stage (**A**) represents a haploid, heterotrichous green alga (such as *Coleochaete,* Ulotrichales), with fertilization occurring in the oogonium, and the 2n zygote dispersing before undergoing meiosis. In **B,** the 2n zygote is retained within the oogonium, and meiosis occurs within the zygote. In stage **C,** the 2n generation is still retained on the haploid plant, but has become multicellular; the water is shallower, producing intermittent drying on the upper branches of both the haploid and diploid generations. In **D,** the diploid generation has developed sporangium-bearing branches, which are emergent, cutinized, and adapted for dispersing meiospores through the air. Further emergence and elaboration of the sporophyte (**E**), accompanied by reduction of the gametophyte, increases the efficiency of spore dispersal on the land. **F** indicates selection of these early emergents; some remain haploid dominant (**F1**) and are selected for wet, shaded habitats. Others (**F2**) become diploid dominant, and are selected for drier areas. Continued selection (**G**) results in two distinct land-plant stocks: a gametophyte-dominant bryophytic line (**G1**) adapted for gamete dispersal in damp, wet/shaded regions, and a sporophyte-dominant vascular plant line (**G2**) adapted for spore production in drier, more upland sites.

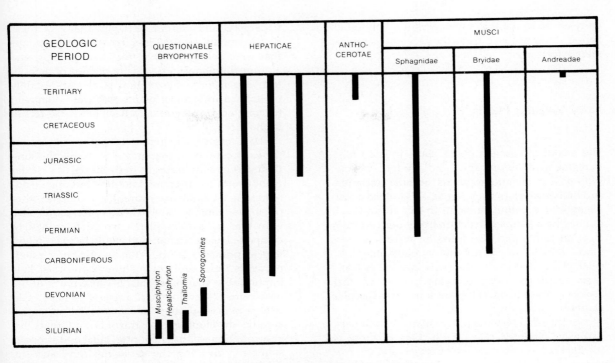

Figure 27-17 Geological record of the main lines of bryophytes. (After Lacey, 1969.)

and complexity in the gametophyte, and increasing size and complexity in the sporophyte.

Bryophyta

The bryophytes, with a fossil record from the Devonian, have three main evolutionary lines: Hepaticae, Anthocerotae, and Musci. All three of these are recorded before the end of the Paleozoic Era, indicating that the main phyletic lines became differentiated early in the history of bryophyte evolution. The fossil record of the Hepaticae is the oldest and also contains the most fossil representatives, but there is no evidence to document that it was ancestral to the other lines (Fig. 27-17). Although the Anthocerotae are often credited with being an ancestral link between bryophytes and chlorophytan precursors, the fossil record of Anthocerotae is very short and late and hence provides no support for the hypothesis. The fact that the main evolutionary lines show well-established characteristics of their respective groups also suggests that the bryophytes evolved earlier than their first record in the Devonian. The record may be foreshortened because they were not preserved, have not been found, or have not been recognized conclusively as bryophytes (e.g., *Sporogonites*, Fig. 27-17).

There are two opposing views on the evolution of the two main morphologic patterns in bryophytes, the thalloid and the leafy. Adherents of the **upgrade theory** believe that the thalloid habit was the first to evolve and the leafy habit was derived. Proponents of the **downgrade evolutionary pathway** hold that the leafy habit evolved first and became flattened and thalloid either by fusion of the leaves or by loss of the leaves and subsequent expansion of the gametophytic axis. The oldest fossil *Hepaticites devonicus* from the Upper Devonian of the United States is thalloid and hence tips the scales in favor of the upgrade viewpoint. However, this is only a single discovery and can hardly clinch the case firmly. Moreover, leafy bryophytes occur in the succeeding Carboniferous Period; their absence from the Devonian may be merely fortuitous.

Early Vascular Plants

The record of vascular plants starts in latest Silurian time with the very small and slender *Cooksonia*. With its thin dichotomous axis, **annular tracheids**, and bulbous terminal sporangia, it is indeed a simple as well as primitive plant. In fact, it is difficult to imagine a simpler structure being derived from algal ancestors and also fitted for terrestrial existence. To date, we do not know of any other vascular plants that live in the same interval as *Cooksonia*, and it is indeed tempting to conclude that *Cooksonia* was the ancestral stock for later vascular plants.

In the early Devonian, two main evolutionary pathways emerged (Fig. 27-18). One of these is the *Rhynia* type, exemplified by *Rhynia* and *Horneophyton*, in which the sporangia are terminal on branch tips. The other is the *Zosterophyllum* type, in which the sporangia are arranged laterally along the axis. The plants of both groups are small, simple, leafless, and rootless and have protostelic conducting strands with annular tracheids. These two types appear to form the early stages of the two evolutionary pathways, possibly from intermediate forms such as *Renalia*. The *Rhynia* type with terminal sporangia could have been ancestral to the *Psilophyton* type of slightly younger stages of the Devonian. The *Zosterophyllum* type, on the other hand, appears as a likely ancestor to the later plants with lateral, **reniform** sporangia, such as *Asteroxylon* and the still-later lycopod types (Fig. 27-18).

Major Evolutionary Trends

During the long interval through which vascular plants were evolving after the early Devonian, major developments in structure and function were taking place as adaptations for a terrestrial mode of life. As the natural by-products of gene changes, variability, and natural selection, many of these developments resulted in successive plant groups that were able to survive, reproduce, adapt, and assume dominance for varying periods of time. In fact, our record of evolution is almost certainly prejudiced in favor of plants that were dominant for long enough periods to have left some record in the rocks.

It is appropriate here to reemphasize some of the basic conditions that prevailed during this long period of evolution. In the early stages of vascular plant evolution, there was nothing but open landscape. As far as we are aware, there were no other organisms already established on terrestrial sites with which vascular plants would have had to compete. Hence competition would have been limited to primary factors, such as obtaining water, nutrients, and sunlight for photosynthesis. It is important to remember this, because much of the evolutionary development appears to have been related to these primary factors. Notable examples are the development of rhizoids and roots for absorption as well as anchoring, taller and stronger stems for invading the aerial medium, and the development of leaves for more efficient photosynthesis. These and other important developments during the evolution of vascular plants are discussed in the following sections.

Evolution of the Axis Throughout the history of the evolution of vascular plants, there have been many modifications in the axis. A few of the more important include evolutionary changes in branching, the character of the stem, the nature of the leaf, and the reproductive organs.

One of the most comprehensive attempts to explain evolutionary changes in the axis is that of the **telome concept.** Although based on the ideas and hypotheses of several notable botanists of earlier times, the telome concept was organized and

Figure 27-18 Probable evolutionary pathways of the terminal sporangial line leading to the trimerophytes and progymnosperms, and of the lateral sporangial line to lycopods, from the earliest vascular plant *Cooksonia*.

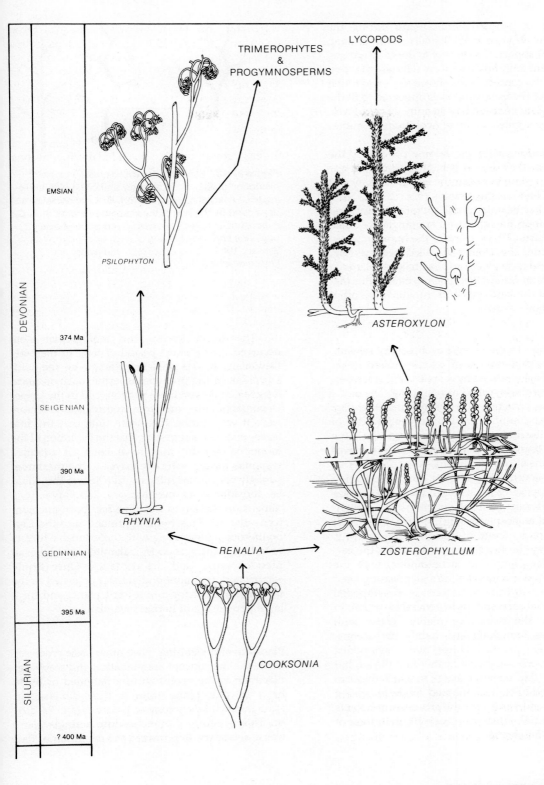

developed by Walter Zimmermann during the period from 1930 to 1965. Initially, Zimmermann (1952, 1965) applied the concept to the early groups. At first, plant morphologists were reluctant to accept and apply the concept. Later, however, it was found that some of the basic proposals appeared to fit the evidence. The concept has steadily gained favor, and is now applied in some evolutionary studies of plants.

The basic unit in the telome concept is the **telome,** described as that part of the dichotomous branching system between the last dichotomy and the distal end of the branch (Fig. 27-19). Such branches that bear terminal sporangia are called **fertile telomes;** those without sporangia are called **sterile telomes.** Those parts of the branching system connecting the telomes are called **mesomes** (Fig. 27-19). These terms were applied to such plants as the *Rhynia* and *Psilophyton* types, which in this theory formed the basic group from which the other vascular plants evolved.

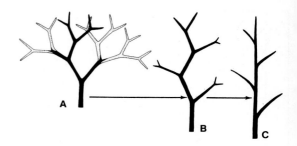

Figure 27-19 Stages in the evolutionary process of overtopping. **A,** a dichotomously branching axis, showing telomes (*t*) and mesomes (*m*). **B,** a sympodial axis developed by the loss of the stippled branches in **A. C,** a monopodial axis developed by reduction of side branches and straightening of main axis. (After W. N. Stewart, 1964, with permission of the author and *Phytomorphology.*)

Overtopping   In the telomic evolutionary scheme, the telomes and groups of telomes called *telome trusses* become modified by several basic processes. In the first of these, called **overtopping,** one of the dichotomous branches grows beyond the other (Fig. 27-19). When combined with the second process of **reduction,** the first stage of this series results in the sympodial branching characteristic of fern leaves and the stems of some lycopods, coenopterid ferns, and pteridosperms (Fig. 27-19B). Further reduction and overtopping of the axis produces the familiar **monopodial** axis of most sphenopsids, gymnosperms, and angiosperms (Fig. 27-19C). In general, this evolutionary trend of branch development is supported by the fossil records, in which the earliest vascular plants are dichotomous, with the sympodial and monopodial forms appearing later. The progression from dichotomous to sympodial to monopodial corresponds in general to the height attained by the respective plants. Those with dichotomous branching are mainly herbaceous, ground-cover plants. Those with sympodial branching reach somewhat farther into the air, but appear to be limited in the extent to which they can be supported. The best-adapted mode for giving the strength necessary for the **arborescent** habit is the monopodial, which can reach 100 m in trees of *Sequoia* and *Eucalyptus.*

The trend toward the monopodial stem occurred over a short period of time in the early Devonian. It was well established by the mid-Devonian in the progymnosperms, such as *Aneurophyton,* and even more pronounced in the Upper Devonian in the progymnosperm tree *Archaeopteris.* It was during this same time that the first leaves evolved, suggesting that the evolution of the monopodial habit may have been an adaptive response for exposing the leaves to the maximum sunlight for photosynthesis. In a sense, this could be considered an evolutionary adaptation to a dimension that up to that time had been uncolonized—the air. This trend continued into the Carboniferous, culminating with the extensive forests of arborescent lycopods, calamites, medullosan pteridosperms, and the cordaites. Once firmly established, the monopodial habit persisted as the main stem habit for arborescent plants and for a large percentage of **herbaceous** plants.

Planation and Webbing   Two other basic processes in the telome concept are **planation** and **webbing.** Planation is the evolution of a flattened or blade habit in one plane from a three-dimensional arrangement of dichotomous branches (Fig. 27-20A, B). This twisting of branches into a single plane was a necessary step toward the evolution of flat-

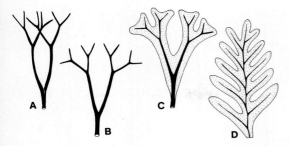

Figure 27-20 Stages of planation and webbing in the evolution of a megaphyll. **A,** three-dimensional branching dichotomous axis. **B,** branches have twisted into one plane. **C,** webbing by flattened leaflike blade extensions of axis. **D,** formation of a pinnately veined megaphyll by overtopping. (After W. N. Stewart, 1964, with permission of the author and *Phytomorphology*.)

tened **megaphyllous leaf** blades, such as those of *Archaeopteris*. There were numerous plants of the middle and upper Devonian with planated branch systems, including the progymnosperm *Protopteridium*, *Cladoxylon* (Cladoxylopsida), and the coenopterid fern *Rhacophyton*.

The second basic process involved in the evolution of the flattened megaphyllous leaf blade is **webbing.** This is the filling-in of the space between planated branches with parenchyma (Fig. 27-20C, D). Examples of webbing can be seen in extinct plants with fernlike foliage, including *Archaeopteris*, and in the Carboniferous pteridosperm genus *Rhacopteris*, various species of which show degrees of webbing resembling a telomic evolutionary progression. These and many other examples indicate that planation and webbing were evolutionary trends in several distinct lineages of vascular plants, including both megaphyllous groups, such as ferns, cycadophytes, and ginkgophytes, and those with **microphyllous leaves,** such as Sphenophyllales.

Syngenesis The last important basic process of the telome theory is **syngenesis.** This is the evolutionary fusion of axes and branches, resulting in the increased size of the axis and included vascular tissues (Fig. 27-21). This fusion of axes results in a single **polystelic axis,** of which there are numerous examples among Devonian and Carboniferous

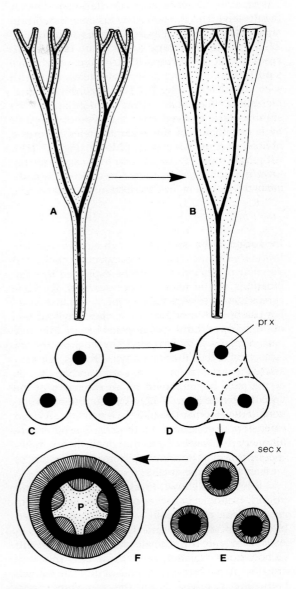

Figure 27-21 Syngenesis of axes. **A, B,** two early stages, showing fusion of separate steles in longitudinal section. **C–F,** showing fusion in transverse sections to form a polystele (**D, E**) and a siphonostele (**F**): *pr x,* primary xylem; *sec x,* secondary xylem; *p,* pith. (After W. N. Stewart, 1964, with permission of the author and *Phytomorphology*.)

Evolution and Phylogeny

plants. One of the best examples of a likely telomic syngenesis series is in the medullosan pteridosperms. As shown in Fig. 27-21C–E, the first stage has separate branches, each with its own **protostele** (**haplostele**). This is followed by fusion of the cortical and epidermal tissues, as in the genus *Sutcliffia* (Fig. 27-21D). The third stage is that found in Carboniferous and Permian stems of *Medullosa*, which have a **polystele** consisting of separate steles, each with secondary xylem (Fig. 27-21E). According to Zimmermann, the main stelar configuration, the siphonostele, evolved from this type of polystele by lateral fusion of the separate steles to form a hollow ring of xylem and phloem (Fig. 27-21F). Although this may be one evolutionary pathway toward a siphonostele, there was at least one additional development, as described in the next section.

Evolution of the Leaf   Although many of the evolutionary changes that have occurred since the first vascular plants evolved can be explained by combinations of the telome processes, not everyone agrees that these were the only mechanisms at work. For example, the origin of the microphyllous leaf of the lycopods and sphenopsids has traditionally been explained by invoking the **enation theory**. According to this, the microphyll with its single vein started as an enation of the cortex and epidermis of the axis, accompanied by an extension of the vascular tissue (Fig. 27-22A–D). Continued evolutionary development resulted in a small, one-veined extension, or enation, from the axis. The other evolutionary pathway based on the telome theory shows the microphyll developing from a reduction of telomes to a single enation (Fig. 27-22E–F).

It has been shown (Beck, 1976) that the compound, frondlike branches of the progymnosperm *Archaeopteris* are actually a system of planated branches of several orders. The only true leaves in *Archaeopteris* are the ultimate bladed segments with dichotomous venation, and these also probably arose by planation and webbing of ultimate telomic branching systems. From the progymnosperm ancestors, leaves apparently evolved in two directions. One path led to the development of the frondlike compound leaves of the pteridosperms and cycads. The other resulted in many variants, ranging from the needle-leaves of the conifers to the strap-shaped leaves of the cordaites and

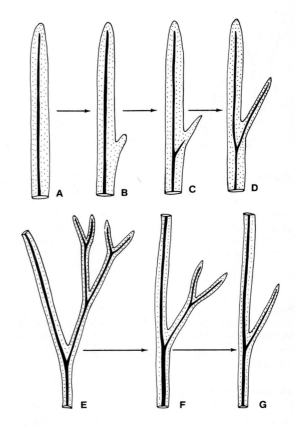

Figure 27-22   Two pathways of evolution of the microphyllous leaf. **A–D,** by progressive extension of outer stem and vascular bundle (enation theory). **E–G,** by progressive reduction of a branched telomic system (telome theory). (After W. N. Stewart, 1964, with permission of the author and *Phytomorphology*.)

the lobed, dichotomously veined leaf of the ginkgophytes.

The general picture of the evolution of the megaphyllous leaf is one of gradual elaboration of terminal branching systems by planation, overtopping, and webbing (Fig. 27-20). The photosynthetic function was hence transferred from the axis to the more specialized bladed areas, or leaves. Although we do not have any concrete answers as to why this happened, it is interesting to speculate that the evolution of the several types of leaves was an adaptation for more efficient photosynthesis. It may also have occurred concomitant with the devel-

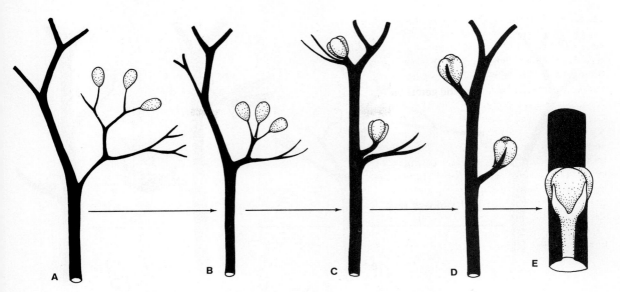

Figure 27-23 Proposed series (**A–E**) for the evolution of the sporangium positioning in *Psilotum* (**E**) by reduction (**A–D**) of a branching fertile telomic axis. (After W. N. Stewart, 1964, with permission of the author and *Phytomorphology*.)

opment of secondary vascular tissues and bark in the stem and branches. These structural tissues were necessary for greater support and water conduction capacity and for protection against wind damage and water loss. At the same time, those features probably prevented the stem from maintaining the main photosynthetic activity, and hence favored selection toward the formation of leaves in the younger, nonwoody portions of the stem. In this way, stronger and taller stems would have evolved to elevate the leaves into positions of maximum sunlight.

Reproductive Systems The evolution of the various patterns of sporangial positioning can also be explained within the framework of the telome concept. Each pattern has evolved through combinations of basic processes, but in different directions and to different extents. The evolution of the sporangial positioning of the *Psilotum* type is interpreted in Figure 27-23A–E. Starting with a sympodial branching system in which overtopping has taken place (Fig. 27-23A), the fertile and sterile telomes become foreshortened by reduction (Fig. 27-23B). Continued shortening of both sterile and fertile segments and fusion of three sporangia (Fig. 27-23C, D) produces the condition of the modern *Psilotum* (Fig. 27-23E). It has a three-lobed sporangium tucked into the axil of the sterile telome, which has undergone extreme shortening and syngenesis with the main axis.

A second pattern of sporangial arrangement is found in *Tmesipteris, Lycopodium,* and the Carboniferous sphenopsid *Sphenophyllum*. This shows a series (Fig. 27-24A–D) similar to that of the *Psilotum* type of foreshortening of both fertile and sterile telomes. From the stage in Figure 27-24B, there appear to be two main paths of reduction and shortening. In the first path, two sporangia are maintained as the fertile telome shortens to give the *Sphenophyllum* type (Fig. 27-24C), with the sporangia attached above and below, and the *Tmesipteris* type (Fig. 27-24D), with two sporangia fused on a very short stalk, which in turn is fused to the very reduced sterile telomic segment.

The telomic series producing the characteristic whorled and recurved pattern of the *Equisetum* and *Calamites* types is shown in Figure 27-25. From the

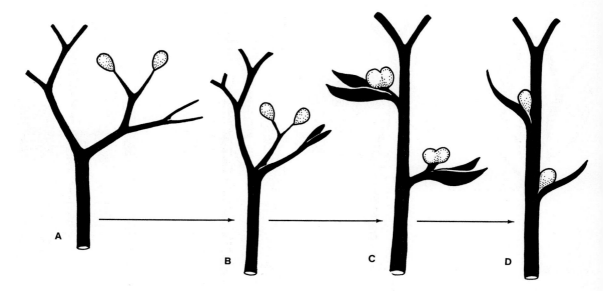

Figure 27-24 A–D, evolution of sporangial positioning in *Lycopodium* (**B**), *Sphenophyllum* (**C**) and *Tmesipteris* (**D**) by reduction of fertile and sterile telomes. (After W. N. Stewart, 1964, with permission of the author and *Phytomorphology*.)

three-dimensional sporangial trusses shown in Figure 27-25A, planation accompanied by another telomic process called *recurvation of sporangial stalks* results in the positioning shown in Figure 27-25B. Further reduction, foreshortening, and fusion of sporangial axes produces the positioning of sporangia in *Equisetum* and *Calamites* (Fig. 27-25C, D). The fact that the branches supporting the discs and sporangia have resulted from the shortening of a telomic axis supports the use of the term **sporangiophore** (*lit.* sporangial-branch) for this structure, rather than any name suggesting homology to a sporophyll.

The last evolutionary series of sporangia is that leading to the marginal positioning of some pterophyte megaphylls, such as the ferns and the Devonian *Cladoxylon*. The series shown in Figure 27-26 starts with three-dimensional telome trusses of the *Rhynia* type (Fig. 27-26A), and proceeds by planation to a flattened system similar to that of *Cladoxylon* (Fig. 27-26B). The last step is webbing of the planated axis to give a leaflike blade with marginal sporangia (Fig. 27-26C). An example is the Carboniferous genus *Acrangiophyllum*, which may be an evolutionary stage leading to the filicalean ferns.

From the above examples it is clear that the telome concept has done much to clarify some of the main evolutionary developments in vascular plants, at least up to the level of angiosperms. To the present time, the telome concept has been difficult to apply to angiosperms. Perhaps another level of fundamental processes has been at work during the evolution of form and structure in angiosperms.

Evolution of Vascular Tissue The changes that took place during the evolution of the stem, leaf, and reproductive structures were accompanied by some significant changes in vascular tissues. The pattern of the stele was modified in many different ways from the simple protostele of the early vascular plants, such as the *Rhynia* type. The major stele types and their apparent evolutionary sequence are shown in Figure 27-27. Note in this series that the plant groups mentioned are examples of groups having that type of stelar structure only; no real phylogenetic relationships whatever are intended.

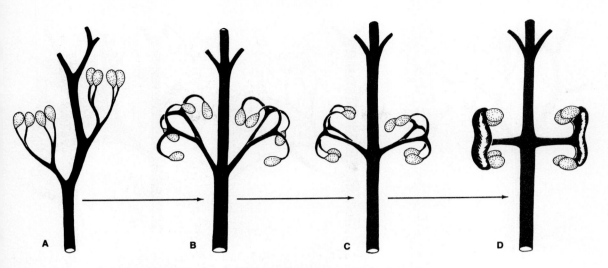

Figure 27-25 **A–D,** evolution of sporangial positioning of the *Equisteum* (**C**) and *Calamites* (**D**) type by reduction and recurvation of fertile telomes. (After W. N. Stewart, 1964, with permission of the author and *Phytomorphology*.)

A significant event was the evolution of the siphonostele. This event appears to be correlated with the development of the arborescent, woody habit. It occurred in the lycopod (*Lepidodendron*), sphenopsid (*Calamites*), and progymnosperm (*Archaeopteris*) lines. It is interesting to speculate that the hollow configuration of the siphonostele may have been an adaptation that provided greater strength in stems that grew to greater heights; a hollow tube is stronger than a solid rod. Hence the evolution of cylindrical primary vascular tissue, together with the formation of secondary xylem, would provide the strength necessary to support tall stems and branch systems and to protect the plants against physical forces such as wind. The production of secondary xylem would also give greater water-carrying capacity for the extensive leaf systems that were developed by early arborescent plants such as *Archaeopteris*.

There were also changes occurring in the xylem cells. The earliest tracheid type in such early plants as *Cooksonia* and *Rhynia* had annular thickenings; still later, the **scalariform** and **bordered-pitted tracheids** developed (Fig. 27-28). Thus the evolutionary trend in tracheid development was toward a greater amount of secondary thickening. Again, it is interesting to speculate that the increase of secondary wall material would have made individual tracheids stronger; hence the xylem as a tissue would have greater strength.

Vessels (Fig. 27-28F), the most efficient water conduits of all tracheary elements, are known to occur in several vascular plant groups, including lycopods (*Selaginella*), sphenopsids (*Equisetum*), ferns (*Pteridium*), and gymnosperms (*Ephedra, Gnetum*), as well as in angiosperms, where they are the predominant water-carrying cell. Hence it is assumed that vessels evolved from tracheids in different lineages at different times but did not become the dominant tracheary element until the evolution of the angiosperms.

Evolution of Heterospory, the Seed Habit, and Pollen Grains These very significant developments in the history of vascular plant evolution were discussed in Chapter 20 (pp. 493–504) but bear summarizing here. The early *Rhynia* type plants

Evolution and Phylogeny   **685**

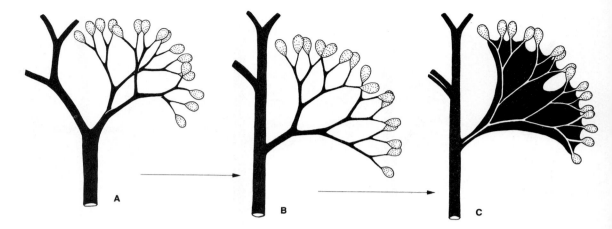

Figure 27-26 **A–C,** evolution of a pterophyte megaphyll by planation and webbing of a three-dimensional fertile telomic truss. **A,** *Rhynia;* **B,** *Cladoxylon;* **C,** *Acrangiophyllum.* (After W. N. Stewart, 1964, with permission of the author and *Phytomorphology.*)

with terminal sporangia contained spores of one size; they were homosporous. From the record of early and middle Devonian spores plus the knowledge that some species of *Archaeopteris* were heterosporous in the Upper Devonian, we assume that heterospory was evolving relatively quickly in the interval between the early Devonian and the early Upper Devonian. Moreover, there appears to have been evolutionary pressure toward the enveloping of megaspores and megasporangia in vegetative tissue, as evidenced by the occurrence of the first seed *Archaeosperma* in the late Devonian (see p. 502).

The evolution of the seed appears intimately tied to the increase in the arborescent habit that was occurring in the lycopods, sphenopsids, and progymnosperms. Although we do not know for certain, it seems reasonable to speculate that the development of the seed habit was an evolutionary response that protected the female gametophyte, while allowing fertilization to occur on the plant but divorced from the necessity for free surface water for fertilization. Significantly, it was approximately at this time that the pollen grain was evolving, assuring similar protection for the male gametophyte inside the pollen wall. By developing elaborations of the wall for floating, the early prepollen appear to have been adapted to reach the ovule by wind dispersal and hence effect fertilization in aerial regions of the plants. This would also allow the young proembryo a chance to develop and mature within the protection of the megaspore, megasporangial wall, and integument before being shed.

Another significant evolutionary trend accompanied the development of the seed and pollen habits, namely gametophyte reduction. Once the trend was started toward the development of the gametophytes *within* the spore, there would have been selection pressure in favor of smaller gametophytes that could be easily and efficiently retained inside the protective tissues and in lighter spores for better wind dispersal. This trend of gametophyte reduction appears to have continued with evolution, reaching the limits of reduction in the angiosperms, where the female gametophyte consists of several cells in the embryo sac, and the male is limited to one or two cells in the pollen grains.

The main evolutionary pathways for the free-sporing vascular plants are shown in Figure 27-29; those for the seed-plant groups are shown in Figure 27-30.

Evolution of Angiospermy The last major event in the evolution of vascular plants was the origin

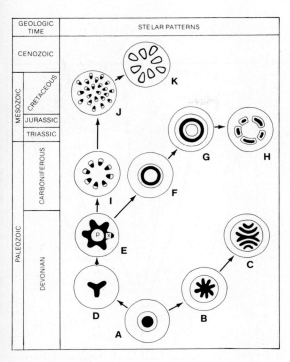

Figure 27-27 Sequence and probable pathways in the evolution of steles.

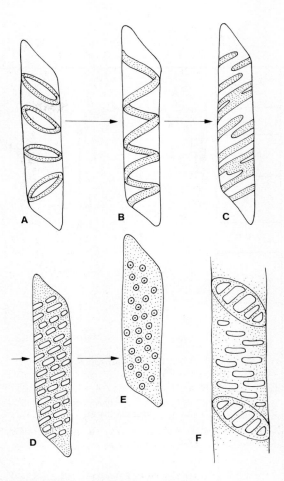

Figure 27-28 Evolutionary series of tracheary elements. **A–E,** tracheids: **A,** annular; **B,** helical; **C,** scalariform; **D,** reticulate; **E,** bordered. **F,** vessel element with scalariform pitting and perforated end plates.

of angiospermy and the subsequent evolution of the angiosperms, which became the dominant group during the last 110 million years. Although they are the most numerous in terms of numbers of taxa and numbers of individuals, we know relatively little about their origin, phylogeny, and the evolutionary patterns that have accompanied their history.

Features that are characteristic of angiosperms are (1) vessels in the xylem; (2) seeds retained in a **carpel;** (3) **netted venation** in leaves; (4) **double fertilization** by nonmotile gametes delivered to the embryo sac by a pollen tube that penetrates the carpel; (5) pollen with radial symmetry of apertures and an exine differentiated into **endexine, columellate** layer, and (in some instances) a **tectum;** and (6) a flower specialized for pollination by various agents. In hunting for the origins of angiosperms, we usually attempt to find those fossils that have the most angiosperm features, or features that could have evolved into those of angiosperms. This is no easy task. It has frustrated many, who would readily second the observation made many years ago by no less a scientist than Charles Darwin that the origin of the angiosperms is an "abominable mystery."

The group of plants generally credited with being the most likely ancestors of angiosperms are the pteridosperms. Features such as seeds, cupules,

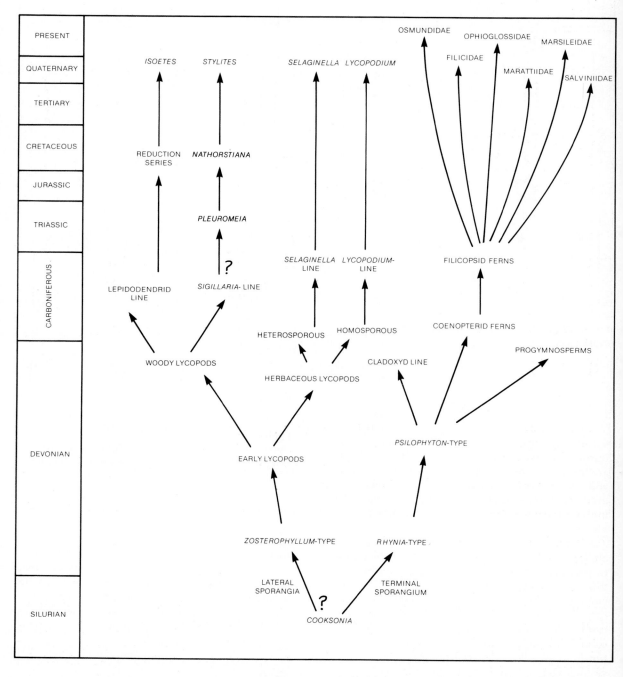

Figure 27-29 Evolutionary pathways in free-sporing vascular plant lineages.

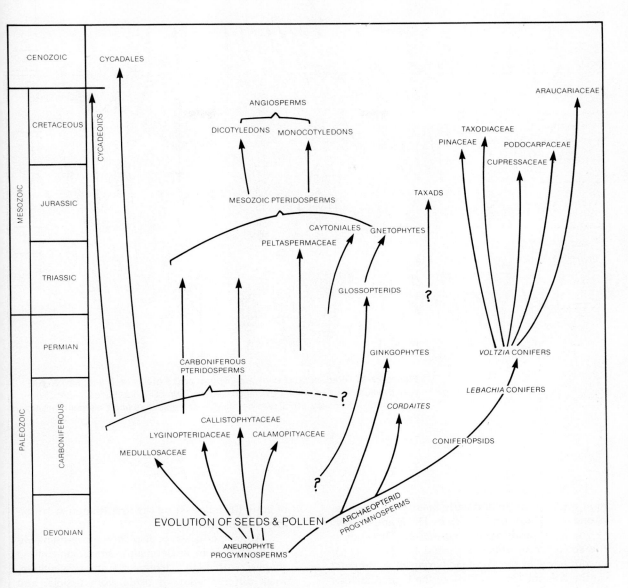

Figure 27-30 Evolutionary pathways in seed-plant lineages.

synangiate pollen organs, monosulcate pollen, and scalariform tracheids are considered to be possible stock from which angiosperms could have evolved. Some of the Mesozoic pteridosperms, such as the Caytoniales, Corystospermaceae, and Peltaspermaceae, with seeds borne inside almost-closed cupules, are considered groups that approached but did not attain angiospermy.

In the last decade and a half, there have been concerted efforts among some paleobotanists and **paleopalynologists** to search for and discover older and older fossils of angiosperms and hence shed

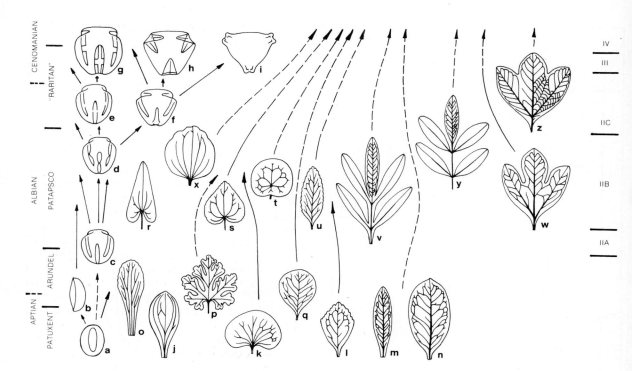

Figure 27-31 Sequence of early angiosperm pollen grains (**a–i**) and leaves (**j–z**) from the Potomac Group of the Atlantic Coastal Plain of United States. (After J. A. Doyle and L. J. Hickey, 1976, with permission of the authors and Columbia University Press.)

light on their origin and early history. The results of one such study, undertaken by Doyle and Hickey (1976), are reproduced in Figure 27-31. This shows the sequence of pollen (*a–i*) and leaf types (*j–z*) encountered in the Potomac Group of sediments on the Atlantic Coastal Plain of the United States. The Potomac leaf and pollen assemblages are the best represented and among the earliest angiosperm assemblages, spanning the interval of *ca.* 125–95 Ma (late Aptian–early Cenomanian), when the angiosperms were undergoing rapid evolution, diversification, and specialization. The pollen sequence progresses from the early **monocolpate** type in zone I, to monocolpates and early **tricolpates** in zone II, and early **tricolporates** (with very weak germinal pores) in the late part of zone II and zone III. True tricolporates appear in the late part of zone III, followed by the first triangular **triporates** in zone III.

Angiosperm leaves also show progressive diversification from single, mainly entire, generalized shapes, with highly irregular primary venation, weak differentiation of blade from petiole, and multistranded midribs in zone I to more pinnately and palmately lobed shape-classes in II-B. Most leaves in this subzone have more organized venation patterns, and greater regularity in the course and pattern of secondary veins. **Pinnately compound** leaves first appear in the upper part of II-B, with more regular tertiary venation pattern than their assumed ancestors. The **palmately lobed** leaves (and derivatives) in subzone II-C and basal zone III are more diverse in form, with rigid venation patterns extending to the fourth order.

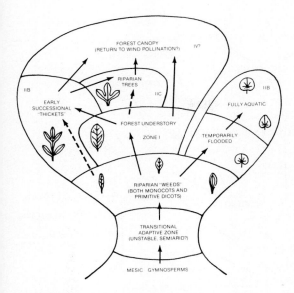

Figure 27-32 Model of early ecological-adaptive evolution of angiosperms. (After J. A. Doyle and L. J. Hickey, 1976, with permission of the authors and Columbia University Press.)

From the results of the Potomac study, Doyle and Hickey concluded that the presence of monocolpate pollen and first-rank leaves of the magnolialian type support the theory that the Magnoliales are closest to the primitive angiosperm stock, and that the monocotyledons appear as early in the record as herbaceous dicotyledons, leaving open the question of which group may have been ancestral to the other, or which came first in angiosperm evolution.

An interesting byproduct of the Potomac study is a model of suggested ecological-adaptive evolution suggested by the leaf and pollen sequence (Fig. 27-32). The sequence has early angiosperms evolving from ancestral gymnosperms (pteridosperms) in unstable, semiarid regions of the Potomac flood plain. Here, primitive monocots and dicots would have colonized disturbed habitats, such as stream banks exposed to flooding, caving, silting, and desiccation. Later, there was divergence into semiaquatic and fully aquatic species, and a forest-understory branch, followed consecutively by the establishment of early successional thickets, **riparian** trees, and finally, in late stages, possibly full forest canopy.

Summary

Throughout this text, we have attempted to show the most likely relationships of the main groups of plants and fungi and, where possible, to depict evolutionary pathways. These are summarized in the flowchart in Figure 27-33. Although in some instances the fossil record supports the suggested relationships, we wish to emphasize that some taxa are necessarily grouped as phenetic series based on the similarities of extant members. This is particularly true in the algae, fungi, and bryophytes, for which the fossil record is too fragmentary to provide phylogenetic series.

In summary, the 4 Ga history of the earth shows progressive evolution of plants and fungi from unicellular prokaryotes to complex, multicellular eukaryotes. The general pattern indicates that the evolution of eukaryotes occurred with a buildup of O_2 in the hydrosphere and eventually in the atmosphere (Fig. 27-34). This was accompanied by the development of the new and significant processes of mitosis, the sexual cycle, and alternation of generations. These set the stage for the subsequent evolution of the multicellular and relatively complex plants of several algal groups, with adaptations to diverse ecological niches in the Paleozoic and subsequent times. The next major evolutionary step was the movement onto land and the development of bryophytes, vascular plants, and fungi as major lines. This was followed by bursts of evolution among the lycopods, sphenophytes, and progymnosperms in late Devonian to early Permian times; the ascendancy of the conifers and other gymnosperms as well as the ferns in the early to middle Mesozoic; and finally the overwhelming predominance of angiosperms from the later Cretaceous to the present. With such a long history, we can only predict that the evolution of plants and fungi will continue for a long time into the future. The challenge is for us to ensure, as potential manipulators, that we don't prevent it from happening.

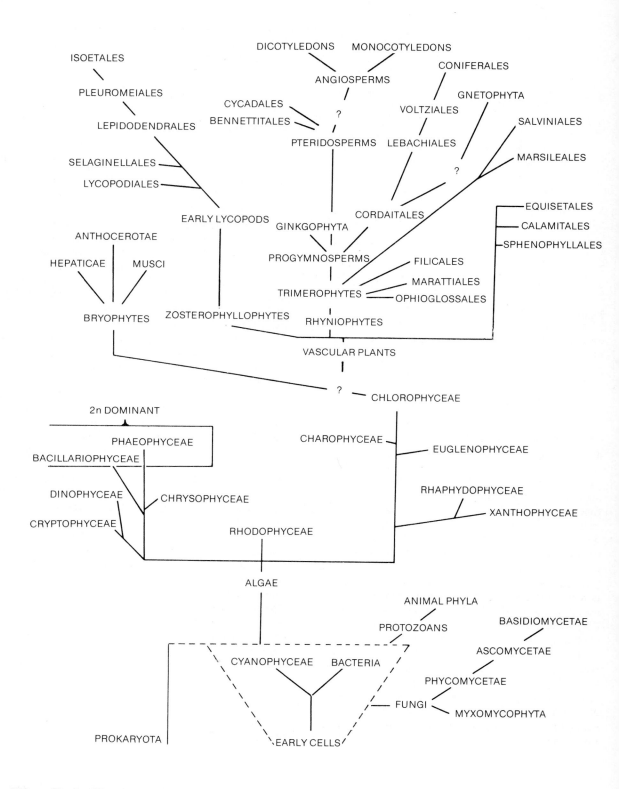

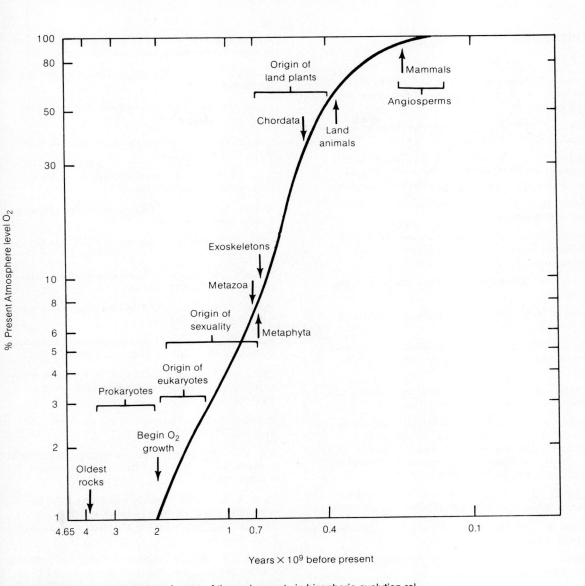

Figure 27-34 Apparent timing of some of the main events in biospheric evolution relative to hypothetical levels of atmospheric oxygen; both scales logarithmic. (Modified from P. Cloud, 1976, with permission of the author and the Paleontological Society, publishers of *Paleobiology*.)

Figure 27-33 Flow diagram of probable evolutionary relationships among main groups of plants.

References

Balch, W. E.; Magrum, L. J.; Fox, G. E.; Wolfe, R. S. and Woese, C. R. 1977. An ancient divergence among the bacteria. *J. Mol. Evol.* 9:305–11.

Beck, C. B. 1976. Current status of the Progymnospermopsida. *Rev. Palaeobot. and Palyn.* 21:5–23.

Brock, T. D. 1973. Evolutionary and ecological aspects of cyanophytes. In Carr, N. G., and B. A. Whitton, eds., *The biology of blue-green algae*. Blackwell, London.

Buetow, D. E. 1976. Phylogenetic origin of the chloroplast. *J. Protozoology* 23:41–47.

Cavalier-Smith, T. 1978. The evolutionary origin and phylogeny of microtubules, mitotic spindles, and eukaryote flagella. *Biosystems* 10:93–114.

―――. 1982. The origins of plastids. *Biol. J. Linn. Soc.* 17:289–306.

Chaloner, W. G. 1970. The rise of the first land plants. *Bio. Rev.* 45:353–77.

Cloud, P. 1976. Beginnings of biospheric evolution and their biogeochemical consequences. *Paleobiology* 2:351–87.

Cronquist, A. 1968. *The evolution and classification of flowering plants*. Nelson & Sons, London.

Doyle, J. A., and Hickey, L. J. 1976. Pollen and leaves from the mid-Cretaceous Potomac Group and their bearing on early angiosperm evolution. In Beck, C. B., ed., *Origin and early evolution of angiosperms*. Columbia Univ. Press, New York & London.

Flugel, E., ed. 1977. *Fossil algae: recent results and developments*. Springer-Verlag, Berlin, Heidelberg, New York.

Fulford, M. 1965. Evolutionary trends and convergence in the Hepaticae. *Bryologist* 68:1–31.

Grambast, L. J. 1974. Phylogeny of the Charophyta. *Taxon* 23:463–81.

Harris, T. M. 1976. The Mesozoic gymnosperms. *Rev. Palaeobot. Palynol.* 21:119–34.

Jeffrey, C. 1962. The origin and differentiation of the archegoniate land-plants. *Bot. Not.* 115:446–54.

Jermy, A. C.; Crabbe, J. A.; and Thomas, B. A., ed. 1973. The phylogeny and classification of ferns. *J. Linn. Soc. Lond.* Supp. 1, Academic Press, London.

Jovet-Ast, S. 1967. Bryophyta. In Boureau, E., ed. *Traité de paléobotanique*. Vol. 2. Masson et Cie., Paris.

Krassilov, V. A. 1977. The origin of angiosperms. *Bot. Rev.* 43:143–76.

Lacey, W. S. 1969. Fossil Bryophytes. *Biol. Rev.* 44:189–205.

Margulis, L. 1970. *Origin of eukaryotic cells*. Yale Univ. Press, New Haven, Conn.

―――; L. To; and D. Chase. 1978. Microtubules in prokaryotes. *Science* 200:1118–24.

Meeuse, A. D. J. 1966. *Fundamentals of phytomorphology*. Ronald Press, New York.

Miller, C. N. 1977. Mesozoic conifers. *Bot. Rev.* 43:218–80.

Muller, J. 1981. Fossil pollen records of extant angiosperms. *Bot. Rev.* 47:1–142.

Pettitt, J. 1970. Heterospory and the origin of the seed habit. *Biol. Rev.* 45:401–15.

Pickett-Heaps, J. D. 1974. The evolution of mitosis and the eukaryotic condition. *Biosystems* 6:37–48.

Pirozynski, K. A., and Mallock, D. W. 1975. The origin of land plants: a matter of mycotrophism. *Biosystems* 6:153–64.

Reijnders, L. 1975. The origin of mitochondria. *J. Mol. Evol.* 5:167–76.

Roe, K. E. 1975. Origin of the alternation of generations in plants: reconsideration of the traditional theories. *The Biologist* 57:1–13.

Savile, D. B. O. 1968. Possible interrelationships between fungal groups. In, Ainsworth and Sussman, eds., *The fungi* Vol. 3. *The fungal population*. Academic Press, New York.

Schopf, J. W. 1975. Precambrian paleobiology: problems and perspectives. *Ann. Rev. Earth Planet. Sci.* 3:213–49.

Schwartz, R. M., and Dayhoff, M. O. 1978. Origins of prokaryotes, eukaryotes, mitochondria, and chloroplasts. *Science* 199:395–403.

Stebbins, G. L. 1974. *Flowering plants*. The Belknap Press, Harvard University, Cambridge, Mass.

Stewart, W. N. 1964. An upward outlook in plant morphology. *Phytomorph.* 14:120–34.

Taylor, F. J. R. 1976. Autogenous theories for the origin of eukaryotes. *Taxon* 25:377–90.

———. 1979. Symbionticism revisited: a discussion of the evolutionary impact of intercellular symbioses. *Proc. Roy. Soc. Lond.* B. 204:267–89.

Taylor, T. N. 1981. *Paleobotany: an introduction to fossil plant biology.* McGraw-Hill Book Co., New York.

Woese, C. R., and Fox, G. E. 1977. The concept of cellular evolution. *J. Mol. Evol.* 10:1–6.

Woese, C. R. 1981. Archaebacteria. *Scientific American,* June, 1981:98–122.

Zimmermann, W. 1952. Main results of the telome theory. *Palaeobotanist* 1:456–70.

———. 1965. *Die telomtheorie.* G. Fischer, Stuttgart.

GLOSSARY

abaxial Situated facing away from the axis of the plant.

abscission layer Cell layer that breaks down or forms cork so that one part of a plant separates from the remainder (usually refers to leaves or fertile organs).

abstrict To separate and discharge part of the plant (applied to release of basidiomycete spores).

accessory pigment Term used in algae to refer to carotenoid and phycobilin pigments, which are usually involved in absorbing light energy that is then made available for photosynthesis.

acellular Term used to designate thalli in which the numerous nuclei and other organelles are not segregated into separate compartments (cells), e.g., myxomycete plasmodium.

acervulus (pl. acervuli) Discoid or pillow-shaped fungal structure in which conidia and conidiophores are formed.

achene Dry, indehiscent, single-seeded fruit.

acidophilic Having the ability to thrive in an acid medium (e.g., the moss *Sphagnum*).

acrasins Substances that are secreted by myxamoebae and that regulate streaming together during the aggregation stage of Dictyosteliomycetes.

acritarch Unicellular organic microfossil with resistant and often highly ornamented outer test. Some considered probably to represent unicellular algae; others may represent various stages of plants and animals.

acrocarpous In mosses, a growth form in which the gametophore is erect, and the sporophyte terminates the main axis.

acrochaetioid Having a branched filamentous form similar to *Acrochaetium* (= *Audouinella*) in Rhodophyta.

acrogynous In Jungermanniales, a condition in which the apical cell of the gametophore produces the female sex organs, thus bearing the sporophyte terminally.

acropetal Developing from the base toward the apex.

actinomorphic In flowers, radially symmetrical; i.e., symmetrical about more than one plane.

actinostele Protostele with vascular tissue arranged in radiating arms interspersed with parenchyma.

adaxial(ly) Situated toward the axis of the plant.

adaxial meristem Dividing tissue on adaxial side of leaf that gives growth in thickness.

adenosine triphosphate See ATP.

adnate Condition in which unlike parts are organically fused.

advanced The state of a certain characteristic (or several) in a more evolved taxon.

adventitious Applied to a structure not arising in its usual place (such as adventitious roots emanating from a stem rather than a root).

adventitious root Root that develops from the stem.

adventive embryony Development of the embryo from nucellar or integument cells rather than cells in the female gametophyte.

aeciospore Dikaryotic spore produced in an aecium of the Uredinales (Heterobasidiomycetes).

aecium (pl. aecia) Structure (often cup-shaped) in which aeciospores are formed.

aeodan Complex phycocolloid substance occurring in cell walls of some red algae.

aeon A period of geologic time of 1×10^9 years; also spelled *eon* (see also Ga).

aerenchyma Tissue containing air spaces within the parenchyma.

aerial root Root arising from an aerial portion of the plant above the soil surface and not penetrating the soil.

aerobe An organism requiring the presence of free oxygen.

aerobic Having the ability to function as an aerobe.

aerola (pl. aerolae) Wall markings in the Bacillariophyceae consisting of thin areas bounded by ridges of siliceous material and having an aggregation of many fine pores.

aethalium (pl. aethalia) A sessile, rounded, or pillow-shaped fructification formed by a massing of all or part of the plasmodium in Myxomycetes.

agamospermy Reproducing asexually.

agamospory Asexual formation of embryo and susequent development of a seed.

agar Complex phycocolloid substance occurring in cell walls of some red algae; also prepared as a commercial product and used to solidify culture media.

agglutination Reaction resulting from union of blood serum proteins with homologous antigens.

aggregate fruit A fruit formed by the fusion of many separate carpels from a single flower.

aggregation Movement of amoebae in Dictyosteliomycetes toward one point prior to pseudoplasmodium (grex) formation.

ahnfeltan Complex phycocolloid substance occurring in cell walls of some red algae.

akinete Thick-walled resting cell or spore in the algae, generally incorporating original vegetative cell wall.

alar (region) cells Cells at basal margin of a moss leaf where it joins the stem, often differing in form or color from other leaf cells.

albuminous Applies to seeds containing endosperm at maturity.

albuminous cells Parenchyma cells in gymnosperm phloem morphologically and physiologically associated with sieve cells but not derived from the same initials.

aleurone (grain) Granules of storage protein in grass seeds.

algal layer In stratified lichen thalli, the zone in which algal cells occur (*see also* phycobiont).

algin Phycocolloid substance occurring in cell walls and intercellular spaces of brown algae; commercially marketed.

alginate Generalized name given to salt of alginic acid.

alginic acid Carboxylated polysaccharide material found in cell walls and intercellular spaces of brown algae.

allele Any one of several alternative forms of a gene at a given chromosome locus.

alloploid Polyploid resulting from an interspecific cross, thus containing two different genomes.

allopolyploidy The formation of a polyploid as a result of chromosome doubling in a hybrid in which there are few (or no) homologous chromosomes.

allotetraploid Tetraploid derived by doubling of the two different genomes from an interspecific cross.

alternation of generations (phases) Alternation of sexual gamete-producing phase with a meiospore-producing stage; usually the alternation of a haploid with a diploid generation.

alveolus (pl. alveoli) Elongated chamber on diatom frustule (wall) from central axis to margin and opening to inside of cell.

ameiotic Reduction of nuclear material by cleavage or splitting without visible chromosome formation.

amitotic Division of nuclear material by cleavage or splitting without visible chromosome formation.

amoeboid Resembling an amoeba; continually changing in shape.

amorphous Without any particular form; uniform throughout.

amphicribral Type of vascular arrangement in which the phloem surrounds the xylem.

amphigastria Ventrally located row of leaves in Jungermanniales (Hepaticae).

amphigenous Growing all around, or over entire surface.

amphiphloic Arrangement of phloem on both sides of xylem.

amphipolyploid See allopolyploid.

amphithecium Outer layer of cells in the early embryonic sporangium of bryophytes.

amphitropous Used with reference to an ovule attached at the middle, with its long axis parallel to the placental surface.

amyloid Starchlike; appearing blue-black after addition of iodine solution, giving a starchlike reaction.

amylopectin Storage polysaccharide starch composed of α-1,6 and α-1,4 glucoside linkages; also known as "branching factor" of starch.

amylose Portion of the storage polysaccharide starch, composed of α-1,4 glucoside linkages; presence gives amyloid (blue-black) reaction.

anacrogynous Condition in hepatics in which the female sex organs are lateral in position, having been formed from subapical cells; thus, the sporophyte is borne laterally.

anaerobe Organism able to grow in the absence of free oxygen.

anaerobic Having the ability to function as an anaerobe.

anagenesis Evolutionary change within a lineage without the lineage dividing into new lineages.

analogy Similarity based on common appearance or function that is not a result of descent from a common ancestor.

anastomose To interconnect or come together, as in the veins of certain leaves.

anastomosis (pl. **anastomoses**) Parts joined or coming together, as in fusion of two filaments.

anatropous Ovule position in carpel, with micropyle facing the placenta (flowering plants).

androecium Collective term for the stamens of a flower.

androgenous cell Cell in pollen grain that gives rise directly to male gametes.

androgenous initial One of the derivatives of the division of the androgenous cell of gymnosperm pollen, which divides to produce a second prothallial cell and a generative cell.

androsporal cell The first cell of gymnosperm pollen following meiosis, which divides to produce a prothallial cell and an androgenous initial.

anemophilous Having anemophily.

anemophily Pollination by wind.

aneuploid Differing from the usual diploid chromosome number by the addition or loss of an extra chromosome.

angiosperm Any vascular plant with seeds covered and protected by a carpel; also termed a *flowering plant*.

angiospermy Seeds contained within a carpel.

anisogamous Having anisogamy.

anisogamy Fusion of gametes of similar form, one of which is always larger; generally the larger gamete is considered the female, and the smaller the male.

anisospory Condition of spores in bryophytes in which the spores of a tetrad are of two different sizes, each size producing a gametophore of a different sex.

annular In the form of rings.

annular thickening Rings of secondary wall thickening of vessel elements and tracheids.

annular tracheid A xylem cell with secondary wall deposited as separate rings of lignocellulose on the inside of the wall; found in the earliest (fossil) vascular plants and in the protoxylem of some living vascular plants.

annulus A ring; in basidiomycetes, remnants of the partial veil on the stipe; in bryophytes, a specialized ring of cells on the sporangium involved in sporangium dehiscence.

anomalous cambium A specialized type of cambium.

anther Terminal portion of stamen bearing the microsporangia or pollen sacs.

antheridial initial A cell in gymnosperm pollen grains resulting from the division of the androsporal cell; the antheridial initial then divides to form a generative cell and a tube cell.

antheridiophore Specialized branch bearing antheridia in members of the Marchantiales (Hepaticae).

antheridium (pl. **antheridia**) Sex organ producing male gametes; in fungi and algae, a single cell; in bryophytes and vascular plants, many cells, including sterile jacket cells.

anthesis Time of flowering.

antibiotic A substance that prevents bacterial or fungal growth or survival.

antibody Protein produced in living body in response to introduction of foreign particles or cells (antigens); the antibody is specific and combines with its inducing antigen.

anticlinal Perpendicular to the circumference or surface (opposite to periclinal).

anticlinal division A cell division with the metaphase plate perpendicular to the nearest surface; applied to cell division in the tunica layers of the apical meristem of angiosperms.

antigen Substance that, when introduced into the body, stimulates the production of the defensive antibodies; bacterial flagella, capsules, cell walls, and many other substances function as antigens.

antipodal cell Vegetative cell in mature megagametophyte of flowering plants, usually located opposite the micropyle.

aperturate Having openings or apertures.

aphlebiae Unusual pinnae on the rachis of some ferns and fern allies.

aphyllous Without leaves.

apical cell A meristematic cell at the apex of a plant; in some more generalized vascular plants, a large meristematic cell at the apex of stems and leaves, from which all succeeding cells are derived.

apical meristem Dividing group of cells (tissue) at the tip of roots and stems, from which all succeeding cell initials are derived, and which gives primary growth.

apical placentation With ovules attached at the top of locule in ovary.

aplanogamete Nonmotile gamete.

aplanosporangium Sporangium in which nonflagellated spores (aplanospores) are produced.

aplanospore Nonflagellate spore.

apocarpous With carpels separated.

apogamous Asexual reproduction in which an embryo develops from an unfertilized cell in the female gametophyte; thus the sporophyte is of the same chromosome number as the gametophyte from which it arose.

Glossary **699**

apogamy Condition in which embryo develops without fusion of gametes.

apogeotropic The growth of roots away from the earth and the force of gravity (i.e., into the air).

apomixis Reproduction without meiosis and/or formation of gametes.

apophysis Swollen basal area of sporangium in some mosses.

aposporous Able to form a gametophyte directly from a sporophyte without the intervention of spores.

apothecium (pl. **apothecia**) Ascocarp (often cup-shaped or discoid) in which the hymenium is exposed at maturity of the ascospores.

appendicular Pertaining to appendices.

arborescent Treelike habit; resembling a tree in structure, appearance, and/or growth.

archegoniophore Specialized branch bearing archegonia in members of the Marchantiales (Hepaticae).

archegonium (pl. **archegonia**) Multicellular sex organ producing single female gamete; generally flask-shaped with elongate neck and swollen portion containing single egg.

archesporial cell In vascular plants, differentiated cell that will ultimately develop into a gametophyte; cell that undergoes meiosis to produce meiospores; spore mother cell.

archesporium Mass of cells from which sporogenous cells originate.

areola (pl. **areolae**) Markings in diatom wall consisting of thin areas bordered by ridges of siliceous material and often containing many finer pores.

areolation Arrangement of network of cells in the bryophyte leaf.

arid Extremely dry (climate).

aril Fleshy covering around seeds; in some conifers, formed as a fleshy outgrowth of the stalk; in flowering plants, of diverse origin and generally associated with tropical plants.

armored Possessing cellulose plates under plasma membrane, as in some dinoflagellates.

aromatic With a distinctive odor.

arthrospore Spore resulting from hyphal fragmentation.

articulate (articulated) Jointed or segmented; divided into portions that may be easily separated.

artificial taxon A taxon containing organisms more closely related to organisms in another taxon at the same level; same as unnatural taxon.

ascidial Having a cup- or pitcher-shaped foliar appendage in flowering plants.

ascidium A cup- or pitcher-shaped organ in flowering plants.

ascocarp Ascus-bearing structure, or "fruiting body" of ascomyetes.

ascogenous hypha In ascomycetes, a hypha that develops from the ascogonial surface after plasmogamy and gives rise to asci.

ascogonium (pl. **ascogonia**) In ascomycetes, female cell that receives nuclei from the antheridium or spermatia.

ascospore Spore formed within an ascus, typically the result of a meiotic followed by a mitotic division.

ascostroma (pl. **ascostromata**) A stroma within which locules and asci develop.

ascus (pl. **asci**) A saclike cell in which ascospores are produced; in Euascomycetes, the cell in which both karyogamy and meiosis occur.

asexual reproduction Production of more individuals identical (except in some fungi) to the parent without syngamy and meiosis.

assimilative Growing and absorbing food; vegetative or nonreproductive.

assimilative system Generalized term used to refer to any system of filaments or organ that enables a plant to manufacture or obtain its chemical source of energy.

asterad In flowering plants, a type of embryo development in which the terminal cell divides longitudinally, and both basal and terminal cells contribute to the embryo.

asynchronous Indicates that events do not occur concurrently or all at the same time.

ATP Adenosine triphosphate: a substance containing energy in the form of high-energy phosphate bonds (this bond energy directly or indirectly drives all energy-requiring processes of life).

autodigestion Process in which enzymes are secreted that digest portions of reproductive structures (e.g., spent portions of gills in *Coprinus* species and the mature gleba in many Gasteromycetes).

autoecious Term applied to rust fungi that require a single host to complete the life cycle.

autogenous (hypothesis) Refers to the origin and differentiation of eukaryotic organelles within the cell, rather than from outside (*see* xenogenous).

autoploid Polyploid resulting from the duplication of a single genome.

autopolyploidy The formation of a polyploid as a result of chromosome doubling in a plant in which many, or even all, of the chromosomes have homologues present.

autotroph Organism that requires for its growth only inorganic substances and light or chemical energy.

autotrophic Having the ability to function as an autotroph.

auxiliary cell In some Florideophyceae, a cell to which the diploid zygote nucleus is transferred and where growth of the carposporophyte is initiated.

auxiliary cell branch In Florideophyceae, the specialized filament of cells in which the auxiliary cell occurs.

auxospore In diatoms, large cell often produced as a result of sexual reproduction.

auxotroph Organism that needs an external supply of some organic substances, especially vitamins.

avirulent Condition in which a strain of a known pathogenic species is unable to establish itself and reproduce in the host.

awn A bristlelike growth or extension.

axenic Term used to refer to a "pure" culture containing a single organism, i.e., it is without contaminating organisms of any kind.

axial Belonging to the axis; around or in the direction of or along an axis.

axil Angle formed between axis and attached appendage.

axile Central in position, as in some algae in which the chloroplast is centrally located in the cell.

axile placentation Attachment of ovules in central area of gynoecium in flowering plants.

axillary Growing from, or pertaining to, an axil.

axillary bud A branch or flowering bud occurring in the axil of a leaf.

bacillus (pl. bacilli) Straight, rod-shaped bacteria.

bacteriochlorophyll Photosynthetic pigment occurring in purple photosynthetic bacteria.

bacteriophages Viruses occurring in bacteria; also termed *phages*.

bark In vascular plants, tissues external to vascular cambium.

basal body Cylindrical body in cell composed of a ring of nine units, each of which is made up of three microtubules; present at base of each flagellum in motile cells (also called a *kinetosome*).

basal cell In Florideophyceae, a cell cut off from the supporting cell.

basal placentation With ovules attached to the bottom of locule in ovary.

basidiocarp Basidium-bearing structure, or "fruiting body," of basidiomycetes.

basidiospore Spore formed exogenously on a basidium, generally following karyogamy and meiosis.

basidium (pl. basidia) In basidiomycetes, cell in which karyogamy and meiosis occur, and upon which basidiospores are borne.

basifixed Attached at the base.

basipetal Development or maturation from the apex toward the base.

benthic (benthonic) Living on and generally attached to the bottom of aquatic habitats.

berry A simple, fleshy, usually indehiscent fruit with one or more seeds (e.g., tomato).

beta-carotene Accessory carotene pigment present in chloroplasts of land plants and most algal groups.

bicollateral bundle Vascular bundle with phloem both external and internal to the xylem.

bifid Forked into two equal parts.

bifurcate Divided into two halves.

binary fission Reproduction occurring when a single cell divides into two equal parts.

biogenesis Evolution of cells or organisms.

biotypes Organisms with the same genotype.

bipod A two-legged stand.

bisaccate Pollen grains with two air sacs or bladders; mainly coniferous pollen (e.g., pine).

biseriate Having two linear rows of cells.

bisexual Having both sexual reproductive structures (male and female) produced by any one individual.

bisporangiate cone A cone containing both megaspores in megasporangia and microspores in microsporangia.

bispore In Florideophyceae, spore produced in a bisporangium (two spores are produced in each sporangium, and they are believed to be meiospores).

blade The broadened, flattened, portion of plant; in large marine algae can be the whole thallus; in vascular plants, part of leaf.

blanc mange Dessert made from milk and gelatinous, mucilaginous, or starchy substances.

bloom Dense growth of planktonic algae giving a distinct color to the water body.

bordered pit A pit in which the secondary wall overarches the pit membrane.

bordered pit pair A pair of opposing pits, both of which are bordered.

bordered pitted tracheid A tracheid with bordered pits, mainly on the radial wall.

bothrosome In Labyrinthulomycetes, organelle responsible for producing the membranes of the ectoplasmic net.

brackish With salinity less than that in the marine environment (generally less than 20 parts per thousand).

bract Leaflike structure subtending one or more flowers or other reproductive organs.

bracteole A small bract.

branch root A root that arises from another, older root.

bryologist One who studies bryophytes (liverworts and mosses).

bud Embryonic shoot with immature stems and leaves.

bud scales Highly modified leaves that surround and protect stem, branch, or reproductive primordia.

budding Type of asexual reproduction in which a small protuberance develops, enlarges, and is separated from the parent cell.

bulb Modified, usually underground, shortened stem enclosed with fleshy leaves.

bulbil A small bulb.

button stage Early developmental stage of basidiocarp in Agaricales when it has the form of a miniature mushroom.

buttress root An adventitious root on a stem that functions as a support; mainly in monocotyledons (e.g., Pandanales).

caducous Deciduous; not persistent.

calcareous Limy; consisting of or containing calcite, or calcium carbonate.

calcite Precipitate of calcium carbonate.

callose Cell wall constituent in some cells of Laminariales (brown algae) and vascular plants.

calyptra The enlarged portion of the archegonium that surrounds and protects the developing sporophyte in bryophytes, forming a sheathing cap over the sporangium in most mosses.

calyx Sterile outer whorl of flower parts composed of sepals external to corolla.

cambium (pl. **cambia**) Lateral meristem in vascular plants that produces either secondary xylem, secondary phloem, and parenchyma, usually in radial rows, or cork.

cap Common name given to pileus of a mushroom.

capillitium In some fungi and Myxomycetes, threadlike strands (often forming a network) interspersed with spores.

capitate Having a head.

capitulum Small head; in Sphagnidae (Musci), dense tuft of branches at apex of gametophyte; in angiosperms, an inflorescence composed of a dense aggregation of sessile flowers.

capsule A case; in bacteria, a nonliving layer of slime material of varying thickness and viscosity, situated immediately outside the cell wall or outer membrane in some prokaryotic cells (may consist of a single polysaccharide or of such material plus amino acids, proteins, etc.); in bryophytes, the spore case containing meiospores and sterile tissue; in angiosperms, a type of dry, dehiscent fruit formed from more than one carpel.

carboxysome Crystalline polyhedral bodies in prokaryote cells involved in carbon dioxide fixation; also called *polyhedral body*.

carinal canal Elongate cavity in sphenophyte stems, thought to function in aeration.

carotene General name for group of orange carotenoid pigments.

carotenoid A yellow, orange, or red pigment that often serves as an accessory pigment in photosynthesis or may serve as a protection against radiation (includes xanthophylls and carotenes; soluble in organic solvents, such as ethanol and acetone).

carpel Floral appendage that encloses ovules; characteristic of angiosperms.

carpellary Bearing carpels.

carpocephalum (pl. **carpocephala**) Specialized, erect, sporangium-bearing branch in Marchantiales (Hepaticae).

carpogonial branch In Florideophyceae, specialized filament of cells, at the end of which is borne the carpogonium.

carpogonium (pl. **carpogonia**) Female gametangium in the red algae.

carposporangium (pl. **carposporangia**) Sporangium produced directly or indirectly as a result of division of the zygote nucleus in red algae.

carpospore Spore produced by a carposporangium in red algae.

carposporophyte Collection of carposporangia occurring in chains on the gonimoblast filaments in Florideophyceae; also referred to as the *gonimoblast*.

carrageenan Complex phycocolloid occurring in the cell wall of some red algae; commercially marketed.

casparian strip Strip of suberin in radial, upper, and lower walls of cells in the endodermis of the root.

cast Preservation of a fossil in the form of a mold that shows the form of the organism but lacks internal structure.

cataphyll In flowering plants, small, scalelike leaf often serving for protection.

catkin Spikelike inflorescence, usually containing scaly bracts, and frequently pendent.

caulid Main shoot or "stem" of the bryophyte gametophore, which supports the photosynthetic and reproductive organs.

cauline Belonging to or arising from a stem.

cauline bundle A vascular bundle that is part of the stem tissue.

cauline placentation Ovules attached to material derived from stem tissue.

cell plate Collection of vesicles forming new membrane and wall in a dividing cell.

cell wall Material produced by and surrounding the protoplast of most plant cells.

cellular endosperm Endosperm in which cell walls form at the first division of the primary endosperm nucleus and all subsequent ones.

cellulose Main cell wall structural component of plants (excluding most fungi and some algae); a polysaccharide made up of many glucose molecules composed of β, 1-4, 1-6 glucoside linkages.

central cell Cell that is axile or central in position in a polysiphonous thallus (e.g., *Polysiphonia*, Florideophyceae), which is surrounded by a series of pericentral cells; large cell in megagametophyte.

centrarch Having solid xylem, with protoxylem elements in the center (see *mesarch*).

centric Central; common name for diatoms with markings centrally arranged.

centrifugal(ly) Developing from the center toward the outside.

centriolar plaque Structure associated with nuclear envelope and lying at the poles of the spindle during nuclear division; believed to be a homologue of the centriole.

centriole Cell organelle consisting of nine units, each composed of three microtubules, lying adjacent to the nucleus and often concerned with the flagella and spindle formation; generally characteristic of animal cells and motile plant cells.

centripetal(ly) Developing from the outside toward the center.

cephalodium (pl. **cephalodia**) Epiphytic lichen growing as a wartlike protuberance on upper surface of host lichen.

chalaza Basal part of ovule, adjacent to the stalk and opposite the micropylar end.

chalazal end The chalazal end of the ovule.

chemoautotroph Autotrophic organism that derives energy from chemical reactions, such as oxidation-reduction reaction of inorganic compounds.

chemotactic The attraction of motile cells by certain chemical substances.

chemotroph *See* chemoautotroph.

chiropterophily Pollination by bats.

chitin Polysaccharide cell wall material composed of glucose units and nitrogen.

chitosan A deacetylated form of chitin found in the walls of many Mucorales.

chlamydomonad Having a morphology similar to the motile unicellular green alga *Chlamydomonas*.

chlamydospore Thick-walled spore produced by rounding up of hyphal cell protoplast and secretion of new wall.

chlorobium chlorophyll Photosynthetic pigment occurring in the green sulfur bacteria.

chlorococcine group Series of green algae that are nonmotile and nonfilamentous in vegetative condition.

chloronema *See* primary protenema.

chlorophyll General name for green photosynthetic pigment, soluble in organic solvents (e.g., ethanol and acetone).

chlorophyllose Containing chlorophyll.

chloroplast Cell organelle (plastid) with membranes (thylakoids) containing photosynthetic pigments (chlorophylls); involved in photosynthesis; gives green color to plants.

chromatid One of two identical longitudinal halves of a chromosome that are joined by a single centromere.

chromatin Nucleoprotein complex of chromosomes.

chromatophore Nonlamellar, pigment-bearing cell organelle occurring in photosynthetic bacteria and animals.

chromoplast Plastid containing yellow or orange pigments.

chromosome Elongate structure in nucleus made of DNA and histone; contains genes in a linear order.

chrysolaminaran Storage polysaccharide composed of β-1,3 1,6 glucoside linkages occurring in chrysophytes and xanthophytes.

cilium (pl. **cilia**) Marginal filamentous hair: in the lichens, hairy extensions on the thallus; in bryophytes, may occur on the leaves or among the inner peristome teeth; also used for motile organelle identical to a short flagellum.

cincinnus A helicoid cyme with short internodes.

circadian rhythm Natural rhythm (of approximately 24-hour cycle): occurs, for example, in the maturation and firing of sporangia in *Pilobolus* (Zygomycetes), in which patterns are repeated over 24-hour cycles.

circinate vernation Characteristic coiling of young leaves (fronds) of true ferns.

circular bordered pit A bordered pit that is circular.

circumscissile Dehiscing by being cut around circumference, as in a capsular fruit.

cisternae Flattened sheets consisting of two unit membranes and the space between them.

cladist One who practices cladistics.

cladistics The erection of a classification based on shared derived traits.

cladode Leaflike small branch.

cladogenesis Phyletic splitting into two lineages.

clamp connection In basidiomycetes, looplike structure connecting adjacent hyphal cells; it is produced during formation of new cells by dikaryotic hyphae.

class A level in the Linnaean hierarchy that includes closely related orders; also, a general term for a group of related objects.

classification The arrangement of objects into classes; the act or process of classifying.

classify To arrange objects into groups.

cleavage furrow The groove that forms around the middle of a cell just prior to splitting into two new cells (e.g., in bacteria).

cleistothecium (pl. cleistothecia) Ascocarp in which the asci are completely enclosed at maturity of ascospores.

clonal With reference to a clone.

clone A population of individuals descended by mitotic division from a single ancestor; all individuals of a clone are of the same biotype or genotype.

cluster analysis Numerical analysis designed to ascertain if clusters of individuals occur in a set of data.

coal ball Concretion of calcium carbonate or silica occurring in certain coal seam layers; often containing abundant, well-preserved plant remains.

coccoid Spherical cell type and growth form or morphological type.

coccolith Scale composed of $CaCO_3$ on organic base occurring in the Prymnesiophyceae.

coccus (pl. cocci) Spherical bacterial cell.

coenobium (pl. coenobia) Colony of algal cells of a definite number, not increasing at maturity; many have definite arrangement of cells.

coenocyte Multinucleate cells or thallus in which nuclei are not separated from one another by septa as in the plasmodia of myxomycetes, hyphae of lower fungi, and some green algae (Bryopsidophyceae) and xanthophytes.

coenocytic Having the condition of a coenocyte.

coenosorus A cluster of sori.

coenozygote Multinucleate zygote in some zygomycetes.

collateral bundle Vascular bundle with phloem only external to the xylem.

collenchyma Primary supporting tissue of vascular plants with unevenly thickened primary walls.

colonial Forming a colony.

colony In bacteria, yeasts, and algae, a mass of individuals or cells of one species living together; in some instances it consists of a few individuals that can be attached to one another in a definite or regular pattern.

colpate Pollen grains with one to several furrows *(colpi)* in the exine.

colporate Pollen grains with both furrows *(colpi)* and pores.

colpus (pl. colpi) In the exine of pollen grain, a furrow running from one pole to another.

columella Small column; in Myxomycetes, often a continuation of stipe into sporangium; in zygomycetes, bulbous septum separating sporangiophore and spore case; in Anthocerotae and Musci, central column of sterile cells in the sporangium surrounded by the sporogenous layer.

columellate Having a columella.

companion cell Specialized phloem parenchyma cell in angiosperms derived from same initial cell as sieve tube element.

compound cone With simple cones clustered or grouped on a secondary axis, as in the gnetophytes.

compound dichasium An inflorescence made up of many dichasia.

compound leaf Leaf with a blade divided into several units or leaflets.

conceptacle In some brown algae, a cavity in the thallus in which gametangia are formed.

conchocelis stage Filamentous stage in the life history of certain members of the class Bangiophyceae (e.g., *Porphyra*).

conchoidal Concave, resembling one half of a bivalve (clam) shell.

conchospore In Bangiophyceae, spore produced on conchocelis stage singly in a conchosporangium.

conchosporangium (pl. conchosporangia) In Bangiophyceae, sporangia produced in a series at the end of a conchospore branch.

conduplicate Folded together lengthwise.

cone A compact strobilus containing sporophylls and sometimes thickened bracts.

cone bract A modified leaf attached to a cone axis.

conidiophore Specialized hypha bearing conidia.

conidiospore See conidium.

conidium (pl. conidia) In bacteria (Actinomycetales), used for chains of endospores formed in hyphal tips; in fungi, essentially a separable portion of a hypha that functions as an asexual propagule.

conjugate division Simultaneous division of the paired nuclei in dikaryotic cells.

conjugation Fusion, usually of isogametes or isogametangia; in bacteria, transfer of genetic material from a donor to a recipient; in phycomycetes and some green algae, fusion of gametangial protoplasts; in yeasts and others, fusion of cell protoplasts following development of a conjugation tube.

connecting filament In Florideophyceae, the nonseptate filament that grows out of the base of the carpogonium and transfers the diploid nucleus to the auxiliary cell (or in some instances, from one auxiliary cell to another).

context Sterile inner part of the cap or pileus in basidiomycetes.

contiguous Adjoining or adjacent.

continental shelf The gradually sloping, submerged (below sea surface) portion of a continential region that occurs in depths less than about 200 m.

contractile root A root that undergoes contraction at some time during its development and thereby effects a change in position of the shoot with regard to the ground.

contractile vacuole Small, pulsating organelle of some algae and fungi; generally in motile cells near the flagellar end.

coralline Refers to the lime-encrusted red algae of the Corallinaceae (Florideophyceae).

coralloid Erect, branched, shrublike, or corallike in form.

cordate Heart shaped.

coriaceous Leathery.

cork In vascular plants, protective tissue impregnated with suberin in the cell walls; derived from the cork cambium.

cork cambium Lateral meristem in stems and roots of some vascular plants that produce cork or phellem cells.

corm Shortened underground stem with scaly leaves; vertical in position.

corolla Sterile whorl of flower parts composed of petals (interior to calyx).

corpus (pl. corpi) The body of a bladdered pollen grain (e.g., in some conifers).

cortex In some brown and red algae, tissue internal to the epidermis but not in the central position; in lichens, the compact surface layer or layers of the thallus; in vascular plants, the portion between the vascular tissue and the epidermis.

corticated Having a cortex; parenchymatous or appearing so.

corymb More or less flat-topped, indeterminate inflorescence.

costa (pl. costae) Ridge: in the diatoms, wall marking formed by well-defined siliceous ridges; in bryophytes, the midrib or multilayered area of leaf or thallus.

cotyledon First leaf produced by embryo of seed plants; also termed *seed leaf.*

cover cells In Florideophyceae, the superifical cells of the thallus that are attached to the supporting cell and cover the tetrasporangium (e.g., *Polysiphonia*).

crassinucellate An ovule in which some of the early products of division of the archesporial cell do not participate in development of the megagametophyte.

crassula (pl. crassulae) Thickening of intercellular material and primary wall along the upper and lower margins of a pit pair in the tracheids of gymnosperms.

crenulate Scalloped.

crista (pl. cristae) Invagination of inner mitochondrial membrane where respiratory enzymes occur.

crossing-over Exchange of parts of two chromatids.

crozier (formation) In ascogenous hyphae, formation of a hook in which conjugate nuclear division occurs, followed by cytokinesis; formation may or may not immediately precede ascus formation; in ferns and allies, the coiled juvenile leaf or stem.

crozier stage Stage of development in a fern when the emerging leaves are tightly coiled into "fiddleheads," or croziers.

cruciate Cross-shaped.

crustose Encrusting: lichen growth form in which the thallus adheres tightly to the substrate.

crystal Body having symmetrically arranged plane surfaces; in plants often formed of calcium oxalate.

culturing Term used with reference to the growing of organisms under controlled conditions, especially bacteria, fungi, and microscopic algae.

cupule Cup-shaped structure surrounding or subtending the ovules in pteridosperms.

cuticle The external, waxy layer covering the outer walls of epidermal cells: in bryophytes and vascular plants, it consists of a waxy compound called *cutin* that is almost impermeable to water; in algae, it may contain other compounds.

cutin Waxy material covering external cell surfaces of some bryophytes and vascular plants (the layer is referred to as a *cuticle*).

cutinization Deposition of cutin.

cyanophycin granule Protein storage body occurring in the bluegreen algae.

cyanophyte starch Storage polysaccharide composed of α-1,4, 1,6 glucoside linkages; considered to be amylopectin.

cyme A branched, flat-topped or convex inflorescence in which the terminal flower on each axis blooms first.

cyphella (pl. **cyphellae**) Cup-shaped depression forming regular opening through the lower cortex of some foliose lichens.

cyst Resistant, sporelike body (often thick-walled), that develops by the rounding up of reproductive cells (lower fungi and Myxomycetes) or vegetative cells (bacteria and some algae); in algae, forms within vegetative cell.

cystidium (pl. **cystidia**) Sterile structure produced in the hymenium of some basidiomycetes.

cystocarp In Florideophyceae, the structure comprising the gonimoblast (carposporophyte) and surrounding pericarp (gametophyte tissue).

cystolith A stalked crystal of calcium carbonate.

cytokinesis Cytoplasmic division, usually following nuclear division.

cytological With reference to the structure of the cell.

cytoplasm Living part of cell, exclusive of nucleus.

cytosis The budding, fusion, evagination or invagination of cell membranes, as in cell cleavage (*see also* endocytosis, exocytosis, and phagocytosis).

deciduous Falling off; not persistent.

decurrent Extending downward.

decussate Pair of opposite appendages occurring alternately at right angles to one another (usually used with leaves, sporophylls, or branches).

decussate branching Branching of stems with pairs of opposite branches alternating at right angles.

dehiscence Method of opening at maturity.

dehiscent Splitting open at maturity.

dendroid Treelike in form.

dentate With teeth, as on the margin of a leaf.

deoxyribonucleic acid (*See* DNA).

derivatives Cells in a meristem that are destined to become part of the mature plant body.

dermatophyte A fungus colonizing keratinized material (hair, skin, nails of man and animals).

detached meristem A patch of meristematic tissue removed some distance from an apical meristem.

determinate (growth) Having a fixed, definite limit of growth.

determinate head A flower head with the oldest flower in the middle.

determinate umbel An umbel in which the branches of the inflorescence do not subdivide.

detritus Particulate organic matter.

dextral Having right-hand helical arrangement; in charophytes, refers to sheath cells around the oogonia.

diarch Having two protoxylem poles.

diaspore Any unit or organ of dispersal.

diatomaceous earth Deposits of the silicified walls of fossil diatoms.

diatomite Diatomaceous earth; fossil deposits of diatoms.

dichasium (pl. **dichasia**) Determinate inflorescence with oldest flower situated between two younger flowers.

dichotomous (dichotomy) Forked into two (more or less) equal parts or branches.

dichotomous branching Branching in which the two arms are more or less equal.

dichotomous venation Branching of veins in leaves into (more or less) equal parts without fusion of veins after they branch.

dicotyledon Group of flowering plants (angiosperms) characterized by having two seed leaves, eustelic stem with cambium, and floral parts in fours or fives or multiples of these.

dictyosome Complex of flattened double lamellae (stacks of cisternae) often with peripheral vesicles; similar to golgi apparatus.

dictyostele Stele with cylindrical arrangement of xylem and phloem in separate vascular bundles; a modified siphonostele.

diffuse (growth) Cell division and elongation occurring throughout the plant.

digitate With broad fingerlike divisions.

dikaryon In ascomycetes and basidiomycetes, hyphae in which each compartment contains a pair of compatible nuclei; initiated through plasmogamy of compatible gametangia or hyphae.

dikaryotic Having the condition of a dikaryon.

dikaryotization Process by which a hypha becomes dikaryotic.

dimorphic In fungi, condition in which the assimilative structure can exist either as budding cells or as hyphae (mainly Hemiascomycetes and Heterobasidiomycetes); in ferns, refers to fronds of a plant that differ conspicuously in morphology.

dioecious Condition in seed plants in which pollen-bearing and ovule-bearing cones or flowers are borne on different plants.

dioicous Condition of bryophyte gametophyte in which the gametophore bears only one kind of sex organ; unisexual.

diploid Having a single set of paired chromosomes (twice the number of chromosomes as in the gametes); $2n$.

diploid apogamy The development of an embryo from a cell of the female gametophyte (other than an egg) that becomes diploid spontaneously.

diploid parthenogenesis The development of an embryo from an egg that becomes diploid spontaneously without fertilization.

diploidization Process in some Florideophyceae in which a specialized haploid cell (auxiliary cell) receives a diploid zygote nucleus or a derivative of it; term used in fungi for *dikaryotization*.

discoid Round and/or flattened with rounded margins; dish-shaped.

discrete Separate or distinct.

dispersed spore genus A genus erected for spores that have been shed and dispersed and recovered from sediments.

dissected Much divided; divided into many slender segments.

disseminule Plant part that gives rise to new plant.

distal(ly) Remote from place of attachment (opposite to *proximal*).

distichous Two-ranked on opposite sides of plant (stem); arranged in two vertical rows; with conspicuous bilateral symmetry.

distromatic Having a thallus two cells thick.

divergent Spread out from a common center.

division A level in the Linnean hierarchy that includes closely related classes, usually restricted to the plant kingdom; identical to phylum in classification of animals.

DNA (deoxyribonucleic acid) Nucleic acid in cells, primarily in nucleus but also in mitochondria and chloroplast; main (genetic) component of chromosomes; transmission of heredity.

dormancy Period of inactivity during which there is no growth; growth is resumed only if certain requirements are met.

dorsal The side opposite an axis; on the back.

dorsal placentation Attachment of ovules to the midrib of carpels in apocarpous gynoecia.

dorsifixed Attached at the back.

dorsiventral Flattened with definite dorsal and ventral surfaces.

dorsiventrality Having a dorsiventral organization or symmetry.

double fertilization Process in flowering plants where there is a fusion of two sperm nuclei, one with an egg nucleus to form the zygote and the other with the polar nuclei (a pair of megagametophyte nuclei) to form the primary endosperm nucleus.

downgrade evolutionary pathway (theory) The hypothesis that the leafy habit was the first to evolve in bryophytes, with the thalloid habit derived later (opposite to *upgrade theory*).

drupe Fleshy, one-seeded, indehiscent fruit.

druse A spherical crystal with many projections.

dulse Dried preparation of *Palmaria palmata* (Florideophyceae); commercially marketed.

dwarf shoots In gymnosperms, branches of an axis that are much smaller than the main branch (e.g., in cones of *Cordaites* and on branches of *Ginkgo* and *Larix*).

ecotype A population, or series of populations, of a species that shows adaptation to a local habitat, and in which the specific adaptations are genetically transferred.

ectexine *See* exine.

ectocarpen A gamone, or sexual attractant, produced by the female gamete in *Ectocarpus* (Phaeophyceae).

ectomycorrhizae Symbiotic association between roots of higher plants and fungi in which the fungus forms a mantle around the root, and hyphae extend inward between cells of the root cortex.

ectophloic Arrangement of phloem external to the xylem.

ectoplasmic net Network of tubules within which gliding cells of Labyrinthulomycetes move; the structurally similar "rhizoids" of *Thraustochytrium* and similar fungi.

ectoplast The outer membrane of the protoplast.

Glossary **707**

effused Referring to type of basidiocarp that is spread out or flattened.

effused-reflexed Referring to a basidiocarp with a resupinate (flat or spreading) portion attached to the substrate and an upper portion extending out like a shelf.

egg Female gamete.

egg apparatus Egg and two synergid nuclei located at the micropylar end of the megagametophyte in angiosperms.

ejectosome In some flagellated algae (dinoflagellates, cryptophytes, raphidophytes), the cytoplasmic organelle ejected when the organism is disturbed; known also as *trichocyst*.

elater Sterile hygroscopic cell among the spores in the sporangium of many Hepaticae; outer part of the meiospore in *Equisetum*.

elaterophore In liverworts, a specialized, elater-bearing mound of cells arising from either the base or roof within the sporangium.

embryo Immature, incompletely differentiated organism; may or may not be retained by parent plant for part of its development; usually diploid, developing from zygote; often dormant.

embryo sac Megagametophyte in angiosperms.

embryogeny Formation of the embryo.

embryonal tube Tier of cells produced by basal tier of suspensor cells in the embryo of vascular plants; also termed *secondary suspensor*.

embryophyte A plant in which an embryo develops *in situ*.

enation Outgrowth from any plant surface; usually referring to a local outgrowth from a stem.

enation theory A theory that proposes that leaves evolved by a progressive outgrowth (enation) from the stem.

encapsulated Development within a capsule.

encrusted Adherent and completely covered with lime as in some of the prostrate coralline red algae.

encrusting Having a prostrate habit, in which the thallus is usually in a flattened, thin layer tightly attached to and spreading over the surface of the substrate.

endarch A type of primary xylem maturation in which the oldest xylem elements (protoxylem) are closer to the center of the axis.

endexine Inner layer of exine of bryophyte spores and pollen grains of vascular plants.

endocarp Innermost layer of the carpel wall, or pericarp of the fruit in angiosperms.

endodermis Inner layer of cortex tissue in vascular plants; often contains cells with thickened walls on the periclinal surface, with or without casparian strips, and often surrounds individual vascular bundles.

endogenous Development from the inside; in the Jungermanniales (Hepaticae), refers to a branch produced by internal stem cells and pushing through mature cells.

endomycorrhizae Symbiotic association between higher plant roots and fungi in which the hyphae grow into the cells as well as between them (characteristically occurs in orchids, Ericaceae, etc.).

endophyte Plant growing within another plant.

endoplasmic reticulum (ER) Internal membrane system found in the cytoplasm, connected with the nuclear envelope; it may have ribosomes attached, in which case it is called *rough endoplasmic reticulum* (*RER*); if ribosomes are not attached, it is called *smooth endoplasmic reticulum* (*SER*).

endosperm Cellular food reserve of angiosperms resulting from double fertilization.

endosperm nucleus The nucleus that will develop into the endosperm.

endosporal(ly) Referring to the development of a gametophyte within the confines of the spore wall.

endospore Spore formed within parent cell; in the bacteria, a thick-walled resistant spore; in bluegreen algae, a thin-walled spore; the term is also used for inner layer of a spore wall.

endosporic Undergoing cell divisions within a spore before the spore coat is ruptured.

endosporic development Development of gametophyte within a meiospore.

endosporous(ly) *See* endosporal(ly).

endostome Inner peristome teeth of bryophytes.

endosymbiosis The ingestion by one cell of another, which forms an integral part of the new cell.

endosymbiotic Organism that lives within another organism (*see* endosymbiosis).

endothecium Inner layer of cells in the early embryonic sporangium of bryophytes.

endozoochory Dispersal of diaspore through ingestion and defecation by animals.

entomophily Pollination by insects.

ephemeral Short lived; transient; persisting very briefly.

epicotyl Portion of seedling above

the cotyledons that will develop into the shoot (includes stems, leaves, etc.).

epidermis Primary tissue forming a protective cell layer on surface of the plant body; outer layer(s) of vascular plants consisting of cuticle, stomata, and trichomes.

epigynous Above the gynoecium in angiosperms.

epiphragm In most Polytrichidae (Musci), a multicellular, parchmentlike membrane closing the mouth of the sporangium after the operculum has fallen; consists of the expanded apex of the columella.

epiphyte Plant growing upon another plant but not necessarily making any nutritional connection.

epitheca (pl. **epithecae**) Outer cell half, or frustule, of diatom cell; often refers only to outer siliceous wall.

epithecium Layer on hymenium surface of apothecium formed by closely packed tips of paraphyses, often forming a brightly colored layer.

epithelial cells Resin-producing cells in the resin canals of some conifers.

ergastic substance Nonliving product of metabolism in plant cells.

erose With an irregular margin that gives the appearance of derivation by erosion or gnawing.

erumpent Swollen.

eukaryotic With reference to organisms (eukaryotes) containing membrane-bound organelles (e.g., nuclei, mitochondria, chloroplasts) within the cells.

eusporangiate Having a sporangium that originates from a multicellular primordium.

eusporangiate sporangium A sporangium that originates from a multicellular primordium; eusporangium.

eusporangium *See* eusporangiate sporangium.

eustele Cylindrical stele composed of anastomosing vascular bundles.

evagination An outgrowth, or unsheathing.

evanescent Of short duration; vanishing or melting away.

evergreen Remaining green during the dormant season, usually due to persistence of leaves.

eversion *See* inversion.

evolution The process of genetic variability and natural selection acting on populations through time; change with time, resulting in biological diversity that can be arranged in a hierarchy.

evolutionary divergence The process of populations becoming different from each other as a result of evolving in different directions.

exalbuminous Referring to seeds lacking endosperm at maturity.

exannulate Without an annulus.

exarch A type of maturation of the primary xylem in which the oldest xylem elements (protoxylem) are located closest to the outside of the axis.

exine Outer thick layer of spores and pollen grains, usually divided into two main layers: an outer ectexine and an inner endexine.

exocarp Outermost layer of carpel wall, or pericarp.

exoenzyme An enzyme secreted or formed externally by the protoplast of a cell and functioning outside the cell proper (chemically breaking down organic materials as a source of energy for growth).

exogenous Developing on the outside.

exosporal The development of the gametophyte outside the confines of the spore wall.

exosporic development Gametophytic development mainly outside the spore.

exosporium The thin, loosely fitting outer layer or coating surrounding the endospore wall of some bacteria.

exostome Outer peristome teeth of bryophytes.

exozoochory Dispersal of diaspores through attachment to the surface of an animal.

extant Living; used in reference to present-day plants.

extracellular enzyme Enzyme acting outside the cell (*see also* exoenzyme).

extranuclear Occurring outside the nuclear envelope.

extranuclear division Nuclear division in which the spindle forms outside the nuclear envelope; the latter structure often disappears during the division.

extrorse Opening of the anther toward outside of flower.

eyespot Red to orange carotenoid-containing organelle in motile cells of many algae.

facultative Being able to live under different conditions.

facultative apomict A plant that sometimes reproduces by apomixis and sometimes by sexual reproduction.

facultative parasite An organism with the ability to exist either saprobically or parasitically, or an organism capable of using one of alternate functional pathways; facultative heteromorph; *see* mixotroph.

fairy ring A circle of mushrooms or other sporocarps, mainly basidiomycetous, formed by an underground mycelium; the mycelium grows radially from its initial starting point, actual extension being restricted to the periphery.

false annulus A discrete group of thick-walled cells on the jacket of the sporangium of some ferns; not directly influencing dehiscence.

false branching In some bluegreen algae, breaking of filament, with one or both ends protruding from the sheath and appearing as a branch.

family A level in the Linnean hierarchy within an order that includes closely related genera.

fascicle A close cluster or bundle.

fascicular Spiral or whorled.

fascicular cambium That part of the cambium that develops within the vascular bundle.

fasciculate In clusters.

female gametophyte The gametophyte that produces the female sex organs; gametophytic cells ($1n$) that develop from divisions of the functional megaspore within an ovule; in flowering plants, the gametophyte within which the female gamete (egg) develops.

fertile telome One branch of an ultimate dichotomy at the tip of a branch; it bears a sporangium or other reproductive structure.

fertilization Union of (male and female) gametes.

fertilization tube Branch from male gametangium that transfers male nuclei or gametes to female gametangium in Oomycetes.

fetid Having an offensive odor.

fiber An elongated support cell with secondary walls.

fibril Threadlike structure; in the moss *Sphagnum*, refers to the fiberlike thickenings on the inner faces of large hyaline cells in the leaf or stem cortex; submicroscopic structure in many cells.

fibrillar Made up of fibrils; fibrous; microscopic or submicroscopic.

fibrous system A root system made up of many more or less equal-sized roots.

filament Threadlike process, structure, or growth form; in the angiosperms, the stalk of the stamen supporting the anther.

filamentous Elongate, threadlike, multicellular growth form or morphological type.

filiform Thread shaped.

filiform apparatus Invagination in the wall of synergids in the female gametophyte of flowering plants.

fission Splitting in two; characteristic especially of bacteria, some yeasts and myxomycetes and bluegreen algae; also termed *binary fission*.

fixation Conversion of N_2 to N-containing compounds by bacteria and bluegreen algae; also, term used for preparing biological material for study (involves preservation of material).

flagellate Referring to cells that have organelles for motility; also a growth form or morphological type.

flagelliferous Whiplike.

flagellin A proteinlike substance; found in muscle; composes the flagella of bacteria.

flagellum (pl. **flagella**) Long, whiplike organelle controlling movement of motile cell: in eukaryotes, external to cell but enclosed within plasma membrane; distinguished from cilium by length and type of movement; in bacteria, not surrounded by membrane.

floral apex Apical meristem that will develop into a flower or, in some plants, an inflorescence.

floridean starch Polysaccharide storage product composed of α-1,4, 1,6 plus 1,3 glucoside linkages occurring in red algae; somewhat similar to amylopectin.

floridorubrin A carmine-purple pigment of unknown function found in certain florideophyte red algae.

floridoside Storage product of some red algae.

flower Characteristic aggregation of sterile and spore-bearing structures in flowering plants.

foliaceous Having leaflike shape.

foliose Leaflike: in lichens, a growth form in which the flattened prostrate thallus may be easily removed from the substrate; in algae, a flattened, usually erect, bladelike thallus.

follicle Dry fruit of one carpel that splits along one suture.

foot In bryophytes, haustoriumlike basal sporophytic cells; in some vascular plants, the basal lobe of the embryo.

foot cell In Laboulbeniomycetes, lowermost cell of thallus, embedded in cuticle of the insect host.

forespore Clear area in cytoplasm of bacterial cell that becomes surrounded by a refractile wall in endospore formation.

form family Family consisting of form genera; i.e., genera based upon a sexual reproductive structure.

form genus In fungi, genus name based on morphology of asexual structures; in fossil plants, a generic name for those fossils with the same form or morphology.

form order In Fungi Imperfecti, order based on imperfect (i.e., asexual) structures.

form species An apparent species, which is really a stage in the life history of a pleomorphic species.

fovea (pl. **foveae**) Pit in the wall of palynomorphs, such as spores, pollen, or dinoflagellate cysts.

fragmentation Breaking apart.

free central placentation Attachment of ovules around central column and free from carpel wall except at base.

free-nuclear division Stage in development in which unwalled nuclei result from repeated division of a primary nucleus.

free-nuclear endosperm Endosperm in which there are many nuclear divisions without cytokinesis before cell walls start to form.

free-sporing Plants that produce spores that are shed before germinating (e.g., ferns, *Equisetum*, and lycopods).

frond Term applied to the usually pinnately compound leaf of ferns and other plants (e.g., palms).

frondiform Leaflike; in some Musci, flaplike, unistratose, and erect.

fructification Reproductive organ or fruiting structure; often used with reference to fungi, myxomycetes and bacteria.

fruit Mature ovary containing seeds, together with other flower parts in some instances.

fruiting (body) The production of fruits; by analogy, the formation of spore-bearing bodies of some nonvascular plants.

frustule Diatom cell wall containing organic material and silica.

fruticose Lichen growth form in which the thallus is shrublike and generally branched.

fucoidan Phycocolloid occurring in cell walls and intercellular spaces of brown algae; commercially marketed.

fucosan vesicles In brown algae, refractive vesicles, usually aggregated about the nucleus, containing a tanninlike substance; also called *physodes*.

fucoserraten A sexual attractant, or gamone, produced by the female gamete (egg) in *Fucus*.

fucosterol A sterol occurring in a number of algal groups (e.g., brown and red algae).

fucoxanthin Xanthophyll pigment characteristic of chrysophytes and brown algae.

funiculus (pl. **funiculi**) Stalk that attaches an ovule to the placenta in an ovary.

funoran Phycocolloid of some red algae; commercially marketed.

furcellaran Phycocolloid of some red algae; commercially marketed.

fusiform Spindle shaped; elongated with tapering, pointed ends.

Ga Term expressing geologic time in years $\times 10^9$ (from L. *gigas*, giant); see gyr.

galactan Polysaccharide containing galactose units.

gametangial contact Sexual reproduction in which, following contact of gametangia, nuclei are transferred from the antheridium to the eggs through the fertilization tube (Oomycetes).

gametangial copulation Sexual reproduction of some lower fungi (Zygomycetes) in which the entire protoplasts of two gametangia fuse.

gametangium (pl. **gametangia**) Structure producing gametes: in fungi and algae, generally a single cell; in bryophytes, a multicellular structure with an outer sterile protective layer.

gamete Sex cell; capable of fusion with another gamete to form a zygote.

gametic meiosis The pattern of alternation of generations in which gametes are either produced directly by meiosis from the multicellular diploid plant, or may produce gametes after several mitotic divisions of individual haploid cells, as in *Fucus* (brown algae).

gametogenesis Formation of gametes.

gametophore The elaborate leafy or thallose structure in bryophytes that bears the sex organs.

gametophyte Gamete-producing plant; generally haploid, and producing gametes by mitosis; usually has half the chromosome number of a sporophyte of the same species.

gamone A chemical substance involved in bringing about sexual fusion of gametes.

gas vacuole Array of cylindrical vesicles, occurring in prokaryotic cells.

gelatinous Composed of proteinaceous (or, in some plants, polysaccharide) material: in lichens, a growth form appearing moist; in fungi and algae, jellylike or slimy; mucilaginous.

gemma (pl. **gemmae**) Specialized structure for vegetative reproduction; in Oomycetes, a cell; in bryophytes, a group of cells produced on the gametophore.

gemmiferous Bearing gemmae.

gene Unit of inheritance, arranged in a linear sequence on the chromosome; specifically, an arrangement of nucleotides in DNA that will code for specific protein.

gene mutation Change in structure of a gene; also known as *point mutation*.

general purpose classification A classification of interest to the greatest number of biologists and upon which nomenclature is based.

generalized The state of a characteristic that is common (general) in a taxon.

generative cell In the pollen grain, cell of male gametophyte that divides to produce sperm.

genetic Pertaining to features resulting from the action of genes.

genome The basic set of chromosomes (*n*) contributed by each parent via gametes.

genotype Genetic constitution of an organism, determined by the assemblage of genes it possesses.

genotypic With reference to the genotype.

genotypic sex determination Determination of sex by single allele difference.

genus (pl. **genera**) A level in the Linnean hierarchy within a family that includes closely related species.

geotropism Growth response of a plant to gravitational pull.

germ tube Hypha produced by a germinating fungal spore.

gill Mycelial tissue on underside of basidiocarp of some basidiomycetes.

girdle Encircling, or middle: in dinoflagellates, transverse groove containing flagellum; in diatoms, region where frustules overlap.

girdle view Side view of diatom or dinoflagellate.

girdling In cycad stems, where leaf traces traverse some distance around the stem before entering a leaf base.

gleba Spore-producing zone in basidiocarp of some Gasteromycetes.

globose Nearly spherical.

globule In charophytes, multicellular male reproductive structure (gametangium), including sterile and fertile cells.

glochidia Elongate, hook-tipped processes on the microsporic massula in the aquatic fern *Azolla*.

glucan Polysaccharide containing glucose units.

glycogen Complex polysaccharide similar to starch but probably with more amylopectin; does not give blue-black color with iodine solution.

glycoprotein Group of enzymes composed of sugar units and amino acids; important in surface reactions of cells (aggregation, clumping, fusion, etc.)

golgi apparatus Collection of cytoplasmic vesicles involved in storage and secondary products; also called *dictyosome*.

golgi body (See **golgi apparatus**)

gonimoblast Collection of gonimoblast filaments and carposporangia occurring in the Florideophyceae; also referred to as the *carposporophyte*.

gonimoblast filaments Cells of carposporophyte (or *gonimoblast*) bearing carposporangia.

gonophyll A theoretical structure consisting of a bract and attached spore-bearing axis.

granum (pl. **grana**) A stack of closely appressed (or fused) thylakoids.

grex In Dictyosteliomycetes, the sluglike structure produced by aggregation of myxamoebae; also called a *pseudoplasmodium*.

groove Longitudinal region present in cryptophytes and euglenophytes; longitudinal or circular region in dinoflagellates.

ground meristem The primary meristem in which the procambium is embedded and which is surrounded by the protoderm; matures to form the ground tissues (pith and cortex).

guard cell In stomata, two cells which open or close the stoma as a result of changes in their turgor pressure.

gymnosperm Seed-bearing vascular plant having integuments only, no enclosing carpel or ovarian wall, and with pollen is deposited directly on the ovule.

gynoecium Female part of a flower consisting of one or more carpels.

gyr Short form for *gigayear* (Ga), or years $\times 10^9$; equivalent to *aeon* and *Ga*.

gyrogonites Name for the lime-encrusted, fossilized oogonia and encircling sheath cells (nucule) of charophytes.

half-bordered pit pair A pair of opposing pits, one of which is simple and one of which is bordered.

half cell One half of desmid cell wall.

halophile An organism that is commonly found in a saline habitat.

halophyte A plant that grows commonly in soil in a saline environment (soil impregnated with salts).

haplocheilic Development of the stomatal apparatus in which the two guard cells develop from a single initial, and the subsidiary cells originate from ordinary epidermal initials.

haploid Having a single set of unpaired chromosomes; the chromosome complement present in the gametes; 1*n*.

haploid apogamy The development of an embryo from a haploid cell (other than an egg) into a haploid sporophyte.

haploid parthenogenesis The development of an embryo from an unfertilized egg on the haploid gametophyte directly into a haploid sporophyte.

haplostele A solid cylindrical stele (protostele) in which a central strand of primary xylem is sheathed by a cylinder of phloem.

hapteron (pl. **haptera**) An attaching structure in some brown algae; usually multicellular, branched, and rootlike.

haptonema Apical coiled organelle in some prymnesiophytes, composed of endoplasmic reticulum extension and surrounded by plasma membrane.

haustorium (pl. **haustoria**) An absorptive structure that derives food from its host by penetrating the cell wall (host cell protoplast is not penetrated).

helical In the form of a helix, or spiral (often used to refer to secondary wall deposition in tracheary elements).

helical thickening Secondary walls of vessel elements and tracheids in the form of a helix or elongated spiral.

helicoid cyme A determinate inflorescence whose lateral branches all develop from the same side of the inflorescence.

helobial endosperm Endosperm in which the first division is accompanied by cytokinesis, but the subsequent development is as in free-nuclear endosperm.

hematochrome Red pigment granules, probably xanthophyll, occurring in some algae.

hemicellulose Structural element in cell walls; a mixture of polysaccharides that are more soluble and less ordered than cellulose.

herbaceous (habit) Having characteristics of herb; with little, if any, secondary growth (and thus not woody).

herbivorous Feeding on plants.

heterocyst Differentiated cell in some filamentous bluegreen algae involved in nitrogen fixation.

heteroecious Applied to rust fungi requiring two hosts to complete their life cycle.

heterogeneous Differing or unlike; heterozygous.

heterokaryosis Condition where mycelium is heterokaryotic.

heterokaryotic Refers to mycelium in which two or more genetically distinct types of nuclei occupy the same cytoplasm; both haploid and diploid nuclei can be present.

heterokont Having flagella of unequal length on a motile cell.

heteromerous Lichen thallus in which algal cells are restricted to a specific layer, creating a stratified appearance.

heteromorphic Morphologically unlike.

heterophylly Production of leaves of different form.

heterosporous Producing meiospores of two different sizes, one of which develops into a female gametophyte and the other into a male gametophyte.

heterospory Having spores of two different sizes; formation of two size classes of meiospores, mega- and microspores, by one plant or one species of plant.

heterothallic (heterothallism) Condition in which sexual reproduction requires contact between two morphologically similar but genetically distinct thalli (e.g., Mucorales).

heterotrichous Having heterotrichy.

heterotrichy Growth form in which erect filaments arise from prostrate portion, as in some algae and bryophytes.

heterotroph Organism that requires an external source of one or more organic compounds as an energy source for growth, etc.

heterotrophic Nutritional regime: organism obtains organic compounds from external sources.

heterozygous Having two different alleles for a given gene at the same locus of homologous chromosomes.

holdfast An attaching, discoid, or rootlike structure of some algae.

holdfast cell An attaching, discoid, elongate or rhizoidal cell of some algae.

homogeneous Having the same nature or consistency.

homokaryotic Refers to a mycelium in which all nuclei have the same genetic makeup.

homologous Refers to structures that have a similar origin.

homologous chromosomes Chromosomes that contain identical sets of loci and pair during meiosis.

homology Similarity due to descent from a common ancestry.

homoplasy A similar feature that appears independently in distantly related taxa.

homosporous Production of meiospores of only one size.

homosporous plant Plant exhibiting homospory.

homospory Having spores of one size; formation of one size class of meiospores in one plant or in one species of plant.

homothallic (homothallism) In zygomycetes and basidiomceytes, lacking distinguishable male and female gametangia; the condition in which a single thallus is able to reproduce sexually without interaction of two differing thalli; in ascomycetes, the condition of species that are hermaphroditic and self-fertile.

homozygous Having the same alleles for a given gene at the same locus on homologous chromosomes.

honeydew Sugary liquid exuded together with pollen; presumably functioning in attracting insects that disperse the pollen.

hormogonium (pl. **hormogonia**) In the filamentous bluegreen algae, multicellular, filamentous segment capable of gliding motion; can be reproductive unit.

host Living organism serving as substrate and/or energy for another.

hyaline Clear or colorless; used with reference to dead cells in leaves and stems of some mosses, especially *Sphagnum*; in fungi, refers to various colorless structures.

hyaloplasm Ground substance of cytoplasm.

hybrid A plant that results from crossing between two different parents.

hybridization The act of fertilization between two different plants.

hydathode Opening opposite vein endings in leaves from which water is exuded under certain conditions; some hydathodes are glandular.

hydroids Specialized, water-conducting cells in bryophytes.

hydrome Water and mineral nutrient conducting (tissue) cells in bryophytes.

hydrophily Pollination by water.

hygroscopic Readily absorbing and retaining moisture; refers to certain cells or structures that respond to changes in humidity.

hymenial With reference to a hymenium.

hymenium Aggregation of asci or basidia and related sterile structures in a continuous layer; also termed *fertile* or *fruiting* layer.

hypanthium Receptacle tube upon which calyx, corolla, and androecium are borne in some angiosperms.

hypha (pl. **hyphae**) One of the tubular filaments composing mycelium.

hypnospore Spore formed inside parental cell and secreting new wall internal to parent wall.

hypocotyl Part of seedling below the cotyledons (may include the root).

hypodermis Layer of cells immediately internal to epidermis.

hypogeous Developing below soil surface.

hypogynous Originating or developing below: in some Florideophyceae, beneath the carpogonium; in angiosperms, below the gynoecium.

hypothallus (pl. **hypothalli**) Thin, shiny, membranous adherent film at base of fructification of myxomycetes.

hypotheca (pl. **hypothecae**) Inner cell half of diatom cell; often refers only to inner, siliceous wall.

hypothecium The layer or zone supporting the hymenium in an apothecium.

hypsophyll Modified leaves associated with flowers.

imbricate Closely overlapping.

imperfect flower Flower lacking either stamens or carpels.

imperfect stage In fungi, stages in which asexual reproductive structures (conidia) are produced.

incomplete Flower lacking one of the whorls of flower parts.

incubous Leaf insertion in Jungermanniales (Hepaticae) in which the upper margin of a leaf lies on top of the lower margin of the leaf directly above it on the same side of the stem.

indehiscent Not opening by a discrete, preformed line.

independent assortment The chance distribution of alleles in meiosis; the distribution of alleles from one pair of homologous chromosomes having no effect on the distribution of alleles from another pair of homologous chromosomes.

indeterminate An inflorescence with the youngest flower at the top.

indeterminate growth Growth that continues indefinitely.

indeterminate head A flat-topped inflorescence with sterile flowers and the youngest flower in the middle.

inducer substance Compound that causes a reaction or another compound to be produced.

indusium (pl. **indusia**) Outgrowth of leaf tissue covering sorus in true ferns.

inferior ovary Ovary situated below the perianth and androecium.

inflorescence Collective term for the grouping of flowers on an axis.

inheritance Traits in offspring that come from the parental generation.

initials Cells in a meristem that do not become part of the mature plant body; the first-formed cells that produce a specific structure or organ.

inner aperture Opening of a pit canal into lumen of a cell.

innovation New branch arising from an old stem, in bryophytes.

inoperculate Opening of a sporangium or ascus by an irregular tear or plug to discharge spores.

insectivore An organism that consumes insects.

integument Outer cell layer or layers of the ovule that covers the nucellus (megasporangium); usually develops into a seed coat or testa.

integumentary With reference to an integument.

integumentary lobe Lobe in the integument; a segment of an integument of early pteridosperm seeds.

intercalary Inserted within or between two cells or tissues.

intercalary band Band of cell wall material on girdle of diatom.

intercalary cell division Cell division among cells within the plant, rather than at a surficial meristem.

intercalary meristem Area of new cell production situated some distance from the apex.

intercellular space Space between cells.

interfascicular cambium That part of the cambium that develops in between the vascular bundles.

internode Portion of axis between two nodes.

interphase Between stages: in mitosis and meiosis, period when there is no chromosomal division, although DNA replication may occur; period between active, assimilative, and aggregation phases in which the myxamoebae cease feeding and undergo internal changes (Dictyosteliomycetes); also used with reference to a nondividing nucleus.

intertidal region Portion of continental shelf periodically exposed between the highest and lowest tide levels.

intine The inner cellulosic layer of the wall of a vascular plant spore or pollen grain.

intranuclear Occurring or remaining inside the nuclear envelope.

intranuclear division Nuclear division in which the spindle lies within the nuclear envelope, which remains intact throughout the division process.

introrse Anthers that open towards the center of the flower.

invagination An ingrowth or ensheathing.

inversion In asexual reproduction in *Volvox*, the process by which the new colony completely turns inside out so that the flagellar ends of cells are externally oriented; sometimes called *eversion*.

inviable gametes Gametes incapable of taking part in the process of fertilization.

involucre In flowering plants, one or more whorls of small leaves or bracts close to and underneath a flower or flower cluster; in thallose bryophytes, a sheaf of tissue that surrounds the female sex organs.

involute Rolled inward toward the top (adaxial surface) along the sides.

Irish Moss Common name given to the carrageenophyte *Chondrus crispus* (Florideophyceae).

irregular Flower that has no line of symmetry.

isidium (pl. isidia) Rigid protuberance of upper part of lichen thallus, which may break off and serve for vegetative reproduction.

isobilateral A tetrad in which the two pairs of spores are oriented perpendicular to each other.

isodiametric Having equal diameters; used to describe cell shape with length and width essentially equal.

isogamous Having isogamy.

isogamy Fusion of gametes that are the same size and are morphologically alike.

isokont Having flagella of same length on motile cell.

isomorphic Morphologically alike.

karyogamy Fusion of two sex nuclei following fusion of protoplasts (plasmogamy).

karyolymph Material in nucleus in which chromosomes are embedded.

keel Canal or cleft in valve of some pennate diatoms.

kelp Common name for brown algae of the order Laminariales (e.g., *Laminaria, Nereocystis*).

keratin Tough, fibrous, sulfur-containing protein forming the outer layer of skin; a major constituent of horn, hooves, scales, hair, claws.

keratinous Composed of keratin.

key A device to aid in the identification of an object by presenting mutually exclusive descriptive phrases with which to compare the object in question.

key classification A key.

kinetochore A region connecting the two chromatids of a chromosome; has a layered appearance at the ultrastructural level.

kombu Edible product of some Laminariales (brown algae).

labeled Marked.

lacuna (pl. lacunae) A gap or space.

lamella (pl. lamellae) Plate, or layer: submicroscopic structure of chloroplast membranes (thylakoids) containing pigments; in bryophytes, refers to thin sheets of flaplike plates of tissue on the dorsal surface of the thallus or leaves; in basidiomycetes, the gills of a mushroom.

lamina (pl. laminae) Blade portion of leaf; in brown algae, expanded leaflike part of thallus.

laminar placentation Attachment of ovules over the surface of the carpel.

laminarin Storage polysaccharide composed of β-1,3, 1,6 glucoside linkages, characteristic of the brown algae.

laminate Flattened and broadly expanded; leaflike.

lamination Flattening of organ into a leaflike form.

lanceolate Narrow and tapering toward each end.

laver General name given to edible dried preparation made from algae, such as *Ulva* (green laver, Chlorophyceae) and *Porphyra* (purple laver, Bangiophyceae).

leaf gap A break in the vascular tissue associated with a leaf trace.

leaflet The unit of a compound leaf.

leaf primordium A structure that will develop into a leaf.

leaf trace Strand of vascular tissue that goes to a leaf.

lenticular Lens-shaped (a double convex).

lenticular plate A biconvex plate.

leptoid Cell specialized for conducting metabolites in bryophytes.

leptoma The thin area in the wall of gymnosperm pollen through which the pollen tube emerges.

leptome Photosynthate-conducting cells in bryophytes.

leptosporangiate With reference to a leptosporangium.

leptosporangium A sporangium formed in ferns, in which a single cell primordium, rather than a group of cells, produces the sporangium.

leucoplast Colorless plastid, generally for starch storage in many green plants.

leucosin See chrysolaminaran.

liana Woody climbing plant, characteristically occurring in the tropics.

libriform fiber Xylem fiber that has thick walls and greatly reduced pits.

lichen Symbiotic relationship of fungus and alga.

lignin Complex molecule made up of phenylpropanoid units associated with cellulose in some cell walls; it serves to harden and strengthen the walls.

ligule A tonguelike outgrowth on the adaxial surface of leaves and sporophylls in heterosporous lycopods and grasses.

limited growth Determinate growth; having a fixed, definite limit.

linear Long and narrow with parallel sides.

linear evolutionary progression Evolutionary sequence of development, where taxa can be placed in a linear series.

Linnean hierarchy The arrangement of organisms into taxa, each subordinate to the one above it, with kingdom as the highest taxon and species as the lowest.

lipid Any group of compounds comprising fats; composed of glycerol and fatty acids.

littoral Belonging to or growing on the seashore.

locule Compartment, cavity, or chamber; in ascomycetes, stromatic chambers containing asci; in angiosperms, cavity in ovules.

locus (pl. loci) A particular location on a chromosome of a gene.

long shoot The main branch that bears short or dwarf shoots in some gymnosperms, e.g., *Pinus*, *Cedrus*, and *Ginkgo*.

lorica (pl. loricae) Surrounding case that is separate from protoplast in some algae.

lumen The inner cavity of a cell or reticulum in which the protoplast is contained.

lutein Xanthophyll pigment.

lyosome Organelle containing digestive enzymes.

Ma Symbol used in geochronology signifying the age of rocks $\times\ 10^6$.

macrocyst Cyst developed by aggregation of myxamoebae in Dictyosteliomycetes; within it, karyogamy and meiosis occur, and myxamoebae are released upon germination.

mamillose Conspicuously bulging.

mannan Polysaccharide material composed of mannose units; occurring in cell walls of some algae and yeasts.

mannitol Saccharide alcohol composed of mannose; part of the polysaccharide laminarin.

mannoglycerate Saccharide storage product in some red algae.

mannose Six-carbon monosaccharide.

mantle Dense mass of fungal hyphae surrounding a root.

marginal meristem Area of dividing cells at the margin of a developing leaf.

marginal placentation Attachment of ovules to the margin of a carpel.

margo Thin portion of the pit membrane between torus and secondary wall.

massula (pl. massulae) Refers to a segment of the periplasmodium that is derived from the sporangia of the fern *Azolla*; the individual massula contains many microspores; in the megaspore, one massula surrounds the megaspore, and the other three sit on top of the megaspore as a cap.

mastigoneme Hairlike thread or process occurring along the length of some flagella; also known as "flimmer" or *tinsel*.

mating type In algae and fungi, the term used to designate a particular genotype with respect to compatibility in sexual reproduction; gametes identical in appearance and referred to as plus (+) and minus (−) rather than as male and female.

medulla Innermost region of thallus in lichens and in some brown and red algae.

megagametophyte Female gametophyte that develops from the megaspore.

megaphyll A large, broad, flattened leaf in which the vascular strand (leaf trace) is much branched (with several to many veins).

megaphyllous Having large leaves with several to many veins.

megaphyllous leaf A leaf that is a megaphyll.

megasporangium (pl. megasporangia) Meiosporangium of heterosporous plants producing usually one to four megaspores.

megaspore Meiospore in heterosporous plants that is formed in the megasporangium and develops into the megagametophyte; it is usually, but not always, larger than the microspore.

megaspore mother cell A cell (meiosporocyte) contained in the megasporangium (nucellus of the ovule) that undergoes meiosis to produce megaspores.

megasporocyte Cell in megasporangium that undergoes meiosis to produce four megaspores.

megasporogenesis Formation of megaspores within megasporangium; in seed plants within ovule.

megasporophyll Leaflike appendage bearing megasporangia.

megastrobilus A strobilus bearing megasporangia, ovules, or seeds.

meiocyte Cell in which meiosis occurs; also termed meiospore mother cell.

meiosis Reduction division in which the number of chromosomes is reduced from the diploid (2*n*) to the haploid (1*n*) state.

meiosporangium Structure in which spores are produced by meiosis (reduction division).

meiospore Spore produced by meiosis; has a haploid chromosome number (spores usually produced in fours).

meristele Individual vascular unit of a dictyostele.

meristem Tissue concerned with formation of new cells.

meristematic Having the characteristics of a meristem.

meristematic activity State of active cell division in a meristem.

meristoderm Outer meristematic cell layer (epidermis) of Laminariales (brown algae).

mesarch A type of maturation of the primary xylem from a central point outward; that is, the oldest xylem elements (protoxylem) are surrounded by the late-formed metaxylem.

mesocarp Middle layer of pericarp or carpel wall.

mesogene development Development of stomata wherein the guard cells and subsidiary cells share a common parental cell.

mesokaryote Term proposed for the dinoflagellate nuclear structure in which histone proteins are lacking and chromosomes remain in condensed state during interphase; also used to denote dinoflagellate evolutionary position between prokaryotes and eukaryotes.

mesophyll Photosynthetic parenchyma in a leaf.

mesophyte Plant of moist habitat, avoiding both extremes of moisture and drought.

mesosome Invagination of plasma membrane in prokaryotic cells; generally contains respiratory enzymes.

metaboly (metabolic) Capable of changing shape, as in the cell of many euglenophytes.

metaphloem Primary phloem formed secondarily after the protophloem.

metaxylem Primary xylem formed secondarily after the protoxylem; generally cell elongation is complete or almost so.

methanogen Producer of methane.

microbe Microscopic organism of any type, whether prokaryote or eukaryote plant, animal, or bacterial.

microbody Cell organelle bounded by a single membrane and containing enzymes.

microfibril Submicroscopic unit.

microfossils Microscopic fossils, including spores, pollen grains, tracheids, pieces of cuticle, small algae, fungi, etc.

microgametophyte In seed plants, gametophyte that develops from the microspore within the wall of the pollen grain; in ferns and allies, the gametophyte produced within a microspore.

microphyll A small leaf in which the vascular strand is simple (has a single vein); there is no leaf gap associated with its trace.

microphyllous Having small leaves with one vein and leaf trace.

microphyllous leaf A leaf that is a microphyll.

micropyle Opening in the integument(s) of the ovule, through which the pollen tube grows.

microsporangium (pl. **microsporangia**) Meiosporangium of heterosporous plants producing many microspores.

microspore Meiospore in heterosporous plants that is formed in the microsporangium and develops into the microgametophyte; it is usually, but not always, smaller than the megaspore.

microsporocyte A cell in a male (micro-) sporangium that undergoes meiosis to produce four microspores.

microsporogenesis Formation of microspores within microsporangia or pollen sacs of the anther.

microsporophyll Leaflike appendage bearing microsporangia.

microstrobilus (pl. **microstrobili**) A strobilus bearing microsporangia or pollen sacs.

microtubular root Microtubules attached to falgellar root in motile cells.

microtubule Elongate, proteinaceous cytoplasmic unit, about 25 nm in diameter; involved in maintaining cell shape; basic component of eukaryotic flagella.

middle lamella Intercellular material that serves to hold contiguous cells together; composed of pectin.

mitochondrion (pl. **mitochondria**) Double membrane-bound organelle in which cellular respiration (and energy release) occurs.

mitosis Equational division in which chromosome number in new cells is same as parental number.

mitosporangium (pl. **mitosporangia**) Structure producing spores by mitosis (equational division).

mitospore Spore produced by mitosis and having same chromosome number as parental cell.

mixotroph Photoautotroph capable of utilizing organic compounds in the environment; may also be termed *facultative heterotroph* or *facultative parasite*.

moniliform Beadlike; sometimes used to describe chromatin in nucleus in dinoflagellates.

monocolpate A pollen grain with one furrow or groove through which the pollen tube emerges.

monocotyledon Flowering plant with one cotyledon (seed leaf).

monoecious Condition in seed plants in which separate pollen-bearing and ovule-bearing cones or flowers are borne on the same plant.

monoicous Condition of bryophtye gametophyte in which the gametophore bears both kinds of sex organs; bisexual.

monokaryotic Hyphal condition in which the compartments contain a single haploid nucleus (e.g., the primary mycelium of basidiomycetes).

monolete Refers to a single suture on a meiospore that is produced in a linear tetrad.

monolete suture A single, linear, thin area on the proximal wall of some spores (e.g., polypod ferns).

monophyletic taxon Evolving from a single ancestral stock.

monophyly Evolution of a taxon from one evolutionary line.

monopodial Having one main axis of growth.

monosaccate Having a single air bladder, such as the pollen of the gymnosperm *Cordaites* or the Lebachiaceae.

monosaccharide A simple sugar.

monosiphonous Having a uniseriate row of cells; not surrounded by pericentral cells (Florideophyceae).

monosporangium (pl. **monosporangia**) Vegetative cell that metamorphoses to produce a single spore; characteristic of some red algae.

monospore Single spore produced by metamorphosis of single vegetative cell, the monosporangium; characteristic of some red algae.

monostromatic Having thallus one cell thick.

monosulcate A pollen grain with one furrow or groove on the distal surface (e.g., cycad or *Ginkgo* pollen).

morphogenesis Development of form; often used to refer to the study of development of form.

mucilage canal Elongate cells in the cortex of Laminariales (brown algae) and cycads that may conduct mucilaginous materials.

mucilage duct A duct containing mucilage or gum or similar carbohydrate.

mucilage hairs Specialized mucilage-producing hairs in bryophytes, most common near leaf axils and growing points of the gametophore.

mucilaginous Jellylike; slimy; gelatinous.

mucocomplex A material composed of cross-linked units constituting the rigid component in prokaryotic cell walls (also referred to as *mucopeptide, murein,* and *peptidoglycan*).

mucopeptide Complex material giving prokaryotic cells rigidity (*see also* mucocomplex and peptidoglycan).

mucronate Terminating abruptly in a distinct spinelike tip.

multiaxial Having a main (central) axis composed of many parallel or almost parallel filaments.

multicellular Having more than one (few to many) cells.

multiflagellate Bearing many flagella.

multiple alleles The presence of several alleles for any one locus within a population.

multiple fruit Fruit that develops from separate ovaries from separate flowers, e.g., pineapple.

multiseriate Having many rows of cells.

multistratose Having many layers.

murein See mucopeptide.

mutation Change in genetic composition; change in gene.

mutualistic Refers to symbiotic associations in which both partners benefit.

mutualistic symbiont One partner in an association (mutualism, mutualistic symbiosis) presumed to be beneficial to both associates, e.g., the fungus or alga in a lichen.

mycelium (pl. mycelia) Mass of hyphae; the thallus of a fungus.

mycobiont The fungus component of a lichen (sometimes also used in reference to mycorrhizal fungi); in most lichens, the mycobiont is an ascomycete, especially apotheciate species.

mycolaminarin Storage product produced by oomycetous fungi; so-called because of its similarity to laminarin.

mycologist One who specializes in the study of fungi.

mycorrhiza (pl. mycorrhizae) Symbiotic association of fungus and root or rootlike structure.

mycorrhizal Containing fungal hyphae within the roots or tissues, in which there is usually a symbiotic relationship.

mycoses Fungus diseases of humans or animals (e.g., ringworm and athlete's foot).

myxamoeba (pl. myxamoebae) Amoeboid cell lacking a cell wall; characteristic of some myxomycetes.

nannoplankton Plankton with dimensions less than 75 μm.

natural classification A classification based on as many characteristics as possible.

natural selection A natural process that results in a change in gene frequency in a population as a result of different degrees of reproduction and survival among individuals of a population.

natural taxon A taxon composed of organisms that are more closely related to each other than they are to organisms in another taxon at the same level.

neck Slender part of archegonium through which the male gamete travels to reach the female gamete.

neck canal cells Inner row of cells in neck region of archegonium; at maturity these cells disintegrate.

neck cells Cells comprising the narrow tube or neck of the archegonium.

nectar Fluid produced by flowers that contains sugars, amino acids, and other nutrients; consumed by pollinators.

nectary The organ in which nectar is secreted.

netted Having a network of interconnected parts, as in the venation pattern of some leaves.

netted venation Leaf venation wherein there is much interconnection between veins.

neutral spore In brown algae, term sometimes used to describe the mitospore produced in a plurilocular mitosporangium.

nitrogen fixation Conversion of free atmospheric nitrogen to bound inorganic or organic forms.

nitrogen fixing Refers to prokaryotes having the ability to fix atmospheric nitrogen.

nodal bract A modified leaflike appendage emanating from a node on stem, cone, or fruiting axis.

node Point on an axis where one or more parts are attached, e.g., leaves.

nodule A small swelling; in pea family, formed on roots in response to nitrogen-forming bacteria.

nomenclature The application of appropriate scientific names to plant taxa following the International Rules of Botanical Nomenclature.

nonseptate Lacking septa.

nori Edible dried preparation of *Porphyra* (Bangiophyceae); also known as *purple laver.*

nucellar cap A multicellular tissue on the tip of the ovule that develops from the protoderm.

nucellus Tissue surrounding the megagametophyte in the ovule of seed plants; the nucellus itself is surrounded by the integument(s) and is usually considered to be the equivalent of the megasporangium.

nuclear body Area in prokaryote cell containing DNA and associated cytoplasm; in bacteria, the area of cytoplasm is devoid of ribosomes (*see also nucleoid*).

nuclear envelope Porate double

membrane enclosing the nucleus.

nucleoid Bacterial "nucleus," consisting of a circular strand of DNA and its associated protoplasm.

nucleolus Body in nucleus containing concentration of RNA.

nucleoplasm Material within nucleus exclusive of DNA and RNA.

nucleus (pl. **nuclei**) DNA- and RNA-containing organelle in eukaryotic cells, bounded by nuclear envelope.

nucule In charophytes, female reproductive structure (gametangium) including oogonium and surrounding sterile vegetative cells; in fossil charophytes, called *gyrogonite*.

numerical phyletics The erection of an inferred phylogeny of extant organisms utilizing numerical analysis of characteristics.

numerical taxonomy The erection of classifications by numerical analysis of characteristics; taximetrics.

nurse cells In some hepatics, sterile cells among the spores that lack any special wall thickenings and frequently disintegrate before the spores mature.

nut An indehiscent, dry, one-celled, one-seeded, hard fruit.

nutlet A small nut.

nutritive cell In some Florideophyceae, a special cell in the carpogonial branch that fuses with the carpogonium following fertilization.

obligate Restricted; generally used for an organism that must lead a parasitic existence; also for bacteria that require aerobic or anaerobic conditions.

obligate parasite An organism that occurs only as a parasite.

odonthalan Mucilaginous galactan occurring in cell walls and intercellular spaces of some red algae.

ontogenetic With reference to the life history or development of an individual organism.

ontogenetic fusion Fusion between different independent structures or the result of them growing together during development to form one unit.

ontogeny Development of an organism or structure in its various stages from initiation to maturity.

oogamous Having oogamy.

oogamy Production of gametes in which the female is large and nonmotile, and the male is small and either motile or nonmotile.

oogonium (pl. **oogonia**) Female gametangium consisting of a single cell (occurring in fungi and algae).

oosphere Egg or female gamete produced in an oogonium (Oomycetes).

oospore Thick-walled resistant spore of oomycetes developing from a fertilized oosphere.

opercular cell A cell that forms a lid or cover.

operculate Refers to a sporangium or ascus that opens by a small lid or cover.

operculum Lid or cover: in the fungi, part of a cell wall; in the Musci, a multicellular apical lid that opens the sporangium.

orbicular Circular in outline.

order A level in the Linnean hierarchy within a class that includes closely related families.

organ Distinct and differentiated part of multicellular organism composed of tissues.

organ system A distinct and differentiated part of a plant, such as root, stem, leaf, or part of flower.

organelle Structure within the eukaryotic cell (e.g., nucleus, chloroplast, etc.); generally membrane bound or composed of membranes.

organographic Pertaining to the origin and structure of organs.

ornithophily Pollination by birds.

orthotropous Upright ovule position in carpel with micropyle away from placenta and on a short funiculus.

osmium organic chemical used for preserving and fixing biological material for microscopic observation.

osmophilic bodies Lipid-containing granules; appear dark when treated with osmium fixation.

ostiole Opening or pore.

outcrossing Cross pollination usually between two different plants of the same species.

outer aperture Opening of a pit canal nearest the external surface of the cell.

outer integument Outer of two coverings on the ovule of an angiosperm.

ovary Ovule-bearing part of the gynoecium: that which matures into a fruit.

overtopping One set of a system of telomes coming to grow longer than the other set at a bifurcation or dichotomy.

overturn Complete mixing of a body of water from surface to bottom resulting from several external factors.

ovoid Egg shaped; shaped like a bird's egg, with one end broader than the other.

ovule Immature seed; it consists of the megagametophyte, nucellus, and integuments.

paleobotanist One who studies fossil plants, their phylogeny, relationships to environment, and age of rocks.

paleopalynologist One who studies fossil spores, pollen, algal cysts, etc.

palisade *See* palisade mesophyll.

palisade mesophyll Parenchyma tissue on the adaxial side of the leaf, made up of tightly packed columnar cells, each bearing many chloroplasts.

palmate Lobed, divided, or ribbed in a handlike fashion in which lobes, divisions, or ribs arise from a single point.

palmately lobed With lobing in a palmate pattern.

palynologist One who studies living and/or fossil spores, pollen grains, algal cysts, etc.

palynology Study of pollen grains, spores, algal cysts, etc.

panicle An indeterminate branching inflorescence in which the branches of the main axis have branched secondary axes.

papilla (pl. **papillae**) Blunt projection or protuberance.

papillate Having papillae.

papillose With small sharp warts, descriptive of sculpturing of spores and cell walls.

pappus In flowering plants, bristlelike or scalelike calyx.

paraflagellar body Swelling near base of flagellum in some euglenophytes; possibly serves as a photoreceptor.

parallel venation Pattern of venation in which veins run parallel throughout the length of the leaf.

paramylan (paramylon) Polysaccharide storage product occurring in euglenophytes composed of β-1,3 glucoside linkages, similar to laminarin and chrysolaminarin.

paraphyletic taxon A taxon that does not include all descendants of one ancestor.

paraphyllium (pl. **paraphyllia**) Filamentous or leaflike chlorophyll-containing outgrowth on stem near leaf base, occurring in some pleurocarpous Bryidae and some Hepaticae.

paraphysis (pl. **paraphyses**) Sterile hair or thread; in the ascomycetes, sterile hypha in the hymenium; in bryophytes and brown algae, unicellular or multicellular hair, generally associated with the sporangia or gametangia.

parasexuality (parasexual cycle) Mechanism of recombination of hereditary material based on the mitotic cycle rather than the sexual cycle occuring in some ascomycetes.

parasite Organism that derives its nutrients and hence energy from a living host.

parasitic Functioning as a parasite (*see* parasite)

paraspore In some Florideophyceae, spore produced in a parasporangium; believed to be a mitospore.

parenchyma Tissue composed of living, thin-walled, randomly arranged cells.

parenchymatous Composed of living, thin-walled, randomly arranged cells.

parichnos Scar on leaf cushion representing parenchyma strands adjacent to leaf vein of *Lepidodendron* (lycopods).

parietal Peripheral in position; as in some algae, the chloroplast is located near the periphery of the cell.

parietal placentation Attachment of ovules in longitudinal rows on carpel wall.

parthenocarpy Fruit development without fertilization.

parthenogenesis The development of a gamete into a new individual without fertilization.

parthenogenetic development Production of new plant from an unfertilized egg.

partial veil Membranous layer covering the developing hymenium in some basidiomycetes.

pathogen Organism causing disease.

pathogenic Disease-causing.

peat Deposits of incompletely decomposed plant material, primarily *Sphagnum* (Musci).

pectic Composed of pectin.

pectin Polysaccharide material (derivative of polygalacturonic acid); commonly found in cell wall and middle lamella.

pedicel Stalk bearing reproductive structures.

peduncle Stalk bearing a strobilus or an inflorescence.

pellicle Proteinaceous covering in euglenophyte cell.

peltate Shield-shaped.

pennate Common name for group of diatoms generally of rectangular shape with markings in parallel rows; winged.

penultimate Next to the last.

peptide Compound involving amino acids; *see* polypeptide.

peptidoglycan Rigid component of cell walls in prokaryotes; composed of cross-linked polysaccharides and polypeptide chains; also called *murein*, *mucopeptide*, and *mucocomplex*.

perennating Surviving from one year to the next.

perennial Lasting several years; refers to a plant that lives for several years.

perfect stage Sexual stage in fungi.

perforation plate End wall of vessel element containing one or more holes (perforations).

perianth Protective organs around reproductive structures; in Jungermanniales (Hepaticae), sheath of fused leaves surrounding archegonia and developing sporophyte; in flowering plants, calyx and corolla of the flower.

pericarp In Florideophyceae, urn-shaped gametophyte tissue surrounding the carposporophyte (sometimes collectively referred to as *cystocarp*); in flowering plants, the mature ovary wall.

pericentral cell Around the central axis, as in *Polysiphonia* (Florideophyceae).

perichaetial (branch) shoot Composite structure, comprising the archegonia and surrounding or subtending leaves of some Hepaticae and most Musci.

perichaetial leaves Leaves surrounding or subtending the archegonia of some Hepaticae and most Musci.

perichaetium Enlarged leaves surrounding archegonia of bryophytes.

periclinal Parallel to the circumference of the surface (opposite to anticlinal).

periclinal division Cell division that results in wall forming parallel to nearest surface.

pericycle Stelar tissue located between endodermis and vascular tissue in many vascular plants.

periderm Outer protective and supportive secondary tissue of some vascular plants, formed by cork cambium.

peridiole Lenticular body in which basidiopores are formed in the Nidulariales (basidiomycetes).

peridium Membranous covering, or outer sterile layer of sporangium of myxomycetes, and some basidiomycetes.

perigene development Development of guard cells in which the guard cell mother cell does not give rise to the subsidiary cells.

perigonial branch The specialized branch that bears antheridia in bryophytes.

perigonial leaves Leaves surrounding the antheridia of some Musci and Hepaticae; also known as *perigonial bracts* in Jungermanniales.

perigonium (pl. perigonia) Antheridium plus associated perigonial leaves or bracts in Hepaticae and Musci.

perigynium (pl. perigynia) In Hepaticae, a sleevelike extension of stem or thallus tissue that surrounds the archegonia.

perigynous Around the gynoecium.

periplast Membrane surrounding the protoplast of cryptophytes.

perisperm Food-storing tissue derived from the nucellus in some seeds.

perispore A wrinkled outer covering of some vascular plant spores, especially in certain ferns.

peristome Sporangium mouth of Musci.

peristome teeth Multicellular toothlike structures ringing mouth (peristome) of sporangium of Musci.

perithecial initial A primordium, typically including the gametangia, that develops into a perithecium (ascomycetes).

perithecium Asococarp in which the hymenium is completely enclosed at maturity of the ascospores except for a small opening or ostiole; generally urn-shaped.

persistent Continuing for an extended period; remaining on a plant; not falling off.

petal Innermost whorl of perianth of flowering plants; often showy and serving to attract insects.

petiole Leaf-stalk bearing lamina or blade of a leaf.

petiolule Stalk of a leaflet.

phaeophyte tannin Tanninlike substance in refractile granules near the nucleus in some brown algae; originally thought to be a polysaccharide and termed *fucosan*.

phage Virus that infects bacterial cells; also called *bacteriophage*.

phagocyte Specialized cell in the body's defensive system capable of ingesting and destroying bacterial cells or other foreign particles.

phagocytosis Ingestion of foreign particles or cells; often involves specialized cells (phagocytes) within an animal body with the foreign material (antigen) first acted upon by antibodies.

phagotroph An organism capable of ingesting solid food particles.

phagotrophic Having characteristics of a phagotroph.

phagotrophically Functioning as a phagotroph; ingesting solid food particles.

phellem (cells) Cork cells forming the outer layers of periderm.

phelloderm Tissue produced by the cork cambium (internal to the cork cambium).

phellogen Meristematic cells that produce cork, also known as cork cambium.

phenetic(s) Referring to the relationships of organisms based on phenotypic characteristics.

phenetic classification A classification based on phenetic characteristics of extant organisms; the arrangement does not reflect

phylogeny because the evolutionary (fossil) history is unknown.

phenetic series (system) An arrangement of taxa demonstrating the degree of relationship believed to exist among them, based on similarity of characteristics, and with no knowledge of their origins, ancestors, or descendants.

phenotype (phenotypic) The external, visible appearance or physical expression of a genetic trait.

phenotypic plasticity Different expressions of one genotype when grown under different environmental conditions.

phenylpropanoid (unit) Repeating C_6, C_3 units in lignin.

phloem Tissue in vascular plants that conducts storage materials, mainly carbohydrates and water, and functions in support.

photic zone Depth of water through which light penetrates.

photoautotroph Autotrophic organism that derives energy for metabolism from visible light.

photophosphorylation Formation of ATP in light involving photosynthesis.

photoreceptor Organelle believed to be light-sensitive; in some euglenophytes an organelle appearing as a swelling near base of flagellum.

photosynthate Compound produced by photosynthetic cells; generally a polysaccharide; also used as a general term for storage product produced by photosynthesis.

phototactic Responding to light by moving toward or away from the stimulus; a locomotory response of motile cells or organisms.

phototropism (phototropic) Growth response toward (positive) or away from (negative) a unilateral light stimulus.

phycobilin Water-soluble proteinaceous pigment, similar to bile pigment, occurring in bluegreen algae, red algae; *see phycobiliprotein.*

phycobiliprotein Water-soluble red or blue bile pigment, occurring in bluegreen algae, red algae, and cryptophytes, serving as an accessory pigment.

phycobilisomes Granules containing biliproteins attached to thylakoid membrane in chloroplast.

phycobiont Algal partner, or algal component of a lichen.

phycocolloid Complex colloidal substance produced by algae, especially some brown and red algae.

phycocyanin Blue phycobilin pigment occurring in bluegreen algae, red algae, and cryptophytes.

phycoerythrin Red phycobilin pigment occurring in bluegreen algae, red algae, and cryptophytes.

phycologist One who studies algae.

phyllid Flattened leaflike appendage in bryophytes.

phyllosporous Spores borne on modified leaves.

phylogenetic classification A classification based on the phylogeny of the organisms classified; also, a classification that reflects the phylogeny of the organisms classified.

phylogenetic series Arrangement of taxa in what is believed to be an evolutionary sequence.

phylogenetic systematics A branch of systematics that classifies organisms using shared, derived traits, and which assumes taxa are the result of evolution.

phylogenetic tree A pictorial representation of a real or inferred phylogeny.

phylogeneticist One who studies phylogeny.

phylogeny An evolutionary history of an organism or group of organisms; includes both extant (living) and extinct (fossil) representatives.

phylum (pl. phyla) One of the main categories used in classification of organisms (often restricted to the animal kingdom); *see* division.

physode *See* fucosan vesicle.

phytoplankton Free-floating or weakly swimming microscopic aquatic plant life; *see* plankton.

pileate Having a pileus.

pileus Cap or structure bearing hymenium in some ascomycetes and basidiomceytes.

pilus (pl. pili) In gram-negative bacteria, proteinaceous filaments protruding through cell wall; smaller than flagella, the filaments are adhesive, and one type is important in conjugation.

pinna (pl. pinnae) Subdivision of compound leaf or frond.

pinnate Branching at uniform angles from different points along a central axis, and all in one plane (as in a feather).

pinnately compound In reference to a compound leaf in which the blade is divided into pinnately arranged leaflets.

pinnule The ultimate subdivision of a pinna or a compound leaf or frond.

pistillate In reference to a flower or plant that has only carpels, but no stamens (i.e., only the female parts are functional).

pit Opening; a cavity usually in the secondary wall where the primary wall is not overlaid by secondary wall.

Glossary   **723**

pit canal The opening between pit chamber and the lumen of the cell.

pit cavity Space in a pit from the pit membrane to the lumen of the cell.

pit chamber The opening in a bordered pit between the outer aperture and the pit membrane.

pit connection In red algae, "plugged" connection between cells.

pit membrane Primary wall and middle lamella at the outside of a pit.

pit pair Opposing pits in contiguous cells.

pith Parenchyma tissue in center of stems and/or roots.

pith ray Parenchyma in between vascular strands.

pitted Containing pits.

placenta (pl. **placentae**) Carpel tissue to which ovules are attached in the flowering plants.

placentation Arrangement of ovules in an ovary.

planated A structure, especially a leaf, which has dichotomous branches arranged in one plane.

planation Dichotomies in two or more planes evolving into one plane; flattening out in one plane, as in a leaf.

plankton Aquatic organisms that are microscopic, and free-floating or weakly swimming; position determined primarily by water currents.

planogamete Motile gamete, generally by means of one or more flagella.

planospore Motile spore with one or more flagella; also termed *zoospore*.

planozygote Zygote motile by means of flagella.

plasmalemma Outer membrane of protoplast; *see* plasma membrane.

plasma membrane Membrane surrounding the protoplast of cell; also called plasmalemma.

plasmids In prokaryote cells, minute self-replicating, extrachromosomal strands of DNA; they contribute to the genotype.

plasmodesma (pl. **plasmodesmata**) A fine cytoplasmic intercellular connection, that passes through the primary cell walls joining the protoplasts of contiguous cells.

plasmodiocarp Sessile sporangium developing from main plasmodial branches in myxomycetes.

plasmodium (pl. **plasmodia**) Naked, acellular assimilative stage in myxomycetes.

plasmogamy Fusion of cytoplasm of gametes prior to nuclear fusion or karyogamy; in higher fungi, fusion of cytoplasm and establishment of the dikaryotic stage.

plastid Double-membrane-bound cytoplasmic organelle in which CO_2 is combined with energy to form carbohydrates; or a storage organelle; or a colored organelle.

plate meristem Actively dividing tissue between the marginal meristem and midrib of leaves.

plectostele A protostele split into many platelike units.

pleomorphism Having more than one form or shape.

pleurocarpous Growth form in Bryidae in which the gametophore is multibranched and creeping (sporophyte is borne on a very short lateral branch).

plume Featherlike tuft.

plumule Terminal bud of an embryo.

plurilocular Having many chambers; used to describe gametangia and mitosporangia occurring in brown algae.

pneumatocyst Hollow area of stipe containing gas; helps keep some brown algae afloat.

podetium (pl. **podetia**) Stiff, erect secondary branch of the thallus bearing the apothecia in some lichens (especially reindeer lichens).

polar nucleus One of two nuclei that migrate to the center of a flowering plant megagametophyte, ultimately fusing with other nuclei to form the endosperm.

polar ring A short, hollow cylinder located at each pole of the spindle of a dividing nucleus (in Florideophyceae).

polarity Condition in which one end of a cell, tissue, or organ is differentiated from the other, usually in both structure and function; usually occurs during development of an organism.

pollen (grain) Immature male gametophyte of seed plants.

pollen chamber Flask-shaped chamber at the top of the nucellus in the ovule; where pollen lands in seed plants other than flowering plants.

pollen drop Exudate of megasporangium in which pollen is held in seed plants other than flowering plants.

pollen tube Tube formed by pollen grain that transports male gametes to vicinity of female gamete in most seed plants.

pollination Transfer of pollen to the vicinity of the ovule.

pollinator Agent that transfers pollen from anther to stigma in angiosperms or from pollen sac to ovule in gymnosperms.

polyarch Root or stem with many primary xylem poles.

polycyclic Series of concentric siphonosteles.

polycyclic solenostele A solenostele consisting of cylinders within cylinders.

polyglucan granules Carbohydrate storage product in bluegreen algae.

polyhedral body *See* carboxysome.

polypeptide Molecule consisting of long chains of amino acids linked together by peptide bonds.

polyphosphate granule Phosphate storage granules in bluegreen algae.

polyphyletic taxon A taxon that includes individuals that do not share a unique common ancestor.

polyphylogeny (polyphyletic) Evolution from more than one ancestral stock.

polyphyly Evolution of a taxon from more than one evolutionary line.

polyploid Having multiple genomes.

polyploidy Having more than two complete sets of chromosomes.

polysaccharide Organic molecule composed of a large number of sugar (saccharide) units linked to one another.

polysiphonous In Florideophyceae, composed of several filaments in tiers of parallel, vertically elongated cells.

polyspore Type of mitospore in which more than four spores are produced from one sporocyte (occurs in some Florideophyceae).

polystele (polystelic) Having more than one stele; consisting of more than one (many) vascular bundles (e.g., as in the stem of the seedfern, *Medullosa*).

polystelic axis An axis that is polystelic (made up of several steles), as in the stem of the seedfern *Medullosa*.

polystromatic Composed of several (many) cell layers.

population The members of one species that occupy one site at a given time.

porate Having one to many pores, as in the exine of pollen grains with pores through the wall.

porphyran Mucilaginous saccharide occurring in cell walls and intercellular spaces of some red algae.

postfertilization In red algae, with reference to stages of development immediately following fertilization.

postical Branches arising in plane of amphigastria (some Hepaticae).

prealga A phagocytic amoeboid prokaryote preceding the first true eukaryote in the (hypothetical) evolution of eukaryotes from prokaryotes.

prepollen Male gametophytes transitional between spores and pollen grains (e.g., seedferns such as the callistophytes).

primary endosperm nucleus Nucleus resulting from fusion of one male (sperm) and the two polar nuclei that will develop into the endosperm.

primary growth Growth in length due to activity of apical meristems.

primary mycelium In basidiomycetes, the mycelium developing from a basidiospore, i.e., the haploid or monokaryotic phase.

primary parietal cell Cell in developing anther that ultimately develops into the anther wall; sister cell of primary sporogenous cell.

primary pit connection In red algae, the plugged aperture that remains following division of a cell.

primary pit fields Thin areas in primary walls where plasmodesmata are concentrated.

primary plasmodium In *Plasmodiophora* (Myxomycetes), plasmodium developing in root hair following infection by haploid swarm cell.

primary primordium First primordium upon which appendages will develop.

primary producers (primary food producers) Photosynthetic organisms that are able to convert light energy into chemical energy for use by other organism.

primary protonema The chlorophyllose protonema produced by a germinating moss spore; sometimes treated as synonymous with *chloronema*.

primary sporogenous cell Cell in developing anther which ultimately forms spores (pollen); sister cell of primary parietal cell.

primary thickening meristem Dividing tissue of restricted distribution distal to the apical meristem that results in growth in girth.

primary tissue Tissue resulting from activity of apical meristems.

primary wall Wall deposited while cell is still enlarging.

primary xylem Formed by procambium; consists of protoxylem and metaxylem.

primary zoospore In Saprolegniales, a planospore with anterior flagella formed in the sporangium; it encysts, and the cyst later releases a secondary zoospore.

primitive Refers to a plant or characteristic that relates to the first-formed (oldest or ancestral) plant or characteristic.

primordium Beginning of an organ; the very earliest cellular level at which an organ can be discerned.

prismatic (crystal) Shaped like a prism.

proboscis In the Fucales, an external membranous structure closely associated with the anterior flagellum.

procambial strand Strand of procambial tissue.

procambium Primary meristem that produces and matures into primary vascular tissue.

procumbent Prostrate; lying down.

progametangial With reference to a progametangium.

progametangium (pl. **progametangia**) In Mucorales, the fertile branch tip in conjugation (the gametangium develops by deposition of a wall in the progametangium).

progenitor Early organism (or ancestors) in a direct evolutionary line.

prokaryote Having cell(s) lacking membrane-bound organelles.

prokaryotic With reference to organisms (prokaryotes) that have nuclear bodies, and lack membrane-bound organelles (nuclear membranes, chromosomes, etc.) in their cells.

promeristem The initiating cells and their most recent derivatives in an apical meristem.

promitochondrion (pl. **promitochondria**) Stage in evolution of mitochondria postulated for certain bacteria that contributed to the formation of the first eukaryotic cells by endosymbiosis.

propagula (pl. **propagulae**) Propagule.

propagule The part of a plant that propagates it; either a vegetative structure or a spore.

propaguliferous Bearing vegetative reproductive structures (propagules).

prophage "Dormant" stage of bacterial virus in which the viral genome is integrated into, and replicates with, that of the host.

prop root A root emerging from stem tissue that helps to support the plant.

prosuspensor Elongate cells in early stage of embryo development.

protein Complex organic molecule composed of one or more polypeptide chains.

prothallial cell Sterile cells formed during development of pollen grains of seed plants other than flowering plants.

protoderm Primary meristem that develops and matures into the epidermis.

protomitochondrion (pl. **protomitochondria**) Stage in evolution of the mitochondrion postulated for certain bacteria that contributed to the formation of the first eukaryotic cell by endosymbiosis.

protonema (pl. **protonemata**) Filamentous gametophyte stage of charophytes and many bryophytes; usually results from spore germination.

protophloem The first primary phloem to mature.

protoplasm Living material of cell.

protoplast That part of a cell contained within the plasma membrane and the cell wall.

protoplastid Photosynthetic body postulated for early eukaryotic cells, representing bluegreen algae that were engulfed during endosymbiosis.

protostele Stele having solid column of vascular tissue.

protostelic Bearing a protostele.

protoxylem Primary xylem that is the first formed before elongation is completed.

protozoa Unicellular organisms, usually microscopic.

proximal(ly) The part nearest the point of attachment.

proximal surface The surface where a structure is joined to others, such as in spores in tetrads following meiosis.

pseudocapillitium (pl. **pseudocapillitia**) Threadlike plasmodial strands in fructification of some Myxomycetes.

pseudocrassinucellate The formation of tissue distal to the megaspore mother cell resulting from periclinal divisions in the nucellar epidermis.

pseudomonopodial Appearing monopodial; refers to branching in which the main leader or branch is not completely dominant (mainly in psilotophytes and lycophytes).

pseudoparaphyllium (pl. **pseudoparaphyllia**) Reduced, chlorophyllose, leaflike or filamenteous structures on the stems of some mosses, arising at branch primordia.

pseudoparenchyma Mass of densely packed filaments, randomly arranged, which may lose their individuality and resemble parenchyma tissue.

pseudoparenchymatous With reference to pseudoparenchyma.

pseudoplasmodium (pl. **pseudoplasmodia**) *See* grex.

pseudopodium (pl. **pseudopodia**) False foot; in Andreaeidae and Sphagnidae (Musci), leafless gametophytic tissue acting as a seta, raising the sporangium above the main part of the gametophore.

pseudoraphe In some pennate diatoms, clear area on valve between rows of striae or costae.

pseudosaccate Having a bladderlike structure around an endospore, as in the spore of the progymnosperm *Tetraxylopteris*.

pseudosaccus The outer cover of a spore that forms a bladderlike structure.

psychrophile A "cold-loving" organism; typically refers to fungi, algae, and bacteria that have maximum growth at low temperatures (below 10 °C).

punctum (pl. **puncta**) Fine pore in the siliceous walls of diatoms; may contain finer pores, but not as complicated as areola.

purple laver Common name given to species of the red alga *Porphyra*, especially when dried; *see also* nori.

pycnidium (pl. **pycnidia**) Flask-shaped structure in which conidia are formed in some ascomycetes and Fungi Imperfecti.

pycniospore Sporelike uninucleate cell produced in pycnia of rusts and functioning in fertilization (i.e., dikaryotization); also called a *spermatium*.

pycnium (pl. **pycnia**) Flasklike to variously shaped subepidermal structure wherein spermatizing agents, the pycniospores or spermatia, develop (Uredinales); also called a *spermagonium*.

pyrenoid Proteinaceous area associated with chloroplasts of some algae and the Anthocerotae; often a center for carbohydrate storage-product formation.

pyriform Pear-shaped.

quiescent center Area at tip of root apical meristem where the rate of cell division is lower than in surrounding tissues.

raceme An indeterminate inflorescence in which secondary axes are unbranched.

rachial (leaves) In the form of a rachis; i.e., a pinnately branched compound axis.

rachis Axis of a compound leaf, or inflorescence.

radial Along a radius; branching out from a central point, like the spokes of a wheel, or arranged evenly around a central axis.

radicle Primary root of embryo below the hypocotyl.

ramentum Sterile, hairlike scales occurring among the seeds of the cone of *Cycadeoidea*.

raphe In pennate diatoms, unsilicified longitudinal groove or split in cell wall; in flowering plants, that portion of funiculus of an ovule that is adnate to the integument.

raphide A needlelike crystal.

ray Tissue initiated by cambium and extending radially in secondary xylem and phloem; consists mainly of parenchyma, but may include tracheids in the xylem.

ray tracheid Cells in some coniferous woods with thick lignified walls and bordered pits flanking top and/or bottom of a ray, or sometimes within the ray parenchyma.

receptacle Part of floral axis supporting floral parts.

receptacular Derived from the receptacle.

receptive hypha In some ascomycetes and Uredinales, specialized filament to which the spermatangial nucleus is transferred.

reciprocal parasitism Partnership between two dissimilar organisms in which both benefit; also referred to as *symbiosis*.

recombination The bringing together of alleles as a result of meiosis followed by sexual reproduction.

red tide Red-colored water usually caused by millions of algal cells containing red pigments.

reduction An elementary process of the telome concept by which structures such as branch systems or fertile trusses have become reduced in size, often accompanied by fusion of parts.

reduction division *See* meiosis.

refractile Capable of reflecting.

refringent Capable of reflecting (light); refractive.

reniform Bean- or kidney-shaped.

repetitive germination Spore germination involving production of a new spore essentially identical to the first.

reservoir Enlarged part of groove in some motile cells such as cryptophytes and eulgenophytes.

resistant sporangium Sporangium-like structure often functioning in overwintering; giving rise to zoospores upon germination.

resupinate Flat or spread on the substrate with hymenium on outer side.

reticulate Netlike.

reticulum Network.

retort cell Flask-shaped cell with an apical pore, occurring on the stem of some Sphagnidae (Musci).

rhizine Bundle of hyphae that attaches the lichen thallus to the substrate.

rhizoid Unicellular or multicellular rootlike filament functioning for attachment to the substratum.

rhizoidal Having the characteristics of a rhizoid.

rhizomatous Possessing a rhizome.

rhizome Underground stem or stemlike structure.

rhizophore Proplike organ produced at a node and forming roots at its tip; present in some lycopods.

rhizoplast Striated microtubular structure connecting basal body to nucleus in some motile cells.

rhizopodial Morphological type of growth form in which the cell is somewhat amoeboid.

rhytidome Outer tissues of bark.

ribosome Granular organelle (nonmembrane bound), composed of RNA and protein; often associated with endoplasmic reticulum of the cytoplasm; site of protein synthesis.

riparian Situated on the banks of a river, lake, or tidewater.

rockweed Common name for brown algae (Phaeophyta) of the order Fucales (e.g., *Fucus*, *Ascophyllum*).

root Underground organ found in most vascular plants; it is usually positively geotropic, develops from a root apical meristem, and functions in absorption.

rootcap Protective tissue distal to the root apical meristem.

root hair An outgrowth from a single root epidermal cell.

root hair cells See root hair.

root scar Scar left after a root has become detached from the stem.

root system All the roots of a plant.

ruminant An animal that chews a cud (e.g., a cow or sheep).

saccate Saclike; refers to pollen grains (mainly conifers) with air bladders or sacci.

saccharide Sugar units.

saccus (pl. sacci) Winglike extensions of the exine in conifer pollen giving buoyancy to the pollen grains.

saprobe (saprobic) Heterotrophic organism deriving its source of energy from dead organisms; also termed *saprophyte*.

saprophyte See saprobe.

sargasterol A sterol occurring in brown algae.

scalariform Ladderlike in form.

scalariform thickening Secondary wall material deposited in a ladderlike pattern in vessel elements and tracheids.

scalariform tracheid A xylem cell with closed end walls and a ladderlike pattern of elongated (perpendicular to the longitudinal axis) pits in the secondary lignocellulose thickening, occurring in many early fossil plants and some extant, particularly ferns.

scale In vascular plants, a platelike outgrowth; in some algae, an external element covering the cell (e.g., some chrysophytes, prymnesiophytes).

sclereid A sclerenchyma cell that is not elongated but somewhat isodiametric and often much ramified.

sclerenchyma Tissue composed of cells with thick, lignified walls; generally dead at maturity.

sclerenchyma fiber Elongate, tapering cell with thick, lignified wall; dead at maturity; important in support.

sclerified Hardened as a result of secondary wall formation and lignin deposition.

sclerotium (pl. sclerotia) In Myxomycetes, a hard plasmodial resting stage; in eumycetes, a resting body composed of a hardened mass of hyphae and frequently rounded in shape.

sclerotized Hardened; hardening of cells through lignification of their walls.

scorpioid cyme Determinate inflorescence with apparently lateral flowers borne alternately, the oldest flowers appearing to be at the base.

secondary growth Growth due to activity of lateral meristems.

secondary mycelium In basidiomycetes, dikaryotic mycelial stage initiated by fusion of primary hyphae or spermatia and receptive hyphae.

secondary nucleus Fusion product of polar nuclei before a sperm nucleus fuses with them.

secondary pit connection In red algae, the plugged aperture remaining following nuclear transfer between one vegetative cell and another.

secondary plasmodium Plasmodium, possibly dikaryotic, developing in root cortex following fusion of motile cells in *Plasmodiophora*.

secondary protonema In mosses, erect chlorophyll-rich branches that arise from the creeping protonema.

secondary spore In basidiomycetes, a spore similar to the basidiospore in its development, but produced by the germinating basidiospore rather than on the basidium.

secondary strobilus A loose aggregate of scales and pollen-producing sacs in the axis of cone bracts that collectively form a compound cone in some gnetophytes (e.g., *Welwitschia*).

secondary suspensor In seed plants, elongated cells derived from divisions of the basal tier of proembryonal cells after the formation of the primary suspensor.

secondary tissue Tissue produced by lateral or secondary meristems; results in growth in diameter and generally provides support.

secondary wall Wall deposited after cell has ceased to enlarge.

secondary xylem Tissue produced by vascular cambium providing conducting and supporting tissues; also referred to as *wood*.

secondary zoospore Zoospore produced from cysts formed by primary zoospore in Saprolegniales.

secretory cells In brown algae, small cells having a secretory function, surrounding mucilage canals.

seed Matured ovule; it consists of an embryo generally in an arrested

seed coat Covering of a seed; see integument.

seed habit Able to produce a seed.

seedling Immature plant that has developed from a seed.

segregation Separation of homologous chromosomes and hence linkage groups of genes at time of meiosis.

selection (pressure) A natural process that results in certain individuals leaving more offspring, thereby changing the nature of the population in which the individuals occur.

semiannular Irregularly ringlike.

semophylesis Partial phylogeny, often used for interpretation of evolutionary development in organs, e.g., sporangia.

sensu In the sense of (in reference to someone).

sepal In flowering plants, outermost whorl of perianth.

separation disc In bluegreen algae, area in a filament formed by death of a cell.

septate Divided by a partition.

septum (pl. septa) A transverse wall, generally perpendicular to the length of the filament.

sereological With reference to a reaction of substances (antibodies) formed in the body with foreign substances (antigens); e.g., components of bacterial cells.

serrate With toothlike margins.

serum Blood plasma minus fibrinogen component.

sessile Without a stalk.

seta (pl. setae) Sporophyte stalk in bryophytes.

sheath Covering external to cell wall.

shell zone A differentiated area at site of bud initiation.

shoot An above ground axis bearing appendages.

shoot system All the shoots of a plant.

sieve area Area in sieve cell or sieve tube element where pores are concentrated.

sieve cell Type of phloem cell characteristic of nonflowering vascular plants, in which pores in lateral sieve areas are equal in size to the pores in sieve areas in end walls.

sieve plate End wall of a sieve tube element, in which pores are larger than pores in lateral sieve area.

sieve tube In Laminariales (brown algae) and angiosperms, a conducting structure composed of tubelike series of sieve tube elements with sieve plate areas in common end walls.

sieve tube element One cell in a series constituting a sieve tube.

silicalemma Membrane in which silica is deposited.

simple fruit A fruit derived from a single carpel or compound ovary of one flower.

simple pit Pit with straight sides, or one that becomes wider near the lumen.

simple pit pair Pair of opposing pits, both of which are simple.

sinistral Having left-hand helical arrangement; used for sheath cells around oogonium in charophytes.

siphonostele (-stelic) A type of stele in which a central pith is surrounded by a cylinder of vascular tissue.

siphonous With reference to morphological type of growth form that is nonseptate and multinucleate, and often elongate; group or series of multinucleate green algae (Bryopsidophyceae).

sirenin A substance produced by female gametes that attracts the male gametes in *Allomyces*.

solenostele An amphiphloic siphonostele (that occurs as a cylinder) with phloem on both sides of the xylem.

soredia Mass of algal cells surrounded by fungus hyphae, extruded through upper or outer cortex of lichen.

sorocarp The simple fruiting body of acrasiomycetes; lacks a containing membrane and often is of irregular shape.

sorophore Stalk holding the sorus in the acrasiomycetes.

sorus (pl. sori) Cluster of spores or spores together with spore-producing structures; may include associated sterile elements.

spadix A thick, elongated, fleshy spike of some sessile flowering plants (as in Araceae) surrounded by a spathe.

spathe A bract or leaf subtending a spadix.

spatulate Spoon-shaped.

spawning In invertebrates and fishes especially, the process of liberating or depositing eggs freely into the water or on the surface of substrates in large numbers.

specialized The state of a characteristic that is rare in a taxon.

special purpose classification A classification based on only a few characteristics and designed to fulfill a special or particular purpose.

speciation The formation of new species.

species Taxonomic unit in Linnean hierarchy within a genus, in which the organisms included possess one or more distinctive characteristics and generally are capable of interbreeding freely.

spermagonium (pl. **spermagonia**) Flask-shaped structure producing the small, sporelike spermatia in some ascomycetes and in the Uredinales (basidiomycetes); also known as *pycnium*.

spermatangium (pl. **spermatangia**) Structure that produces a spermatium in red algae.

spermatium (pl. **spermatia**) Nonmotile cell functioning as male gamete; occurs in red algae and in some ascomycetes and basidiomycetes.

spermatogenous cell One of two cells in the last stage of development of the male gametophyte of gymnosperms that divides to form two male gametes.

sperm cell Male gamete.

spheroidal Approaching the form of a sphere.

spike Usually unbranched, elongated, simple indeterminate inflorescence bearing sessile flowers.

spindle Intracellular microfibrillar structure formed during nuclear division.

spine A strong, sharp-pointed appendage; in vascular plants, a woody emergence (appendage) on a stem.

spinose Having spines.

spiral tracheid Tracheid with secondary wall deposited in the form of a spiral.

spirillum (pl. **spirilla**) Helical or coiled morphological form of bacterial cell; also termed *spiral*.

spongy mesophyll Mesophyll on underside of leaf, consisting of lobed cells with large intercellular spaces.

sporangiophore Special branch bearing sporangia.

sporangiospore Spore produced in sporangium.

sporangium (pl. **sporangia**) Structure in which spores are produced; unicellular in algae, fungi, and bacteria; in bryophytes, multicellular with outer sterile layer of protective cells.

spore General name for reproductive structure, usually unicellular, but multicellular in some fungi and a few bryophytes.

sporeling Young plant produced by germination of a spore.

sporic meiosis Pattern of alternation of generations in which spores are formed by meiosis, and the spores develop into multicellular gametophytes before gametes are produced; occurs in many algae, all bryophytes, and vascular plants.

sporocarp Many-celled structure bearing spores; a fruiting body in fungi.

sporocyte A cell that gives rise to a spore or spores.

sporophyll Leaflike appendage bearing sporangia.

sporophyte Spore-producing plant, generally diploid and producing meiospores.

sporulation Releasing spores.

spur Tubular or saclike projection from a flower or flower part.

spur shoot A much-reduced and slow-growing branch in conifers, as in stems of *Ginkgo*, *Larix*, and *Cedrus*.

squamule Small, loosely attached lobe in certain lichens.

squamulose Lichen growth form similar to foliose type but with numerous, small, loosely attached thallus lobes (squamules).

stachysporous Having spores borne on stems.

stalk cell Cell in the fertile axis of some red algae, e.g., *Polysiphonia*, to which the tetrasporangium is attached.

stamen The male sporophyll in a flower, consisting of an anther and a filament.

stamen tube Tube formed by fusion of stamens.

staminate With reference to a flowering plant having only male parts functional.

starch Storage polysaccharide composed of α-1,4, 1,6 glucoside linkages (amylose and amylopectin), characteristic of green algae, pyrrhophytes, cryptophytes and most embryophytes.

statospore In chrysophytes, type of ornamented resting spore (sometimes a zygote).

stelar Pertaining to a stele, i.e., the vascular cylinder.

stele The vascular tissue and associated internal and adjacent parenchyma in a plant axis.

stellate Star-shaped.

stem-calyptra In some hepatics, a protective sheath formed from both stem tissue and archegonial cells that sheathes the young sporophyte.

stephanokont Having an anterior ring of flagella on a motile cell.

stereids Thick-walled, elongate supporting cells in the gametophores and seta of mosses.

sterigma (pl. **sterigmata**) Minute spore-bearing process in basidiomycetes.

sterile cell In bryophytes, cells within a sporangium that do not produce spores; in red algae (Florideophyceae), a cell closely associated with the carpogonial branch; in gymnosperms, one of two cells in the last stage of male gametophyte development within the pollen grain.

sterile telome One terminal branch of a dichotomy devoid of a reproductive structure (e.g., sporangia).

sterol Type of lipid present in some plants, possibly as a storage product; e.g., ergosterol, fucosterol, sargasterol, and sitosterol.

stigma Portion of angiosperm flower upon which pollen is deposited.

stigmarian system An initially four-forked dichotomizing base of the stem of fossil lycopods in which the roots are borne spirally.

stigmatoid tissue Cytoplasmically dense cells lining stigma, through which pollen tubes grow.

stipe Stalk lacking vascular tissue; may be unicellular or multicellular.

stipitate Having a stipe or special stalk.

stipule Modified part of a leaf arising at point of attachment to the stem at the node.

stolon Aerial runner; in zygomcyetes, aerial hyphae, usually rhizoids and sporangiophores at points of contact with the substrate; in vascular plants, stem usually on surface of substrate.

stoloniferous Having stolons, or propagating itself by stolons.

stoma (pl. stomata) Pore in epidermis formed by two generally kidney-shaped guard cells; in some bryophytes and most vascular plants.

stomatal apparatus The two guard cells and the opening they surround, together with any subsidiary cells.

stomium An opening on the side of a (fern) sporangium, or anther through which dehiscence takes place.

stria (pl. striae) Minute groove; in diatoms, fine row of puncta in wall appearing as lines.

striate Lined.

striated band Flagellar root with many short lines at right angles to length of root.

strobilus (pl. strobili) Collection or loose aggregation of sporophylls and associated bracts.

stroma (pl. stromata) A compact mass of fungus cells, or of mixed host and fungal cells, in or on which spores or sporocarps are formed.

stromatolite Finely layered calcareous formations produced by some bacteria and bluegreen algae.

structures Organized groups or parts; usually refers to groups of tissue.

style The elongated extension of the ovary between the stigma and the base of the ovary.

suberin Fatty substance present in or on the walls of cork and endodermal cells.

submarginal Near or under the margin.

submarginal placentation Attachment of ovule near the margin of the carpel.

subopposite Nearly oppositely arranged.

subsidiary cell Differentiated epidermal cell associated with and flanking the guard cells of stomata, often assisting in the stomatal function.

substrate (substratum) Foundation underlying surface providing point of attachment, or host for plant.

subtend To extend under, or be opposite to.

subtending bract A modified leaf or bract bearing another organ, intimately associated with another structure (e.g., a stem in the axil).

subtending leaf Stretching or extending underneath.

subtidal region That portion of the continental shelf below the lowest low-tide level (never exposed).

succubous Leaf insertion in Jungermanniales (Hepaticae) in which lower margin of a leaf lies on top of upper margin of leaf directly below it on the same side of the stem.

succulent Fleshy tissue with much water; filled with juice.

sudd Masses of vegetation which break away, often blocking the channel of tropical rivers.

sulcus (pl. sulci) Longitudinal furrow; in dinoflagellates, longitudinal posterior groove containing the trailing flagellum. Also a thin area of a pollen wall in the form of a furrow, used mainly in pollen of cycads and *Ginkgo*.

sunken stomata Stomata below the epidermis in which the guard cells are depressed below the surface cells.

superficial On the surface.

superior ovary Ovary in which the floral structures are inserted below it.

supporting cell Specialized cell from which carpogonial branch arises in some Florideophyceae.

sushi A type of Japanese sandwich made with red seaweed (*Porphyra*).

suspensor Multicellular filamentous structure produced by the first divisions of the embryo in seed plants.

swarm cell Flagellated cell; in Myxomycetes resulting from spore germination (also called *swarmer*).

swarmer Motile, flagellated reproductive cell. *See also* swarm cell.

swarming The process of releasing swarmers and subsequent apparently random movements of swarmers.

symbiosis Partnership between two dissimilar organisms, in which both benefit; also referred to as *reciprocal parasitism*.

symbiotic Pertaining to symbiosis, a living together; originally coined for association of two dissimilar organisms, but often used in the same sense as *mutualitic*.

symbiotically Functioning as a symbiont.

sympleisiomorphy Character state that characterizes all taxa in a classification.

sympodial A branch system consisting of alternating branches, the oldest branch being at the base.

sympodium (pl. **sympodia**) A stem made up of a series of superimposed branches, so as to imitate a simple axis.

synandrium (pl. **synandria**) United microsporangia.

synangium (pl. **synangia**) United sporangia.

synapomorphy Shared derived trait that characterizes a taxon.

synaptonemal complex Organelle present during meiosis and linking paired, homologus chromosomes.

syncarpous A gynoecium with fused carpels.

syndetocheilic Stomatal development in which subsidiary cells and guard cells originate from the same initials.

synergids Micropylar nuclei associated with the egg in flowering plants; part of the egg apparatus.

syngamy Fusion of gametes; fertilization.

syngenesis Coming together (fusion) of parts.

tactic With reference to directed movements of motile cells in response to stimuli (e.g., aerotaxis, phototaxis, etc.).

tapetum Nutritive layer of cells within a sporangium.

taproot Single large root.

taximetrics *See* numerical taxonomy.

taxon (pl. **taxa**) General term that can be applied to any taxonomic group or entity.

tectum The portion of the wall on top of the columellae in the wall (exine) of some angiosperm pollen grains; may be solid or perforated.

teliospore Thick-walled resting spore that bears the basidium in some rusts and smuts (basidiomycetes).

telium (pl. **telia**) Structure producing teliospores in some rusts and smuts (basidiomycetes).

telome Single terminal segment of a branching axis; can be either sterile or fertile.

telome concept A theory whereby various vascular plant organs are derived by one to several elementary processes (e.g., planation, webbing, overtopping, and syngenesis of the axis).

tendril A twisting, threadlike extension by which a plant grasps an object; may be modified stem or leaf.

teninucellate Ovule in which the archesporial cell develops into the megagametophyte without any cell divisions prior to meiosis.

tepal In flowering plants, unit of an undifferentiated perianth.

terete Cylindrical and tapering.

testa Seed coat.

tetrad scar The mark or scar on a spore surface where the spore was in contact with three others in the meiotic tetrad.

tetrahedral Arrangement of spores in a meiotic tetrad in which the four spores are arranged as at the four corners of a tetrahedron.

tetrapolar Condition referring to sexual compatibility of some basidiomycetes in which two sets of factors are involved (such as A, a and B, b).

tetrarch Protostele with four protoxylem poles.

tetrasporangium (pl. **tetrasporangia**) Meiosporangium in Florideophyceae (red algae) in which four spores are produced.

tetraspore Meiospore produced in Florideophyceae (red algae).

tetrasporine group With reference to a series of uninucleate green algae with a nonlinear arrangement of cells.

tetrasporophyte Plant producing tetraspores, usually free-living diploid plant in Florideophyceae (red algae).

thallose Having a simple plant body without differentiation into leaves or leaflike structures; type of growth form in some plants, especially Hepaticae and lichens.

thallus The body of a simple organism; one not differentiated into roots, stems, and leaves.

theca (pl. **thecae**) Case; cell wall of diatoms and cell covering of dinoflagellates.

thecate Having thecae (e.g., diatoms, dinoflagellates); *see* theca.

thermoacidophile Organism living in hot, acid conditions.

thermophilic Heat loving; with reference to organisms capable of growth at high temperatures (optimum 40–50 °C for many thermophilic fungi; up to about 90 °C for thermophilic bacteria and 75 °C for bluegreen algae), as in hotsprings.

thorn Sharpened appendage derived from stems, leaves, or roots.

thylakoid Photosynthetic membranes in eukaryotic cells grouped within chloroplast envelope.

thylakoid membrane Cytoplasmic membrane in prokaryotes containing photosynthetic and respiratory pigments and ribosomes.

thyrse Inflorescence with an indeterminate main axis and determinate lateral axes.

tinsel flagellum Flagellum with many fine tubular hairs, or mastigonemes, in one or two rows along the length of the flagellum.

tissue Group of cells organized into a structural and functional unit.

tonoplast Membrane surrounding the vacuole.

toral growth Growth in the form of a ring.

torus (pl. tori) Thickened central portion in the pit membrane of the bordered pits in some gymnosperms.

trabecula (pl. trabeculae) Row of cells bridging an intercellular space.

tracheary element Conducting element in xylem.

tracheid A xylem conducting element in vascular plants lacking perforations in end wall, although pits are abundant throughout the wall; several kinds of secondary thickenings occur, such as annular, bordered, helical, reticulate, or scalariform.

transduction Transfer of genetic material from one bacterial cell to another by bacterial viruses (bacteriophages).

transformation The incorporation of genetic material of dead cells from the medium into the genetic make-up of a living cell, as in some bacteria.

transfusion cells Cells around the vein and in the mesophyll of some cycad leaves with scalariform or reticulate pitting, apparently representing modified tracheids; cells with active cytoplasm and invaginations, functioning to transfer matter from one cell to another.

transition region Region in the hypocotyl of a seedling, where the vascular tissue characteristic of roots changes into that characteristic of the stem.

transition zone Intercalary meristem between lamina and stipe in some brown algae.

triarch Protostele with three protoxylem poles.

trichoblast Simple or branched, often colorless, hairlike branch in some Florideophyceae (red algae).

trichocine group With reference to a series of uninucleate green algae with linear (filamentous) or parenchymatous arrangement of cells.

trichocyst Cytoplasmic organelle in some flagellated algae (cryptophytes, raphidophytes, and dinoflagellates) that can be released upon being disturbed; also known as *ejectosome*.

trichogyne Receptive, hairlike extension of female gametangium in Florideophyceae (red algae) and ascomycetes.

trichome Linear row of cells; in the bluegreen algae, exclusive of sheath material; epidermal outgrowth (hair) in angiosperms.

trichothallic Intercalary growth at base of a hairlike, uniseriate filament in brown algae.

trichothallic growth Growth resulting from activity of a trichothallic meristem.

trichotomy Forked into three branches.

tricolpate Flowering plant pollen with three colpi (germinal furrows).

tricolporate Flowering plant pollen with three colpi (germinal furrows) with a pore in the center of each colpus.

trifurcate Forking into three parts.

trigone Conspicuous corner thickening in cell walls of leaves of Hepaticae.

trilete Spores with a three-armed tetrad scar in the shape of a Y.

trilete mark The Y-shaped scar on a plant spore where the other three spores were attached to it in a tetrahedral meiotic tetrad.

triploid Polyploid having three times the haploid chromosome number.

triporate Flowering plant pollen with three germinal pores.

true branching Branch in a filamentous form produced by a change in direction of cell division.

truncate Appearing as if cut off straight at the end.

tuber Underground storage organ; in the bryophytes, group of nutrient-rich cells that can remain dormant during the unfavorable season; in vascular plants, underground stem containing storage compounds.

tubule Small tube.

tunica layer Layer of apical meristem in which cells divide only anticlinally.

turbinate Shaped like a top.

umbel Umbrella-shaped inflorescence in which pedicels radiate from a common point like the ribs of an umbrella.

umbonate Bearing a convex elevation in the center.

unarmored Lacking specific articulated plates or armor, as in some dinoflagellates.

uniaxial Having a main (central) axis consisting of a single filament of usually large cells.

unifacial leaf Leaf flattened parallel to a radius of the stem to which it is attached.

unilocular Having one chamber; usually refers to the meiosporangium in the brown algae.

uniseriate Having a single linear row of cells.

uniseriate ray A ray of a single row of cells (one cell thick).

unisexual Having only one type of sexual structure (either male or female) produced by any one individual.

unistratose Having one layer.

universal veil Membrane covering the developing basidiocarp in the Agaricales.

unlimited growth Indeterminate growth; growth that occurs indefinitely, or without limit.

unnatural taxon A taxon that includes some organisms that are closely related to organisms in another taxon at the same rank.

upgrade theory The hypothesis that the thalloid habit evolved first in bryophytes, with the leafy habit derived later (opposite to *downgrade theory*).

uredium (pl. **uredia**) Structure producing uredospores in some rusts (basidiomycetes).

uredospore Dikaryotic repeating spore in some rusts (basidiomycetes).

vacuole A membrane-bound, fluid-filled sac within the cytoplasm containing cell sap (water plus solutes) and bounded by tonoplast.

vallecular canal Air-containing canal alternating with the vascular bundles in stem of some sphenophytes.

valve In diatoms, each half of the silicified portion of the cell; in hepatics, parts resulting from bending outward of sporangium wall when sporangium opens by means of regular longitudinal splits.

valve view Surface view of diatom cell.

vascular bundle An elongated strand of primary conducting tissue in vascular plants.

vascular cambium Lateral meristem that produces secondary xylem to the inside and secondary phloem to the outside.

vascular cryptogam Vascular plants that reproduce by spores rather than seeds (e.g., lycopods and ferns).

vascular ray Ray in secondary vascular tissues.

vascular strand *See* vascular bundle.

vascular system The conductive and support system of a plant composed of xylem and phloem.

vascular tissue Conducting and support tissues in tracheophytes.

vector Organism that transmits a disease from one plant or animal to another (e.g., transmission of malaria by mosquito); any means by which a diaspore is transferred from the parent plant to a new site.

vegetative With reference to the nonreproductive or sterile phase of plant structure and growth.

vegetative reproduction Asexual reproduction (progeny have same genetic constitution as parent).

vein Vascular bundle in a leaf.

velamen Multiple epidermis on orchid roots.

velum A protective flap of tissue covering or partially covering the face of the sporangium in *Isoetes*.

venation The pattern of vein arrangement in a leaf.

venter Lower, swollen, egg-containing portion of archegonium.

ventral The surface against the substratum or oriented toward the axis.

versatile Attached near the middle.

vesicle Small, intracellular membrane-bound sac.

vessel Xylem conducting structure of some vascular plants composed of tubelike series of vessel elements with perforations in common end walls; several kinds of secondary thickenings occur, such as annular, bordered, helical, reticulate, or scalariform.

vessel element One cell in a series constituting a vessel.

vestigial Persisting as a fragmentary remnant; rudimentary structure or organ.

vibrio Short, curved, rod-shaped bacterial cell.

violaxanthin Xanthophyll pigment.

volutin Stored food substance in bacteria, often appearing as granules.

volva Cuplike fragment of universal veil at base of stipe of some Agaricales (basidiomycetes).

volvocine group With reference to a series of uninucleate green algae with flagellated vegetative cells.

webbing An elementary process in the evolution of leaves by the telome concept; involves the filling in of tissue between planated, dichotomously branched branch tips, or telomes.

whiplash flagellum Smooth-surfaced flagellum, generally lacking tubular hairs (although other hairs can be present); may have thin distal region.

whorl Three or more leaves, flowers, sporangia, or other plant parts at one point on an axis or node.

wilting Drooping (usually as a result of water loss).

wood Secondary xylem; the xylem of vascular tissue, best developed in lignified plants that have secondary growth.

woody Characterized by presence of wood, or secondary xylem.

xanthophyll General name for group of yellow carotenoid pigments composed of oxygenated hydrocarbons.

xenogenous (hypothesis) Refers to origin of the eukaryotic cell by deriving the mitochondria and chloroplasts through engulfing bacteria and bluegreen algae, respectively (*see* autogenous).

xeric Dry, with reference to a habitat or environmental condition; climatically arid or with limited precipitation.

xeromorphic Refers to morphological or structural features of plants adapted to very dry habitats.

xerophytic Pertaining to xerophytes, or plants adapted to very dry habitats.

xylan Water-soluble polysaccharide xylose units occurring in cell walls of some red algae and green algae.

xylary fibers Fibers in xylem.

xylem Tissue that conducts water and mineral nutrients and functions in support.

xylose Five-carbon monosaccharide.

zeaxanthin Xanthophyll pigment.

zooplankton Free-floating or weakly swimming aquatic animal life; *see* plankton.

zoospore Motile spore with one or more flagella; also termed *planospore*.

zooxanthellae Algal cells (often yellow) living symbiotically in cells of certain invertebrate animals; algae known to be members of the dinoflagellates, cryptophytes, and xanthophytes.

zygomorphic Bilateral symmetry; i.e., symmetrical only about a single axis.

zygophore Special hyphal branch involved in gametangial copulation in the Mucorales (phycomycetes).

zygospore Thick-walled resting spore resulting from the fusion of gametangia (conjugation) in zygomycetes.

zygote Product of syngamy; diploid cell resulting from fusion of two haploid gametes.

zygotic meiosis The pattern of alternation of generations in which meiosis occurs directly in the zygote, restoring the haploid condition; only diploid stage is the zygote.

INDEX

Numbers in **boldface** indicate pages on which illustrations appear.

Abies, **293**, 569, **579**
abscission, 520
abscission scars, 541; *Ginkgo biloba*, **543**
abstricted, 103
acellular slime molds, 118
Acacia, 603, 604, 605, **607**
Acaena, **652**
accessory photosynthetic pigments, 175
accessory pigments, 174
acellular, 39
Acer, 645, 649; *A. saccharinum*, **606**; *A. striatum*, 354
Acetabularia, 204, **207**, 208
achene, 647, 649
Achillea millefolium, **3**
Achlya, **40**, 65
Achromobacter, 22
acorn, 649
Acrangiophyllum, 684, **686**
acrasins, 130
acritarchs, 668
acrocarpous, 321, 332
Acrochaetiales, 253
acrochaetioid stage, 275, **283**
acrochaetioid tetrasporophyte, 273
Acrochaetium, 253
acrogynous, 310
Actaea, 649
actinomorphic, 609, 628
Actinomycetales, 21, 24, **25**, 26
actinomycosin, 122
Actinostachys, 474
actinostele, 407, 413
adaptation series, leaves, 691
adaptations, Potomac ecological niches, 691; pollination, 648
adaxial meristem, 604, 605
Adenostema, 604
adhesion, small disseminules, 653; to fur and feathers, 651
Adiantum, 476, 540
adnate, 612
advanced, 12
adventitious roots, 435, 456, 643
adventive embryogeny, 640
aecial initials, 99
aeciospores, 99
aeons, 658
aerenchyma, 350
aerial hyphae, 58, 65, 67, 72
aerial roots, 606
aerial spore dispersal, 675
aerobe, 19
aerobic, 19, 63
Aesculus hippocastanum, **608**
aethalia, 125, 126, 127
after-ripening, 645
agamospermy, 640
agar, 253, 277
Agaricales, 104, **104**, **108**; basidiocarps, **108**, **110**
Agaricus (see also *Psalliota*), **106**; *A. bisporus*, 104
Agarum, 219, 220
Agathis, 556, 566
agglutination, 20
aggregate fruit, 647
aggregation, 128, 130
aggregation center, 128, 130, 132
Aglaozonia, 220
Agmenellum, 30
agricultural crops, 653
Agropyron repens, 640
Agrostis interrupta, **599**
Ahnfeltia, 253
Ailanthus, 649
air, chambers, 315, 489; currents, 107; dispersal, fruits, **607**
airborne, fruits, **646**; seeds, **646**
Akebia, 593
akinete, 29, 146
alar cells, mosses, 331
Alaria, 220, **223**
Albizzia julibrissen, 645
Albugo, **58**, 67; *A. cruciferarum*, **68**
albuminous, cells, 368, 569, 584; seeds, 644
alcoholic beverages, 78
Aldanophyton, 403, 404
alders, 655
Alectoria, 88
alerone, 292
Alethopteris, 507, **509**
Aleuria, 88
algae, 38, 39, 49, 69, 142-172, 569, 668, 691;
  Bacillariophyceae (see also diatoms), 152-158;
  Chlorophyta (see also green algae), 175-218, 671, 672; Chrysophyceae (see also golden algae), 148-152; Chrysophyta (see also chrysophytes), 142-143, 670; Cryptophyta (see also cryptophytes), 169-172, 670; Euglenophyta (see also euglenophytes), 165-172, 670; Phaeophyta (see also brown algae), 219-251, 670, 672; Prymnesiophyceae (see also prymnesiophytes), 158-159, 671; Pyrrophyta (see also dinoflagellates, Dinophyceae), 159-165, 670; relationships, 663; Rhodophyta (see also red algae), 252-285, 670, 671; Tribophyceae (see also xanthophytes, yellow-green algae), 143-148
algal, ancestors, 115, 674; cells, 107; component, lichens, 214; groups, 49; groups, phenetic schemes, 668; groups, relationships, **671**; layer, 77; partner, 63
algal-fungal association, 75
algin, 249
alginates, 249
alginic acid, 249
Alisma, 622, 649
Alismataceae, 622, 625
alkaloids, 8, 85
Allium, 375, **607**
Allomyces, 70; *A. arbuscula*, **71**
allophycocyanin, 254
allopolyploidy, 5
alpha carotene, 255
Alternaria, 90
alternation of generations (see also alternation of phases), **46**, **47**, 217; evolutionary development, **665**; heteromorphic, brown algae, 225, 237; isomorphic, brown algae, 225, 237, 241, 242; Phaeophyta, 225, 237, 663, 665, 691

alternation of haploid and diploid; generations, 208, 665; phases, 197
alternation of isomorphic phases, 208
alternation of phases, 217
Amanita, 104; *A. phalloides*, 107; *A. verna*, 107,**110**
amino acids, 25, 63
ammonium alginate, 249
amoebae, 39, 139
amoeboid, 119; prealga, 663
amoeboid movement, 119
amorphous pectic layer, 175
amphibians, 65
amphigastria, 310
amphiphloic, 356
amphiphloic siphonostele, 357
Amphipleura, **153**
Amphora, **52**
amphotericin B, **24**
amylopectin, 255
Anabaena, **29**, **31**, 491, 521
Anabaenopsis, **29**
Anachoropteridaceae, 456
anacrogynous, 309
anaerobe, 19
anaerobic, 19, 33; prokaryotes, 659
anagenesis, 12
Anagraecum sesquipedale, **632**
analogous, 12
Analipus, 220, 226, **229**
analogy, 12
Ananas, 647
anastomoses, 29
anatomy, **236**, **238**, 588; Ephedrales, 584; *Ginkgo biloba*, 541; stem, 507
anatropous, 634
ancestors, 115
ancestral algae, 675; Carboniferous pteridosperms, 532
ancient angiosperms, 642
Andreaceae, 325
Andreaea, 325-327, **327**; spores, 326; *A. rupestris*, **327**
Andreaeidae, 325; anatomy, **327**; morphology, **327**
Andreaeobryum, 325, 326; *A. macrosporum*, **327**
androecial primordia, 622
androecium, 610, 622, 623, 626, 629
androgenous initial, 544; *Ginkgo biloba*, **548**
androsporal cell, 544; *Ginkgo biloba*, **548**
Anemia, 474, **475**
anemophilous, flowers, 631, **633**, 656; pollen, 570
anemophily, 626, 629, 656
Aneurophytales, 493, 494, **495**, **496**, 499, 505, 511; relationships, 494
Aneurophyton, 494, 495, 680
Aneurospora, 494
Angiopteris, 465; *A. andrewsii*, **470**; *A. evecta*, **469**
angiospermous embryogeny, 569, 592
angiosperms, 97, 293, **294**, 295, **297**, 298, **298**, 299, **299**, 300, **300**, 355, 371, 373, 374, 377, 582, 585, 592, 593-656, 680, 691; adaptive evolution, **691**; adhesion of small disseminules, 653; ancestry, 518; ancient, 642; androecium, 622; anemophily, 629; animal dispersal, 645, 651; anther, 638;

737

angiosperms (continued)
  asexual reproduction, 640; carpels, 610; chiropterophily, 631; classification, 655; classification schemes, 655; companion cells, 368; complete flowers, 616; diptera flowers, 628; dispersal by adhesion to fur and feathers, 651; dispersal by ingestion, 651; dispersal mechanisms, 644; distinctive features, 593; embryo, 639, 643; endosperm, 639; entomophily, 626; evolution, 515, 690, 691; evolutionary lines, 606; floral development, 620; floral function, 626; floral interpretation, 623; flower, 606, 655; fruits, 647; general morphology, **347, 348**; groups, 596; gynoecium, 622; hydrophily, 631; hymenoptera flowers, 628; imperfect flowers, 616; incomplete flowers, 616; inflorescences, 617; integuments, 634; leaf development, 604; leaf, morphology, 605; leaves, 689; likely ancestors, 687; long range dispersal, 653; major groups, 596; man as dispersal agent, 653; mechanical dispersal, 653; megagametophyte, 636; microgametophyte, 638; nucellus, 634; nutritive tissue, 644; ocean dispersal, 651; origins, 687; ornithophily, 629; ovule, 631, 634; perfect flowers, 616; petals, 607, 622; plumed fruits and seeds, 649; pollen, flowers, 628; pollen, grains, **642**; pollen, sequence, **690**; pollination, 626; rain dispersal, 649; reproductive, apex, **375**; reproductive, plant body, 606; reproductive, structures, 641; root, system, 605; root, variation, 606; roots, 374; seeds, 643; seeds, dormancy, 645; seeds, habit correlation, 645; sepals, 607, 620; shoot system, 596; stamens, 610; stem variation, 598; stems, 598; stream dispersal, 651; sudd dispersal, 651; testa, 634; tumbleweeds, 648; vegetative, apex, **375**; vegetative, plant body, 596; vessels, 582; water dispersal, 645, 649; wind dispersal, 644, 648; winged fruits and seeds, 649; xylem, 361
angiospermy, 515; characteristics, 687; evolution, 686; origin, 687
angular collenchyma, 350
animal, 653; dispersal, 645, 651; pathogens, 22; pollinators, 609
animals, 13, 190, 448
anisogametes, 70, 163, 208, 241, 242
anisogamous, 43, 201
anisogamy, 45, 69, 225, 235, 241, 242, 671; brown algae, 235, 241, 242
annual bluegrass, 639
annuals, 237, 648
annular, 360; secondary walls, 293; thickenings, 308; tracheids, 678, **687**
Annularia, 435
annulus, 104, 333, 460, 481, 489
anomalous cambia, 606
anther, 297, 610, 626, 631, 638
antheridia (see also spermatangia), 64, 65, 79, **234, 239, 247, 252, 296, 298**, 304, 305, 309, 310, 311, 314, 315, 321, 327, 331, **400, 402, 463**, 463
antheridial initial, 522, 571; *Ginkgo biloba*, **548**
antheridiophore, 315
anthers, 638
Anthoceros, 317, **320**; *A. punctatus*, **320**
Anthocerotaceae, 315
Anthocerotae (see also hornworts), 301, 315-317, 677
Anthocerotales, anatomy, 320, **320**
anthrax, 21
antibiotics, 15, 75, 81, 82, 277
antibodies, 20
anticlinal divisions, 226, 373; brown algae, 226
antigens, 19, 20
antipodal cells, **595**
antipodals, 636, 638
Antithamnion, **267**, 273

Antitrichia curtipendula, **336**
Aphanizomenon, **31**, 33, 34
aphlebiae, 456
Aphyllophorales, 103, 104; basidiocarps, **104**; hymenia, **106**
aphyllous, 342
Apiaceae, 649
apical, cell, 226, 266, 269, **281**, 306, 309, 311, 321, 373, 374, 376; depression, 214; dominance, 305; flagellation, 163; groove, 214; growth, 212, 226, 227, **232, 234**, 262, **281**; growth, brown algae, 226; initial, 373; intrusive growth, 354; lid, 321; meristem, 346, 347, 371, 373, 374, 521, 567, 596, 622, 623, 634, 643; meristem cells, cytology, 371; meristems, angiosperms, 373; shoot meristem, 342; sporophyte, 53
Apistonema, **159**
Apium, 598
aplanosporangia, **45**, 260
aplanospores (see also carpospore, monospore, polyspore, tetraspore), 40, **45**, 60, 62, 145, 149, 179, 235, 260, **263**
apocarpous, 612
apogamous, 486
apogamy, 640
apogeotropic roots, 521
apomictic reproduction, 640
apomictically, 640
apomixis, 640
aposporous, 486
apothecia, 87, 88
apothecial stage, 88
appendages, 376
appendicular, 623
apple scab, 90
apples, 649
aquatic, animals, 69; fungi, 43, 64; plants, 651; seed plants, 214; vascular plants, 342
Aquilegia, 622, 649
Araucaria, **372**, 373, 566, 567, 569, 570, 580; *A. heterophylla*, **556**
Araucariaceae, 561, 562, 571, 572; embryo development, 571
arborescent, habit, 502, 680; lycopods, 407; plants, 685
Archaebacteria, 14, 25, 36, 660
Archaeocalamites, 430, **434**, 443, 444
Archaeocycas, **531**, 532
Archaeolithophyllum, 279
Archaeopteridales, 493, 494, 499
Archaeopteris, 493-495, **496**-498, 499, **499**, 500, **500**, 502, 513, 540, 549, **553**, 576, 577, 578, 625, 680-682, 685, 686; *A. macilenta*, **499**
Archaeosperma, 499, 502, 686
archegonial, chamber, 314; initials, 530; jacket, 531; tubes, 587, **589**
archegoniophore, 315
archegonium, 298, **299**, 304, 305, 309, 310, 312, 314, 315, 317, 321, 333, **400, 402**, 460, 463, **463, 501**, 506, 530, 530, 548, 571, **574, 583**, 585, 592; *Ginkgo biloba*, **551**
archesporial cells, 636, 640
archesporium, 638
Archidium alterniflorium, **335**, 339
Arctium minus, **652**
Arcyria, 126, **129**
areola, 152
areolation, 332
aril, 645
Aristolochia, 629, **353, 357**; *A. sipho*, **615**
armored, 159, 163
aromatic, 628
arthropods, 90
articulated, coralline algae, 279; stems, 429
artificial taxon, 12
Arum maculatum, **627**, 628
Asclepias, 644, **646, 649**; *A. speciosa*, **615**
asci, 75, 78, 80, 81, 85, **94**
ascidial, 622

ascocarps, 81, 82, 85, 89, 96; development, Euascomycetes, **83**
ascogenous hyphae, 80, 92; Euascomycetes, **83**
ascogonia, 79, 80, 85
ascogonium, tip, 80
ascomycetes (see also Ascomycotina), 668
ascomycetous fungi, 279
Ascomycotina (see also ascomycetes), 56, 62, 64, 75, 78-92, 95, 116; ascocarps, 81, 82, 85, 89, 96; ascus, 75, 78, 80, 81, 85, **94**; Clavicipitales, 85; Discomycetes, 87; Erysiphales, 82, 85; Euascomycetes, 78, 79, 80, 81, 115, 116; Eurotiales, 81, 82, **87, 88**; Hemiascomycetes, 78, **80**, 95, 114, 116; Laboulbeniales, 64, 90, 95; Loculoascomycetes, 89; parasexuality, 62; Pezizales, **81**, 88, 89, **93**; Plectomycetes, 81, 82; Pleosporales, 89, **94**; Pyrenomycetes, 85; Taphrinales, 78; Xylariales, 85
Ascophyllum, 220, 226
Ascoseira, 220
Ascoseirales, 220
ascospores, 62, 75, 78, 80, 81, 82, 85, **94**, 115; formation, 78; germination, 79; release mechanism, 81; two-celled, 90
ascostromata, 90, **94**
ascus, 81, 116
asexual reproduction, 40, **42**, 60, 62, 64, 72, **74**, 75, 78, 79, 95, 149, 163, 167, 170, 179, 201, 235, 260, 266, 306, 311, 315, 640, 659; Florideophyceae, 266; hepatics, **307**; mosses, 306; Phaeophyta, 235; yeasts, 114
asexually, 640
ash, 649
Asparagus, 599, **600**, 649
Aspergillus, 82, **87**, 114
Aspidiales, 481, **483**
Asplenium, 481
assimilative, phase, 127; body, 47; cell, 139; myxamoeba, **132**; phase, 60, 62, 128, 139; stage, 60; system, 190; thallus, 64
assimilatory starch, **289, 290**
Astasia, 167, 169
astaxanthin, 169
Asteraceae, 342, 607, 622, 639, 644, 649
asterad, 639; embryo, 639
Asterionella, **157**
Asterocystis, 253, **259**
Asterophyllites, **435**
Asteroxylon, 406, 407, 678; *A. mackiei*, **406**
astrosclereid, 351, **351**
asynchronous, 58
Athalamia hyalina, **319**
athlete's foot, 75, 82
atmosphere, 691; evolution, 660
atmospheric oxygen, 659, **660**; evolution, **693**
aucanten, 223
Audouinella, 253
Aulacomnium androgynum, **307**
aureomycin, 24
autodigestion, 104, 107
autogenous, evolution, eukaryotic cell, **662**; hypotheses, 661; school, 661
autopolyploidy, 5
autotrophic, 14, 39, 501
autotrophic bacteria, 15
autotrophs, 659
auxiliary cell, 253, 273
auxospore, 155
Avena, 633
axial, 59
axile placentation, 612, 616, 623
axillary, bud, 378, 617; sporangia, 416
axils, 377, 617
axis, embryonic, 643; evolution, 678, **681**
Azolla, 489, 491; *A. filiculoides*, **490**
Azollopsis, 491
Azotobacter, 21, **23**

bacillar bacteria, 658, **659**
Bacillariophyceae (see also diatoms), 142, 143, 149, **151**, 152-158, **152**, **153**, 172; cell structure, 152; classification, 155; ecology, 155; fossils, 152, 158, 172; frustule, 152, 155; fucoxanthin, 149, 152, 172; hypotheca, 152, 155; morphological variation, 155; pennate, 155, 158; reproduction, 155
bacilliform, 21
Bacillus, 15, **19**, 21; *B. anthracis*, 21
bacteria (see also Schizomycophyta), 13, 14-25, 36, 39, 128, 659, 660; Actinomycetales, 16, 21, 22, 24, **25**, **26**; anaerobic, 19, 33; autotrophic, 15; bacillar, **659**; Beggiatoales, 21; biochemistry, 33; cell, form, 15; cell, structure, **16**, **27**, **29**; characteristics, 21; classification, 19; DNA, 16, 18, 19, 20, **21**, 27; Eubacteriales, 21; fission, 18; flagella, 16, 18, **24**; fossil record, 660; fragmentation, 29, 33; genetic recombination, 18; gram-negative, 21, 27; gram-positive, 21; heterotrophic, 14, 15, 16, 19; lipids, 27; Myxobacteriales, 21, 24, **25**, **26**; nitrogen fixing, 21, 33, 34, 36; nucleoid, 16, 18, **18**, **20**, **23**; orders, 21-24; peptidoglycan, 15, 27; phage, 18; pili, 16; plasma membrane, 15, **16**, **17**, 18, 22; plasmid, 16; polyglucan granules, 27; polyhedral bodies, 27; Pseudomonadales, 22, 27, 33; pseudomonads, **17**, 22; reproduction, 16, 18; Rickettsiales, 21; Spirochaetales, 21; structure, 15, **16**; true, 21
bacterial, chromosome, **21**; conjugation, 18, **20**; ecology, 24; endospore, 18, 19, **19**; fission, **18**; flagellum, 16; form, **15**; recombination, **21**; structure, **17**
bacteriochlorophyll, 33; *a*, 22; *b*, 22; *c*, 22; *d*, 22; *e*, 22
bacteriophage, **18**
Baiera, 540, **541**, 549
baker's yeast, 63, **78**
bakeries, 85
baking, 75, 78
bald cypress, 561
Balsaminaceae, **654**
Baltic amber, 308, 311
banana, 647
banded iron formation, 659, **660**
baneberry, 649
Bangia, 253, **259**, 260, 262, **265**
Bangiales, 253
Bangiophyceae (see also red algae), 33, 252-254, **255**, 256, 257, 259, 260-262, **261**, **262**, **263**, **264**, 265, 282, 668, 671; classification, 260; conchocelis stage, 262; life histories, 260; morphological diversity, 260; *Porphyra*, 260; reproduction, 260
Baragwanathia, 404, **404**
Barbados gooseberry, 593
barberry, **101**, **648**, 649; leaves, 99
bark, 569
barley, 105, 596, 648
barnyard manure, 249
basal bodies, 178, 214, 225
basal sterile cell, 271, **278**
basidia, 75, 92, 95, 96, 97, 101, 103, 104, 107
basidiocarps, 92, 95, 96, 97, **98**, 103, 104, 107, 111
basidiolichen, 107
basidiomycetes, 62, 92, **92**, 95, 96, **97**, 102, 668; earliest fossil, 668; hyphae, **96**; lichens, 102, 107
basidiomycetous fungi, 95
Basidiomycotina, 56, 62, 64, 75, 92-114, 116; Agaricales, 104; Aphyllophorales, 103; Gasteromycetes, 107; Heterobasidiomycetes, 97; Homobasidiomycetes, 96, 97, 103, **103**; Hymenomycetes, **103**; Lycoperdales, 107, **112**, 113; Nidulariales, 107; Phallales, 107, **113**; Tremellales, 97, **98**; Uredinales, 97, **99**; Ustilaginales, 97, 101

basidiospores, 62, 75, 92, **95**, 96, 99, 101, 103, 104
basidium, 75, 96, **96**, 101; development, 116
basifixed, 610
basswood, 647, 649
bat pollinators, 609
Batrachospermum, 253, **267**, 283
beans, 643
beech, 102
bees, 628
beetles, 82, 628
beets, 606
Beggiatoa, 63, 64
Beggiatoales, 21
Bennettitales, 532
bennettites, 642
benthic algae, 249
Berberis, **648**, 649; *B. vulgaris*, **99**
berry, 647, **648**, 649
Bertholletia excelsa, 644
Beta, 606; *B. vulgaris*, **613**, **641**
beta-carotene, 220, 255
beta-glucan, 64
beta-sitosterol, 255
bicollateral bundles, 598
Bidens alba, **604**
bifid, 401
biflagellate, motile cells, 139; sperm, 188; swarm cell, 136; zoospores, 64, 65
Bifurcariopsis, 220, 239
bilateral, symmetry, **276**; tinsels, 170
bilin pigments, 254
biliproteins, 254, 255
binary fission, 18
binucleate cells, 60
biochemical features, 49; embryophytes, 287
biochemistry, 33
biogenesis period, 658
biological diversity, 655
bioluminescent, 130
biosphere, evolution, **660**
biospheric evolution, atmospheric oxygen, **693**
biotin, 62
bipod, 631
bipolar pili, **17**
bird, 629, 645, 651; baths, 214; flowers, 629; pollinators, 609
bird's nest fungi, 103, 107
Biscalitheca, 458
biseriate, 190
bisexual, 252, 262, 448; flowers, 656
bisporangiate cones, 535, 537
black bread mold, 73
bladder kelp, 8
bladdered pollen grains, 515
blade (see also lamina), 601
Blasia pusilla, 337
Blastocladiales, 69, 70, **71**
blister rust, 101
blooms, 36, 172
blue cheese, 82
blue molds, 82
bluegreen algae (see also Cyanophyta), 13, 14, 21, 27-37, 39, 172, 190, 279, 317, 659, 660; bloom, **35**, 36; cell structure, 27; cells, 521; Chroococcales, 30; classification, 29; colonization, 34; ecology, 34; endosymbiont, 36; endosymbiotic, 521; morphological variation, 29; movement, 27; nitrogen fixation, 491; Nostocales, 33; physiology, 33; reproduction, 29; Stigonematales, 33
body, cell, 544; nuclei, 585; scales, 217
Boletus, 104, **108**
Bonnemaisonia, 253, 275, **284**
Bordeaux mixture, **67**
border, **294**
bordered, 361; pit, 293, 294, 505, 506, 569; pit pair, 293; pitting, 535
bordered-pitted tracheids, 493, 576, 685
Bossiella, 253, **254**

Botrychium, 458, 473, **464**; *B. multifidum*, **464**
Botryopteridaceae, 456
Botrytis cinerea, 88
botulism, 21
Bowenia, 520
Bowmanites, **432**
B-phycoerythrin, red algae, 254
brachysclereid, 351, **351**
bracken fern, **351**, 363
bracket fungi, 55, 92, 103
brackish, habit, 169; water, 148
bracteoles, 582, **583**, 584, 585, 586, 589, **590**, 592; inner, **591**; outer, **591**
bracts, 537, 558, **574**, 585, 589, 617, 620, 625, 649
branch, 376, 584; evolution, 680; sclereids, **351**; traces, 495
branched sclereids, **351**
branching systems, 379
Brassica, **604**
Brazil nut, 644
bread, 82; molds, pink, 85; molds, red, 85
breaking dormancy, 645
brewer's yeast, 78
brewing, 75
Brie cheese, 82
bristles, 628
broad crook, 80
brome grass, 5
Bromus, 5; *B. inermis*, **642**
Brotherella roellii, **336**
brown algae (see also Phaeophyta), 8, 51, 64, 115, 165, 219-251; algin, 249; alginates, 249; alginic acid, 249; alternation of generations, 225; alternation of heteromorphic generations, 237; alternation of isomorphic generations, 225, 237; ammonium alginate, 249; anisogamy, 225, 235, 241, 242; annual sporophyte, 237; antheridia, **234**, 239, **247**; anticlinal divisions, 226; apical cell, 226; apical growth, 226, **227**, **232**, **234**; aplanospores, 235; aucanten, 223; basal bodies, 225; calcification, 220; calcium alginate, 249; callose, 235, **238**; carotenoid pigments, 220; cell structure, 219; cell wall, 219; cellular differentiation, 231; cellulose, 220; Cenozoic record, 242; centromeres, 223; chlorophyll *c*, 242; chloroplast, 220, **221**, **222**; classification, 9; conceptacles, 239, **247**; cortex, 231, **238**; corticated thallus, 226; Cyclosporideae, 219, 220, 225; D-glucose, 221; D-glucuronic acid, 220; diatomite deposits, 242; dictyosomes, 221; diffuse growth, 231; distribution, 242; D-mannitol, 221; DNA, 220, 223; D-xylose, 220; ecology, 242; ectocarpalean line, 223, 225, 226, 235, 242; ectocarpen, 223, 237; egg, 235, 239, **246**, **247**; embryo, 239; endoplasmic reticulum, 220, 221; evolutionary lines, 242, **672**; eyespot, 223, 225; female gametophyte, 239, **244**, **246**; fertilization, 239; flagella, 223, **223**, **232**; floats, 235; fossils, 242, **248**, **250**; free-nuclear division, 235; fucalean line, 219, 225, 235, 242; fucosan vesicles, 223; fucoserraten, 223, 241; fucoxanthin, 220, 242; galactose, 220; gametes, 223, 235, 237, 239, **245**, 249; gamone, 223, 237, 241; growth types, filamentous, **227**; growth types, parenchymatous, **230**; growth types, pseudoparenchymatous, **227**; hairs, **244**; haptera, 231; heteromorphic generations, 237; heterotrichy, 226; importance, 247; intercalary cell division, **231**; intercalary meristems, 226, 231; iodine salts, 249; isogametes, 241; isogamy, 225, 235, 241; kelps, 219; lamina, 221; laminaran, 221; life histories, 235; life history complexities, 241; lines of evolution, 241; male gametangia, 235; male gametophyte, 239; mastigonemes, 223; medulla, 231, **238**; meiosis, 239, 665;

Index **739**

brown algae (*continued*)
meiosporangia, 237; meiospores, **234**, 235, 237, 239, **245**; meristematic activity, 226; meristoderm, 231, **238**; mitospores, 235, **245**; motile cells, 223; mucilage, canals, 231; mucilage, ducts, **238**; multifiden, 223; multinucleate, 223; nuclear envelope, 220; nucleus, oogamy, 235, 239, 241; organelles, 223; parenchymatous, **232-234, 238**; pectic acid, 220; perennial, 237, 239; phenetic relationships, **672**; phycocolloids, 219; physodes, 223; pits, **238**; plants, 219; plasmodesmata, 235; pneumatocysts, 235; polarity, 231, 239; proboscis, **223**, 225; propagules, **232**, 235; pseudoparenchymatous, 226, **227, 229, 236**, 242; pyrenoid, 223; receptacles, 239, **247**; relationships, 241; reproduction, **232**, 235, **238**; rhizoplast, 225; ribosomes, 221; secretory cells, 231, **238**; sex determination, 223; sexual reproduction, 225, 235; sieve, cells, **238**; sieve, plates, **238**; sieve, tube elements, 235; sodium alginate, 249; sperms, 235, 239, **246, 247**; storage products, 221; syngamy, **245**; thylakoids, 220; transition zone, 231; trichothallic growth, 226, **227, 236**; unilocular sporangium, 219, 228, 235, 237, **238**, 239; uninucleate, 223; uses, 247; vacuoles, 221; zoospores, 223, 235, 239, 249; zygote, 237, 239, **245, 246**

brown rot, 87
Bruchia brevifolia, **339**; *B. drummondii*, **339**
brussels sprouts, 67
Bryidae, 330-333, 336; acrogynous, 333, **334**; alar cells, 331, **334**; annulus, 333; calyptra, 305, 337; cell surface ornamentation, 333, **334**; gametophore, 332; gemmae, 306, **307**; leaf areolation, 332; morphology of leaves, **334**; paraphyllia, 333; peristome, 336, **338**; pleurocarpous, 332, 333, **385**; rhizoids, 333; seta, 336; size, 331; sporangium, 333, **337**; spores, 315, 336, **339**; sporophyte, 326, 333, **335, 337**; stomata, 328, 333, 337; stomata, of sporangia, **337**; variety, **336**; variety, gametophytes and sporophytes, **335**; variety, leaves, **335**; variety, peristomes, **338**; variety, size and spore morphology, **339**; variety, sporangia and calyptrae, **337**
Bryophyta (see also bryophytes), 38, 39, 51, 53, 287, 295, 296, 301
bryophytes (see also hepatics, hornworts, mosses), 33, 40, 49, 51, 53, 209, 298, 300, 301, 303-339, 675, 691; Andreaeidae, 325-327, **327**; antheridia, 298; Anthocerotae, 315-317, **320**; Bryidae, 331-340, **333-339**; Buxbaumiidae, 329-331, **332**; Calobryales, 308-310, **308**; calyptra, 305, 308, 309, 322, 328, 329, 337; conducting tissue, 321, **323**, 329; fossil record, 677; gametophyte, 303; gemmae, 306, **307**; geological record, **677**; Hepaticae, 306-315; hornworts, 315-317; Jungermanniae, 308-312; Jungermanniales, 309-311, **309-312**; life history, 305, **305, 306**; liverworts, 306-315; Marchantiae, 312-315; Marchantiales, 314-315, **318, 319**; Metzgeriales, 311-312, **313, 314**; Monocleales, 314, **318**; morphological variation, 204; mosses, 317-339; Musci, 317-339; nutrient uptake, 301; origin, 672; paraphyllia, 331, 333; paraphyses, 327, 328, 331, 335; peat, 325; perianth, 310; perichaetium, 322, 327, 328; perigonia, 310, 312; perigynium, 312; peristome, 308, 336; peristome teeth, 322, 329, 331, 336; phylogenetic history, 672; pleurocarpus, 321, 332, 333; Polytrichidae, **323**, 328, 329, **330**, 333, 336; propagula, 306; protonema, 303, 305, 306, 321, 325, 327, 329, 337; protonemal flaps, **327, 328**, 331; reproduction, 201; rhizoids, 308, 310, 313,

bryophytes (*continued*)
314, 315, 321, 322, 325, 329, 333; sex organs, 305; sperm, **304**; Sphaerocarpales, 312-314, **316**; Sphagnidae, 322-325, **324, 326**; spore germination, 311; sporophyte, 304; Tetraphidae, 327, 328, **328**
Bryopsidophyceae, 174, 175, 200-208, **203, 205**, 216, 217, 672; cell structure, 200; Cladophorales, 204; classification, 204; Dasycladales, 208; Siphonocladales, 204
Bryopsis, 200, **203**, 204, **207**, 208, 217
Bryum, 335; *B. violaceum*, **307**
buckwheat, 649
bud, formation, 78; scales, 604; scar, 78
budding, 40, **45**, 56, 62, 75, 78, **80**, 95, 96; fungi, 114
buds, 376
Bulawayan rock series, 659
bulb, 599
bulbils, 460
Bumilleria, **144**, 145, **147**
buttercup, 622, 649
button stage, 104, 107, **107**
buttress roots, 606, **610**
butyric acid, 548
Buxbaumia, 329, 331; *B. aphylla*, **332**
Buxbaumiidae, 329; anatomy, **332**; morphology, **332**

cabbage, 67, 132, 643
Cachonia, **165**
Cactaceae, 623
cacti, 593, 596, 599, 604, 644
caducous leaves, 307
Calamites, 407, 430, **432, 435, 436**, 444, 683-685, **685**
calamites, 680
Calamophyton, 450, **452**
Calamopityaceae, 511
Calamostachys, **437**
Calandrinia compressa, **613**
calcareous, formation, 36; red algae, 279
calcification, 220, 254
calcified fossils, 214
calcium, 216; alginate, 249; carbonate, 125, 143, 175, 200, 209, 216, 220; carbonate, deposition, 216; carbonate, scales, 158
calcium oxalate, 521
California redwood, 561
calla lily, 605
Callandrium, 510, 511
Callistophytaceae, 510, 511, 513, **514**, 515, 516
Callistophytales, 509, 518
callistophyte pteridosperms, 515
Callistophyton, 510, 511, **512-514**
Callithamnion, 675
Callixylon, 493; tracheids, **496**
Callophyllis, 253, **272**
callose, 235, **238**, 364, **365**, 368; brown algae, 235, **238**
Callospermarion, 510, 511, **514**
Calobryaceae, 308, 310
Calobryales, 308; morphology, **308**
Calopityaceae, 513
Caltha, 651
Calypogeia fissa, **309**
calyptra, 305, 308, 309, 322, 329, 489
calyx, 606, 607, 609, 623, 655; tube, **616**
cambium, 423, 588, 606
Cambrian, 216, 217, 665, 668, 672
Camembert, 82
canal, 167
cancer, 72
Candida, 114; *C. albicans*, 114, **116**
Cannabis, 7, 291
cap, 104; cell, 304; differentiation, 208
capillitium, 123, 125, 126, 127
capsule, 15, **53**, 647, 649
carbohydrate, 25, 40, 142, 158

carbonate, 216; rocks, 659
Carboniferous, 403, 411, 460, 473, 478, 494, 502, 504, 505-507, 510, 511, 513, **514**, 534, 540, **553**, 554, 555, 576-578, 626, 668, 680-682, 684; age, 315; gymnosperms, **500**; Period, 678; pteridosperms, 549, 681; lycopod forests, 405, 407
Cardamine hirsuta, **654**
Cardiomanes, **479**
Carex, **642**; *C. magellanica*, 653
carinal canal, 438, **440**
carnation, 596
carotenes, in algae, 254, 255
carotenoid pigments, 195, 214, 220, 254
carotenoids, 22, 33, 142, 143, 149, **152**, 159, 165, 169, 170, 172, 175, 200, 217, 287
carpel, 516, 593, 610, 612, 622, 623, 633, 634, 655, 687; flowering plants, **623, 624**; leaflike, 623; number, 612
carpellary, 622; tissue, 623, 625; units, 610, 612
Carpinus caroliniana, **650**
carpogonial branch, 271, **278, 280**; initial, 278
carpogonium, 252, 262, 267, 271, 273, 275, **278, 280, 283**
Carpomitra, 220
Carpophyllum, 220
carposporangia, red algae, 253, **267**, 273, 275, **280, 282-284**
carpospores, red algae, 253, 262, 275, **278, 282, 283**
carposporophyte, 253, **267**, 271, 273, 275, **280, 282-284**; development, **267**; initial, 278
carrageenan, red algae, 254, 277
carrion flies, 628
carrot, 606
Caryophyllaceae, 622, 623, 626, 640
caryophyllad embryo, 639
caryopsis, 649
casparian strips, 356
Castilleja, **646**, 649
castor bean, 643
Catalpa, **646**, 649
cataphyll, 604, 605
cattails, 655
cattle, 214
cauline, 495; bundles, 584; placentation, 622, 623, 625
Caytonanthus, 516, **517**
Caytonia, 516, **517**
Caytoniales, 516, **517**, 518, 689
caytonias, 516
Cedrus, 561, **565, 566, 579**
cell, arrangement, Chlorophyceae, 181; cleavage, 167; cleavage, furrow, 40; components, Phaeophyta, 221; covering, 40, 49, 51, 142, 143; division, 169, 197, 200, 217, 254, 368; division, fungi, **60**; division, red algae, **254**; enlargement, 368; form, 15; organelles, red algae, 254; plate, 178, **179**, 292; propulsion, 40; reproduction, 27; structure, 15, 27, **29**, 40, 143, 148, 152, 159, 165, 169, 175, 200, 219, 253; wall, 40, 49, 58, 143, 146, 152, 175, 204, 217, 219, 287, **289, 290**, 292; wall, chemistry, 116, 178, 193, 208; wall, composition, 253; wall, embryophytes, 292
cells, embryophytes, 287; xylem, **362**
cellular, components, classification, 287; differentiation, Phaeophyta, 231; endosperm, 639; slime molds, 46, 127
cellulose, 25, 46, 49, 55, 56, 58, 64, 70, 102, 119, 128, 130, 139, 143, 149, 159, 163, 175, 193, 195, 200, 208, 217, 219, 220, 253, 292; plates, 159; rings, 197
Cenacrum, 253
Cenozoic, 172; records, 242
Centaurea, **646**
central, cells, **280**, 638, 639; mother cell, 373; pore, 56
centrarch, 389

centric diatoms, 155, 158
centrifugal, 356
centrifugally, 223
centrioles, 58, 254, 279, 663
centripetal, 356; maturation, 356
centromeres, 223
centrosomes, 223
Cephalotaxaceae, 561, 562, 571
Cephalozia bicuspidata, 312
Ceramiales, 253, 269, 273, 275, **278, 280, 281**
Ceramium, 253, **257**
Ceratiomyxa, 125; *C. fruticulosa*, 125, **126**
Ceratiomyxales, 125
Ceratium, **162**, 163, **166**
Ceratocystis ulmi, 82
Ceratophyllum, 631
Ceratopteris, 476
Ceratozamia, 520
cereal grains, 644
Chaetoceros, **157**, 158
Chaetophorales, 181, **184, 185**, 190, 193, 195, **195, 196**
chalazal, 639; chamber, 639; end, 548, **595**, 634, 636
Champia, 253
Chara, 45, **210**, 212, **212**, **217**
characteristics, bluegreen algae, 33; Cyanophyceae, 33; fungi, divisions, 56; fungi, subdivisions, 56; used in classification, 12
Charales (see also Charophyceae), **210**, 212
Charophyceae, 174, 175, 208-214, **210, 212, 216, 217**; diversity, 212; fossil record, 672, **673**; morphological variation, 212; reproduction, 209; structure, 209
charophytes, 209, 210, 676; gyrogonites, distribution, **673**
cheeses, 82
Cheilolejeunea clausa, **311**
Cheiroglossa, 463
chemise, 604
Chenopodiaceae, 640
chenopodiad embryo, 640
cherry, 628, 649
cherts, 659
Chilomonas, 172
chiropterophily, 626, 631
chitin, 46, 56, 58, 64, 69, 70, 72, 116; walls, 116
chitosan, 56, 58, 72
Chlamydobacteriales, 21
chlamydomonad form, 188
Chlamydomonas, **177**, 181, 183, **187**, 188, 193, 195, 197, 200, 214, 216, 217
Chlorella, 181, 190, **191**
chlorenchyma, 350
chloroamphenicol, 24
Chlorococcales, 155, 181, **184**, 188, 190, **191**, 204, 214
Chlorococcum, 181, 188, 190, **191**
Chlorophyceae (see also green algae), 146, 149, 155, 163, 174, 175-201, **176, 177**, 181, **182, 184**, 204, 209, 212, 214, 217; cell structure, 175; Chaetophorales, 190; classification, 181; filamentous, 190-200; morphological variation, 181; motile forms, 181-188; nonfilamentous, 188; nonmotile, 188; Oedogoniales, 195; reproduction, 179; Ulotrichales, 190; Ulvales, 190; Zygnematales, 197
chlorophyll, 49, 142, 143, 152, 159, 287
chlorophyll *a*, 33, 49, 143, 149, 159, 165, 170, 174, 175, 217, 255, 287
chlorophyll *b*, 165, 175, 217, 287
chlorophyll *c*, 143, 149, 159, 170, 242; c_1, 152; c_2, 152, 159
chlorophyll *d*, red algae, 255
chlorophyllose cells, 308
Chlorophyta (see also green algae, Chlorophyceae, Bryopsidophyceae, Prasinophyceae, Charophyceae), 39, 51, 53,

Chlorophyta (*continued*)
142, 159, 172, 174, 175-218, **216**, 287; Bryopsidophyceae, 200-208; characteristics, 175, 665, 668, 671, 672, 674; Charophyceae, 208-214; Chlorophyceae, 175-200; ecology, 214; phenetic relationships, **674**; Prasinophyceae, 214; relationships, 216
chlorophytan fossils, 671
chlorophytes, 216, 671
chloroplasts, 40, 49, 50, 59, 143, 145, 149, 152, 154, 156, 159, 166, 170, 172, 175, **176**, 177, 178, 188, 216, 220, **221**, 254, 288, **289, 290**, 309, 310, 312, 315, 317, 354, 662, 671; electron micrograph, **222**; envelope, 254; Phaeophyta, 220; red algae, 254; thylakoids, 49
Chlorosaccus, 144
cholesterol (see also sterol), 255
Chondrus, 253, 277
Chorda, 42
Chordaria, **223**, 225; zoospore, **224**
Chordariales, 220, 241, 242
Choreocolax, 253
Christensenia, 465
chromatin, 170, 172
Chromatium, 24
chromatophore, 49
chromoplasts, 288
chromosome, 2, 40, 58, 62, 159, 165, 254, 313; red algae, 254; structure, 663
Chromulina, 149, 151
Chroococcales, 30, 33
Chroomonas, **169**, **170**, 172
Chrysochromulina, **159**, **161**
chrysolaminaran, 142, 143, 149, 152
Chrysophyceae (see also golden algae), 142, 143, 145, 146, 148-152, **151**, 158, 163, 172; cell structure, 149; classification, 149; ecology, 149; morphological variation, 149; reproduction 149
Chrysophyta (see also Bacillariophyceae, Chrysophyceae, chrysophytes, Eustigmatophyceae, Prymnesiophyceae, Raphidophyceae, Xanthophyceae), 51, 142, 158, 159, 172, 242; characteristics, 143
chrysophytes, 51, 142, 158, 159, 172, 242; Bacillariophyceae, 142, 143, 149, 152, 155, 158, 172, **151-153, 157**; Chrysophyceae, 142, 143, 145, 146, 148, 149, **151**, 158, 163, 172; Eustigmatophyceae, 142; Prymnesiophyceae, 142, 143, 158, 159, **161**, 172; Raphidophyceae, 142; Tribophyceae, 142, 143, **144**, 145, 146, 149, 172
Chytridiales, 69, **69**, 70, 115
Chytridiomycota, 39, 56, **57, 61**, 64, 69, 70, 115, 118
chytrids, 69, 70
Cibotium, **461**
ciliates, 172
cincinnus, 620, **621**
Cinclidium stygium, **338**
circinate vernation, 407, 456
circinately vernate axis, 392
circular, 361; bordered pit, **294**, 494, 495; chromosome, 663
cisternae, 287, 288
citric acid, 82
Citrus, 640, 645
citrus fruits, 82
cladistics, 10; analysis, 10
cladists, 12
cladogenesis, 12
cladogram, 11, **11**
Cladonia, **51**, 77, 88; *C. rangiferina*, 88
Cladophora, 204, **207**, 208, 214, 216; *C. glomerata*, 216
Cladophorales, 200, 204, **205, 207**, 208
Cladostephus, 220
cladoxylids, 450-455
Cladoxylon, 450, 681, 684, **686**; *C. scoparium*, 450, **451**

cladoxylons, 494
cladoxylophytes, 450-455
Cladoxylopsida, 346, 450-455, 681
clamp connections, 92, **97**, 116, 668
clamps, 115
Clarkia, 4
class, 6
classical flower theory, 625, 655
classification, 6, 7, 12, 19, 49, 64, 146, 149, 155, 163, 169, 170, 174, 181, 204, 225, 260, 344, 345, 655; angiosperms, 655; Bangiophyceae, 260; biological, 7; brown algae, **9**; features used, 12; Florideophyceae, 262; fruits, **647**, 649; general, 6; general purpose, 8; interpretation, 12; key, 8; kinds, 8; natural, 10; ovules, 634; Phaeophyta, 225; schemes, angiosperms, 655; special purpose, 8; system, 55
Claviceps, 87; *C. purpurea*, 85, **91**
Clavicipitales, 85
Clavulina, **104**
cleavage, furrow, cell division, **663**; spores, 123
cleistothecia, 82
Clematis, 598, **599**, 607, **646**
Climacium dendroides, **335**
clone, 460
Closterium, 21, 181, 197, **201**
Clostridium botulinum, 21
club, moss, 356, 377; root, 132
cluster analysis, 10
coal ball, 456, 510
coalescence, 122
coccoid, 21, 29; bluegreen algae, 658; form, 146; types, 149
coccolithophorids, 172
coccoliths, 158
Coccolithus, **161**
coccus, 15, **16**
coconut, 649, 651
Cocos, 649, 651
Codiales, 200, 204, **205**, 208, **210**, 216, 672
Codium, 204, **207**, 208; *C. fragile*, 208; life history, 210
coenobial, **191**; forms, **193**
coenobium, 179
coenocyte, 146, 204
coenocytic, 56, 72, 146, **162**, 163, 208; mycelia, 64; stage, angiosperms, 593
coenopterid fern, 680, 681
Coenopteridopsida, 346, 450, 456-458
coenopterids, 456-458, 494
coenosorus, 484
Coleochaete, 181, **181**, 193, 195, **195, 196**, 212, 214, 217, 675
Coleus, **348, 349**, 353
collateral bundles, 598
collenchyma, 346, 350, 389, 584; angular, 350; cell types, **350**; cells, 350; lacunar, 350; lamellar, 350; tissue, 350
colonial, 29, 30, **30**, 146; Chlorococcales, 179; flagellates, 149
colonies, 30, 149, 188
colonizing terrestrial habitat, 668, 674, 675
colorless, 214
colpae, 642
colpate, 640
Colpodexylon, 406
colporate, 642
columbine, **622**, 649
columella, 123, 126, 321, 322, 336
columellate, 687
columnar, cell, 350; sclereid, **351**, 354
commensalistic relationships, 21
companion cell, 368, **370**, 589, 596; angiosperms, 368
compatible, hyphae, 92; nuclei, 62; thalli, 65
competition, 644
complete flower, 616

Compositae, **642**
Composopogon, 253
compound, cones, 571, 582; cyme, **621**; diachasium, 620, **621**; leaf, 506, 601, 604, 605; male cone, **590**; ovary, 253; strobili, 592; zoospore, 146
conceptacles, brown algae, 239, **247**
conchocelis, **258, 264,** 282; stage, 262, 265, 626, 671
Conchocelis, 253, 262; filaments, red algae, 262
conchoidal plates, 569
conchosporangial branch, 262, 264
conchosporangium, 262
conchospores, **264**
condensed chromosomes, 159, 172
conducting tissues, 675
conduplicate, 622
cone-bracts, 586
cones, 430, 435, 522, 535; bisporangiate, 535, 537; conifers, **562,** 569, **570, 573;** *Cordaites,* 555, 558; *Cycadeoidea*, 535; *Lebachia*, 558, **560**
congenital fusion, 672
conidia, 56, 62, 64, 72, 75, 79, 82, 85, 87, 92, 95, 113, 114
conidial, chain, **26**; stages, 85
conidiophores, 82, **87,** 88, 97
conifer, 102, 691; evolution, 580; ovules, 571; seeds, 299; sperm, **300**
Coniferales, **494, 504,** 554, 555, 560-573; Araucariaceae, 561, 562; Cephalotaxaceae, 561, 562, 571; Cupressaceae, 561, 562, 569, 571, 572, 580; distribution, 561; embryo, 573; families and distribution, 560; Lebachiaceae, 556, 560, 571; life history, 561; morphology, 561; morphological characteristics, 561; Pinaceae, 561, 562, 569, 570, 571, 580; Podocarpaceae, 561, 562, 570, 571, 580; reproduction, 569; stem anatomy, 567; Taxaceae, 561, 562; Taxodiaceae, 561, 562, 570, 577
Coniferophyta, 554-580
coniferophyte line, 592
coniferophytes, 549, 554-580; Coniferales, 554, 555, 560-562; Cordaitales, 554, 556-558, 560, 571; evolution, 576; evolutionary pattern, **580**
coniferopsid, 554
Coniferopsida, 346, 493, 554-580
coniferous bordered pit, **294**
conifers, **294,** 299, **494,** 500, 511, 540, 554, 556, 560, 569, 571, 578, 655; apical meristem, 567; embryonic development, **575**; fossil history, 560; leaves, 566; ovules, 571; periderm, 569; seeds, 563, **573**; sperm, **572;** stem morphology, 561
conjugation, **20,** 43, 101; amoeboid, gametes, 200
connecting filament, 253
Conocephalum conicum, **319**
continental shelf, 249
contractile, roots, 606; vacuoles, 166, **177,** 178
Cooksonia, 344, 384, 678, **679,** 685; *C. caledonica,* **384**
copepods, 169
Coprinus, **49,** 104, **110**
coral, fungi, 103; reef area, 277
Corallina, 253
Corallinaceae, 253, 668
coralline algae, 254, 277; articulated, 277; encrusting, 277
coral reef formation, 200
corals, 172
Cordaianthus, **553,** 554
Cordaitaceae, 554
Cordaitales, 504, 549, 554, 556, 558, 560, 571
cordaiten ancestors, 585
Cordaites, 407, 430, 549, 554, 555, **555, 557,** 558, 577, **578,** 585, 680, 682; female cone, **558**; leaves, 555, **557**; male reproductive organs,
Cordaites (*continued*)
557; pollen grain, **502;** reproductive organs, 555; stem, **556**; stomata, 555
Cordyline, 381
cork, 295, 296, 380; cambium, 347, 368, 380, 494, 569
corky ribs, 651
corm, 380, 423, 599, 606, 620
corn, 74, 596, 644
Cornus nuttallii, 647
corolla, 607, 609, 610, 622, 623, 628, 655
corpus, 373, 555, 588; wall, **502**
cortex, **297,** 346, 606; brown algae, 231, **238;** *Ginkgo biloba*, **544**
cortical, cells, 321; canals, 438; system, 598
corticated thallus, brown algae 226
corticating filaments, **267**
Corylus, 649
Corynebacterium diptheriae, 21
Corynepteris, 458
Corystospermaceae, 689
Corystosporales, **514,** 515, 518
Coscinodiscus, **153, 157,** 158
Cosmarium, **176,** 181, 197, **201, 203**
costa, 152, 306
cotyledonary node, 643
cotyledons, 532, 549, **549,** 573, **599,** 640, 643
couch grass, 640
cover cell, red algae, 273, **280**
c-phycocyanin, red algae, 254
crassinucellate, 636
Crassula, **609**
Crassulaceae, 664
crassulae, 543
Crataegus, 599, **600**
Cretaceous, 519, 534, 554, 626
cristae, **290**
Crocus, **600**
Cronartium ribicola, 101
crop losses, 75
crossing over, 2
Crossotheca, 506, **508**
crowberry family, 641
crowfoot, 649
crown cells, 212
crozier, 80, 115, 116; development, Euascomycetes, **83**; stage, 484
Crucibulum, 107, 111, **114**
Cruciferae, **654**
crucifers, 132
Cruoria, 253
Cruoriopsis, 253
crustose, 76, 88; lichens, 76, **77,** 78; tetrasporophyte, 275
Cryptocolea imbricata, **311**
Cryptomonas, **169,** 170, 172
Cryptonemiales, 253
Cryptophyceae (see also cryptophytes), 169, **169,** 170, **170**
Cryptophyta (see also Cryptophyceae, cryptophytes), 39, 51, 142, 169-172; cell structure, 169; classification, 170; ecology, 170; morphological variation, 170; reproduction, 169
cryptophytes, 51, 169, 172; characteristics, 143
Cryptosiphonia, 253
crystals, embryophytes, **291**
cucumber, **648**
Cucumis sativis, **648**
Cucurbitaceae, 598, **654**
cud, 25
culmination, 130, **134**
Cumagloia, 253, **268**
Cuneiphycus, 279
Cupressaceae, 561, 562, 566, 569, 571, 572, 580
cupule, 506, 515, 687
cuticle, 296, 354, 399, 644, 675
cutin, 293, 295, 354
cutinization, 675
cutinized, 309, 315; spores, 384
Cutleria, 220, 223
Cutleriales, 220
Cyanidium, **671**
Cyanobacteria (see also bluegreen algae, cyanophytes), 13, 14, 27, 30, 38
cyanocobalamin, 169
Cyanophyceae (see also bluegreen algae), 27, 29, **29, 30,** 34, 163; developmental lines, **32;** morphological diversity, **32**
cyanophycin granules, 27, **29**
Cyanophyta (see also bluegreen algae), 14, 27, 38, 170, 172
cyanophytes (see also bluegreen algae), 29
Cyathea cunninghamii, **344;** *C. elegans,* **482**
Cyatheales, 481, **482**
Cyathus, 107, 111, 481, **482**
cycad, 504, 505, 506, 509, 520, 522, 537, 543, 544, 585, 642, 682; adaptation, 520; embryogeny, **530**; extant, 520; female cone, **527;** female gametophyte, **529;** frond, 576; gum, 519; leaves, 520; life history, 521, **524**; male gametophyte, **526**; morphology, 519; ovulate sporophylls, **528**; ovule, **528;** pollen cone, **525**; propagation, 519; reproduction, 522; roots, 521; sperm, **300, 527**; stem, anatomy, 521; stem, vascular pattern, **523**; xerophytic, 520
Cycadales, 535, **539**
Cycadeoidaceae, 534
Cycadeoidea, **533, 534, 534,** 536, 537, **534, 535, 539**; ovulate receptacle, **536**; reproductive cones, 535
Cycadeoidales, 532, 534, 539; phylogeny, 538
Cycadeoidopsida, 346, 532-539; Cycadeoidales, 532; Bennettitales, 532; Williamsoniales, 537
cycadeoids, 519, 521, 532-539, 585; axillary fertile branch, **539**; cone, evolutionary development, **538**
cycadophytes, 519-539, 540, 585, 588; leaves, detached, 537; stomatal pattern, **537**
Cycadopsida, 346, 493, 519-532
Cycas, 519, 521, 522, 526, 527, **528, 529, 531,** 532; *C. circinalis,* **527;** *C. revoluta,* 507, 519, 521, **521, 522, 527, 528**
Cyclamen europaeum, **613**
cyclic AMP, 130
Cyclosporideae, 219, 220, 225
Cydonia, 649
Cymathere, 220
Cymbella, **153, 157,** 158
cyme, 620
Cynoglossum officinale, **652**
Cyperaceae, 622, **642**
cypsela, 649, **650**
cystidia, 104
cystocarp, red algae, 271, **280, 282, 284**
Cystodinium, **162,** 163
cystoliths, 291, **291**
Cystophora, 220, 242
Cystoseira, 220, 242
Cystoseiraceae, **250**
Cystoseirites, 242
cysts (see also statospore), 18, 70, 119, 146, 149, 155, 163, 169, 170, 172, 668; germination, 132; production, 155
cytokinesis, 178, 217, 292
cytology, apical meristem cells, 371
cytoplasm, 287, **290**; embryophytes, 287
cytoplasmic, connections, 217; continuity, red algae, 254, 273; DNA particles, 662; streaming, 79
cytosis, 662, 663

Dacrymyces, **98**
dairy products, 249
daisies, 596
damping off, 65

dandelion, 607, 649
dasyclad algae, phylogeny, 666, 667
Dasycladaceae, 668
Dasycladales, 200, 204, **205**, 208, 216, 665, 672; phylogenetic relationships, 668
Dasycladus, 204
data matrix, 10
date, 649; palms, 626
Datura, 626
Daucus, 606; *D. carota*, **610**
dawn redwood, 561
Dawsonia, 328; *D.superba*, **330**
DDT, 4
death angels, 107
decay, stored fruits, 72; vegetation, 72
decaying, animals, 72; plants, 72
deciduous, 40, 447; leaves, 403
decussate branching, 494
deer, 214
definitive callose, 364
Degeneria, 625
dehiscence slits, 392
dehiscent, 647
Delphinium, 649
Dendroceros, 317
Dennstaedtia, 357, **358**
Derbesia, 200, 204, **207**, 208, 217; *D. neglecta*, 208; *D. tenuissima*, 208; life history, **209**
derivatives, 371
derived trait, 12
dermatophytes, 82
Desmarestia, 220, 226, 231, **236**
Desmarestiales, 220, **236**, 237, 242
desmids (see also Zygnematales), **52**
destroying angels, 107
detached, meristem, 376; leaves, cycadophytes, 537
deterioration of stored foods, 82
determinate, 271; head, 620, **621**; inflorescence, 620; type, 271; umbel, 620, 621
detritus, 277
Deuteromycetes, 113
developing, basidium, 96; plant, **348**
development, gametophyte, **246**
Devonian, 308, 383, 384, 392, 395, 403, 404, 407, 493-499, 500, 502, 504, 511, 540, **553**, 576, 659, 671, 672, 677, 678, 680, 681, 684, 686, 691; plants, 493; Rhynie chert, 668; rocks, 494; spores, 502
diabetes, 72
Diachea, **131**, 216
diarch, 521, 606
diatom, deposits, 242; frustules, 158
diatomite deposits, 242
diatoms (see also Bacillariophyceae), **52**, 152, **163**, 172
dichasium, 620; compound, 620
dichotomous, branch, **379**; branching, 377, 680; venation, 447
dichotomously branched system, **379**
Dicnemon calycinum, **339**
dicot, 97
Dicotyledonae, 598
dicotyledons, 346, 363, 447, 596, 598, 605, 636, 639, 691; herbaceous, 691; leaf, 605
Dicranopteris, **480**
Dicranum scoparium, **335**, **336**
Dictydium, **121**, 123, **124**, 125, 127, **127**
Dictyosiphonales, 220, 241
dictyosome (see also golgi apparatus), 221, 254, 287, **289**, **290**, 365, 636; vesicle, **290**
dictyostele, 357, 360, 481
dictyostelids (see also Dictyosteliomycetes), 118, 127, 128, 130, 132, 139
Dictyosteliomycetes, 47, 118, 127-132; aggregation, 126; culmination, 130; life history, 128; migration, 130; myxamoebae, 126; sexual reproduction, 130; spores, 126
Dictyostelium discoideum, 128, 130, 132, **132**, 134

Dictyota, 220, 226, **234**, 237, 241; sperm, **223**, **224**
Dictyotales, 220, 225, 226, **234**, 238, 242
Didymosphaenia, **153**
diffuse, 231, 260
digestive vacuoles, 119
Digitalis purpurea, **630**
dikaryon, **47**, 49, 62, 99, 665
dikaryotic, 46, 62, 101; aeciospores, **100**; ascogenous hyphae, 81; cells, 62; hyphae, 62, 78, 79, 81; mycelium, 62, 92, 95; phase, 116; spore, 99, 101; stage, 99, 115; uredospores, **100**
dikaryotization, 92, 99
Dimorpha, **52**
dimorphic, 484; fungi, 58
Dinobryon, 149, **151**
dinoflagellates, 51, 159, 163, 663; blooms, 165
Dinophyceae (see also dinoflagellates), 159, **162**, 163, 165, **165**, **166**, 170, 172
Dinothrix, 163
dioecious, 522, 569, 584, 617, 631
dioicous, 448
Dioon, 520, 521, 526; *D. spinulosum*, **522**; pollen cone, **525**
Diphysciaceae, 329, 331
Diphyscium, 329, 331; *D. foliosum*, **332**
Diplasiolejeunea rudolphiana, **311**
diploid, apogamy, 640; parthenogenesis, 640; phase, inception, 665; teliospores, **100**; zoospores, 62
diplontic life history, 43
diptera flowers, 628
Dipteris, 484
dipterocarps, 649
Dipterocarpus, 649; *D. retusus*, **650**
dipterous flowers, **627**, 628
diptheria, 21
disaccate pollen, **502**, **503**, 555, 558, **561**, **571**; evolution, 561
discharge tube, 70
Discomycetes, 87, 88
discomycetous types, 115
discussion, Filicidae, 484
diseases, 136; hair, 82; skin, 82
dispersability, change, 655
dispersal, adhesion to fur and feathers, 651; enclosed seeds, 648; ingestion, 651; man, 653; mechanisms, 644; seed, 644
dispersed, 107
dispersed-spore genus, **494**
disseminules, 655
distally, 226
distinctive features, embryophytes, 301; vascular plants, 593
distribution, 242, 277; brown algae, 242; red algae, 277; vascular plants, 342
distromatic, 260
divergent, 447
diversity, 181; Pterophytina, 447; vascular plants, 342
DNA, 2, 16, 18, 19, 20, 21, **21**, 27, 40, 220, 223, 254, 288, 374, 662, 663; fibrils, 663; strand, 36
DNA-containing mitochondria, 662
dodder, 647
dogs, 214
dogwood family, 641
Dolerotheca, 508, **509**, **510**, 511
dormancy, 296, 645; breaking, 644; seed, 644
dorsal placentation, 612
dorsifixed, 610
dorsiventrality, 407
double fertilization, 301, 593, 639, 687
Douglas fir, 347, **569**, **573**, 574
downgrade evolutionary pathway, 677
Downingia, 623
downy mildew of grape, 67
downy mildews, 67
Draparnaldia, 181, 193, **195**

Drepanophycus, 404; *D. spinaeformis*, 392, **405**
Drimys, 593, 622, 623
drupe, 647, 649
druse, 291, **291**
dry, 647
dry-rot fungi, 63
duckweeds, 342
Dudresnaya, 253
dulse, 277
dung, 72, 78, 102, 107
dung-inhabiting fungi, 72, 104
Durvillaea, 220, 246
Durvilleales, 220
Dutch elm disease, 82
Dutchman's pipe, 629
dwarf shoot, 555, 561, **565**
D-glucose, 220
D-glucuronic acid, 220
D-mannitol, 221
D-xylose, 220

earliest, conifers, 577; evolution, 658; fossils, 658; gymnosperms, 494; lycopods, 403; records, 511; vascular plants, 678
Early, Cretaceous, **553**; Devonian, 502, 678; Permian, 553
Ecballium elaterium, 653, **654**
Echinosteliales, 126
Echinostelium, 126; *E. minutum*, **128**
ecological adaptative evolution, 691
ecology, 34, 146, 149, 158, 163, 169, 170, 242, 277; Chlorophyta, 214; Phaeophyta, 242
ecotypes, 3
ectocarpalean evolutionary line, 242
ectocarpalean line, 223, 225, 226, 235; brown algae, 225
Ectocarpales, 220, 235, 237, 239, 241, 245, 671
ectocarpen, 223, 237
Ectocarpus, 220, **221**, 223, 226, **228**, 237, 241, 242; life history, **245**
ectophloic, 356; siphonostele, 356
ectoplasmic net, 139
ectoplast, 287
eelgrass, 342; disease, 139
effused, 104
egg, 43, 146, 188, 197, 212, 235, 239, **246**, **247**, **249**, **299**, **300**, 506, 544, 548, 636, 638, 639, 640
egg, apparatus, 298, **299**, 636, 639; cells, **591**; nucleus, 531, **595**, 638
Egregia, 220
ejectosomes, 170
Eklonia, 220
Elaphoglossum, **481**
elaters, 308, 310-312, 314, 442
electron micrograph, flagella, **224**
Eleocharis tuberosa, **600**
elm, 649
embryo, 53, 101, 239, 299, 447, 502, 530, 532, 573, 588, **599**, 639, 640, 643, 644; development, 532; development, Araucariaceae, 571; *Ginkgo biloba*, **549**; *Gnetum*, 588-591, **590**, **591**; sac, 298, 593, **599**, 636, 639, 686, **687**; types, asterad, 639; types, caryophyllad, 639, 640; types, chenopodiad, 640; types, onagrad, 639; types, piperad, 639; types, solanad, 639, 640
embryo-producing plants, 174
embryogeny, 252, 285, 548
embryonal tier, 572
embryonic, axis, 643; development, **589**; initial, 585; root, 643
embryophytes, 286-656; biochemical features, 287; cell wall, 292; cells, 287; crystals, 291; cytoplasm, 287; distinctive features, 301; ergastic substances, 291; female structures, 298; fertilization, 300; general features, 287; male structures, 296; morphological trends,

embryophytes (continued)
295, 296; nonliving protoplast, 288; nucleus, 288; protein, 291; protoplast, 287; relationships, 301; reproductive trends, 296; starch, 291; vacuoles, 291; vegetative trends, 295
Empetraceae, 641
enation, 392, 397, 682
enation theory, 682
Encalypta, **339**; *E. affinis*, **339**; *E. alpina*, **339**; *E. ciliata*, **339**; *E. rhaptocarpa*, **339**
Encephalartos, 519, 520, **527**
encrusting coralline algae, 277
encyst, 64, 169, 170
encysting, 64
endarch, 356, 535, 554; xylem, 588
endexine, 687
endocarp, 647
endodermis, 295, 356, 357, **357**, 566
endoparasites, 132
endoparasitic slime molds, 118, 132
endoplasmic reticulum, 58, 60, 142, 145, 149, 158, 170, 178, 200, 220, 221, 254, 287, 288, **289, 290,** 463, 371, 636, 661
endosperm, 301, 585, 590, 638, 639, 642, 644; cellular, 639; free-nuclear, 639; helobial, 639; nucleus, 593, **595, 599**
endosporally, 522
endospores, 18, 19
endosporic, 416, 447; development, 286
endostome, 331
endosymbiosis, 662, 663
endosymbiotic, 172; bluegreen algae, 521; prokaryotes, 36
endothecium, 638
endozoically, 169
endozoochory, 651
Enteromorpha, 181, 190; *E. intestinalis*, 8
entomophilous flowers, 628, 629, 656
entomophily, 626
Entomophthora musae, 75
Entomophthorales, 74
environmental requirements, 62
enzymes, 103, 130, 178
Eomycetopsis, 668
Ephedra, 363, 373, 582, **583,** 584, 585, 588; anatomy, 584; leaves, 591; morphology, 584; reproduction, 584; *E. sinica*, 584
Ephedrales, 582-592; anatomy, 584; gametophyte, 584; morphology, 583, 584; relationships, 585, 588; reproduction, 584, 586
ephedrine, 584
Ephemerum minutissimum, **339**
Epichrysis, 149, **151**
epicotyl, 532, 643
epidermal, cells, 354; hairs, 507
epidermis, 67, 295, 354, 543; cuticle, **297,** 354; root system, 354, 355; shoot system, 354
epigynous, 612, **618,** 643; flowers, 628
Epilobium, 644; *E. angustifolium*, **642**
epiphragm, 329
Epiphyllum oxypetalum, 632
epiphytes, 146, 310, 463, 606
epiphytically, 214
Epipyxis, 149, **151**
epitheca, 152, 155
epithelial cells, 569
Equisetum, 356, **358,** 363, **372, 375,** 376, 377, **377, 378,** 429, 438, **439-441,** 444, 521, 581, 582, 584, 683-685, **685**; *E. arvense*, 438; *E. hymenale*, **439**, 443; *E. myriochaetum*, 438, 442; *E. scirpoides*, 438, **440**; *E. sylvaticum*, **439**; *E. temateia*, **441**
ergastic substances, embryophytes, 287, 291
ergot of rye, 85
ergots (see also *Claviceps*), 85
Ericaceae, 641, 645
Eristophytaceae, 554
Erysiphales, 82

Erysiphe, **89**; *E. graminis*, **89**
Erythraea centaurium, 613
Erythrocladia, 253, **259**
Erythrodermis, 253
Erythropeltidales, 253
Erythrotrichia, 253, **259**
Escherichia, 21; *E. coli*, 21, **23**
Etapteris, 457
Euascomycetes, 78, 79, 80, 81, **81, 83,** 115, 116
euascomycetous fungi, 81
Eubacteria, 14, 21, 660
Eubacteriales, 21, 22, **23**
Eucalyptus, 347, 596, 680
Eucheuma, 253, 277
Eudorina, 8, 181, **187**
Euglena, **167-168,** 169
euglenoid, 51, 165, 169
Euglenophyceae (see also euglenids, euglenophytes), 165, 167, **167, 168,** 169, 170
Euglenophyta (see also euglenids, euglenoids, euglenophytes) 39, 51, 142, 165-169; cell structure, 165; characteristics, 143; classification, 169; ecology, 169; reproduction, 165
euglenophytes, 165, 172
eukaryotes, 658, 660-663, 691; diversification, **664**; evolution, 660, 691
eukaryotic, 175; algae, 38, 39, 40, 49, 662, **662**; cells, 663; divisions, geological ranges, 670; flagellar types, **42**; heterotrophs, 55; nucleus, 663; organelles, 58, 216; organisms, 38; plants, 43
Eumycota, 39, 56, 64, 75, **75,** 76, 115, 118
Eupatorium, 640
Euphorbia, **345,** 610; *E. pulcherrima*, **608**
European spotted arum lily, 628
Eurotiales, 81, 82; ascocarps, **88**; conidiophores, **87**
Eurotium, **88**
eusporangiate, 383, 401, 460; development, 522
eusporangium, 460
eustele, 357, 358, **360**
Eustigmatophyceae, 142
evanescent, 271
evergreen, 447
evolution, 4, 12, 342, 658, 659; angiosperms, 691; angiospermy, 686; atmosphere, 660; atmospheric oxygen, **693**; axis, 678, **681**; biosphere, 660; bluegreen algae, 659; conifers, 576; dichotomous to sympodial to monopodial, 680, **680**; early life forms, 658; eukaryotes, 660, 663, 691; fungi, 115; heterospory, 685; hypothetical model, 675; land plants, 672; leaf, 682; line to land plants, 214, 675, **676**; lithosphere, **660**; megaphyllous leaf, 680, 681, **681,** 682; microphyllous leaf, **682**; Paleozoic pteridosperms, 511; pollen grains, 504, 685, 686; progymnospermous ancestors, 511; prokaryotes, 658; pterophyte megaphyll, **686**; seed, 500, 502, 686; seed habit, 685; siphonostele, 685; sporangial position, 683, **684, 685**; spores, **503**; steles, **360, 687**; tracheary elements, 685, **687**; vascular tissues, 684
evolutionary, branches, 504; change, 658; development, alternation of generations, **665**; development, presumed, 663; fusion, petals, 623; fusion, sepals, 623; fusion, stamens, 623; interpretation vascular plants, 344; line to vascular plants, 217, **676**; lines, angiosperms, 344, 606; loss of parts, 623; origin, 209; pathways, 549, 675; pathways, free-sporing plants, 686, **688**; pathways, sporangial line, **679**; pathways, vascular plants, 678, **689**; patterns, 665; process, overtopping, **680**; progression, linear 502; relationships, green algae, 178; relationships, groups of plants, **691**; sequence, brown algae, 242; series, algae to

evolutionary, branches (continued)
early land plants, 217, **676**; series, tracheary elements, **687**; stages, 663; trends, 660; trends, major, 678
exalbuminous seeds, 644
exanulate sporangia, 505
exarch, 356, 420; xylem, 399
exine, 494, 504, 555, 584, 642, 687
exocarp, 647
exoenzymes, 15, 46
exosporic, 321, 447, 460; development, 286; gametophyte, 415
exostome, 331
exozoochory, 651
extant, 242; cycads, 520; genera, 519
extracellular enzymes, 55
extranuclear, 223
extrorse, 610
eyespot, 146, 149, 159, 166, 167, 170, 178, 223, 225

Fabaceae, **642,** 644, 654
facultative, 19; apomicts, 640; parasites, 75
Fagaceae, **642**
Fagopyrum, 649
Fagus sylvatica, **642**
fairy ring, 92
false branches, 33
Farlowia, 253
fascicular cambium, 378, 599
fats, 40, 143, 159
fatty acids, 25, 143
Fauchea, 253
feathers, 82, 653
feet, 653
female, cone, 526, 555, 584, 586, 587, **588,** 589; flowers, 628; gametangia, 64, 70, 90, 212, 237, 675; gametophyte, 239, **244, 246,** 253, **282, 283,** 286, 500, **501,** 502, 521, 530, 548, 561, 571, **574,** 585, 587, 591, 593, **595,** 631, 686; gametophyte, conifers, **563,** 571, **572, 574**; gametophyte, cycad, **528, 529,** 530, 531; gametophyte, development, 571; gametophyte, evolution of reproductive units, 578; gametophyte, generation, 530; gametophyte, *Ginkgo biloba*, **550**; gametophyte, *Gnetum*, 589; gametophyte, plants, 195; gametophyte, prothallus, 587; gametophyte, reproductive units, 577; gametophyte, structure, embryophytes, 298; gametophyte, types, **637**; trichoblast, 271
fenestrations, 254
ferns, 97, 217, 295, **299,** 300, 377, 391; distribution, 460; embryo, 460; fertilization, 460; fossils, 460
fertile telome, 680
fertilization (see also syngamy), 70, 188, 212, 239, 269, 271, 327, 525, 526, 531, 544, 585, 587, 588, 592, 593, 610, 626, 638, 639, 647, 674, 675, 686; angiosperms, **300**; embryophytes, 300; medulla, **238**; tubes, 65, 589
fertilizer, 249
fetid odor, 628
fiber, cells, 506; sclereids, **351,** 354
fibers, 351, 352, **352,** 354, 359, 363, 505, 588
fibrillar wall, 175
fibrils, 122, 178
fibrous, root, **609**; system, 605
Ficus, 291, 354, 647
fig, 291, 354, 647
filament, 610; unbranched, 193
filamentous, 29, 33, 146, 260; brown algae, 226, **227, 228**; Chlorophyceae, 190, 195; conchocelis stage, 282; Cyanophyceae, **31**; forms, 149; medulla, **270**; organization, 225; species, 64
filaments, limited growth, **276**; unlimited growth, **276**

744   Index

filicalean ferns, 684
Filicidae, 461, 473-484; Aspidiales, 481; Cyatheales, 481; discussion, 484; fossil record, 473; Gleicheniales, 478; Hymenophyllales, 476; Polypodiales, 484; Pteridales, 476; Schizaeales, 474; sporangium dehiscence, **483**
Filicopsida, 346, 361, 450, 460-491; Filicidae, 473; general features, 460; Ophioglossidae, 462; Marattiidae, 465; Marsileidae, 487; Osmundidae, 471; Salviniidae, 489
filicopsids, 460-491
filiform apparatus, 636, 638, 639
filmy ferns, 342, 476
filter feeders, 181
filter-feeding animals, 249
fine lenses, 82
fireweeds, 644
firs, 569
first, atmosphere, 659; photoautotrophs, 659
fish, eggs, 65; aquarium hatcheries, 65
Fissidens adianthoides, **336**
fission, 18, 60, 118, 119, 128, 139
flagella, 16, **17, 24,** 39, 40, **42,** 64, 119, 142, 145, 146, 149, 158, 163, 165, 166, 167, 169, 170, 178, 416, 544, 661; insertion, 143
flagellar, apparatus, 178; arrangement, **43, 182;** characteristics, 116; groove, 170; root, 178, 200, 217; swelling, 166, 170
flagellated, 159, 172; cells, 118, 223; sperms, 531
flagellation, 51
flatworms, 169
fleshy, 647; pericarp, 651; sporophylls, 626
flies, 167, 628
floats, brown algae, 235
floral, apex, 620; apical meristem, 622; appendages, 620; appendages, initiation, 620; cup, 610, 612, 616; development, 620; function, 626; interpretation, 623; morphology, 620, 625; nectaries, 628; structure, 626
floridean starch, red algae, 255
Florideophyceae, 115, 252-254, **257,** 262-277, **270, 276, 278, 280-284,** 671; asexual reproduction, 266; auxiliary cell, 273; classification, 262; growth types, **276**; life histories, 266, 269; morphological diversity, 262; multiaxial growth types, **274**; *Nemalion*, 275; *Polysiphonia*, 269; vegetative reproduction, 266
Florinites, 557
Florinites-type cordaite, **502**
flower, bats, 631; modified stem, 623; nectar, 628; theory, classical, 655
flowers, **297,** 593, 606, 607, 609, 622, 623, 625, 647
flowering plants (see also angiosperms), 342, 593-656
fly, 629
foliar, fungicide, 67; traces, 584
foliose, 76, 88, 260; lichen thallus, **77**; plant, 195; thallus, 260; type, 190
follicle, 649
Fontinalis antipyretica, **338**
food, reserves, 40, 58, 152; vacuoles, 128
foot, 305, 317
foot cell, 90
form, class, 111; genus, 430; orders, 114
fossil, 158, 172; Chlorophyta, 216; ferns, 460; forests, Sphenophytina, 443; forms, 152; gymnosperms, 576; history, 216, 217, 279; leaves, ginkgophytes, **541**; lycopods, 424; Marattidae, **470**; Marsileidae, **470**, 489; material, 308; Phaeophyta, 242, **248, 250**; pollen, 592, 642; pollen, grains, 592; record, 152, 172, 279, 315, 504, 532, 656; record, algal divisions, 668, **670**; record, Filicidae, 473; record, Marattidae, 465; Rhodophyta, 270; sphenophytes, 444

Fossombronia, 312
Fragaria, 640, 647
Fragilaria, **157,** 158
fragmentation, 29, 33, 40, 145, 179, 200, 201, 235, 260, 315
Fraxinus, 649, **650**
free-central placentation, 612
free-nuclear, division, 235, 239, 548, 585; endosperm, 639
free-nucleate stage, 589
free nuclei, 530
free-sporing, 530; evolution of groups, **688**; plants, 548
freshwater, 172
Fritschiella, 181, **182,** 190, 193, 195, **195, 196,** 217, 287
frond, 456
fructification, 123, 125
fruit, 593, 610, 626, 647; classification, 647, 649; types, 647
fruit-eating bats, 631
fruiting bodies, 47, 62, 92
fruiting structures, 47
fruit rots, 65
Frullania dilatata, **312**; *F. kunzei*, **311**
frustules, 152, 155
fruticose, 88; lichens, 76; structure, 88; thallus, 88
fucalean, evolutionary line, 219; line, 225, 235, 242
Fucales, 220, 226, 231, 235, 239, 241, 242, **247, 248,** 665, 671
fucoidan, 219
fucosan vesicles, brown algae (see also phaeophyte tannin), 223
fucoserraten, brown algae, 220, 223, 241
fucosterol (see also sterol), 255
fucoxanthin, 149, 152, 172, 220, 242
Fucus, 220, 223, **223,** 225, 226, 231, 239, 241, 242, 246; early development, **238, 240**; life history, **247**; reproduction, **248**; sperm, 224
Fuligo, **120, 121**; *F. septica*, **125,** 127
functional, eggs, *Fucus*, 239; megaspores, 500, **501,** 502, 530, 548, 571, 636, 638
fungal, groups, 668; hyphae, 668; partner, 63; phenetic relationships, **669**; structure, electron microscope, **58**; zoospores, 60
fungallike filaments, fossil, 668
fungi, 13, 38, 39, 46, 47, 55-118, 132, 139, 165, 190, 647, 668, 691; Ascomycotina, 78-92; Agaricales, 104; Aphyllophorales, 103; Basidiomycotina, 92-114; cells, 58; characteristics, 55; Chytridiomycota, 69; classification, 64; Clavicipitales, 85; Discomycetes, 87; Erysiphales, 85; Euascomycetes, 79; Eumycota, 75-114; form class, 111; Gasteromycetes, 107; general characteristics, 55; Helotiales, 87; Hemiascomycetes, 78; Heterobasidiomycetes, 97; Homobasidiomycetes, 103; Hymenomycetes, 103; Hyphochytridiomycota, 69, 70; Imperfecti, 111; Laboulbeniomycetes, 90; Lecanorales, 88; Loculascomycetes, 89; Lycoperdales, 107; Nidulariales, 107; nutritional requirements, 62; Oomycota, 64-68; Peronosporales, 65; Pezizales, 88; Phallales, 107; Plectomycetes, 81; Pleosporales, 89; Pyrenomycetes, 85; relationships, 114; reproduction, 60; Saprolegniales, 64; thallus, 55; Tremellales, 97; Uredinales, 97; Ustilaginales, 101; Xylariales, 85; Zygomycetes, 72; Zygomycota, 72-74
Fungi Imperfecti, 56, 62, 64, 68, 75, 82, 90, 111, 113
funiculus, 610, 634, 644, 645
funoran, 277
fur, 653

Fusarium, **60**
fusiform rays, **568,** 569
fusion, 72; gametangia, 62; hyphae, 62; protoplasts, 70

Ga, 685, 691
galactans, 58, 253
galactose, 220
Galinsoga, **650**
gametangia (see also antheridium, archegonium, carpogonium, oogonium, spermagonium, spermatangium), 43, **45,** 60, 65, 70, 72, 73, 136, 146, 197, 295, 665
gametangial, contact, 56, 79; conjugation, **45**; fusion, 56, 81
gamete production, 675
gametes, 43, **46,** 60, 69, 70, 136, 145, 146, 149, 155, 163, 169, 179, 188, 223, 235, **245,** 249
gametic life history, 665, **665**
gametophore, 303, **317,** 321, 322, 325, 327, 329, 333
gametophytes, 43, 53, 239, 287, 295, **297,** 303, 321, **464,** 570, 584, 589, 593, 675, 686; bryophytes, 303; conifers, **563,** 570-572, **572**; cycad, **528, 529,** 530, 531; decreasing, complexity, 677; decreasing, size, 677; Ephedrales, 584; *Ginkgo biloba*, 543, 544, **547, 548, 548, 550, 551;** Gnetales, 589, **589**; Musci, 321; Pterophytina, 447; reduction, 686; structure, 304; Welwitschiales, 587
gametophytic, generations, 237; phases, 208
gamone, 223, 237, 241
gangrene, 21
garden pea, 604
Garrya elliptica, **633**
Garryaceae, 641
gas vacuoles, 27, **29**
Gasteromycetes, 103, 107
gelans, 254
Gelidium, 253
gemmae, 306, 311, 315, 327, 399
gene frequency, 2
general, characteristics, 55; features, embryophytes, 287; features, Filicopsida, 460; morphology, 403; morphology, lycopods, 403; morphology, Sphenophytina, 429; purpose classification, 8
generalized, characteristics, 12; vascular plants, 374
generative, apospory, 640; cell, 544, 571, 588, 589, 638; cell, *Ginkgo biloba*, **548**; nucleus, 588
genes, 1
genetic, differences, 2; mutation, 62; recombination, 18, 644; senescence, conifers, 561; sources of variation, 2; tool, 85; uniformity, 640; variability, 62, 665
geological ranges, eukaryotic divisions, **670**; prokaryotic divisions, **670**
Geothallus, 312, 313
Geotrichum, **116**
Geraniaceae, 649, **654**
Geranium bicknellii, **654**
germ tubes, 65, 67, 70, 96
germinal furrow, 525, 584
germination, 119, 645; ascospores, 78; seed, 636; spores, 336
giant kelp, 8
Gigartina, 253, 275
Gigartinales, 253, **282**
gills, 104
Ginkgo, 300, 541, **542,** 544, 549, **553,** 561, 567, 569, 570, 571; androgenous initial, **548**; androsporal cell, **548**; antheridial initials, **548**; distribution, 540; epidermis, **546**; female reproductive organs, **549**; generative cell, **548**; leaf, **546**; leaf, anatomy, **543**; leaf, blade, **546**; leaf, morphology, 543; male gametophyte, **548**; megasporangium, **551**; megasporocytes, **551**; morphology, 541;

Index 745

Ginkgo (continued)
palisade, **546**; pollen grain, **548**; tube cell, **548**; *G. biloba*, 8, 540, **542**, 543, **543**, 549, **552**, **553**, 567, 584: abscission scars, **545**; anatomy, 541; cortex, **545**; distribution, 541; female gametophyte, 548, **550, 551**; long shoot, **545**; male gametophyte, 543, 544, **547, 548**; morphology, 541; ovule, **549**; oxalate crystals, **546**; periderm, **545**; phylogeny, 549, 553, **553**; pollen grains, **548**; prothallial cell, **548**; reproduction, 543; secondary phloem, **545**; seeds, **549**; spongy mesophyll, **546**; spur shoots, **545**; uses, 541; vascular cambium, **545**; young embryo, 549
Ginkgoales, 8
Ginkgoites, 540, **541**
ginkgophytes, 540-553, 682; fossil leaves, **541**; phylogeny, 549
Ginkgopsida, 346, 493, 540-553
girdle, 152, 155; thylakoid **258**, 282
girdling, 521
glandular tapetum, 638
gleba, 107
glebal mass, 107
Gleichenia, 356, **358**
Gleicheniales, 478, **480**
Glenodinium, **165**
globular glands, 506
Globularia cordifolia, **613**
globule, 212

Gloiosiphonia, 253
Glomisporites, 491
Glossophora, 220
Glossopteris, 516; flora, 516; reproductive features, 516
glucans, 46, 56, 58, 69, 72, 116
glucose, 49, 62
gluten, 291
glycogen, 46, 58, 64
glycoprotein, 27
gnat, 629
Gnetales, 518, 588-591; reproduction, 589
Gnetophyta, 582-592
gnetophytes, 582-592; characteristics, 582; phylogeny, 591; vessels, 582
Gnetopsida, 346, 493, 516, 582-592
Gnetum, 363, 585, 588, 589, **590**, 591, **591**, 592, 685; embryo, 590; female gametophyte, 589; fertilization, 591; leaves, 588; relationships, 591; *G. gnemon*, 588
golden algae (see also Chrysophyceae, chrysophytes), 115, 148, 149
golgi (see also dictyosome), 200; apparatus, 170; body, 661
Gomphonema, **157**, 158
gonimoblast (see also carposporophyte), 253, **267, 268**, 269, 273
gonimoblast filaments, 253, 275, **283**
Goniotrichum, 253, **259**
Gonium, 4, 8, 181, **187**
gonophylls, 625
gonorrhea, 21
Gonyaulax, **162**, 163
gooseberry plants, 101
Gorgonzola cheese, 82
Gosslingia, 392; *G. breconensis*, 392, **393**
Gracilaria, 253, 273, 275; life history, **282**
grains, 63; cereal, 644
Gramineae, **642**
gram-negative, 21; bacteria, 27
gram-positive, 15, 21; bacteria, 27
gram stain, 15, 19
grana, 288, **290**
grape, 67, 88, 364, **648**, 649; industry, 67
grapefruit, 640
grass family, 363, 616, 641, 649

grasses, 64, 99, 596, 622, 643, 645; flower, 625; ovule, 625
green algae, 5, 51, 53, 174, 179, 214, 217, 287, 342, 668; evolutionary relationships, 178, 671, 672, 673, 674; filamentous, 174; heterotrichous, 217; motile, 174; multicellular, 174; nonfilamentous, 174; nonmotile, 174; parenchymatous, 174, 217; siphonous, 174; unicellular, 174
green, molds, 82; sulfur bacteria, 22
grex, 130, 132; migration, 132
Griffithsia, 675
Grimmia pulvinata, **335**
groove, 163, 165, 166, 170
ground, meristem, 346, 374, 571; pine, 416
growth, 368, 622, 623; inhibitors, 645; filamentous, **227**; Florideophyceae, **276**; multiaxial Florideophyceae, **274**; parenchymatous, **230**; phenomena, 368; plant body, 346; plants, 368; pseudoparenchymatous, **227**; types, brown algae, **227**, 230, 230; uniaxial Florideophyceae, 266
guard, cells, 354, 355, 521, 537, 584; mother cell, 355
gum, 347
Gymnodinium, **162**, 163
Gymnogongrus, 253
gymnosperm, 97, 295, 300; apex, 373, 500; ovule, 500; pits, 293; sperm, **300**; tracheids, 493
gymnospermous, characteristics, 494; groups, 495; relationships, 499; seed, 499; trees, 342
gymnosperms, 295, 298, 299, 300, 355, 363, 368, 371, 373, 374, 377, 493, 500, 511, 539, 625, 629, 647, 680, 691; classes, 493; evolutionary progression in leaf development, **576**; fossils, 576; pollination, 629; stomata, 355
gynoecium, **297**, 610, 611, 612, 622, 623, 625, 626, 640; variation, 622
gyrogonites, 214, 672, **673**

habitat, 51, 146; nonvascular eukaryotes, 39
Haematococcus, 181, 214
hairs, 82, 628, 644, 651
Halarachnion, 253
half-bordered pit pair, **294**
half-cells, 200
half-inferior ovary, 612
Halicystis, 204, 208
Halidrys, 220, 242
Halimeda, 204, **207**
Halogeton, 649
halophiles, 25
halophytes, 312
Halopteris, 220
Halosaccion, 253
Hamamelidaceae, **654**
Hamamelis virginiana, **654**
Hapalosiphon, 33, **34**
haplocheilic, 355, 538, 555, 584; stomata, 537, **537**, 538, 585
haplodiplontic, 43
Haplogloia, 220
haploid, 43, 146, 181
haploid gametophyte, 195
Haplomitrium, 308, 309; *H. hookeri*, **308**
haplontic life history, 43
haplostele, 682
haptera, 231
haptonema, 146, 158, 159
Harpagophytum, **652**
harvesting barge, kelp, **251**
Harveyella, 253, 263
haustoria, 58, 65, 67, 76, 85, 90, 97, 101, 525; endosperm, 639
hawkweed, 640

hawthorn, 599
hay-scented fern, 357
hazel nut, 649
heath family, 641, 645
Hecistopteris, 476
Hedera helix, **601**
helical, 360; tracheids, 399, **687**
helicoid cyme, 620, **621**
Helminthostachys, 462, 463
helobial endosperm, 639
Helotiales, 87, 88, 89, **92**
hematochrome (see also astaxanthin), 169
hemiascomycetes, 78, **80**, 95, 114, 116
hemicellulose, 292
Hemitrichia, 126, **129**
hemp, 291
Hepaticae (see also hepatics), 301, 306-315, 677
Hepaticites devonicus, 312, **314**, 677; *H. kidsoni*, 314
hepatics (see also Hepaticae), 306, 308, 317, 321, 342; asexual reproduction, 307; evolutionary lines, 308; gametophyte structure, 308; geological age, 308, 677; life history, **306**; oil bodies, **311**; sporophyte structure, 308
herbaceous, 403, 446, 648; habit, 403, 448; plants, 649, 680
herbivores, 88
Hernandia ovigera, **613**
heterobasidiomycetes, 75, 96, 97, 102, 114, 116
Heterochloris, **144**, 220
heterocyst, 29, **31**, 34, **34**, 35
heterokaryosis, 62
heterokaryotic, mycelia, 62; spore, 62
heteromorphic, 43, 193; generations, 237
heterosporous, 295, 298, 416, 447, 499, 502, 686; ancestors, **501**; genera, 403
heterospory, 286, 403, 500; evolution, 685
heterothallic, 73, 78, 81
heterotrichous, 217, 674; form, 217; protonema, 329
heterotrichy, 190, 226
heterotrophic, 14; bacteria, 15; eukaryotes, 118; prokaryotes, 659; species, 19
heterotrophs, 1, 5, 15, 165
heterozygous, 2
Hibiscus, **614**
Hicklingia, 392; *H. edwardi*, **393**
Hieracium, 640
hierarchy, 12
higher fungi, 55, 56, 58, 75
higher plants, 216
Himanthalia, 220
Himantothallus, 22
histones, 16
holdfast, 195, 231, 260; cell, 193, 197
homobasidiomycetes, 96, 97, 103, **103**
homologous chromosomes, 2
homology, 12
homoplasy, 12
homosporous, 295, 383, 401, 407, 460, 502, 686; ancestor, **501**; ferns, 461; plants, 286
homospory, 447
homothallic, 73
honey guides, **629**
honeydew, 85, 99
honeysuckers, 629, **629**
hooks, 651
hop tree, 649
Hormidium, 193
hormogonia, 29
hormones, 65, 130
Hormosira, 220, 247
Horneophyton, 317, 386, 678; *H. lignieri*, 386
hornworts, 53, 301, 315-317, 386
horsetails, 356, 363, 377, 429-443; modern, 435, 438
Hosta, 605, **606**
housefly, 75

hummingbirds, 629
Huperzia, 401, 406, 411, 412, 413, 416, **417**; *H. gnidioides*, **417**; *H. nummularifolia*, **417**; *H. phlegmarioides*, **417**; *H. selago*, **414**
hyaloplasm, 287
hybrid, inviability, 4, 5; sterility, 4, 5
hybrids, 443
hybridization, 5
hydathodes, 441
Hydnellum, **104**
Hydnora africanum, **345**
hydrocarbons, 223
Hydroclathrus, 220
Hydrodictyon, 181, 190, **191**, 204
hydroids, 321, 329, 333, 336
hydrophily, 626, 631
hydromes, 296
hydrosphere, 691
Hydrurus, 149, 151
Hygrohypnum, 53; *H. smithii*, **336**
hygroscopic, 653; peristome teeth, 321
Hylocomium splendens, **333, 335**
hymenial layer, 103
hymenium, 104
Hymenomonas, **159, 161**
Hymenomycetes, 103; cystidia, **107**
Hymenophyllales, 476, 478, **479**
Hymenophyllum, 476; *H. affine*, **479**
Hymenophytum flabellatum, **313**
hymenoptera flower, 628, **630**
hypanthium, 610, 612
Hypericaceae, 612
hyphae, **25**, 55
hyphal, fusion, 92; remains, 668; vesicles, 668
Hyphochytridiomycota, 39, 56, 64, 69, 70, 115, 116, 118
Hypnum lindbergii, **334**
hypocotyl, 532, 643, 644
hypodermal layers, 520
hypodermis, 566
hypogynous, 275, 612, **618**
hypothallus, 123
hypotheca, 152, 155
hypsophylls, 604, 605

ice, island, 342; plant, 604
Impatiens, 653; *I. aurella*, **654**
imperfect flower, 616, 617
imperforate, 360
importance, brown algae, 247; red algae, 277
inception, meiosis, 663; syngamy, 663
incomplete flower, 616, 617
independent assortment, 2
indeterminate, growth, 315; head, 620, **621**; inflorescence, 620; umbel, 620, **621**
Indian pipe, 102
indicators, physical-chemical characteristics, 158
inducer substances, 188
indusium, 460, 476, 478, 481
inferior, 647; ovary, 612, 623, 647, 648
inflorescences, 556, 617, 620; determinate, 620; indeterminate, 620; types, **621**
ingest, 118
ingesting, 130
ingestion, rods, 169; dispersal, 651
initials, 371, 374
initiation, floral appendages, 620
inky cap, 104
inner, aperture, 293; peristome, 321
inoperculate, 87
insect, dispersal, 73, 107; eggs, 628; flowers, 629; pollinators, 609, **615**
insectivores, 631
insects, 62, 75, 85, 90, 99, 107, 609, 626, 628
integument, **297**, 299, **299, 501**, 502, 506, 508, 522, 527, 548, 582, 583, **583**, 585, 586, 589, **591, 595**, 634, 636, 644, 645; second, 584

integumentary beak, **591**; lobe, **501**, 502
intercalary, bands, 152, 155; cell, 33; cell, division, 226; growth, 605; meristems, 315, 317, 444; meristems, brown algae, 226, 231; oogonia, 65
intercellular space, **289, 290, 292**, 350
interfascicular cambium, 378, 598
intergranal lamellae, 288
International Code of Botanical Nomenclature, 113
internode, 212, 430, 606
intine, 555, 570
intranuclear division, brown algae, 223
introrse, 610
inversion, 188, **189**
invertebrates, 169, 214
inviable, gametes, 5; seeds, 647
involucre, 583-585
involute, 622
iodine, 249
Ipomoea, 606
Iridaea, 253
Irish famine, 67
Irish moss, 277
irregular, 609
isidia, 78, **79**
isobilateral, 638
isodiametric, 347
Isoetes, 378, 380, 411, 412, 416, 420, 423, 424, **424, 426**, 427; *I. nuttallii*, **424**
isogametes, 70, 193, 241; brown algae, 241
isogamous, 43, 146, 149, 193; species, 183
isogamy, **45**, 69, 225, 235, 241, 671
isomorphic, 43, 70, 193, 195, 237; generations, 237, **665**
Isothecium, **51**

jacket, 317
Jamesonia, 476
jams, 82
Jatropha gossypifolia, **604**
jellies, 82
jelly fungi, 97
joint pine, 584
Juglans regia, **613**
Julescraneia, 242; *J. grandicornis*, **250**
Juncaceae, 609
Juncus, 350
Jungermanniae, 308
Jungermanniales, 310, 311; leaf arrangement, **310**; morphology, **309**; oil bodies, **311**; spore, discharge, **312**; spore, germination, 311
Juniperus, 561
Jurassic, 515, 537

Kalanchoe, 640
karyogamy, 47, 49, 62, 75, 78, 101, 136
karyolymph, 288
Katodinium, **165**
Kaulangiophyton, 392, 404; *K. akana*, **394**
kelp (see also Laminariales), 82, 219
Kentosphaera, **183**
keratin, 69, 82
keratinous materials, 82
key classification, 8
Kigelia, **634**
kinetochore, 254
Kingdom, Fungi, 38; Plantae, 38; Protista, 38
kingdoms of organisms, 13
Klebsormidium, 181, **182**, 190, 193, **193**, 195, 212, 214, 217
knot-weed family, 622

Laboulbenia, 90, 95
Laboulbeniales, 90, **95**
Laboulbeniomycetes, 90
Labyrinthula, 139, **140**; *L. macrocystis*, 139;

Labyrinthula (*continued*)
L. minuta, **140**; *L. vitellina*, **140**
labyrinthulids, 118, 139
Labyrinthulomycetes, 46, 118, 139
Lactobacillus, 21
Lactuca, **350**, 623
lacuna, 438
lacunar collenchyma, 350
Lagenidiales, 64
lake sediment, 158
lamellae, mosses, 329
lamellar collenchyma, 351
Lamiaceae, 649
lamina, 231, 601
laminar, placentation, 612; stamens, 623
laminaran, 49, 143, 165, 172, 221
Laminaria, 220, **223**, 231, 249; *L. digitata*, 2; *L. groenlandica*, 4
Laminariales, **9**, 219, 220, 231, 235, 237, **238**, 239, 242, **244**, **246**, 249, **250**, 665; sperm, 245; zoospore, 245
lamination, 193, 195, 209, 217, 495
land plants, algal ancestors, 674; evolution, 672, **676**; origin, 672, **676**
Lantana camara, **622**
Larix, 561, 566, **579**, 580
larkspur, 649
larvae, 628
Lastraea, 481
late blight of potato, 65
late, Cambrian, 279; Cretaceous, 489; Devonian, 392; Permian, 471
lateral, flagellar insertion, 163; meristems, 368; sterile cell, 271, **278**
Lathyrus, **614**
Laurentian Great Lakes, 216
leadwort, 638
leaf, 346, 376, **541**, 584, 585, **586**, 601, **641**, 656, 678; adaptation, 691; anatomy, 566; angiosperms, 601; bases, 535, 601; compressions, 540; development, angiosperms, 604; evolution, 682; gap, 376, 521; initiation, 376; margins, flowering plants, **605**; morphology, angiosperms, 605; palmately lobed, 690; pinnately compound, 690; primordium, 604, 605; stalk, 601; traces, 404, 521, 535
leaflet, 601, 605
leaf-like carpels, 623
leafspot diseases, 78
leafy gametophyte stage, **53**
leather goods, 82
Leathesia, 220, 226, **229**, 231, 690
Lebachia, 556, 558, **559**, 560, 561, 577, **578**; anatomy, 556; dwarf shoot, **560**; foliage twigs, **559**; monosaccate pollen, 558; pollen, **560**; pollen grain, 558; pollen-bearing cones, **559**; seed-bearing cones, **560**
Lebachiaceae, 556, 560, 571
Lecanorales, 88
Leclerquia, 406
legumes, 649, 653
Lejeunea flava, **309**; *L. minutilobula*, **311**
lemon, 640
lenticular, 569
Lentinus edodes, 104
Lepidocarpon, 412
Lepidodendron, **407**, **408**, 410, 411, **412**, 424, 430, 685
lepidoptera flowers, 628, **632**
Lepidopteris, 516, 517; morphology, 516; seeds, 516
Lepidostrobus, **412**
Lepidozia reptans, **309**
Leptocycas, 532, **532**
leptoids, mosses, 321, 329
leptoma, 570
leptome, 296
Leptomitales, 64

Leptopteris, 471
leptosporangiate, 460; fern gametophyte, **463**
leptosporangium, 460
Lescunaea incurvata, **333**
Lessonia, 220
lettuce, 67, 88, 623
Leuconostoc, 21
leucoplast, 288
liana, 588, 589, 598
libriform fibers, 363
Liceales, 125
lichens, 49, 63, 75, 76, 88, 102, 190, 214; fungi, 49; growth, 76; nature of association, 49; reproduction, 79; thalli, 77
life histories, 46, 53, 118, 136, 146, 174, 183, 201, 204, 208, 235, 260, 266, 269, 305, 665; Bangiophyceae, 260; brown algae, 235; bryophytes, 305; Chlorophyceae, 181; *Codium*, **209**, **210**; complexities, Phaeophyta, 241; Coniferales, 561; Cycadopsida, 521; euascomycetes, 81, **87**; Florideophyceae, 266, 269; *Gracilaria*, **282**; leafy hepatic, **305**; mosses, **306**; *Nemalion*, 275, **283**; Nemalion-type, 275; Phaeophyta, 235; *Polysiphonia*, 269, **278**; *Polysiphonia*-type, 269; *Porphyra*, 260
light, 645; for assimilative growth, 64; for reproductive stages, 64
lignified, cells, 341; secondary walls, 357
lignin, 293, 295, 301
ligules, 403, 406, 409
Lilaea scilloides, **613**
Liliaceae, 647
lilies, 610, 640, 659
Liliidae, 596
Lilium, 640; *L. canadense*, **613**; *L. martagon*, **632**
lima bean, **643**
lime, 125
lines of evolution, brown algae, 241
Linnaea borealis, **601**
Linnaean hierarchy, 8
lipid, granules, 178; reserves, 643
lipids, 27, 46, 58
Liriodendron tulipifera, **602**
lithocyst, **291**
Lithophyllum, 279
lithosphere, evolution, **660**
Lithothamnium, 253, 254, 279
litter, 102
liverworts, 53, 301, 306–315
livestock, 651
lobar pneumonia, 21
Lobelia cardinalis, **596**
loci, 1
locule, 610, 612, 638
Loculoascomycetes, 89
long shoot, 541, 543, 566; *Ginkgo biloba*, **543**
long-range dispersal, 653
Lophozia ascendens, **307**; *L. silvicola*, **307**, **311**
lorica, 149, 165
loricate, 149, 169
low temperature, 63
Lower, Cretaceous, 516; Devonian, 386, 391, 392, 499
Lower Permian, 434, 532
LSD, 85
lumen, 293, 351
lutein, 255
Luzula, 622
Lycogala, 125, 126; *L. epidendrum*, 126, **127**
Lycoperdales, 107; basidiocarps, **112**, **113**
Lycoperdon, 107, 112, 113
Lycopersicum, **648**
lycophytes, 403–427
Lycophytina (see also lycopods), 345, 403–427
Lycopodium, **298**, 356, **358**, **372**, 377, 403, 407, 411, 412, 413, **415**, 416, **416**, **418**, 419, 420, 423, 675, 683, **684**; *L. clavatum*, **415**; *L. complanatum*, **419**; *L. inundatum*, **415**; *L.*

Lycopodium (continued)
obscurum, **415**
lycopods (see also Lycophytina), 295, 298, 403–427, 680, 682; fossils, 424; general morphology, 403; modern, 412
Lyginopteridaceae, 511, 513
Lyginopteridales, 505
Lyginopteris, 506, **506**, **507**, **508**, 511; seeds, 506; *L. oldhamia*, **506**
Lygodium, 474; *L. palmatum*, **475**
Lyngbya, 30
lysine, 116
lysosomes, 661

macrocyst, 130, 132; development, 130; germination, 130
Macrocystis, **52**, 220, 231, 242, 247, 249; *M. angustifolia*, 8; *M. integrifolia*, 8; *M. pyrifera*, 8, **251**
Macromitrium comatum, **339**
Macropiper, 598
Macrozamia, 519, 520, 527, **527**; *M. denisonii*, 522
magnesium, 63
Magnolia, 610, **612**, 645, 647
magnolia family, 609
Magnoliaceae, 609
Magnoliales, 656, 691
magnolias, 610
Magnoliidae, 596
Magnoliopsida, 346, 355, 357, 593–656 subclass characteristics, 598
Mahonia, **157**, 158
maidenhair tree, 540
maize, 67, 101
major groups, 596
male, branch, **278**; cone, 522, 555, 569, 576, **583**, 584, 586, 587, 589; flowers, 628; gametangia, 64, 70, 146, 235; gametes, 195, 197, 201, 296, 300, **300**, 587, **591**; gametophyte, 239, **282**, 286, 295, 298, 504, 521, 522, 544, 561, 584, 586, 588, **591**; gametophyte, *Ginkgo biloba*, **548**; reproductive units, 577; strobili, **557**, 589; strobili, *Cordaites*, **557**; trichoblast, 271, **278**
Mallomonas, 149, **151**
mallow family, 620, 644
Malus, 649; *M. plumula*, **616**
Malvaceae, 622, 644, 649
mammals, 651
man, as dispersal agent, 653
mannan, 56, 58, 116, 175, 200, 208
Mannia rupestris, **319**; *M. sibirica*, **319**
mannitol, 255
mannose, 200
mantle, 102
maple, 645, 649
Marasmuis oreades, **97**
Marattia, **469**
Marattiidae, 465, **469**; fossils, **470**
Marchantia polymorpha, **307**
Marchantiae, 312-315
Marchantiales, 314, 315, **319**; anatomy, **319**
marginal, meristem, 604; placentation, 612, 616
margo, 290, **294**
marine, algae, 46; angiosperm, 139; habitat, 169; Rhodophyta, 277; marsh gas, 25; plants, 653
marsh marigold, 651
Marsilea, 363, 486, 489, 582; *M. quadrifolia*, **489**
Marsileidae, 461, 487, 489
Marsupella sprucei, **309**
massulae, 491
mastigonemes (see also tinsel flagella), 223
mating, gametes, **197**; strains, 78; types, 43, 85, 132, 183, 188
Matonia, 357, **358**
Matteuccia, 484
maturation of seed, 634
mature pollen, 584

mechanical dispersal, 653
mechanically dispersed seeds, **654**
mechanisms preventing hybridization, 4
medicines, 82
medulla, 77, 78, 231, **238**
Medullosa, 507, 508, **509**, **510**, 511, 682; *M. noei*, 507, 508
Medullosaceae, 509, 511, 513, 538
Medullosales, 507
medullosan pteridosperms, 535, 680, 682
Megaceros, 317; *M. endivaefolius*, **320**
megagametophyte, **297**, **299**, 593, 631, 634, 636, 638, 640, 641; development, 636; walls, 636
megaphyllous, 342; leaf, evolution, 681, **681**, 682, 686
megaphylls, 376
megasporangium, **297**, 411, 420, 489, 499, 500, 502, 522, 527, 530, 548, 593, 634
megaspore, 286, **297**, 403, 420, 491, **501**, 504, 530, 548, 586, 589, 636, 638, 686; mother cell, 530, 586, 589, 636; wall, **299**
megasporocytes, 548; *Ginkgo biloba*, **550**
megasporophyll, **530**, **531**, 532, **426**, 610, 623, 625
meiosis, 3, 40, 43, 60, 62, 72, 80, 123, 130, 146, 155, 239, 275, 306, 638, 663, 665; zygote, 56
meiosporangia (see also tetrasporangium, unilocular sporangium), 237, 273, **282**
meiospores, 40, 43, 179, 188, 195, 197, **234**, 235, 237, 239, **245**, 273, **278**, **283**, **284**, 675
meiotic division, 60, 223
Melanconiales, 114
Melobesia, 253
Melosira, **157**, 158
meningitis, 21
Merismopedia, 30, **30**
meristele, 357, 465
meristematic, activity, 226; cell, **289**, **300**
meristems, 368
meristoderm, 231, **238**
Mesambryanthemum, 604
mesarch, 438, 494, 506, 508, 510; primary xylem, 495; protoxylem, **496**, 508
mesocarp, 647
mesogene development, 355
mesokaryotes, 663
mesophyll, 350, 409, 507
Mesophyllum, 253
mesosome, 15, **16**, 18, **18**, 680, **680**; membrane, 17
Mesozoic, 172, 506, 513, 515, 516, 518, 537, 539, 540, 544, **553**, 556, 561, 580, 592, 691; angiosperms, 656; Era, 279, 500, 532; floras, 532; phylogeny, 518; pteridosperms, 515, 689; rocks, 540; strata, 172
metaphase, 223
metaphloem, 355, 363
Metasequoia, 561, 566, 576
metaxylem, 355, 357, 360, **497**
methanogens, 25
Metzgeria, **307**
Metzgeriales, 311, **313**, 442, 476; fossils, 314
Mexican tea, 651
Micrasterias, 181, 197, **201**
microbes, 24
microbiology, 277
microbodies, 661
Microcycas, 519, 520
Microcystis, 30, **30**, 34
microfibrils, 200, 216, 354, 355
microfossils, 36, 668
microgametophyte, 286, 295, **297**, 298, **298**, **300**, 593, 638
Microgonium, 478; *M. tahitiense*, **344**
microphyll, 376, 682
microphyllous leaf, 681, 682; evolution, **682**
micropylar, cell, 639; chamber, 639; end, 639; tube, 584, **588**, 589
micropyle, **300**, 342, 376, **501**, 502, 506, 508,

748 Index

micropyle (continued)
  525, 527, 535, 582, 590, 634, 636
Microspora, **176**
microsporangia, 411, 420, 489, 499, 543, 593; Ginkgo biloba, **547**
microspore, mother cells, 638; nucleus, 638
microspores, 286, **297**, 403, 420, 491, **501**, 543, 570, 584, 638
microsporocytes, 584
microsporogenesis, 584, **596**
microsporophyll, 535, 558, 569, 610, 623, 682; Ginkgo biloba, **547**
microstrobili, 543; Ginkgo biloba, **547**; Cordaites **557**
microtubular, arrangement, 178; roots, 178
microtubules, 40, **42**, 158, 159, 165, 170, 172, 178, **179**, 181, 195, 214, 217, 661
Mid-Devonian, 502
Middle, Carboniferous, 465; Devonian, 493, 494, 499, 502
middle lamella, 292, **292**
Mid-Pennsylvania, **514**
midrib, 604
migrating grex, 130
migration, 130; period, 130
mildew, 67, 78
milkweed, 144, 610, 649
Mimosa, 610
Mischococcus, **144**, 146
Mitella, 651, **652**
mitochondria, 119, **181**, 200, 254, 266, 288, **289**, **290**, 368, 371, 636, 660; DNA, 288
mitosis, 40, 178, 663; evolutionary stages in development, 663
mitosporangia (see also plurilocular sporangium), 237
mitospores, 235, **245**, 273
mitotic, crossing-over, 62; cell division, 223, 663; spindle, 178
mitrewort, 651, **652**
Mnium marginatum, **334**
mode of nutrition, 257
modern, lycopods, 411; horsetails, 435
modified, leaves, 623; stems, 623
Mohria, 474
molds, 63, 75
Monera, 13, 14, 36
monerans, 14-37, 38; relationships, 36
Moniliales, 114, **116**; conidia, **115**
Monilinia, 57, 88; M. fruticola, **92**
monkey puzzle, 373
monoblepharidales, 69, 70, **72**
Monoblepharis, **72**
Monoclea, 314
Monocleales, 314, **318**
monocolpate, 642, 690
monocot, 97
Monocotyledonae, 598
monocotyledons, 357, 363, 364, 374, 378, 447, 596, 598, 639, 642, 691; apical meristem, **375**; roots, 606; stems, **359**
monoecious, 569, 584, 611, 631
monoicous, 448
monokaryotic, 92; hyphae, 81, 101; mycelia, 92, 102
monolete, 461; scar, 461; suture, 508
monophyletic, 626; taxon, 11
monophyly, 11
monopodial, 377, 391; axis, 406, 680; branch developoment, 680; branching system, **379**; habit, evolution, 680, **680**
monosaccharides, 49
Monoscalitheca, 458
monosiphonous, 271
monosporangia, 260, **267**
monospores, 260, 262, **262**
Monostroma, 181, 190, **193**, 217
monostromatic, 260, **262**
monosulcate, 511; pollen, 535, 689

Monotropa, 102
Morchella, 88, **93**
morels, 15, 88
mormon tea, 363, 373
morphological, classification, 29; diversity, **31**, 225, 260, 262, **272**; diversity, Bangiophyceae, 260; diversity, Florideophyceae, 262; diversity, Phaeophyta, 225; trends, embryophytes, 295; variation, 29, 146, 149, 155, 163, 169, 170, 174, 181, 204, 212
morphology, 230, **236**, 238, 267, **268**, 270; Coniferales, 561; Cycadopsida, 520; Ephedrales, 584; Ginkgo biloba, 541, 542; lycopods, 403
mosses (see also bryophytes), 53, 212, 217, **298**, **299**, 301, 317, 328, 331, 317-339; asexual reproduction, 306; conducting systems, **323**; geological record, 677, **677**; life history, **306**; sperm, **300**
motile cells, 69, 223; Chlorophyceae, 181; gametes, **45**, 62, 69; isogametes, 56
Mougeotia, **176**, **177**, 181, 197, **201**, 214
mountain ash, 649
movement, 155
mucilage, 254, 315; canals, 231, 521; cavities, 543; ducts, **238**, 465; hairs, 308
mucilaginous, pads, 159; tubes, 158
mucopeptides, 27
mucopolysaccharide, 128
Mucor, 72
Mucorales, 72, 73, 74, **74**, **75**
multiaxial, **270**; construction, **268**; Florideophyceae, **268**, **272**, **276**
multicellular sporangium, **45**
multifiden, 223
multiflagellate zoospores, 146
multinucleate, 119, 200, 204, 223, 245, 254; cells, 200, 204; hyphae, 64; zoospore, 146
multiple, alleles, 2; fruit, 647; epidermis, 354
multiseriate, 260; bluegreen algae, 33
multistratose, 304
Musci (see also mosses), 301, 317-339, 677, **677**; Andreaeidae, 325-327; Bryidae, 331-339; Buxbaumiidae, 329-331, **332**; gametophyte, 321; Polytrichidae, 323, 328-329, **330**; Sphagnidae, 322-325; sporophyte, 321; Tetraphidae, 327, 328
Muscoflorschuetzia, 329
mushrooms, 55, 62, 75, 92, 103, 104, **108**
mustard, alga, 14; family, 67
mutation, 2, 677
mutational changes, 659
Mutinus, **113**
mutualistic, associations, 74; relationships, 21, 25; symbionts, 55, 63; symbioses, 75
Mycelia Sterilia, 114
mycelium, 56, 62
Mycetozoa, 118
Mycobacterium, 24, **25**, **26**
mycobiont, 49, 76, 78
mycolaminaran, 115
mycologist, 47, 62
Mycoplasmatales, 21
mycorrhizae, 74, 102, 103, **103**, 384, 463
mycorrhizal, 416, **464**; associations, 647; associations, forest trees, 102; associations, orchids, 102; fungi, 102, 104, 115, 416, 675
mycoses, 72
Mycota, 46
Myeloxylon, **509**
Myrionema, 220, 226, **228**
Myristica, 649
myxamoebae, 118, 119, 128, 130, 139; aggregation, 128; phagotrophic, 130
Myxobacteriales, 21
myxomycetes, 40, 46, 47, 118; classification, 125; fructifications, 125; life history, 118, 119, **119**; myxamoebae, 119; plasmodium, 119; sclerotium, 122; spores, 119; sporulation,

myxomycetes (continued)
  122; swarm cells, 119; zygote, 119
Myxomycota (see also Dictyosteliomycetes, Labyrinthulomycetes, Myxomycetes, Plasmodiophoromycetes), 39, 46, 118; relationships, 139

Naiadita, 312; N. lanceolata, **316**
Najas, 342
nannoplankton, 148, 149
Narcissus, **618**
Nathorstiana, 424, **427**
natural, classification, 10; order, 6; selection, 3; taxa, 12
Navicula, 158
neck, **299**; canal cell, 304, 305; cells, 530, 548, **551**, 571
nectar, 626, 628, 631; guides, 628
nectaries, 626
needle grass, 4, 7
Nelumbo, 645; N. nucifera, 645
Nemaliales, 253, 254, 275, **283**, **284**, 285
Nemalion, 253, **268**, 275, 277; life history, **283**; type of life history, 275, 285
nematodes, 169
Neoagardhiella, 253, **270**
Neohodgsonia mirabilis, **319**
Neomeris, 204
Nereocystis, 220, 237, **238**, 239, 241, 242, 260; life history, **244**, **246**
Nerium oleander, **601**
net, formation, 139
netted venation, 447, **606**
Neurospora, 85; N. crassa, **83**, 85; N. sitophila, 85
neutral spores, 235
Nicotiana, 375
Nidulariales, 107; basidiocarps, **114**
nightshades, 640
Nitella, 5, 212, **212**; N. furcata, 5
Nitophyllum, 253
Nitrobacter, 22
nitrogen, 63, 216, 246
nitrogen fixation, 21, 33, 34, 36; bluegreen algae, 491
nitrogenase, 34
Nitrosococcus, 22
Nitrosomonas, 22
Nocardia, **25**, **26**
Noctiluca, 163
nodal bract, 586
node, 212, 429, 584, 601, 620
nodules, 450
nomenclature, 8
nonarticulated coralline algae, 279
nonfilamentous Chlorophyceae, 188
nonliving part of protoplast, embryophytes, 288
nonmotile, female gamete, 252; gametes, **45**
nonphotosynthetic eukaryotes, 46
nonrecurrent apomixis, 640
nonseptate hyphae, 56, 64
nontubular hairs, 51
nonvascular eukaryotes, 38-54; algae, 49; bryophytes, 53; cell structure, 40; fungi, 47; habitat, 39
nori, 277
Nostoc, **29**, **31**, 33, 34, 317, 521
Nostocales, 30, 33
Notothylas, 317
Nowakowskiella, **69**
nucellar, cap, 636; cell, 640
nucellus, 299, **299**, **501**, 506, 508, 511, 522, 525, 527, 561, 571, **583**, **589**, 631, 634, 636, 644
nuclear, 172; divisions, 119; endosperm, 639; envelope, 40, 220, 223, 254, 663; fusion, 46; membrane, 58, **289**; pair, 74; transfer, 56
nuclei, embryophytes, 288; Phaeophyta, 223; Rhodophyta, 254

Index **749**

nucleoid, 16, **18**, **23**, 279
nucleoli, 16, 58, 159, 165, 167, 170, 172, 254, 288
nucleoplasm, 16, 27
nucleus, 38, 165, 167, 170, 254, 279, 288, **289**
nucule, 212
numerical taxonomy, 10
nurse cells, 312
nut, 647, 649
nutrition, 257
nutritional requirements, 62
nutritive tissue, 530, 644
Nymphaea, 610, 645
nystatin, 24

oak, 102, 363, 596, 601, 645, 655; wilt, 62
oats, 101, 644
obligate, 19; mycorrhizal association, 342; parasites, 67, 78, 90, 101; parasites, vascular plants, 85
oblong bordered pit, **294**
ocean dispersal, 651
oceans, 651
Ochromonas, 149, **151**
Octoblepharum albidum, **339**
Odonthalia, 253, 257
Oedipodium griffithianum, **336**
Oedogoniales, 178, 181, **185**, 190, 195, 197, **199**, 208
Oedogonium, **177**, 181, 197, **199**, 214
oil, 40, 143, 159
oil bodies, 309, 310, 321; hepatics, **311**
oil deposits, 642
Old Red Continent, 395
oligochaetes, 169
olive, 649
Olpidium, **69**, **70**
Omphalina ericetorum, 107, **111**
Onagraceae, 639, **642**
onagrad embryo, 639
onions, 67
Onoclea, **378**, 484
ontogenetic, 622; fusion, sepals, 620; origins, 648
ontogeny, 231, 399
oogamous, 43, 188, 195, 197, 235, 675
oogamy, **45**, 69, 70, 155, 225, 235, 241, 671
oogonia (see also carpogonia), 64, 65, 239, 252, 672, 675
oomycetes, 64, 70
oomycetous fungi, 115
Oomycota, 39, 56, **57**, **60**, **61**, 62, 64, 65, 69, 72, 115, 116, 118; electron micrograph of zoospore flagella, **61**
oospore, 65
operculate asci, 89
operculum, 81, 308, 322, 329, 333; mosses, 308, 322, 329, 333
Ophiocytium, **144**, 146
Ophioglossidae, 461-463, 465, 471
Ophioglossum, 462, 463, **466**, **467**; *O. palmatum*, 463, **466**; *O. pendulum*, 463; *O. vulgatum*, **466**
Opuntia, **600**, 644
oranges, 640
orchid, family, 610, 644, 645; fruit, 648; seedlings, 102
Orchidaceae, 644, 645, 647
orchids, 102, 342, 596, 643
Ordovician, 279
organ, derivation, Ginkgoales, **553**; systems, 383
organelles, 36, 223
organic, acids, 82; scales, 172
organographic differences, embryos, 640
organs, 383; evolutionary pathways, 549
origin of prokaryotes, 658
ornithophily, 76, 629
Orthothecium rufescens, **336**
Orthotrichum ohioense, **337**

orthotropous, 634
Oscillatoria, 33, **34**
Osmunda, 356, **358**, **372**, **377**, 471, **472**, 473; *O. regalis*, **472**
Osmundaceae, 471, 473
osmundaceous fern, 471
Osmundidae, 461, 471
ostiole, 85, 271, 273, **280**, **282**, **284**
Ostrya virginiana, **650**
outer, aperture, 293, **294**; integument, 500; peristome, 331
ovary, 610, 612, 613, 626, 638, 643, 647; inferior, 612; position, **618**; superior, 612; types, **619**; wall, 623, 647
overtopping, 680, **680**
ovulate, flowers, 631; receptacle, *Cycadeoidea*, **536**
ovules, 101, **297**, 299, 502, 525, 526, 527, 540, 544, 548, 555, 558, 569, 571, 576, **583**, 589, **591**, 593, 610, 623, 625, 631, 634, 643; classification, 634; conifer, 571; development, **595**, 636, 641; *Ginkgo biloba*, **549**; grasses, 625; gymnosperm, 500; initiation, 620; primordia, 610; structure, 527
ovuliferous scale, 571, **574**
oxalate cells, 543
oxalic acid, 291
Oxalis, 653
Oxycoccus, 613
oxygen, 49, 659; levels, 216
oxygenated atmosphere, buildup, 659
ozone layer, 659

Pachytesta, 508, **510**
Padina, 220, **221**, 226, 242
Paeonia, 649
Paleocystophora, 242; *P. delicatula*, **250**
Paleohalidrys, 242; *P. lompocensis*, **250**
paleopalynologists, 689
Paleozoic, 505, 511, 516, 518, 540, 554, 556, 668, 671; Era, 279, 677; fungi, 668; pteridosperms, 505, 515, 516, 519, 532, 538, 539; pteridosperms, characteristics, 511; pteridosperms, phylogeny, 511
palisade, 605; layer, 604; mesophyll, 350
Palmaria, 253, 277; *P. palmata*, 277
Palmariales, 279
palmate, 601, 603; leaf form, **604**; pattern, 603
palms, 364, 596, 606
Pandanus, 610, **611**
Pandorina, 8, 181, **183**, **187**
panicle, 620, **621**
Papaver, 628
Papilionaceae, 649
pappus, 607; scales, **650**
Parachaetetes, 279
paramylon, 143, 165
paraphyletic taxon, 11
paraphyllia, 331, 333; Bryidae, **333**
paraphyly, 11
paraphyses, 81, **92**, **238**, 321, 327, 328, 331, 335, 476, 484; bryophytes, **304**; 321
parasexuality, 62
parasites, 1, 69, 139, 257, 645; on algae, 64; on higher plants, 69
parasitic, 70, 72; flowering plants, **345**; fungi, 58; relationships, 21; species, 72
Parazolla, 491
parenchyma, 181, 260, 346, 356, 359, 363, 368, 380, 506, 508, 598, 599, 606; cell, 368; cell, types, **349**; tissue, 347, 598
parenchymatous, 193, 217, 260, **261**; form, **232**, **233**, **234**, **238**, 671; growth, 226; kelps, 231; organization, 225
parichnos, scars, 409; system, 409
parietal, 636; placentation, 612, 616, 623
parthenocarpic fruit, 647
parthenocarpy, 647
parthenogenesis, 640; diploid, 640

parthenogenetic, 260
partial veil, 104
pattern of root growth, **610**
pea family, 610, 644, 649
peach leaf curl, 78
peaches, 649
pear, 649
pearlwort, 651, **652**
peat (see also *Sphagnum*), 325; deposit, 645
pectic, acid, 220; layer, 253
pectin, 175, 292
Pediastrum, 181, 190, **191**
pedicels, 617, 620
Pedinomonas, **215**, **216**, 217
peduncle, 387, 617
Pelagophycus, 220, 242, **251**
Pellaea, 476; *P. glabella*, **477**
Pellia neesiana, 313
pellicle, 165, 167
Peltaspermaceae, 689
Peltaspermales, 515, **517**
peltate, 622
Peltigera, **51**
Pelvetiopsis, 220
Penicillium, **63**, 75, 82, 114
Peniophora, **104**
pennate, 155, 158; diatoms, 155, 158
Pennsylvanian Period, 279
peony, 649
Peperomia, 639
peptidoglycan, 15, 27
Peranema, **167**, 169
Percursaria, 181, 190, **193**
perennials, 237, 239
Pereskia, 593
perfect flower, 616
perforation plate, 360, 361, 582, 685, **687**
perianth, **309**, 310, 589, 610, 628, 631, 655
pericarp, 271, **280**, **282**; angiosperms, 647, 649; wall, **284**
pericentral cells, 280, **281**
perichaetia, **306**, 322, 327-329, 333
perichaetial branches, 322
periclinal, cell division, 373; divisions, 226
pericycle, 356, 380, 399
periderm, 411, 493, 506, 508, 510, 515, 521, 543, 567, 569, **569**; *Ginkgo biloba*, **544**
Peridinium, 163, 165, **166**
peridiole, 107, 111; cord, 111
peridium, 107, 111, 123, 125, 126
perigene development, 355
perigonia, 310, 312, 333
perigonial, branch, **309**, 322; leaves, 321, 336
perigynous, 612, **619**
periplast, 170
perisperm, 644
peristome, 309, 336; teeth, 322, **328**, 329, 330, 331, **332**, 336, **338**
Perithalia, 220
perithecia, 85, 89, 90; development, Euascomycetes, **83**
perithecial initial, 85
perivascular fibers, 354
Permian, 515, 516, 532, 534, 540, 554, 578, 592
Peronospora, 67
Peronosporales, 64, 65, 67
persistent nuclear membrane, 58
Pertica, 389, 391, 450; *P. quadrifaria*, **390**
Pestalotia, 115
Petalonia, 220
petals, 593, 607, 609, 610, 612, 622, 623, 625, 628; evolutionary fusion, 623; initiation, 622; primordium, 622
petiole, 446, 601, 605
petiolule, 601
petrified stems, 510, 511
Petrocelis, 253, 275
Peziza, **81**, 88, 93
Pezizales, 88, 89; ascocarps, **93**
Phacus, **167**, **168**, 169

Phaeoceros, 317; *P. laevis*, **320**
Phaeocystis, **161**
Phaeophyceae (see also brown algae), 219, 220
Phaeophycidae, 219, 220, 225
Phaeophyta (see also brown algae), **9**, 39, 51, 142, 145, 172, 217, 219–251, **227**, **233**, 671; alternation, heteromorphic generations, 237; alternation, isomorphic generations, 237; apical growth, 226; asexual reproduction, 235; cell components, 221; cellular differentiation, 231; chloroplast, 220; classification, 225; distribution, 242; ecology, 242; ectocarpalean line, 225; filamentous organization, 225; flagellate cells, 223; importance, 247; intercalary meristem, 226; life histories, 235; life histories, complexities, 241; lines of evolution, 241; morphological diversity, 225; nucleus, 223; orders, 220; parenchymatous growth, 226; parenchymatous, organization, 225; phenetic relationships, **672**; phenetic series, **672**; relationships, 242; reproduction, 235; sexual reproduction, 235; storage products, 221; structural differentiation, 231; uses, 247; vegetative reproduction, 235
phaeophyte tannin, 223
Phaeostrophion, 220, 226, 231, **233**
Phaeothamnion, 149, **151**
phage, 18; particles, **21**
phagotrophic, 149, 159, 165, 169
phagotrophically, 60, 122
phagotrophy, 132, 139
Phallales, 107; basidiocarps, **113**
Phallus, **113**
Phaseolus lunatus, 643; *P. vulgaris*, **608**
Phasmatocycas, 532
phellem, 380, 569
phelloderm, 380, 569
phellogen, 347, 380, **382**, 569
phenetic, algal groups, 668; analysis, 11; brown algae, **671**; fungi, 668, **669**; red algae, **671**; relationships, brown algae, **672**; relationships, Chlorophyta, **674**; schemes, 217, 668
phenetic series, 216, 217, 668, **672**; fungi, 668, **669**; Phaeophyta, **672**; Rhodophyta, **671**
phenetics, 10
phenolics, 609
phenotypes, 3
phenotypic plasticity, 2
Philonotis fontana, **337**
phloem, 38, 296, **297**, 301, 363, 380, **569**, 596, 598, 599, 606; fibers, 363; parenchyma, 368, 380, 569
phloemlike elements, 231
Phoenix, 649
Phoma, **115**
phosphorous, 63, 216, 249
photoautotrophic forms, 159
photoautotrophs, 172, 659
photophosphorylation, 33
photosynthate, 49, 235
photosynthesis, 159, 355; algae, 38; prokaryotes, 31; versus heterotrophy, 115
photosynthetic, animals, 142; cell, **289**; eukaryotes, 49; forms, 49; lamella (see also thylakoids), 220, 254; nonvascular eukaryotes, 39; organs, 601
phragmoplast, 287
phycobilin pigments, 254, 279
phycobiliproteins (see also biliproteins), 27, 33, 49, 143, 170, 172
phycobilisomes, 36, 255
phycobiont, 49, 76, 214
phycocolloids, 49, 219
phycocyanin, 49, 254
phycocyanobilin, 254
phycoerythrin, 49, 254, 277
Phycomyces (see also lower fungi), 64
Phylloglossum, 411, 412, 413, 416, **420**

Phyllospadix, 260
phyllosporous, 622, 625; ovules, 622
phylogenetic, history, Bryopsidophyceae, 672; relationships, 668; series, 665; systematics, 10
phylogeneticist, 242
phylogenies, 655
phylogeny, 216, 279, 658; Chlorophyta, 216; Cycadeoidales, 538; Cycadopsida, 532; dasyclad algae, **666**, **667**; ginkgophytes, 549; Gnetophyta, 591; Mesozoic pteridosperms, 518; Paleozoic pteridosperms, 511; Rhodophyta, 279
Physarales, **121**–**123**, 126, 127
Physarum, **121**–**123**, 126, **127**; *P. bivalve*, **125**; *P. polycephalum*, 126
physiology, 33
Physocarpus opulifolius, **616**
physodes, brown algae, 223
phytoflagellates, 172
Phytophthora, 65; *P. infestans*, 65, 67, **68**
phytoplankton, 149, 158, 163, 214
phytosterols, 255
Picea, 560, **564**
pigmentation, 208, 217; red algae, 254
pigments (see also pigmentation), 49, 51, 58, 143, 204, 254
Pilayella, 220, **221**–**223**, 226, **228**
pileate, 104
pileus, 104
pili, 16
Pilularia, 489
Pinaceae, 561, 562, 569, 570, 571, 572, 580, **584**
pine, 300, 565–567, 569; leaf, **566**; life history, 563; male gametophyte, 570, **572**; pollen, 570, **571**; stem, 567, **567**, 568
pineapple, 596, 647
pink, bread mold, 85; family, 622
pinnae, 450, 451, 456, **461**, **465**
pinnate, 601, 603, **604**
Pinnularia, **152**, 158
pinnule, 450, 460
Pinus, 300, **349**, **353**, 560, 561, **561**, 563, 566, 567, 569, 578, **579**; cotyledons, **575**; dwarf shoot, 565; female gametophyte, **574**; generative cell, **572**; life history, 563; male, cone, **570**; male, gametophyte, **572**; microspore nucleus, **572**; pollen, **571**; pollen, sac, **570**; secondary xylem, **568**; stem, **57**; tube cell, **572**; *P. aristata*, 554
Piperaceae, 639
piperad embryo, 639
pistil, 612; compound, 612; simple, 612
pistillate, 617; flower, 617, 631
Pisum, 604, **623**; *P. sativum*, **602**
pit, canal, 293; cavity, 293; chamber, 293, **294**; connection, **257**; membrane, 293, 360; pair, 293
pith, 346, 356, 399, 495, 510; ray, 346, 378, 599
pits, **238**, 293, 360, 363, 495; in secondary walls, **294**
pitted, 360; elements, 231; tracheids, 399
placenta, 610, 612, 623
placentation, 612; axile, 612; dorsal, 612; free-central, 612; laminar, 612; marginal, 612; parietal, 612; submarginal, 612; types, **619**
Plagiochila tridenticulata, **307**
planated, 494
planation, 495, 680; stages, **681**
plane tree, 647
plankton, 148, 158, 214
planospore (see also zoospores), 40, 60, 179
planozygote, 163
plant, body, 346; cell wall structure, **292**; cells, 287; parasites, 92, **92**, 114
Plantae, 38
plantain lily, 605
plants, 13, 39; evolutionary relationships, 688, 689, **692**

plasma membrane, 15, **16**, **17**, 18, 22, 40, 58, 122, 123, 128, 178, **289**, **290**, 292, **292**, **294**, 666
plasmalemma, 287
plasmid, 16
plasmodesmata, 217, 235, 368, 531
plasmodia, 119, 132, 136, 139
plasmodial stages, 139
plasmodiocarps, 123, 125, 126
Plasmodiophora brassicae, 132, 136, **137**, **139**
Plasmodiophoromycetes, 46, 47, 118, 132, 139
plasmodium, 119, 122, 125, 126, 136
plasmogamy, **47**, 62, 78, 80, 136
plastids (see also chloroplasts), 200, 288
Platanus, 647
plate meristem, 604
plates, 163
Platymonas, 215
Plectomycetes, 81, 82
plectostele, 416
Pleodorina, 181, 187
Pleospora, 90, **94**
Pleosporales, 89, **94**
pleurocarpous, 321, 322; Bryidae, 333
Pleurochloris, **144**, 146, 159
Pleuromeia, 424, **427**
Pleurotus, 104, **108**
plum, 649
Plumbago, 638; *P. capensis*, 639
plume, 649
plumed fruits, **649**
plumed seeds, 649
plumule, 643
plurilocular, **228**, 235, 237; sporangia, **245**
pneumatocysts, brown algae, 235
Poa annua, 639; *P. bulbosa*, 640
Poaceae, 599, 616, 622, 641, 644, 649
Podocarpaceae, 561, 562, 570, 571, 580
Podocarpus, 566, 571
Pohlia annotina, 307
polar, flagellum, 21; gaps, 223; nuclei, 300, **300**, 599, 638; ring, 254
polarity, 231, 239; *Fucus*, 231
pollen, 511, 544, 610, 626, 628, 631, 656, 689; chamber, **501**, 506, 508, 510, 511, 531, 544, 585, 589, 590; *Cordaites*, 557; cycad, **526**; dispersal, 570; drop, 525, 585; drop, mechanism, 590; *Ephedra*, **583**; evolution, 558; evolution, coniferalean saccate, 561; flowers, 628; formation, 638; fossil, 592, 642; *Ginkgo biloba*, **547**, **548**; grain, 295, 504, 506, 510, 511, 516, 521, 522, **557**, 561, 570, **571**, 576, **583**, **587**, 590, 591, 638, 641, 642, 686; *Lebachia*, 560; mechanisms, 578; monocolpate, 691; monosulcate, 535; *Pinus*, **571**; rigid, 586; sacs, 522, 525, 555, 558, 569, **583**, 589, **590**; structure, 578; tube, 300, 504, 525, 529, 587, 590, **599**, 634, 638, 639, 687; wall, 504; *Welwitschia*, **587**
pollen-bearing, cones, 555, 558; organs, 515
pollinating, adaptations, 648; agents, 631; gymnosperms, 629; mechanisms, 579, 610
pollination, 8, 300, 448, 502, 526, 544, 571, 584, 610, 626, 631
pollinators, 629
pollution, 158
polyarch, 465, 606
polycyclic solenostele, 357
polyembryonic, 591
polyglucan, bodies, **29**; granules, 27
Polygonaceae, 622
polyhedral bodies, 27
Polykrikos, **162**, 163
Polyneura, 253, **272**
polypeptides, 15, 20
polyphosphate, bodies, 27; granules, 27, **29**
polyphyletic, 625; taxon, 11
polyphyly, 4, 11
polyploid species, **6**
polyploids, kinds, **3**

polyploidy, 4
Polypodiales, 484, **486**
Polypodium, **358**, 537; *P. vulgare*, **486**
Polyporus, **49**, **104**
polysaccharides (see also chrysolaminaran, cyanophyte starch, floridean starch, laminaran, starch), 15, 40, 49, 58, 143, 149, 158, **162**, 200, 255
polysaccharide carbohydrates, 49
Polysiphonia, 253, 269, 273, **280**; life history, **278**; reproduction, **280**; structure, 280
Polysiphonia-type life history, 269, 273, 275, **278**, 285
polysiphonous, 269, 271; structure, **280**
Polysphondylium, 132, **136**
polystele, 507, 682
polystelic, 420; axis, **507**, 681, **681**; stems, 507
Polystichum lonchitis, **485**
polystromatic, 260
Polytrichaceae, 328
Polytrichidae, 328, 329, 333, 336; anatomy, **330**; morphology, **330**
Polytrichum, **53**; *P. commune*, **323**, **330**
pome, 649
Pomoideae, 649
Pongola rock series, 659
poplar, 645
poppy, 628
poptillo, 584
population genetics, 1
populations, 1
Populus, 645, **645**
porate, 642
pore, 104
Porphyra, 253, **255**, **257**, **258**, 260, **261–265**, 262, 277, 282, 285, 671; life history, 262; *P. nereocystis*, 260, **265**; *P. perforata*, 7
Porphyridiales, 253
Porphyridium, **256**, **259**, 277
Porphyrodiscus, 253
Postelsia, 220; *P. palmaeformis*, 8
postfertilization, 263; processes, 252; stage, 285
Postilla, **532**
potassium, 63, 249
potato, 67, 640, 649; crop of Ireland, 67; plants, 675
Potentilla, 649
Potomac, leaf assemblages, 690; pollen, 690
Potomac Group, 690
powdery, scab of potato, 132; mildew, 85
Prasinophyceae, 174, 175, 214, **215**
prasinophytes, 214
Precambrian, 115, 172, 216, 658, 659, 660, 663, 665, 668, 672; sediments, 36
preferns, 450–458
prepollen, 504, 509, **514**, 686
prespore cells, 130
prestalk cells, 130
presumed evolutionary developments, 663
prickles, 651
primary cysts, 64
primary, cell wall, 292; endosperm nucleus, 300, 639, 643; growth, 346, 596; mycelium, 95; nucleus, 208; parietal cell, 638; phloem, 356, 363, 368, 371; pit, connections, 254, **255**, 262, 263, 279; pit, field, **292**, 293, **293**; plasmodium, 136; primordium, 622; producers, 247; sporogenous cell, 638; suspensors, 572; thickening meristem, 347, 374; tissues, 346; vascular bundles, 598; vascular bundles, system, 598; vascular bundles, tissues, 355, 643; wall, **292**; xylem, 357, 360, 508, 510, 543, 555; xylem, secondary wall deposition, **363**; zoospores, 64
primitive, 12; angiosperm stock, 676, 691; ascomycetes, 58; basidiomycetes, 58; earth, 658; flower, 623; primordium, 622, 638
Primula, **614**
Primulaceae, 648

Proboscidea, 652
proboscis, **223**, 225, 628
procambial strands, 598
procambium, 346, 357, 363, 373, 374, 378, 598, 604
process of evolution, 1
proembryo, 532, **551**, 571, 585, 587, 591, **599**, 686
progenitors, 214; of land plants, 216, 673–675, **676**
Progymnospermopsida, 346, 493-504
progymnospermous ancestors, 577
progymnosperms, 391, 493–504, **505**, 577, **679**, 680, 681, 691; ancestors, 682; evolutionary relationships, **500**; leaf, 576; spore, **502**
prokaryotes, 14, 25, 39, 658, 659, 660, 691
prokaryotic, ancestors, 661; cells, 216; divisions, 668; divisions, geological ranges, **670**; fossils, 36; microfossils, 668; nucleoid, 663
prop root, 606, **611**
propagula, bryophytes, 306, **307**
propagules, 114, **232**, 235, 299, 412, 423, 644, 655; brown algae, **232**
Prorocentrum, **162**, 163
protein, 40, 58, 122, 170, 291, 292; embryophytes, 291; fibrils, 155; reserves, 643
proteinaceous, 170; strips, 165
prothallial cell, 522, 544, 561, 570, 586; *Ginkgo biloba*, 548
prothallus, **463**, 590
Protista, 13, 38, 46, 142, 165
protistan, algae, 142; ancestry, 115; relationships, 172
protistans, 115, 118–173; acellular slime molds, 118; algae, 142–173; dictyosteliomycetes, 127–132; endoparasitic slime molds, 132–136; labyrinthulids, 139; Labyrinthulomycetes, 139; Myxomycota, 118–141; myxomycetes, 118–127; Plasmodiophoromycetes, 132–136; true slime molds, 118
protists, 39, 115
protoalga, **664**
protoderm, 346, 354, 357, 374, 636, 638
Protolepidodendron, 404, **405**, 406, **406**, 409
protomitochondria, 662
protonema, bryophytes, 212, 303, 305, 306, 321, 325, 327, 329, 337
protonemal flaps, 331
protophloem, 355, 356, 363
protoplasm, 122; embryophytes, 287
protoplast, 58, 287, 364; embryophytes, 287
protoplastids, 662
Protopteridium, 681
protostele, 356, 357, 376, 383, 438, 494, **496**, **497**, 682
protostelic, 399
protostelid amoebae, 139
Protostelida, 139
protostelids, 139
Prototheca, 181, 214
protothecosis, 214
prototype, Florideophyceae, 282; land plants, 217
protoxylem, 606
Protozoa, 38, 139, 165, 214
proximally, 226
Prunus, **594**, **618**, 628, 649
Prymnesiophyceae (see also prymnesiophytes), 142, 143, 158–159, **159**, **161**, 172
Psalliota campestris (see also *Agaricus*), 287
Psaronius, 465, **470**
Pseudobornia, 429, 430; *P. ursina*, 429, **431**
pseudocrassinucellate, 636
Pseudoctenis, **537**
pseudoelaters, 317
pseudofilamentous, brown algae, 226
Pseudogloiophloea, 253

Pseudomonadales, 21, 22, **24**, 27, 33
pseudomonads, 22
Pseudomonas, **17**, 22
pseudomonopodial, 377; axis, 387; branch system, **379**
pseudoparenchymatous, 226, 242; cortex, 270, **270**; forms **227**, **229**, **236**, 671; organization, 226; precursors, 226, **230**, **233**
pseudopodium, 322, **324**, 327
pseudosaccate, 494, **497**
pseudosaccus, 494, **497**
Pseudosporochnus, 450, **453**; *P. nodosus*, 450
Pseudotsuga, 347, **569**, 571, **573**, **574**, **579**, 580
Pseudovoltzia, **578**, **579**
psilophytes, 397, 499
Psilophyton, 387, 389, 391, 678, 680; *P. dapsile*, 389, **389**; *P. microspinosum*, 388, **389**; *P. princeps*, 387, **388**, 389
psilotophytes, 397-402
Psilotophytina, 345, 346, 397, 401, 474
Psilotum, 296, 356, **358**, 376, 377, **377**, 397, 401, 683, **684**, 685; gametophyte, **402**; *P. complanatum*, 401; *P. nudum*, 397, **398**, 399, **400**
psycrophiles, 63
Ptelea, 649
Pteridales, 476
Pteridium, **357**, **358**, 363, 582, 685
Pteridospermopsida, 346, 357, 493, 505-518
pteridospermous ancestors of vascular plants, 655
pteridosperms (see also seed ferns), 493, 505–518, **514**, **516**, **531**, 532, 549, 642, 656, 680, 682, 687, 691; ancestors, 522; callistophyte, 515; Carboniferous, 505, 549, 681; Carboniferous, ancestral link, 532; characteristics, 511; concept, 505; detached organs, 510; frond, 576; medullosan, 535, 680, 682; Mesozoic, 515, 689; Paleozoic, 505, 511, 515, 516, 519, 522, 538, 539; phylogeny, 511, 518; prepollen, 511
Pterigoneurum ovatum, **336**
Pteris, 476
Pterophyllum, 353, **537**
pterophyte megaphyll evolution, 686
pterophytes, 446-448
Pterophytina, 346, 397, 446-448; classes, 346; diversity, 447; gametophyte, 447; modification, 448; sporangia, 448; sporophyte, 446
Pteropsida, 346
Pteruchus, 515, **517**, 518
Pterygophora, 220, 237
Puccinia graminis, 97, **99**, 101, **101**
puff balls, 55, 78, 92, 103, 107
puncta, 152
Punctaria, 220
purple laver, 277
purple, photosynthetic bacteria, 36; sulfur bacteria, 22
pycnia, 99
pycnidia, 114
pycniospores, 99
Pylaiella (see also *Pilayella*), 220
pyrenoids, 143, 149, 153, 159, 165, 170, 175, **176**, **177**, **193**, 200, 212, 220, 223, 254, 255, 315, 317
pyrenomycete, 85
Pyrocystis, 163, **166**
Pyrrhophyta (see also pyrrhophytes, Dinophyceae, dinoflagellates), 39, 51, 142, 159–170, 172, 242; cell structure, 159; classification, 159; ecology, 159; morphological variation, 159; reproduction, 159
pyrrhophytes (see also dinoflagellates), 159, 172
Pyrus, 353, 649
Pythium, **61**, 65

quack grass, 640

quadriflagellate zygotes, 195
Quercus, 353, 363, 645, 647, 649, 655;
   Q. macrocarpa, 2
quiescent center, 376
quillwort, 378
quince, 649

raceme, 620, **621**
rachial leaves, 495
rachis, 456, 460, 476, 601
radial, 359; symmetry, **276**
radicle, 532, 573, 643
radiocarbon dating, 645
radiolarian, 172
rain, 649
rainwash, 649
Ralfsia, 220, 226, **228**, 241
ramentum, 535
Ranalean, 623; plants, 623, 625
Ranales, 642, 655
Ranunculaceae, 612
Ranunculus, 357, 622, 649; *R. repens*, **600**
raphe, 155, 634; channel, 155
raphides, 291
Raphidophyceae, 142
raspberry, 647
ray, 495, **496**
ray tracheid, 495, **568**, 569
Reboulia hemisphaerica, **319**
receptacles, 315, 606, 610, 623, 628, 647; brown algae, 239, **247**
receptacular, 623
receptive hyphae, 99
recombination, 2, 188
recurrent apomixis, 640
red algae (see also Rhodophyta), 33, 36, 51, 115, 172, **255**, 277, 279, 668
red, bread molds, 85; clover, 628; snow, 214; tide, 163, 165, 172
reduction, 495, 680; division, 43
redwood, 347, 561
Regnellidium, 463, 489
reindeer moss, 88
reinfecting stage, 79
relationships, 111, 172, 585, 588; algal groups, biochemical, **671**; algal groups, morphological, 691; algal groups, ultrastructural, **671**; brown algae, 241; Chlorophyta, 216; embryophytes, 301; Ephedrales, 585, 588; fungi, 114; Monera, 36; Phaeophyta, 235, 242
Renalia, 387, 678; *R. hueberi*, **384**, 387
repeating stage, fungi, 79
reproduction, 27, 60, 143, 149, 155, 159, 165, 169, 170, 174, 179, 201, 209, **232**, 235, **238**, 260, **267**, 569, 586, 588; Bangiophyceae, 260; bluegreen algae, 29; Coniferales, 569; Cycadopsida, 522; Ephedrales, 584, 586; *Ginkgo biloba*, 543; Gnetales, 589; Phaeophyta, 242; Welwitschiales, 586
reproductive, trends, embryophytes, 296; plant body, 265; plant body, angiosperms, 606; stages, **265**; structures, 516, 641; system, 683
reserve lipids, 643; products, 49, 53; protein, 643
reservoir, 166, 167, 169
resin, ducts, **566**, 567; canals, **568**, 582
resistant, 70; cysts, 119; sporangia, 70; spore, **45**, 62, 96, 118
reticulate, 360, 361
Rhabdosporites, 494, **497**
Rhacomitrium canescens, **338**; *R. lanuginosum*, **337**
Rhacophyton, 681; *R. ceratangium*, **455**, 456
Rhacopteris, 681
Rhizanthella, 344
Rhizidiomyces, **72**
Rhizobium, 21, **23**
Rhizochloris, **144**

Rhizoglossum, 463
rhizoidal, 212, 260; processes, 260
rhizoids, 55, 58, 70, 190, 193, 296, 308, 310, 313, 314, 315, 321, 322, 325, 329, 333, 384, 400, 678
rhizomatous, 383
rhizome, 399, 484, 584, 599, 640
rhizophore, 423, 424, **424**
Rhizophydium, 57, **69**, 70
rhizoplast, brown algae, 225
Rhizopus, 72, 73, **74**, **75**; *R. stolonifer*, 73
Rhodochaete, 253, 260
Rhodochaetales, 253
Rhododendron, **613**
Rhodomela, 253, 257
Rhodomelaceae, **281**
Rhodophyceae (see also red algae, Rhodophyta)
Rhodophyta (see also red algae), 39, 51, 142, 170, 172, 217, 252–285, 668; Bangiophyceae, 260–262; cell, division, 254; cell, organelles, 254; cell, structure, 253; chloroplasts, 254; chromosomes, 254; distribution, 277; importance, 277; life histories, *Nemalion*-type, 275; life histories, *Polysiphonia*-type, 269; mode of nutrition, 257; nucleus, 254; orders, 252; phenetic series, **671**; phylogeny, 279; pigments, 254; polysaccharides, 255; storage products, 255; uses, 277
rhodophytes, 36
rhodopsin, 25
Rhodymenia, 253; *R. palmata*, 277
Rhodymeniales, 253
Rhynia, 384, 386, 391, 678, 680, 684, 685; *R. gwynne-vaughnii*, **385**, **386**; *R. major*, 384, **385**
rhyniophytes, 383-387, 499; evolution, 678, **679**; fossils, 317
Rhyniophytina, 301, 345, 383-387, 397; classification, 344
rhytidome, 380
rib meristem, 567
ribbon grass, 631
ribonucleic acid, 14
ribosomes, 16, 60, 211, 221, 254, 288, **290**, 365, 368, 630
Riccardia multifida, **313**; *R. palmata*, **307**
Riccia, 315; *R. beyrichiana*, **319**
Ricciocarpus natans, **319**
rice, 644, 596; beverage, 72
Ricinus, 643
Rickettsiales, 21
Riella, 312; *R. americana*, **316**
ringworm, 75, 82
riparian trees, 691
RNA, 14, 25, 27, 60, 288, 374, 376; polymerase, 25
Robinia pseudoacacia, **603**, **643**
rockweeds, 219
Rodeites, 489
role in vegetation, vascular plants, 312
root, 75, 296, **297**, 346, 463, 678; adventitious, 643; apical meristem, 374, 640; appendages, 378; contractile, 606; cortex, 374; embryonic, 643; hair, 136, 354, 355, 606; hair, cell, 355; meristem, 342; nodules, 15; rots, 65, 78; scars, 409; storage, 606; system, 346, 605; system, angiosperms, 354, 605, **609**; variation, 606; variation, angiosperms, 606
rootcap, 374; mother cells, 374
Roquefort cheese, 82
Rosa, 628
Rosaceae, 647
rose, 628
rotifers, 169
rotten meat, 628
rough endoplasmic reticulum, **290**
royal fern, 356
R-phycocyanin, 254
R-phycoerythrin, 254

Rufusiella, **162**, 163
rumen, 25
ruminant, 15, 25; animals, 25
rush family, 609
rust, 97, 101, 115; haustorium, **99**
rye, 596, 644; flower, 85

sac fungi, 78
saccate prepollen, **502**
Saccharomyces, 63, **80**; *S. cerevisiae*, 78
saccus, 504, **514**, 555
Sagenopteris, 516, **517**
Sagina, 651
St. Anthony's fire, 85, 87
sake, 72
Salicaceae, **642**
salinity, 214
Salix, **623**, 645, 655; *S. scouleriana*, **603**; *S. sitchensis*, **642**
salmon, 65
Salmonella enteritidis, 21; *S. typhi*, 21
Salvia officinalis, **613**; *S. pratensis*, **630**
Salvinia, 489, **490**, **491**
Salviniales, 489
Salviniidae, 461, 489, **490**
salt marsh, 149, 158; pools, 172
samara, **648**, 649
saprobes, 75, 101, 102, 114, 214
saprobic, 70, 72, 675; aquatic fungi, 104; species, 63, 69
saprobically, 165
Saprolegnia, **49**, 64, 65, **67**, 115
Saprolegniales, 64, 65, 67
saprophytes, 645
saprophytic, 102; flowering plants, 345
Sarcodiotheca, 263, 271
Sarcothamnus scoparius, **630**
Sargassum, 220, 235, 242, 247
sauerkraut, 15
sausage tree, **634**
Sawdonia, 387, 392, 407; *S. ornata*, 387, **394**
sawn timber, 102
scalariform, 360, 361; pattern, 535, 685, **687**; sclereids, 351; tracheids, 399, 685, **687**, 689
scale, mosses, 310; trees, **409**
scales, 40, 143, 149, 212, 315, 628; calcium carbonate, 158; inorganic, 149; organic, 158; siliceous, 149
scaly flagella, **42**, 212, 214
Scapania nemorosa, **309**
Scenedesmus, 181, 190, **191**
Schizaea, 474; *S. pusilla*, **475**
Schizaeales, **472**, 474, 476, 478
schizocarp, 649
Schizomycetes, 21, 22
Schizomycophyta (see also bacteria), 14, 25; characteristics, 21
Schizoneura, 443, **444**; *S. gondwanensis*, **444**; *S. manchuriensis*, 444
Schizothrix, 30, **34**
Sciaromium tricostatum, **336**
Scirpus, 349
sclereids, 351, 586, **587**; types, 351
sclerenchyma, 346, 351, **440**; fibers, **352**; tissue, 351; types of sclereids, **351**, 357
sclerified, 584; epidermal cell, **351**; layer, 644
sclerotia, 85, 87, 119, 122
Scoleopteris incisifolia, **470**
scorpioid cyme, **612**, 620
Scouleria aquatica, **337**
Scytosiphon, 220, **221**, 226, 231, **233**, 241
Scytosiphonales, 220, 241
sea palm, 8
secondary, basidiospores, 96; cyst, 64; female strobilus, **588**; growth, 363, 378, 521, 596, 598; growth, dicotyledons, 381, 598; male strobili, 589; mycelium, 95; phloem, 363, 569, **584**; phloem, *Ginkgo biloba*, **544**; pit connections, **255**; plant body, 598;

secondary (continued)
plasmodium, 136; primordia, 622; strobilus, 586, **587**; suspensor cells, 572; tissue, 347; tracheids, **497**; vascular tissue, 355; walls, 292, **293**; walls, annular, 363, 685, **687**; walls, deposition, 361, **363**; walls, helical, 363, 685, **687**; walls, pitted, **363**; walls, sclariform, **363**, 685, **687**; wood, 495; xylem, 347, 357, 360, 494, 496, 521, 543, 555, 584, 606; secondary, wall deposition, **363**; zoospores, 64
secondary zoospores, 64
secretory cells, brown algae, 231, **238**
sedges, 622
seed, 295, **297**, 299, 448, **501**, 508, 516, 522, 532, 571, 573, **574**, 590, 593, 610, 643, 647, 687; coat, 502, 585, 636, 644, 645; correlation with habitat, 645; dispersal, 626, 644; dispersal, air, 607; dispersal, enclosed seeds, 648; dormancy, 644, 645; evolution, 500, **501**, 502, 685, 686; ferns, 493, 502, 505–518, 538, 588, 626; ferns, lineage, 511; germination, 636; habit, 500; habit, evolution, 685; leaves, 640; plants, 447, 502, 626; plants, flowering, **643**; plants, lineages, evolutionary pathways, **688**, **689**; plants, maturation, 634; plants, sections, **641**; viability, 645
seed-bearing, branches, 515; cones, 555, 558; organ, 516
secretory, cavities, 510, 511; cells, 508; glands, 506; tapetum, 638
seedless, grapes, 647; oranges, 647
seedlings, 604
segmental polyploids, 5
Selaginella, 296, **298**, **299**, 363, 376, 377, 378, **378**, **380**, 411, 412, 416, **420**, 530, 582, 685; *S. kraussiana*, **421**
selection, 675; advantages, 645; pressures, 4, 675
self-sterile, 79, 81
semophylesis, 549
sensitive plant, 610
sepals, 593, 606, 607, 609, 610, 612, 620, 622, 623, 628; evolutionary fusion, 623; initiation, 622
septa, 27, 56, 204, 254; formation, 64
septal pores, 79
septate, 72; basidium, 96; hyphae, 56, 79
Sequoia, 347, 554, 561, 571, 680
Sequoiadendron, 554, 567; *S. giganteum*, **343**, 561, **564**
serum, 20
sessile, 601
seta, mosses, 305, 308, 310, 314, 321, 322, 328, 329, 333, 336
sex, 663; chromosomes, 542; determination, brown algae, 223; differentiation, 541
sexual, cycle, 691; reproduction, 18, **45**, 60, 62, 64, 65, 69, 70, 73, **75**, 79, **83**, 95, 114, 130, 146, 155, 163, 169, 208, 225, 235, 262, 675; stage, 62
Shanorella, **88**
sheath, 30, 33, 212
sheep, 214; wool, 651
shellfish, 165
shell zone, 376
shoot, apex, 549; apical meristem, **372**, 373, 374, 640; system, 346, 354, 596
short shoot, 541, 585
shrubs, 596
Sibbaldia procumbens, **613**
sieve, areas, 363, 364, 364, **365**, 416, 569; cells, **238**, 363, 364, 510, 569, 580, 584; elements, 363, 364, 365, 368, **372**; plate, **238**, 364, **364**; plate, development of pores, 366; tube, 364, 588, 596; tube, development, **369**; tube, elements, 235, 321, 329, 364; tube elements, protoplast, 367
Sigillaria, **407**, **410**, 411, 424
silica, 143, 149, 152, 155, 165, 216

silicalemma, 155
silicified, 149
Silurian, 279
simple, dichasium, **621**; fruits, 647; pistil, 612; pit, 293, **294**, 569; pit pair, 293
Siphonocladales, 200, 204, **205**, **207**, 208
Siphonocladus, 204
siphonostele, 356, **359**, 411, 420, 438, 465, 682, 685; amphiphloic, 359; endophloic, **359**; evolution, 685
siphonous forms, 200
skin diseases, 214
slime, 122, 365; bacteria, 127, 128; molds, 13, 39, 46, 115, 118, 125, 132, 165; plug, 365, 372; trail, 130
slime bacteria, 127, 128
Smilax laurifolia, **602**
Smithora, 253, 260
smooth endoplasmic reticulum, **290**
smuts, 97, 101
snow mold, 63
sodium alginate, 249
soil, 69, 72, 102, 107, 159, 214; fungi, 65; saprobes, 82
soil-inhabiting, fungi, 72, 103, 104; species, 92
Solanaceae, 8, 598, 640
solanad embryo, 640
Solanum, 640, 649; *S. tuberosum*, 640
Solenopora, 279
Solenoporaceae, 279
Solenostoma hyalinum, **309**
somatic apospory, 640
Soranthera, 220
Sorbus, 649
Sordaria fimicola, **90**
soredia, 78
Sorghum, **641**; *S. sorbifolia*, **616**
sorocarp, 130, 132
sorophore, 130
sorus, 130, 132, 139, 460, 473, 522
soybeans, 74
spadix, 628
spathe, 628
Spathodea, **646**
spawning, 249
special-purpose classification, 8
specialized, characteristics, 12; tracheophytes, 360
spermatangia, 90, 92, 99, 252, **267**, 273, **278**, **280**, **282**
spermatia, 252, 262, **264**, 271, 275, **278**, **282**, 283
spermatial transfer, 115
spermatogenous cell, 526, 589
sperm, 43, 146, 195, 197, 212, 235, 239, **246**, **247**, 300, **300**, 304, 416, 460, 506, 526, **530**, 532, 544, 571, **572**, 585, 639; cells, 593, 648; colonies, 188; cycad, **526**; Laminariales, 245; multiflagellate, 429; nuclei, **300**, 585, 586, 639
Spermopteris, **531**, **532**
Spermothamnion, 253
Sphacelaria, **40**, 220, **221**, 226, **232**, 235, 237
Sphacelariales, 220, 226, **232**, 235, 242
Sphaerocarpales, 312, 314, **316**
Sphaerocarpos, 312, 313, **316**; genetics, 313; *S. texanus*, 316
Sphaeropsidales, 114
Sphagnidae, 322, 333; anatomy, **324**; morphology, **324**
Sphagnum, 149, 322, **324**, 325, **326**, 327; bog, 146, 148; peat, 325; spore dispersal, **326**; spores, 325; *S. magellanicum*, **324**; *S. palustre*, **324**; *S. papillosum*, **324**; *S. squarrosum*, **324**; *S. tenellum*, **324**
Sphenophyllales, 681, 683, **684**
Sphenophyllostachys, 430, **432**
Sphenophyllum, 430, **432**, **433**, 444
sphenophytes, 429-443; early, 429; fossil, 443
Sphenophytina (see also sphenophytes, sphenopsids), 345, 429–443
sphenopsids, 680, 682

spike, 620, **621**; moss, 377
spindle, 223; fibers, 254; formation, 195; microtubules, 663
spines, 103, 599, 601, 651
spiral, thickenings, 308; tracheids, 399
spirilla, 15, **15**
Spirochaetales, 21
Spirogyra, **176**, **177**, 181, 197, **201**, **203**, 214
Spirulina, **34**
Splachnum luteum, **337**
splash cups, **652**
splashing rain, 62
sponges, 172
Spongospora subterranea, 132
spongy mesophyll, 350, 605; *Ginkgo biloba*, **547**
sporangia, **40**, **45**, 60, 62, 64, 65, 67, 70, 72, 122, 125, 126, 295, 305, 308, 310, 315, 321, 329, 333, 458, 494, **500**, **501**, 506, 522, 625, 665, 678, 680, 683, 686; evolution, positioning, **683-685**; evolutionary series, 684; marginal positioning, 684, **686**
sporangial, arrangement, evolution, 683; development, **462**; positioning, evolution, 683, **684**, 685; stalk, 684; wall, 321, **501**
sporangiophores, **40**, **45**, 56, 60, 64, 72, 75, 429, 430, 442, **464**, 684
sporangiospores, 72
spore, dispersal, 331; evolution, **502**; halo, 75; mother cell, 311, 522, **595**; patterns, 311
spores (see also akinete, aplanospore, arthrospore, ascospore, basidiospore, carpospore, chlamydospore, cyst, hynospore, monospore, oospore, planospore, polyspore, statospore, tetraspore, zoospore, zygospore), 40, **45**, 55, 72, 90, 118, 119, 123, 125, 127, 128, 130, 163, 172, 179, 190, 201, 204, 311, 315, 317, 336, 400, 506, 668, 675
sporic life history, 43, **47**, 665
sporocarps, 58, 62, 64, 75, 78, 81, 119, 122, 123, 125, 126, 489, **489**; formation, **134**
Sporochnales, 220
Sporochnus, 220
sporocytes, 491
Sporogonites, 677
sporophylls, 403, 447, 526, 527, 532, 538, 582
sporophyte, 43, 53, 287, 295, **297**, 304, 314, 317, 322, 333, 675, **676**; adaptation, 675; Anthocerotae, 317; bryophytes, 304, 675, **676**; budding, 640; generations, 237; hepatics, 308; hornworts, 317; increasing, complexity, 677; increasing, size, 677; Musci, 321; plants, 195
sporulation, 75, 122, 136
spur shoot, 541, 543, 561; *Ginkgo biloba*, **543**
squash family, 598
squirting cucumber, 653
stachysporous, 625
stalk, 104, 155, **280**, 309, 327; cell, 130, 273, **280**, 544, 588; formation, 132; nuclei, 585
stamen, 610, 612, 622, 623, 625, 628, 631, 655; arrangement, **614**; diversity, **613**; evolutionary fusion, 623; tube, 622
staminate, 617; flower, 617
Stangeria, 520, 521
Stapelia, **615**, 628; *S. grandiflora*, **615**
starch, 49, **55**, 58, 143, 159, 170, 172, 178, 200, 204, 212, 217, 254, 279, 287, 291; chloroplast, 178; embryophytes, 291; grains, 291, **292**, 521; sheath, 356
statospore, 149
Staurastrum, **52**, 181, 197, **201**
Stauropteris, 456; *S. burntislandica*, 456; *S. oldhamii*, **454**, 456
stele, 356, **357**, **400**, 446, **464**, 495, 507, 684; evolution, **359**, 687; evolutionary sequence, 685; types, **358**, 685
stem, 346, **361**, 376, 535, 584, 598, 678; anatomy, 543, 567; anatomy, Coniferales, 567; anatomy, Cycadopsida, 520; cankers, 78;

stem (*continued*)
  tissue, 625; variations, angiosperms, 598
stem-calyptra, **308**, 310
Stemonitales, 126
Stemonitis, **131**, 136
stereid cells, **323**, 332, 336
Stereum, **104**
sterile, cell, 526; hybrids, **6**; telomes, 680
sterols, 65, 143
sticky fruits, 651
Stigeoclonium, **176**, 181, 190, 193, 195, **195**, **196**, 217
stigma, 5, 300, 610, 612, 623, 626, 628, 631, 638
stigmarian, rhizophore, 424; system, 409, **409**
stigmatoid tissue, 638
Stigonema, 33, **34**
Stigonematales, 33
stilton cheese, 82
stinkhorns, 103, 107
Stipa, 4; polyploid series, **7**
stipe, 104, 107, 231
stipules, 463, 601, **603**
stolons, 599, **600**, 640
stomata, 65, 317, 329, 333, 354, 407, 409, 446, 507, 543, 566, 584, 585, 588, 675
stomatal, apparatus, 537; rows, 521
stone, cells, 351; fruits, 87
stonecrop family, 604
stoneworts (see also Charophyceae), 4, 209
storage, organs, 643; products, 40, 49, 51, 53, 142, 143, 221, 255; products, Phaeophyta, 221; products, Rhodophyta, 255; roots, 606, **610**
strawberry, 88, 640, 647
stream, 651; dispersal, 651
streaming, 122
Streblonema, 220, 241
Streptococcus, 21; *S. pneumoniae*, 21
Streptomyces, 24, **26**
streptomycin, 24
striae, 152
striated band, 163
striped maple, 354
strobili, 403, 411, **555**, 585, **586**
stromata, 85, 89, 288, **290**
stromatolites, 36, 659, **661**
stromatolitic formation, **661**
structural differentiation, Phaeophyta, 231
structure, 209
style, 101, 610, 612, 623, 638
Stylites, 411, 412, 416, 423, 424, 427, **427**
suberin, 293, 295, 356
submarginal placentation, 612
submerged hyphae, 58, 72
subsidiary cells, 355, 521, 537, **537**, 538
subtending bract, 620
subterranean ascomycetes, 89
succulents, 604
sudd, 651; dispersal, 651
sugar, alcohols, 46; beets, 67; cane rot, 82; granules, **24**
sulcus, 504, 511, **514**, 584
sulfur bacteria, 22
sulfuric acid, 24
sun birds, 629
sunflower family, 607, 649
sunflowers, 596
sunken stomata, 520, **522**, 543, **546**, 555, **557**, 566
superior, 612, 647; ovary, 612, 647
supporting cell, 271, 273
suspensor, 572, 585, 586; tier, 572
Sutcliffia, 682
swarm cells, 118, 119, 125, 132, 136, 139
swarmers (see also gametes, zoospores), 149, 155, 163, 181
swarming, 195
Swaziland Supergroup, 36
sweet potato rot, 82, 606
sycamore, 647

symbiosis (see also endosymbiosis), 662
symbiotic, 21; mycorrhizal association, 102; relationships, 33
symbiotically, 22, 190, 214
symplastic growth, 354
sympleisiomorphy, 12
sympodial, 377; axis, **680**; branch development, 680; terminal branches, 379
synandrium, 586
synangia, 399, **400**, 471, 505, 508, 511
synangiate pollen organs, 513, 689
synapomorphy, 12
syncarpous, 612; nuclear division, 119
synchronous nuclear divisions, 119
syndetocheilic, 355, 537, **537**, 585; stomata, 537, 588, 591
synergids, 298, **299**, **595**, 636, 639
syngamy (see also sexual reproduction), 43, 47, 62, 119, 181, **245**, **283**, 663
syngenesis, 681; axes, **681**
Synura, 149, **151**
Syringoderma, 220, 226, **233**
systematics, 7

Tabellaria, **153**
tadpoles, 169
Taeniocrada, 386; *T. decheniana*, 386, **387**
Takakia, 308, 309, 310; *T. lepidozioides*, **308**
Talaromyces, **88**
tannin, 543; cells, 465
tap root, 585, 605, **609**
tapetum, 522, 636, 638; glandular, 638; secretory, 638
Taphrina, 79; *T. deformans*, **78**
Taphrinales, 78
Taraxacum, **642**, 649
Targionia hypophylla, **319**
Taxaceae, 561, 562; foliage seeds, **576**
taxad evolution, 580
Taxales, 554, 561, 562, 573-576; distribution, 561; morphological characteristics, 561
taxic responses, 40
Taxodiaceae, 561, 562, 570, 577, **580**
Taxodium, 561, 566
taxon, 7, 165, 169
Taxopsida, 573
Taxus, 300, 573, 575, 576; morphology, 573
Tayloria splachnoides, 338
tectum, 680, 687
Tedeleaceae, 456
telia, 101
teliospore, 101
telome, 495, 502, 680, **680**; concept, 678; processes, 680-684; theory, 682; trusses, 680
telomic process, 576, 680
tendrils, 601, 604
teninucellate, 636
tentacle, 163
tepals, 610
terrestrial, 214; plants, 69
Tertiary, 311, 519, 540, 561, 592, 626; age, 308
testa, 643, 644
tetanus, 21
tetracycline, 24
tetrad scar, 504
tetrads of meiospores, 306
tetrahedral, 638
Tetragonidium, **157**, **169**, 172
Tetraphidae, 327, 328; anatomy, **328**; morphology, **328**
Tetraphis, 327; *T. pellucida*, **328**
tetrarch, 521, 606;
Tetraspora, 181, **182**
Tetrasporales, 181
tetrasporangia, 253, **278**, **280**, **282**, 283, **284**
tetraspores, 253, 273, 275, **278**, **282-284**
tetrasporophyte, red algae, 253, 273, **278**, **280**, 283
Tetraxylopteris, 494, **496-498**; *T. schmidtii*, **495**

Tetrodontium, 327
textiles, 75
Thalassiophyllum, 220
Thalassiosira, **157**
thallophytes, 38
thallus, 55, 79
theca (see also epitheca, hypotheca), 152
thecate, 159
Theriotia, 329
thermal pollution, 36
thermoacidiophiles, 25
thermophilic, 63
thiamin, 62, 216
Thiobacillus, 22
Thismia neptunis, **345**
Thlaspi, **614**
thorns, 599, 606
Thuidium delicatulum, **333**
Thuretellopsis, 253
Thursophyton, 406, 407
thyamine, 25
thylakoid, 27, **29**, 142, 143, 159, 165, 170, 175, **176**, 178, 200, 220, 254, **256-258**, 279, 282; girdle, 282; membranes, 662
thyrse, 620
tide, 179; pools, 149, 179
Tilia, 647, 649, **650**
Tilopteridales, 220
Tilopteris, 220
timber, 75
Timmia bavarica, **338**
tinsel, **42**, 60, 146, 170; flagellum, 40, 70, 139, 149, 155
tissue, differentiation, 346; collenchyma, 350; parenchyma, 347; sclerenchyma, 351; vascular, 355; vascular, plants, 347
Tmesipteris, 397, **397**, **398**, **400**, 401, 683, **684**; *T. tannensis*, **400**
tobacco family, 598
Todea, 471, 473
Tolypella, 15
Tolypothrix, **31**, 33
tomato, **648**, 649
tonoplast, 287, **289**, 290, **290**, 368
tooth fungi, 103
toral growth, 620, 622
Torreya, **576**
Tortula bistratosa, **334**; *T. fragilifolia*, **307**; *T. papillosa*, **334**; *T. princeps*, **337**
torus, **294**, 361
touch-me-not, 653
toxin production, 34
toxins, 34, 36, 82, 104, 165
trabeculae, 420, 535
tracheary, elements, 301, 360, 363; evolution, 685; evolutionary series, **687**
tracheid, 359, 361, 363, 423, 494, 495, **496**, 505, 506, 510, 535, 543, 556, 567, 576, 586, 685; pitting, 521; wall, 495
Trachelomonas, **167**, **168**, 169
Tracheophyta, 49, 286, 287, 295, 301, 329, 341, **343**, **344**; subdivisions, 345
tracheophytes, 46, 49, 341, 342, 368, 371, 392, 593; aquatic, 342
Trailliella, 253
Trailliella-stage, 275, **284**
trained, dogs, 89; pigs, 89
transfusion, cells, 521; tissue, 567; tracheids, 567
transition, aquatic to terrestrial mode, 674; region, 643; zone, 373; zone, brown algae, 231
transverse, flagellum, 163; groove, 163; wall formation, **18**; walls, 146
Trebouxia, 181
tree ferns, 460, 520
tree-of-heaven, 649
Tremella, **98**
Tremellales, 97, **98**
Trentepohlia, **177**, **183**

triarch, 606
Triassic, 430, 515, 516, 532, **553**
Tribonema, 144, **145**, 146, 148
Tribophyceae (see also xanthophytes, yellow-green algae), 142-148, **144**, **146**, 149, 163, 172; cell structure, 143; classification, 146; ecology, 146; morphological variation, 146; reproduction, 143
Triceratium, **153**, 158
Trichiales, 126
trichoblast, red algae, 269, **278**, 280
trichogyne, 80, 90, **95**, 252, 262, 271, **283**
Trichomanes, 476, **479**; *T. reniforme*, **479**; *T. saxifragoides*, **479**; *T. tahitiense*, **479**
trichome, 30, **31**, 355
Trichomycetes, 72
Trichopitys, 540, **541**, 549, **553**
trichothallic, growth, 226, **227**, **236**; hairs, **244**; meristem, **236**
tricolpate, 690, **690**
tricolporate, 690, **690**
Trifolium, 628, **630**
trigones, 310
trigonous corners, 321
trilete, 494, 506; mark, 494
trimerophytes, 383, 387-391, 499, **679**
Trimerophytina, 345, 383, 387-391, 395
Trimerophyton, 389, 391; *T. robustius*, **390**
Triplaris, **650**
triporates, triangular, 690, **690**
Tristachya, 430, **434**
tRNA, 25
Tropaeolum, **629**
true, ferns, 460-491; slime molds, 46, 118, 139
true slime molds, 46, 118, 139
truffles, 89
Tsuga, **579**, 580
tube, cell, *Ginkgo biloba*, **548**; nucleus, 571, 589
Tuberales, 89
tubers, 313, 416
tubular, carpels, 622; tinsels, 145
tubules, 104, 139
Tulipa, **618**
tumbleweeds, 648
tunica, 373, 588; layer, 373, 567; organization, 373, 374
Turnerella, 253
turnip, 67
tylosoids, 569
type, life histories, 43; steles, 357
Typha, 655
typhoid fever, 21

Ullmannia, 558, **561**
Ulmus, 649; *U. americana*, **606**
Ulota megalospora, **339**; *U.phyllantha*, 307
Ulothrix, 193, **193**, 214, 217
Ulotrichales, **185**, 674, 675
ultrastructure, **256**, **257**, **258**; eukaryotic cells, **40**
ultraviolet light, 609
Ulvales, 181, 190, 193
Ulvaria, **183**
umbel, determinate, 620
Umkomasia, **517**
unarmored, 159
unbranched filaments, 193
Uncinula, **89**; *U. salicis*, **89**
underground buds, 584
uniaxial, construction, **267**; Florideophyceae, **267**, **274**
unicell, 39
unicellular, 29, 30, 260; Chlorophyceae, 178
unifacial leaves, 603, 605
uniflagellate zoospores, 69
unilateral hairs, 166
unilocular, **228**, 235, 239; sporangia, brown algae, 219, 237, **238**

uninucleate, 223, 254; primary zoospores, 64; thalli, 55
uniseriate, 226, 260; filament, 190, 247; rays, 494
unisexual, 252, 262, 448; flowers, 656
unistratose, 304, 401
universal veil, 104
unnatural taxa, 12
upgrade theory, 677
Upper, Carboniferous, 513; Cretaceous, 484; Devonian, 386, 493, 494, 499, 502, 677, 680, 686; Triassic, 532
uredia, 99, 101
Uredinales, 97, **99**
uredospores, 99, 101
urns, 214
Urospora, 204, **207**
Ursinia, 646
uses, brown algae, 247; red algae, 277
Usnea, **88**
Ustilaginales, 97, 101, **101**
Ustilago, 101, **101**, 102; *U. hordei*, **102**; *U. maydis*, **102**

vacuoles, 58, 200, 221, 287, **289**, 290, **290**, 291; embryophytes, 291
vallecular canal, 438, 440
Vallisneria, 631; *V. americana*, **635**
Valonia, 204, **207**
valve, markings, 158; mold, 155; view, diatoms, 152
valves, 653
variability, sexual process, 663
variation, 2, 8; branching pattern, 377; genetic sources, 2
vascular bundles, 346, 357, 438, 566, 598, 603, 623, 634; monocotyledons, **359**; primary, **357**; types, **357**
vascular cambium, dicotyledons, **381**
vascular cryptogams, 502
vascular plants, 46, **53**, 132, 287, 296, 341-382, 456, 675, 678, 680, 686; angiosperms, 347; appendages, 376; branches, 376; cambium, 342, 357, 363, 368, 378, 494, 521, 584; cambium, *Ginkgo biloba*, **544**; cells, 347; classification, 344; collenchyma, 350; cytology of apical meristem cells, 371; derivatives, 371; distribution, 342; diversity, 342; early, 678; epidermis, 354; evolutionary interpretation, 344; evolutionary pathways, 678; fibers, 354; free-sporing, evolution, 686; growth, 346, 368; growth, of plant body, 346; initials, 371; leaves, 376; lineages, evolutionary pathways, **688**; origin, 672; parenchyma, 347, 368; phellogen, 380; phloem, 363; plant body, 346; protostele, 356; role in vegetation, 312, 342; root, apical meristem, 374; root, system, 354; sclerenchyma, 351; secondary growth, 378; shoot, apical meristem, 373; shoot, system, 354; sieve, cells, 363; sieve elements, 363; sieve, tube elements, 364; siphonostele, 356; stele, 356; tissues, 347, 355; xylem, 357
vascular, rays, 359; strands, 404, 494, 505; structure, 585; system, 341; tissues, 301, 355
vascularization in flower, **615**
Vaucheria, 115, 145, 146, **147**, 148, **148**
vegetative, cell division, 155; hyphae, 79; plant body, angiosperms, 596; propagation, 321; propagules, 313, 326; reproduction, **45**, 163, 179, 200, 212, 235, 260, 478; reproduction, angiosperms, 640; reproduction, Florideophyceae, 266; reproduction, Phaeophyta, 235; stages, **265**; trends, embryophytes, 295; units, 208
veil, 104
veins, **272**, 603
velamen, 354

velum, 423
venation, 430, 460; angiosperm leaves, **605**; netted, 603; pattern, 521, 601, 603; reticulate, 516, 603
venter, 304, 305
ventral canal, cell, 548; nucleus, 531, 585
Venturia, 94; *V. inaequalis*, **90**
Veronica, **627**
versatile, 610
Verticillium, **81**, 114
Vesicaspora, **510**, 511, **514**, 516
vesicles, 22, 58, 159, 223
vessel elements, 359, 360, 361, 363, 596, 687
vessels, 363, 582, 592, 685; angiosperms, 582; gametophytes, 582
vibrio, 15, **15**
Vicia, 653; *V. gigantea*, **654**
Viola, **606**, **629**, 645
violet, 645
virgin's bower, 598
virus, **21**
viscid fruits, 651
vitalized, protostele, 356, 357, **358**; siphonostele, 357
vitamins, 25, 62, 78, 172, 277
Vitis, 364, **382**, **648**, 649
Vitreisporites, 516
Voltziaceae, 556, 558, 560
Voltziales, 554, 556
volva, 104
Volvariella volvacea, 104
Volvocales, 179, 181, **184**, **187**, **189**, 190, 217
Volvox, 8, 42, 181, **184**, 188, **189**, 197

wall formation (see also cell wall), **179**, **181**
water, 62; dispersal, 73, 645, 649
water lily, 610
water plantain, 622
water plants, 653
water-plantain family, 649
wattle, 603, 604
webbing, 495, 680, 681; stages, **681**
weeds, 640, 653
Weeksia, 253
Welwitschia, 586, **586**, 587, **587**, 588, **588**, 589, **589**, 590-592; *W. bainesii*, 585
Welwitschiales, 585
wheat, 63, 64, 99, 101, 596, 644; rust fungi, 97
whiplash, **42**, 60, 146, 163, 172; flagellum, 40, 69, 304; type, 178
whisk ferns, 377, 397
white pine blister rust, 101
white pines, 101
white rusts, 67
whooping cough, 21
Williamsonia, 537; reconstruction, **537**
Williamsoniaceae, 534, 536
Williamsoniales, 537
willow, 645, 655
wilting, 74
wind, dispersal, 306, 644, 648; dissemination, 648; pollinated, 631; pollinated, flowers, 629, 631; pollination, 629
wine industry, 67
winged, achene, 649, **650**; fruits, 649; seeds, 649
Wolfia punctata, **343**
Woloszynskia, 163
wood, 102, 103, 107, 347; rush, 622
wood-inhabiting fungi, 103, 104
Woodsia ilvensis, **485**
woody, 446
wool, 651
wound parasites of fish, 65

X- and Y-chromosomes, 541
Xanthophycomonas, **144**

xanthophyll pigments, 220
xanthophylls, 49, 149, 152, 169, 172, 208, 254, 255; Tribophyceae, 142, 143
xanthophytes (see also Tribophyceae), 143, 148
xenogenous, hypothesis, 662; model, eukaryotic cells, **662**
xeric, 513
xeromorphic characteristics, 507
xylan, 175, 200, 208, 253
Xylariales, 85
xylary fibers, 363
xylem, 38, 296, **297**, 301, 355-357, 363, 380, 554, 598, 599, 606; cells, **362**; parenchyma, 359, 558; maturation, 411; ray, 569
xylose, 200

yeasts, 58, 75, 78, 115
yellow-green algae (see also Tribophyceae, xanthophytes), 142, 143, 145, 146
yogurt, 15

young, root, 549; sporophyte, **246**

Zamia, 521, 522, **523, 527**; pollen grain, **526**; *Z. pygmaea*, 527
Zanardinia, 220
Zantedeschia, 605
Zea, 357, **358**, 375; *Z. mays*, 610
zeaxanthin, 255
Zebrina, **353**
Zonaria, 226
zoosporangium, **45,** 70, 146
zoospore, 40, **42, 45,** 56, 60, 62, 64, 65, 67, 69, 70, **71, 73,** 145, 146, 149, 179, 188, 190, 193, 195, 197, 208, 212, 223, 235, 239, 249; encystment, 70; Laminariales, 245; release, 65
Zostera, 260, 342
zosterophyllophytes, 383, 391-395
Zosterophyllophytina, 345, 391-395, 407
Zosterophyllum, 392, 404, 678; *Z. myretonianum*,

Zosterophyllum (continued)
  391, 392; *Z. rhenanum*, **391,** 392
Zygnema, **176, 177,** 181, 197, **201,** 214
Zygnematales, 115, 179, 181, **185,** 190, 195, 197, **201,** 203, 214
zygomorphic, 609; corolla, 622; flower, 628
zygomycetes, 72, 668
Zygomycota, 39, 56, 64, 72, 74, 115, 116, 118
Zygopteris, 456, 457; *Z. berryvillensis*, 457, **457**; *Z. primaria*, 457
zygospores (see also zygote), **183**
zygote, 43, 62, 64, 65, 70, 74, 119, 146, 163, 183, 188, 195, 201, 208, 212, 237, 239, **245, 246,** 300, 305, 532, 571, 585, 587, 639, 643, 675; germination, 188; nucellus, 273, 275, **278,** 548
zygotic life history, 43
zygotic meiosis, 665

Kingdom Plantae (continued)
(Vascular Plants)

SUBDIVISION	CLASS	SUBCLASS	ORDER	FAMILY
Rhyniophytina (*Rhyniophytes*)	Rhyniopsida		Rhyniales	
Trimerophytina (*Trimerophytes*)	Trimerophytopsida		Trimerophytales	
Zosterophyllophytina (*Zosterophyllophytes*)	Zosterophyllopsida		Zosterophyllales	
Psilotophytina (*Psilotophytes*)	Psilotopsida		Psilotales	
Lycophytina (*Lycopods*)	Lycopsida		Lepidodendrales	
			Pleuromeiales	
			Isoetales	
			Selaginellales	
			Lycopodiales	
Sphenophytina (*Sphenophytes*)	Sphenopsida		Sphenophyllales	
			Calamitales	
			Equisetales	
Pterophytina (*Pterophytes*) (*Preferns*)	Cladoxylopsida		Cladoxylales	
	Coenopteridopsida		Coenopteridales	
		Ophioglossidae	Ophioglossales	
		Marattiidae	Marattiales	
		Osmundidae	Osmundales	
		Filicidae	Schizaeales	
			Pteridales	
			Dicksoniales	
(*True Ferns*)	Filicopsida		Hymenophyllales	
			Gleicheniales	
			Cyatheales	
			Aspidiales	
			Blechnales	
			Matoniales	
			Polypodiales	
		Marsileidae	Marsileales	
		Salviniidae	Salviniales	
Plantae — "Tracheophytes" (Division TRACHEOPHYTA) (*Gymnosperms*)	Progymnospermopsida (*Progymnosperms*)		Aneurophytales	
			Archaeopteridales	
	Pteridospermopsida (*Seed Ferns*)		Lyginopteridales	Lyginopteridaceae
			Medullosales	Medullosaceae
			Callistophytales	Callistophytaceae
			Corystospermales	Corystospermaceae
			Peltaspermales	Peltaspermaceae
			Caytoniales	Caytoniaceae
	Cycadopsida (*Cycadophytes*)		Cycadales	Cycadaceae
	Cycadeoidopsida (*Cycadeoids*)		Cycadeoidales	Cycadeoidiaceae
			Williamsoniales	Williamsoniaceae
	Ginkgopsida (*Ginkgos*)		Ginkgoales	Ginkgoaceae
	Coniferopsida (*Coniferophytes*)		Cordaitales	Cordaitaceae
			Volziales	Lebachiaceae, Volziaceae
			Coniferales	Pinaceae, Cupressaceae, Podocarpaceae, Taxodiaceae Araucariaceae, Cephalotaxaceae
			Taxales	Taxaceae
	Gnetopsida (*Gnetophytes*)		Ephedrales	
			Welwitschiales	
			Gnetales	
(*Angiosperms*)	Magnoliopsida (*Angiosperms*)	Magnoliidae (*Dicots*)	Magnoliales	Magnoliaceae, Degeneriaceae, Annonaceae
			Laurales	Calycanthaceae, Lauraceae
			Piperales	Piperaceae
			Aristolochiales	Aristolochiaceae
			Rafflesiales	Rafflesiaceae
			Nymphaeales	Nymphaeaceae, Ceratophyllaceae
			Illiciales	Illiciaceae, Schizandraceae
			Nelumbonales	Nelumbonaceae
			Ranunculales	Ranunculaceae, Lardizabalaceae, Berberidaceae
			Papaverales	Papaveraceae, Fumariaceae
			Sarraceniales	Sarraceniaceae
			Trochodendrales	Trochodendraceae
			Cercidiphyllales	Cercidiphyllaceae
			Eupteleales	Eupteleaceae
			Didymelales	Didymelaceae
			Hamamelidales	Hamamelidaceae, Platanaceae
			Eucommiales	Eucommiaceae
			Urticales	Ulmaceae, Moraceae, Cannabaceae, Urticaceae